The Routledge Handbook of Technology, Crime and Justice

Technology has become increasingly important to both the function and our understanding of the justice process. Many forms of criminal behaviour are highly dependent upon technology, and crime control has become a predominantly technologically driven process – one where 'traditional' technological aids such as fingerprinting or blood sample analysis are supplemented by a dizzying array of tools and techniques including surveillance devices and DNA profiling.

This book offers the first comprehensive and holistic overview of global research on technology, crime and justice. It is divided into five parts, each corresponding with the key stages of the offending and justice process:

- Part I addresses the current conceptual understanding of technology within academia and the criminal justice system;
- Part II gives a comprehensive overview of the current relations between technology and criminal behaviour;
- Part III explores the current technologies within crime control and the ways in which technology underpins contemporary formal and informal social control;
- Part IV sets out some of the fundamental impacts technology is now having upon the judicial process;
- Part V reveals the emerging technologies for crime, control and justice and considers the extent to which new technology can be effectively regulated.

This landmark collection will be essential reading for academics, students and theorists within criminology, sociology, law, engineering and technology, and computer science, as well as practitioners and professionals working within and around the criminal justice system.

M. R. McGuire is Senior Lecturer in Criminology at the University of Surrey, UK.

Thomas J. Holt is Professor of Criminal Justice at Michigan State University, USA.

WITHDRAWN

The Routledge Handbook of Technology, Crime and Justice

Edited by M. R. McGuire and Thomas J. Holt

Routledge
Taylor & Francis Group

LONDON AND NEW YORK

First published 2017
by Routledge
2 Park Square, Milton Park, Abingdon, Oxon OX14 4RN

and by Routledge
711 Third Avenue, New York, NY 10017

Routledge is an imprint of the Taylor & Francis Group, an informa business

British Library Cataloguing-in-Publication Data
A catalogue record for this book is available from the British Library

Library of Congress Cataloging in Publication Data
Names: McGuire, M. R., editor. | Holt, Thomas J., 1978- editor.
Title: The Routledge handbook of technology, crime and justice / edited by M.R. McGuire and Thomas J. Holt.
Description: Abingdon, Oxon ; New York, NY : Routledge, 2017. | Series: Routledge international handbooks | Includes bibliographical references and index.
Identifiers: LCCN 2016030191| ISBN 9781138820135 (hardback) | ISBN 9781315743981 (ebook)
Subjects: LCSH: Criminology. | Technological innovations. | Criminal investigation--Technological innovations. | Criminal justice, Administration of--Technological innovations.
Classification: LCC HV6030 .R687 2017 | DDC 364--dc23
LC record available at https://lccn.loc.gov/2016030191

ISBN: 978-1-138-82013-5 (hbk)
ISBN: 978-1-315-74398-1 (ebk)

Typeset in Bembo
by Saxon Graphics Ltd, Derby

Printed and bound in Great Britain by
TJ International Ltd, Padstow, Cornwall

Contents

Contents

Contents

Figures

Images

Tables

Tables

Contributors

Kimberly L. Barrett is an Assistant Professor of Criminology in the Department of Sociology, Anthropology, and Criminology at Eastern Michigan University. Her research interests include green criminology, environmental justice, schools and crime, and youth public health issues. Kimberly's recent work has been published in the *British Journal of Criminology*, *Journal of Crime and Justice*, and *Journal of School Violence*.

Susan W. Brenner is the Samuel A. McCray Chair in Law at the University of Dayton School of Law. She specializes in two distinct areas of law: grand jury practice and cyberconflict, i.e., cybercrime, cyberterrorism and cyberwarfare. Professor Brenner is a member of the American Academy of Forensic Sciences and has published widely in the area of technology, especially in relation to cybercrime. Her most recent book is *Cybercrime and the Law: Challenges, Issues, and Outcomes* (2012). She also writes a blog, CYB3RCRIM3.

Philip Brey is Professor of Philosophy of Technology at the University of Twente in the Netherlands and Scientific Director of the 3TU.Centre for Ethics and Technology. He investigates philosophically how technology is interacting and co-evolving with human beings, society and culture, and he investigates social, political and ethical issues in the development, use and regulation of technology in society. His focus is on new and emerging technologies, especially information technology, biomedical technology and technologies for sustainability. He has contributed to major approaches in his field, such as the embedded values approach for analysing values in technology, the structural ethics approach for studying ethical implications of social and technological structures, and the anticipatory technology ethics (ATE) approach for assessing ethical implications of new and emerging technologies. He is president of the International Society for Ethics and Information Technology and past president of the Society for Philosophy and Technology.

Patrick Bishop is a Senior Lecturer in the College of Law and Criminology, Swansea University. He is predominantly an environmental lawyer and his research interests encompass the enforcement of environmental offences, regulatory approaches and the role of science in the formulation of environmental policy. His publications in a number of peer-reviewed journals reflect these interests. He is the co-editor of *Environmental Law and Policy in Wales: Responding to Global and Local Challenges* (2013).

Albert Borgmann is Regents Professor Emeritus of Philosophy at the University of Montana, Missoula where he has taught since 1970. His special area is the philosophy of society, culture and technology. Among his publications are *Technology and the Character of Contemporary Life*

(1984), *Crossing the Postmodern Divide* (1992), *Holding On to Reality: The Nature of Information at the Turn of the Millennium* (1999), *Power Failure: Christianity in the Culture of Technology* (2003), and *Real American Ethics* (2006).

Avi Brisman is an Associate Professor in the School of Justice Studies at Eastern Kentucky University in Richmond, KY (USA). He received a BA from Oberlin College (Oberlin, OH), an M.F.A. from Pratt Institute (Brooklyn, NY), a JD with honors from the University of Connecticut School of Law (Hartford, CT), and a PhD in anthropology from Emory University (Atlanta, GA). He is widely published in green criminology on linkages between armed conflict and environmental degradation, representations of environmental crime and harm in film and literature, and individual and collective resistance to environmental crime and harm. He is co-editor, with Nigel South, of the *Routledge International Handbook of Green Criminology* (2013), co-editor, with Nigel South and Rob White, of *Environmental Crime and Social Conflict: Contemporary and Emerging Issues* (2015), and co-author, with Nigel South, of *Green Cultural Criminology: Constructions of Environmental Harm, Consumerism, and Resistance to Ecocide* (2014).

Jo Bryce is a Senior Lecturer at the School of Psychology, University of Central Lancashire, UK. Her research interests focus on the psychological, social and forensic aspects of the Internet and related technologies, with a specific focus on their use by young people, associated risk exposure and Internet safety. Her work has also focused on the use of ICT to facilitate sexual offending and other forms of cybercrime (e.g., cyberstalking, filesharing). These research areas have developed from her work as a former Coordinator of the UK National Awareness Node for Child Safety on the Internet (now known as Safer Internet Centres), and the project lead for the completed ISCA and INSAFE projects (both funded by the European Commission Safer Internet Plan). She has extensive experience of researching young peoples' online behaviours and experiences using qualitative and quantitative methodologies, and has undertaken reviews of associated online technologies and services for the European Commission and Council of Europe. She is a member of the Evidence Group of the UK Council on Child Internet Safety, and regularly consults on associated issues with a variety of stakeholders in government, industry, education and enforcement.

Mark Button is Director of the Centre for Counter Fraud Studies at the Institute of Criminal Justice Studies, University of Portsmouth. Mark has written extensively on counter fraud and private policing issues, publishing many articles and chapters and completing eight books. His latest book (co-authored with Martin Tunley, Andrew Whittaker and Jim Gee) is titled the *Accredited Counter Fraud Specialist's Handbook*. Some of the most significant research projects include leading the research on behalf of the National Fraud Authority and ACPO on fraud victims; the Department for International Development on fraud measurement; Acromas (AA and Saga) on 'Cash-for-Crash fraudsters'; the Midlands Fraud Forum, Eversheds and PKF on 'Sanctioning Fraudsters'. Mark has also acted as a consultant for the United Nations Offices on Drugs and Crime on developing international standards for Civilian Private Security Services and the United Nations Development Programme/European Union on enhancing civilian oversight of the Turkish private security industry. Mark also holds the position of Head of Secretariat of the Counter Fraud Professional Accreditation Board. He is also a former Director of the Security Institute. Before joining the University of Portsmouth he was a Research Assistant to the Rt Hon Bruce George MP specialising in policing, security and home affairs issues. Mark completed his undergraduate studies at the University of Exeter, his Masters at the University of Warwick and his Doctorate at the London School of Economics.

James Byrne is a Professor in the School of Criminology and Justice Studies at the University of Massachusetts, Lowell and an Adjunct Professor at Griffith University in Australia. Professor Byrne is the author/editor of several books, including *The New Technology of Crime, Law and Social Control* (with Donald Rebovich, 2007), and *The Culture of Prison Violence* (with Don Hummer and Faye Taxman, 2008). Since 2008, he has served as the Editor-in-Chief of the journal *Victims and Offenders: An International Journal of Evidence-based Research, Policy, and Practice*, and he serves on the editorial boards of several professional journals, including *Criminology and Public Policy, Health and Justice, Federal Probation*, and the *European Journal of Probation*. In 2011, he was the recipient of both The Distinguished Scholar Award and The Marguerite Warren and Ted B. Palmer Differential Intervention Award from the American Society of Criminology. In 2012, Professor Byrne was selected to serve as a member on the Panel of Experts – Correctional Services Advisory and Accreditation Panel, Ministry of Justice, United Kingdom and in 2014 he served as the External Inspector of Prisons, Office of the Inspector General, Queensland Correctional Services. His forthcoming book *Technology and Criminal Justice: A Global Perspective,* co-authored with Don Hummer, examines the impact of technology in police, courts, and corrections systems globally.

Lisa Claydon is a Senior Lecturer in Law at the Open University. Her research centres upon the overlap between neuroscience and the criminal law, in particular the use of neuroscience to inform decisions by the courts when it is used in evidence in the courtroom. She is currently involved in an AHRC funded project entitled *A Sense of Agency* which is examining neurocognitive and legal approaches to a personal sense of agency. She is a Programme Committee member of the International Neuroethics Society and Secretary to the European Association for Neuroscience and Law.

Simon A. Cole is Professor of Criminology, Law and Society and Director of the Newkirk Center for Science & Society at the University of California, Irvine. He received his PhD in Science and Technology Studies from Cornell University. Dr. Cole is the author of *Suspect Identities: A History of Fingerprinting and Criminal Identification* (2001), which was awarded the 2003 Rachel Carson Prize by the Society for Social Studies of Science, and *Truth Machine: The Contentious History of DNA Fingerprinting* (2008, with Michael Lynch, Ruth McNally & Kathleen Jordan). He is Co-Editor of the journal *Theoretical Criminology*.

Cassandra Cross is a Senior Lecturer in the School of Justice, Queensland University of Technology. Previously, she worked as a research/policy officer with the Queensland Police Service, where she commenced research on the topic of online fraud and was awarded a Churchill Fellowship in 2011. Since taking up her position at QUT, she has published in this area across several journals and continued her research into online fraud focusing across the prevention, victim support and policing aspects of this crime. She has received two highly competitive Criminology Research Grants, the first in 2013 to conduct the first Australian study into the reporting experiences and support needs of online fraud victims, and the second in 2015 to examine the restoration of identity for identity theft victims. She is co-author (with Professor Mark Button) of the book *Cyber Frauds, Scams, and their Victims*, which will be published in 2017.

Andrew S. Denney is currently an Assistant Professor in the Department of Criminology and Criminal Justice at the University of West Florida. His research focuses on institutional corrections, offender re-entry, faith-based programming, sex offences and offenders, and sexual deviance. His most recent publications have appeared in *Children and Youth Services Review, Criminal Justice Review,* and the *Journal of Qualitative Criminal Justice and Criminology.*

Adam Edwards is Reader in Politics and Criminology at Cardiff University, School of Social Sciences and editor of the Routledge Series on *Crime, Security and Justice.* His research interests include the impact of emergent technologies on crime, security and justice including the uses of Big Data in predicting crime and civil unrest. He co-founded the Collaborative Online Social Media ObServatory (COSMOS) and is author of *Crime Control and Community* (with G. Hughes, 2002), *Transnational Organised Crime* (with P. Gill, 2003) and *Policing European Metropolises* (with E. Devroe and P. Ponsaers, 2017). Adam is also a member of the Foundation for Responsible Robotics (FRR).

Paul Ekblom was originally a social researcher in the UK Home Office and worked on diverse crime prevention projects, horizon-scanning and developing the discipline of crime prevention. He then spent a decade at the Design Against Crime Research Centre, Central Saint Martins, University of the Arts, London where he remains Professorial Associate. He is also Visiting Professor at the Applied Criminology Centre, University of Huddersfield, and the Department of Security and Crime Science, UCL. He has collaborated with the EU Crime Prevention Network, Europol, Council of Europe, Australian Institute of Criminology, Abu Dhabi Government, and the UN. His current research covers design, counter-terrorism, evolutionary approaches to arms races, and developing practice-knowledge frameworks, viewable at www.designagainstcrime.com/methodology-resources/crime-frameworks and http://5isframework.wordpress.com.

Roy Fenoff is an Assistant Professor in the Department of Criminal Justice at The Citadel (The Military College of South Carolina). He is also a Board Certified Forensic Document Examiner (through testing by the Forensic Specialties Accreditation Board) and an expert in forgery detection. Dr. Fenoff provides scientific advice, offers training, and conducts forensic examinations for individuals, law enforcement, and law firms throughout the United States and abroad. He earned a BS in Entomology and a BA in Criminal Justice from the University of Georgia in 2004, an MS in Medical/Veterinary Entomology from the University of Wyoming in 2007, and a PhD in Criminal Justice from Michigan State University in 2015. Dr. Fenoff specializes in forgery and document fraud, crime prevention, food fraud, and the use of science in the courts. His current research focuses on understanding the myriad of document fraud issues faced by the food industry. Dr. Fenoff is a published author who has presented his work at a variety of criminal justice, forensic science, and food safety conferences. In addition to his current position at The Citadel, Dr. Fenoff is a research collaborator with Michigan State University's Food Fraud Initiative. (www.RoyFenoff.com)

Steven Furnell is a Professor of Information Systems Security and leads the Centre for Security, Communications and Network Research at Plymouth University. He is also an Adjunct Professor with Edith Cowan University in Western Australia and an Honorary Professor with Nelson Mandela Metropolitan University in South Africa. His research interests include usability of security and privacy technologies, security management and culture, and technologies for user authentication and intrusion detection. He has authored more than 270 papers in refereed international journals and conference proceedings, as well as books including *Cybercrime: Vandalizing the Information Society* (2001) and *Computer Insecurity: Risking the System* (2005). Professor Furnell is the BCS representative to Technical Committee 11 (security and privacy) within the International Federation for Information Processing, and is a member of related working groups on security management, security education, and human aspects of security. He is also a board member of the Institute of Information Security Professionals, and chairs the academic partnership committee and southwest branch. Further details can be found at www. plymouth.ac.uk/cscan, with a variety of security podcasts also available via www.cscan.org/ podcasts. Steve can also be followed on Twitter (@smfurnell).

Jérôme Goffette is *Maître de conférences* (Lecturer) in Philosophy of Medicine at the University Claude Bernard Lyon 1 and a member of the University's research unit on Science and Society: History, Education, and Practices (S2HEP). His research focuses primarily on anthropotechnics and human enhancement, a topic on which he has written more than 20 articles, as well as a book: *Naissance de l'anthropotechnie* [*Birth of Anthropotechnics*] (Paris 2006). He co-directed with S. Bateman *et al. Inquiring into Human Enhancement – Interdisciplinary and International Perspectives* (London 2015). His second area of research concerns the body and its imaginary. Among other publications on this topic, he co-edited a collective volume with Lauric Guillaud *L'imaginaire médical dans le fantastique et la science-fiction* [*Medical Imagination in Fantastic and Science Fiction*] (Paris 2010), and published with Jonathan Simon 'The internal environment: Claude Bernard's concept and its representation in *Fantastic Voyage* (R. Fleisher)' in Landers and Munoz (eds), *Anatomy and the Organization of Knowledge, 1500–1850.*

Keith Guzik is Associate Professor of Sociology at the University of Colorado, Denver. He is the author of *Making Things Stick: Surveillance Technologies and Mexico's War on Crime*; *Arresting Abuse: Mandatory Legal Interventions, Power, and Intimate Abusers*; and co-editor with Andrew Pickering of *The Mangle in Practice: Science, Society, and Becoming.*

Simon Hallsworth is Professor and Pro Vice Chancellor of the University of Suffolk. Though a sociologist by training, his work crosses the boundaries between sociology, social policy, political economy and political theory. Simon has written extensively on punitivity and penal change in modern societies and is also one of the UK's leading commentators on street violence and violent street collectives. His most recent book, *The Gang and Beyond: Interpreting Violent Street Worlds*, was published in 2013.

Mireille Hildebrandt is a lawyer and a philosopher. She holds the research chair of 'Interfacing Law and Technology' at the Law and Criminology Faculty of Vrije Universiteit Brussels, and the Chair of 'Smart Environments, Data Protection and the Rule of Law' at the Science Faculty of Radboud University Nijmegen. Her research is focused on the implications of data-driven intelligence for the grammar of democracy and the rule of law.

Thomas J. Holt is Professor in the School of Criminal Justice at Michigan State University specializing in cybercrime, policing, and policy. He received his PhD in Criminology and Criminal Justice from the University of Missouri-Saint Louis in 2005. He has published extensively on cybercrime and cyberterror in outlets such as the *British Journal of Criminology*, *Crime and Delinquency, Deviant Behavior, Sexual Abuse*, the *Journal of Criminal Justice*, and *Terrorism and Political Violence*. He is also a co-author of the book *Cybercrime In Progress: Theory and Prevention of Technology-Enabled Offenses.*

Chris Jay Hoofnagle is Adjunct Full Professor of Information and of Law at the University of California, Berkeley. An elected member of the American Legal Institute, he is author of the recent Federal Trade Commission Privacy Law and Policy (2016).

Ibrahim Altaweel, Jaime Cabrera, Hen Su Choi, and **Katie Ho**, are, respectively, students at Diablo Valley College, California State University Fullerton, University of California San Diego, and Mount Holyoke College, South Hadley. **Nathaniel Good** is a member of the Professional Faculty, School of Information at the University of California, Berkeley.

Don Hummer is a faculty member in the School of Public Affairs at the Pennsylvania State University, Harrisburg. In addition to research interests in institutional corrections, his writing focuses on offender re-entry, parole, and community violence. He is co-author/editor of *The Culture of Prison Violence* (2008) and *Handbook of Police Administration* (2008). His work has appeared in peer-reviewed outlets such as *Aggression & Violent Behavior, Probation Journal, Law & Policy,* and *Journal of Crime & Justice*. He is a former University senator at Penn State and is a past recipient of teaching awards presented by the National Consortium for Academics and Sports & the University of Massachusetts-Lowell. Dr Hummer formerly served as Vice Chairperson of the Herbert A. Schaffner Youth Center Advisory Board, Dauphin (PA) County's secure youth detention facility.

Richard Jones is Lecturer in Criminology at the School of Law, University of Edinburgh. His research and publications are in the areas of the use of new technologies in policing and criminal justice (including surveillance, security, the electronic monitoring of offenders, and cyber security); cybercrime; criminal justice and penal politics; and theoretical criminology. He was involved in the IRISS (Increasing Resilience in Surveillance Societies) FP7 project which examined the role of surveillance and resilience in democratic societies. At Edinburgh, Richard is a member of the Centre for Research on Information, Surveillance and Privacy (CRISP), CeSeR (Centre for Security Research), and SIPR (Scottish Institute for Policing Research). He is a member of the Editorial Board of the *British Journal of Criminology*.

Maria Kaspersson is a Senior Lecturer in Criminology at Greenwich University. Her doctoral thesis explored homicide in Stockholm between the sixteenth and twentieth centuries. Maria's current research interests include the history of homicide, domestic homicide and domestic violence, honour-related violence, the Swedish prostitution law, The Dangerous Dogs Act and 'status' dogs.

Max Kilger is a Senior Lecturer in the Department of Information Systems & Cyber Security at the University of Texas at San Antonio (UTSA). Dr Kilger received his PhD in Social Psychology from Stanford University. He has more than fifteen years of experience in the area of information security concentrating on the social and psychological factors motivating malicious online actors, hacking groups and cyberterrorists. Max has written and co-authored a number of journal articles and book chapters on profiling, the social structure of the hacking community, cyberviolence and the emergence of cyberterrorism. He is a founding and board member of the Honeynet Project, a not-for-profit information security organization with 54 teams of experts in 44 countries working for the public good.

Victoria Knight is a senior research fellow in the Faculty of Health and Life Sciences, De Montfort University. She has expertise across three core areas: 1) digital technologies in prisons, 2) emotion and prison, and 3) offenders and education. Victoria has extensive qualitative research experience for engaging 'hard to reach' groups in a range of settings; both community and prison. Victoria has written a book about her extensive sociological study of in-cell television in prison – *Remote Control: Television in Prison* (2016). Victoria is a member of the

Independent Monitoring Board at HMP Leicester and the editorial board for the *Prison Service Journal*. Victoria continues to research, publish and consult on digital technologies in secure settings. She is also the founder and convenor of the Emotion and Criminal Justice Cluster at De Montfort University.

Christopher Lawless is Lecturer in Applied Social Sciences at Durham University, UK. He has been researching the sociology of forensic science for more than ten years and has published numerous articles and books on this subject. Dr Lawless continues to study emerging forensic technologies and their potential social, sociological, ethical, legal and scientific impacts. Dr Lawless also continues to pursue research interests relating to the broader relations between technology, law and policy.

Fredric I. Lederer is Chancellor Professor of Law and Director of the Center for Legal and Court Technology (CLCT), formerly the Courtroom 21 Project, at William & Mary Law School. Professor Lederer graduated from the Polytechnic Institute of Brooklyn, now the NYU School of Engineering, and received his JD from Columbia University School of Law, where he was a member of the Columbia Law Review and recipient of the Archie O' Dawson Prize. As Founder and Director of CLCT, Professor Lederer is responsible for the McGlothlin Courtroom, the world's most technologically advanced trial and appellate courtroom, and the Court Affiliates, an organization of federal, state, and foreign courts. Professor Lederer's areas of specialization include legal technology, evidence, technology-augmented trial practice, electronic discovery and data seizures, criminal procedure, military law, and legal skills. He is the author or co-author of 11 books, numerous articles, and two law-related education television series.

Stuart Macdonald is Professor in Law at Swansea University. He is co-director of the University's £7.6m EPSRC CHERISH-Digital Economy Centre (www.cherish-de.uk) and the multidisciplinary Cyberterrorism Project (www.cyberterrorism-project.org), author of *Text, Cases and Materials on Criminal Law* (2015) and co-editor of *Violent Extremism Online: New Perspectives on Terrorism and the Internet* (2016), *Terrorism Online: Politics, Law and Technology* (2015) and *Cyberterrorism: Understanding, Assessment and Response* (2014). He has previously received research funding from the British Academy and NATO's Emerging Security Challenges Division, and has held visiting scholarships at Columbia University Law School, New York, the Institute of Criminology at the University of Sydney and the Faculté de Droit, Université de Grenoble. In 2016/17 he will be the holder of a Fulbright Cyber Security Award.

Gary T. Marx is Professor Emeritus MIT. He has taught at the University of California, Berkeley (from where he received his PhD), Harvard and the University of Colorado. He is the author of *Protest and Prejudice*; *Undercover Police Surveillance in America*; *Undercover: Police Surveillance in Comparative Perspective*; and *Windows into the Soul: Surveillance and Society in an Age of High Technology* among other books. Additional information is at www.garymarx.net.

Carole McCartney is a Reader in the School of Law, Northumbria University. Previously senior lecturer in criminal law and criminal justice at the University of Leeds, and Bond University, Queensland, Australia, Carole has written on Australian justice, Innocence Projects, miscarriages of justice, international policing cooperation, and DNA, forensic science and criminal justice more widely. She established an Innocence Project at the University of Leeds in 2005, and was project manager for the Nuffield Council on Bioethics report 'The Forensic Uses of Bio-information: Ethical Issues' and the Nuffield Foundation project 'The Future of Forensic Bioinformation'. She has run projects on forensic science education and forensic regulation and completed an EU Marie Curie international research fellowship (2009–2012) on 'Forensic Identification Frontiers'. She currently teaches and researches in the areas of criminal law, criminal evidence, and forensic science.

M. R. McGuire has developed an international profile in the critical study of technology, crime and the justice system, in particular issues around cyber-offending and cybercrime. His first book *Hypercrime: The New Geometry of Harm* (2008), critiqued the notion of cybercrime as a way of modelling computer-enabled offending and was awarded the 2008 British Society of Criminology runners-up Book Prize. His most recent publication, *Technology, Crime & Justice: The Question Concerning Technomia* (2012) was the first book in the field of Criminology and Criminal Justice to provide an overview of the implication of technology for the justice system and complements a range of applied studies in this area, including the comprehensive *UK Review of Cybercrime* conducted for the Home Office. He is joint editor of the *Handbook of Technology, Crime and Justice* (2016) and is currently preparing a monograph *The Organisation of Cybercrime*, which will provide one of the first detailed studies of the use of digital technologies by organised crime groups.

Andrew Newton is Reader in Criminology and Associate Director of the Applied Criminology Centre, University of Huddersfield. His research interests include crime and place; policy analysis and evaluation; crime prevention; detecting and explaining crime patterns; crime, technology and society; crime and the built environment; police decision-making; measuring feelings of safety; and research methods. He is specifically interested in crime and its relationship with alcohol, violence and the Night-Time Economy (NTE), acquisitive crime, safer travel, and corridors of crime. His research has been funded by a range of organisations including the Home Office, the Department for Transport, Alcohol Research UK (formerly the AERC), the ESPRC, JISC, the Railway Safety and Standards Board (RSSB), the European Regional Development Fund, the Government Office for the North West, Merseyside Police and Merseytravel Passenger Transport Authority, and Liverpool and Manchester CitySafe Partnerships.

Ugo Pagallo is a former lawyer and current Professor of Jurisprudence at the Department of Law, University of Turin (Italy). He is author of ten monographs, numerous essays in scholarly journals and book chapters, chief editor of the *Digitalica* series published by Giappichelli in Turin and co-editor of the *AICOL* series by Springer. He is a member of the European RPAS Steering Group (2011–2012), of the Group of Experts for the Onlife Initiative set up by the European Commission (2012–2013), of the Ethical Committee of the CAPER project, supported through the Seventh Framework Programme for Research and Technological Development (2013–2014), and Expert for the evaluation of proposals in the Horizon 2020 robotics program (2015). Ugo is Faculty Fellow at the Center for Transnational Legal Studies in London, UK (2013, 2008); Vice-President of the Italian Association of Legal Informatics, NEXA Fellow at the Center for Internet & Society at the Politecnico of Turin, and is now collaborating with the Joint International Doctoral (PhD) degree in Law, Science and Technology, part of the EU's Erasmus Mundus Joint Doctorates (EMJDs). His main interests are Artificial Intelligence and law, network theory, robotics, and information technology law (specially data protection law and copyright).

Andrew Puddephatt is Executive Director of Global Partners Digital and leads their work on internet human rights issues and communications policy. (www.gp-digital.org). He provides support for public interest groups in Africa, Latin America and South and South East Asia and leads the Secretariat for the inter-governmental Freedom Online Coalition (www.freedomonlinecoalition.com). He has published widely on different aspects of freedom of expression, and co-authored UNESCO's (2012) global study of privacy and freedom of expression online. He developed a major scoping study of global digital communication trends on the behalf of the Ford Foundation and acts as the facilitator for the steering committee of Best Bits, a civil society network on Internet governance and Internet rights that offers an open space for groups to discuss how to advance human rights online (http://bestbits.net). He is currently Chair of International Media Support in Denmark, Deputy Chair of the Sigrid Rausing Trust; a management board member of the European Council on Foreign Relations, and Chair of the human rights charity Global Dialogue. He was awarded an OBE for services to human rights in January 2003.

Paddy Rawlinson is an Associate Professor in International Criminology at Western Sydney University, having previously taught at the London School of Economics, Leicester and Bangor Universities in the UK. Her research has covered organised crime in Russian and Eastern Europe, with particular emphasis on political economy, corruption and child trafficking. In her current research she brings a criminological perspective to medical violence, for example, unethical clinical trials, and how this manifests within the state-corporate context.

Marcus K. Rogers, Ph.D., CISSP, CCCI, DFCP, is the Dept. Head in the Computer & Information Technology, Purdue University. He is a Professor, University Faculty Scholar, Fellow of the Center for Education and Research in Information Assurance and Security (CERIAS), Fellow of the American Academy of Forensic Sciences (AAFS) and Chair of the Digital & Multimedia Science Section of the AAFS. Dr Rogers is a former police officer with the Winnipeg Police Service, serving in the High Tech Crime Unit. His areas of research and interest cover the behavioural aspects of the deviant use of technology, cybercriminal behavioural analysis and understanding cyber terrorism.

Nigel South is Professor in the Department of Sociology and Director, Centre for Criminology at the University of Essex, England. He is an Adjunct Professor in the School of Justice at Queensland University of Technology, Brisbane. In 2013 he received a Lifetime Achievement Award from the American Society of Criminology, Division on Critical Criminology and serves as European Editor for *Critical Criminology*. With Avi Brisman, he is co-editor of the *Routledge International Handbook of Green Criminology* (2013), co-author of *Green Cultural Criminology* (2014), and both are co-editors (with Rob White) of *Environmental Crime and Social Conflict: Contemporary and Emerging Issues* (2015).

Peter Squires is Professor of Criminology and Public Policy at the University of Brighton and the author of several books, two of which directly examine gun-enabled crime and firearms control issues: *Gun Culture or Gun Control?* (2000) and *Gun Crime in Global Contexts* (2014). He has also written on the development of police armed-response policy: *Shooting to Kill? - Policing Firearms and Armed Response* (2010). He has been an advisor on firearms controls and a member of the National Council of Police Chiefs Independent Advisory Group on the Criminal Use of Firearms.

John Spink is Director of the Food Fraud Initiative at Michigan State University (USA). His 2009 MSU Packaging PhD work was on Anti-Counterfeit Strategy and his broad research expands from Food Fraud to product fraud-related business risks (including Enterprise Risk Management ERM and COSO), and a range of outreach activities that cover policy and trade issues. While conducting his research and outreach he has a full teaching load with graduate courses 'Packaging for Food Safety', 'Anti-Counterfeiting and Product Protection (Food Fraud)', and 'Quantifying Food Risk'. He is widely published in leading academic journals with important works such as 'Defining the Public Health Threat of Food Fraud'. He was a co-author on 'Food Fraud' in the first comprehensive Chinese language Food Safety textbook. His leadership positions include product fraud-related activities with ISO, GFSI Food Fraud Think Tank, and US Pharmacopeia (USP). Global activities include engagements with the European Commission, INTERPOL and Operation Opson, New Zealand MPI, and serves as Advisor on Food Fraud to the Chinese National Center for Food Safety Risk Assessment (CFSA). Outreach includes MSU's biannual 'Food Fraud MOOC' (Massive Open Online Course) that offers free training and certificates online. (www.FoodFraud.msu.edu).

Victoria Sutton is the Director of the Center for Biodefense, Law and Public Policy, the only center at a law school in the US to focus solely on issues of law and biosecurity. She is the founding editor-in-chief of the *Journal for Biosecurity, Biosafety and Biodefense Law*, an official journal of the Texas Tech University School of Law. She was awarded the Paul Whitfield Horn Professorship in 2010 and is a permanent visiting faculty member of the Vytautas Magnus University School of Law, Lithuania. She has served two Presidents, President George H.W. Bush in the White House Science Office, and President George W. Bush as Chief Counsel for the Research and Innovative Technology Administration, US Department of Transportation. She was a member of the Texas Task Force on Infectious Diseases during the Ebola events and currently serves on the Texas Task Force on Infectious Diseases addressing the Zika threat.

Richard Tewksbury is Professor of Criminal Justice at the University of Louisville. His research focuses on issues of sexual deviance, alternative sexual expressions, sex offending and societal reactions to sexual deviance/crime. His most recent publications have appeared in *Corrections: Policy, Practice and Research*, the *American Journal of Criminal Justice*, and the *International Journal of Offender Therapy and Comparative Criminology*. He is also the current editor of *Criminal Justice Studies*.

David S. Wall is Professor of Criminology at the Centre for Criminal Justice studies, School of Law, University of Leeds, UK where he researches and teaches cybercrime, identity crime, organised crime, policing and intellectual property crime. He has published a wide range of articles and books on these subjects. He also has a sustained track record of interdisciplinary funded research in these areas from the EU FP6, FP7, H2020, ESRC, EPSRC, AHRC and other funders, such as the Home Office and DSTL. David has been a member of various Governmental working groups, such as the Ministerial Working Group on Horizon Planning 2020–25, the Home Office Cybercrime Working Group, looking at issues of policy, costs and harms of crime and technology to society, and the HMIC Digital Crime and Policing working group. He is an Academician of the Academy of Social Sciences (FAcSS), a Fellow of the Royal Society of Arts (FRSA). He re-joined Leeds University in August 2015 from Durham where he was Professor of Criminology and Head of the School of Applied Social Sciences.

Martin Wasik is Emeritus Professor of Law at Keele University. He was formerly Professor of Law at Manchester, and then at Keele. As chairman of the Sentencing Advisory Panel from 1999 to 2008, he was closely involved in the development of sentencing guidelines in England and Wales. In recognition of this work he was appointed CBE. He is a recorder (part-time circuit judge) on the Midland Circuit since 2008, and he works with the Judicial College in the delivery of judicial training. He is the author of much published work including *A Practical Approach to Sentencing* (5th edn, 2014) and is co-author of *Blackstone's Criminal Practice* (annual editions since 1991).

Rob White is Professor of Criminology in the School of Social Sciences at the University of Tasmania, Australia. He has written extensively in the areas of youth studies, criminology and eco-justice. Among his recent books are *Environmental Crime and Collaborative State Intervention* (co-ed. with Grant Pink); *Environmental Harm: An Eco-Justice Perspective*; *Innovative Justice* (with Hannah Graham); and *Environmental Crime and Social Conflict* (co-ed. with Avi Brisman and Nigel South).

Introduction

M. R. McGuire

The increasingly significant role of technology within the crime and criminal justice field has been only incompletely discussed. Where it has, it has almost always been digital technologies which have served as the primary focus of analysis. Even the most preliminary scoping of the field supports this conclusion, for example, enter 'technology crime' into a search engine and the *only* results produced will be those relating to cybercrime. This 'digital myopia' has tended to obscure many other varieties of technology equally worthy of criminological and socio-legal research. In particular, it has caused us to downplay the role of tools which utilise biological or chemical processes – tools which represent both some of technology's most deadly risks and its most coercive potentials. At a time when technology is transforming almost every aspect of the crime and justice process the limited attention paid to the very wide range of technologies which can be implicated in crime and control and the failure to grasp the continuities and discontinuities here in any kind of joined up way is surprising. Technologies rarely work in isolation, but as a composite set of artefacts, processes and practices which must be understood holistically if they are to be understood at all.

There are of course many challenges which must be confronted in developing any kind of framework for understanding technology crime and control in general terms. One objection might be that this is a subject area which is simply too broad to be amenable to any kind of general overview. Technologies may be pervasive to everyday life, but they are so varied in their impacts that each variety can only be dealt with individually and any theory of them can only be limited at best. In response to this it could be argued that the ubiquity of technology within contemporary life – and in particular within the crime and criminal justice process – is so striking that to say nothing about this does not just leave us with an incomplete understanding. It also makes us vulnerable to criminal threats or coercive uses of technology which we lack the conceptual resources to understand. Another worry is that any study of technology risks being out of date before it is published. Given the seemingly relentless pace of technological change, those devices which are of interest today may well be irrelevant tomorrow. One way to address both these concerns and to bring greater order into our thinking about technology within the crime and justice field is to develop a framework which focusses upon the very basics – what (and how) technology *does* and where in the crime and justice process it does this. Secondly, by avoiding becoming too concerned about the impact of every new technological development (though some inevitably are discussed) and focusing more upon the structural constants of technologisation upon harm, offending and justice we hope to show just how far these are being radically transformed. We also hope to indicate how poorly prepared our legal and justice systems are for this creeping and all but invisible sea-change and to outline the risk we now face which is that the crime and justice process are becoming the province of specialists we no longer understand.

To this end the *Handbook* is divided into five parts, with each part corresponding to the key stages of the offending and justice process – from an original crime/act of harm, through to the institutional responses to this, to their culmination in punishment and the implications of this for future developments in technology and for justice itself. These five sections and the more specific chapters within them are detailed below.

Part I Technology, crime and justice: theory and history

Part I addresses the generally poor conceptual understanding of technology which has plagued academic thinking as much as it has the work of the policy makers and criminal justice practitioners with more immediate responsibility for managing or responding to the use of technology in the crime and criminal justice process.

In Chapter 1, 'Theorizing technology and its role in crime and law enforcement', Philip Brey addresses what has often been overlooked in discussions of technology within the CJS – how best to *define* technology. He analyses the strengths and weaknesses of one of the most important and useful theories of technology – the idea that it *extends* or, as Jérôme Goffette puts this in a later chapter, *enhances* the body and what it can do. Extensional theories of technology provide an important bridge between informal ideas of it as a 'tool' and a growing consensus around the 'post-human' approach – where technology and the body are linked together as an assemblage (an 'actant' in Latour's terminology) to form a composite human-technical agent or 'cyborg' (in Haraway's seminal formulation). Extensional theories provide an important general framework which helps clarify relations between high and low technologies. For example, they help us see that a stick 'extends' the force a hand can exert in similar (albeit less spectacular) ways that a foodmixer or a nuclear missile can do. Brey's chapter introduces some new and important distinctions for extensional theory – in particular how extension can function *collectively* as well as individually (for example, extending the power of agencies) – and provides some introductory examples of how this idea might be applied to the use and misuse of technology in criminological contexts.

Aside from conceptual gaps, another problem for the way technology crime and control has been approached within the criminal justice field (and beyond) is the assumption that this is something largely specific to high technology societies – in particular those dependent upon information technology. This misplaced assumption is challenged in Chapter 2, 'Technology crime and technology control: contexts and history', in which M. R. McGuire argues that such a narrow focus profoundly undermines our understanding of how criminals and control agents use technologies. By considering a wider historical context for technological use and misuse a more profound history begins to emerge, one that demonstrates how technology crime posed similar problems in Ancient China, the Roman Empire, Medieval Europe or indeed any other pre-modern civilisation as it now does for us. Equally well, the misuse of technology by the powerful is not just relevant in the context of contemporary challenges like surveillance but has always been central to control, whether in the Greek phalanx, interception of communications within postal systems or the use of trained State poisoners.

Part II Technology, crime and harm

In Part II a comprehensive overview of the current relations between technology and criminal behaviour is set out.

Section 1 examines our most familiar variety of technology-enabled crime – the misuse of digital devices and networks. Under its usual term of reference – 'cybercrime' – this variety of offending has often appeared to be a concept which is all but equivalent to technology crime in general. Indeed, it could be said that it is with cybercrime that much of the contemporary rationale for associating technology with crime can be derived – together with many of the recurring problems in defining how far technology can be taken to be agentic to criminal acts. A standard distinction here between *cyber-dependent* and *cyber-enabled* crime has attempted to address this by distinguishing between crimes which cannot be committed without digital technology (cyber-dependent crimes) and those which are extended or made more easy by this technology.[1]

In Chapter 3, 'The evolving landscape of technology-dependent crime', Steven Furnell sets out an up-to-date overview of cyber-*dependent* offending. Furnell considers the various ways in which familiar examples of cyber-dependent crime like malware generation and distribution or DDoS attacks have evolved over the last 20 years. He then goes on to analyse how the emergence of wholly new digital technologies – from mobile networks to the Internet of Things – may change this offending landscape. Chapters 4–11 further develop this framework by shifting attention to cyber-*enabled* offending, beginning with one of the most common examples – the use of digital networks to commit property crime, specifically fraud. In Chapter 4, 'Technology and fraud: the 'fraudogenic' consequences of the Internet revolution', Mark Button and Cassandra Cross suggest that digital technology has had a decisive impact upon fraud-related offences, one which our criminal justice institutions have failed to keep abreast of. Instead, the technology has been used with remarkable success by a range of offenders, many atypical of previous types. Part of the problem they identify is not just the new means, but the inability of our existing protections to manage it. The fault lies as much with the failures of systems for recording fraud as the technical means for committing it, with many examples simply not being recorded in crime figures. Chapters 5, 6 and 7 explore what has been the other main area of concern raised by digital technology – the problem of cyber-enabled violence (whether psychological or speech based), sexual abuse and other forms of sexual misconduct. In Chapter 5, 'ICTs and child sexual offending: exploitation through indecent images', Jo Bryce looks at a particular area of concern – the way digital technologies may have encouraged paedophiliac abuse of children by enabling the making and distribution of illegal sexualised imagery. Bryce provides an extensive overview of current research in this area and emphasises the importance of understanding the role of ICT in facilitating and detecting this form of sexual offending against children and young people.

Just as the development of printing encouraged a wider distribution of pornography, the internet has facilitated unprecedented access to sexual materials of all kinds. Aside from the more obvious criminalities attached to exploitative sexual imagery, the widened access to sexual materials this has brought has posed new challenges in deciding where we are to draw the line between harm and criminality. In Chapter 6, 'ICTs and sexuality', Andrew Denney and Richard Tewksbury examine a range of sexual practices which have been transformed by digital technology, from pornography to prostitution. They evaluate some of the typical harms which have been argued to result, such as sex addiction or the promotion of sexually inappropriate behaviours, and set out some of the emerging forms of criminality which have come with the greater access to sexual materials and behaviours provided by ICTs – such as 'sexting' and revenge porn. In Chapter 7, 'ICTs and interpersonal violence', Thomas J. Holt takes up some of the wider areas of concern about the impacts of ICT upon personal safety and well-being. Holt sets out detailed evidence around the way information technology has influenced bullying, harassment, abuse and other forms of violence which, though largely psychological or speech based, can be just as distressing as more traditional and immediate forms of physical violence.

Beyond these more direct physical concerns are other criminalities which digital networks appear to have enhanced. One emerging problem is the way that new networks have also created new kinds of marketplaces, many of them illicit, such as new drug markets, forums for the illicit exchange of body parts or the weapons trade. These in turn often enable other kinds of criminality. In Chapter 8, 'Online pharmacies and technology crime', Chris Jay Hoofnagle *et al.* look at a particular example of this – online sales of pharmaceuticals. Their research into online pharmacies found clear evidence of exploitation of the vulnerable and links to criminal groups. For example, many of the pharmacies in their sample were run by small numbers of individuals and searches for them often led to hacked websites. In Chapter 9, 'The theft of ideas as a cybercrime: downloading and changes in the business model of creative arts', David S. Wall examines one of the most frequently raised (and commonly prosecuted) examples of digital crime – the problem of intellectual and copyright theft seen in activities like piracy or illicit downloading. Wall considers the many anomalies which have arisen in attempting to control this – for example the emergence of technologies designed to automatically censor IP content in live streams which at the same time perversely incentivise deviance. Or the emergence of 'speculative invoicing' by legal firms which is corrupting the very process that they seek to protect. He highlights the curious paradox of circulation and control here, one which demands that successful management of digital IP lets ideas roam free enough for consumers to buy into it and incorporate into the next generation of popular culture whilst also needing to restrict it in order to maintain it as a commodity for profit.

The unique nature of interactions across digital networks has created an explosion of data – much of it highly personal such as birthdates, banking details, records of social life and so on. In Chapter 10, 'ICTs, privacy and the (criminal) misuse of data', Andrew Puddephatt examines how this data explosion has not only generated a range of new kinds of harms and criminalities around the use and misuse of data, but has also raised fundamentally new questions about what privacy *is* and how far it should be legally protected. Puddephatt argues that the current business model underpinning the delivery of online services, which involves data acquisition and trading at a huge and global level (what Zuboff has termed 'surveillance capitalism'[2]), poses significant challenges to the protection of privacy. Creating a new economic paradigm where personal data is 'owned' and licenced by its bearer might offer one kind of solution.

Section 2 extends the scope of technology crime by considering two technological artefacts and processes which, though far less discussed, might equally be seen to enable or enhance criminal behaviours. These involve the kind of offending enabled by chemical or biological tools. Like ICTs, such technologies can be characterised in terms of the spatial ranges at which they operate, but in contrast with them these involve 'very close' or micro-level forms of interaction – a converse set of enablings to the 'distance shrinking' forms of agency which are the hallmark of digital networks. Though their scope for criminal misuse is significant, bio-chemical related technologies represent only an entrée into the ways that the micro-realm can be misused. Of particular concern is our increasing capacity to manipulate forces and entities at the atomic and sub-atomic levels which has been only minimally discussed by criminologists to date. Some useful work has been done on the misuse of nuclear materials – whether by terrorists seeking to construct 'dirty bombs' or worse, by States and companies whose malpractice (or incompetence) have led to significant radioactive pollution at sites like Chernobyl and Fukushima. This however is only part of a much larger story. The prospect of 'nanocrime' discussed later by Susan Brenner outlines a further example of how the atomic level might be useful to criminals and emphasises why, given the potential power of technologies which allow us to manipulate the very structure of matter, there is an urgent need for better understanding of their criminological consequences.

In Chapter 11, 'Crime and chemical production', Kimberly Barrett looks at the huge range of synthetic chemicals now being produced by the chemical industry and evaluates the harms and criminalities which can be associated with their production. She considers the limited regulatory frameworks which cover potentially harmful outcomes of the chemical industry, such as increased circulation of carcinogens, toxic spills or the creation of widespread birth defects. In Chapter 12, 'Pharmatechnologies and the ills of medical progress', Paddy Rawlinson considers another variation of what might be called 'chemical crime' – the often dubious practices and products of the global pharmaceutical industry. She focusses upon two areas where the status of pharmatechnology can be considered in criminological terms – the clinical trial and the use of vaccines. Situating pharmatechnology as an increasingly powerful mode of governance, she concludes that it has also taken on a pathological element, one which often directly contradicts its perceived status as a technology which enhances public well-being.

An equally, if not more deadly set of technologies is now emerging around the use and manipulation of micro-organisms and in Chapter 13, 'Bioengineering and biocrime', Victoria Sutton suggests that the harmful potentials of bio-engineering mean that we are at the dawn of a new kind of criminal era, one where 'bio-crime' becomes as much an option for the ill-intentioned as more traditional misuses of technology. Sutton sets out a catalogue of the varied criminal potentials of bio-crime, including biological terrorism and the misuse of biological weapons. Particularly worrying are the possibilities around 'bio-hacking' where individuals armed with relatively limited (high school or university) biological knowledge can engage in basic forms of DIY bio-engineering to create wholly new bio-organisms – with wholly unpredictable consequences. It seems inevitable that some will seek to exploit this 'synthetic biology' for their advantages. Targeted crop destruction, personalised DNA weapons or the theft of genomic trade secrets of the type recently obtained in the cornfields of Iowa by Chinese agents are just some of the emerging scenarios around bio-crime.

In reflecting upon the huge range of harms, risks and criminalities which can be generated by this class of technologies it is striking how many fall within the domain of what has come to be called 'green' criminology. What is also striking is how little green criminologists have had to say about technology *per se* – even though it has often been so central to their concerns. In Chapter 14, 'Technology, environmental harm and green criminology', Rob White provides a keynote discussion for this section which aims to address this theoretical lacuna directly. White sets out a series of contexts where technology impinges upon green criminological thinking, most obviously as a source of harm to the environment which, at its most catastrophic, can extend to 'ecocide'. More positively, he also notes that technology can make significant contributions in *detecting* green crimes. He identifies a central paradox of technology – that something so often proposed as a solution to societal problems can end up generating more problems than it solves. White concludes by stressing the need for green criminology to look more carefully at the questions concerning technology and its interpretations. He suggests that a Heideggerian-style focus upon technology as *techne* may be useful for green criminologists, for this emphasises that technology has its origins in craft and natural understanding as much as it does in material, artificial devices. In particular by learning how traditional societies draw upon folk skills and folk knowledge in the use of tools, a better understanding of *techne* can contribute to a better understanding of how best to *interpret* environmental harms as well as to manage them.

Section 3 of Part II completes the survey of technologies as criminal tools by considering a still wider range of possible examples. It is telling that none of these are ever considered to count as instances of 'technology crime' even though they arguably pose as many – if not more risks than the technologies in previous sections. Weapons technologies figure as one of the more prominent examples here. As one of the most deadly of all technologies – whether the missiles dispensed by State military agencies or the knives which lie behind so much youth violence – weapons have rarely been categorised in terms of crimes and harms which are ultimately *technological*. In Chapter 15, 'Guns, technology and crime', Peter Squires examines gun crime – one of the most familiar examples of this variety of technologically enabled offending. Taking the example of the UK, a jurisdiction where stringent gun-controls are in place, Squires details how a very mixed economy of gun access has emerged. On the one hand this includes sophisticated military-issue assault weapons which have heightened the risks posed by targeted assassinations or mass shootings perpetrated by 'lone wolf' fanatics or terrorists. At the other end of the technological spectrum he points to the way in which 'junk' guns can be used to further crime – whether these are converted firearms, 'antique', collectors' weapons or simply replicas used to intimidate victims. The prospect of 3D printed handguns – whilst not practicable at present – presents a longer-term scenario for weapons technology harm within public space. In Chapter 16, 'Crime, transport and technology', Andrew Newton considers an equally fascinating (and under-researched) area of technology-related offending – the misuse of transportation and its networks. Newton notes the huge range of technologies which could be discussed under this category – from land transport, water transport, rail transport, air transport through to spaceflights, and argues that we need to think of transport *systems* as much as specific technologies in this regard. Newton notes certain parallels with cybercrime observing that there can be transport *dependent* crime – where transport itself is the target (as in damage to trains, vehicle theft or hijacking), transport *enabled* crime, where transport technologies help further certain offences (such as sexual assaults) or transport *enhanced* crime (such as the use of getaway vehicles in robbery, or the use of drones). He criticises the lack of reflection about the criminal potentials of new transport products and warns that the criminal justice system needs to keep pace with these – especially as we move into an era of 'intelligent' transport technologies. In Chapter 17, 'Food fraud and food fraud detection technologies', Roy Fenoff and John Spink draw attention to another rapidly emerging issue of technological crime and its control – the problem of 'food fraud' and the technologies being developed to combat this. Fenoff and Spink detail how traditional notions of 'food protection risks' involving food safety, food quality, and food defence have now had to be extended to include 'food fraud'. Though passing off foodstuffs as something which they are not is an age-old problem, food fraud crimes have been significantly enhanced by the increased role of technologies such as chemical processing, global transportation and new production methods; these have widened food fraud to incorporate illegal practices around additives, counterfeits, misbranding, the masking of product origins, and even intentional distribution of hazardous substances. They detail the various strategies being used to address food fraud and the way that improved technologies of detection has enhanced a focus upon prevention. Technologies here have largely focussed upon food authentication or tracing food supply chains and there is a sophisticated range of technological processes now involved in these goals, from spectroscopic analysis to the use of GPS. Fenoff and Spink concede that continued innovation in detection technologies will be necessary as criminals learn to adapt to existing controls. Consumer demand for information about the food they eat will be one driver of innovation and will enable us all to take a more central role in food authentication.

In Chapter 18, 'Consumer technologies, crime and environmental implications', Avi Brisman and Nigel South focus in more detail upon the technologies which sustain consumption practices and the crimes and harms which can result. They argue that consumption has tended to be overlooked in favour of production in discussions of technology – in spite of the fact that the demands placed by consumers upon the market drives technological development in crucial – and often harmful – ways. They examine two examples of 'technologies of consumption' –

the case of bottled water (together with the widespread use of plastics) and so-called 'hybrid' vehicles (like the Prius) which, far from promoting 'sustainable' consumption practices, often produce entirely the opposite outcome. They challenge assumptions about 'miraculous' technical fixes which simply maintain consumption practices which are, ultimately, unsustainable. They conclude by reflecting back on a paradox noted by Jock Young – a growing disillusionment with technology which is at odds with consumer societies' growing dependence upon it.

Part II of the *Handbook* concludes with Max Kilger's keynote discussion which considers how the previous chapters have defined what it is for a technology to be used in criminal ways. In Chapter 19, 'Evaluating technologies as criminal tools', Kilger assesses the kinds of continuities and discontinuities which hold when associating criminal acts with the very wide range of technologies which can produce them. The capacity of technology to generate harms on mass as well as individual scales leads him to a surprising conclusion. If traditional forms of law, justice and policing – which have centred upon individual victims – have to be revised to incorporate this 'mass' element more completely, then criminal acts as we have traditionally understood them may need to become increasingly blurred with terrorist acts. This has the benefit of making it clear that culpability for mass technological harms perpetrated by 'legitimate' (i.e. state or corporate) agents can be equated with that perpetrated by more extremist elements. However, given the current folk-devil status of terminology used in connection with terrorism, this is also an ominous prospect. For as Kilger suggests, the tendency on the part of control agents to blur organised crime with terrorism or to use technologies like mass surveillance to secure public safety suggests that such a blurring may benefit coercive social control as much as it does accountability.

Part III Technology and control

Contemporary crime control – and social control in general – have both been decisively shaped by the increased capacities provided by technology. Whether this involves the origins of forensics in the nineteenth century, new communications and transport innovations such as police radios and patrol cars or contemporary developments like CCTV, DNA profiling and anti-virus protection, there are scarcely any aspects of crime control that are not now dependent upon technological artefacts and practices. This section provides an overview of these and the many and various ways in which technology underpins contemporary formal and informal social control.

In Chapter 20, 'Crime, situational prevention and technology', Paul Ekblom draws upon his influential work in the field of crime prevention and design to address a surprising gap – a failure on the part of crime-prevention theorists to engage with any in-depth analysis of technology's function within this. Ekblom observes how the classic components of situational crime prevention (SCP) theory – opportunities, problems and solutions – are all ineluctably intertwined with technological concepts and responses and sets out a comprehensive schema for understanding the many levels at which technology is crucial. He concludes by warning that 'narrow and linear technological determinism' is of no use if crime prevention through technology is to prove effective. Rather, he counsels that it is only where there is a proper understanding of the complex social, physical and informational systems in which technology is situated that SCP will be able to move to a 'next level' in utilising it.

In Chapter 21, 'Technology, innovation and twenty-first-century policing', James Byrne and Don Hummer consider the long association between law enforcement and the use of technology. They argue that contemporary policing is almost unimaginable without the support of technologies like cars, radio communication, access to criminal record data, crime analysis units, body cameras and so on. The crucial critical challenge then is whether technology and technological innovation improves police performance or not. Whilst conceding that a full overview of this question is beyond the scope of the chapter, Byrne and Hummer utilise evaluation research around recent innovations like gunshot location software, cameras (i.e. dashboard mounted cameras) and the use of social media as an investigative tool to arrive at some provisional conclusions. They suggest that such innovations do seem to enhance the operational efficiency of police forces. They also point to evidence indicating that such innovations may enhance police legitimacy by making what they do more transparent to the public. Though potential police misuse of such technologies – whether to obscure wrongdoing or to enact it – can undermine these improvements in police–public trust, the 'fix' to this is not ultimately a technological one. Instead, Byrne and Hummer argue that it is only by addressing deep malaises within policing cultures that public concerns about police misconduct can be effectively addressed.

In Chapter 22, 'Contemporary landscapes of forensic innovation', Christopher Lawless looks at one of the key technologies used within crime control – the use of forensics. As one of the oldest and most relied upon crime control technologies, forensics has been subject to continual development and innovation. In this chapter, Lawless questions what drives such innovation. Is it the certainties of science and technology and the drive for greater 'truth' in obtaining evidence? Or is it, as Lawless suggests, a far more 'messy' process than this, something shaped by the (often competing) interests of policymakers, funders, law enforcement officials, academic researchers, forensic practitioners and commercial manufacturers. Especially problematic has been the increasing role of commercial imperatives in developing forensic technologies, for this has resulted in a focus upon costs and profits which is not always conducive to effectiveness. Such pressures threaten to impact upon public expectations around forensic technology – in particular how juries perceive their authority in criminal trials.

In Chapter 23, 'Technology and digital forensics', Marcus Rogers discusses the extension of traditional forensics into a whole new field – the digital environments where our lives are increasingly conducted. Rogers sets out a history of digital forensics from an early 'pre-science' era when digital evidence was poorly understood and overly trusted by the courts, through to a 'second generation phase when the volume of digital forensic data increased so dramatically that a need for automated tools was created'. A third generation of these tools has now emerged which does not just use automation to enhance the speed of forensic data analysis, but which can 'make decisions' about what the data means by finding patterns and associations within it. Whilst this may improve digital forensic analysis, it will also mean judges will increasingly be forced to accept the verdicts of algorithmic rather than human experts whose fallibility cannot be known (because of the lack of any defined error rate). The result has been an increased scepticism on the part of legal officials towards convicting on the basis of digital evidence, but this is unlikely to signal any diminished importance for the technology. Rather it will simply increase pressures to develop more robust and testable digital forensic tools, together with greater education for legal officials.

In Chapter 24, 'DNA and identification', Carole McCartney guides us through the pros and cons of DNA profiling – a technology which has become so fundamental to identifying the guilty (and clearing the innocent) that it is sometimes called 'the technology of justice'. McCartney acknowledges the many successes of DNA profiling technologies but also reminds us that the kind of infallibility now associated with them was once granted to fingerprint evidence. She details a number of cases where profiling has failed spectacularly and discusses the role of various national DNA databases in furthering social control. One indicator of this she points to are the attempts to retain the DNA of innocent citizens on such databases (often long after they have been cleared of any suspicions of criminality) and the associated 'function creep' which occurs when criminal justice technologies are extended beyond their normal brief. Whether the persuasive power of DNA in convincing trial juries of guilt or potential developments such as 'DNA photofits' and universal databases which retain the DNA of every citizen will ultimately culminate in a situation where 'genetic justice' outweighs traditional varieties of justice remains to be seen. But McCartney reminds us that it is essential to be clear, not only that criminal justice agents (such as the police) are not scientists, but also that DNA remains a marginal tool when considering the full spectrum of criminal investigation techniques.

Technology also shapes social control in more informal ways, and in Chapter 25, 'Visual surveillance technologies', Richard Jones provides a detailed survey of the way that contemporary surveillance and control have been hugely expanded by the advent of visual technologies, in particular CCTV. Jones traces the origins of such technologies from early advances in optics and developments in photography to the burgeoning field of contemporary technologies which enable visual scrutiny such as ANPR, satellite imaging or body-worn cameras. He considers the convergence of these technologies and the impacts upon policing and social control this is having. He concludes by observing that, whilst such technologies also enable what has been called 'sousveillance' – the capacity for ordinary citizens to record activities of social control agents – this capacity remains asymmetric, with the State and power elite still very much in control of who can observe whom.

In Chapter 26, 'Big data, predictive machines and security: the minority report', Adam Edwards further develops our understanding of the emerging structures of extra-legal control by examining some of the impacts the transition to a data-rich 'big data' society is having upon them, in particular the belief that better, more integrated data will ultimately underpin a development once envisaged by the writer Philip K. Dick – systems which can 'predict' future crimes. Edwards counsels against such overly simplistic assumptions. Instead he suggests that a more non-linear approach such as that adopted within the UK Justice Matrix (which models criminalisation via interactions between social attitudes and criminal justice data) might offer a more fruitful direction. He also notes how 'big data' is not just likely to contribute to wider projects of social control than crime prevention, but may well enhance criminal activity itself – a point which seems rather to have been forgotten by big data enthusiasts.

In Chapter 27, 'Cognitive neuroscience, criminal justice and control', Lisa Claydon considers how the development of neuroscience and brain scanning technologies such as MRI are beginning to challenge the process of law as it has been traditionally construed. She considers the various contexts in which neuroscientific evidence might be permissible in the courtroom and goes on to consider the wider social control aspects of developments here, in particular, the claims advanced by 'neurocriminologists' that brain structures might predispose individuals towards criminality and so could be used to intervene even before crimes have been committed. This version of the 'pre-crime' scenario is mirrored in other potentially disturbing control scenarios – for example, research being conducted by the US Defense Agency DARPA in memory retrieval which could have implications beyond any use as an investigative crime tool. She concludes that the benefits of neuroscience and its technologies will need to be balanced by appropriate ethical frameworks.

This section culminates in a seminal discussion of the continuities between technology and social control, by Gary T. Marx, one of the most important contemporary theorists in this field. In Chapter 28, 'The uncertainty principle: qualification, contingency, and fluidity in technology and social control', Marx, together with Keith Guzik, sets out a masterly overview of the many ways in which technology can be deployed for the purposes of social control. They start their account with the tale of Odysseus and the Sirens – a tale they interpret as demonstrating our faith that technical interventions (the wax and rope Odysseus uses to block communication and movement) will succeed in controlling external risks. Marx suggests that this faith has now become manifested in a 'maximum security society' where varying modalities of (technical) hard and soft control determine who, what, where, and how we can be. The problem is that this faith in rational-technical solutions overlooks a fundamental uncertainty principle which is endemic to technology and which will always undermine such solutions. Marx and Guzik offer a simple example of this – the repeated problems encountered by the Mexican government when implementing its REPUVE programme – an attempt to combat crimes involving automobiles (such as car thefts or kidnappings) by using RFID technologies to create a national car registry. They meticulously detail the various factors which undermined this programme and derive a taxonomy of uncertainty which encompasses uncertainties of function; consequence; context and environment. These reflect a wider range of 'fallacies of technology' which Marx details at the end of his chapter.

Part IV Technology and the process of justice

Perhaps one of the most significant areas where technology is transforming the criminal justice process lies at its culmination – in the courtroom and the possible sanctions which follow. In this part, the *Handbook* sets out some of the fundamental impacts technology is now having upon the judicial process – from the trial through to sentencing and punishment. Can we trust technology to 'deliver truth' in the way the trial demands? Do the regulatory powers of technology now go beyond that of the law?' Could it even be said that technology is now eroding enlightenment concepts of justice altogether?

In Chapter 29, 'Establishing culpability; forensic technologies and justice', Simon Cole looks at the contribution of technology to establishing culpability. He describes the historical ideal of 'mechanical objectivity', and uses this to evaluate a number of technologies which have been thought to offer more reliable, less subjective forms of 'truth' within the trial process. Cole begins with photographic evidence which, in spite of the apparently neutral basis it offers for establishing facts, has been involved in some controversial rulings, for example, the acquittal of police officers involved in the Rodney King beatings, in the face of detailed video evidence of the assault. He highlights similar questions about the level of faith placed in fingerprinting technologies. Drawing upon his own widely cited work which has challenged its reliability, Cole reminds us that any identification made upon the basis of fingerprints is not categorical, but probabilistic – and therefore subject to error. And, like all forensic science techniques, it is subject to the problem of 'confirmation bias' – the tendency to select those pieces of evidence which support or confirm what one already believes. He casts similar doubts upon polygraph (lie detector) testing, observing that in spite of the illusion of 'mechanical truth' that polygraph devices convey, they always ultimately require subjective human actors to operate them and to interpret their results. Similar qualms apply to neuroimaging technologies which, as Lisa Claydon details in Chapter 27, many believe offer still more powerful ways of 'reading' the mind. Cole concludes by echoing Carole McCartney's challenge to DNA profiling discussed in an earlier chapter, suggesting that no matter how elevated its status as our latest 'truth machine', interpretation remains a central and unavoidable problem. Interpretation, together with the role of context will continue to delimit the infallibilities of any future truth machine we come to rely upon. In Chapter 30, 'Technology-augmented and virtual courts and courtrooms', Fredric Lederer shifts focus towards the courtroom itself and details the huge number of ways in which its operations are now being transformed by our new technological order. On one level this relates to the court's administrative functions – for example, the use of electronic filing and case management systems to manage caseloads. On another level, such transformations go to the very heart of the court's central purpose – to conduct criminal and civil trials. Here we see an increasing need to cope with advances in electronic communications and an ever greater dependence upon data-oriented forms of litigation. Thus there has been a surge in technology-based evidence such as mobile-phone recordings, drones, body cameras, forensic materials or social media data. As a result, the technology-augmented courtroom has evolved to facilitate new ways of displaying such evidence – such as enhanced visual displays which may encompass 3D (holographic) or immersive virtual reality displays of crime scenes. A longer-term outcome of this is the potential emergence of the virtual courtroom, where there are no humans present at all and forms of remote testimony and representation become the norm.

Chapters 31 to 34 deal with the aftermath of the trial process and the impacts of technology upon how those determined to be guilty of offences are disposed of. In Chapter 31, 'Computer-assisted sentencing', Martin Wasik looks at the significant impacts digital technology is having upon the sentencing process, such as enhanced storage and retrieval of case notes, in particular the idea of an 'algorithmically' driven sentencing process, which could – in principle – be determined by a computer, rather than a judge. Though he accepts that sentencing follows guidelines which can resemble an algorithmic process, Wasik is sceptical as to whether sentencing could ever be fully automated, as too many of the steps require human judgement. In Chapter 32, 'The technology of confinement and quasi-therapeutic control: managing souls with in-cell television', Victoria Knight looks at the ways in which technology has changed the way in which confinement functions as a punishment, focusing specifically upon its use to pacify or control prisoners. Drawing upon data from a detailed study of television use in prisons, Knight sets out an ethnography of the way that television has been used as a multi-layered form of reward, punishment, therapy and control within the contemporary penitentiary. She concludes by considering how information technology is now being used to augment this complex matrix – with the result that the very nature of confinement itself may now be changing.

In Chapter 33, 'Punitivity and technology', Simon Hallsworth and Maria Kaspersson set out a general framework for interpreting the relationship between punishment and technology. Hallsworth develops his earlier work on punitivity by considering how punishment has always been a practice enhanced by technological artefacts and processes. Together, they trace a historical trajectory of this relationship beginning with its origins in the pre-modern order, where technology not only served as a way of maximising pain delivery but also as a method for delivering 'punitive spectacles'. They go on to outline the enlightenment shift towards a use of technology to 'civilise' (and sanitise) punishment – a shift perfected as much in the gas chambers of the Holocaust or devices which dispose of the excrement of US execution victims more efficiently as it has been in more 'humane' punitive methods. Hallsworth and Kaspersson conclude by suggesting we may be entering a new order of punitivity where contemporary communications technologies are uneasily conjoined with the resurgence of pre-modern punitivities which celebrate torture or which turn brutal beheadings into globalised spectacles.

Given the faith in technology as a regulatory tool a vital question which then arises is how should technology *itself* be regulated? In Chapter 34, 'Public and expert voices in the legal regulation of technology', Patrick Bishop and Stuart Macdonald discuss some of the major difficulties here. They explore such problems in the context of three regulatory regimes directed specifically at technology: UK law around Human Fertilisation and the use of embryos; EU law around chemicals; and the scope of international law (specifically the Basel Convention) in controlling hazardous waste. They draw particular attention to the increased dependence upon a new kind of (technically based) authority within such regimes – the need to use 'expert witnesses' (who understand the technology in a way juries and judges cannot). They conclude by suggesting that there is little in the way of public participation in the regulation of technology at present. Instead, each of their case studies indicates that such regulation is often discretionary and issues largely from within the regulated industry. This inevitably brings the legitimacy of existing legal controls over technology into question.

In Chapter 35, 'The force of law and the force of technology', Mireille Hildebrandt inverts the questions raised in the previous chapter by asking, not how *we* should regulate technology, but how technology now regulates *us*. In the keynote discussion for this section she describes how the norms which drive our legal order are becoming subsumed within what might be called 'technological norms'. Such norms stand outside the function of the legislator, but emerge, unexpectedly, from interactions between devices, technical infrastructures and humans. On the plus side, this means that regulation is loosened from coercive (human) power. On the less positive side it means that such regulation is harder to discern and therefore harder to contest. Maintaining the current system of 'technology neutral' law (i.e. laws which ignore or overlook the direct role of technology), simply means that further (technology specific) laws must then be introduced to maintain legal authority. It also means that, unless legal agents like barristers rethink how they conduct themselves within legal-technic frameworks they will simply become redundant. The prospect of improvised, poorly conceptualised technology law, or legal experts becoming beholden to the dictates of technology, does not bode well for better justice.

Part V Emerging technologies of crime and justice

As suggested above, one of the fundamental problems in developing any kind of analysis of technology – whether in relation to crime, criminal justice or beyond – is the constant emergence of new technologies. Not only do these challenge our certainties, they also present continual new opportunities for harm. In this section, two examples of some of the more significant emerging technologies for crime, control and justice are discussed. This is supplemented by more general considerations about the extent to which *any* emerging technology and the enhancements it offers can be effectively regulated.

In Chapter 36, 'Nanocrime 2.0', Susan Brenner examines the emergence of nanotechnology – a technology which utilises atomic or molecular structures for various ends – and the 'second industrial revolution' many have predicted this will bring. The huge scope of its possible applications are likely to bring equally varied criminal risks – risks we are singularly unprepared for. Though she warns that we should not stretch the analogy too far, Brenner imagines an unfolding of 'nanocrime' which resembles what happened with cybercrime, a key precedent for how new technologies can create consequences for which we are unprepared. For example, like networked computers, nanosystems might be the subject of 'target crimes' where the technology is used to attack itself. One scenario here is where nanoparticles injected into a body to perform a medical task are 'hacked' by other nanoparticles for the purpose of harm. Or, in a parallel with DDoS attacks, nanoparticles might be used to 'deny' service – i.e. to block other nanoparticles from performing their legitimate function. She also outlines 'tool' crimes where, just as computers can facilitate offences like theft, nanotechnology is used to facilitate crimes against the person or property. For example, nanoparticles might be used to deliver a fatal poison, or as a way of counterfeiting goods. And because forgery would happen from the atomic level upwards, not only could documents or designer items be perfectly copied, so too could unique items like the Mona Lisa. At the extreme we can even imagine the creation of 'intelligent nano-entities' which, once unleashed, commit the kinds of autonomous offences discussed by Ugo Pagallo in the following chapter. In Chapter 37, 'AI and bad robots: the criminology of automation', Ugo Pagallo confronts the increasing challenges being posed by robotics to criminal law. At present the possibility that machine agents may become so sophisticated that they can, autonomously, 'commit crimes' stands outside existing legal norms since they do not possess consciousness and cannot therefore exhibit 'mens rea' or intention to commit crime. Pagallo considers a range of potential 'robotic' offences and details how the ongoing Japanese 'Tokku' experiment – the creation of 'living zones' for testing problems in human and (simple) robotic interaction – could be extended to manage such offences. Pagallo concludes by conceding that the existing interests of sovereign states means it is unlikely that we will see the emergence of any effective international system of robotic law in the near future. And, whilst it also remains unlikely that we will be able to attribute mens rea to robots anytime soon, other scenarios – such as the loss of control of robotic agents like those used by the military, need to be taken more seriously.

In Chapter 38, 'Technology, body and human enhancement: prospects and justice', Jérôme Goffette considers a way around the challenges posed to our socio-legal structures by the constant emergence of new technologies. His suggestion – which echoes themes proposed earlier by Philip Brey – is to characterise the impacts of any new technology in terms of the way that it *enhances* existing human capacity. Goffette discusses how more familiar enhancements like cosmetic surgery, the use of mood modifiers or 'performance enhancement' in athletics have already posed problems for law and regulation. As human-technical enhancement becomes increasingly prevalent, a new regulatory field defined by what he terms 'anthropotechnics' will develop. However, whether we are ready as a society to effectively regulate our enhancements remains to be seen.

The *Handbook* concludes with a keynote discussion by Albert Borgmann, one of the most important contemporary theorists of technology, whose work has developed the seminal Heideggerian 'question concerning technology', to ask whether technologies can not only be 'useful' – but good. In Chapter 39, 'Technology and justice', Borgmann questions whether we have developed sufficient conceptual resources to deal with the kinds of challenges to justice being posed by technology seen in this *Handbook* and elsewhere. His conclusion is that we have not. He notes (with honourable exceptions) the absence of any developed body of thought about technology until the 1970s and a parallel failure to build consistently upon key theories of justice – in particular the normative view developed by Rawls in his ground-breaking *Theory of Justice*. One result of these epistemic lacunae has been what he calls 'shallow justice' – a world where we have lost sight of what justice might be and where we have allowed our conceptions of it to become colonised by machinery. A world of shallow justice is one where 'glamorous surfaces' prevail and technological 'miracles' like electric power or running water are so normalised as to have become mundane. In such a world we expect machines to provide us with security and questions of justice only arise when our technologies malfunction or fail. Borgmann warns that we cannot have deep justice in a shallow world – yet nor can we dismantle technology to achieve this. What can be done is to relegate technology to the background of our lives in favour of practices which promote engagement over the indifference which so often comes with shallow justice.

Notes

1 See McGuire and Dowling (2013) *Cybercrime: A Review of the Evidence*, HOS/11/047, Home Office. Reference is also sometimes made to 'cyber-assisted' crime, though how ICT is agentically different here than with cyber-enabled crime is not always clear.
2 See Zuboff (2015) Big Other: Surveillance Capitalism and the Prospects of an Information Civilization. *Journal of Information Technology* 30, 75–89. doi:10.1057/jit.2015.5.

Part I

Technology, crime and justice

Theory and history

Theorizing technology and its role in crime and law enforcement

Philip Brey

1. Introduction

Virtually any human practice or institution that exists in society has been seriously affected by the use of technology. However, the role of technology in society is little understood. Decision-making processes concerning technology are often driven by false and naïve images of technology and its interaction with society. A sound understanding of technology and its role in human practices and institutions requires specialist knowledge. Such knowledge is developed in several fields in the social sciences and humanities, including science and technology studies (STS), philosophy of technology, ergonomics, management of technology, and others. In this chapter, I will review these fields, with particular reference to approaches and theories in STS and the philosophy of technology, and I will report on their main findings regarding the role of technology in society. This review will be carried out in Section 2.

A second aim of this chapter is to introduce a particular theory of technology and to apply it to an analysis of the role of technology in crime and law enforcement. The theory in question, *extension theory*, provides a powerful perspective for understanding the role of technology in society, in particular the powers that technological artifacts give to their users (individuals, groups and organizations) and how these powers help them further their aims. After introducing extension theory in Section 3, I will use it in an analysis of the use of technology in crime and law enforcement in Section 4. This analysis intends to show how new technologies have extended the powers of both criminals and law enforcement, and how these new powers play out against each other. The chapter ends with a concluding section that reviews the chapter's main findings.

2. Understanding technology

Technology is a key driver of social change. The industrial revolution and the technologies it has yielded have radically transformed a great number of societies. Modern societies are shaped by modern technology, which has caused revolutionary changes in work, the economy, social organization and the way of life. The ever increasing pace of technological innovation promises more such changes for the future. Because of the profoundly influential role of technology in every sector and institution of society, whether it is healthcare, education, government, law, business, the arts, or any other, an adequate understanding of these sectors nowadays requires an understanding of the role and impact on them of technology.

Several academic fields have emerged to study technology and its role in society, including the nature of technology, its design, development, and historical evolution, its dependence on social and economic developments, its impact on individuals, society, and the environment, and the governance of technology. Most centrally, technology is studied in the interdisciplinary field of science and technology studies (STS), sometimes also called science, technology and innovation studies (STIS) (Hackett *et al.* 2007; Sismondo 2009). This field aims to study the evolution of the institutions and practices of science and technology in society, their impacts on society, and their governance and regulation. STS emerged in the 1970s, and has established itself as a field with departments and programs around the world, as well as specialized conferences and journals.

STS is a loosely knit field, with a wide variety of contributing disciplines, such as sociology, history, cultural studies, anthropology, policy studies, urban studies and economics. It is fair to say, however, that it is dominated by sociological, historical and governance approaches. Sociological approaches are concerned with an understanding of institutions, modes of organization and practices of science and technology, and their interactions with other sectors of society. Historical approaches study the historical development of science and technology and the social context in which its institutions and actors operated.[1] Governance approaches, finally, investigate how science and technology and its impacts can be governed or managed in society, by government agencies as well as by the private sector and other relevant actors (Smits *et al.* 2010).

Several other fields exist that study some aspect or dimension of technology and its role in society. They are either closely associated with STS, or have developed along a largely separate trajectory. They include, amongst others, economics of technology, ergonomics, material culture studies, internet psychology, design studies, technology management, innovation management, and the philosophy of technology. I will now further introduce one of these fields, the philosophy of technology. The philosophy of technology is a field that, next to STS, aims to develop theories of technology and society of a general nature. It is also the source of the theoretical perspective on technology that I will be developing in Sections 3 and 4 of this chapter.

The philosophy of technology emerged in the course of the twentieth century, largely as a response to the major impacts of technology on society during that century, many of them perceived to be negative by philosophers. These include the unprecedented destruction and carnage during the First and Second World Wars, the rationalization and automation of work which brings alienation, the rise of a consumer culture that promotes materialism and shallow consumption, and the destruction of the environment due to modern technology. Philosophers like Martin Heidegger, Jacques Ellul, Herbert Marcuse and Günther Anders developed extensive critiques of modern technology based on these perceived social consequences.

Since the 1990s, the broad critiques of these classical authors have made way for more pragmatic and empirically oriented philosophical studies that aim to better understand specific technologies and their role in society. This new orientation has been called the empirical turn in the philosophy of technology (Kroes and Meijers 2000; Brey 2010a). Studies after the empirical turn tend to be more descriptive, more informed by empirical results from their own case studies and from empirical studies in other fields, more focused on concrete technologies and social problems, less pessimistic about technology, and less determinist about technological change and impacts of technology. In addition to this descriptivist empirical turn, the 1990s also saw the emergence of ethics of technology as a major field of study, including new specializations like computer ethics, nanoethics and ethics of design. Due to the empirical turn, the philosophy of technology has moved closer to STS, a field from which it has incorporated methods and ideas. In spite of this alignment, the philosophy of technology has largely retained a status separate from that of STS.

What is technology?

Let us now turn to the insights into technology and society that can be gained from these fields. A proper understanding of these matters should begin with an adequate understanding of the concept of technology. Unfortunately, definitions of technology in dictionaries and professional textbooks differ widely in their meaning and scope. Technology is variably defined as a process of making things, a type of knowledge for making things, or the very things that are made (Reydon 2012; Li-Hua 2009). In addition, many definitions hold that technology involves the application of science to useful ends, but there are also definitions that make no reference to science and define technology as the application of knowledge for practical ends, where presumably this knowledge can be of any kind (Mitcham and Schatzberg 2009). Moreover, some definitions hold that technology involves the manipulation of matter for practical ends, whereas others merely hold that it involves the creation of practical value, so that presumably applications of the social sciences would also qualify as technology ("social technology" or "soft technology") (Jin 2011).

Most commonly, however, technology is related to engineering, and definitions incorporate three elements: technology is something that involves the manipulation of matter, has an orientation towards practical ends, and has a basis in the sciences. Technology can, according to this conception, be defined as the development, through the application of science and mathematics, of physically defined or implemented means (devices, systems, methods, procedures) that can serve practical ends. This assumes that technology is the mere application of science, whereas convincing arguments have been made that, although science is applied in engineering, engineering also involves the application and creation of unique engineering knowledge, which is a highly formalized, evidence-based and systematic type of knowledge (Vincenti 1990). So probably the definition should be amended to hold that technology is developed through the application of science, mathematics and engineering knowledge.[2]

This definition of technology characterizes it as a process that is carried out by engineers and applied scientists. However, in other contexts, "technology" appears to refer instead to the means created in this process, as when it is said that a store sells office technology, or to a form of knowledge, namely knowledge of the techniques for the production of artifacts and other useful means. Technology is hence defined as a process, a set of objects, or form of knowledge, depending on context (Mitcham 1994).

Subclasses of technology are often defined in relation to engineering branches. The main branches of engineering are chemical engineering, civil engineering, electrical engineering and mechanical engineering, and they produce chemical, civil, electrical and mechanical technologies, respectively. More specialized branches, such as software engineering and biomedical engineering, also yield corresponding technologies. Note that engineering branches are sometimes defined in terms of technological features (e.g. mechanical engineering), and sometimes in terms of application areas (e.g. environmental engineering).

For a consideration of consequences of technology for society, the most relevant aspects of it are not technological knowledge or technology development processes carried out by engineers, but rather the technological products that result from engineering: devices, systems, procedures and methods that are used for practical purposes and through this use, impact society. The products of technology development are of two basic kinds: objects (tools, devices, systems) and instructions for carrying out processes (procedures, methods). The latter includes methods and procedures for processes like food irradiation, iron smelting and automated assembly of laser diodes. The focus of theorists has, however, mostly been on the objects produced by engineers: *technological artifacts*, which are the physical products of technology.

The notion of a "technological artifact" or "technological product" is usually associated with electronic and technologically complex devices, such as computers, mobile phones and automobiles. However, there are many human-made products that are the result of science-based technological production processes but that are not themselves machines or electronic devices. A nylon jacket is not generally seen as a piece of technology; it is not a device with mechanical parts and does not run on electricity. However, it is a technological artifact in the sense that it is a product of advanced technological processes of production of nylon polymers out of chemical compounds which are subsequently mechanically melted, spun and fused together into wearable items. Most products that people buy in stores are technological artifacts in this sense, including hammers, cups, tables and even many foods. It is often difficult to distinguish between technologically produced artifacts and craft-based artifacts; a wooden chair could be of either type. For this reason, theorists sometimes for convenience refer to any human-made artifact as a technological artifact, although in other contexts it may be appropriate to restrict the scope to engineering-based products, machines or electronic devices.

The most exemplary class of technological artifacts consists of (electronic) machines. A *machine* is a tool that consists of parts with different functions and that receives energy from a source and transforms it, through the interoperation of its parts, into useful actions. In the evolution of technology, one sees the evolution from simple tools and devices to mechanical machines (machines powered by natural forces like wind and water, or human and animal labor), followed by electrical machines (powered by electricity, or converting mechanical to electric energy), electronic machines (electric machines with active electrical components) and computing machines (electronic machines in which the store and flow of electrons is interpreted as information).

New technologies and fields of engineering emerge regularly, and new technological products and methods, often incorporating multiple technologies, are developed at an ever increasing rate. The constant emergence of new technologies means that technologies exist in society at different stages. A rough distinction can be made between entrenched and emerging technologies. *Entrenched technologies* are established technologies with long histories that have yielded many products that are broadly used in society. Automotive technology is an example. *Emerging technologies* are new technologies that have not yet yielded many concrete products and applications, if any, and that are still mostly defined in terms of fundamental research. Examples are nanotechnology (although it is yielding more applications every year) and synthetic biology. This distinction is important because in assessing future impacts of technology on society, one should not only consider innovations within entrenched technologies, but also potential applications of emerging technologies.

The mutual shaping of technology and society

Majority support has emerged in STS and philosophy of technology for a number of positions concerning the way in which technology is developed and interacts with society, which I will now attempt to spell out. A first point of agreement is that technology does not evolve in a deterministic fashion, but rather evolves along a path that is strongly influenced by a wide variety of social factors. Technology is *socially shaped*, meaning that technological change is conditioned by social factors, and technological designs and functions are the outcome of social processes rather than of internal standards of scientific-technological rationality (Mackenzie and Wajcman 1999).

The social shaping thesis denies the technological determinist idea that technological change follows a fixed, linear path that can be explained by reference to some inner technological 'logic,' or perhaps through economic laws. Instead, technological change is radically underdetermined by such constraint, and involves technological controversies, disagreements and difficulties, that involve different actors or relevant social groups that engage in strategies to shape technology according to their own insights. The thesis implies that technologies will evolve differently in different societies (assuming that they are developed separately in them), and that society can exert strong influence over the way in which technologies develop.

Against a linear path model of technological change, proposals have been made for a *variation and selection model*, according to which technological change is multidirectional: there are always multiple varieties of particular design concepts; some die, and others, which have a good fit with social context, survive (e.g. Pinch and Bijker 1987; Ziman 2000; Brey 2008). Many studies have been performed to show how users, regulators, civil society organizations and other actors affect the development and design of technology and the way in which technological artifacts are interpreted and used (Bijker *et al.* 1987; MacKenzie and Wajcman 1999; Oudshoorn and Pinch 2003).

Theorists also agree that the meaning and use of technologies is not pre-given, but that instead technology has *interpretive flexibility*, meaning that technological products can be interpreted and used in different ways (Pinch and Bijker 1987). Different actors in society will attribute different meanings, functions and uses to technologies, and engage in processes of social negotiation. If such negotiations are successful, they come to a close and interpretive flexibility diminishes because the technology is said to stabilize, along with its co-evolved meanings and social relations. This implies the embedding of the technology in a stable network consisting of humans and other technologies, and the acceptance of a dominant framework on how to interpret and use the technology. It implies that the contents of the technology are 'black-boxed' and no longer a site for controversy, although black boxes can be opened at any time. The history of technology shows how artifacts like the telephone, the Internet and the automobile take on particular functions and societal roles that vary from time to time and from place to place.

A second major point of agreement between scholars of technology is that *society is shaped by technology*, meaning that technologies shape their social contexts. This is converse to the earlier claim that technology is socially shaped. The claim is not just that particular uses of technology have an influence of society, but also that technological innovations and products themselves shape their social contexts. This claim goes against a naïve conception of technology as a neutral means that does not itself affect society, and according to which it is the choices made by its users that determine its impacts on society (Feenberg 2002). The technological shaping thesis holds that technological innovations and products engender a multiplicity of functions, meanings and effects that always, often quite subtly, accompany their use. Technologies become part of the fabric of society, of its social structure and culture, and have a deep and lasting impact on social structure and human behavior (Callon 1987; Latour 1992). Naturally, such an impact could also involve effects on criminality and responses to this.

The idea that society is technologically shaped means, according to many scholars, that technology seriously affects social roles and relations, political arrangements, organizational structures, and cultural beliefs and practices. Scholars have argued that technical artifacts sometimes have built-in political consequences (Winner 1980), that they may contain gender biases (Wajcman 1991), that they may subtly guide the behavior of their users (Sclove 1995; Latour 1992), that they may support or hinder the realization and implementation of moral values and norms (Brey 2010b), that they may presuppose certain types of users and may fail to accommodate non-standard users (Akrich 1992) and that they may modify fundamental cultural categories used in human thought (Akrich 1992; Turkle 1984).

Let us consider some examples. Bruno Latour (1992) has aimed to show how mundane artifacts, like seat belts and hotel keys, induce their users towards certain behaviors. A hotel key, for example, has heavy weights attached to them in an attempt to compel hotel guests to bring their key to the reception desk upon leaving their room. Langdon Winner (1980) has argued that nuclear power plants require centralized, hierarchical managerial control for their proper operation, unlike, for example, solar energy technology. In this way, nuclear plants shape society by requiring a particular mode of social organization for their operation.

Richard Sclove (1995) has pointed out that modern sofas with separate seat cushions define distinct personal spaces, and thus work to both respect and perpetuate modern Western culture's emphasis on individuality and privacy, in contrast to Japanese futon sofa-beds, for example. Sherry Turkle (1984), finally, discusses how computers and computer toys affect conceptions of life, since due to their intelligent behavior, they lead children to reassess the traditional dividing lines between 'alive' and 'not alive' and hence to develop a different concept of 'alive'. Most authors would not want to claim that technological artifacts have inherent powers to affect such changes. Rather, it is technologies-in-use, technologies that are already embedded in a social context and have been assigned an interpretation, that generate such consequences.

To conclude, many scholars have proposed that the development of technologies, and therefore these technologies themselves, is shaped by society, and that technological innovations and products shape social and organizational structures and arrangements, and human practices and beliefs. Taken together, these claims can be summarized to say that society and technology co-shape or co-construct each other. Technological change is moreover held not to be linear but to proceed through a quasi-evolutionary process of variation and selection. In addition, technologies have interpretive flexibility, implying that their meanings, functions and capabilities are constantly open to renegotiation by users and other social actors.

3. The extension view

Extension theories of technology provide a perspective on technology and its relation to human beings that has significant explanatory power and that can be usefully applied in the analysis of the role of technology in society. Extension theories are based on the idea that technological artifacts can be understood as means that build upon, and extend, capabilities of the human body and mind; they are extensions of the human organism. An analysis of technology and its function in society should take into account how technological artifacts extend, enhance, augment, supplement or replace human body parts and organs and their capabilities. Classical versions of extension theory have been put forward in the nineteenth and twentieth centuries by thinkers like Marshall McLuhan, Arnold Gehlen, Henry Bergson and Ernst Kapp. In recent years, philosophers, including myself, have attempted to further systematize and improve extension theory (Brey 2000, 2005; Lawson 2010; Heersmink 2012; Steinert 2015).

The most influential classical version of extension theory has been put forward by Marshall McLuhan. In his famous *Understanding Media*, subtitled *The Extensions of Man*, McLuhan depicts technologies as extensions of human beings that extend beyond the skin and various body parts and organs of humans (McLuhan 1964). McLuhan made a basic distinction between two kinds of extensions: those of the body and those of cognitive functions, including the senses, the central nervous system, and the brain. Extensions of the body are technologies that augment or aid parts of the body that are used for acting on or protecting oneself from the environment, or regulating bodily functions. These include our limbs, teeth, skin, and bodily heat control systems. He theorized that during the mechanical age, technologies were developed that extended our bodies: bows, spears and knives extended our hands, nails and teeth; the wheel extended our feet; and clothing extended the function of bodily heat control and protection of the skin.

During the electrical age, McLuhan claimed, our senses and central nervous system were extended. He analyzed media as extensions of the senses, particularly those of sight and sound. The radio and telephone are long-distant ears, and visual media are extensions of vision. Electric media, in particular, are extensions of the information processing functions of the central nervous system in that they take over functions of information management, storage and retrieval normally performed by it. McLuhan foresaw another age of digital computer technology in which creative cognition and higher thought, or "consciousness" is also extended and taken over by technology. He envisioned that these cognitive functions would be automated and translated into information functions performed by digital computers.

McLuhan's account provides an interesting perspective on technologies as the externalization and partial replacement of human bodily and mental faculties. However, when considered as comprehensive theory of technology, there are problems with it. The main problem is that McLuhan holds that every technological artifact is functionally similar to some human organ and serves to augment or complement this organ's function. There are, however, many technological artifacts that appear to have functions that are not similar to any found in human organs. For example, electric lighting serves to emit light, but there is no human organ that has a similar function. Likewise, magnets, ionizers, roads and nuclear reactors do not seem to have functional analogues in the human body or mind.

If not all technological artifacts extend human organs, can extension theory still be maintained? I have proposed a version of it which categorizes all technological artifacts as extensions of the person, but not necessarily in a one-on-one relation with human faculties (Brey 2000). In my account, persons come equipped with a set of means, or capabilities, that is provided to them by the unaided body and mind. Technological artifacts add to, or extend, this set of means. Some do this by replicating, strengthening or enhancing means that are already present in the human body and mind. Others do this by providing qualitatively new means to people, such as the means to illuminate one's environment, to fly, or to magnetize iron.

I have also emphasized, in my account, that by extending our means, technologies extend our ability to attain ends, that is, to realize our intentions. Linking technology to human intention is important because it helps explain the evolution and application of technology: technological artifacts are developed and used because they are believed to help unaided persons realize their intentions. My extension theory can therefore be summed up as follows: all technological artifacts extend the set of naturally given means (i.e., human bodily and mental faculties) by which human intentions are realized, either by replicating or augmenting human faculties, or by introducing qualitatively novel capabilities.

I have since come to believe that this account, though true, is not complete, because it ignores the fact that people are always situated in an environment, and in virtually any environment, people have more means available to them than just their bodily and mental faculties. Environments contain various elements that can also serve as means: natural objects such as rocks and plants, technological artifacts and infrastructures, and people and animals that could serve as means to one's ends. So in any given situation, the total set of means available to a person does not just consist of those means provided by her body and mind and any technological artifacts that she has in possession, but also includes the means that she has access to in her environment. Therefore, any technological artifact that is provided to a person does not merely extend the means provided to her by her faculties, but should rather be seen as extending the set of means that is defined by the human faculties and environmental elements that are available to her for use. The added value of a technological artifact for a person depends on its added functionality relative to the already available set of means, and relative to the intentions that the person wants to realize.

The fact that things in the environment other than technological artifacts can serve as means to realize our intentions also serves to broaden the notion of extension. Many things can serve to extend the abilities of the unaided body and mind, not just technological artifacts. Natural objects are frequently used as means to realize our intentions: a rock is used to crack a nut, a tree is used to provide us with wood for building a home. Other people can serve as a (temporary) means to ends, when we ask someone to pass the salt or when we hire someone to do work for us. In addition, we have devised modes of social organization and social expression that help us realize our intentions. A language, for example, is a shared system of symbols and conventions that help us communicate. Social roles and relations confer means to people that help them reach ends. For example, someone's appointment as teacher or accountant gives him the ability to contribute to society with his skills and to earn an income. The establishment of a friendship between persons provides them both with means to have rewarding interactions and to get emotional and practical support.

Organizational and institutional structures, procedures, and regulations can also directly or indirectly provide people with means that they otherwise would not have, including means to reach joint ends, and to help direct one's actions and make them more effective. In addition, acquired knowledge and skills also serve to extend the means or abilities that are available to persons. Acquired physical or social skills, like maintaining good balance, typing, driving, speaking French, solving differential equations, or greeting people, can help people further one's intentions. The acquisition of (factual) knowledge has a similar role, as it can help people make decisions, guide actions, and function as a useful resource for others. It can be concluded that there are many things other than technological artifacts that can serve to extend the set of means available to people to realize their intentions: natural objects and structures, people and animals, social conventions and procedures, social and organizational structures, and acquired knowledge and skills. Any consideration of the function of a technological artifact for a person should consider how it adds to all of these means that are already available to him or her.

Extension theory also needs to take into account that although the primary role of technological artifacts is to serve as extensions for human users to further their intentions, technological artifacts are also found in different roles. Notably, sometimes technological artifacts are mere elements in the environment that are not used by persons to serve their ends. For a particular person, a car or computer system owned by someone else may merely be an encountered object that does not serve any ends for her, although it may indirectly benefit the realization of her ends or harm them. Technological artifacts can also serve as (part of) the target or goal of actions, rather than as a means by which goals are reached. They may be objects that someone seeks to observe, study, damage, dispose of, repair, develop or acquire. In such actions, the artifact does not serve as a means for realizing goals, but rather as an object at which one's means are directed.

Types of technological extensions

So far, I have only considered cases of technology use in which artifacts are used by one person in order to further his or her intentions. However, artifacts can also be used by collectives and organizations. For example, the roof, central heating system and sewer system of an apartment building do not serve as means for one particular person, but rather serve as means for a group or collective: the inhabitants of apartments in the building. Likewise, a computer network or website used by an organization is not in place to serve the needs of any particular person, but rather to serve the needs and objectives of the organization as a whole. The organization should be seen as an agent with its own intentions and objectives, and technologies are used by it to further these objectives.

A distinction therefore needs to be made between technological artifacts that serve as *individual extensions* and those that function as *collective* and *organization extensions*, of groups and organizations. Collective and organizational extensions are also used by individuals, and help further their intentions. For example, the sewer system in an apartment building is also used by its inhabitants considered as individuals, and employees of a firm make individual use of its computer network, even when their intention in doing so is to further the objectives of the firm. However, such individual uses derive from the collective or organizational function of the artifact or system.

Note, also, that artifacts may serve different functions for different persons, collectives or organizations at the same time, simultaneously extending them in different ways. For example, an organization's computer network may simultaneously serve the individual ends of employees when they use it for private messaging. Seat belts protect drivers from physical harm, but simultaneously also help protect insurers from financial loss. A pair of jeans extends the skin of the wearer, but the logo it displays also furthers the intentions of the manufacturer by serving as an advertisement of the brand. The latter two examples also show that an artifact can further an agent's intentions without the agent being an end-user of the artifact. Both the insurance company and the jeans manufacturer have however taken prior actions to ensure that these artifacts serve as means that further their ends, even if used by others.

Another useful distinction is between *physical* and *social extensions*, where physical extensions extend physical capabilities of agents and social extensions extend social capabilities. A physical capability is one that enables non-social physical actions such as cutting, transporting, repairing and cooking. A social capability enables social actions, which are actions that have as their primary intent to impact social phenomena, such as communicating, becoming friends, starting an organization, taking up a new profession, firing someone, and purchasing something. Social actions may centrally involve physical objects, as when someone purchases a dishwasher, but what makes such action social is that they are about altering the socially constituted status of such objects, which, in the case of the dishwasher, is its property relation to an owner.

Technological artifacts may either function as physical extensions, as social extensions, or as both. Mechanical tools and devices, such as drills, automobiles, heaters, and binoculars, generally serve as physical extensions. They extend physical capabilities of persons, including those of physical manipulation and alteration, movement, temperature regulation and perception. Technologies that function as social extensions depend on socially constructed meanings and statuses that have been granted to them. As philosopher John Searle (1995) has argued, artifacts can acquire powers that do not (solely) derive from their physical properties but rather from collectively imposed *status functions*. Status functions are socially agreed meanings or statuses that are bestowed upon things, events or persons in virtue of which they are able to perform certain social function or roles. Dollar bills, for example, are able to function as a medium of exchange not because they have inherent physical powers to perform this function, but because people have collectively assigned the function of being a medium of exchange to them. Without this assigned function, dollar bills could not function as such. Likewise, a text editor can only function as a text editor because people have agreed that the squiggles manipulated by them constitute words and sentences that have meanings. Otherwise, it would only be a device for manipulating meaningless squiggles.

Technological artifacts that constitute social extensions include information and communication technologies, as well as other artifacts that are used to serve social, cultural, organizational and institutional roles. A clock, for example, serves to display time, where time is itself a social convention. Technological artifacts can also serve multiple physical and social functions simultaneously. For example, an expensive car can have both a physical function of transportation, and a social function of signaling status and wealth.

Within the class of physical and technological extensions, further distinctions can be made between several subclasses. Within the class of physical extensions, artifacts can be distinguished that have a function for physical action (e.g. cutting, welding, refining, heating), for transportation (e.g. automobiles, airplanes), for perception (e.g. microscopes, telescopes), for regulation (e.g. thermostats, cooling systems), and for protection (e.g. clothing, walls, landmines) against others. Within the class of social extensions, some artifacts serve a communication function (e.g. telephone), others an information function (e.g. web browser), and yet others may serve functions of play, friendship, work collaboration, and creative expression, as well as various organizational and institutional functions.

A special class is constituted by so-called *cognitive artifacts* (Brey 2005; Norman 1991) that extend cognitive capabilities of persons and organizations. A cognitive artifact is an artifact capable of displaying, storing, retrieving or operating upon information, and having this as its primary function. Computers are, of course, cognitive artifacts, but so are thermometers, newspapers, clocks, and measuring rods. Cognitive artifacts mainly belong to the class of social extensions, because most of them depend on socially defined conventions to assign meanings to signs and symbols. Arguably, one could also include devices like telescopes and hearing aids that enhance perceptual information and that have been categorized as physical extensions. Cognitive artifacts extend human cognitive function by extending human cognitive abilities like memorization and information storage, interpretation, reasoning, calculating, and conceptual thought.

In Brey (2005), I argued that computers initially qualified as cognitive artifact, since their function in the 1960s and 1970s was to process and produce information within scientific research, education, administration, and defense, amongst other fields. Since the introduction of computers with graphical user interfaces and multimedia capabilities in the late 1970s and early 1980s, and the World Wide Web in the early 1990s, computers acquired functions that are noncognitive and non-informational, such as playing computer games, listening to and composing music, watching movies, and communicating with other users. I call these activities noncognitive and non-informational even though they involve cognition and information processing. I do so because their primary function is not to process or display of information, but rather to enable activities in which performance of these information functions is not the central purpose.

So what is extended by computers when they adopt these noncognitive functions? All kinds of abilities. Computers have become extremely versatile, multifunctional devices that can perform or support many very different actions and processes, such as drawing, playing music, banking, chatting, operating external devices like heating systems and refrigerators, and live streaming of camera images from remote places. Almost any social action can be performed with or mediated by a computer, like chatting, paying, becoming friends, and signing a contract, and some physical actions, like playing music, watching movies, and turning one's heater up, can also be performed with computers. Computers have therefore become very powerful technological extensions that replicate in virtual, informational form the capabilities of ordinary technological artifacts and other extensions.

4. Crime and technological extensions

Extension theory can help illuminate the role of technology in crime and law enforcement. I will now perform a general analysis of this role from the point of view of extension theory. The analysis proceeds as follows. First, I will identify the main classes of actors in crime and law enforcement, their intentions, and the relation between these intentions. Next, I will survey and classify technologies used by these actors and their role in realizing the actor's intentions. Finally, I will provide a historical analysis of the role of technology in crime and law enforcement that aims to assess how technologies strengthen the powers of criminals and law enforcement.

For the first step in the analysis, we will consider the main classes of actors in crime and law enforcement, their intentions, and the relations between these intentions. The dialectic of crime and law enforcement involves three main classes of actors: criminals, human crime targets and law enforcement officers (cf. Cohen and Felson 1979). Let us consider them in turn. Criminals are individuals that habitually or incidentally commit crimes. Crimes are acts committed in violation of law and punishable by the state. A criminal normally has two kinds of intentions relative to an impending criminal act. First, he normally has the intention to commit the act in question. I say "normally", because there are crimes in which the intent to commit the act in question was not there, for example in involuntary manslaughter. I will not consider such cases here. I will also restrict my discussion to cases in which the perpetrator knows that the act he intends to commit is unlawful, even though there are cases of criminal behavior in which the perpetrator is unaware that the act he commits is unlawful. In most cases, however, the perpetrator has an intention to commit an act that he knows to be unlawful. This intention is normally accompanied by a second intention, which is to escape punishment for committing this act.

The second class of actors, within which I will make no further differentiations, is that of law enforcement officers, by which I mean individuals sanctioned by the state to enforce the law, particularly criminal law. This includes anyone with a state-sanctioned duty to prevent, detect and investigate crimes and apprehend and persecute criminals. The intentions of law enforcement officers directly oppose those of criminals. They aim to prevent unlawful acts from being committed, and once such acts have been committed, they intend to ensure that the perpetrator does not escape punishment for committing the act.

A third and final class that should be considered is that of people and organizations that are targets of crime, and therefore its potential victims. Persons are potential targets of so-called crimes against persons, such as murder, aggravated assault, rape and kidnapping. Both persons and organizations are potential targets of crimes against property, such as burglary, larceny, robbery, vandalism, arson and fraud. Crime targets have an interest and therefore normally an intent to prevent themselves from being a victim of crimes against persons and against property. In other words, they have an intent to prevent criminal acts from taking place with them or their property as the target. They may also have an interest in the punishment of the perpetrators once a criminal act has been committed against them, but this is not invariably the case.

In comparing the intents of criminals, law enforcement officers and human crime targets, we see, first of all, that they have opposing intentions regarding the perpetration of criminal acts: criminals have an intent to commit criminal acts, law enforcement officers have an intent to prevent criminal acts from being committed, and crime targets have an intent to prevent criminal acts from being committed against them and their property.[3] Second, criminals and law enforcement officers have opposing intents regarding punishment: criminals have an intent to escape punishment for criminal acts committed by them, whereas law enforcement officers have an intent to ensure that such punishment is meted out.[4]

These intentions by the three classes of actors induce them to acquire and use technological and non-technological extensions that are believed by them to make these intentions come true. These extensions may include technological artifacts, knowledge, skills, persons who provide assistance and support, social conventions and procedures, and social and organizational structures. For example, to ensure a successful robbery of a liquor store, a criminal may mobilize and use relevant technological artifacts (for example, a gun and a getaway car), knowledge (of the layout of the liquor store, its security features, and its contents), skills (in using a gun, or in breaking open a cash registry), accomplices (other persons who help in the holdup or drive the getaway car), and social conventions and procedures (for example, hand signals that can be made to accomplices, and agreed-on procedures for carrying out the robbery).

Technological extensions for and against crime

The technologies sought by criminals for their criminal pursuits come in two main types: technologies for committing crimes, and technologies for escaping punishment.[5] (Some technologies may fall into both categories.) What the relevant technologies are in these categories depends on the type of crime. Let us first consider technologies for committing crimes. These are technological artifacts and processes that enable a criminal act to be committed at all or that enable it to be committed more reliably, effectively and safely. (Premeditated) crimes against persons, such as murder, aggravated assault and rape, may include technological artifacts to help perpetrators get near the victim, to threaten or restrain them, to hide the crime from bystanders and observers so that it is not interrupted, and to commit the physical act that constitutes the crime. Crimes against property, such as burglary and robbery, include technologies to get close to the crime target and get access to it, to avoid detection while doing so, to coerce or restrain persons on the premises, and to transport the property to a safe place. Crimes against society, finally, such as prostitution, narcotics violations, gambling and weapons law violations, constitute a diverse group that involves a variety of technologies for committing the crime.[6]

Artifacts for committing crimes include the following:[7]

- Artifacts for getting information about crime targets and for planning a crime: cameras, eavesdropping equipment, hacking software, etc.
- Artifacts for getting access to crime targets: burglary tools like crowbars, lock picks, bolt cutters, torches, and explosives.
- Artifacts to disable or block security systems and cameras: wire cutters, signal jammers, hacking tools, etc.
- Artifacts for threatening, restraining or coercing persons: knives, guns, handcuffs, sedatives, artifacts such as nude photos that are used for blackmail, etc.
- Artifacts for maintaining communication between criminals: mobile telephones, computers, etc.
- For cybercrime: software tools and services for breaking into computers, causing damage, and stealing data, such as exploit kits, botnets, fake fingerprint kits and Trojan horses.

Artifacts for escaping punishment include the following:

- Artifacts for preventing, altering, destroying, concealing or falsifying physical evidence (fingerprints, tire tracks, bodies, DNA, camera images, physical evidence of break-in, etc.): gloves, masks, cleaning products, disposable phones, data-erasing software, encryption software, web anonymizers, artifacts that are used to manufacture or constitute false evidence (e.g. fabricated DNA evidence, false Internet alibis, a murder weapon planted on a victim to suggest suicide), etc.
- Artifacts for a secure escape from a crime scene or from a police pursuit: fast cars, weapons, bullet-proof vests, mobile phones, door reinforcements (in hideouts), alarm systems, false digital trails, etc.

Criminals have increasingly powerful technologies at their disposal to effectively commit crimes and escape punishment for them. The technologies used in law enforcement have exactly the opposite aim: preventing crime and ensuring that criminals are punished for their crimes. Their first intent, that of preventing crime, is aided by artifacts that include the following:

- Databases with information on crime targets, (potential) criminals, and the possession and the distribution of (illegal) means for committing crimes, which help law enforcement determine where to mobilize its resources to prevent crime.
- Artifacts that aid live surveillance and investigation of (potential) criminals, their preparatory activities, and crime targets: surveillance cameras, wiretaps, satellite feeds, live maps, helicopters, etc.
- Artifacts for hardening crime targets against crime (to the extent that this is seen as the responsibility of law enforcement): locks, fences, alarm system, firewalls, etc.

The intent of punishing crime is aided by artifacts such as the following:

- Artifacts to support the detection, processing and matching of physical evidence: fingerprint kits, cameras, databases (for convicts and their fingerprints and DNA data, licensed drivers, stolen and found property, vehicles, and so forth), etc.
- Artifacts for effectively locating and interrogating witnesses and suspects: GPS trackers, databases, lie detectors, etc.
- Artifacts for the pursuit and apprehension of suspects: fast police cars, weapons, bullet-proof armor, two-way radios, arrest cars, thermal imaging cameras, battering rams, handcuffs, etc.

Crime targets, finally, invest in artifacts and systems that prevent them from becoming the victim of crimes. They do not normally engage in surveillance of potential criminals and criminal activity, but focus on the immediate protection of themselves and their property. Artifacts used therein may include the following:

- Artifacts for physically protecting oneself and one's property and for deterrence: weapons, locks, fences, firewalls, reinforcements, safe rooms, etc.
- Artifacts for surveillance of crime targets and for signaling criminal activity: surveillance cameras, alarm systems, etc.

The employment of these technological artifacts by criminals, law enforcement and human crime targets should be understood as part of their strategies to optimize the set of means (extensions) that they have available to realize their intentions. Since these intentions are diametrically opposed between criminals on the one hand and law enforcement and human crime targets on the other, we see that an increase in one party's capability of realizing its intentions corresponds with a decrease in the opposing party's capability of realizing its intentions. This decrease is either achieved by employing means that can damage or destroy the opposing party's means (e.g. wire cutters that incapacitate an alarm), means that leave opposing means intact but that create conditions to render them less effective or irrelevant (e.g. a bullet-proof vest that partially neutralizes the advantage of a gun, a signal-jamming device, gloves that make fingerprint collection irrelevant), and means that otherwise add new capabilities that cannot be effectively countered at the time (e.g. a Trojan horse that can pass through any firewall).

The process of adding better and better means that iteratively negate the advantage of the opposing parties or exploits their weaknesses makes the process of extension resemble an arms race (Ekblom 1997). As criminals equip themselves with better, more powerful technological and other extensions to support their criminal intent, law enforcement and human crime targets respond in kind, and vice versa. This includes the iterative development and deployment of technologies to thwart the advantage of the opponent: criminals build Trojan horses to infect computers, which are then countered by firewalls, which are then in turn countered by Trojan horses that can disable or circumvent firewalls, etc. What one sees in countries in which crime is rampant is that criminals have won the arms race between criminals and law enforcement: they have the best technologies, modes of organization, information, training for skills, and other extensions. This, however, is not to say that systemic corruption, social deprivation, or high levels of inequality might not also contribute to high crime societies.

The evolving impact of technology on crime and law enforcement

To expand the above analysis, I will now briefly consider the evolution of technologies for crime and law enforcement in historical perspective, from the point of view of extension theory. Throughout history, technological advances have altered the targets of crime, the kinds of crimes that are committed against them, the ways in which they are committed, and the way in which law enforcement operates to prevent and respond to these crimes. Prior to the late nineteenth century, the means for protecting crime targets mostly consisted of physical devices and structures like locks, fences, and weapons, and surveillance was limited to observation by law enforcement officers walking the streets and limited keeping of physical records that were difficult to search and reproduce. There was no developed forensic science, and so persecution of crimes based on physical evidence was often difficult, which frequently made it difficult to persecute crimes at all. In this context, societies tended to choose harsh methods of punishment, often in public, as a deterrent and used torture to extract confessions. In this way, the lack of other means of law enforcement to prevent crime and persecute criminals could be partially redressed.

Crime and law enforcement entered a new stage in the late nineteenth and twentieth centuries, with the introduction of electric and electronic devices and the establishment in the early twentieth century of forensic science. Electricity and electronics gave electronic communications, which benefited both criminals and law enforcement, and electronic surveillance, which mainly benefited the latter. Advances in forensics gave law enforcement powerful new tools for apprehending and persecuting criminals. On balance, the new technologies benefited law enforcement more than they did criminals. However, the increasing complexity of society, especially of cities, and the rise of organized crime (which constitutes a type of social extension of the individual criminal) partially offset this advantage. Many of the technological products of the twentieth century also became targets of crime themselves, because of their value and expense. Products like cars, televisions and stereos became prime targets of theft.

With the introduction of computers in society in the twentieth century, computer crime became an issue. Its real breakthrough came in the 1990s, with the emergence of the Internet as a mass medium. The emergence of the Internet and computer networks made it possible for criminals to break into computers from a distance and therefore made computer crime much easier than it was before their emergence. Computer crime, or cybercrime, is crime in which the computer is used as a target or tool. The emergence of information technology has not only resulted in computer hardware becoming a target of crime (e.g. computers, smartphones, tablets) but also software and data have become a frequent crime target (e.g. digitally encoded credit card numbers, electronic money, downloadable digital content, personal information records and personal identities). When persons and organizations are the target of cybercrime, or when society is the target, the computer functions as a tool rather than the target (Wall 2007; Brenner 2010). Computers are nowadays used as tools for a wide variety of crimes, such as harassment, extortion, identity theft, drug trafficking, and child pornography.

As argued in Section 3, computers reproduce in electronic form many social actions, objects, practices and institutions, from communicating to listening to music to banking. The Internet has become itself a kind of society, an electronic environment in which people work, play, form social relations, and engage in many of the practices that they also engage in in physical environments. It is for this reason that cybercrime should not be understood as just a new type of crime on top of all the others. Because the institutions and structures of society have moved for a significant part online, the Internet has become the new frontier for a broad range of previously existing criminal pursuits, including theft, fraud, extortion, harassment, vandalism, prostitution, child pornography, drug trafficking, and many more.

Cybercrime is more difficult to combat than ordinary crime. In cybercrime, criminals are able to maintain a large distance between themselves and the crime scene (McGuire 2009; Yar 2013). They have strong means for remaining undetected and erasing their tracks, such as anonymizers, strong encryption software, and darknets. Criminal law is usually running behind the latest technological developments and cybercrime often crosses international legal borders, which requires strong international cooperation between law enforcement agents. Internet service providers often do not want to cooperate to protect the privacy of their clients, and laws that protect civil liberties may limit the abilities of law enforcement to do surveillance. Law enforcement has in recent years made heavy investments in means to fight cybercrime, which include setting up monitoring systems and databases to track and investigate suspicious activity over digital networks; strong collaborations between jurisdictions and the private sector; gaining access to national and global surveillance programs for espionage and combating terrorism such as ECHELON; and efforts to outlaw software tools and practices that aid cybercrime. Nevertheless, cybercriminals are still at an advantage.

Although law enforcement is still at a disadvantage on the playing field for cybercrime, information technology has given it powerful tools to combat ordinary (non-computer) crimes. The creation of large, searchable databases for individuals, past crimes, and forensic evidence, and the introduction of data mining techniques have provided law enforcement with powerful tools to track and sometimes even predict crime. Security cameras, sometimes equipped with facial recognition and other tracking software, have provided new sources of evidence and detection. Information technology has also made new forensic techniques possible, such as odor-detecting devices that can find buried human bodies and imaging technologies for examining hidden features in documents. On the whole, information technology may have given law enforcement the advantage in ordinary crime, but it has opened up a new frontier of cybercrime where criminals have the advantage.

5. Conclusion

This chapter had two objectives: to review the major theories and approaches for the analysis of the role of technology in society, and to introduce a particular theory within this category – extension theory – and apply it to an analysis of the role of technology in crime and law enforcement. The review was carried out in Section 2. There, I introduced and reviewed the fields of science and technology studies (STS) and the philosophy of technology, which are arguably the two fields most concerned with providing general theories of technology and its relation to society. I reviewed discussions of the nature of technology and technological artifacts, and of the ways in which society impacts the development of technology, and technology impacts social practices and institutions. Some main findings were that technology development is not a linear, deterministic process but involves an evolutionary process of variation and selection that is shaped by social factors and interests, that technological artifacts have interpretive flexibility and have different meanings and uses for different actors, and that technologies are not neutral but shape their social context, including human behaviors and practices, social and political arrangements and institutions, and human culture.

In Sections 2 and 3, I introduced the extension theory of technology, as a powerful account for the analysis of the role of technology in society, and I applied it to the analysis of crime and law enforcement. The extension theory holds that technological artifacts should be analyzed as means that extend the ability of persons, groups and organizations to carry out their intentions or goals, along with non-technological extensions such as skills, knowledge, human resources, and organizational structures. Analyses of the role of technological artifacts as extensions, or added means, for realizing intentions can provide for assessments of the added value of technology in certain practices and of assessments of the way in which technologies transform and improve these practices.

In the analysis of the role of technology in crime and law enforcement, it was found that criminals on the one hand and law enforcement and human crime targets on the other have diametrically opposed objectives. Technological artifacts used by these groups, such as guns, Trojan horses, surveillance databases and alarm systems, were analyzed as means that are added to an existing set of means and that, in relation to this set, may or may not enhance the ability of one group to reach its objectives and weaken the ability of the opposing group. I ended with a brief analysis of different stages in the evolution of the role of technology in crime and law enforcement, including the introduction of electric and electronic technology and the establishment of forensic science in the late nineteenth and twentieth centuries, and the introduction of information technology and cybercrime in the late twentieth century. I hope to have shown that extension theory provides a powerful tool for the analysis and assessment of the role of technology in crime and law enforcement that invites further studies within its framework.

Notes

1 In Chapter 2 of this volume, M. R. McGuire offers one of the few socio-legal/criminological accounts of the historical development of technology crime and control.
2 Another problem with the definition is that it does not allow for the creation of technological means that are not based in science, even though there seem to be cases of technological invention that are not science-based, such as James Watt's invention of the steam engine, which was hardly based on scientific principles. This example shows that it is difficult to arrive at a definition of technology that

captures all of our intuitions. However, even if Watt's invention did not, twentieth- and twenty-first-century technology invariably finds its basis in science, mathematics, and formalized engineering knowledge.

3 McCord (2004) offers some good discussions of the role of intentions in offending.

4 There are additional, less influential, intentions that shape the extensions of the three classes of actors. For example, it is an intention of society as a whole that law enforcement does not abuse its powers. The extensions of law enforcement are to some extent shaped to prevent such abuse. For example, some jurisdictions nowadays require police to wear bodycams, motivated in large part for their ability to expose police abuses.

5 There is not an extensive literature on the technologies of crime and control, but notable works include Byrne and Rebovich (2007), McGuire (2012), and Marx (2016).

6 The classification of crimes into crimes against persons, against property and against society is taken from the classification system used in the United States (National Incident Based Reporting System).

7 The notion of artifact is conceived broadly to include both technological and non-technological products.

References

Akrich, M. (1992). The De-Scription of Technical Objects, in Bijker, W., and Law, J. (eds), *Shaping Technology/Building Society: Studies in Sociotechnical Change*. Cambridge, MA: MIT Press.

Bijker, W., Hughes, T., and Pinch, T. (eds) (1987). *The Social Construction of Technological Systems: New Directions in the Sociology and History of Technology*. Cambridge, MA: MIT Press.

Brenner, S.W. (2010). *Cybercrime: Criminal Threats from Cyberspace*. Santa Barbara, CA: Praeger.

Brey, P. (2000). Theories of Technology as Extension of Human Faculties, in Mitcham, C. (ed.), *Metaphysics, Epistemology, and Technology. Research in Philosophy and Technology*, vol. 19. London: Elsevier/JAI Press.

Brey, P. (2005). The Epistemology and Ontology of Human-Computer Interaction. *Minds and Machines*, 15, 3–4: 383–98.

Brey, P. (2008). Technological Design as an Evolutionary Process, in Vermaas, P., Kroes, P., Light, A. and Moore, S. (eds), *Philosophy and Design: From Engineering to Architecture*. Springer (Netherlands).

Brey, P. (2010a). Philosophy of Technology after the Empirical Turn. *Techné: Research in Philosophy and Technology*, 14, 1: 36–48.

Brey, P. (2010b), Values in Technology and Disclosive Computer Ethics, in Floridi. L. (ed.), *The Cambridge Handbook of Information and Computer Ethics*. Cambridge: Cambridge University Press, 41–58.

Byrne, J.M., and Rebovich, D.J. (2007). *The New Technology of Crime, Law and Social Control*. Monsey, NY: Criminal Justice Press.

Callon, M. (1987). Society in the Making: The Study of Technology as a Tool for Sociological Analysis, in Bijker, W., Hughes, T. and Pinch, T. (eds), *The Social Construction of Technological Systems: New Directions in the Sociology and History of Technology*. Cambridge, MA: MIT Press.

Cohen, L.E. and Felson, M. (1979). Social change and crime rate trends: A routine activity approach. *American Sociological Review*, 44, 4: 588–608.

Ekblom, P (1997). Gearing up against crime: A dynamic framework to help designers keep up with the adaptive criminal in a changing world, *International Journal of Risk, Security and Crime Prevention*, 2, 4: 249–65.

Feenberg, A. (2002). *Transforming Technology: A Critical Theory Revisited*. Oxford: Oxford University Press.

Hackett, E., Amsterdamska, O., Lynch, M. and Wajcman, J. (eds) (2007). *The Handbook of Science and Technology Studies*, 3rd edition. Cambridge, MA: MIT Press.

Heersmink, R. (2012). Defending extension theory: A response to Kiran and Verbeek, *Philosophy & Technology*, 25, 1: 121–8.

Jin, Z. (2011). *Global Technological Change. From Hard Technology to Soft Technology*, 2nd ed. Bristol, UK and Portland, USA: Intellect Books.

Kroes, P. and Meijers, A. (eds) (2000). *The Empirical Turn in the Philosophy of Technology*. Amsterdam: JAI.

Latour, B. (1992). Where are the missing masses? The sociology of a few mundane artifacts, in Bijker, W., and Law, J. (eds), *Shaping Technology/Building Society: Studies in Sociotechnical Change*. Cambridge, MA: MIT Press.

Lawson, C. (2010). Technology and the extension of human capabilities, *Journal for the Theory of Social Behaviour*, 40, 2: 207–23.

Li-Hua, R. (2009): Definitions of Technology, in Olsen, J.K.B., Pedersen, S.A., and Hendricks, V.F. (eds), *A Companion to the Philosophy of Technology*. Chichester: Wiley-Blackwell, 18–22.

MacKenzie, D., and Wajcman, J. (eds) (1999). *The Social Shaping of Technology*, 2nd edn, Open University Press.

Marx, G.T. (2016). *Windows into the Soul: Surveillance and Society in an Age of High Technology*. Chicago: University of Chicago Press.

McCord, J. (2004). *Beyond empiricism: Institutions and Intentions in the Study of Crime*. New Brunswick, N.J: Transaction Publishers.

McGuire, M. (2009). *Hypercrime: The New Geometry of Harm*. Oxford: Routledge-Cavendish.

McGuire, M. (2012). *Technology, Crime & Justice: The Question Concerning Technomia*. Oxford: Routledge, Taylor & Francis.

McLuhan, M. (1964). *Understanding Media: The Extensions of Man*. New York: The New American Library, Inc.

Mitcham, C. (1994). *Thinking through Technology: The Path Between Engineering and Philosophy*. University of Chicago Press.

Mitcham, C., and Schatzberg, E. (2009). Defining Technology and the Engineering Sciences, in Gabbay, D.M., Meijers, A., and Woods, J. (eds), *Philosophy of Technology and Engineering Sciences*, Vol. 9. Amsterdam: Elsevier, 27–65.

Norman, D. (1991). Cognitive Artifacts, in Carroll, J. (ed.), *Designing Interaction: Psychology at the Human-Computer Interface*. Cambridge: Cambridge University Press.

Oudshoorn, N., and Pinch, T. (eds) (2003). *How Users Matter: The Co-Construction of Users and Technology*. Cambridge, MA: MIT Press.

Pinch, T., and Bijker, W. (1987). The Social Construction of Facts and Artifacts: Or How the Sociology of Science and the Sociology of Technology Might Benefit Each Other, in Bijker, W., Hughes, T., and Pinch, T. (eds), *The Social Construction of Technological Systems: New Directions in the Sociology and History of Technology*. Cambridge, MA: MIT Press, 17–50.

Reydon, T.A.C. (2012). Philosophy of Technology, in Fieser, J. and Dowden, B. (eds), *Internet Encyclopedia of Philosophy*.

Sclove, R. (1995). *Democracy and Technology*. New York: Guilford Press.

Searle, J.R. (1995). *The Construction of Social Reality*. New York: Free Press.

Sismondo, S. (2009). *An Introduction to Science and Technology Studies*, 2nd ed. Toronto: Wiley-Blackwell.

Smits, R., Kuhlmann, S., and Shapira, P. (eds) (2010). *The Theory and Practice of Innovation Policy. An International Research Handbook*. Cheltenham, UK: Edward Elgar.

Steinert, S. (2015). Taking stock of extension theory of technology, *Philosophy & Technology*, 1–18. doi: 10.1007/s13347-014-0186-3.

Turkle, S. (1984). *The Second Self: Computers and the Human Spirit*. New York: Simon & Schuster.

Vincenti, W. (1990). *What Engineers Know and How They Know It: Analytical Studies from Aeronautical History*. Baltimore: Johns Hopkins University Press.

Wajcman, J. (1991). *Feminism Confronts Technology*. Cambridge, UK: Polity Press.

Wall, D.S. (2007). *Cybercrime: The Transformation of Crime in the Information Age*. Cambridge, UK: Polity Press.

Winner, L. (1980). Do artifacts have politics? *Daedalus*, 109: 121–36.

Yar, M. (2013). *Cybercrime and Society*. London: Sage Publications.

Ziman, J. (ed.) (2000). *Technological Innovation as an Evolutionary Process*. Cambridge: Cambridge University Press.

2

Technology crime and technology control

Contexts and history

M. R. McGuire

Just as productive uses of technology are all but contiguous with (if not directly responsible for) the emergence of human culture and society so – presumably – are its misuses. Yet whilst terms such as 'technology-dependent crime' or 'technology-enabled crime' have become increasingly widespread, the concept of technology itself remains rather under-discussed within the criminal justice field. Indeed the idea of technology crime seems only to have emerged in the post-1990s period and has come to refer, almost exclusively, to one very specific variety of offending – that perpetrated via digital networks like the Internet (so-called 'cybercrime'). Our understanding of the ways in which technology has been used to impose forms of social control has, by contrast, been more developed. Here key contributions such as Marx's account of the role of technical-production methods in shaping social order (2007), Heidegger's assault upon technological instrumentalism (1949) or Marcuse's claim that 'technological rationality' underpins contemporary governance (1982) have all helped shape our awareness of this uneasy relationship.

Of course, any well-founded historical overview of technology crime or control will be dependent upon the history of technology and its influence upon society, and here there have been more structured lines of thought. We are all familiar with the idea that socio-historical change can be ordered in terms of technological development by way of periodisations such as the 'stone age', 'iron age', 'industrial age' or our more recent 'information age'. Elsewhere, commentators like Singer *et al.* (1954), or Pacey (1991), have attempted more developed schema but it is Lewis Mumford (1934) who set out the most sophisticated version of technological history, defining three key 'eras': the Eotechnic (1000–1750, handicrafts, low technologies), the Paleotechnic (1750–1900, industrialisation) and the Neotechnic (1900–2000).

Perhaps the most common (and simple) way of ordering socio-technological development has been to distinguish between 'pre-industrial' and 'industrial' societies.[1] Though this distinction is one that I will occasionally invoke in the following discussion, it comes with a health warning. For it risks fostering an assumption that technology crime and control are largely features of industrial (i.e. 'technologised') societies. Worse, it also implies that *concepts* of technology crime emerged with the onset of industrialisation – even though it has only been during the post-industrial, 'information society' stage that a consistent association between technology and crime has been suggested. In this chapter, I will argue that there is a far longer and livelier engagement with technology by lawbreakers and lawmakers than current thinking might suggest. And whilst it is undoubtedly true that the dramatic technological shifts of the eighteenth and nineteenth centuries had a decisive impact both upon crime (from urban disorder to speeding offences) and our contemporary criminal justice systems (from professional police forces to forensic science methods), there is also a rich, much longer history of technology crime and control waiting to be excavated.

Pre-industrial technological crime: patterns and portents

Associations between the use of technology and criminality are arguably as intuitive as they are ancient. In my book *Technology, Crime and Justice: The Question Concerning Technomia* (2012) for example, I noted how mythologies across the world have often seen the very *acquisition* of technology as a criminal or deviant act – usually manifested in 'fire-theft' tales. Conversely, Lewis Mumford argued that crime itself could be seen as a kind of technology – a 'tool' which makes illicit desire fulfilment far easier.

> Robbery is perhaps the oldest of labor saving devices [...] to obtain women without possessing charm, to achieve power without possessing intelligence, to enjoy the rewards of consecutive and tedious labor without lifting a finger in work or learnt a single useful skill.
>
> *(Mumford 1934: 83)*

But in spite of such apparently pervasive associations, definitive examples of technological crime, technological regulation or technological misuse in the pre-industrial era are hard to come by. There seem to be several reasons for this. One immediate and obvious problem relates to the many cultural variations in how societies have sought to control technology, or to punish its misuse. For example, the kind of technology crime and control that might be witnessed in Han-dynasty China or in Ancient Rome seems unlikely to be comparable to that experienced in medieval Paris or eighteenth-century London. Significant complications also arise from the wide historical disparities in what constituted criminal offences. Take, for example, slavery, which is now (almost) universally treated as a criminal act. Given that many pre-modern societies were implicated in some form of this, to what extent should the technologies used to sustain it – such as slave-ships or manacles – now be associated with pre-modern technology crime? Similarly, could the use of axes in furthering deforestation during the pre-industrial period really be defined in terms of 'green crime', given that forestry clearance was often a matter of survival (cf. Winner 1978: 17)? There are also conceptual and methodological challenges. For example, since the sense in which technology can be defined as 'enabling' or 'assisting' with criminal/control behaviours remains undeveloped for contemporary technology we lack clear guidance when attempting to define earlier examples of it. The seeming lack of any very adequate historical sources for technology crime further complicates meaningful analysis.

In fact, there are certain solutions to all of these problems. Cultural variations pose no more of a decisive obstacle to a history of technology crime than to histories of crime in general, so must simply be accepted as they are presented. Similarly, explaining how technology enables or assists in criminality – whether in the contemporary or pre-modern sense – becomes a lot easier if we can draw upon definitions of technology which cover more than complex machinery or devices. For example if, as Philip Brey has argued in the previous chapter, we follow Kapp, McLuhan and others in viewing technology as something which *extends* human capacity (cf. Kapp 1877, McLuhan 1964) then there need be no big mysteries about the causal role of technology in effecting crime or control. Technology 'enables' simply by extending the range of what we already do, whether this involves complex contemporary technologies like a computer or more simple, traditional examples like a waterwheel. In turn, with just a little digging, there is a useful range of documentary evidence around technology crime and control available in sources from the ancient and classical periods. For example, the technical records kept by Chinese Imperial court bureaucrats (cf. Needham 1954); legal codes such as the Roman book of twelve tables or the Babylonian code of Hammurabi (cf. Watson *et al.* 2001) and travellers' tales and historical accounts of technology such as those provided by Herodotus and Thucydides. More technical sources can be found in texts like Vitruvius' *On Architecture*, Frontinus' work on aqueducts (see Humphrey *et al.* 1997; Cuomo 2008), early Indian mathematical texts (Yadav and Mohan 2010) or Islamic works on astronomy and medicine (Hassan and Hill 1986). And whilst detailed sources from the early medieval period are more lacking, a wealth of legal documents, personal narratives and other more technology-specific texts begin to become more widely available from the late fifteenth century onwards (see Singer *et al.* 1954; White 1962; Gimpel 1976). From this point, as the scientific revolution gathered pace, there is an increasingly rich range of documentation for tracing licit and illicit uses of technology. Clearly, much work remains to be done in analysing such diverse material, but even with the limited research tools which are available a surprising amount can be said about technology crime and control in the pre-industrial world.

Technology, violence and theft in the pre-industrial world

One immediate way to bring some order into any proto history of technology crime is to focus upon an obvious fixed pattern – those basic constancies in human (mis-)behaviour across history. It is perhaps not surprising then that it is our two most familiar categories of criminality – violence and property crime – which provide some of the richest available evidence for the role of technology in enabling harm. As one of our earliest technological innovations, weapons figure strongly here since they significantly enhanced our capacity to inflict violence, irrespective of how primitive such tools might be. In fact, archaeological records of weapons use enable us to trace technology-enabled violence to at least 230,000 years ago (Gill 2015) and as civilisation developed, weapon technologies became ever more integral to the acquisition and maintenance of power. The famous Narmer Palette dating from the thirty-first century BCE, which shows the unification of Upper and Lower Egypt, represents this by one of the oldest depictions of technology-enabled violence – King Narmer, smiting his opponent with a sophisticated mace or club.

The prevalence of violence throughout the pre-modern period cannot just be attributed to the greater social acceptability of using it to attain goals, but also to the far greater number of individuals likely to be carrying or using weapons technologies of various kinds. These usually took the form of basic stabbing, smashing, or choking devices (whether specifically designed for violence or, like a pike, for other tasks). However, more sophisticated weaponry also contributed to violence from a very early stage, in particular those devices which enabled violence 'at a distance', such as spears, slings, the bow and arrow or military devices like the siege catapult (ballistae) which aided military conquest for over 2,000 years (van Creveld 1989; O'Connell 1989). By the sixteenth century Chinese inventions like the cross-bow and, of course, gunpowder had further increased the potential range, accuracy and destructive power of the technologies of violence (Kelly 2004).

It is clear enough then that there is a long association between technology and violence, but more developed conclusions – for example the *volume* of violent crime that can be correlated

Image 2.1   Narmer Palette, thirty-first century BC
Source: Wikimedia Commons

with specific technologies – are less easy to come by. It is usually accepted that levels of violence were higher in the ancient world, a trend which continued into the late medieval period, when most historians and criminologists concur that a gradual decline began (see, amongst many here, Johnson and Monkkonen 1996; Emsley 1987). By the twelfth/thirteenth century, the percentage of those who fell victim to violence in England may have been as high as 23 per cent per 100,000 population (see Sharpe 1996; Eisner 2003), a figure which by the 1600s fell to around 5.7 per cent per 100,000, and still further to 2.3 per cent by the beginning of the nineteenth century (see Beattie 1986, chapters 3–4; Eisner 2003). Since such estimates are highly provisional, any conclusions about the extent to which certain kinds of tools or weapons *enabled* historical violence are, at best, speculative. However, some inferences seem plausible enough. Legal restrictions upon who was permitted to carry swords meant that 'technologically enabled' violence was more likely to involve improvised varieties of weapon. Slaves were forbidden to carry any kind of weaponry in most ancient societies – unless in the strict service of their masters (Mendelsohn 1932; Westermann 1984) and there were similar restrictions in the medieval period when only lords, knights and privileged commoners were entitled to bear arms (Kaeuper 1988). Weapon control could also be determined by location, as in the prohibitions around sacred, or high security zones. In Rome, no-one was permitted to carry weapons into the inner city of Rome proper – the pomoerium – and even the Praetorian guards, the military elite responsible for protecting the Emperor, were compelled to wear civilian dress when entering this part of the City (Beard *et al.* 1998). It is also true that access to more sophisticated stabbing and cutting technologies like swords would have been expensive, so that casualties resulting from such 'advanced' weaponry were more likely to be associated with war or 'elite', ritualised forms of combat like duelling rather than opportunistic crime or crimes of passion. Duelling was not uncommon, even after it was criminalised, with one estimate suggesting that in France alone around 4,000 deaths between 1589 and 1607 could be attributed to it (Baldick 1970). Around a third of the 172 duels recorded during the reign of George III of England (1760–1820) produced fatalities and two-thirds, (96) serious woundings (Norris 2009).

One demonstration of the level of harm which can be associated with improvised weapons technologies is provided by a study of violence in East Anglia during the period 1422–42. This suggested that over a quarter of murder cases were facilitated by everyday tools like turf shovels or hedge stakes (Maddern 1992). This pattern is further supported by an analysis of 364 killings investigated at Assize courts in Hertfordshire and Sussex during the 1590s which indicate that "the vast majority were not the result of calculated violence, rather they occurred during acts of sudden, unpremeditated aggression [involving] [...] a variety of knives and blunt instruments" (Cockburn 1977: 57).

More concentrated or higher volume forms of violence (such as genocides or massacres) have often been associated with a greater dependence upon technological means and the historical record appears to bear this out. In one of the earliest records we have of mass killing – a slaughter of 27 individuals which took place around 10,000 years ago near Lake Turkana in North Kenya – tips from stone arrow or spear projectiles found in two male victim's bodies, along with multiple blunt trauma marks on skulls and broken limbs (presumably inflicted by clubs) provide compelling evidence of technology's early role in supporting mass killing (Handwerk 2016). Sources from advanced ancient civilisations like the Assyrians also provide examples of the ways in which technical means aided their conquests and subsequent genocides on mass scales. Inscriptions left by King Ashurbanipal depict how rebellious cities were quelled by sophisticated siege engines, giant shield walls and earth ramps – with technology equally central to the mass slaughters which followed. Following the siege of Suru, Ashurbanipal proudly describes his technical ingenuity in constructing a giant

> pillar at the city gate (where) I flayed all the chief men who had revolted and I covered the pillar with their skins; some I walled up inside the pillar, some I impaled upon the pillar on stakes.

> *(cited in Anglim 2013: 185)*

In fact, the more this theme is pursued, the more we are forced to conclude that technological power was no less integral to supporting genocide in the pre-modern world than it has been with contemporary mass killings such as those conducted by the Nazi regime. The use of disciplined cavalry by Mongol hordes to devastate central Asia – which even on conservative estimates is reckoned to have left around 28 million dead (Rummel 1998: lines 441–535) – or the way the Roman military machine was directed at annihilating over 2,000,000 Jewish citizens during revolts between 66 and 136 CE (Josephus 1987: ix 3) provide typical examples of this. But conquest was only one motivation for technologically enabled mass killing in the pre-modern era. The sophisticated technology-driven spectacles enacted in the Roman Colosseum (including naval battles and special cages and elevators for delivering animals into the arena) contributed to over a million individuals estimated to have died in the gladiatorial spectacles which took place between the advent of Imperial Rome (around 20 BCE) to the outlawing of combat as entertainment by the Emperor Constantine in around 325 CE (Grant 1967).

Weapons technology also played a predictable role in enabling the second historical constant in behaviour under consideration here – property-related crime, though details are again largely sketchy. In Rome, the problem of violent robbery, burglary and theft appears to have been so serious that, until the second century AD, it was lawful to kill any thief bearing a weapon. Roman law generally carried higher sentences for 'manifest theft' (where a thief was caught in the act) and many interpret this as a way of ensuring thieves did not carry weapons which might result in injury to victims (Garoupa and Gomez 2005). By the early 1300s we know that robbery and theft comprised around three quarters of offences tried before assize courts – a proportion which seems to have remained fairly constant until the eighteenth century when, like violent crime, it too began to fall (Sharpe 1996: 23ff). The use of tools like clubs, garottes or stabbing instruments to effect such offences seems likely and from the sixteenth century, the advent of firearms began to dramatically widen the technological options. Around 7 per cent of robbery cases investigated in assizes during the 1540s involved the use of 'pocket dags' or firearms – a trend which caused the Privy Council to lament that it was now "a common thing for [...] thieves to carry pistols" (cited in Cockburn 1977: 57).

But there is good evidence to suggest that force or the use of weapons were not the only technological options for enabling property crime. Indeed, there are numerous accounts of specialised uses of technology or 'thieves' tools' to effect this. For example, some of the earliest recorded varieties of property crime – the tomb thefts conducted in Ancient Egypt – were often very sophisticated technological operations. Records of a builder called Amenpenofer, who worked for Amenhotep, High Priest of Amen-Re Sonter, at the time of Ramesses the Great, describe how he formed a prototypical organised crime syndicate with seven other builders, woodworkers, farmers and a boatman. They deployed their formidable battery of technical skills (and presumably the specialised tools they possessed) to break into the pyramid of Sobekmesef by cutting passages and successfully removing all the traps and obstacles which had been set. Their impressive spoils, which included golden burial masks, amulets with precious stones and jewellery were then divided amongst them (Montet 1958). Elsewhere, Exodus 22.2 refers to thieves 'breaking in' to houses of the time, probably with elementary digging tools, given that many houses would have been mud or clay based.[2]

The technological sophistication of many forms of domestic security in Rome indicates the skill of Roman thieves in using tools to conduct burglaries. In Roman towns such as Pompeii for example, the homes of the wealthy had extra thick walls with doors made of very heavy wood. These were not only bolted at night but also used what we would call a 'fox lock' today, that is, large timbers or metal bars angled against the door and inserted into indentations carved in the floor or walls to brace it just for this purpose (Conzémius 2013). The specialised uses of technology by thieves can sometimes be discerned in the jargon and slang used by perpetrators. For example, the fifteenth-century term 'hooker' or 'hooksman' refers to criminals who made use of hooked staffs to steal line or clothes through windows. This contrasts with the 'angler' who used hooks to steal from shop windows, the 'betty' who could use a variety of devices to pick locks, or even the use of 'cleymans' – sophisticated fake sores used by beggars to defraud the charitable (Harman 1573; Taylor 2010). Elizabethan tinkers of the sixteenth century became especially notorious for their use of specialised burglars' tools. One source tells of a knight who spotted a set of skeleton keys amongst a tinker's goods and tricked him into carrying a letter to the next gaol, where he was promptly detained (Salgado 1992). Enhancements to a thief's capacity might also be attained through relatively 'low-tech' options like animal, or even human extensions to the body. There is evidence, for example, of a dog who was trained to pick pockets by one Tom Gerrard, a housebreaker executed at Newgate in 1711 (Rayner and Crook 1926) and the use of young boys or 'divers' who could climb through windows too small or high for adult thieves was another common technique (Hallsworth 2005). We even find evidence of more sophisticated chemical technologies like drugs being used for theft. In 1675, for example, George Clerk and John Ransay, thieves executed at Newgate, were found to have administered various concoctions to their victims, including potions containing opium poppies, in order to put them to sleep whilst they were robbed (Rayner and Crook 1926). Enhanced organisation as a technology for theft should also not be overlooked. The gangs of 'cutpurses' who plagued cities in the seventeenth and eighteenth centuries augmented their already effective techniques of illicit acquisition by using the simple but effective technology of working together in close organisation. There is even evidence of technical skills like these being passed on in special underworld 'schools' (Salgado 1992).

In an echo of contemporary trends around fraud, property crime seems generally to have been a more fertile area for technological innovation than violent crime. Indeed, some of the most telling examples of the ongoing technological struggle between crime and its prevention can be seen in historical subversions of what Simmel called "the purest example of the tool" (1978: 210) – money itself (see McGuire 2007). The shift in monetary systems from simple barter towards coinage (Davies 2002) required a number of supporting technologies such as metallurgy and coin making devices, or more precise measurement tools like scales to ensure parity of exchange. And just as the emergence of digital currencies or online banking has provided fertile opportunities for cybercriminality, the increased centrality of technology in sustaining historical monetary systems also created new options for technological innovations around theft. The emergence of counterfeiting techniques from around 700 BCE – very soon after the first coins – serve as an early example of this (Davies 2002). In Rome, forgery became one of the seven main categories of crime during Sulla's pre-Imperial reorganisation of the courts (Dillon and Garland 2005: 529). Execution was the normal penalty for counterfeiting in China – especially after the introduction of paper money in the thirteenth-century Yuan dynasty, when warnings about fraud were printed on the notes. Like contemporary forms of hi-tech theft, the tactic of stealing money simply by copying it was the province of the criminal specialist, who had access to the necessary resources. 'Insiders', in particular those responsible for producing coinage, were especially likely to exploit their access privileges. The Emperor Nero perfected this form of State-driven technological theft by continually debasing Roman coinage – replacing up to 14 per cent of the silver in the denarius coin and 11 per cent of the gold in the aureus coin with base metals (Butcher and Ponting 2005). The fact that there was up to 60 per cent devaluation of the denarius (and its effective termination as a trading item) in early Imperial Rome indicates how widespread the practice was (ibid.). Even by the medieval period it remained a kind of economic policy for monarchs to raise capital in this way (Spufford 1989: 289ff).

The lucrative possibilities raised by counterfeiting continued to stimulate technological innovations in copying into the medieval period. Venetian records indicate a rash of fake coins during the thirteenth and fourteenth centuries and one renowned coin forger there even set up a mini-mint in his own home. Here he was able to beat out copper ingots and to then cut and stamp them with dies stolen from the Mint (Stahl 2008). Even with these fairly limited technical resources be was able to manufacture more than 13,000 fake coins (ibid.). And just as inadequate methods of authenticating identity have hugely inflated digital frauds, the success of pre-modern counterfeiting was closely associated with the inadequate technologies for detecting fake coinage (Singman 2000).

Pre-industrial 'hi-tech' crime (I) – communications technologies

Most of the artefacts considered in the previous section have involved criminality effected by way of relatively simple or 'low-tech' tools. The history of technology crime starts to become a good deal more interesting if we can find any kinds of precedent for misuses of more sophisticated kinds of technology. This points us towards another kind of structuring device for bringing the as yet fragmented history of technology crime into some kind of clearer focus, one which centres more directly upon the *kind* of technology involved. To this end, I will consider how far two of the key classes of 'hi-tech' criminality considered in this *Handbook* – those effected by 'distance-shrinking' technologies like communications tools, or by 'proximal', micro technologies such as those centred upon chemical and biological technologies (see McGuire 2012) – might have figured in pre-industrial crimes. Prima facie, the use of such tools to conduct crime or impose control in the pre-modern era seems highly unlikely. However, if we accept a definition of technology in terms of the way it *extends* human capacity, then plausible examples of how our communicative capacities or our capacities to access or use micro-phenomena were extended by certain tools begin to emerge. The fact that such extensions were often deployed without any developed scientific understanding of what was involved does not count as an objection to this idea. For one does not need to know what a tool is made of, or to what end to be able to use it in some way. After all, phishing emails are still thought of in terms of cybercrime, even though the senders are unlikely to understand how an email is coded, or transmitted across digital networks.

If then we turn to the first example of more sophisticated technologies – those based upon communication and information – we can quickly find examples which offered a capacity to enact harm from a distance. Though communications were certainly not instantaneous, early distance messaging technology, such as the prototype postal networks – which we know existed from at least the ancient period onwards – could certainly be used to facilitate assassinations, conspiracies, frauds, sexual stalking and other harms (McGuire 2012; Poe 2011). Such crimes just took longer to arrange and to execute. Alternatively, the arrival of the magnetic compass from China during the thirteenth century (a technology which could also be argued to extend the range of potential social interaction), eventually led to a general increase in pillage and piracy – especially following the discovery of the gold-rich Americas (see Lewis 1937; Garner 2002). But perhaps the locus classicus for an association between communications technology and crime in the pre-modern era came with the advent of the printing press. Following its rapid adoption in Europe from the mid-fifteenth century onwards a surge in intellectual property theft involving unauthorised reproductions of original texts and pamphlets began to develop. This was complemented by a rise in the reproduction of fake official documents such as letters of authentication, court orders and so on (see Hiatt 2004 for some examples). As with cybercrime, legal responses and criminalisation followed with the development of copyright law throughout the 1600s. This emerged from a combination of monopolies, privileges and rights granted to printers and culminated in the world's first dedicated copyright law – the UK's Statute of Anne in 1710 (Deazley *et al.* 2010).

An awareness that communications technology offered criminal possibility can be seen in the attempts to restrict access to it from very early times. For with the advent of printing a range of issues around censorship and freedom of expression came into a new focus. Thus, the fears of the cultural power-elite that communications technology might threaten the security of their power meant that blasphemy and sedition joined copyright theft as the most common offences associated with communications technology. The flood of new texts led to Pope Paul IV issuing the first 'Index of Prohibited Books' in 1559 (a list maintained until 1948) whilst monarchs such as Charles IX attempted to criminalise the publication of any printed document without express royal permission (Polastron 2007). In England, an explosion of seditious pamphlets around the Civil War period highlighted the challenges to the political status quo posed by the new printing technologies. There were just five printers working in London in 1500, but the advent of cheap moveable-type technologies led to a surge in new 'start-up' printing businesses (Seibert 1952) which significantly increased possibilities for public information and expression. Cromwell's Printing Act of 1649 attempted to stem this tide, restricting the printing of pamphlets to London and imposing a requirement for licences (ibid.). But significant arrests and imprisonments for 'seditious pamphleting' continued well into the eighteenth century, extending even to influential commentators such as the author Daniel Defoe (Thomas 1969).

Pre-industrial 'hi-tech' crime (II) – micro technologies

There is then, a plausible sense in which a form of communications technology offending (a 'proto-cybercrime' even) can be identified in the pre-industrial period. However there seems far less likelihood of being able to identify any harms/offences associated with the second class of 'high' technologies discussed in this book – those centred upon chemical or biological phenomena. One problem is that the use of such phenomena was based more upon practical, folk knowledge than genuine scientific understanding. Another is the (apparent) lack of any obvious tools that could be used to interact with micro-phenomena. However, a focus upon technology as something which 'adds capacity' again suggests another possible interpretation.

One immediate example relates to the use of chemicals – most obviously poisons of various kinds. No very complex chemical processes were required to obtain these since they were readily available in a variety of natural sources like plants (such as aconite), minerals (such as arsenic) or animals (most obviously snake venom). The Egyptians were experts in distilling poisons, including cyanides extracted from peachstones (Blyth and Blyth 1906) whilst use of arsenic was common in early China – both as a remedy *against* poison, as well as a means of inflicting death (Wexler 2015). The scale of poisoning in the ancient world is evidenced in the fact that the world's first law specifically prohibiting it – the *Lex Cornelia* – was passed by the Roman lawmaker Sulla in around 82 BCE (Wax 2006). The widespread penchant for using poison as an assassination tool can also be seen in the emergence of the 'food taster', who became an essential technology of security in royal courts from China to Persia and beyond (Emsley 2006). By the middle ages, the need for security technologies of this kind became still more pressing as poison-related crime became an almost commonplace social/political harm. This was perhaps most notoriously seen in the habit of the Borgias and other leaders of the time for engaging in spectacular 'dinner-party' executions of their rivals (Collard 2008). Accordingly, silversmiths began to construct cups made from materials like agate, rock-crystal or ostrich shells – substances believed to indicate the presence of poisons in liquids (ibid.). Seventeenth-century Venice and Rome were even reputed to host 'schools of poisoning' where students studied specialist texts such as Porta's *Natural Magic* and were instructed in differing poison types and methods of administering them (Emsley 2006).

Poison presents a relatively straightforward instance of pre-modern misuses of chemical tools, but can anything similar be said for their biological counterparts? Once again, there do seem to be certain precedents. One example might relate to crude attempts to manage the gene pool seen in traditional practices like infant exposure. This served to dispose of unwanted genetic variants like females or disabled infants (Patterson 1985; Boswell 1988) and was a practice especially tolerated in both Greece and Rome. Indeed, Roman commentators of the time were puzzled by the failure of other ancient civilisations to engage in the practice – Strabo for example thought Egyptians most odd because they 'raised every child that was born' (Strabo 1932: 17.2.5).

Overall, perhaps the best evidenced examples of the use of bio-chemical technology to enact harm can be found within the context of war. From China and Iran to India or Europe ingenious methods for enhancing military capacity by way of biological or chemical substances can be found. There is, for example, good physical evidence for the use of chemical weapons dating from as far back as AD 256. Recent excavations at the Persian city of Dura-Europos have revealed that Roman soldiers who had attempted to tunnel under the walls died as a result of the Persian tactic of flooding the tunnel with toxic gases obtained from burning bitumen and sulphur crystals (Tharoor 2009). Other examples can be found in the use of so-called 'Greek fire' (an incendiary weapon made from a chemical formula now lost) which was used as a form of defence by the Byzantine Empire against the Turks for over 500 years. Similarly, the suggestion by Sun Tzu, the Chinese general, that what he called 'fire weapons' were useful devices on the battlefield (2009) appears to have been heeded given records of Chinese armies using incendiary bombs (Roland 1992; Mayor 2003). Examples of aggressive uses of pathogens and toxins can be seen in the deployment of so-called 'scorpion bombs' which were sometimes catapulted into enemy cities and troop formations (Mayor 2003). There also seems to have been a widespread awareness that infections could be used as military technologies, with early forms of biological warfare, usually involving attempts to spread diseases, complementing the use of incendiary chemicals on the battlefield. Ancient swordsmen were reputed to smear human faeces on their swords in order to create infections and there are also stories of arrows dipped into putrefied human blood for the same end (ibid.). Prototypical biowarfare was most dramatically extended in the notorious Siege of Caffa in 1346, when the Tartar armies catapulted corpses of plague victims into the city (Wheelis 1999), but attempts to contaminate water sources by the use of diseased animal carcasses were probably more common than such tactics. As early as 1675 the Strasbourg Agreement (between the French and Holy Roman Empire) had made the use of 'odious toxic devices' on the battlefield the subject of explicit legal prohibitions (Zanders 2003).

The development of alchemy – the predecessor of modern chemistry – provides a fascinating example of the way changing cultural-historical contexts impacted upon public perceptions of science-technology as 'deviant'. For alchemy was both sponsored and suppressed, depending upon the interests at work. The received view – that alchemy was forbidden because its methods involved 'consorting with demons' is certainly not false – the literary figure of Dr Faustus reflects widespread public unease that the alchemist's aim of transmuting base metals into gold, or finding an elixir of life was somehow 'demonic' (Haynes 2006). Other concerns centred upon whether the creation of alchemical gold could result in a currency that was 'legal' for exchange (Nummedal 2007: 110). Suspicions of fraud were also raised in texts like Maier's *Examen Fucurom Pseudo-Chymicorum*, which detailed common tricks used to simulate the 'transmutation' of gold. Pope John XXII's ruling of 1317 charged alchemists with '*Spondent quas non exhibent*' (promising what they do not produce) and specified punishments such as branding or immolation, whilst Charles V of France banned the use of alchemical equipment altogether in 1380. Other monarchs were more pragmatic – Edward III of England, for example, employed two alchemists in the hope they might be able to produce gold at will, and subsequent monarchs followed this line (Nummedal 2007). But fears of devil-worship meant that public attitudes were less tolerant. The private home of the celebrated alchemist Dr Dee – private astrologer to Elizabeth I – was ransacked by a mob and John Lambe, a confidant of the Duke of Buckingham was stoned to death by an angry mob in London after being exposed by the Royal Society in 1627 (Salgado 1992). The suspicions provoked by alchemical experimentation with chemical substances set the scene for a more general antipathy towards the emerging scientific method. Indeed, in one of the most famous examples of this – Galileo's development and use of the telescope to map the night sky – the undesirable implications of this new science and technology resulted in overt attempts to criminalise and suppress it. For Galileo's insight that what he saw through this new distance-compressing technology provided evidence for the 'heretic' Copernican heliocentric model of the solar system was not lauded as any scientific breakthrough. Instead he was tried and convicted by the papacy and placed under house arrest until his death (Reston 1994).

A further important historical context for associations between the use of bio-chemical processes and criminality can be seen in attitudes towards early industrial processes and controls aimed at protecting public health. In his third satire, Juvenal complained about the "smoke pouring out of buildings" in Rome (1992), whilst the foul smells generated by the use of urine in the fulleries meant that the dyeing of cloth was restricted to specially designated areas of the city (Wacke 2002). There were also Roman by-laws controlling drains and the production of effluent (Wacke 2002; see also Bauman 1996; Harries 2007). The prototypical industrial technologies of the medieval world generated similar complaints about public nuisance and threats to health. Amongst the earliest known attempts at emission controls were those enacted against the burning of sea coal in the City of London in 1306, with certain offenders even executed for breaching these rules (Isaac 1953). However, such controls seem to have done little to stem the growth of industrial pollution in the capital. By the 1600s the essayist John Evelyn was complaining about the "hellish smoke of (London) town … [which] … impairs the health of its inhabitants" (Evelyn 1995). Following the Great Fire of 1666 Evelyn suggested that the King impose a variety of early 'zoning controls' over industry during the rebuilding of London, expressing the hope that

> the necessary evils of brewhouses, bakehouses, dyers, salt, soap and sugar boilers will now (be dispersed) to some other parts about the river towards Bow and Wandsworth.
>
> *(Evelyn 1995/1998: 154)*

Some of the most demanding controls around bio-chemical processes and the environment that we know of were directed at the technologies associated with agriculture and food production. There are of course well known religious codes governing food hygiene and consumption found in the Middle East, India and elsewhere but wherever food production became more industrialised, formal State controls seem also to have emerged. For, as today, the success of pre-modern civilisations depended heavily upon the technical means for ensuring adequate supplies of food and water to large population centres. Concerns about the potential misuse or failure of these are evidenced by the many regulatory codes in existence from very early times. In sixth-century Athens, the laws of Solon set out concise prescriptions on where and how to extract water (Plutarch 1914) just as effective management of the complex systems of pumps, water wheels and aqueducts which supplied Rome required a specialised cadre of technocratic police – the *curatores acquarum* (Berger 1953). They had the power to punish 'water violators', with fines imposed upon anyone who polluted a public fountain and liabilities of up to 100,000 sesterces for wilful damage to an aqueduct structure (Aicher 1995: 26). In Egypt, meat inspectors were employed to check quality, whilst in Medieval York, laws were imposed prohibiting the sale of meat kept for more than 24 hours and there are numerous records of innkeepers and shopkeepers who were prosecuted for selling pies with undercooked or tainted meat (Carlin 1998).

Closely related to concerns about the misuse of bio-chemical processes in food production and distribution were those related to health and medicine. We have substantial evidence of medical malpractice and attempts to regulate this stretching back as far as ancient Egypt. Aristotle describes the legal penalties faced by Egyptian doctors for irregular courses of treatment (Carrick 2001), whilst in Babylon, the Code of Hammurabi contained at least ten laws specifically directed at medical practice. Some stipulated extremely harsh sanctions for failure – for example:

> If a doctor has treated a man with a metal knife for a severe wound, and has caused the man to die, or has opened a man's tumor with a metal knife and destroyed the man's eye, his hands shall be cut off.
>
> *(King 2004: 218)*

The regulatory standards imposed upon ancient medical practitioners are perhaps most familiar to us from the writings of the Greek doctor Hippocrates, with the 'Hippocratic oath' (do no harm) still recited as a symbolic rite of passage by doctors in many jurisdictions (Temkin 2002). Standards of medical practice were especially fastidious within the Islamic world where texts such as Ali Ruhawi's tenth-century *Adab al-Tabib* (*Ethics of a Physician*) or Abu al-Hasan Al-Tabari's ninth-century work *Firdous al-Hikmat* (*The Paradise of Wisdom*) echoed Hippocrates' commitment to developing rules for 'good practice'. Some suggestions – such as the advice that doctors should 'wear clean clothes, be dignified, and have well-groomed hair and beard' (cited in Al-Ghazal 2004) were more about hygiene and presentation than medical expertise per se, but there were also restrictions on certain medical practices. In the *al-Tabib* for example, doctors were recommended not to 'give drugs to a pregnant woman for an abortion unless necessary for the mother's health' (Al-Ghazal 2004).[3] The preference for professional codes and self-determining standards in the Greek and Arabic worlds indicate a familiar sounding desire to keep central government out of medical regulation – one maintained in the modern era by bodies such as the UK General Medical Council (GMC), or the American Medical Association (AMA).

Religious, as well as State authority played its part in criminalising certain types of medical practice, in particular the restrictions placed upon anatomical dissection. The belief that damage to the physical body might undermine chances of resurrection was not limited to Christian contexts – pre-christian Rome also held dissection to be sacrilegious (Toynbee 1996) and the restrictions which resulted significantly impacted upon the work of early physicians like Galen. He was forced to leave Rome and conduct his research in Pergamum, where wounds sustained in a nearby gladiator school provided him with legitimate opportunities to further anatomical knowledge (Hankinson 2008).

The fourth-century *Codex Theodosianus* indicates the existence of a highly formalised hierarchical system of medical control in late Rome, overseen by the *Comes Archiatorum* – a kind of 'chief surgeon' who presided over a group of around 14 medical officials known as *archiater* (Matthews 2000). Elsewhere, the Justinian code of AD 529 was one of the earliest legal codes requiring education and proof of competence for doctors, with penalties for malpractice and limitations upon the numbers of practitioners. In Europe as a whole, the hiatus in centralised control of medical practice and technology which came with the fall of Rome did not last for too long. By 1140 Roger II of Sicily had passed edicts which required properly organised medical teaching with 'set courses, examinations and qualifications', a precedent followed soon afterwards in (1224) by the Holy Roman Emperor Frederick II (Gradwohl 1976).

Pre-industrial technology control and regulation by technology

Evidence within the pre-modern era that technology was not only a criminal tool but one integral to control is most obviously seen in its use as an enhancer of power. Access to superior military technologies – iron swords over bronze swords for example – is a clear example of how these could extend the capacity to impose control over enemies or criminals. But control could also be enhanced by less forceful means, with early communication technologies often central to this.

Control and early ICTs

The use of written records as a technology for bolstering sovereign power has been widely noted (Innis 1950). Successful management of the new agriculturally based empires like Egypt or Sumer depended heavily upon the capacity to maintain records, such as the annual flood levels of the Nile, or the amount of grain kept in the royal storehouses. With this technology of data-storage came a new social class – the scribe – a prototypical information elite who wielded significant power. In Sumer their role in overseeing legal contracts and processes gave them unique access to the workings of the State (Sjöberg 1975) and scribes were even more powerful in Egypt, since they were also often members of the priesthood.

The development of State messaging-networks complemented writing and record keeping in the creation of distinctive 'communication'-based forms of power. Early optical or sonic messaging technologies (like fire beacons or drums) may have been rudimentary, but organised properly, could still be highly effective ways of extending political reach. King Darius of Persia for example reputedly created a system where messengers stood on top of hilltops and shouted to each other – thereby delivering messages over distances that might take up to 30 days by foot (Diodorus Siculus 1947: 19.17.5–6). Networks based upon couriers working on foot or by horse could be even more effective. For example, the power of Inca emperors had much to do with their postal system which was operated by relay runners called 'chasqui', stationed every 1.4 kilometres on the mail trail. The system reputedly provided a one-day delivery for every 150 miles of road and permitted emperors to enforce their dictates quickly and efficiently (Hyslop 1984).

In the UK, the first integrated postal system was created by Henry VIII – and was immediately associated with extensions to surveillance power (Brayshay et al. 1998) serving as it did as 'the government's mouthpiece, eyes, and ears', (Ellis 1958). A 'secret office' which existed alongside the official service was authorised (by Royal warrant) to open and intercept any mail and pass on material of interest to the Secretary of State (Desai 2007). By the late 1700s warrants had also been issued which required that the mail of political opposition or foreign diplomats could be opened as a matter of course (ibid.). The reputation of the British Post Office for surveillance became so bad that citizens in the US colonies were determined to establish a parallel 'constitutional post' – first set up by Goddard during 1774, and adopted as the model for the US system as a whole in the Second Continental Congress of 1775. Goddard's messaging system was founded on new, more ethical principles of communication which stated that, 'mails shall be under lock and key, and liable to the inspection of no person but the respective Postmasters to whom directed' (cited in Desai 2007: 50).

The widespread disrespect for privacy rights in the UK and the philosophy that 'secrecy made legality unimportant' (ibid.: 25) was manifested within every succeeding shift in communications technology. For example, the surge in public communication which followed the introduction of the uniform penny post in the 1840s significantly extended surveillance possibilities. This pattern – enhanced communication technology, followed by enhanced scrutiny – has some obvious parallels with later technologies, such as the Internet, especially in the lack of accountability or transparency. Following a letter-opening scandal in 1844, a Commons Parliamentary committee outlined the direction in which communications control was heading, remarking that, "to leave it a mystery whether or not his power is ever exercised, is the way best calculated to deter the evil-minded from applying the post to improper use" (cited in Pedersen 2006).

Pre-modern bio-control: disciplining the body

Early forms of chemical control in the pre-modern world might be seen in the use of poison, which was as useful for technological regulation as it was for crime. A particularly novel variety of this was found in Ancient India where the secret services used an ingenious form of early bio-control in the form of female agents (*visakanyas* or 'poison damsels') who saturated their bodies with poisons in order to execute enemies of the State by tempting them with sex (Penzer 1980).[4] In general, however, pre-modern technologies of bio-chemical control were centred more upon the macro-level, with the body figuring as a central disciplinary subject. For example, trials by 'ordeal' were commonplace throughout the ancient and pre-modern world, with rudimentary technologies of fire and water used to decide the guilt of an accused (Lea 1973; Leeson 2010). Suspects might be required to retrieve a ring from a cauldron of boiling water or to carry a piece of red hot metal a predetermined distance, with innocence or guilt then established by the extent to which the body could subsequently 'regenerate' itself within a set time period.

But it was with the practice of torture where control by pain dispensation reached its highest levels of technical sophistication within pre-modern justice systems. Chinese authorities displayed a particular ingenuity in this regard – for example in the use of the 'Pao-Luo' (human grill) seen during the Zhou dynasty, where suspects were forced to walk across heated metal beams, or the later 'death by a thousand cuts' where considerable bio-medical expertise was deployed in inflicting the maximum possible pain upon victims (van Gulick 2008). The classical world was no less accomplished in the creation of sophisticated torture devices – for example, the 'brazen bull' involved placing victims inside a hollow bronze cast bull which was slowly heated over an open fire. A sophisticated technology of tubes and airstops then amplified the dying screams of the victim(s), generating a sound which resembled a braying bull, thereby providing additional amusement for spectators (Dubois 1991). By the medieval period an impressive array of torture technology attested to the ongoing development of a 'science' of pain delivery (see Hallsworth and Kaspersson, Chapter 33 in this volume, for one kind of history of this). One of the best known artefacts, the rack, may have been a fairly straightforward piece of engineering but in the hands of a skilled operator could be used to attain very precise correlations between the turning of the handles and targeted pain delivery (not too much or too little). And of course the rack was often merely an 'entree' into a far wider field of pain delivery technologies. There was for example the 'Judas Cradle' which required a victim to be seated on a triangular-shaped seat where he or she was slowly impaled. Or the 'Chair of Torture', a formidable piece of engineering incorporating anything between 500 to 1,500 spikes which covered the back, arm-rests, seat, leg-rests and foot-rests. As the victim's wrists were tied to the chair any struggles made the spikes penetrate the flesh even further. Similarly fiendish ingenuity was at work in the (largely self-explanatory) function of devices such as the 'pear of anguish', the 'breast ripper', the 'thumbscrew' or the 'knee splitter' (see Innes 1998 and Kellaway 2002 for these and other examples). Executions could also be fairly developed technological spectacles – as an expert technician of death a hangman could provide instant death through a broken neck, or the long, lingering and painful variety associated with the infamous 'hangmans dance'.

The pre-modern world was not totally lacking in evidentially based forms of crime control – with precedents for the use of chemical or biological tools in shaping this. Almost every history of forensic science begins with tales of the Chinese judge, Song Ci, who worked around the thirteenth century (McKnight 1981). In his memoirs, *Collected Cases of Injustice Rectified*, Song Ci discusses a number of cases over which he presided where 'forensic'-style techniques were used. One famous example involved a man who was murdered with what appeared to be a sickle. In order to detect the murderer, the investigator instructed everyone in the village to bring their sickle to a pre-arranged location where flies, attracted by the lingering smell of blood, eventually gathered on a particular one – so forcing the owner to confess to his crime. Elsewhere, the Caroline Code of 1533 (proclaimed by the Germanic Emperor, Charles V) was one of the first regulatory systems to suggest that expert medical testimony might be required in order to guide judges in cases of 'murder, wounding, poisoning, hanging, drowning, infanticide, and abortion and in other circumstances involving injury to the person' (Polsky and Beresford 1943).

Though science had clearly not yet developed sufficiently for the contemporary 'gold standard' in establishing guilt – genetic information – to be available, the body could be used to provide other physical indicators. For example, heart and pulse rates, or the extent to which a suspect blushed during questioning were all advanced as methods for 'reading' evidence of deviant actions or intentions at various points. In his 1730 essay 'An Effectual Scheme for the Immediate Preventing of Street Robberies and Suppressing all Other Disorders of the Night', the writer Daniel Defoe suggested how an investigator might apply such techniques, arguing: "take hold of (an offender's) wrist and feel his pulse, there you shall find his guilt", for it is impossible for "the most firm resolution of even the most harden'd offender (to) conceal and cover it" (cited in Matté 1996: 11). But Defoe was also concerned about the kind of justice involved in 'making a man an evidence against himself' (ibid.) – a concern that has become no less pressing for subsequent generations.

Early industrialisation and technology crime/control

Though the early industrial period clearly enhanced the range of technologies available for crime or control no specific concept of 'technology crime' became associated with this – either at the time, or by subsequent commentators. What was more debated was whether the very process of industrialisation itself violated societal norms in some deep, perhaps even criminal way. For the huge shift in productive power brought by new technologies like Hargreaves' Spinning Jenny also brought a range of harmful impacts upon social mobility, income and leisure in their wake. An increasing concentration of the population into factory production centres, the disruption of families through long working hours and child labour, or the creation of widespread poverty in the midst of plenty (Hobsbawm 1969) all fuelled an emerging realisation that with technology came harms as well as benefits. Such views were not restricted to intellectuals like Morris, Ruskin or Carlyle but were also found in the workforces within the new mills and factories. For of course the harms of technology were far more immediately apparent to them and so provoked a more concrete challenge. Their resistance has subsequently (and some might say unfairly) been interpreted as simple 'anti-technologism', though it could equally be interpreted as a response to a new kind of dilemma – which technological innovations are to be regarded as socially acceptable and which are not.

The 'Luddite' uprising, as this early phase of resistance has come to be characterised (even though its leader 'Ned Ludd' was almost certainly a fiction and it was never an 'uprising')[5] was initially provoked by the introduction of a new kind of spinning/weaving machine for making stockings more cheaply (though at a lower quality). This led to a series of disturbances and factory break-ins at stocking manufacturers in Nottinghamshire between 1811 and 1812, and the destruction of more than 200 of the new machines. Elsewhere, mills were besieged or torched and unrest began to spread to other parts of the countryside. The response of the governing elite set another precedent for this new technological order – a readiness to ignore the harms that came with it and to treat mere *resistance* to technology as criminal. Over

12,000 troops were quickly dispatched to Nottingham – a force larger even than that used by Wellington to fight the Peninsular War against Napoleon (Thomis 1970). By June of 1812 an intimidating army of 30,000 soldiers was camped out on the moors above Manchester ready to crush any further signs of protest. The use of military force was quickly complemented by draconian new forms of criminalisation like the 1812 Frame-Breaking Act which stipulated the death penalty for anyone daring to damage this new device.

Urbanisation, crime and control: the new spaces of crime

Of all the new technological forms that emerged with industrialisation, perhaps the most significant was the modern city itself. Of course, as technological 'complexes', cities have always enhanced our species' capacity to live in closer proximity, in greater numbers and in greater comfort – offering as they do more effective sewage and waste disposal technologies, enhanced provision of shelter and better access to food. But the power of the contemporary industrial city is witnessed in the sheer scale of the demographic reorganisations they facilitated. It took over 200 years between 1500 and 1750 for the UK population to double (from 3 to 6 million) but less than a hundred (between 1780 and 1860) for the figure to double again, a doubling process then repeated in less than 40 years, between 1860 and 1900 (Jeffries 2005). The exponential growth in population facilitated by the shift towards urban centres meant that, by 1900, most of the UK population was living in towns and cities such as London (where the population trebled from around 750,000 to around 3 million by 1900), Liverpool (which rose from 22,000 to 450,000) or Manchester (up from 18,000 to 376,000) (ibid.).

The shifts in the nature of crime and control which followed can be argued to have been significantly 'enabled' by the development of densely populated urban centres. A first, and obvious set of factors here related to a seeming increase in the *volume* and *variety* of criminal opportunities. These arose in part from higher levels of wealth – particularly amongst the newly affluent middle classes – which created larger numbers of potential victims. At the same time this new wealth accelerated social division, with the proximity of the very poor to the very rich (evidenced in Booth's poverty maps of London (Booth 2010)) producing obvious temptations for illicit acquisition. A further factor was seen in the densely packed structures and streets of the modern city which offered greater anonymity to would-be criminals, and more places to 'disappear' (Emsley 1987).

Rapid urbanisation certainly created contemporaneous perceptions of a rise in crime – especially in relation to property offences. Engels, for example, reported that crime was "higher in Britain than any other country in the world" (see Phillips 1993: 158) and figures for crime, published for the first time in 1810 by the UK government suggested that the 4,000 or so individuals indicted for trial in 1805 had risen to over 31,000 by 1842 (ibid.). Later commentators have also interpreted such figures to imply a 'vast increase in crime' (Perkin 1969: 162) during the early industrial period – whether in terms of volume, or organisation – such as the emergence of new street gangs (Hobbs 1994: 444–5). But some caution is also necessary, especially as it has since become clearer how readily technological change tends to be associated with perceptions of 'new crime waves'. The rise in the number of new offences created over this period may have inflated figures somewhat – in 1833 alone, for example, the list of serious crimes was increased from 50 to 73 (Taylor 2010). However, it cannot be claimed that rises in crime can be attributed to rises in the population (rather than any increase in offending itself) since the rise in recorded crime rate was nearly double the percentage increase in population rate (Taylor 2010). How best to interpret crime data over this period has been the subject of lengthy debates (see for example Tobias 1972; Sindall 1990; Hudson 1992; and Taylor 2010).

If technological processes like industrialisation or urbanisation *did* cause a rise in crime, there were equally many new technologies which arose to control it – most obviously what Ellul called the 'technical apparatus of the police' (1992). Thus, given the new capacities for detection and arrest this brought, any 'rise in crime' may simply have been a rise in *prosecutions* (see Emsley (1987) for one version of this argument). Officials at the time were well aware of this possibility, as an 1827 cabinet report indicates:

> only a portion of cases committed for trial [...] is to be deemed indicative of a proportionate increase in crime and that even of that proportion, much may be accounted for in the more ready detection and trial of culprits.
>
> *(cited in Phillips 1993: 159)*

Certain qualifications therefore seem required before any conclusion that crime 'rose' over the period of industrialisation can be accepted, still less that it was technology that was responsible. For example, the fact that indictable offences *fell* by over a third between 1850 and 1900 – at a time when a plethora of new technologies from electricity to the automobile emerged (Taylor 2010: 18) – suggests a different conclusion.

Whatever the verdict on levels of crime, the contribution of industrialisation to contemporary technological justice seems indubitable. For with it came a variety of tools now central to any modern criminal justice system. In addition to the police themselves, these might also include specialised, technology-driven crime experts – the 'detectives'; prototypical crime science methods such as blood analysis, fingerprinting, or the use of statistical mapping; and shifts in the technology of penality and social control, towards new, more 'humane' forms of control, centred on the prison or institutions like the hospital or the asylum.

High-tech crime and control in the nineteenth century

With the more advanced technologies that gradually emerged in the nineteenth century – such as the telegraph, or railways – more obvious continuities with what we might now think of as 'technology crime' began to emerge. But no-one ever spoke of 'telegraph' crime or 'railway crime', and any new offending patterns that did emerge cannot be wholly separated from the new class of offences designed to manage misuse of such technologies. For example, in the ten years after 1895 there was a 50 per cent rise in the number of cyclists prosecuted, along with a sharp rise in motoring offences – though this was clearly connected more to the new laws around transportation than any specific upsurge in technology crime (Taylor 2010: 21; see also Andrew Newton, Chapter 16 of this volume). As has been typical with the introduction of any new technology, there were also less rational concerns. In 1862 for example, the *Railway Traveller's Handy Book* warned passengers that: "In going through a tunnel it is always as well to have the hands and arms disposed for defence so that in the event of an attack the assailant may be immediately beaten back or restrained". Such advice followed upon a series of scares about an influx of thieves and con-men onto the new train system (Kalla-Bishop 1977), just as the advent of the telephone provoked worries that it might help extend the power of organised crime (Pool 1983, McGuire 2007). Such concerns seem minor compared to contemporary hyperbole – even though the impression made upon the public by these new communication information and transport technologies was equal to anything we have seen with the Internet. Huge crowds flocked to demonstrations of the new telegraphic technology and the completion of the Atlantic cable in 1866 even led to spontaneous street celebrations (Leblow 1995: 11).

The rapid exploitation of communications technology for control purposes has stronger resonances with contemporary experience. The precursor to the electronic telegraph – the optical signalling system developed by Claude Chappe – was under French state control from the outset and the very first message it carried – a communique reporting French success in retaking the city of LeQuesnoy from the Austrians and the Prussians – was a military one (Coe 1993: 6). It was not long before the new criminal justice agencies also took advantage of it – as early as the late 1840s UK police had used the telegraph to apprehend two high-profile suspects (Standage 1998; McGuire 2007). By 1856, police forces in Boston, Philadelphia and New York had created telegraph networks which linked precinct offices to operational headquarters and other forces quickly followed their lead (Tarr 1992: 10). Riots and outbreaks of urban unrest in the emerging US cities between the 1830s and the 1860s provided further motivations for enhanced police communication technology. Municipal authorities established large reserve forces ready to intervene at the first signs of disorder and telegraphic alarm systems connected to precinct headquarters were sited along the beats of police officers. As the Mayor of Philadelphia commented in 1855, "Now the police force has but one soul, and that soul is the telegraph" (cited in Tarr 1992: 10–11). Citizens were not slow to recognise the function of this 'soul' in enabling the suppression of their protests. During the draft riots of the 1860s in New York, rioters attacked telegraph poles and pulled down lines to impede police responses (ibid.).

The advent of the telephone enhanced police communications power still further. The technology was quickly used to link police stations in Albany NY (Thomas 1974) and by the early 1880s experiments were being conducted in Chicago on the use of the telephone as an operational device (ibid.). Combined telegraph/telephone signalling points emerged, with patrol wagons situated at each precinct office ready to respond to calls for assistance. 'Success' was both immediate and dramatic – in the USA arrests facilitated by these new call boxes rose from 6 to 44 per cent of the total between 1881 and 1885 (Tarr 1992: 12; Stewart 1994). As the *New Scientist* presciently observed, civic organisation would now become

> sensitive at every point, and the transmission of intelligence from there to the brain and subordinate nervous ganglia – that is the central and district police stations – will be practically instantaneous.
>
> *(cited in Tarr 1992: 12)*

In addition to enhanced operational interconnectivity, these new networks inevitably offered authorities a chance to build upon their already considerable capacity for surveillance. A telegraph network could often be 'tapped' very simply – either by attaching a feeder wire to the main cable and then reading off the signals as they were fed through (Petersen and Zamir 2001) or, more simply, by influencing the telegraph operators upon whom the system depended. If an operator couldn't be bribed or coerced into handing over messages, there was always the option of acquiring the paper tapes upon which the dot-dash messages were printed out – by fair means or foul. The ease in intercepting telegraphic communications worked both ways – creating some rich opportunities for financial and other frauds. Wheatstone, who helped develop the UK telegraph network, designed a 'pocket cryptograph' machine or cipher machine to keep messages secret – particularly from intrusive postal officials who caused complaints from the public to rise from 1 in 2,000 to 1 in 600 following their takeover of the system (Roberts 2011).

The industrialisation of chemical and biological science from the seventeenth century generated other new contexts for 'hi-tech' styles of offending. The discovery of new techniques for soda-ash production in 1773 or the patenting of the Deacon process a few years later (which enabled the separation of chlorine from chlorine gas) (Jackson 1996) were just some of the many milestones which had shifted chemistry away from laboratory curiosity to industrial tool. But the industrialisation of chemical knowledge brought, in its wake, harms that no-one had properly anticipated – not least a surge in pollution emanating from the new factories and disease and death for huge numbers of individuals living and working around them. Rapid and large scale damage to local environments provided one of the more obvious indicators of this – by 1877 for example, a Royal Commission survey of the River Tawe in South Wales found that a river which had been all but pristine less than 60 years before was now a toxic brew of copper alkali, sulphuric acids, iron sulphates, slag, cinders and coal (Markham 1994: 162). Sharp rises in respiratory and intestinal diseases along with a succession of epidemics like cholera and typhoid were some of the more deadly outcomes (Haley 1978).

The limited controls upon pollution imposed by earlier regulators were quickly made redundant by this upsurge of toxicity in the environment. As early as 1819, the first Parliamentary Select Committee on the effects of Steam Engines indicated an awareness that something more needed to be done about industrial emissions (Parliamentary Papers 1819). The resulting conclusion, that the public had a 'right' to live in clean environments, or to be protected against sickness and illness by governments, indicated profoundly new concepts in how to determine technological harm and its regulation. But the fact that there was little understanding of *how* air and other pollutions might be injurious to health meant little pressure was placed upon manufacturers to do very much about it (see Beck 1959: 479). Thus, whilst seminal research like Chadwick's *Report on the Sanitary Condition of the Labouring Population of Great Britain* (1842) noted 'atmospheric impurities', there was little attention paid to the responsibility of industry in generating them. Aside from a few obvious observations (for example that flowers in gardens were now often covered in soot), government largely deferred to business in regulating industrialised science. Whilst further standing committee reports of 1843 and 1845 addressed the issue of industrial control more keenly and set out ways in which legislation might help reduce 'smoke nuisance' (Parliamentary Papers 1843 and 1845) six bills on clean air were thrown out between 1844 and 1850 (Ashby and Anderson 1981).

The arguments against emission controls made by factory owners and industrialists sound eerily familiar. They were, reported various committee members 'only concerned with the immediate outlay … [so that] … the advantages [of controls] were too remote to be taken into consideration' (Beck 1959: 482). Claims that there was 'no proof' that devices for reducing emissions would work, or that the pressures of foreign competition meant British manufacturers could not 'afford' the expense of emission controls might almost have been advanced yesterday. In the end however, the manifest effects of industrialisation made it impossible to completely defer to the demands of commerce. Thus the 1847 Improvement Clauses Act (which focused on factory smoke) was further complemented by the 1866 Sanitary Act and the 1875 Public Health Act which created new regulatory authorities, directed to take action against smoke nuisances and to set new standards for housing, sewage and drainage, water supply and contagious diseases (ibid.). But new laws were one thing, enforcement of them quite another. Though factory inspectors had been appointed under the Nuisance Removal Act of 1856, their powers were very limited – for example, they could only enter factory premises under limited circumstances between 9 am and 6 pm. The Alkali Act of 1863 created an 'Alkali Inspector' with four assistants who were appointed to stop hydrochloric gas being dispatched into the atmosphere from 'alkali works'. These emissions included hydrogen chloride, which became hydrochloric acid when exposed to the atmosphere and severely damaged vegetation. The act required that 95 per cent of emissions should be stopped and was successful at first – resulting in a drop from 45,000 tonnes of annual emissions, to less than 45 tonnes (Russell 2000). But as a sign of what was to come the Inspector soon became responsible for setting standards, as well as enforcing them, a move which inevitably led to an overly cosy relationship with the industry and a resulting lack of accountability and transparency in decision making. And whilst the second Alkali Act of 1874 required industry to always use the best available technologies for reducing pollution problems, enforcement and penalties for offenders remained grossly inadequate. There were only around three prosecutions between the early 1900s and 1970s (Russell ibid).

The harmful impacts of the emerging chemical industries were also seen in a range of commercial by-products, in particular food and pharmaceutical items. The increasing tendency by food manufacturers to add chemical substances like phosphorus to medicines, tonics or even breakfast cereals from the mid- to late-1800s helped focus public attention on new issues around technology-driven criminality (Crellin 2004: 76). In 1861, *Reynolds Weekly* highlighted how public scandals like the 'Bath-bun poisoning and the Bradford poisoning cases [have] brought the issue of food adulteration to the attention of various public bodies' (07 April 1861). New medicines and their ingredients were another area where technological controls were weak, but the significant rise in chemically induced deaths which resulted brought the activities of new pharmaceutical corporations like Burroughs Wellcome in the UK or Parker Davis in the USA into new focus (Crellin 2004). For such deaths could only be partly attributed to public misuse of the more widely available, commercially produced chemicals. Thus, whilst the 1851 Arsenic Act may have been a response to a surge in poisoning (Bartrip 1992), it is also worth noting that over 500 deaths between 1837 and 1838 were a direct result of the widespread *commercial* use of arsenic – in soap, wallpaper, or even as an aphrodisiac (Whorton 2010).

Biological science offered fewer opportunities for harmful exploitation since it was not yet as advanced or as commercially developed as chemistry. As a result, what we might think of as 'biological' crimes were, as in the pre-industrial period, still largely restricted to military or medical contexts. For example in the 1760s, British forces engaged in colonial Indian wars in North America were reported to have advanced the possibilities of biological warfare by distributing blankets contaminated with smallpox to the local population – though evidence for this has been disputed (Fenn 2000). Instead, some of our best examples of wrongdoing involving biological technology and knowledge come with the development of modern medicine which was often assisted by practices which did not just violate ethical and moral considerations, but which were (in terms of contemporary law) unquestionably criminal. The lurid crimes of those who answered medical science's need for bodies for anatomical dissection by murdering to order (like Burke and Hare) has often deflected scrutiny away from the organised and widely tolerated trade in corpses between criminal gangs and the medical elite (Frank 1976). And whilst the Anatomy Act of 1832 helped increase the supply of cadavers for medical research, it did so at the expense of any justice for the poor whose bodies could now be routinely transferred from the workhouses direct to the dissection chamber (Richardson 1987).

Supplementing the grey legal areas around dissection were a range of questionable medical practices involving the use of untried drugs, deliberate infection, experimentation and other forms of exploitation. The Apothecaries Act of 1815 had attempted to introduce some degree of professionalism in the dispensing of drugs, but conflicts of professional interest between the Society of Apothecaries and the Royal College of Surgeons created regulatory inconsistencies between those working in the medical profession and those selling remedies (Coley 2000). The steep rise in numbers of doctors practising in the UK (from 14,415 in 1861 to 35,650 in 1900 – Robinson 2011) forced the medical establishment to take the question of medical regulation more seriously – though the formation of the British Medical Association in 1856 and the General Medical Council in 1858 ensured that, as elsewhere with technology, self-regulation was to be the preferred model (Bynum 1994).

An inherent weakness of the self-regulatory approach was quickly revealed when none of the new professional bodies failed to intervene to prevent the growing exploitation of marginalised or vulnerable members of society for research purposes. One notorious example of this can be seen in a well-documented series of shocking biological research programmes carried out in the USA on plantation slaves and working-class black citizens during the nineteenth century (see Schiebinger 2004). Contemporary accounts relate stories of slaves being placed in boiling hot ovens to study the effects of heat stroke, parts of their bodies amputated to test analgesics and many other distressing experiments (Washington 2007). What qualms were raised about this tended to have more to do with whether results from black or other 'subhuman' classes could be extended to the whole (i.e. white) population than with the ethics per se. Respected surgeons like Dr J. Marion Sims of Carolina conducted numerous gynaecological experiments on black female slaves, usually without any anaesthesia (Axelson 1985). Yet, like many other medical researchers of this time, Sims was widely honoured (even today the Medical University of South Carolina maintains a J. Marion Sims Chair in Obstetrics-Gynecology).

Summary

This chapter has been able to offer only the most limited introduction to the associations between technology, crime and control in the pre-modern and early industrial periods. However, what (I hope) has been made clear is the rich history of this which is waiting to be deciphered, one which gives the lie to assumptions that technology crime somehow 'begins' with the Internet, or is something found mainly within late industrial, information societies. It is also worth stressing that in attempting to interrogate the roots of this phenomenon we do more than engage in mere historical description. By developing a better understanding of how the impacts of technology upon social control and crime have evolved over time it also becomes easier to identify some of the key features which underpin this relationship. Not only does this provide a more complete set of tools for interpreting the wide range of technology crimes we see later in the *Handbook*, it also suggests more effective ways of responding to them. For it is only where we can appreciate the complex and subtle forms of technological artefacts and processes which exist that a more informed socio-legal or criminological approach to their possible misuse can be developed.[6]

Notes

1 A distinction that has been challenged on various historical grounds (cf. Hartwell 1990). See also Landes (2003).
2 Gill's *Exposition of the* Old Testament (1748) discusses the Hebraic commentators who suggested the kinds of tools which might be used for such an operation.
3 Note that the original version of the Hippocratic Oath also contains proscriptions against abortion.
4 Bio-control for similar ends was also exerted by using women infected with venereal diseases.
5 The name comes from a series of letters sent to factory owners signed by one 'General Ned Ludd and his Army of Redressers' which demanded justice for sacked workers.
6 Some sections of this chapter were adapted and revised from McGuire 2012, Chapter 2.

References

Aicher, P. 1995. *Guide to the Aqueducts of Ancient Rome*, Wauconda, Illinois: Bolchazy-Carducci Publishers.
Al-Ghazal, S. 2004. Medical ethics in Islamic history at a glance, *Journal of the International Society for the History of Islamic Medicine*, 3, 4: 12–14.
Anglim, S. 2013. *Fighting Techniques of the Ancient World*, London: Amber Books.

Ashby, E. and Anderson, M. 1981. *The Politics of Clean Air*, Oxford: Oxford University Press.

Axelsen D. E. 1985. Women as victims of medical experimentation: J. Marion Sims' surgery on slave women, 1845–1850, *Sage*, 2, 2:10–13.

Baldick, R. 1970. *The Duel: A History of Duelling*, New York: Spring Books.

Bartrip, P. 1992. A 'pennurth of arsenic for rat poison'. The Arsenic Act 1851 and the prevention of secret poisoning, *Medical History*, 36: 53–69.

Bauman, R. 1996. *Crime and Punishment in Ancient Rome*, London and New York: Routledge.

Beard, M., North, J. and Price, S. 1998. *Religions of Rome, Volume 1: A History*, New York: Cambridge University Press.

Beattie, J. M. 1986. *Crime and the Courts in England, 1660–1800*, Princeton, NJ: Princeton University Press.

Beck, A. 1959. Some aspects of the history of anti-pollution legislation in England 1819–1954, *Journal of the History of Medicine*, XIV, 10: 475ff.

Berger, A. 1953. Encyclopedic dictionary of Roman law, *Transactions of the American Philosophical Society*, New Ser., V. 43, Pt. 2.

Blyth, A. and Blyth, M. 1906. *Poisons: Their Effects and Detection*, London: Charles Griffin and Company.

Booth, C. 2010. *Poverty Maps of London*, LSE Charles Booth Online Archive.

Boswell, J. 1988. *The Kindness of Strangers: The Abandonment of Children in Western Europe from Late Antiquity to the Renaissance*, New York: Pantheon.

Brayshay, M., Harrison P. and Chalkley, B, 1988. Knowledge, nationhood and governance: The speed of the Royal Post in early modern England, *Journal of Historical Geography*, 24, 3: 265–88.

Butcher, K. and Ponting, M. 2005. The Roman denarius under the Julio-Claudian emperors: Mints, metallurgy and technology, *Oxford Journal of Archaeology*, 24, 2: 163–97.

Bynum, W. 1994. *Science and the Practice of Medicine in the Nineteenth Century*, Cambridge: Cambridge University Press.

Carlin, M. 1998. Fast food and urban living standards in medieval England, in Carlin, M. and Rosenthal, J. (eds) *Food and Eating in Medieval Europe*, London, Rio Grande, Ohio: Hambledon Press, pp. 27–52, pp. 27–28.

Carrick, P. 2001. *Medical Ethics in the Ancient World*, Washington: Georgetown University Press.

Cockburn, J. S. 1977. The nature and incidence of crime, 1559–1625, in Cockburn, J. S. (ed.), *Crime in England 1550–1800*, London: Methuen.

Coe, L. 1993. *The Telegraph: A History of Morse's Invention and its Predecessors in the United States*, Jefferson, NC: McFarland and Company.

Coley, N. 2000. Forensic chemistry in 19th-century Britain, *Endeavour*, 22, 4: 143–7.

Collard, F. 2008. *The Crime of Poison in the Middle Ages*, (trans. Nelson-Campbell, D.), Westport, CT: Praeger.

Conzémius, M. 2013. *Private Security in Ancient Rome*, Pétange (see Private Security in Ancient Rome - Education.lu).

Crellin, J. 2004. *A Social History of Medicines in the Twentieth Century: To Be Taken Three Times a Day*, London: Haworth Press Inc.

Cuomo, S. 2008. Ancient written sources for engineering and technology, in Oleson, J. (ed.) *The Oxford Handbook of Engineering and Technology in the Classical World*, Oxford: Oxford University Press, pp. 15–34.

Davies, G. 2002. *A History of Money from Ancient Times to the Present Day*, 3rd. edn, Cardiff: University of Wales Press.

Deazley, R. Kretschmer, M. and Bently, L. (eds) 2010. *Privilege and Property: Essays on the History of Copyright*, Cambridge: Open Book Publishers.

Desai, A. C. 2007. Wiretapping before the wires: The post office and the birth of communications privacy, *Stanford Law Review*, 60: 553ff.

Dillon, M. and Garland, L. 2005. *Ancient Rome: From the Early Republic to the Assassination of Julius Caesar*, London: Taylor and Francis.

Diodorus Siculus, 1947. *The Library of History*, Loeb Classical Library Cambridge, MA: Harvard University Press.

Dubois, P. 1991. *Torture and Truth*, New York: Routledge.

Eisner, M. 2003. Long term historical trends of violent crime, *Crime and Justice: A Review of Research*, 30: 83–142.

Ellis, K. 1958. *The Post in the Eighteenth Century: A Study in Administrative History*, Oxford: Oxford University Press.

Ellul, J. 1992. *Betrayal by Technology*, Amsterdam, Netherlands: Stichting ReRunProdukties. Videocassette.

Emsley, C. 1987. *Crime and Society in England, 1750-1900 (Themes In British Social History)*, Harlow: Longman.

Emsley, J. 2006. *The Elements of Murder: A History of Poison*, Oxford: Oxford University Press.

Evelyn, J. 1995. London Redivivum, in De la Bedoyere (ed.) *The Writings of John Evelyn*, Woodbridge: Boydell Press (sections also in Chant, 1998).

Fenn, E. 2000. Biological warfare in eighteenth-century North America: Beyond Jeffery Amherst, *Journal of American History*, 86, 4: 1552–80.

Frank, J. 1976. Body snatching: A grave medical problem, *Yale Journal of Biology and Medicine*, 49: 399–410.

Gardner, P. L. (1994). The relationship between technology and science: Some historical and philosophical reflections. Part 1, *International Journal of Technology and Design Education*, 4, 2: 123–54.

Garner, R. 2002. The forensic polygraph, in Levinson, D. (ed) *Encyclopedia of Crime and Punishment*, London: SAGE Publications, pp. 725–727.

Garoupa, N. and Gomez, F. 2005. Paying the price for being caught: The economics of manifest and non-manifest theft in Roman criminal law, *Indret Revista Para el Analisis del Derecho*, I 2005, www. indret.com/pdf/276_en.pdf.

Gill, J. 1748–1763. *An Exposition of the Old Testament* (vols 1–3), London.

Gill, V. 2015. 'Evidence of 430,000-year-old human violence found,' BBC 28 May 2015.

Gimpel, J. 1976. *The Medieval Machine: The Industrial Revolution of the Middle Ages*, London: Penguin.

Gradwohl, R.B.H. 1976. *Gradwohl's Legal Medicine*, Boston, MA: Year Book Medical Publications.

Grant, M. 1967. *Gladiators*, New York: Delacorte Press.

Haley, B. 1978. *The Healthy Body and Victorian Culture*, Cambridge, MA: Harvard University Press.

Hallsworth, S. 2005. *Street Crime*, Cullompton: Willan.

Handwerk, B. 2016. An ancient, brutal massacre may be the earliest evidence of war, www.smithsonian. com, 20 Jan 2016.

Hankinson, R.J. 2008. *The Cambridge Companion to Galen*, Cambridge: Cambridge University Press.

Harman. T. 1573. *A Caveat Or Warening for Common Cursetors Vulgarely Called Vagabones*, London: Reeves and Turner.

Harries, K. 2007. *Law and Crime in the Roman World*, Cambridge: Cambridge University Press.

Hartwell, R.M. 1990. Was there an Industrial Revolution, *Social Science History*, 14, 4, 567–576.

Hassan, A. and Hill, D. 1986. *Islamic Technology: An Illustrated History*, Cambridge: Cambridge University Press.

Haynes, R. 2006. The alchemist in fiction: The master narrative, *Hyle*, 12, I.

Heidegger, M. 1949/1977. The question concerning technology, in *The Question Concerning Technology and Other Essays*. William Lovitt, editor and translator, New York: Harper & Row, 1977, pp. 3–35.

Hiatt, A. 2004. *The Making of Medieval Forgeries: False Documents in Fifteenth Century England*, London: British Library.

Hobbs, D. 1994. Professional and organized crime in Britain, in M.R. Maguire, R. Morgan and R. Reiner (eds) *The Oxford Handbook of Criminology*, Oxford: Clarendon Press, pp. 441–468.

Hobsbawm, E.J. 1969. *The Pelican History of Britain, Volume 3: Industry and Empire*, Harmondsworth: Penguin.

Hudson, P. 1992. *The Industrial Revolution*, London: Edward Arnold.

Humphrey, J., Oleson, J. and Sherwood, A. 1997. *Greek and Roman Technology: A Sourcebook – Annotated Translations of Greek and Latin Texts and Documents*, New York: Routledge.

Hyslop, J. 1984. *The Inka Road System*, New York: Academic Press.

Innes, B. 1998. *The History of Torture*, New York: St. Martin's Press.

Innis, H. 1950/2007. *Empire and Communications*, Toronto: Dundurn Press.

Isaac, P. 1953. Air pollution and Man's health, *Public Health Reports* (1896–1970), 68, 9: 868–70.

Jackson, T. 1996. *Material Concerns: Pollution, Profit and Quality of Life*, London: Routledge.

Jeffries, J. 2005. The UK population: Past, present and future, Chapter 1 in *Focus on People and Migration*, London: Office for National Statistics.

Johnson, E. and Monkkonen, E. 1996. *The Civilization of Crime: Violence in Town and Country since the Middle Ages*, Urbana: University of Illinois Press.

Josephus 1987. *The Works of Josephus*, Translated by William Whiston, A.M., Peabody, MA: Hendrickson Publishers, Inc.

Juvenal, 1992. *The Satires* (trans. Rudd, N.), Oxford: Oxford University Press.

Kaeuper, R. 1988. *War, Justice, and Public Order: England and France in the Later Middle Ages*, Oxford: Clarendon Press.

Kalla-Bishop, P. 1977. *The Golden Years of Trains, 1830–1920*, London: Phoebus.

Kapp, E. 1877. *Grundlinien einer Philosophie der Technik*, Braunschweig, Verlag George Westermann.

Kellaway, J. 2002. *The History of Torture and Execution: From Early Civilization through Medieval Times to the Present*, New York: The Lyons Press.

Kelly, J. 2004. *Gunpowder: Alchemy, Bombards, & Pyrotechnics: The History of the Explosive that Changed the World*, New York: Basic Books.

King, L. (trans) 2004. *The Code of Hammurabi*, Kila MT: Kessinger Publishing.

Landes, D. 2003. *The Unbound Prometheus: Technical Change and Industrial Development in Western Europe from 1750 to the Present* (2nd ed.), New York: Cambridge University Press.

Lea, H. 1973. *The Ordeal*, Philadelphia: University of Pennsylvania Press.

Leblow, I. 1995. *Information Highways and Byways: From the Telegraph to the 21st Century*, New York: IEEE Press.

Leeson, P. 2010. *Ordeals*. Available at SSRN: http://ssrn.com/abstract=1530944.

Lewis, E. 1937. Responsibility for piracy in the middle ages, *Journal of Comparative Legislation and International Law*, 19, 1: 77–89.

Maddern, P. 1992. *Violence & Social Order: East Anglia 1422–1442*, Oxford: Oxford University Press.

Marcuse, H. 1982. Some social implications of modern technology, in Arato, A. and Gebhardt, E. (eds), *The Essential Frankfurt School Reader*, New York: Continuum, pp. 138–62.

Markham, A. 1994. *A Brief History of Pollution*, New York: St. Martin's Press.

Marx, K. 2007. *Capital: A Critique of Political Economy*, Vol. I, New York, Cosimo.

Matté, J.A. 1996. *Forensic Psychophysiology Using the Polygraph: Scientific Truth Verification, Lie Detection*, Williamsville, NY: J.A.M. Publications.

Matthews, J. 2000. *Laying Down the Law: A Study of the Theodosian Code*, New York: Yale University Press.

Mayor, A. 2003. *Greek Fire, Poison Arrows & Scorpion Bombs: Biological and Chemical Warfare in the Ancient World*, New York: Overlook.

McGuire, M. R. 2007. *Hypercrime; The New Geometry of Harm*, Abingdon: Routledge.

——2012. *Technology, Crime and Justice: The Question Concerning Technomia*, London: Routledge.

McKnight, B. (tr.) 1981. The washing away of wrongs; Sung Tz'u: forensic medicine in thirteenth-century China, *Science, Medicine and Technology in East Asia, 1, xv*, Ann Arbor: University of Michigan Center for Chinese Studies.

McLuhan, M. 1964. *Understanding Media: The Extensions of Man*, New York: McGraw Hill.

Mendelsohn, Isaac, 1932. *Legal Aspects of Slavery in Babylonia, Assyria and Palestine*, Williamsport, PA: Bayard Press.

Montet, P. 1958. *Everyday Life in Egypt in the Days of Ramesses The Great*, (trans Maxwell Hyslop, A. and Drower, M.), University of Pennsylvania Press.

Mumford, L. 1934. *Technics and Civilization*, New York: Harcourt, Brace & Company, Inc.

Needham, J. 1954. *Science and Civilisation in China Vol I*, Cambridge: Cambridge University Press.

Norris, J. 2009. *Pistols at Dawn: A History of Duelling*, Gloucester: History Press.

Nummedal. T. 2007. *Alchemy and Authority in the Holy Roman Empire*, Chicago: University of Chicago Press.

O'Connell, R. 1989. *Of Arms & Men: A History of War Weapons and Aggression*, Oxford: Oxford University Press.

Pacey, A. 1991. *Technology in World Civilisation: A Thousand Year History*, Cambridge MA: MIT Press.

Parliamentary Papers, 1819. *Report from the Select Committee on Steam Engines*, Preface.

——1843 *Report from the Select Committee on Smoke Prevention, VII.*

——1845 *Second Report of the Commissioners of Inquiry into the State of Large Towns and Populous Districts.*

Patterson, C. 1985. 'Not worth the rearing': The causes of infant exposure in Ancient Greece, *Transactions of The American Philological Association*, 115: 103–23.

Pedersen, S. 2006. Spies in the Post Office: Sovereignty, surveillance, and communication in 18th Century Denmark, Paper presented at XIV International Economic History Congress, Helsinki 2006.

Penzer, N. 1980. *Poison Damsels*, Manchester, NH: Ayer Co Pub.

Perkin, H. 1969. *Origins of the Modern British State 1780-1880*, London: Routledge Kegan and Paul.

Petersen, J. and Zamir, S. 2001. *Understanding Surveillance Technologies: Spy Devices, Their Origins & Applications*, New York: CRC Press.

Philips, D. 1993. Crime, law and punishment in the Industrial Revolution, in O'Brien, P.K. and Quinault, R. (eds), *The Industrial Revolution and British Society*. Cambridge: Cambridge University Press, pp. 156-182.

Poe, M. 2011. *A History of Communications: Media and Society From the Evolution of Speech to the Internet*, Cambridge: Cambridge University Press.

Polastron, L. 2007. *Books on Fire: The Destruction of Libraries throughout History*, Rochester, VT: Inner Traditions.

Polsky, S. and Beresford, S. 1943. Some probative aspects of the early Germanic codes, Carolina and Bambergensis, *Boston University Law Review*, 23: 183.

Pool, I. 1983. *Forecasting the Telephone: A Retrospective Technology Assessment*, Norwood, NJ: Ablex Publishing.

Plutarch, 1914. *Lives* (Vol I), (trans. Perrin, B.) Loeb Classical Library, Cambridge MA: Harvard University Press.

Rayner, J. and Crook, G. (eds) 1926. *The Complete Newgate Calendar*, London: Navarre Society, 5 vols.

Reston, James Jr., 1994. *Galileo: A Life*, New York: Harper Collins.

Richardson, R. 1987. *Death, Dissection and the Destitute*, London: Routledge Chapman and Hall.

Roberts, S. 2011. Distant Writing, see: http://distantwriting.co.uk/default.aspx.

Robinson, B. 2011. Victorian medicine – from fluke to theory, BBC History, *Victorians*.

Roland, A. 1992. Secrecy, technology, and war: Greek fire and the defense of Byzantium, *Technology and Culture*, 33, 4: 655–79.

Rummel, R. 1998. *Statistics of Democide: Genocide and Mass Murder since 1900*, Rutgers: Transaction Press.

Russell, C. (ed.) 2000. *Chemistry, Society and Environment: A New History of the British Chemical Industry*, Royal Society of Chemistry: Print on demand edition.

Salgado, G. 1992. *The Elizabethan Underworld*, New York: St Martins Press.

Schiebinger, L. 2004. Human experimentation in the 18th century; Natural boundaries and valid testing, in Daston, L. and Vidal, F. (eds) *The Moral Authority of Nature*, Chicago: University of Chicago Press, pp. 384–407.

Seibert, F. 1952. *Freedom of the Press in England 1476–1776*, Urbana: University of Illinois Press.

Sharpe, J. 1996. Crime in England, long term trends and the problem of modernization, in Johnson, E. and Monkkonen, E. (eds) *The Civilization of Crime: Violence in Town and Country since the Middle Ages*, Urbana: University of Illinois Press pp. 17–34.

Simmel, G. 1978. *The Philosophy of Money*, London: Routledge Kegan & Paul.

Sindall, R. 1990. *Street Violence in the 19th Century: Media Panic or Real Danger?* Leicester: Leicester University Press.

Singer, C., Holy, E.J., Holmyard, E.J., and Hall, A.R., (eds) 1954. *A History of Technology*, Oxford: Clarendon Press.

Singman, J. 2000. *Daily Life in Medieval Europe*, Westport. CT: Greenwood Publishing Group.

Sjöberg, A. 1975. The Old Babylonian Eduba, in Liebermann, S. (ed.) Sumerological Studies in Honour of Thorkild Jacobsen, *Assyriological Studies*, 20: 123–57.

Spufford, P. 1989. *Money and its Use in Medieval Europe*, Cambridge: Cambridge University Press.

Stahl, A. 2008. Coin and punishment in medieval Venice, in Karras, R., Kaye, J., and Matter, E. (eds) *Law and the Illicit in Medieval Europe*, Philadelphia: University of Pennsylvania Press, pp. 164–79.

Standage, T. 1998. *The Victorian Internet*, New York: Walker and Co.

Stewart, R. 1994. The Police Signal Box: A 100 year history, available at www.britishtelephones.com/police/boxes.pdf.

Strabo. 1932. *Geography*, Loeb Classical Library, Cambridge MA: Harvard University Press.

Suler, J. 2004. The online disinhibition effect, *CyberPsychology & Behavior*, 7, 3:

Sun Tzu. 2009. *The Art of War* (trans. Giles, L.), United States: Pax Librorum Publishing House.

Tarr, J.A. 1992. The municipal telegraph network: Origins of the fire and police alarm systems in American cities, *FLUX Cahiers scientifiques internationaux Réseaux et Territoires*, 8, 9: 5–18.

Taylor, D. 2010. *Hooligans, Harlots & Hangmen: Crime and Punishment in Victorian Britain*, London: Praeger.

Temkin, O. 2002. What does the Hippocratic Oath say?, in *On Second Thought and Other Essays in the History of Medicine*, Baltimore: Johns Hopkins University Press, pp. 21–8.

Tharoor, I. 2009. Why chemical warfare is ancient history, *Time*, 13 Feb 2009.

Thomas, D. 1969. *A Long Time Burning: The History of Literary Censorship in England*, New York: Praeger.

Thomas, M. 1974. *Police Communications*, Illinois: CC Thomas.

Thomis, M. 1970. *The Luddites: Machine-breaking in Regency England*, Hampton, Conn.: Archon Books.

Tobias, J. 1972. *Crime and Industrial Society in the 19th Century*, London: Penguin.

Toynbee, J.M. 1996. *Death and Burial in the Roman World*, Baltimore: Johns Hopkins University Press.

van Creveld, M. 1989. *Technology & War: From 2000 BC to the Present*, New York: Free Press.

van Gulik R. (ed.) 2008. *Crime and Punishment in Ancient China*, Tang-Yin-Pi-Shih: Orchid Press.

Wacke, A. 2002. Protection of the environment in Roman Law? *Roman Legal Tradition*, Vol 1, pp. 1–24.

Washington, H. 2007. *Medical Apartheid: The Dark History of Medical Experimentation on Black Americans from Colonial Times to the Present*, New York: Doubleday.

Watson, A., Cairns, J. and Robinson, O. 2001. *Critical Studies in Ancient Law, Comparative Law and Legal History*, Oxford: Hart.

Wax, P. 2006. Historical principles and perspectives, in Goldfrank, L., Flomenbaum, N., Lewin, N., Howland, M., Hoffman, R., and Nelson, L.(eds) *Goldfranks Toxological emergencies*, New York: McGraw Hill, pp. 1–22.

Westermann, W. 1984. *The Slave Systems of Greek and Roman Antiquity*, Philadelphia: DIANE Publishing.

Wexler, P. 2015. *Toxicology and Environmental Health, Vol II: Toxicology in Antiquity*, London: Elsevier.

Wheelis, M. 1999. Biological warfare before 1914, in Geissler, E., and van Courtland Moon, J.E. (eds), *Biological and Toxin Weapons: Research, Development and Use from the Middle Ages to 1945*, Oxford, UK: Oxford University Press, pp. 8–34.

White, L. 1962. *Medieval Technology and Social Change*, Oxford: Oxford University Press.

Whorton, J. 2010. *The Arsenic Century: How Victorian Britain was Poisoned at Home, Work, and Play*, Oxford: Oxford University Press.

Winner, L. 1978. *Autonomous Technology: Technics-out-of-Control as a Theme in Political Thought*, Cambridge MA: MIT Press.

Yadav, B. and Mohan, M. 2010. *Ancient Indian leaps into Mathematics*, New York: Birkhauser.

Zanders, J.P. 2003. International norms against chemical and biological warfare: An ambiguous legacy, *Journal of Conflict and Security Law* 8, 2: 391–410.

Part II
Technology, crime and harm

Part II

Technology, crime and harm

Section 1

Information communication technologies (ICTs) and digital crime

3

The evolving landscape of technology-dependent crime

Steven Furnell

This chapter examines the evolving relationship between information technology and its technology-specific modes of harm and criminality. These so-called 'cyber-dependent' offences can involve a broad spectrum of activities, everything from the creation and distribution of malware, to distributed attacks and the targeted disruption of networks. The discussion examines the current situation (and aspects of the path that has led us towards it), with particular reference to the technologically driven and technologically based nature of the offences involved.

Technological growth: a foundation for future attack

When we look at technology-dependent threats, one of the first things to realise is that there is a great deal of technology out there to be targeted and potentially exploited. To illustrate the point, statistics from the International Telecommunications Union (ITU 2015) reveal the following trends in terms of ICT growth since the millennium:

- A seven-fold increase in global Internet penetration (i.e. individuals using it) from 2000 to 2015, up from 6.5 per cent to 43 per cent. In terms of the raw numbers, Internet users have increased from 400 million to 3.2 billion in the same period.
- Domestic Internet access increased from 18 per cent in 2005 to 46 per cent in 2015.
- 97 per cent penetration of cellular phones, up from 738 million in 2000, to over 7 billion in 2015.
- Mobile broadband subscriptions increased 12-fold since 2007, reaching 47 per cent penetration in 2015 (noting that, by comparison, fixed-line broadband had only reached 11 per cent penetration by the same point). In parallel, 69 per cent of the worldwide population were within coverage for 3G mobile data, compared to less than half just four years earlier.

In all of these cases, the figures are based on the worldwide averages, and there are of course notable variations in specific cases. As may be expected, the penetration within developed countries is substantially higher than developing or least developed countries, with over 80 per cent of the population online in the former, as against approximately a third and a tenth in the latter cases.

A consequence of these advances, especially when looking at the population in the developed world, is that we expect to be online, and we expect to use the related devices and services. Unfortunately, however, we are often not so prepared when it comes to an expectation to secure and protect them. Indeed, the increased usage has not necessarily been accompanied by a corresponding growth in the associated security, and so there is arguably a far greater population of users and devices that have the potential to fall victim to cybercrime and other forms of online attack.

Classifying the crimes

Part of the challenge of examining cyber-dependent crimes (and indeed other security breaches such as insider abuse and frauds) is that the domain itself gives rise to variations in the use of terminology. This in turn can lead to confusion over what should be counted, and the potential for resultant misrepresentation has existed for some time. Even taking a look at just the subset of threats relating to malware serves to reveal the challenge of the situation. For example, some terms (e.g. virus, worm and Trojan) are technical definitions, insofar as they group programs according to how the code functions. In this respect at least, there is common agreement across the industry. However, sometimes other, non-technical, terms are used to describe malware and malware-related programs. The term 'spyware' is one such example of this. As the name suggests, it refers to software that monitors activity on a computer. Unfortunately, this could include programs that are malicious and those that are not (such as monitoring agents that may be deployed on end-systems within an organisational network).

In 2001, the author's book *Cybercrime: Vandalizing the Information Society* (Furnell 2001) aimed to provide an introduction to the cybercrime problem and to explore some of the main dimensions in which problems could be encountered. As such, there was coverage of core themes such as hacking and malware, along with various examples of (then) relatively new problems such as website defacement and Distributed Denial of Service (DDoS) attacks. However, looking at the book today, it really does seem a reflection of a more innocent age (although it did not seem like it at the time). The coverage predates a whole range of topics that one would now consider fundamental to the topic of cybercrime, and the wider threat landscape. As examples of just a few things that the book does *not* cover, one can list phishing, mobile malware, and Advanced Persistent Threats – three issues that it would be inconceivable to omit from a discussion of cybercrime in today's context. However, one aspect of the book that appears to remain relevant is the way in which it presented a top-level stratification of the cybercrime problem, splitting it into those crimes that are assisted by the presence of computers, and those that have emerged as a direct result of them. This distinction is maintained in much of today's discussion of the problem, with relatively recent definitions from the UK's *Serious and Organised Crime Strategy* (HM Government 2013) maintaining similar descriptions but with more up-to-date labels. The related definitions are presented and contrasted in Table 3.1.

Viewed within this terminology, this chapter is focusing upon the subset of cyber-dependent (computer-focused) crimes. However, while the discussion will continue to use the terms, it is worth briefly considering the validity of this distinction, because the distinction is arguably less meaningful now, given that the wider crime can involve many methods. For example, in the case of a banking Trojan, do we view the malware as the crime, or what it does, or both? With malware being created and distributed to serve underlying criminal motives of computer-enabled fraud and theft, the boundary can blur quite easily. Another contributor to this blurring is that some of the channels for cyber-dependent crimes can also support cyber-enabled crime as well, for example, both malware (cyber-dependent) and phishing (cyber-enabled) can be

Table 3.1 Similarity across the years – top-level categorisations of cybercrime

From Cybercrime (2001)	From Serious and Organised Crime Strategy (2013)
Computer-assisted crimes. Cases in which the computer is used in a supporting capacity, but the underlying crime or offence either predates the emergence of computers or could be committed without them.	**Cyber-enabled crimes** are traditional crimes that are increased in their scale or reach by the use of computers, computer networks or other ICT. Unlike cyber-dependent crimes, they can still be committed without the use of ICT.
Computer-focused crimes. Cases in which the category of crime has emerged as a direct result of computer technology and there is no direct parallel in other sectors.	**Cyber-dependent crimes** are offences that can only be committed by using a computer, computer networks, or other form of ICT … Cyber-dependent crimes are primarily acts directed against computers or network resources, although there may be secondary outcomes from the attacks, such as fraud.

encountered by email. Thus, from the victim perspective, the distinction between them is arguably less meaningful in practice; they can both be seen as email-based threats. As such, while the naming potentially fits for the purposes of developing a taxonomy, it is relevant to consider how much utility it actually has in practice (e.g. is it aligned with how individuals and businesses think of cybercrime?).

Framing things in terms of cyber-dependent and cyber-enabled is focusing upon the means of attack, rather than the motivation and intended outcome – and it is arguably the latter aspects that will interest people in terms of understanding why they might fall victim and the consequent case for protection. Moreover, if we look at a particular attack in more detail, we can often find multiple means being employed in pursuit of a single outcome. For example, distribution of malware may be used to establish a botnet, which may in turn be used to launch DDoS attacks against nominated targets. While there are several affected parties here (e.g. anyone who receives the malware, and particularly those whose systems get infected and become part of the botnet), the main impact is arguably felt by those that find themselves on the receiving end of the resulting DDoS attack (for whom the resulting outcome could range from system outage and disruption, through to loss of revenue and reputational damage). It is also possible to imagine a scenario that intermixes cyber-dependent and cyber-enabled methods within the conduct of a single attack. For example, a phishing attack could be used to acquire a user's login credentials, which in turn lead to a hacking incident, that results in data theft from the victim organisation (with the data itself potentially going on to be used in other ways, e.g. to commit fraud). Indeed, a key part of the evolution over the past few years has been the interplay between the categories. For example, Ransomware (as exemplified by cases such as CryptoLocker[1]) introduces a clear overlap between a technically dependent form of attack and the rather more long-standing financial motive of extortion. The latter is certainly not a cyber-dependent crime, but doing it in this manner, underpinned by malware-style techniques for propagation and payload, certainly depends upon the technology to achieve it.

Ultimately, therefore, a cyber-dependent technique can result in the same impacts as a cyber-enabled one, or they can be used together to the same effect. This view was very much reflected within a 2015 study funded by the UK Home Office, which sought to examine the appropriate means of understanding the scale, trends and measurement of cyber-dependent crime (Furnell et al. 2015). As part of the investigation, views were canvassed from a variety of professionals and practitioners from the anti-malware and wider Internet security industry, and one of the observations emerging was that the provided examples of cyber-dependent crime (e.g. malware, DDoS and hacking) were in fact crime *tools* rather than crime *types*. While this viewpoint clearly contradicts the definition from the *Serious and Organised Crime Strategy* (and indeed legislation that would classify these activities as criminal acts in their own right), it also serves to further highlight the significant perceptual differences that exist around the topic.

One of the factors contributing to the inconsistency of the terminology and vocabulary is the dynamic nature of the domain. The appearance of new threats leads to new names being introduced, and further potential for confusion arises from the industry itself seeking to differentiate its product and service offerings. This point is well-illustrated by the following quote from one of the respondents in the aforementioned Home Office study:

> Most vocabulary seems to come from Vendors' Marketing teams … as new vendors think of better ways of dealing with security they need to change the view of security professionals to fit in with their paradigm.
>
> *(Furnell et al. 2015: 9)*

Although this could be taken to be an implicit criticism of the vendors concerned, the reality is that both they and the victims of the attacks are sitting within an environment that will not stay still. In fact, new names are often *required* in order to enable a distinction to be made between new approaches and those that preceded them. As such, while a high-level classification of crimes as cyber-dependent and cyber-enabled can be valid as a conceptual distinction, it is often less meaningful in practical terms. If a distinction is to be made between *-dependent* and *-enabled* categories, then it is perhaps more meaningfully applied to the underlying methods, leaving the *crimes* to be considered as cyber-related, regardless of how they happen (recognising that as time goes on, more and more criminal activity is likely to have a cyber component involved in it).

Whatever we call it, it's getting worse

Setting aside the specific debate around naming the cyber-crimes, one thing that is clear is that there can be lots of them. There is now a much greater diversity in how the problems are categorised. To illustrate the point, Table 3.2 lists the various categories used by three long-standing and widely cited survey series, all of which give attention to cyber-dependent crimes alongside other types of security incidents (with the table presenting only those categories with potential to relate to the former).[2] While it is possible to identify some points of direct comparability between the lists (e.g. all three have a distinct category relating to malware), there is also a very clear variation in the nature and granularity of the groupings across the set. At the time of writing all of the categories would remain valid, but the picture of cybercrime that emerges would have the potential to look quite different depending upon the lens that is used. It is also worth noting that even here, some of the categories (e.g. 'exploit of user's social network profile') could end up capturing both cyber-dependent and cyber-enabled crimes, depending upon what exactly was done in a given attack.

So, while we do not have a definitive list of cybercrime and attack types, we do nonetheless have a growing one. Even then, however, the snapshot of categorisations provided by Table 3.1 only gives part of the picture, and in order to get a feel for the evolution of the problem, it is worth looking at how the *level* of reporting against the different types has changed over time. As an example, we can look in more detail at the related numbers from the Information Security Breaches Survey (ISBS) series, which has been carried out amongst UK organisations since the early 1990s. Figure 3.1 tracks the related survey categories across the last decade, based

Table 3.2 Cyber-dependent crime categorisations from leading survey series

2010/11 Computer Crime and Security Survey [a]	Global Information Security Survey 2014 [b]	Information Security Breaches Survey 2015 [c]
• Malware infection • Bots/zombies within the organisation • Password sniffing • Denial of Service • Website defacement • Other exploit of public-facing website • Exploit of wireless network • Exploit of DNS server • Exploit of client web browser • Exploit of user's social network profile • Instant messaging abuse • Insider abuse of Internet access or email (i.e. pornography, pirated software, etc.) • Unauthorised access or privilege escalation by insider • System penetration by outsider	• Cyber attacks to disrupt or deface the organisation • Cyber attacks to steal financial information (credit card numbers, bank information, etc.) • Cyber attacks to steal intellectual property or data • Internal attacks (e.g. by disgruntled employees) • Malware (e.g. viruses, worms and Trojan horses) • Zero-day attacks	• Infection by viruses or malicious software • Actual penetration into the organisation's network • Denial of Service attack • Attack on Internet or telecommunications traffic

Notes: (a) Richardson (2010); (b) Ernst & Young (2014); (c) HM Government 2015

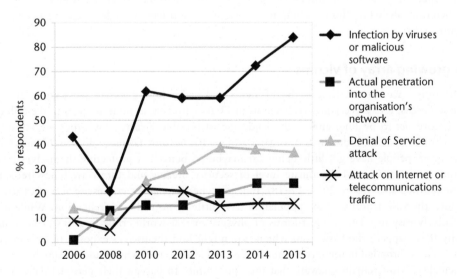

Figure 3.1 Tracking the cyber-dependent crime categories from the Information Security Breaches Survey series

upon the reported data from large organisations (those with 250+ employees). The surveys all included responses from small and medium organisations as well, but differed in whether they reported these distinctly, or included them within an overall figure (whereas the data for large organisations was consistently separated across all editions of the report). The chart focuses upon categories that were consistently reported across the surveys (another ostensibly cyber-dependent category, 'significant attempt to break into the organisation's network', also appeared in the surveys from 2006 through to 2013, but was dropped in the later versions). It should be noted that the frequency of the survey changes partway through the period, because it began as a biennial publication and then became an annual study from 2012/13 onwards.

As indicated earlier, we cannot be definitively sure that all of the incidents are exclusively related to cyber-dependent crimes. For example, some of the reports relating to the penetration of the organisation's network could have resulted from staff disclosing login credentials in response to phishing messages, and the penetration could in turn have led to data theft (so the predominance of the activity would be cyber-enabled, with the penetration just being the means to an end).

When looking at these figures in the context of the wider ISBS findings, it should be noted that they are not necessarily the most prominent categories of attack, nor necessarily the ones considered to be the worst or most costly to the organisations concerned.[3] However, when assessing the evolution of cyber-dependent crime, the key thing to note is basically the trend over time – and the fact that *nothing* appears to have got better when considered on balance across the decade, and reported incidence in all of the categories has ended the period at a significantly higher level than it began. This occurs in spite of organisations reporting increased use of controls, and serves to evidence the challenge of keeping up with the problems. This will also have been further complicated by the broadening of the technology landscape to be protected, with the increased use of mobile devices, the emergence of cloud computing, and increased IT outsourcing, all being trends that were observed across the period and adding to the complexity of safeguarding IT security. Basically, the advancements can open up new opportunities for attack and exploitation, bringing advantages for the cybercriminals, while those seeking to ensure protection have to fight harder to keep pace with change.

A growing array of victims

Of course, we do not have to be IT nerds and technology geeks to get caught out by these threats; today everyone has become an IT user (in fact, as more services move online, it is hard *not* to be one) and so the landscape of possible victims – whether we consider these to be systems, individuals or organisations – has become far more diverse. Meanwhile, however, people have got little better (or maybe no better!) at protecting themselves from other perspectives, and the innate level of security literacy and threat awareness remains low (Furnell and Moore 2014, 12–18). Indeed, cyber-dependent crimes are also ones about which the vast majority of the population have little or no real comprehension. While it is relatively easy to relate to most forms of physical or real-world crime in terms of how and why they happen, the basis from which one might find themselves exposed to hacking or malware is shrouded in mystery for most users, and leads to the 'why would it happen to me?' mentality. Most remain unaware that the opportunity to harvest their personal data, or to enlist their computer as a participant in a botnet, makes their system as valid a target as many others. Moreover, if the system is vulnerable to the type of exploit that a given attacker (or malware) has got within their arsenal, then it makes the system a preferable target to a system that has been better protected.

The exposure is also likely to increase as more systems, applications and devices rely upon us to update them in order to ensure that they remain protected. This is something that we have been pretty poor at doing, even with one or two things to manage, and so the problem is only going to be amplified as there is more technology to be looked after. As an example of previous bad practice in this regard, we can consider the slow and lingering death of Windows XP – an operating system launched in 2001, superseded in 2007, withdrawn from support in April 2014, and yet still in wide scale use well over a year later (with figures in early July 2015 suggesting that it still accounted for around 13 per cent of Windows-based installations, as against only slightly – 17 per cent – using the a version of Windows 8, which was the latest version at the time of writing) (NetMarketShare 2015). The significant risk here is the ongoing exploitability of the systems concerned. Even during its supported lifetime Windows XP was hardly a stranger to vulnerabilities being discovered and exploits being released. As such, it became a routine requirement (at least amongst dutiful system administrators in organisations and diligent end-users at home) to keep the systems regularly patched. Those that were not updated found themselves at far greater risk, both from malware and other forms of online attack. So, as these systems continue to be used beyond the officially declared end-of-life, and no longer receive any ongoing support from Microsoft in terms of security updates, they represent a lingering risk to both their owners and other Internet users who may find themselves indirectly affected as a result (e.g. thanks to a compromised XP system being harnessed into a botnet and sending out spam to all and sundry). Although some might argue that, with such a significant user base still remaining, Microsoft ought to have further extended support, the key requirement here is really for users to be sufficiently aware that their technology poses a risk and be prepared to take action. However, while we have readily adopted the technology, we do not have such an established culture of protecting it.

It does not help when things go from being safe to becoming a danger. Mobile devices are a case in point here, as we have transitioned from basic mobile phones (with limited processing and storage, and a communications capability that typically just covered voice calls and text messaging) to smartphones (with their myriad apps, gigabytes of storage, and always-on broadband data connectivity). We have gone from having very little to worry about aside from the loss or theft of the device itself (in which the main value came from the cost of the hardware) to a point where mobile platforms are now exposed to a whole range of specific threats, with the key value now almost certainly being related to the data/content rather than the physical devices.

These examples highlight the fact that tackling the cybercrime threats requires more than just installing a security technology and expecting the problem to be solved. Potential victims, be they organisations or individuals, need to recognise the routes through which exposure can occur, as well as the fact that something that was safe today could be vulnerable tomorrow. This becomes further apparent when looking at the nature of the threats themselves. As such, the remainder of the discussion in the chapter is focused upon some examples that illustrate this evolution. Rather than attempting to consider all the different manifestations of cybercrime, attention is instead directed towards a specific example of the problem – namely the threat of malicious software, or malware. This is not a difficult choice to make, given that this is typically the one that many people most readily associate with computers, and indeed perhaps best exemplifies the way in which technology has given rise to an offence that was previously without parallel in the traditional crime space.

Malware: the threat that keeps on giving

The problem posed by malicious software – as traditionally represented by viruses, worms, and Trojan horses – provides a good illustration of the changing nature of the cyber-dependent threats. In today's world, these threats can be readily encountered across many devices and online activities, and so malware is basically an issue about which all users ought to have a level of awareness … and ideally some associated protection.

Of course, malware has been a long-standing problem, posing a growing threat to systems ever since the original PC viruses of the mid-1980s.[4] In the early days they spread primarily via the exchange of removable media (specifically floppy disks), and in many cases the payload effects were fairly innocuous (although infection was still not something that anyone particularly welcomed). As time passed, the techniques evolved and viruses soon found their way to infecting executable programs, documents, and various other carriers in addition to boot sectors in which they had originally concealed themselves on the disks. The most significant development in terms of amplifying the potential impact of the problem was when malware strains emerged that started to leverage our Internet connectivity. Of course, the first network-based malware actually predated this by some years, with the Internet (or Morris) Worm of 1988 representing the first large-scale incident.[5] However, the Internet in 1988 was a very different thing, with only around 60,000 host systems, residing mostly within scientific and educational establishments. By the time that malware harnessed it en masse in the late 1990s, things had changed significantly, with growing public adoption by organisations and home users meaning that there were now millions of systems and users to fall victim. Indeed, this was early evidence that our IT practices would directly influence the nature of the threats – in short, where we go, the malware will follow.

Reading the security surveys of today, malware typically takes the top spot in terms of reported incidents. For example, looking again at the Information Security Breaches Survey series, the detail in Figure 3.1 has already illustrated that it was the most frequently encountered category amongst the cyber-dependent categories. Moreover, in the most recent releases of the survey it was also the most frequently encountered of all breach categories amongst both large and small organisations, placing it ahead of 'Attacks by an unauthorised outsider', 'Theft or fraud involving computers', and 'Other incidents caused by staff'. For small businesses in particular, it was significantly ahead of other breach categories, suggesting that while the potential for staff-related incidents and more specific targeting by external attackers may increase with organisation size, the indiscriminate nature of malware can mean that size is far less of a factor. So, while it does not necessarily end up being the most costly category, malware is clearly amongst the most prevalent, and therefore something that organisations would ignore at their peril.

Part of the reason for this prevalence is that malware has managed to find its way to us via every conceivable channel, plus some that we would not naturally expect. As successive online services have proven popular (from email and instant messaging through to social networking and mobile apps), so the malware has followed along and attempted to use them as a channel for infection. Its chances here have often been aided by the fact that public awareness of the threat tends not to keep pace with the reality, and while many users may ultimately have got the message that unsolicited emails and attachments may be unsafe, they may remain totally vulnerable to malware reaching them through an unexpected route such as their social network.

As time has gone on, the malware problem has basically become more problematic in all dimensions:

- The threat has increased in *volume*. For example, in 2014 Kaspersky Lab reported that it was identifying 325,000 new malicious files per day (Kaspersky Lab 2014a). This represented an increase of 125,000 per day compared to 2012, and is a world apart from 2008, when we were talking in terms of about 500,000 known malware strains in total and 8,000 new ones being discovered per month (and back then at least, even these numbers seemed significant).
- Current malware has a greater degree of *sophistication* than that of the past. In addition to the technical complexity of the code (as exemplified in landmark cases such as Stuxnet, Duqu and Flame, all of which were noted for their complexity (Kushner 2013)), where malware might once use a specific technique to get into a system and do a specific thing once it got there, today's malware can employ a multitude of methods and will often just steal everything that is available, and allow the attackers to work out what to do with it later (Lee 2011).
- There is now a greater range of infection *routes*, meaning that potential victims can encounter the problem through a wider range of services and devices. Gone are the days when malware was just a concern for PCs, or when the most likely way to get it was as an email attachment. It can now reach us across various devices, and via all manner of applications, and more particularly via websites and other online routes if our systems are hosting vulnerabilities that leave them exploitable.

Moreover, all of this has been accompanied by a notable change in the intended behaviours, with today's malware having more overtly criminal intentions than that of the past. Put another way, many attacks have now gone from being the *motive* to being the *means*. Rather than simply releasing malware in order for it to be seen and make a mark through media attention, the end objective is now often an organised criminal activity rather than predominantly the realm of individuals or groups actuating in pursuit of challenge, mischief or other non-financial motives. It also ends up being used in support of state-sponsored attacks, corporate espionage, and as an extension of real-world conflict.

There has also been a significant shift in the participants and their roles. The creator of the malware or exploit code is no longer necessarily going to be the ultimate user, and is often developing it specifically for others to use – often for a fee. This in turn brings the concept of a marketplace, where attacks – both general and targeted – can be bought by those that desire and require them. A good example here is ZeuS, a highly prevalent banking Trojan, which was originally identified in 2007 but became most prominent from 2009 onwards, leading to millions of infected machines in over 190 countries. ZeuS offered a variety of mechanisms for stealing data, including the use of webinjects, which were used to inject rogue content into banking websites (e.g. changing the page from asking for selected login information to asking for full details of the user's secret information, which would then enable theft from their account later). Such webinjects could be found for sale in online forums, with targets including American, British, Canadian, and German banks, and priced according to the scope of the targets concerned (e.g. the cost of one webinject pack was $60, whereas a UK webinject pack cost $800, and updating or modification of webinjects was $20 each) (Klein 2011). Clearly the ability for attacks to be purchased in this way, by parties that would otherwise not have direct access to them, affects the potential scale of both the problem and the victim base, with attackers having a more specific motive and their targets facing a more specific risk.

A further change, and arguably the most notable one from the victim perspective, is the fact that the malware can now reach us in contexts that were previously immune or unrelated to it. This links to the point about the range of infection routes in the list above, and is particularly illustrated by the rise of malware on mobile devices such as smartphones. While this threat had long been foreseen by many security professionals, by the time it actually arrived mobile phones were an established technology that millions of people had owned for years, and were accustomed to using without any concern about malware. With this in mind, the change of circumstance is a good example of how the cybercrime landscape can evolve in new directions and catch us out. As such, this is the focus of the penultimate section below.

Mobile malware – an old problem in a new guise

As highlighted by the ITU statistics at the start of the chapter, mobile devices have been the big growth area for IT in global terms, and so following some of the observations from earlier discussion, that alone makes them an attractive target for attack. And, as also mentioned earlier, mobile phones are a technology that have moved from being relatively 'safe' in security terms (and completely safe from the perspective of malware) to being an area in which users need to be explicitly aware of the threat and take steps to protect against it. Moreover, this has happened within a very short space of time, with the almost inevitable consequence that those using the technologies have often been caught off-guard. For example, if the numbers are to be believed, then we have gone from a situation where there was an average of over 800 new mobile malware discoveries per month during 2011 (Kaspersky Lab 2013), to an average approaching 5,000 per *day* just three years later (G DATA 2015).

While there are a variety of mobile devices and operating systems to choose from, they have proven to be far from equal in their attraction to malware writers. Indeed, although various mobile platforms, including Symbian, Windows Mobile and even iOS (or iPhone OS as it was originally called in the pre-iPad days) had been around in the market for some time before, it is Google's Android operating system that has proven to be by far the most attractive target in this space. The reason for this is based upon two significant factors. Firstly, Android quickly amassed a large user base, running on products from multiple manufacturers, spanning a full spectrum of budget and high-end devices. Secondly, the sources from which users can download apps were not regulated and restricted in the same way as those on some of the other platforms. For example, while Apple operated a walled-garden approach with iOS – only allowing users to download apps from its own App Store, and requiring apps to be submitted to a formal approval process before being made available (which in turn involves some level of code verification in order to guard against malicious content) – the Android approach was far more permissive, with Google Play (formerly the Android Market) allowing open and relatively unrestricted submission of new apps. As a consequence, apps with malicious functionality were able to pass through unchecked, giving them ready access to a large community of potential users; as a result of this, Android was quickly able to develop an unwanted monopoly position in its share of the mobile malware market. Indeed, looking at figures published by Kaspersky Lab in late 2014, 98 per cent of all mobile malware detections during 2013 were on Android (Kaspersky Lab 2014b).

In parallel, however, the *awareness* of the threat and the accompanying use of malware protection on mobile devices has simply not kept pace. For example, in a survey that we conducted amongst 1,222 users in the UK, the US, Malaysia and South Africa,[6] we discovered that while 91 per cent reported having antivirus protection on their desktop or laptop PCs, only 10 per cent claimed to have similar protection on their phone. Of course, as readers familiar with the market during this period will know, users on iPhones would not have had the option to do so in the first place, as no such apps were actually available.[7] With this in mind, we also looked at antivirus usage amongst the specific subset of respondents that were Android users.

This yielded 688 eligible respondents, but the level of antivirus usage remained very low, with only 14 per cent reporting having it.

This lag between the emergence of new crimes and our awareness and preparedness for them does not bode well for the future. The evidence of the past suggests that we will only see a further broadening of the attack surface – in terms of both devices and services – and so new technological opportunities ought to be approached with more readiness from the outset, rather than recognised retrospectively. Unfortunately, we have yet to see signs that this is the case...

At the time of writing we would appear to be standing on the brink of a new era of problems thanks to the emergence of the Internet of Things and a whole new landscape of online devices. Our drive for technological innovation appears to show a remarkable capacity for ignoring the security lessons of the past. In this respect, the IoT is following the well-trodden path of PCs, wireless networking, mobile devices, with all the attention going towards innovating and deploying the technology, while the risks are given little or no consideration. As such, the devices have significant potential to further increase the breadth of exploitable technologies – and to do so on a large scale. As an example, one can now readily purchase IP-enabled surveillance cameras, promoted as a solution for home security and remote monitoring of other premises. However, many such systems are used with default passwords, meaning that anyone willing to do a tiny bit of research to find out what the default is can tap into the device and watch the video that is being captured (NetworkWorld 2014). It is, of course, somewhat ironic that devices specifically installed to improve security in the physical world should find themselves fundamentally vulnerable to being exploited on the cyber side. However, the same problem has previously been seen with unsecured wireless access points, with devices shipped without security enabled and with default passwords to administer them, and so it is not hard to foresee that a similar potential for misuse would exist here.

Conclusions

Whether or not we agree with their specific categorisation as cyber-dependent crimes, it is clear that the various technology-centric methods of attack have become an increasing threat to the security of our systems and data. The very nature of our technology-dependent society means that this situation is highly unlikely to be reversed. Unfortunately, however, the evolution of these threats is significantly outpacing our ability to recognise and respond to them. Indeed, from a cynical perspective, one of the main advancements over the years appears to be that we have now got more names for the things that can harm us; our ability to protect against them still leaves a lot to be desired.

In reading the chapter, it may have been noted that it has devoted little attention to describing the specific workings of the malware or other attacks. The basic reason is that, if we have an eye to the future, many of these underlying details do not matter. The specifics will change, and will depend upon the technologies and opportunities of the day. What is notable is the broader trend towards greater exploitation and sophistication. And what is therefore important is to realise that the only way to manage the situation is through ongoing vigilance. Old threats will persist for as long as the opportunity exists to use them, and new ones will emerge where fresh opportunities can be found. Meanwhile, criminal activities will continue to be drawn towards whatever openings the technology can offer – and this tends to follow the technologies that we ourselves find the most attractive for our own reasons. As such, we need to recognise that the attackers are on the journey with us, and take steps to ensure that they cannot enjoy the ride at our expense.

Notes

1 First appearing in September 2013, CryptoLocker was PC-based cryptoware/ransomware that encrypted the contents of the user's drive(s) and demanded payment in order to get the decryption key. Over half a million computers were reportedly infected with it between September 2013 and May 2014, with FBI estimates in June 2014 suggesting that $27 million had been paid out by victims in order to recover their data.

2 In addition to being taken from established survey series, the three selected examples present perspectives from the United States, the United Kingdom, and a global sample group.

3 For example, while malware infection was actually the most commonly encountered incident in the 2015 survey (with 84 per cent of respondents reporting it), only 11 per cent considered it to have been the cause of their worst security incident. By contrast, a third attributed their worst incident to a category not shown in Figure 3.1, namely 'Theft or unauthorised disclosure of confidential information' – which could arguably result from deliberate cyber-enabled or cyber-dependent attacks, or indeed from accidental incidents.

4 The first reported example was the MS-DOS-based Brain virus in January 1986. In this case the payload effect was essentially harmless, simply changing the name of the disk (the volume label) to become "©Brain".

5 Written as an experiment by Cornell University student Robert Morris, the Internet Worm spread far faster than he had intended, and quickly ended up infecting around 10 per cent of the hosts on the entire Internet. Moreover, the volume of traffic that these systems then generated, as the worm sought to find other systems to infect, served to bring the network to a standstill. The incident was the direct catalyst for the formation of CERT – the Computer Emergency Response Team – in order to ensure that the growing Internet community was ready to deal with such incidents in the future.

6 The data was collected in different phases, collectively spanning the period between September 2013 and November 2014.

7 The background here is that because App Store content is subject to verification before it is approved for release, this is considered to remove (or at least minimize) the potential for malware to slip through and find its way onto user devices. With this safeguard implicitly providing protection by default, Apple does not see a valid market for antivirus tools and does not approve them for inclusion in the App Store. There is, however, a risk for users that have chosen to *jailbreak* their iPhone/iPad, in order to enable apps to be installed from non-approved sources. In these cases, malware has a route onto the device, and may exploit the additional access rights that jailbreaking will have made available.

References

Ernst & Young. 2014. *Get ahead of cybercrime – EY's Global Information Security Survey 2014*. Available at: http://www.ey.com/Publication/vwLUAssets/EY-cyber-threat-intelligence-how-to-get-ahead-of-cybercrime/$FILE/EY-cyber-threat-intelligence-how-to-get-ahead-of-cybercrime.pdf (accessed 31 Aug 2016).

Furnell, S. 2001. *Cybercrime: Vandalizing the Information Society*. London: Addison Wesley.

Furnell, S., Emm, D. and Papadaki, M. 2015. The challenge of measuring cyber-dependent crimes, *Computer Fraud and Security*, October 2015: 5–12.

Furnell, S. and Moore, L. 2014. Security literacy: The missing link in today's online society?, *Computer Fraud & Security*, May 2014: 12–18.

G DATA. 2015. *Mobile Malware Report. Threat Report: Q1/2015*. G DATA Software AG. Available at: https://secure.gd/dl-us-mmwr201501 (accessed 1 July 2015).

HM Government. 2013. *Serious and Organised Crime Strategy*. October 2013. Available at: https://www.gov.uk/government/publications/serious-organised-crime-strategy (accessed 31 Aug 2016).

HM Government. 2015. *Information Security Breaches Survey – Technical Report*. Department for Business, Innovation and Skills. Available at: https://www.gov.uk/government/uploads/system/uploads/attachment_data/file/432412/bis-15-302-information_security_breaches_survey_2015-full-report.pdf (accessed 31 Aug 2016).

ITU 2015. *ICT Facts and Figures – The world in 2015*. ICT Data and Statistics Division, International Telecommunication Union. May 2015.

Kaspersky Lab. 2013. 99% of all mobile threats target Android devices, 7 January 2013. Available at: www.kaspersky.com/about/news/virus/2013/99_of_all_mobile_threats_target_Android_devices (accessed 31 Aug 2016).

Kaspersky Lab. 2014. Kaspersky lab is detecting 325,000 new malicious files every day, *Virus News*, 3 December 2014. Available at: www.kaspersky.com/about/news/virus/2014/Kaspersky-Lab-is-Detecting-325000-New-Malicious-Files-Every-Day (accessed 2 July 2015).

Kaspersky Lab. 2014. Mobile cyber-threats: A joint study by Kaspersky Lab and INTERPOL, 6 October 2014. Available at: https://securelist.com/analysis/publications/66978/mobile-cyber-threats-a-joint-study-by-kaspersky-lab-and-interpol/ (accessed 31 Aug 2016).

Klein, A. 2011. Webinjects for sale on the underground market, Trusteer Blog, 2 November 2011. Available at: www.trusteer.com/cn/node/355 (accessed 21 July 2015).

Kushner, D. 2013. The real story of Stuxnet, *IEEE Spectrum*, 26 February 2013. Available at: http://spectrum.ieee.org/telecom/security/the-real-story-of-stuxnet (accessed 31 Aug 2016).

Lee, D. 2011. 'Steal everything' era of hacking, BBC News, 27 April 2011. Available at: www.bbc.co.uk/news/technology-13213632 (accessed 31 Aug 2016).

NetMartketShare. 2015. Desktop operating system market share, Available at: www.netmarketshare.com/operating-system-market-share.aspx (accessed 2 July 2015).

NetworkWorld. 2014. Peeping into 73,000 unsecured security cameras thanks to default passwords, *Network World*, 6 November 2014. Available at: www.networkworld.com/article/2844283/microsoft-subnet/peeping-into-73-000-unsecured-security-cameras-thanks-to-default-passwords.html (accessed 1 July 2015).

Richardson, R. 2010. *15th Annual 2010/2011 Computer Crime and Security Survey*. Computer Security Institute.

4

Technology and fraud

The 'fraudogenic' consequences of the Internet revolution

Mark Button and Cassandra Cross

Introduction

Recorded crime in many industrialised countries over the last twenty years has been falling and this has stimulated much debate over whether this is really happening and if so, what explains it (Farrell *et al.* 2011; Knepper 2015). One common proposition is that enhanced security and crime prevention measures have reduced opportunities for crime and contributed to a decline (Van Dijk 2008; Farrell *et al.* 2011). However, this is really only one half of the story. This chapter will argue that juxtaposed to declining crime trends and reduced opportunities for 'traditional crimes' to occur, there has been a technological revolution, which over the last 20 years, has centred around the Internet and related technologies. This has created a vast pool of new opportunities for crime, particularly centred around fraud. Therefore, the chapter will argue that this technological evolution has created a 'fraudogenic' environment, in which the 'criminal justice complex', by which we mean criminal justice bodies, politicians and many criminologists have been 'behind the curve'. There are now in many industrial countries huge levels of fraud victimisation resulting from cyber-enabled and dependent attacks, but which traditional accurate measures of crime, such as crime surveys, have been ill-equipped to capture, resulting in an attenuation of the problem. This is starting to change, but much more attention and resources need to be dedicated to the problem.

This chapter will largely use the UK as the basis for discussion, but examples where relevant from other industrialised countries will also be used. The chapter will begin by exploring the concept of fraud and the technological aspects of it. It will then examine the technological changes which have occurred, before examining the impact these have had for fraud. The chapter will use the themes of opportunity, globalisation, risk and responsibilisation to illustrate some of the major changes. The chapter will end with a more theoretical discussion arguing the changes have resulted in a more 'fraudogenic' environment and that current responses to it have effectively attenuated the problem.

Defining fraud and its relationship with technology

Fraud covers a broad range of activities united by some form of misrepresentation by a party to secure an advantage for that party or cause a disadvantage to others. It covers criminal, civil and regulatory acts of deviance. Fraud is not a new phenomenon and its relationship to technology is well established. While current conceptualisations of fraud are strongly linked to new technologies (such as the Internet), it is important to note that older technologies provided the means for deception in previous centuries. For example, in medieval times seals were seen as "objects denoting both identity and authority" (Bedos-Rezak 2000: 1531) and therefore the target for earlier criminals to perpetrate both fraud and identity theft offences. The evolution of technology has not changed the desire or willingness of offenders to engage in fraud, rather it has simply altered the means through which these offences are executed.

Unlike many other volume crimes such as burglary or theft, it is not always clear with frauds whether they actually are frauds and if so whether they meet criminal, civil or regulatory thresholds of evidence. In many countries the criminal acts of fraud are often covered by a complex mix of legislation. England and Wales is one of the few countries to have codified the criminal act of fraud with the *Fraud Act 2006* (although many criminal acts of fraud are still dealt with by other legislation, such as tax frauds, social security frauds etc.). Under the 2006 Act there are a number of ways in which fraud can be committed:

* Fraud by False Representation (this could cover the submission of a false invoice for services for a payment).
* Fraud by Failing to Disclose information (this could be a person is paid for 40 hours per week, but in fact only works 30 and fails to disclose this or a prospective employee is asked for certain information on the application form but doesn't provide it).
* Fraud by Abuse of Position (this is where a person in a position of trust abuses their position such as an accountant diverting funds to their own personal account).

The legislation also set out a series of other offences such as:

* Possession of articles for use in frauds and making or supplying articles for use in frauds (this is very wide ranging and could include catching someone at home with a paper or electronic copy of a false invoice which could be submitted to a company).
* Participating in a fraudulent business (this could be a car dealership founded on enhancing the value of cars by turning back the mileage clocks).
* Obtaining services dishonestly (this could be securing an insurance policy by providing false or inaccurate information) (Farrell *et al.* 2007).

It is interesting to note that the 2006 legislation in England and Wales recognised the growing role of technology with the offences around the possession, using and making articles to enable frauds to take place. Theoretically, this encompasses the role of technology to a certain degree. As will be noted shortly in this chapter, technology has increased exponentially the opportunities to commit fraud. In the United Kingdom the growing technological element to crime has led the police to divide such crimes between cyber-enabled and cyber-dependent. McGuire and Dowling (2013) define these as:

Cyber-dependent crimes (or 'pure' cyber crimes) are offences that can only be committed using a computer, computer networks or other form of information communications technology (ICT).

Cyber-enabled crimes are traditional crimes, which can be increased in their scale or reach by use of computers, computer networks or other forms of information communications technology (ICT). Unlike cyber-dependent crimes, they can be committed without the use of ICT. Two of the most widely published instances of cyber-enabled crime relate to fraud and theft.

Cybercrime is much broader than fraud, covering illegal pornography, online harassment and hacking, to name but some. It would therefore be useful to illustrate the two types of cybercrime vis-à-vis fraud in Table 4.1.

Table 4.1 shows the wide diversity of cyber-related frauds. There is much interest and focus upon cyber-dependent crimes and frauds. These clearly are a major problem, but the vast majority of cyber-related frauds arise from the cyber-enabled frauds. Indeed, at a conference in 2015, Commander Head of the City of London Police (drawing upon Action Fraud data) estimated 70 per cent of frauds reported to Action Fraud were cyber-enabled (Head 2015) (the other 30 per cent also includes non-cyber related fraud). Clearly the skills necessary to perpetrate cyber-enabled fraud are much less than for cyber-dependent. Setting up websites, sending emails etc., do not require significant information technology skills, whereas most cyber-dependent frauds do. Before we examine in more depth the opportunities technology has provided for fraud it would be useful to examine the extent of fraud.

Table 4.1 Selected examples of cyber-enabled and dependent frauds

Cyber-dependent frauds	*Cyber-enabled frauds*
Hacking The computer systems of organisations and individuals are hacked to secure personal information to perpetrate frauds or to actually divert monies.	*Card Not Present Fraud* Fraudsters secure the relevant numbers from plastic cards and then use them to purchase goods and services online.
Monitoring Aids such as key-sniffers/spyware placed on computers to enable passwords and personal details to be identified to perpetrate frauds. This also includes computer viruses targeting machines in order to gain passwords and personal information, or to enable control to facilitate other actions (such as Dedicated Denial of Service (DDoS) attacks.	*Fraudulent Sales* Goods and services which are non-existent or not as described sold online. *Phishing Scams* Emails sent purporting to be from an official body which persuade the respondent into supplying personal information which can then be used for a fraud. *Mass Marketing Frauds* (also known as *Advanced fee fraud*) A wide variety of frauds which seek to secure a payment from the victim, ranging from fake lottery prizes, to inheritance notifications to business investment opportunities. *Romance Frauds* The method in which offenders use dating websites and other social media sites under the guise of legitimate relationships to trick 'lovers' into sending (often recurring amounts of) money.

Source: Adapted from McGuire and Dowling (2013).

Note: Identity frauds where a person's personal information is used to commit a fraud such as apply for a loan in their name or use their credit card details to purchase goods and services may be either the result of cyber-dependent or cyber-enabled means and often the victim will not know how their personal details were gained.

Extent of fraud

Fraud is one of the, if not the most, voluminous crimes in most industrialised countries with financial losses that eclipse all other crimes (Button and Tunley 2015). It is therefore a gigantic problem, but we do not know the true extent because of the following reasons:

- The ambiguity of fraud: many acts are open to interpretation whether they are crimes, civil issues, or regulatory transgressions (Levi 2008; Button and Tunley 2015).
- Unknowing victims: many acts of fraud lie hidden as legitimate transactions such as false invoices which have passed accounting controls, been paid and will remain as legitimate until exposed (if at all) (Button and Gee 2013; Button and Tunley 2015).
- Non-reporting individual victims: many victims are often reluctant to report victimisation due to the shame and stigma associated with this type of victimisation and a strong victim blaming discourse (Button et al. 2009a; Cross 2013, 2015).
- Non-reporting organisational victims: many organisations are unwilling to report to authorities for fear of damaging their reputations and potential legal liability (Furnell 2002). This is exacerbated by a lack of mandatory data breach reporting in countries like Australia, whereby organisations who detect a compromise to security and details of customers/ employees are not required to report this to authorities (Burdon et al. 2010).

Table 4.2 The extent of fraud

The extent of fraud against individuals	The extent of fraud against organisations
England and Wales Mass marketing fraud, survey found 48% targeted, 8% a victim (OFT 2006)	*England and Wales* 11% of retailers, 8% accommodation/ food sector, 7% recreation sector, 3% agriculture, fishing and food experienced a fraud (ONS 2014)
Identity fraud, 8.8% victims of identity fraud in previous 12 months (National Fraud Authority 2013)	
Card fraud, 5% victims in previous year (ONS 2015a)	
Australia Survey indicated that 1.2 million Australians over the age of 15 were a victim of at least one personal fraud in the last 12 months. This equates to a national victimisation rate of 6.7% (ABS 2012)	*Australia* 57% of organisations surveyed experienced economic crime in the past 24 months (PwC 2015)
Almost $82 million was lost by Australians to fraud in 2014 (ACCC 2015)	Stemming from this, 47% of organisations experienced in excess of 10 fraudulent incidents in this time period (PwC 2015)
USA Online fraud losses over $525 million in 2012 (IC3 2013) 11.7 million persons (5% of all persons aged 16 or older in the USA) experienced at least one type of identity theft in a 2-year period (Langton and Planty 2010)	*USA* Forty-five per cent of US organisations suffered from economic crime (including fraud) in the past 2 years (PwC 2015)
Canada Over $74 million was lost to mass marketing fraud alone in 2013 (CAFC 2015)	*Canada* Thirty-six per cent of organisations reported being victims of economic crime (PwC 2015)

- Reporting fraud is not always straightforward: in some countries there are multiple bodies to report fraud to, meaning some victims often find it difficult to identify the right body and are sent on 'merry-go-rounds'. Sometimes agencies question whether a report can be accepted. Organisations suffering large volumes of fraud sometimes are unable to file bulk reports so do not report (Button *et al.* 2009a; Cross *et al.* 2016).
- Measures of fraud are poor: there are a variety of measures of fraud which have suspect methodologies and crime statistics – because of the issues above, they are particularly poor in gauging the size of the problem (Levi and Burrows 2008; Fitzgerald 2014).

The latter issue is particularly important and will be developed further here. Crime statistics are well known for their flaws, but these are exacerbated by the issues identified above. There are lots of industry barometers published which use varying methodologies of rigour to estimate the extent of organisational fraud (Levi and Burrows 2008). Victimisation surveys, despite being perceived as the gold standard of crime measurement, do not generally cover fraud in any depth (Fitzgerald 2014). For example, the England and Wales crime survey only seeks information on the experience of plastic card fraud in the past 12 months. It does not cover the many other types of fraud a person could have suffered. The other major flaw in the survey format is its exclusive focus on individual victimisation. The most accurate measures of organisational fraud, fraud loss measurement exercises, are not carried out by a large number of organisations (Button and Gee 2013) nor do they feature in crime victimisation surveys. Nevertheless, the table above notes some of the most accurate measures of fraud for individuals and organisations from around the industrialised world.

A technological explosion of connectivity

> A scam artist is committing malpractice if he's not using the Internet.
>
> *(Danner 2000 cited in Langenderfer and Shrimp 2001: 764)*

The above quote by Danner sums up the importance of the technological changes which have occurred over the last two decades. Technology, particularly the Internet, has opened up new ways to perpetrate frauds and to industrialise old ones. It is worth illustrating some of the technological changes which have occurred to give a context to the substantial changes which have occurred around the world, before we look more specifically at its impact upon fraud.

The expansion of the Internet

In December 1995 across the globe there were estimated to be 16 million users of the Internet (Jewkes and Yar 2010). By 2000, there were around 361 million users of the Internet worldwide (Internet World Stats 2012). This had grown to almost 2.5 billion in 2011, amounting to 35 per cent of the world's population using the Internet, compared to 18 per cent in 2006 or just over 1.1 billion. While the overall number of users has increased dramatically, there are still distinct lines of disadvantage across the globe. For example, in the developed world in 2015, there were 80.8 (per 100) households with a computer compared to 32.9 (per 100) in the developing world (International Telecommunications Union (ITU) 2015). The disparity continues between developed and developing nations for households with Internet access at home (81.3 compared to 34.1) and individuals using the Internet (82.2 compared to 35.3) (ITU 2015). The bandwidth for Internet access has also been growing exponentially during this period, enabling a greater range of uses for the Internet to be pursued. Prices have also been dropping for access and in the developing world in the two years before 2011 average prices had dropped by over 50 per cent (ITU 2012).

While there are significant differences between developed and developing countries regarding households with computers and Internet access, the advancement of mobile computing devices has increased access across many countries. Access to the Internet has expanded from desk-based computers, to laptops, to tablets, to televisions to mobile phones. In 2015, mobile-cellular telephone subscriptions were 120.6 (per 100) in the developed world compared to 91.8 (per 100) in developing countries. This highlights less of a disparity in terms of access to mobile devices which then has an impact in terms of access to Internet networks. Indeed, global mobile phone subscriptions in the four years prior to 2011 had been growing at the rate of 45 per cent annually with 87 per cent of the global population or 5.9 billion having one and within that a doubling of those subscriptions with access to mobile broadband to 1.2 billion (ITU 2012). Smartphones, which have advanced computing and Internet access facilities, have also been growing substantially. Prevalence research for Google found rates of growth for smartphone usage from February 2011 to February 2012 to have grown from 26 per cent to 47 per cent in the USA, 39 per cent to 51 per cent in the UK, 27 per cent to 38 per cent in France, 18 per cent to 29 per cent in Germany and 6 per cent to 20 per cent in Japan (Think Insights with Google 2012). These statistics show the growing opportunities for not only those in industrialised countries to have access to the Internet and related technologies, but also significant parts of the developing world too.

The growth of social networking

Completely new forms of social interaction have emerged through social networking sites such as Facebook, Tinder and LinkedIn. Indeed Facebook, which was only launched in 2004 in the USA, has grown to almost a billion users, with 936 million daily active users recorded in March 2015 (Facebook 2015). Social networking sites dedicated to people seeking love, sex and friendship have also emerged. Research by YouGov in the UK found that 1 in 5 relationships now begin online (Nolan 2012).

The growth of online shopping and banking

The trading site eBay, formed in 1995, now has a global customer base of 233 million enabling millions to trade globally with one another on a daily basis (eBay, n.d.). Research conducted in the UK in 2009 found 66 per cent of 16- to 74-year-olds had ordered goods or services on the Internet in the previous 12 months (Randall 2010). New forms of undertaking shopping have grown from nothing to the norm in less than a decade. Amazon went online in 1995 selling books and now sells virtually every consumer item and had global revenues of $61 billion in 2012 (Amazon 2013). The Internet has also changed the way that financial transactions and business are conducted. By 2009, over 22 million people in the UK were banking online for their main account, which was over half the Internet users at that time (UK Payments Administration 2009). Online payment portals (such as PayPal) have enabled business and consumer activity to move into an online context.

Technology and commerce

Further to this, the use of technology (such as computers, the Internet, mobile phones) has also fuelled completely new ways of doing business and providing services. Take applying for jobs –

many organisations advertise opportunities online and provide websites where applications are completed online. Many payments by both organisations and individuals are done so through online banking or through automated telephone systems, rather than paper cheques. Loans, social security benefits, grants, tax returns, and passports are also increasingly secured via online and automated telephone technology. Even participation in electoral voting is shifting to an online environment. Methods of communication, between companies, their staff and customers have also shifted from traditional face-to-face interactions to that of email and other modern forms of communication. There is no doubt that the Internet has revolutionised and somewhat disrupted traditional methods of business, communication and knowledge production. This of itself is not a negative thing, as these new methods can provide new and more efficient means to provide services. However, they can also create new threats and more opportunities for fraud to occur, as the following section will demonstrate.

The impact of technology on fraud

These changes in technology and the way things are done have had a significant impact on fraud. The remainder of this chapter will note these changes under the following key themes, most of which are underpinned by the proliferation of opportunities for fraud to occur. Second, the transformation of technology has fuelled the globalisation of fraud, where frauds increasingly take place at a global cross-border level. Third, these changes have fuelled an increased exposure and likelihood of becoming a victim of fraud for both individuals and organisations, through the course of conducting legitimate business or other activities. Traditional understandings of crime risk and exposure are no longer adequate to explain the potential targeting and victimisation of an individual in the virtual environment. Indeed, there are now persons who 20 years ago were designated at a very low risk of crime victimisation who are now regularly exposed to the risk of crime in the form of cyber-enabled fraud. Fourth, the changes have fuelled a responsibilisation strategy for dealing with the problem as the state infrastructures have been unable to cope. Each of these will now be explored in more depth.

Opportunities for fraud

There is extensive research illustrating the links between crime and opportunity (Mayhew *et al.* 1976; Cohen and Felson 1979; Clarke 1980). The importance of opportunity has also been noted in fraud (Cressey 1973; Newman and Clarke 2003). The technological changes which have occurred have increased the potential opportunities for fraud exponentially. First and foremost, completely new opportunities for frauds have occurred. The cyber-dependent frauds for the technologically skilled provide new opportunities to secure sensitive information which can be used to perpetrate a fraud or be sold to someone else to do so. However, as noted earlier, these types of frauds only account for a small proportion of frauds, with the majority emanating from cyber-enabled means. Much larger are the cyber-enabled frauds. Some of these new opportunities will now be explored.

New types of fraud

Advances in technology have created opportunities for new types of fraud, which either did not exist before or were much more difficult to successfully pursue. The Internet needs to be viewed as a set of 'social practices' and it is through these legitimate social practices that some people will create distinct opportunities for offending (Yar 2013: 6). Therefore, illegitimate online activities must be viewed in conjunction with their legitimate counterparts. For instance, the expansion of online dating and social networking sites has created the opportunity for romance fraud. Typically, the modus operandi of such scams involves a person either establishing a false profile or stealing the legitimate identity of another person, enticing potential suitors and then developing what is perceived to be a genuine online relationship. Once trust is established between the victim and offender, the fraudster creates fictitious scenarios where the victim is groomed and pressured into handing over money (Whitty 2013). For example, there might be a business deal where a short-term loan is required or an accident which requires money to pay for hospital treatment for a relative. The Internet has not exclusively created these types of fraudulent scenarios; such frauds were possible and occurred before the Internet, however they were much more difficult to successfully pursue as they required face-to-face contact. Dating and romance fraud is now one of the most prevalent and lucrative categories of online fraud victimisation, recording victim losses close to $28m in Australia alone in 2014 (ACCC 2015). One of the recurring themes of many cyber-enabled frauds is the 'at-a-distance' aspect, which makes the frauds easier to perpetrate (Duffield and Grabosky 2001), and makes the use of false and stolen identities a more attractive option for offenders, which in turn, is harder to detect by victims. This is also evident in phishing emails, whereby potential victims receive an authoritative-looking email from a legitimate organisation (such as a bank or other service provider). This email generally claims there has been a security breach and requests potential victims to confirm their identity or 'reset' their passwords. Again such scams were possible before email, but would have required much more effort and resources to perpetrate. These two examples illustrate how the emerging technologies have fuelled new opportunities for fraud from legitimate activities which are now dominant within a virtual environment. Indeed, it is estimated globally there are 29 billion spam emails daily and that the email virus rate is 1 in 196 and phishing emails are 1 in 392 (Symantec 2014).

Adapting to new technology

A second strand is the ease and convenience through which technology has created opportunities to undertake frauds via new methods. For example, the advent of online sales platforms such as eBay and Gumtree has made it simpler for fraudsters to sell non-existent and fake goods. Indeed Treadwell (2012) has noted how traditional criminals have moved from car boot sales to eBay to sell counterfeit goods. The proliferation of technology which is relatively easy to use has also enabled many ordinary people (and criminals) to use computer software such as Photoshop and related technologies to create documents which can be used to perpetrate fraud. As an example, in 2015 a man in Bedfordshire, UK, was convicted of three fraud-related offences for displaying a false 'blue badge' (parking passes for disabled motorists that enable free parking and to park in a restricted area). The driver had used Photoshop to create a new pass from a legitimate one which had expired (*Bedford Today* 2015). Twenty years ago it would have been very difficult for an ordinary person to find low-priced technology to perpetrate such offences.

Industrialisation of 'old frauds'

Perhaps the biggest increase in opportunities has come from the ability for fraudsters to 'industrialise' old frauds. Just as the advent of modern technology has increased opportunities for businesses to reach larger markets to sell goods and services, this is also evident with fraudsters. At one level there are a number of frauds which were already perpetrated on a mass scale through traditional communications such as mail and telephone, which have become even easier to perpetrate using the Internet, email and also to be done so more cheaply. Advanced fee fraud schemes (also known as Nigerian 419 scams), where a potential victim is presented with a scenario where they are asked to pay a small amount of money up-front in return for a promised larger amount down the track, have moved from paper letters to email (Smith *et al.* 1999). Such scenarios can include inheritance notifications, business/investment opportunities and corrupt officials seeking assistance. The fraudsters can send out millions of emails at little cost hoping to hook a tiny few. The same applies to the very common fake lottery scams, where emails can replace paper notifications of a win in a foreign lottery which requires a 'fee' to release the substantial prize (although paper versions of this fraud are still common, particularly for targeting pensioners). The Internet, email and cheap telecommunications have enabled fraudsters to target potential victims on an industrial scale. For example, in 2015 the City of London Police in the UK reported over 1,500 Britons had reported being conned by purchasing bogus holidays online (Mailonline 2015). Another relevant example is that of computer prediction software and sports investment schemes. While betting and gambling is certainly not a new phenomenon, the introduction of technology has the proliferation of scams which offer software packages that supposedly guaranteed to deliver strong profits for unsuspecting victims. However, victims report receiving software which does not work or does not deliver the returns as promised, despite them having outlaid substantial amounts of money (ACCC 2014).

'Normalising' fraud

Technological changes have also created new opportunities for fraud, which could almost be described as the 'normalisation' of fraud. By this we mean the 'dark web', which now contains websites where individuals can purchase credit card details to undertake fraud. In the same way that eBay and Gumtree engage in legitimate trade, there are websites on the dark web which are dedicated to the sale and trade of stolen credentials (such as credit card numbers/passwords/personal information) which facilitates further criminal activity. Indeed some of the websites even 'ape' legitimate website sales techniques such that there are 'free tests for regular customers' and 'discounts for regular clients' (RSA 2015). Thus anyone with access to the Internet and knowledge of where to find such websites can purchase the tools to conduct fraud, just as easily as they can buy any other goods and services.

Again this demonstrates the need to conceptualise cybercrime in the context of legitimate activities carried out through the Internet. This is further driven by the value of big data and an increasing amount of personal information being shared and uploaded by individuals into the cloud. This wealth of data is a lucrative target for offenders, who can use such information to perpetrate a range of fraudulent offences, including phishing and identity theft.

Fraud entrepreneurs

The substantial number of opportunities that have arisen have given rise to the growth of fraud entrepreneurs or 'scampreneurs' (Button *et al.* 2009b). These are persons who have realised the potential of the technological changes and used these to invent scams to target victims and make large sums of money. Many are 'traditional' criminals who have adapted to the new opportunities, but some are a new breed of persons lured into criminality who see the easy money to be made and the low risks. In 2011, a teenager who had attended an exclusive British public school (elite fee paying school) was convicted of running an £18 million Internet scam, which involved a social networking site for crooks and the selling of stolen credit card details amongst others (Mailonline 2011). In addition, in many countries worldwide there has been a sharp decline in the number of armed robberies of banks and other financial institutions across the past decade. While increased security and crime prevention measures are undoubtedly a strong contributing factor (Brown 2015; Weatherburn and Holmes 2013), many also assert the notion that offenders have turned their efforts to online crime, as it is seen to offer lower risk and higher rewards compared to its offline counterpart (Nicas 2013).

Globalisation of fraud

As previously stated, one of the unique characteristics of cybercrime more broadly is its transnational nature, whereby offenders will seek to exploit geographic jurisdictional boundaries to avoid police detection and prosecution. Technological changes have enabled those inclined to fraud to broaden their horizons and target victims across the world. There are a number of dimensions to this globalisation. First there are those from the poorer parts of the world who now have the capacity and opportunity to target those in the richer parts of the globe. Second, there are fraudsters based in industrialised countries who can reach a global audience with their scams. Third, there are more organised fraudsters who move their operations to a third country to target other countries to avoid detection. Each of these dimensions to globalisation will now be explored.

Technological changes have opened up opportunities for poor but skilled persons in developing nations to target citizens in richer countries. In 2014, a report by investment bank Credit Suisse found that 1 per cent of the world's population owned 48 per cent of global wealth (Credit Suisse 2014). The same report found the bottom 50 per cent of the world's population owned less than 1 per cent of global wealth. This demonstrates a huge inequality in the distribution of wealth in the world. For individuals in the poorer parts of the world equipped with appropriate technological skills and the increasing ease and lower costs of connecting with the richer parts of the world, this has provided the opportunity to target them with frauds. A report on advanced fee frauds estimated there were over 800,000 active perpetrators of such frauds globally with 'too many to count' in Nigeria, but also substantial numbers operating in countries such as Ghana, India, South Africa and Cameroon, to name a few (Ultrascan Global Investigations 2014). Many attribute cyber-fraud to predominantly West African nations such as Nigeria and Ghana, and while historically this may have been accurate (evidenced by terms such as Nigerian fraud and 419 fraud, being the section of the Nigerian Criminal Code which refers to these criminal acts), it is now a worldwide phenomenon (Australian Crime Commission (AUSTRAC) 2011; see also Schoenmakers et al. 2009).

Globalisation has also enabled some scammers to establish a scam that lures victims from around the globe. The sale of bogus goods and services have been particularly open to this. In 2009, a group were convicted in the UK of selling counterfeit premium brand golf clubs on eBay, which had made them 'millions'. The counterfeit clubs were manufactured in China and marketed on eBay. Victims were drawn from the UK, China, Thailand, Australia, Germany, the USA, Singapore and Hong Kong (This is Money 2009).

The other aspect to globalisation is that is has enabled some organised fraudsters to locate in one country and target another. This can be done at much lower costs using telephones, the Internet and related technologies. Location in other countries is often associated with local law enforcement who are unlikely to be interested due to their own citizens not being targeted. For example, in Europe many boiler rooms selling bogus investments have been based in Spain, around Marbella and have targeted the UK. In North America there have been groups that have located in Canada to target the USA (Button 2012). In Australia, victims have been targeted from countries all over the world, and continue to send money to West African countries (Cross and Blackshaw 2015) as well as Europe, North America and Asia.

Risk of fraud

The impact of technology has ultimately led to the increased risk of an individual being targeted and becoming a victim of fraud. As noted in Table 4.2 earlier there is much evidence from around the world to illustrate the large numbers of victims. On an individual level most people who use modern technology run the daily gauntlet of fraud attempts upon them. Most people regularly receive phishing emails, notifications of lottery wins, offerings of bogus products to name some. When we scan the Internet for bargains there is the risk of stumbling across bogus websites and legitimate sites too, selling counterfeit and non-existent products and services. Seeking love and friendship online runs the risk of Internet fraudsters with fake identities befriending you. Downloading a potential job opportunity or 'amusing video' sent to us could also mean putting malware on our computers and smartphones which could be used to steal our identity. Our computers and smartphones have become the windows through which a potentially global community of fraudsters can target and defraud us. Many segments of society which traditionally have been at very low risk of crime victimisation, have in a short period of time, become much more high risk. For example, many crime surveys show young men as the most likely to be victims of crime (theft and violence) and older people to be the least likely victims of crime (ONS 2013). However, the technological changes that have emerged have opened up many older people to be high risk of a variety of frauds (OFT 2006). Thus the pensioner safe in their house at very low risk of suffering a property related crime, as soon as they switch on their computer or smartphone, check their email, surf the web or answer the phone is suddenly connecting to a much higher risk world. There are many who readily accept that older people are attractive targets for potential fraud offenders (Reiboldt and Vogel 2001), arguably through factors such as access to life savings, superannuation, ability to access additional lines of credit and their likely ownership of property. Many offenders will seek to target older people specifically based on these circumstances and use the Internet to perpetrate this.

To develop this point further the impact of technology has created a completely new habitus in which we live, work and play; but unlike the traditional world many have not yet learnt to appreciate the risks. Many people would know the risks and how to manage them of purchasing goods from a 'carboot' sale, of visiting certain locations at night and of talking to strangers. The Internet, however, has created a completely new habitus in which we operate and many are still learning how to gauge and manage the risks of this new technology. The application of traditional crime prevention measures (such as locking the front door, keeping valuables out of sight in a vehicle, and shredding personal documents) are not always as readily applicable to the virtual environment and therefore, need to be modified to ensure safety. Digital networks, including the Internet, also pose new threats to our safety that we are yet to comprehend or effectively combat and this remains a challenge into the future.

Responsibilisation of fraud

The substantial rise in cyber-frauds has also been confronted with a law enforcement response in many countries which is not commensurate with the growing risk. There are a number of dimensions to this. First, the police have not shifted enough resources towards cyber-fraud. Second, the priority and commitment given to cyber-fraud has been low. Third the complexity and global nature of many frauds has meant many police forces have been unable to pursue many cases. The consequences of these changes is a very high rate of attrition for fraud-related offences. The City of London Police estimate 80 per cent of frauds are not reported (Head 2015). Button *et al.* (2012) drawing on an even wider range of fraud and using a crude method estimated only 1.5 per cent of fraud is reported, with only 0.4 per cent resulting in criminal sanctions or detections.

Recent research by Button *et al.* (2015) found that of 230,599 police officers and staff in the United Kingdom, only 652.3 are dedicated to fraud. This compares in England and Wales to 828 police officers dedicated to complaints and discipline (Home Office 2014). Thus there are in England and Wales more police officers dedicated to investigating allegations of misconduct against their own staff, than there are police officers and staff dedicated to combatting the huge problem of fraud. Added to this are resources in a number of other state bodies such as the National Crime Agency (NCA) and Serious Fraud Office (SFO), but these add only hundreds more and they tend to focus on a few cases which are large and complex. Included in the NCA is also the National Cybercrime Unit, but again there are relatively small numbers of staff and a significant proportion of their work focusses upon child pornography. Amongst many police staff their commitment to fraud is low and even some specialist fraud police are rooted more in traditional frauds (London Assembly 2015). The added dimension of cross-borders also makes many police reluctant to investigate (Button 2012). There are signs in the UK this is beginning to change with the establishment of FALCON (Fraud and Linked Crime Online) in the Metropolitan Police dedicated to fraud and online related crime with several hundred officers (London Assembly 2015).

The consequence of the limited state response has been what Garland (1996) describes as 'responsibilisation'. This is a process whereby the state seeks to encourage individuals and organisations to take responsibility for their own protection. This is particularly evident in the context of identity crime and fraud which stems from the misuse of personal information (Monahan 2009; Whitson and Haggerty 2008). Countering cyber-dependent frauds (and crimes) for organisations have largely become the preserve of large companies specialising in this area who have the expertise, capacity and resources to offer extensive services for healthy fees, such as Kaspersky and EMC, for example. For many cyber-enabled frauds that target organisations and individuals, victims face challenges securing police interest and are often left to investigate themselves (if they have the capacity) or pay corporate investigators to do so (Button *et al.* 2015; Cross *et al.* 2016). There has been much encouragement from various state bodies of measures for organisations to pursue to give greater resilience to cyber-fraud attacks as has been pursued under the auspices of the National Cybercrime Strategy (Cabinet Office 2014). Individuals have also been targeted with numerous campaigns to raise awareness of scams and to encourage the pursuit of certain basic protections. Action Fraud, in the UK, regularly conducts awareness campaigns around different types of scams (this is also evident in other countries such as Australia through the work of SCAMWatch). There is clearly more research required on the effectiveness of such strategies and it is clear from the scale of the threat that more of such work is required. Nevertheless, what is clear is that the state in many countries cannot cope with cyber-fraud and has had to relinquish the lead responsibility for dealing with it to organisations and individuals.

Conclusion: The 'fraudogenic' consequences of technological changes

The discussion above has illustrated how technological changes have contributed to a variety of new types of opportunity to commit fraud, combined with a globalisation of the problem resulting in increased risk of victimisation, combined also with a response so far from the state which has not kept pace with the challenge. In seeking to prevent crime (particularly property crime) Clarke (2005) amongst others has noted the importance of factors and theories relating to opportunity, routine activities and rational choice. Clarke along with other collaborators has sought to construct 25 techniques of crime prevention rooted around five themes of increasing the effort, increasing the risks, reducing the rewards, reducing provocations and reducing excuses. The impact of technological change has effectively created huge gaps in three of these areas in society. Committing fraud using the latest technology has thus become easier to do and also enables the targeting of greater numbers of potential victims. The risks of getting caught and punished in an online environment are less than for many traditional crimes and frauds. The potential rewards for many cyber-frauds are huge. The consequences of the technological revolution can therefore be seen as criminogenic and because of the unique opportunities for fraud, perhaps the special term 'fraudogenic' should be used. As will now be discussed, the 'falling behind the curve' of what could be described as the 'criminal justice complex' (the politicians, criminal justice agencies and researchers) has contributed further to the inadequate response of the United Kingdom (and other countries) to this problem.

The 'criminal justice complex': turning the 'supertanker'

The huge expansion of fraud fuelled by the technological revolution has seriously challenged what could be described as the 'criminal justice complex'. The focus here will be the UK, but clearly there is resonance elsewhere in the world. Politicians, criminal justice agencies and some criminologists have been locked into a crime problem defined by the crime statistics and in the UK context the crime surveys of England and Wales and of Scotland. These have shown downward trends in volume property crimes since the 1990s (Knepper 2015). The police have been happy to claim the credit of their activities for reduced crime. Politicians of all parties have been happy to note such trends, pointing to the policies they have implemented to tackle crime. Although there have been a few criminologists that have pointed to the glaring exception of fraud from these figures (Fitzgerald 2014), many have embraced the idea that crime has been falling (Aebi and Linde 2010; Farrell et al. 2011). Those organisations involved in the 'criminal justice complex' have all had a vested interest in the traditional crime paradigm. This chapter is not arguing that such crimes have not fallen, because clearly they have. What this chapter argues is that crime has shifted to new forms which are harder to measure and the established indicators have not yet developed measures to capture their true extent. Until more research and accurate measurement is conducted, it is difficult to determine the true size of the problem and to what extent traditional criminals have moved to cyber-frauds and how much of an impact the globalisation of fraud has had.

The lack of accurate measurement has contributed to what Button and Tunley (2015) describe as deviancy attenuation. This is a process which is the opposite to the well-established concept of deviancy amplification (Becker 1963; Wilkins 1964). Button and Tunley (2015) illustrate the concept in Figure 4.1. At '1 o'clock', the process begins with the narrowing of behaviours defined as the crime of fraud. As was noted earlier fraud is frequently relabelled as error, a civil dispute, or a consumer issue, for example. Added to that is limited measurement, which leads to statistics showing fraud is not a major problem. This leads to a lower priority and resources dedicated to the problem, which means detections and sanctions are lower, which then reinforces to decision-makers that fraud is not a problem. There is an exception to this, which is benefits/social security fraud, where there are clear definitions, accurate measurement, commitment to tackle it, media interest fuelling perception of it as a problem and as a

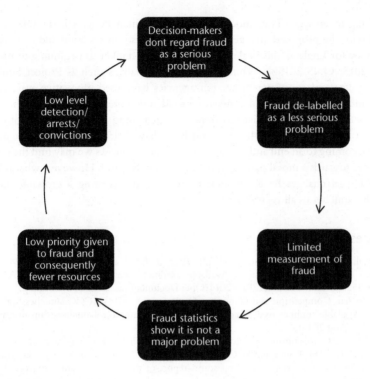

Figure 4.1 Deviancy attenuation
Source: Button and Tunley (2015: 13)

consequence, deviancy amplification. However, with cyber-frauds on the whole, a process of deviancy attenuation has been occurring over the last decade or so .

Button and Tunley (2015) argue that the process of deviancy attenuation is underpinned by an immoral phlegmatism – the opposite to a moral panic – amongst the key decision-makers in government, law enforcement and organisations, which attenuates the problem. For the authors this immoral phlegmatism amongst most of those involved in the 'criminal justice complex' is not down to a conspiracy to hide the true level of crime, although there are clearly politicians and police leaders who have been happy to benefit from it. Rather this complex can be viewed as a 'large supertanker'. It has been heading in the same direction for some time with slight tweaks here and there. However, in a relatively short period of time there has been a technological revolution. This has probably impacted upon traditional volume crimes, reducing them, while at the same time increasing opportunities for cyber-frauds (Aebi and Linde 2010; Knepper 2015). This has required a change in direction, which many parts of the 'criminal justice complex' have been slow to adapt to. Police forces have been slow to shift resources to a crime where they do not know how large it is or understand it. Knowledge and resources in law enforcement have been slow to grapple with the challenges and complexities of the new fraud problems (London Assembly 2015). Politicians have been reluctant to accept new ways to measure, because of the likely gigantic leap in levels of crime which could at best cause political embarrassment and worst end their careers (Fitzgerald 2014). For some criminologists with careers built on traditional crimes, their measures and scepticism of declining crime have been slow to appreciate the significant shifts in crime, how to measure them accurately and how to understand them. All of these will take time to change and there is clearly evidence that changes are beginning to emerge. The launch of the Metropolitan Police's FALCON unit in 2015 combined with the proposed changes to extend the questions on fraud and cybercrime in the Crime Survey for England and Wales are signs the 'supertanker' is beginning to turn (London Assembly 2015; ONS 2015b). While in Australia, initiatives such as Project Sunbird where Australian police and consumer protection agencies have used innovative ways through the pro-active monitoring of financial transactions to identify potential victims and then contacting them to seek to prevent further victimisation, show increasing signs of change in other countries (Department of Commerce 2014; Cross and Blackshaw 2015; Cross 2016). Attenuation may therefore be coming to an end and as interest (and concern) grows we may find the environment shifting more towards a moral panic and deviancy amplification. However, what is clear is that the technological changes have left their mark on fraud, creating a growing and common problem that touches us all regularly.

References

ABS (Australian Bureau of Statistics) (2012) *Personal Fraud 2010-2011*, Cat 4528.0, Canberra: Australian Bureau of Statistics. Available online: www.abs.gov.au/AUSSTATS/abs@.nsf/Lookup/4528.0Main+Features12010-2011?OpenDocument (accessed 31 Aug 2016).

ACCC (Australian Competition and Consumer Commission) (2014) ACCC launches Scam Disruption Project. Available online: www.accc.gov.au/media-release/accc-launches-scam-disruption-project (accessed 31 Aug 2016).

ACCC (Australian Competition & Consumer Commission) (2015) *Targeting scams: Report of the ACCC on scams activity 2014*. Canberra: ACCC. Available online: www.accc.gov.au/publications/targeting-scams-report-on-scam-activity/targeting-scams-report-of-the-accc-on-scam-activity-2014 (accessed 31 Aug 2016).

Aebi, M. and Linde, A. (2010) Is there a crime drop in Western Europe? *European Journal on Criminal Policy and Research*, 16: 251–277.

Amazon (2013) Press Release. Available online: http://phx.corporate-ir.net/phoenix.zhtml?c=97664&p=irol-newsArticle&ID=1779040&highlight (accessed 5 February 2013).

Australian Transaction Reports and Analysis Centre (AUSTRAC) (2011) AUSTRAC typologies and case studies report 2011. Available online: www.austrac.gov.au/sites/default/files/documents/typ_rpt11_full.pdf (accessed 31 Aug 2016).

Becker, H. (1963) *Outsiders: Studies in the Sociology of Deviance*. London: Macmillan.

Bedford Today (2015) Man is fined for using fake blue badge he made using photoshop. Available online: www.bedfordtoday.co.uk/news/local/man-is-fined-for-using-fake-blue-badge-he-made-using-photoshop-1-6788247 (accessed 14 July 2015).

Bedos-Rezak, B. (2000) Medieval identity: A sign and concept, *The American Historical Review*, 105: 1489–1533.

Brown, R. (2015) Explaining the property crime drop: The offender perspective, *Trends and Issues in Crime and Criminal Justice*, no 495.

Burdon, M., Lane, B. and von Nessen, P. (2010) The mandatory notification of data breaches: Issues arising for Australian and EU legal developments, *Computer Law and Security. Review*, 26 (2): 115–129.

Button, M. (2012) Cross-border fraud and the case for an 'Interfraud'. *Policing: An International Journal of Police Strategies & Management* 35: 285–303.

Button, M., Blackbourn, D. and Tunley, M. (2015) 'The not so thin blue line after all?' Investigative resources dedicated to fighting fraud/economic crime in the United Kingdom, *Policing*, 9: 129–142.

Button, M. and Gee, J. (2013) *Countering Fraud for Competitive Advantage*. Chichester: Wiley.

Button, M., Lewis, C., Shepherd, D., Brooks, G., and Wakefield, A. (2012) *Fraud and Punishment: Enhancing Deterrence through More Effective Sanctions*. Portsmouth: Centre for Counter Fraud Studies.

Button, M., Lewis, C. and Tapley, J. (2009a) *A Better Deal for Victims*. London: National Fraud Authority.

Button, M., Lewis, C. and Tapley, J. (2009b) *Fraud Typologies and the Victims of Fraud Literature Review*. London: National Fraud Authority.

Button, M. and Tunley, M. (2015) Explaining fraud deviancy attenuation in the United Kingdom, *Crime, Law and Social Change*, 63: 49–64.

Button, M., Wakefield, A., Brooks, G. and Lewis, C. (2015) Confronting the 'fraud bottleneck': private sanctions for fraud and their implications for justice, *Journal of Criminological Research, Policy and Practice*, 1: 159–174.

Cabinet Office (2014) The UK Cyber Security Strategy: Report on Progress and Forward Plans. Online. Available online: www.gov.uk/government/uploads/system/uploads/attachment_data/file/386093/The_UK_Cyber_Security_Strategy_Report_on_Progress_and_Forward_Plans_-_De____.pdf (accessed 14 July 2015).

CAFC (Canadian Anti-Fraud Centre) (2015) *Annual Statistics Report 2014*. Available online: www. antifraudcentre-centreantifraude.ca/reports-rapports/2014/ann-ann-eng.htm (accessed 31 Aug 2016).

Clarke, R. V. (1980) Situational crime prevention: Theory and practice, *British Journal of Criminology*, 20: 136–147.

Clarke, R. V. (2005) Seven misconceptions of situational crime prevention. In Tilley, N. (ed.) *Handbook of Crime Prevention and Community Safety*. Cullompton: Willan.

Cohen, L. E. and Felson, M. (1979) Social changes and crime rates: A routine activities approach, *American Sociological Review*, 44: 588–608.

Credit Suisse (2014) Global Wealth Report. Available online: https://publications.credit-suisse.com/tasks/render/file/?fileID=60931FDE-A2D2-F568-B041B58C5EA591A4 (accessed 14 July 2015).

Cressey, D. (1973) *Other People's Money*. Montclair, NJ: Patterson Smith.

Cross, C. (2013) 'Nobody's holding a gun to your head...' Examining current discourses surrounding victims of online fraud. In Richards, K. and Tauri, J. (eds) *Crime, Justice and Social Democracy: Proceedings of the 2nd International Conference*. Queensland: Crime and Justice Research Centre, Queensland University of Technology, pp. 25-32.

Cross, C. (2015) No laughing matter: Blaming the victim of online fraud, *International Review of Victimology*, 21(2): 187–204.

Cross, C. (2016) Policing online fraud in Australia: The emergence of a victim-oriented approach. In Berents, H. and Scott, J. (eds) *Crime, Justice and Social Democracy: Proceedings of the 3rd International Conference 2015*, Brisbane: Crime and Justice Research Centre, pp. 1–8.

Cross, C. and Blackshaw, D. (2015) Improving the police response to online fraud, *Policing: A Journal of Policy and Practice*, 9 (2): 119–128.

Cross, C., Richards, K. and Smith, R. (2016) *Improving the response to online fraud victims: An examination of reporting and support*, Criminology Research Grant Final Report. Available online: http://www.crg.aic. gov.au/reports/201617.html#1314-29 (accessed 31 October 2016).

Department of Commerce (2014) Project Sunbird flying high. Available online: www.scamnet.wa.gov. au/scamnet/Types_Of_Scams-Advanced_fee_frauds-OperationProject_Sunbird.htm (accessed 13 August 2015).

Duffield, G. and Grabosky, P. (2001) *The Psychology of Fraud. Australian Institute of Criminology: Trends and Issues in Criminal Justice*. Canberra: Australian Institute of Criminology.

eBay (n.d.) *The Company*. Available online: http://pages.ebay.co.uk/aboutebay/thecompany/company overview.html (accessed 5 February 2014).

Facebook (2015) Stats. Available online: http://newsroom.fb.com/company-info/ (accessed 14 July 2015).

Farrell, G., Tseloni, A., Mailley, J. and Tilley, N. (2011) The crime drop and the security hypothesis, *Journal of Research in Crime and Delinquency*, 48: 147–175.

Farrell, S., Yeo, N. and Ladenburg, G. (2007) *Blackstone's Guide to the Fraud Act 2006*. Oxford: Oxford University Press.

Fitzgerald, M. (2014) Falling Crime or Flawed Statistics? Presentation at BSC/Mannheim seminar, LSE, 15 January 2014.

Furnell, S. (2002) *Cybercrime: Vandalising the Information Society*. London: Addison-Wesley.

Garland, D. (1996). The limits of sovereign state: Strategies of crime control in contemporary society, *British Journal of Criminology*, 35, 445–71.

Head, S. (2015) Presentation to the 2015 Annual Counter Fraud and Forensic Accounting Conference, 2 June 2015. Available online: https://vimeo.com/130322766 (accessed 14 July 2015).

Home Office (2014) Tables for 'Police Workforce', England and Wales 31 March 2014. Available online: www.gov.uk/government/statistics/tables-for-police-workforce-england-and-wales-31-march-2014 (accessed 14 July 2015).

IC3 (2013) 2012 Internet Crime Report. Richmond: National White Collar Crime Centre.

International Telecommunications Union (ITU) (2012) *The World in 2011*. Available online: www.itu. int/ITU-D/ict/facts/2011/material/ICTFactsFigures2011.pdf (accessed 14 July 2015).

International Telecommunications Union (ITU) (2015) Key 2005-2015 ICT data for the world, by geographic region and by level of development. Available online: www.itu.int/en/ITU-D/Statistics/ Pages/stat/default.aspx (accessed 13 August 2015).

Internet World Stats (2012), *Internet Usage Statistics*. Available online: www.internetworldstats.com/stats. htm (accessed 14 July 2015).

Jewkes, Y. and Yar, M. (2010) Introduction: The internet, cybercrime, and the challenges of the 21st century. In Jewkes, Y. and Yar, M. (eds) *Handbook of Internet Crime*. Cullompton: Wiley, pp. 1–8.

Knepper, P. (2015) Falling crime rates: What happened last time? *Theoretical Criminology*, 19: 59–76.

Langenderfer, J., and Shrimp, T. A. (2001). Consumer vulnerability to scams, swindles, and fraud: A new theory of visceral influences on persuasion, *Psychology & Marketing*, 18, 763–783.

Langton, L. and Planty, M. (2010) *National Crime Victimisation Survey Supplement, Victims of Identity Theft 2008*. Bureau of Justice Statistics, US Department of Justice Available online: www.bjs.gov/content/ pub/ascii/vit08.txt (accessed 13 August 2015).

Levi, M. (2008) Organized frauds and organising frauds: Unpacking the research on networks and organisation, *Criminology and Criminal Justice*, 8: 389–419.

Levi, M. and Burrows, J. (2008) Measuring the impact of fraud in the UK: A conceptual and empirical journey, *British Journal of Criminology*, 48: 293–318.

London Assembly (2015) *Tightening the Net. The Metropolitan Police Service's Response to Online Theft and Fraud*. London: London Assembly.

McGuire, M. and Dowling, S. (2013) *Cybercrime a Review of the Evidence*. Home Office Research Study 75. Available online: www.gov.uk/government/uploads/system/uploads/attachment_data/file/248621/ horr75-chap2.pdf and www.gov.uk/government/uploads/system/uploads/attachment_data/file/ 246754/horr75-chap3.pdf (accessed 14 July 2015).

Mailonline (2011) Pictured with Piles of Cash: The Public Schoolboy Jailed for Five Years for Masterminding £18 million Internet Scam. Online. Available online: www.dailymail.co.uk/news/article-1362207/ Public-school-boy-masterminded-18million-Internet-scam-jailed-years.html (accessed 14 July 2015).

Mailonline (2015) British Holidaymakers Conned out of £2.2 million by fake holiday letting sites, hackers and bogus airline tickets. Available online: www.dailymail.co.uk/travel/travel_news/article-3035595/ British-holidaymakers-conned-2-2million-fake-holiday-letting-websites-hackers-bogus-airline- tickets.html (accessed 14 July 2015).

Mayhew, P., Clarke, R. V., Sturman, A. and Hough, J. M. (1976) *Crime as Opportunity*. London: HMSO.

Monahan, T. (2009) Identity theft vulnerability: Neoliberal governance through crime construction, *Theoretical Criminology*, 13(2): 155–176.

National Fraud Authority (2013) *Annual Fraud Indicator*. London: National Fraud Authority.

Newman, G. R. and Clarke, R. V. (2003) *Superhighway Robbery*. Cullompton: Willan.

Nicas, J. (2013) Crime that no longer pays? *Wall Street Journal*. Available online: www.wsj.com/articles/ SB10001424127887323926104578274541161239474 (accessed 13 August 2015).

Nolan, S. (2012) Huge Spike in Online Dating After Christmas as Holiday Spirit Encourages Thousands to Cure their Loneliness. Available online: www.dailymail.co.uk/news/article-2254968/Online- dating-statistics-Huge-spike-Christmas-holiday-spirit-encourages-thousands-cure-loneliness. html#axzz2K7uODPCt (accessed 5 February 2013).

OFT (Office of Fair Trading) (2006) *Research on Impact of Mass Marketed Scams*. London: Office of Fair Trading.

ONS (2013) The Likelihood of Becoming a Victim of Crime. Available online: www.ons.gov.uk/ons/ rel/crime-stats/crime-statistics/period-ending-march-2012/sty-a-victim-of-crime.html (accessed 14 July 2015).

ONS (2014) Crime Against Business: Headline Findings from the 2013 Commercial Victimisation Survey. Available online: www.gov.uk/government/uploads/system/uploads/attachment_data/file/284818/ crime-against-businesses-headlines-2013-pdf.pdf (accessed 14 July 2015).

ONS (2015a) Crime in England and Wales, Year Ending September 2014. Available online: www.ons. gov.uk/ons/dcp171778_392380.pdf (accessed 14 July 2015).

ONS (2015b) *Update: Extending the Crime Survey for England and Wales (CSEW) to Include Fraud and Cybercrime*. London: ONS.

PriceWaterhouseCoopers (PwC) (2015) *Global Economic Crime Survey 2014*. Available online: www.pwc. com/gx/en/economic-crime-survey/downloads.jhtml (accessed 13 August 2015).

Randall, C. (2010) *E-Society*. London: Office for National Statistics.

Reiboldt, W. and Vogel, R. (2001) A critical analysis of telemarketing fraud in a gated senior community, *Journal of Elder Abuse and Neglect*, 13(4): 21–38.

RSA (2015) *Cybercrime 2015 An Insider Look at the Changing Threat Landscape*. London: RSA.

Scamnet (n.d.) *Project Sunbird*. Available online: www.scamnet.wa.gov.au/scamnet/Fight_Back-Project_Sunbird.htm (accessed 13 August 2015).

Schoenmakers, Y., de Vries, R. and van Wijk, A. (2009) *Mountains of Gold: An Exploratory Research in Nigerian 419 Fraud*. Netherlands: SWP.

Smith, R. G., Holmes, M. N. and Kaufmann, R. G. (1999) Nigerian advanced fee fraud, *Trends and Issues in Crime and Criminal Justice*, No 121, Canberra: Australian Institute of Criminology.

South Australian Police (2014) *Annual Report 2013-2014* Available online: www.police.sa.gov.au/__data/assets/pdf_file/0008/58616/Annual-report-2013-2014_INTERNET.pdf (accessed 13 August 2015).

Symantec (2014) Internet Security Threat Report 2014. Available online: www.symantec.com/content/en/us/enterprise/other_resources/b-istr_main_report_v19_21291018.en-us.pdf (accessed 14 July 2015).

Think Insights With Google (2012) *Key Market Report: Trends in Digital Device and Internet Usage*. Available online: www.thinkwithgoogle.com/insights/library/studies/trends-in-digital-device-and-internet-usage-2012/ (accessed 5 February 2013).

This is Money (2009) Ebay Fraudster Made Millions from Fake Clubs. Online. Available online: www.thisismoney.co.uk/money/bills/article-1685712/Ebay-fraudster-made-millions-from-fake-golf-clubs.html (accessed 5 February 2013).

Treadwell, J. (2012) From the car boot to booting it up? eBay, online counterfeit crime and the transformation of the criminal marketplace, *Criminology and Criminal Justice*, 12: 175–191.

UK Payments Administration (2009) *Number Of Internet Users Now Banking Online Exceeds 50% For The First Time Ever*. Available online: www.ukpayments.org.uk/media_centre/press_releases/-/page/871/ (accessed 5 February 2013).

Ultrascan Global Investigations (2014) 419 Advanced Fee Fraud Statistics 2013. Available online: www.ultrascan-agi.com/public_html/html/pdf_files/Pre-Release-419_Advance_Fee_Fraud_Statistics_2013-July-10-2014-NOT-FINAL-1.pdf (accessed 14 July 2015).

Van Dijk, J. (2008) *The World of Crime*. London: SAGE.

Weatherburn, D. and Holmes, J. (2013) The decline in robbery and theft: Inter-state comparisons. *Crime and Justice Statistics: Bureau Brief*, no. 89.

Whitson, J. and Haggerty, K. (2008) Identity theft and the care of the virtual self, *Economy and Society*, 37(4): 572–594.

Whitty, M. (2013) The scammers persuasive techniques model: Development of a stage model to explain the online dating romance scam, *British Journal of Criminology*, 53: 665–884.

Wilkins, L. (1964) *Social Deviance: Social Policy, Action and Research*. London: Tavistock.

Yar, M. (2013) *Cybercrime and Society*, 2nd edition. London: Sage.

5

ICTs and child sexual offending

Exploitation through indecent images

Jo Bryce

Introduction

The dynamic nature of information communication technologies has created a unique and complex online environment which can be exploited by offenders to facilitate a variety of criminal behaviours related to child sexual offending (Bryce 2015; Long *et al.* 2013). The sexual exploitation of children through the production, possession and distribution of indecent images (IIOC) is one category of cybercrime which has emerged as an important enforcement and safeguarding issue (Long *et al.* 2013). Advances in technology and the widespread adoption of the Internet have created more opportunities for individuals to access and disseminate this material (Al Mutawa *et al.* 2015; Beech *et al.* 2008). This has resulted in action by enforcement, government and industry to address the issue. Theoretical and empirical research examining offender and victim characteristics, as well as the relationship between possession or downloading and contact offending (Beech *et al.* 2008; Bourke and Hernandez 2009; Long *et al.* 2013), has also informed evidence-based approaches to prevention and response at the national and international level.

This chapter provides an overview of the currently available literature on IIOC offending, with a specific focus on offences associated with possession and downloading of this material. It starts by briefly outlining offence definitions and prevalence, before reviewing the available evidence about victim experiences and impacts. It then examines offender motivations and psychological characteristics, as well as the relationship between possession and potential escalation to contact offending. The final section considers the investigative utility of IIOC, offender collections and related digital forensic evidence. The overall aim of the chapter is to identify the current knowledge gaps and research challenges associated with this category of cybercrime, as well as the implications of the intersection between technology and this form of sexual offending for the criminal justice system.

Definitions

Indecent images are a form of sexual exploitation which depict children being sexually abused and exploited (CEOP 2013). This content is illegal in the UK, USA, Europe and many other countries. At the European level, two Conventions of the Council of Europe relate to the production, circulation and possession of IIOC. Article 9 of the Budapest Convention on

Cybercrime (2001) specifies IIOC related offences, and associated definitions of content and the child. Articles 20 and 21 of the Lanzarote Convention on the Protection of Children against Sexual Exploitation and Sexual Abuse (2007) also address these issues, as well as the need for international cooperation, education of children and corporate liability. In the UK, the Sexual Offences Act (2003) extended the Protection of Children Act (1978) to create new offences specifically focusing on the sexual exploitation of children through indecent images (Sentencing Guidelines Council 2007), though these remained within the Protection of Children Act (1978) and Criminal Justice Act (1988).

Section 1 of the UK Protection of Children Act (1978) specifies the offences of 'making', 'taking/permit to take', and 'distributing' IIOC. Offences relating to 'taking/permit to take' refer to the production of IIOC. Police and NGO reports suggest that this material is produced in a number of different contexts. These include intra-familial (e.g. parents, carers) and extra-familial sexual abuse (e.g. family friends, teachers), as well as online grooming processes and commercial child prostitution (CEOP 2013; IWF 2013). Section 1 of the Protection of Children Act (1978) also specifies the offence of 'distributing' IIOC. This refers to the dissemination of content though online communities/networks and technologies (e.g. Peer-2-Peer, TOR), commercial online subscription sites, and between individuals using a variety of online communications networks (CEOP 2013; IWF 2013). The final offence specified by Section 1 of the Protection of Children Act (1978) is that of 'making' IIOC. This offence refers to accessing or downloading IIOC from the Internet. This is similar to the possession offence specified by Section 160 of the Criminal Justice Act (1988). As both these offences relate to possession or downloading of these images, they do not require the offender to have direct physical contact with the victim (CEOP 2013). For the purpose of this chapter, the term 'possession' will be used to cover both offences due to the similarity of the offending behaviour involved.

The offence categories described above map onto official criminal justice system data sources for examining the prevalence of IIOC related offences in the UK. These figures suggest a general increase in all three offence categories in recent years. For example, Crown Prosecution Service data on the number of offences charged and reaching the Magistrates Courts indicates an increase in possession offences from 2,768 in 2006–7 to 3,849 in 2012–13 (McGuire and Dowling 2013; Smith et al. 2013). The same data source indicates a higher proportion of offences for making IIOC, increasing from 10,761 in 2006–7 to 14,033 in 2011–12 (McGuire and Dowling 2013; Smith et al. 2013). Figures relating to the number of offenders found guilty for take, make or distribute offences have also increased from 768 in 2006–7 to 1,248 in 2012–13 (Ministry of Justice 2013). This trend of increases in charges and sentences for IIOC offences are likely to reflect increasingly pro-active and successful enforcement investigations, rather than a direct increase in levels of offending (McManus and Almond 2013; Wolak et al. 2012). However, there are a number of difficulties associated with the use of these figures to estimate the prevalence of IIOC offences as they reflect different levels of reporting and legislative frameworks, making their interpretation and comparison problematic (McManus and Almond 2013). For example, the CPS figures refer to the number of offences charged during the specified period, which creates difficulties comparing them with MOJ data which reflect the number of offenders (McGuire and Dowling 2013).

It is also important to recognise that as the majority of offences are likely to remain undetected, official figures can only provide a partial indication of the scale of the problem (Bryce 2014; McGuire and Dowling 2013). This is related to the awareness among offenders of the legal and social sanctions associated with their activities, and associated actions to hide their identity and behaviour (e.g. file encryption, use of anonymous email and proxy services) (Seto *et al.* 2015). As a result, inexperienced or less skilled offenders are more likely to be detected and prosecuted than those who are more technologically sophisticated, as they may be more effective at evading detection (Seto *et al.* 2015). This has implications for interpretation of the results of empirical studies using offender populations as they may only provide data about the motivations, psychological characteristics and deficits of those individuals who are unsuccessful in their attempts to evade detection. It also demonstrates the role of ICT in allowing IIOC offences to be committed more successfully through their use to evade detection, and is a further enabling characteristic of technology in this category of crime.

Differences in the dynamics of victim-offender relationships between the online and offline environment also have implications for the reporting, investigation, detection and recording of IIOC offences. For example, the ability for ICT to facilitate offending remotely without direct physical contact has altered the nature of the victim-offender relationship. This has investigative implications as reports to the police for contact offending or production offences are generally initiated as a result of disclosure by the victim or the concerns of a parent or other adult (Palmer 2005). In these instances, victim awareness and disclosure may be the trigger for an investigation. In the online environment, detection may occur as a result of law enforcement investigations, intelligence and monitoring of online offender networks. It may also be initiated on the basis of suspicion or image discovery by family members, or in the course of device repairs. As a result, the investigation of possession and dissemination offences may not directly involve or identify the victims depicted in the images, and this is not required for detection, recording or a successful prosecution. This highlights the influence of the characteristics of the online environment and communication (e.g. perceived anonymity, disinhibition) on the dynamics of offending behaviour and related investigative issues. It also has implications for the process by which victims come to the attention of law enforcement and safeguarding agencies.

Challenges associated with victim awareness, recognition and reporting of their experiences also have implications for estimating the number of victims of IIOC offending (Bryce 2010). In cases where images are produced as a result of victimisation through contact offending, and the victim is aware that this has occurred, they may disclose their experiences and a subsequent report be made to the police. However, if the victim is very young or IIOC are produced during the online grooming process, they may lack awareness that images have been produced and/or are being disseminated online. They may also perceive the behaviour involved in production of images to be part of a romantic relationship (Bryce 2010; Webster *et al.* 2012). Victims may also be reluctant to report their experiences as the result of direct or implied threats by offenders, the experience of self-blame and shame, or concerns that their reports will not be believed or taken seriously (Bryce 2010). Given these reporting barriers, as well as the difficulties associated with victim identification in possession and distribution offences, it is difficult to estimate the number of children and young people victimised in this way.

An alternative source of data for estimating the number of victims, prevalence of offending and the volume of IIOC in circulation is law enforcement and databases of IIOC recovered during the course of investigations. For example, the COPINE Project image database contained 700,000 images in 2004, which researchers estimated to represent between 7,000 and 70,000 victims[1] (Holland 2005). The ChildBase database of the Child Exploitation and Online Protection Centre (CEOP) contained 807,525 unique still abusive images in 2009 (Quayle and Jones 2011). However, as these images were recovered during investigations or the online monitoring of offender networks, they can only provide a partial indication of the amount of images being produced, disseminated and downloaded at any given time. It is also important to note that media reporting of criminal cases provides evidence that some offenders have larger numbers of images in their collections when investigated by the police. For example, an offender found guilty of 20 offences of making and possessing indecent images of children in May 2015 was found to have over a million digital images in his collection (BBC 2015).

Regardless of the challenges associated with estimating the prevalence of offending and victimisation, the available data indicates that there is a high level of demand for IIOC, and a significant number of offenders involved in the production, dissemination and possession of this material (McManus and Almond 2013). This further suggests that a high number of children and young people are victimised in this way, the majority of whom are unknown to the relevant police and safeguarding agencies. It also demonstrates the central role of digital technology in producing and disseminating such material.

Victim characteristics

This relates to the general lack of data about the characteristics of the victims of IIOC offending. The main source of information about the children and young people depicted in this material is law enforcement databases and IIOC examined through reporting agencies. The IWF produce yearly reports which identify the gender, age and content of the victim images they examine. These generally suggest the dominance of female victims in images (80 per cent), with a large proportion being aged under 10 years old (80 per cent) (IWF 2014). The IWF have also reported an increase in the severity of the content of images from 58 per cent depicting penetrative abuse in 2008 to 65.6 per cent in 2010 (CEOP 2013; IWF 2010). This trend in severity of content and a decrease in the age of victims has raised concerns over the associated potential for an increase in the likelihood of progression from accessing images to non-contact and contact offending (CEOP 2013).

There are few studies which have examined the relationship between victims and offenders in relation to the production of IIOC. However, data from the US National Centre for Missing and Exploited Children (NCMEC) Statistical Report (2012) examined data for the 4,638 victims identified between 1998 and June 2012. The study found that a high percentage of offenders had a prior offline relationship with the victim before they were involved in the production of IIOC. For example, 18 per cent of cases involved parents/guardians and 27 per cent involved neighbours/family friends. Fifteen per cent of images were produced as the result of the victims meeting a perpetrator online and transmitting self-produced images, or where victims and offenders met online and images were produced as the result of an offline meeting. The finding that the majority of offenders were known to victims prior to production of images is consistent with UK research suggesting that the majority of young people are sexually abused by someone already known to them (Radford et al. 2011). This indicates that the offending typically involved in the production of IIOC frequently occurs in established family or community contexts, and suggests that police should examine whether this has occurred when investigating cases of contact offending. This is important as the online dissemination of material produced in this context represents a form of ongoing victimisation as other offenders continue to access and download IIOC. This further demonstrates the importance of ICT as a facilitator of the production and dissemination of IIOC. It also creates opportunities for offending where the victim and offender do not have a prior offline relationship as the result of providing channels by which offenders can select and approach potential victims.

Victim impacts

Although a small number of studies have been conducted with victims of online grooming and sexual exploitation (e.g. Whittle et al. 2013, 2014), there is little published research on the experience and consequences of victimisation associated with IIOC offending (Quayle *et al.* 2008). This reflects difficulties in victim identification associated with the role of technology in mediating victimisation after the point of production which does not involve direct contact between the victim and offender(s). However, there is an established body of literature examining the effects of sexual victimisation experienced by children and young people in offline settings which suggests that it is a significant risk factor for behavioural and psychological problems in childhood and adulthood (Fergusson *et al.* 2008; Marriott *et al.* 2014). There are also a small number of studies which examine the psychological and social impacts of victimisation through the production of IIOC in offline contexts (e.g. Scott 2001; Svedin and Back 2003) which have identified similar social and psychological impacts to other forms of offline sexual abuse (Quayle *et al.* 2008). Based on this evidence, it is clear that similar effects will be associated with sexual victimisation in which IIOC are produced and circulated online.

In addition to the ongoing anxiety, guilt and shame associated with the original victimisation experiences, practitioners involved in counselling victims have identified other factors associated with the impact of IIOC offending (Palmer 2005; Quayle *et al.* 2008). These relate to the additional stress and shame which result from knowledge that a digital record of their victimisation is potentially circulating online indefinitely (Palmer 2005; Quayle *et al.* 2008). As each download represents another instance of victimisation, distribution of IIOC online creates opportunities for continued victimisation beyond the initial circumstances of production. The associated experience of loss of control, feelings of helplessness and lack of closure can further intensify the psychological, social and physical effects of this form of sexual exploitation (Palmer 2005; Quayle *et al.* 2008). This demonstrates the role of ICT and the online environment in changing the dynamics of this form of offending by facilitating further victimisation and exacerbating the related impacts of sexual exploitation.

Quayle *et al.* (2008) also suggest that the effects of being a victim of sexual abuse and the associated production of IIOC are likely to differ according to victim awareness, the nature and length of victimisation, the relationship with the offender, as well as victim life experiences and resilience. Similar results have also been identified in research examining the impacts of online grooming on young people (Marriott *et al.* 2014; Whittle *et al.* 2013). These factors can further influence the psychological impacts of victimisation, and should be explored by practitioners when assessing associated support requirements for victims and their families (Quayle *et al.* 2008).

Offender demographics

This section of the chapter reviews current understanding of offender demographics, motivations and explanations for offending behaviour in order to identify current gaps in the literature and associated investigative challenges. The evidence suggests that there is no clear demographic profile for those who use ICT to commit IIOC offences[2] other than the majority being male and of white ethnicity (Babchishin *et al.* 2014). There is some evidence that IIOC offenders are younger than contact offenders, but figures vary between different studies (Babchishin *et al.* 2014). The research evidence also suggests that IIOC offenders are less likely to be married or have children than contact offenders, or to have access to children through employment, and this has been identified as a situational determinant of offending behaviour (Long *et al.* 2012; McManus *et al.* 2014). However, the ability of ICT to provide offenders with a channel to interact with young people and facilitate offending should also be examined when investigating their potential involvement in contact offending (Long *et al.* 2012; McManus *et al.* 2014). This would provide a more comprehensive risk assessment which more fully recognises the technological dimensions of this form of offending.

Motivations and explanations for possession/access

Research has identified a number of offender motivations and explanations for the possession of IIOC. These include curiosity, sexual gratification, replacement for unsatisfying offline relationships, as a coping strategy for personal or psychological problems, and as the result of the addictive properties of the Internet (e.g. Quayle and Taylor 2002; Seto et al. 2010). Many of these are not directly focused on sexual interest in children, suggesting that this motivation may not be central to offending for some individuals. If this behaviour is not motivated by sexual interest in children in many offenders, it is important to understand why IIOC are used to meet other needs (e.g. coping mechanisms for interpersonal difficulties) when they could access adult sexually explicit materials online which do not carry the strong legal, moral and social sanctions associated with IIOC (Bryce 2010).

However, understanding in this area is based on the analysis of offender accounts obtained through research, police or clinician interviews. As a result, it is difficult to assess the extent to which they are accurate reflections of behavioural motivations at the time of offending, or represent post-offence rationalisations (Howitt and Sheldon 2007). Winder and Gough (2010) identified a number of strategies used by offenders to distance themselves from their behaviour and minimise its impact. This included denying active involvement in creating victims as the result of accessing IIOC, as well as claiming that victims did not appear to be harmed or were enjoying the activities depicted. This potential for offender bias, self-presentation and justification in explanations for their behaviour must be recognised when drawing conclusions from empirical evidence in this area.

Offender psychological characteristics and deficits

Quantitative research has also examined offender motivations, and their relationship with psychological deficits, pro-offending attitudes and deviant sexuality (Babchishin et al. 2014). The utility of the 'pathways model' of contact offending (Ward and Siegert 2002) for understanding IIOC possession offending has been examined in order to determine potential similarities in pro-offending attitudes and psychological dysfunction with contact offenders (Henry et al. 2010; Middleton et al. 2006). The model specifies that offending is not motivated by the same set of factors for all offenders. Instead, there are a number of different psychological variables and deficits which combine at different levels to create different pathways into offending (Henry et al. 2010; Middleton et al. 2006).

Henry et al. (2010) identified three specific pathways in IIOC offenders. The inadequate pathway was characterised by socio-affective difficulties, low self-esteem, high levels of loneliness and a lack of pro-offending attitudes. The deviant pathway was associated with empathy deficits, pro-offending attitudes, low self-esteem and high levels of loneliness. These pathways have also been identified in contact offenders, suggesting the suitability of established risk assessment and treatment strategies for use with IIOC possession offenders demonstrating these characteristics (Henry et al. 2010). The normal pathway was characterised by greater emotional stability and a lack of pro-offending attitudes, deficits or sexual interest in children. This suggests that there may also be a group of offenders without these characteristics who experience pathways into offending not specified by the model (Henry et al. 2010).

This may reflect differences in cognitive distortions[3] between IIOC and contact offenders (Henry et al. 2010). These beliefs have been identified in both categories, though contact offenders have been found to have significantly higher scores for this factor than IIOC offenders (Babchishin et al. 2011; Elliott et al. 2012; Merdian et al. 2014). It is possible that the cognitive distortions of IIOC offenders have greater offence-specificity and differ from those which characterise contact offenders (Merdian *et al.* 2014). For example, these may relate to beliefs about the lack of victim harm and lack of personal responsibility as the result of not being involved in the production of images (Sheehan and Sullivan 2010). These may also be reinforced by the distancing nature of technology and the online environment, the availability of a wide variety of IIOC content, the ability to connect with other offenders, and the expectation that technology will enable offenders to evade detection. The development of the measures of cognitive distortions in samples of contact offenders suggests that their reliability and validity may be limited for IIOC offenders, and the need to develop a more specialised assessment tool for this group which recognises the technological contexts of offending (Henry *et al.* 2010; Merdian *et al.* 2014).

There is also some disagreement among researchers over the function of cognitive distortions and the specific focus of related psychometric measures (Merdian *et al.* 2014; Ó Ciardha and Gannon 2011). Cognitive distortions have been described as functioning in three different ways. They may represent pre-offence beliefs which facilitate offending (e.g. Abel *et al.* 1984), or post-offence rationalisations to explain and justify the behaviour (e.g. Gannon and Polaschek 2005). An alternative perspective views them as reflecting the distorted early experiences of offenders, particularly early childhood experiences of sexual abuse by adults or sexual activity with other children (Howitt and Sheldon 2007). The qualitative studies of offender motivations reviewed earlier suggest that the evidence is more consistent with their being post-offence rationalisations.

It is also important to recognise the potential for offender response bias when completing psychometric tests to lead them to respond in ways which present themselves more positively or underestimate their psychological deficits (Henry *et al.* 2010). When offenders participate in research during or after a prison sentence and treatment, their responding may reflect learned clinical explanations for their behaviours, and obscure the motivational and emotional states experienced at the time of offending (Bryce 2010; Henry *et al.* 2010; Howitt and Sheldon 2007). This suggests that some offenders may deny or minimise their pro-offending attitudes and sexual interest in children, consistent with the higher scores for socially desirable responding in the normal group in the Henry *et al.* study. It is also consistent with the explanations or strategies utilised by offenders to distance themselves from responsibility for their behaviour in the previously described qualitative research (e.g. Winder and Gough 2010).

Despite these potential limitations, the evidence suggests that IIOC offenders are a heterogeneous group who demonstrate different levels of sexual interest in children and psychological deficits (Henry *et al.* 2010). Further research is required to develop greater understanding of cognitive distortions, pathways to offending, and the reliability and validity of existing measures for this specific group of offenders. There is also a need to examine the implications of the lack of direct victim interaction for reinforcing the fantasy related aspects of IIOC use, objectification and related strategies for maintaining psychological distance from victims. This includes a consideration of the role of ICT and the online environment in the escalation from accessing IIOC to contact offending, and the potential to reinforce cognitive distortions which enable offenders to rationalise their behaviour.

It is also likely that there will be differences in offending behaviour and cognitive distortions between these different groups of offenders, and that these may also be implicated in varying levels of risk of progression to contact offending, However, this has yet to be addressed in the research literature, and the key focus for empirical studies has been to examine factors which potentially influence the nature of the relationship between possession and contact offending.

Relationship between IIOC possession and contact sexual offending

The relationship between possession of IIOC and contact offending is a central issue for enforcement, risk assessment and treatment (McManus *et al.* 2014; Seto *et al.* 2011). The identification of patterns of offending behaviour which can predict the probability that an offender may progress to committing sexual offences against children can enable the police and other relevant agencies in suspect prioritisation, development of risk management strategies, and allocation of resources (HMIC/HMCPSI 2012).

The existing research evidence suggests that IIOC offenders are a heterogeneous group with varying risk of subsequent contact offending (Middleton, Beech, and Mandeville-Norden 2006; Long *et al.* 2012; Webb *et al.* 2007). A number of potential relationships between access to/use of technology in acquiring IIOC and contact offending have been identified in the literature (Long *et al.* 2012). It has been suggested that some offenders use IIOC as a diversion from contact offending which prevents them from acting on their deviant sexual fantasies (Babchishin *et al.* 2011; Elliott *et al.* 2009). Other researchers have suggested that use of IIOC represents an extension of existing offending behaviour, with offenders using content as part of victim grooming to normalise sexual activity between adults and children (e.g. Bourke and Hernandez 2009). This further demonstrates the role of ICT in creating opportunities for extending the scope and nature of sexual offending against children and young people, as well as criminal behaviour more generally (see Chapter 1 for further discussion of the ability of technology to extend criminal behaviour and capacity).

However, it is the potential that ICT enabled access and use of IIOC may escalate to contact offending that has received the greatest empirical attention (Buschman *et al.* 2010). IIOC may provide a script which offenders follow when progressing to contact offending (Quayle and Taylor 2002). It has also been suggested that the sexual arousal associated with accessing IIOC may lead to desensitisation and offenders to seek more violent and degrading images (Bourke and Hernandez 2009). This 'fantasy escalation effect' is potentially associated with an increase in the likelihood of contact offending as the use of IIOC may be unable to provide sexual gratification in the long-term (McManus and Almond 2014; Sheehan and Sullivan 2010). Empirical research has addressed the potential escalation of offending by examining criminal justice system data relating to the criminal history of IIOC offenders, as well as rates of recidivism.

Previous criminal history of IIOC offenders

Based on empirical evidence that previous convictions for sexual offences are a strong predictor of future risk of recidivism (Hanson and Bussière 1998; Robilotta *et al.* 2008), research examining the criminal history of convicted IIOC possession offenders has been a key focus in the literature (Seto 2009). For example, some studies suggest that many IIOC offenders had no convictions or contact with the criminal justice system prior to their index offence charge (Aslan and Edelmann 2014; Frei *et al.* 2005). Comparative research also suggests that this category of offender have fewer previous convictions than contact offenders (Elliott *et al.* 2009; Webb *et al.* 2007). However, other studies found that 22 per cent of IIOC offenders had previously been arrested, with 11 per cent having prior convictions for offences against children (Wolak *et al.* 2005). Seto and Eke (2005) found that 24 per cent of their sample had a criminal record for contact offences and 15 per cent for possession of IIOC.

The figures presented in these studies suggest that a relatively small proportion of IIOC offenders have previous general, contact or IIOC convictions. This could be argued to indicate a relatively low risk of recidivism and progression to contact offending (Seto 2013). However, these data sources may not provide a full account of previous criminal behaviour as official records only provide information on detected and prosecuted offences (Babchishin et al. 2014). It is likely that some IIOC offenders will have been offending for some time before being caught and have undetected offences in their history, whilst contact offenders may have more previous convictions as a result of their behaviour being more easily detected (Aslan and Edelman 2014). This is consistent with studies indicating that many IIOC offenders disclose unreported crimes during treatment (Galbreath et al. 2002; Webb et al. 2007). For example, the meta-analysis conducted by Seto et al. (2011) found that 12 per cent of IIOC offenders disclosed a previously undetected contact offence against a child, with the proportion rising to 55 per cent in the self-report studies included in their meta-analysis.

This further complicates the development of a clear understanding of the link between criminal history and future risk of IIOC or contact offending, although it has been argued that the evidence is not sufficient to claim that many offenders will commit further crimes as the recidivism rates for contact offenders are also low (Hanson and Bussière 1998; Hanson and Morton-Burgon 2005; Seto 2009). However, it is also important to examine the criminal history of contact offenders for prior IIOC convictions and associated indications of risk of escalation. This is less frequently reported in the literature, though one recent study found no evidence that the contact offenders in their sample had previous convictions for IIOC offences (Aslan and Edelmann 2014). As previously described, the data sources used may not provide a full account of previous criminal behaviour due to the potential for undetected offences to be present in offenders' prior history (Babchishin et al. 2014).

Reoffending rates for IIOC offences and contact offending

Research has also examined the reoffending rates of IIOC offenders as a way of establishing risk of escalation to contact offending (Eke et al. 2011; Seto and Eke 2005). Seto and Eke (2005) found that 6 per cent of IIOC offenders were subsequently convicted of a new IIOC and 17 per cent of a general offence. A more recent 4-year follow-up found a similar recidivism rate of 7 per cent, with 4 per cent of the sample being subsequently convicted of a contact sexual offence (Eke et al. 2011). Another meta-analysis found a recidivism rate of 3.4 per cent for new IIOC and 2.0 per cent for new contact offences (Seto et al. 2011). Wakeling et al. (2011) found a lower rate of recidivism of 2.1 per cent one year after release for IIOC offenders, which increased to 3.1 per cent after two years. Seventy-four per cent of those who were subsequently convicted were charged with an IIOC and 19 per cent with a contact offence.

These figures suggest that recidivism in this group of offenders is more likely to reflect non-sexual than IIOC offences or escalation to contact offending. However, recidivism studies have similar limitations related to undetected offences as those examining criminal history, suggesting the potential underestimation of actual reoffending rates (Aslan and Edelmann 2014; Babchishin et al. 2014). As a result, they may not necessarily provide an accurate estimate of the level of reoffending or possible escalation. Studies vary in the definitions of recidivism used, sampling and data sources, as well as the follow-up periods examined which potentially explain identified variation in rates between studies (Lussier 2005; Lussier and Cale 2013). It is also important to note that these studies treat IIOC offenders as a homogeneous group, and do not consider how risk of recidivism and escalation may vary according to the different subgroups of offenders identified in relation to the pathways model. This represents an important area for further empirical investigation, and can inform the development of appropriate risk assessment strategies and associated offender interventions.

The research reviewed in this section suggests that a relatively small proportion of IIOC offenders have previous convictions for possession offences, and associated risk of escalation to contact offending. This is consistent with claims that the relationship between IIOC possession and contact offending is complex and not necessarily directly causal (McCarthy 2010). However, the available evidence suggests that some offenders do pose an increased risk of contact offending. This indicates the need to identify factors which predict escalation, and to develop typologies of the different relationships between possession and risk of contact offending (Eke *et al.* 2011). One recent approach to addressing this has considered the utility of examining the content of IIOC collections to determine risk of escalation (McManus *et al.* 2014; Long *et al.* 2012). This is examined in further detail in the next section of the chapter which examines the investigative, evidential and behavioural utility of examining the content of IIOC and associated forensic evidence.

IIOC as crime scenes, victim identification and forensic data analysis

As IIOC depict a record of criminal offence, they can be analysed as digital crime scenes and enable police investigations, the detection of offenders and identification of victims. There are a number of different levels of analysis which can be applied to IIOC and related forensic data. These include the content of images and collections, image metadata, and wider digital forensic evidence recovered from offender devices during investigations. Analysing these different sources of data can enable the identification of investigatively and evidentially relevant information (Glasgow 2010; Long *et al.* 2012; McManus *et al.* 2014). Analysis of the content of images can provide details of victim and offender characteristics (e.g. age, gender, physical description), and assist in offender and victim identification (Holland 2005). The processes by which IIOC and related forensic data can be analysed in this way are described by Holland (2005), and demonstrate the involved resourcing and investigative challenges. This is reflected in the comparatively low number of successful identifications in comparison to the number of victims depicted in IIOC as described earlier. In the UK, for example, CEOP's Victim Identification Team made 47 identifications in 2009–2010 (CEOP 2013). At the international level, over 7,800 victims from more than 40 countries had been identified using the International Child Sexual Exploitation Image Database by November 2015 (Interpol 2015). However, these figures demonstrate that victim identification is possible, and fulfils an important safeguarding role, as well as having an investigative and evidential function.

Analysis of IIOC collections can also enable the identification of offence-related characteristics and preferences (e.g. nature and severity of sexual activity depicted), and inform risk assessment and the classification of offences for charging and sentencing (McManus *et al.* 2014). It can also contribute to the development of further understanding of offender motivations and behaviours, as it has been argued that offenders choose content that is consistent with their sexual fantasies and interests (e.g. Lanning 1992; Glasgow 2010). As a result, these interests are likely to be reflected in the victims depicted in the content of IIOC collections, including the age and stage of victim development that is most sexually arousing to an offender (Quayle and Taylor 2002). This suggests that an examination of image collection can potentially enable the identification of victim and offence preferences (e.g. age, gender, specific sexual activities). This has been examined empirically in comparative studies of the image collections of IIOC possession and dual offenders[4] (Long *et al.* 2012; McManus *et al.* 2014). McManus *et al.* (2014) found that 63.3 per cent of the image collection examined indicated a preference for female victims, 12.4 per cent for males, and 23.3 per cent for both. Both studies found that there were no victim age or gender differences between IIOC and dual offenders based on analysis of image collections.

However, they did find evidence that the victims in the collections of dual offenders had a smaller victim age range. The authors suggest that this may represent greater victim age specificity in image collections in dual offenders which may subsequently be reflected in victim choice when contact offending (Long *et al.* 2012; McManus *et al.* 2014).

These studies also found that there were no differences between offenders in the proportion of their collection across different SAP levels.[5] However, Long *et al.* (2012) found that dual offenders had greater specificity on images at SAP levels 3–4, whilst IIOC offenders focused on images at SAP level 1. The authors suggest that these results indicate preferred fantasies and sexual activities which may be related to escalation and implications for the type of contact offences committed (Long *et al.* 2012; McManus *et al.* 2014). As a result, possession of IIOC should be viewed as indicating a risk of contact offending, with specific investigative priority placed on offenders with greater specificity in their samples at higher SAP levels, access to children and a prior criminal history as factors associated with escalation (Long *et al.* 2012).

Forensic analysis of offender hardware may also lead to the identification of other data which is relevant to understanding offender motivations and behaviour, as well as risk of progression to contact offending (Glasgow 2010). This includes evidence of image labelling and organisation, involvement in production and dissemination, attempts to contact young people online, actions to evade detection, and association with other offenders (Glasgow 2010; Al Mutawa *et al.* 2015). Initial exploratory research has examined the utility of analysing this type of data for developing knowledge of offender motivations and behaviour, as well as improving investigative procedures (Al Mutawa *et al.* 2015). For example, evidence of IIOC related queries in P2P client software and web browser search engines can indicate specific victim and activity interests, as well as provide evidence that behaviour was intentional (Al Mutawa *et al.* 2015). Use of anti-forensics tools, peer2peer applications and encryption (e.g. TOR, Darknet) suggest intentional activities to conceal IIOC and associated behaviour, offender awareness of the existence of files and their legal status. This evidence can be used when interviewing suspects to challenge claims about accidental access or virus infection as an explanation for the discovery of IIOC on their devices (Al Mutawa *et al.* 2015). This is a new area of empirical research that demonstrates potential investigative and behavioural relevance, and can further contribute to understanding this form of offending.

Conclusion

The review of the literature provided by this chapter indicates the utility of a combined approach to understanding the dynamics of this form of online sexual exploitation of children and young people. It has examined the available evidence relating to demographics and motivations, pro-offending attitudes and psychological deficits, as well as the nature of IIOC collections and offence-related behaviours identified based on analysis of digital forensic evidence. A combined examination of these factors can contribute towards developing further understanding of the characteristics of IIOC offenders, and the potential risk of escalation to contact offending. Both of these aspects of offending require further empirical research to address the existing knowledge gaps and further inform evidence-based approaches to investigation and victim safeguarding.

This should include the development of further understanding of the role of ICT and technological contexts in facilitating and detecting this form of sexual offending against children and young people. This includes the influence of the characteristics of the online environment and mediated communication on the dynamics of offending behaviour (Bryce 2015). It also raises the question of whether a new category of child sexual offenders has emerged as a result of the expanded opportunities for the production, dissemination and access to IIOC afforded by the online environment, or whether this represents a new medium for facilitating offending (Seto *et al.* 2011). It is unlikely that the presence of IIOC online alone is sufficient to encourage individuals to become involved in this form of offending without the presence of the other psychological characteristics and deficits discussed earlier in the chapter. It has, however, expanded access to this material, and altered the situational determinants of production and contact offending by providing opportunities for victim contact for those offenders who do not have access to children and young people in the offline environment. This may further reinforce pro-offending attitudes and fantasies, as well as offender minimisation of the seriousness of their activities by enabling further distancing from victims and the impacts of their behaviour.

There is also a need for greater consideration of the links between IIOC offending and other forms of online sexual exploitation. The reporting by law enforcement of instances where sexually explicit material produced by young people and distributed in online peer networks has subsequently been identified in offender image collections (CEOP 2013) indicates the need to develop a more detailed understanding of the intersection between the normative online behaviour of young people and opportunities for offending (Bryce 2014). The tendency in the literature to examine IIOC offending and other forms of online sexual exploitation as distinct categories is problematic as it does not enable sufficient consideration of its intersection with online grooming and non-contact offending. Greater consideration of their inter-relationship is required, as well as the associated investigative and victimisation implications.

The existence of online offender networks and their involvement in distribution of IIOC represents an additional impact of digital technology/ICT on the offending process which has yet to be fully explored empirically. The small number of studies which have examined their structure and function demonstrate their role in allowing communication between individuals with sexual interest in children (e.g. Durkin and Bryant 1999; Holt *et al.* 2010; Quayle and Taylor 2002). This may lead to the validation and reinforcement of offence-supportive beliefs and deviant sexual scripts, as well as the potential for the planning and performance of offending behaviour in both online and offline environments (e.g. Bourke *et al.* 2012; Holt *et al.* 2010; Quayle and Taylor 2002). It is also possible that networking with other offenders may be an additional risk factor for contact offending which has yet to be fully examined empirically.

Despite the challenges associated with investigation and offender detection examined in this chapter, the online environment makes offending visible as well as facilitating it. This highlights the investigative and evidential utility of content, collections and related digital forensic evidence, as well as their ability to further inform understanding of offender motivations, characteristics and the dynamics of the behaviour.

Finally, it is also important to develop further understanding of victimisation processes and impacts in the context of the online environment and mediated communication. This area of research is underdeveloped due to the difficulties of victim identification, as well as the ethical and safeguarding issues associated with participation in empirical studies. It has recently been argued that victimisation should be conceptualised as a complex process which involves the dynamics and impacts of the offence, and involves a variety of stakeholders in addition to the victim and offender (Bryce *et al.* 2016; Fohr 2015). This is particularly relevant to understanding victimisation through the production and dissemination of IIOC as the affordances of technology, the online environment and associated influences on interaction and behaviour are implicated in facilitating offending and victimisation (Bryce 2015). The process of victimisation is extended beyond the point of the production of IIOC to the wider and indefinite circulation of images online, and the related opportunities for unlimited numbers of offenders to access the content. This additional source of ongoing victimisation may interfere in recovery and coping with the associated traumatic impacts. The role of the police, practitioners and the wider criminal justice system is assisting victims and their families through investigations and the legal process also requires further consideration to ensure that the associated potential for revictimisation and further trauma are minimised.

Each of the areas examined in this chapter represent a specific aspect of the offending and victimisation process, and demonstrate the influence of digital technology in shaping the generation, dissemination and collection of IIOC. It is important to examine each of these dimensions, as well as their inter-relationships, in order to develop further understanding of the offence process, detection and investigation, as well as the safeguarding and support of victims. This will further contribute to the developing evidence base which can inform the implementation of appropriate prevention and response strategies by different stakeholders.

Notes

1   This difference reflects the difficulties of victim identification given potential appearances in multiple images over periods of time in which changes in appearance and body characteristics may occur.
2   This refers specifically to those offenders who have been charged with accessing or possession of IIOC, but do not evidence any indication of engagement in contact offending.
3   Cognitive distortions refer to the extent to which offenders perceive children as being able to consent to sexual contact, and to engage in such activities without any associated harm (Ward and Keenan 1999).
4   Convicted for both IIOC possession and contact offences.
5   The content and severity of IIOC in the UK was rated using the Sentence Advisory Panel (SAP) classification system which specified five SAP levels until 2014. This system is used by the courts when sentencing IIOC offences, based on the examination of offender collections by police officers to identify the proportion of images at different SAP levels. This classification scale was recently reduced from five to three categories (See Sentencing Guidance Council 2014). The original five SAP levels which are reported in empirical studies are:

    1.   Images depicting erotic posing with no sexual activity.
    2.   Non-penetrative sexual activity between children, or solo masturbation by a child.
    3.   Non-penetrative sexual activity between adults and children.
    4.   Penetrative sexual activity involving a child or children, or both children and adults.
    5.   Sadism or penetration of, or by, an animal.

References

Abel, G. G., Becker J. V. and Cunningham-Rathner, J. (1984) Complications, consent, and cognitions in sex between children and adults, *International Journal of Law Psychiatry*, 7: 89–103.

Al Mutawa, N., Bryce, J., Franqueira, V.N.L. and Marrington, A. (2015) Behavioural Evidence Analysis Applied to Digital Forensics: An Empirical Analysis of Child Pornography Cases Using P2P Networks, in *Availability, Reliability and Security (ARES), 10th International Conference*, pp. 293–302. 24–27 Aug. 2015. doi: 10.1109/ARES.2015.49.

Aslan, D. and Edelmann, R. (2014) Demographic and offence characteristics: A comparison of sex offenders convicted of possessing indecent images of children, committing contact sex offenders or both offences. *The Journal of Forensic Psychiatry & Psychology*, 25: 121–134.

Aslan, D., Edelmann, R., Bray, D. and Worrell, M. (2014) Entering the world of sex offenders: an exploration of offending behaviour patterns of those with both internet and contact sex offences against children, *Journal of Forensic Practice*, 16: 110–126.

Babchishin, K. M., Hanson, R. and Hermann, C. A. (2011) The characteristics of online sex offenders: A meta-analysis, *Sexual Abuse: Journal of Research and Treatment*, 23: 92–123.

Babchishin, K. M., Hanson, R. and Van Zuylen, H. (2014) Online child pornography offenders are different: A meta-analysis of the characteristics of online and offline sex offenders against children, *Archives of Sexual Behavior*, Advance online publication. doi: 10.1007/s10508-014-0270-x.

BBC (2015) Court told paedophiles may have escaped prosecution. BBC News. http://www.bbc.co.uk/news/uk-northern-ireland-32551334. 01.05.16 Accessed 11.09.16.

Beech, A. R., Elliott, I. A., Birgden, A. and Findlater, D. (2008) The Internet and child sexual offending: A criminological review, *Aggression and Violent Behavior*, 13: 216–228.

Bourke, M. L. and Hernandez, A. E. (2009) The 'Butner Study' redux: A report of the incidence of hands-on child victimization by child pornography offenders, *Journal of Family Violence*, 24: 183–19.

Bourke, P., Ward, T. and Rose, C. (2012) Expertise and sexual offending: A preliminary empirical model, *Journal of Interpersonal Violence*, 27: 2391–2414.

Bryce, J. (2010) Online sexual exploitation of children and young people, in Jewkes, Y. and Yar, M. (eds), *Handbook of Internet Crime*. London: Willan.

Bryce, J. (2014) The technological mediation of leisure in contemporary society, in Elkington, S. and Gammon, S. (eds), *Leisure in Mind: Meanings, Motives and Learning*. London: Routledge.

Bryce, J. (2015) Cyberpsychology and human factors, *IET Engineering and Technology Reference*, 1–8. doi: 10.1049/etr.2014.0028.

Bryce, J., Brooks, M., Robinson, P., Stokes, R., Irving, M., Lowe, M., Graham-Kevan, N., Willan, V. J., Khan, R. and Karwacka, M. (2016) A qualitative examination of engagement with support services by victims of violent crime, *International Review of Victimology*, 22(3): 1–17.

Buschman, J., Wilcox, D., Krapohl, D., Oelrich, M. and Hackett, S. (2010) Cybersex offender risk assessment. An explorative study, *Journal of Sexual Aggression*, 16: 197–209.

CEOP (Child Exploitation and Online Protection Centre) (2013) *A Picture of Abuse: A Thematic Assessment of the Risk of Contact Child Sexual Abuse Posed by Those Who Possess Indecent Images of Children*. London: CEOP.

Council of Europe (2001) Convention on Cybercrime, CETS 185: 1–27.

Council of Europe (2007) Convention on the Protection of Children against Sexual Exploitation and Sexual Abuse, CETS 201: 1–21.

Criminal Justice Act, c. 33 O.P.S.I. (1998).

Durkin, K. F. and Bryant, C. D. (1997) Misuse of the Internet by pedophiles: Implications for law enforcement and probation practice, *Federal Probation*, 61: 14–18.

Durkin, K. F. and Bryant, C. D. (1999) Propagandizing pederasty: A thematic analysis of the on-line exculpatory accounts of unrepentant pedophiles, *Deviant Behaviour*, 20: 103–127.

Eke, A. W., Seto, M. C. and Williams, J. (2011) Examining the criminal history and future offending of child pornography offenders: An extended prospective follow-up study, *Law and Human Behavior*, 35: 466–478.

Elliott, I. A. and Beech, A. R. (2009) Understanding online child pornography use: Applying sexual offense theory to internet offenders, *Aggression and Violent Behavior*, 14: 180–193.

Elliott, I. A., Beech, A. R. and Mandeville-Norden, R. (2012) The psychological profiles of internet, contact and mixed internet/contact sex offenders, *Sexual Abuse: A Journal of Research and Treatment*, 25: 3–20.

Elliott, I. A., Beech, A. R., Mandeville-Norden, R. and Hayes, E. (2009) Psychological profiles of internet sex offenders: Comparisons with contact sexual offenders, *Sexual Abuse: A Journal of Research and Treatment*, 21: 76–92.

Fergusson, D. M., Boden, J. M. and Horwood, L. J. (2008) Exposure to childhood sexual and physical abuse and adjustment in early adulthood, *Child Abuse Negl*, 32: 607–619.

Fohring, S. (2015) An integrated model of victimisation as an explanation of non-involvement with the criminal justice system, *International Review of Victimology*, 21(1): 45–70.

Frei, A., Erenay, N., Dittmann, V. and Graf, M. (2005) Paedophilia on the internet – a study of 33 convicted offenders in the Canton of Lucerne, *Swiss Medical Weekly*, 135: 488–494.

Galbreath, N. W., Berlin, F. S. and Sawyer, D. (2002) Paraphilias and the internet, in Cooper, A. (ed.) *Sex and the Internet: A Guidebook for Clinicians*. Philadelphia: Brunner-Routledge, pp. 187–285.

Gannon, T. and Polaschek, D. (2005) Do child molesters deliberately fake good on cognitive distortion questionnaires? An information processing-based investigation, *Sexual Abuse: A Journal of Research And Treatment*, 17: 183–200.

Gannon, T. A. and Polaschek, D. L. L. (2006) Cognitive distortions in child molesters: A reexamination of key theories and research, *Clinical Psychology Review*, 26: 1000–1019.

Glasgow, D. (2010) The potential of digital evidence to contribute to risk assessment of internet offenders, *Journal of Sexual Aggression*, 16: 87–106.

Hanson, R. K. and Bussière, M. T. (1998) Predicting relapse: A meta-analysis of sexual offender recidivism studies, *Journal of Consulting and Clinical Psychology*, 66: 348–362.

Hanson, R. K. and Morton-Burgon, K. (2005) Predictors of Sexual Recidivism: An Updated Meta-Analysis 2004-02, *Public Works and Government Services Canada* Cat. No.: PS3-1/2004-2 ISBN: 0-662-68051-0.

Henry, O., Mandeville-Norden, R., Hayes, E. and Egan, V. (2010) Do internet-based sexual offenders reduce to normal, inadequate and deviant groups? *Journal of Sexual Aggression*, 16: 33–46.

HMIC/HMCPSI (2012) *Forging the links: Rape investigation and prosecution. A joint review by HMIC and HMCPSI.* ISBN: 978-1-84987-688-9.

Holland, G. (2003) Identifying victims of child abuse images: An analysis of successful identifications, in Taylor, M. and Quayle, E. (eds) *Child Pornography: An Internet Crime.* New York: Brunner-Routledge.

Holland, G. (2005) Identifying victims of child abuse images: An analysis of successful identifications, in Taylor, M. and Quayle, E. (eds), *Viewing Child Pornography on the Internet: Understanding the Offence, Managing the Offender, Helping the Victims.* Lyme Regis, UK: Russell House Publishing Ltd.

Holt T., Kristie, B. and Burker, N. (2010) Considering the pedophile subculture online, *Sexual Abuse*, 22: 3–24.

Home Office (2011) *Crime in England and Wales 2010/11. Findings from the British Crime Survey and Police Recorded Crime*, 2nd edition. London: Home Office Statistical Bulletin.

Howitt, D. and Sheldon, K. (2007) The role of cognitive distortions in paedophilic offending: Internet and contact offenders compared, *Psychology, Crime and Law*, 13: 469–486.

Interpol (2015) *Crimes Against Children*, available at: www.interpol.int/Crime-areas/Crimes-against-children/Victim-identification.

IWF (Internet Watch Foundation) (2010) *Internet Watch Foundation Annual and Charity Report.* London: IWF.

IWF (Internet Watch Foundation) (2012) *Internet Watch Foundation Operational Trends Report 2012.* London: IWF.

IWF (Internet Watch Foundation) (2013) *Internet Watch Foundation Annual and Charity Report.* London: IWF.

IWF (Internet Watch Foundation) (2014) *Internet Watch Foundation Annual Report 2014.* London: IWF.

Lanning, K. (1992) *Investigators Guide to Allegations of 'Ritual Child Abuse'*, Quantico, VA: Behavioral Science Unit, National Centre for the Analysis of Violent Crime, FBI Academy.

Long, M. L., Alison, L. A. and McManus, M. A. (2012) Child pornography and likelihood of contact abuse: a comparison between contact child sexual offenders and non-contact offenders, *Sexual Abuse: A Journal of Research and Treatment*, 25: 370–395.

Lussier, P. (2005) The criminal activity of sexual offenders in adulthood: Revisiting the specialization debate, *Sexual Abuse: A Journal of Research and Treatment*, 17: 269–292.

Lussier, P. and Cale, J. (2013) Beyond sexual recidivism: A review of the sexual criminal career parameters of adult sex offenders, *Aggression and Violent Behavior*, 18: 445–457.

Marriott, C., Hamilton-Giachritsis, C. and Harrop, C. (2014) Factors promoting resilience following childhood sexual abuse: A structured, narrative review of the literature, *Child Abuse Review*, 23: 17–34.

McCarthy, J. A. (2010) Internet sexual activity: A comparison between contact and non-contact child pornography offenders, *Journal of Sexual Aggression*, 16: 203–217.

McGuire, M. and Dowling, S. (2013) *Cyber Crime: A Review of the Evidence, Home Office Research Report 75.* London, UK: Home Office.

McManus, M. and Almond, L. (2014) Trends of indecent images of children and child sexual offences between 2005/2006 and 2012/2013 within the United Kingdom, *Journal of Sexual Aggression*, 20: 142–155.

McManus, M., Long, M. L., Alison, L. and Almond, L. (2014) Factors associated with contact child sexual abuse in a sample of indecent image offenders, *Journal of Sexual Aggression*, Advance online publication. DOI: 10.1080/13552600.2014.927009.

Merdian, H. L., Curtis, C., Thakker, J., Wilson, N. and Boer, D. P. (2014) The endorsement of cognitive distortions: Comparing child pornography offenders and contact sex offenders, *Psychology, Crime & Law*, 20(10): 971–993.

Middleton, D., Elliott, I. A., Mandeville-Norden, R. and Beech, A. R. (2006) An investigation into the application of the Ward and Siegert Pathways Model of child sexual abuse with Internet offenders, *Psychology, Crime and Law*, 12: 589–603.

Ministry of Justice (2013) Unpublished data. London: Ministry of Justice.

NCMEC (2012) NCMEC Statistical Report, available at: www.saferinternet.org.uk/content/childnet/safterinternetcentre/downloads/Research_Highlights/UKCCIS_RH62_NCMEC_Statistics_Report.pdf.

Ó Ciardha, C., and Gannon, T. (2011) The cognitive distortions of child molesters are in need of treatment, *Journal of Sexual Aggression*, 17: 130–141.

Palmer, T. (2005) Behind the screen: Children who are the subjects of abusive images, in Quayle, E. and Taylor, M. (eds), *Viewing Child Pornography on the Internet*. Lyme Regis, UK: Russell House Publishing.

Protection of Children Act, c. 37 O.P.S.I. (1978).

Quayle, E. and Jones, T. (2011) Sexualised images of children on the internet, *Sexual Abuse: A Journal of Research and Treatment*, 23(1): 7–21.

Quayle, E., Lööf, L. and Palmer, T. (2008) *Child Pornography and Sexual Exploitation of Children Online*. Bangkok: ECPAT International.

Quayle, E. and Palmer, T. (2010) Where is the harm? Technology mediated abuse and exploitation of children, *Ontario Association of Children's Aid Societies Journal*, 55: 35–40.

Quayle, E. and Taylor, M. (2002) Child pornography and the Internet: Perpetuating a cycle of abuse, *Deviant Behavior*, 23: 331–61.

Quayle, E. and Taylor, M. (2003) Model of problematic Internet use in people with a sexual interest in children, *CyberPsychology and Behavior*, 6: 93–106.

Quayle, E., Vaughan, M. and Taylor, M. (2006) Sex offenders, Internet child abuse images and emotional avoidance: The importance of values, *Aggression and Violent Behavior*, 11: 1–11.

Radford, L., Corral, S., Bradley, C., Fisher, H., Bassett, C., Howat, N. and Collishaw, S. (2011) *Child Abuse and Neglect in the UK Today*. London: NSPCC.

Robilotta, S., Mercado, C.C. and DeGue, S. (2008) Application of the polygraph examination in the assessment and treatment of Internet sex offenders, *Journal of Forensic Psychology Practice*, 8: 383–393.

Scott, S. (2001) *The Politics and Experience of Child Sexual Abuse: Beyond Disbelief*. Open University Press: Buckingham.

Sentencing Guidelines Council (2007) *Sexual Offences Act 2003: Definitive Guideline*. London.

Seto, M. C. (2013) *Internet Sex Offenders*. Washington DC: American Psychological Society.

Seto, M. C. (2009) Pedophilia, *Annual Review of Clinical Psychology*, 5: 391–407.

Seto, M. C. and Eke, A. W. (2005) The criminal histories and later offending of child pornography offenders, *Sexual Abuse: A Journal of Research and Treatment*, 17: 201–210.

Seto, M. C., Hanson, R. and Babchishin, K. M. (2011) Contact sexual offending by men with online sexual offenses, *Sexual Abuse: Journal of Research and Treatment*, 23: 124–145.

Seto, M. C., Hermann, C. A., Kjellgren, C., Priebe, G., Svedin, C. G. and Långström, N. (2015) Viewing child pornography: Prevalence and correlates in a representative community sample of young Swedish men, *Archives of Sexual Behavior*, 44(1): 67–79.

Seto, M. C., Reeves, L. and Jung, S. (2010) Explanations given by child pornography offenders for their crimes, *Journal of Sexual Aggression*, 16: 169–180.

Seto, M. C., Wood, J., Babchishin, K. M. and Flynn, S. (2012) Online solicitation offenders are different from child pornography offenders and lower risk contact sexual offenders, *Law and Human Behavior*, 36: 320–330.

Sexual Offences Act, c. 42 O.P.S.I. (2003).

Sheehan, V. and Sullivan, J. (2010) A qualitative analysis of child sex offenders involved in the manufacture of indecent images of children, *Journal of Sexual Aggression*, 16: 143–167.

Smith, K., Taylor, P. and Elkin, M. (2013) *Crimes detected in England and Wales 2012/13*. Available at: http://socialwelfare.bl.uk/subjectareas/services-activity/criminal-justice/homeoffice/152950hosb0213.pdf.

Svedin, C. G. and Back, K. (2003) *Why Didn't They Tell Us? Sexual Abuse in Child Pornography*. Stockholm: Save the Children Sweden.

Von Weiler, J., Haardt-Becker, A. and Schulte, S. (2010) Care and treatment of child victims of child pornographic exploitation (CPE) in Germany, *Journal of Sexual Aggression*, 16: 211–222.

Wakeling, H. C., Howard, P. and Barnett, G. (2011) Comparing the validity of the RM2000 scale and OGRS3 for predicting recidivism by internet sexual offenders, *Sexual Abuse: A Journal of Research and Treatment*, 23: 146–148.

Ward, T. and Keenan, T. (1999) Child molesters' implicit theories, *Journal of Interpersonal Violence*, 14: 821–838.

Ward, T. and Siegert, R. J. (2002) Toward a comprehensive theory of child sexual abuse: A theory-knitting perspective, *Psychology, Crime and Law*, 9: 319–351.

Webb, L., Craisatti, J. and Keen, S. (2007) Characteristics of Internet child pornography offenders: A comparison with child molesters, *Sexual Abuse: A Journal of Research and Treatment*, 19: 449–465.

Webster, S., Davidson, J., Bifulco, A., Gottschalk, P., Caretti, V. and Pham, T. (2012) *European Online Grooming Report*. European Commission, Safer Internet Plus Programme.

Whittle, H. C., Hamilton-Giachritsis, C. E. and Beech, A. B. (2014) 'Under his spell': Victims' perspectives of being groomed online, *Social Sciences*, 3: 404–426.

Whittle, H. C., Hamilton-Giachritsis, C., Beech, A. and Collings, G. (2013) A review of young people's vulnerabilities to online grooming, *Aggression and Violent Behavior*, 18: 135–146.

Whittle, H. C., Hamilton-Giachritsis, C., Beech, A. and Collings, G. (2013) A review of online grooming: Characteristics and concerns, *Aggression and Violent Behavior*, 18: 62–70.

Whittle, H. C., Hamilton-Giachritsis, C. and Beech, A. (2013) Victims' voices: The impact of online grooming and sexual abuse, *Universal Journal of Psychology*, 1: 59–71.

Winder, B. and Gough, B. (2010) 'I never touched anybody – that's my defence': A qualitative analysis of internet sex offender accounts, *Journal of Sexual Aggression*, 16: 125–141.

Wolak, J., Finkelhor, D. and Mitchell, K. (2005) *Child Pornography Possessors and the Internet: A National Study*. Arlington, VA: National Center for Missing & Exploited Children.

Wolak, J., Finkelhor, D., Mitchell, K. J. and Ybarra. L. (2008) Online 'predators' and their victims. Myths, realities, and implications for prevention and treatment, *American Psychologist*, 63: 111–128.

Wolak, J., Finkelhor, D. and Mitchell, K. J. (2011) Child pornography possessors: Trends in offender and case characteristics, *Sexual Abuse: A Journal of Research and Treatment*, 23: 22–42.

Wolak, J., Finkelhor, D. and Mitchell, K. J. (2012) Trends in Arrests for Child Pornography Possession: The Third National Juvenile Online Victimization Study' (NJOV-3). Crimes Against Children Research Center.

6

ICTs and sexuality

Andrew S. Denney and Richard Tewksbury

It seems so obvious: If we invent a machine, the first thing we are going to do – after making a profit – is use it to watch porn. When the projector was invented roughly a century ago, the first movies were not damsels in distress tied to train tracks or Charlie Chaplin-style slapsticks; they were stilted porn shorts called stag films. VHS became the dominant standard for VCRs largely because Sony wouldn't allow pornographers to use Betamax; the movie industry followed porn's lead. DVDs, the Internet, cell phones. You name it, pornography planted its big flag there first, or at least shortly thereafter.

(Brown 2006)

The advancement and increased use of various information communication technologies (ICTs) has had a substantial impact on numerous aspects of everyday life. These impacts range from the way individuals communicate with one another to how individuals explore and develop various personal qualities. As ICTs have significantly influenced the development of general aspects of everyday life, so too have they impacted how one might explore and act upon various aspects of one's sexuality. ICTs have drastically influenced new ways in which sexuality and sexual deviance is learned, expressed, modified, and even created. Consequently, both informal and formal mechanisms of social control have begun reacting to this increased use of ICTs in regard to the discovery, exploration, and enactment of various sexual behaviors.

From the period of approximately 1940 to the time of this publication, many advanced civilizations throughout the world have seen a monumental increase in the development and use of ICTs in everyday life. Notable examples of ICTs that have had a substantial impact on the way we live are televisions, beepers, home entertainment video equipment (e.g. VCRs, DVDs), personal computers (PCs), cellular phones, and computer tablets. Such ICTs can be used primarily for entertainment purposes (e.g. televisions) or as a central piece of the global economy (e.g. mobile banking on a cellular phone). However, with an estimated 3.17 billion fee-based users worldwide (Statista 2015) (i.e. approximately half of the world's population), the Internet has united many of the ICT devices individuals use in everyday life (e.g. computers, cellular phones). *Fee-based* refers to Internet users who pay a subscription fee for Internet access, not including those who access it through other, typically free, sources (e.g. libraries, Internet cafes). Thus, the number of individuals with Internet access is potentially much higher. As such, the influence of the Internet and its ability to connect a multitude of ICT devices has by far been the most impactful technology for sexuality in how it is learned, expressed, modified, and even created.

To date, the Internet has helped normalize many sexual behaviors that were once considered to be extremely taboo, perhaps even illegal, due to the Internet's interactive capabilities that expanded around the turn of the twenty-first century (Månsson and Sodërlind 2013; Ross 2006). The end result at this point is a crowd of virtual communities where individuals with likeminded interests – sexual and other – are able to connect globally in real-time, an ability not previously feasible at any prior point in history (Durkin *et al.* 2006; Månsson and Sodërlind 2013). As such, the ways in which various aspects of sexuality are learned, expressed, and modified – for deviant and non-deviant ways – has drastically changed.

The purpose of this chapter is four-fold. First, to explore how the development of ICTs – primarily through the Internet – has impacted the ways some individuals discover and explore various aspects of their sexuality. Second, to investigate how ICTs have increased the ease of the discovery, enactment, and modification of various forms of sexual deviance. Third, how such exposure to various mediums of sexual deviance (e.g. Internet pornography, online subcultures) are believed to lead to "real-world" consequences (e.g. addiction, sexual compulsivity, sexual offending). The fourth and final aim of this chapter is to discuss various media, legislative, and law enforcement responses to the dangers of new ways of viewing/expressing sexuality on these mediums. The chapter will close with a brief discussion of the broader implications of ICTs on sexuality for both criminal justice and society as a whole.

The evolution of sexuality through ICTs

Overview of sexuality

Sexuality is a commonly used term that can describe a wide range of emotions, identities, and behaviors. Prior to the discussion of the main thrust of this chapter, it is imperative that a relatively brief overview of *sexuality* is provided for a stronger context of the role that ICTs have played in the modification and expansion of the expression of sexuality. Specifically, there are three key ways in which sexuality can be defined and applied. These three primary ways are (1) general sexual feelings, (2) an individual's overall sexual preference, and (3) the sexual activity in which an individual chooses to engage.

One general way that the concept of sexuality is defined pertains to one's sexual feelings, either private or overt. For example, sexuality can refer to a female going through puberty who realizes she is sexually attracted to boys or girls her age. Additionally, sexuality, in this manner, can refer to the same individual coming to the realization that certain clothing styles may make themselves more attractive to potential partners, thus recognizing and enacting the concept of *sexual attractiveness*.

A second way in which the concept of *sexuality* is often applied is to an individual's preference for sexual partners (e.g. heterosexual, homosexual, bisexual). That is, one who identifies as heterosexual is stating, perhaps directly or indirectly, that they are only sexually interested in a member of the opposite sex. However, an individual that identifies as bisexual is stating that they are sexually interested in both sexes. Therefore, by accepting and demonstrating this aspect of one's sexuality, one is establishing a general boundary for potential sexual partners.

A third and final primary way that the concept of *sexuality* can be applied is through the physical manifestation(s) of one's emotions and/or sexual preferences. In essence, sexuality in this manner can pertain to any physical expression of human sexuality. This can range from two individuals engaging in consensual sex to an individual becoming sexually aroused by dressing in a clothing material that might be an object of their fetish (e.g. leather, latex, rubber). As such, sexuality can apply to a diverse range of emotions, preferences, and physical actions.

Traditionally, one's sexuality has been argued to develop through various social interactions with others that are either closely known to one or even in relations with complete strangers (Gagnon and Simon 1967). Moreover, *sexuality* has been argued to be scripted (see Simon and Gagnon 1986). That is, one learns, modifies, and enacts various aspects of one's sexuality through interaction with one's environment. Specifically, Simon and Gagnon (1986) argue that there are three distinct ways in which sexuality is scripted, these being via (1) *cultural scenarios*, (2) *interpersonal scripts*, and (3) *intrapsychic scripts*.

The first sexuality script, *cultural scenarios*, refers to both historical and culturally shared standards for sexuality (Simon and Gagnon 1986). Moreover, this loosely means that which is considered *normal* sexual behavior in a particular society. For example, in Western civilization, historically, it has been the general norm that one should be heterosexual and engage in heterosexual behavior(s).

The second sexuality script, *interpersonal scripts*, refers to the ability of individuals to engage in complex context-specific behavior(s) with similar expectations and understandings with their chosen sexual partner(s) (Simon and Gagnon 1986). Generally, such agreed upon standards and expectations for all individuals involved are derived from what is considered *normal* in society. Therefore, this allows individuals to both agree to a desired sexual behavior and decide that certain behaviors are allowed/forbidden throughout the commission of a sexual act (e.g. hitting, name calling). That is, some behaviors are developed through verbal and non-verbal cue interpretations as ICTs oftentimes indirectly force users to replace verbal context – generally present in face-to-face communication – with text-based detail.

The third and final sexuality script argued by Simon and Gagnon (1986) to be key in the formation of one's sexuality is the *intrapsychic script*. This refers to individual qualities that contribute to one becoming sexually aroused and why one might become involved in sexual activities in the first place (Simon and Gagnon 1986). This can also refer to individuals deviating from the expected social norm when it comes to sex, thus helping form a sexual subculture that differs from typical sexuality expressed in society (i.e. sexual deviance).

Although sexuality has traditionally been argued to be formed and enacted via the three above scripts, relatively recent developments in ICTs – primarily with the Internet as the glue that unites various devices – have been the basis for debate on how these longstanding sexuality scripts might be impacted (see Doring 2009; Kvalem *et al.* 2014). Specifically, it has been argued that sexually explicit media (SEM), portrayed largely in mainstream pornography, deviates from the *cultural scenario* script by typically focusing on the sex act itself and avoiding other key components of the *cultural scenario* script (e.g. courtship). Moreover, one might view pornography that involves extreme violence (e.g. depictions or actual commissions of sexual assault) to where they perceive such behavior as normal.

Most individuals who report watching pornography report viewing mainstream pornography that focuses primarily on *hardcore sex* (e.g. oral, vaginal, and/or anal penetration) (Hald 2006; Paul 2009). As such, it has been argued that this increased viewing may have an impact on how sexuality is conceptualized and enacted in "real-life" (Daneback *et al.* 2012; Daneback *et al.* 2009; Paul 2009). This concern has gained even more credence with Internet pornography, deviant sexual communities, and a variety of other potential influences being readily available through a multitude of ICTs (e.g. cellular phones, PCs, tablet computers) that can be accessed relatively anywhere as long as an Internet connection is available.

The expression of sexuality through ICTs

Technology and various expressions of sexuality have gone hand-in-hand throughout history. Moreover, various technologies have long been considered by researchers to have a significant influence on how humans interact with one another with a term used to describe this impact known as *technicways* (Bryant 1984; Durkin and Bryant 1995; Odum 1937). For example, some of the earliest forms of technology, such as the Gutenberg Press, have been used to demonstrate various expressions of sexuality with the production of some form of pornography or *obscene* material. This is a trend that has held true in relatively recent developments in ICTs over the past several decades.

Throughout the past approximate half century, there have been multiple key ICTs developed and used to demonstrate various forms of sexuality, both legal and illegal expressions. Some of the notable ICTs used to express both sexuality and sexual deviance are televisions (e.g. subscription-based pornography channels, viewing pornography DVDs), beepers (e.g. used by some prostitutes to connect with pimps and johns), and Citizen's Band (CB) radio (e.g. used by some prostitutes to connect with truck drivers) (see Forsyth and Quinn 2014). Although the above-mentioned ICTs have had a significant impact on the expression/consumption of various forms of sexuality and even sexual deviance, it can be argued that no such technology has had near the initial and/or sustained impact upon sexuality and deviance as the Internet and ICTs. That is, the ability of the Internet to connect billions of individuals across the world at any given time simultaneously across a multitude of devices (e.g. PCs, laptop, cell phone, tablet computer) (Durkin and Bryant 1995). As such, the evolution of the Internet has significantly impacted how sexuality is learned, expressed, and even modified (Doring 2009; Gauthier and Forsyth 1999).

Internet sexuality

The Internet has revolutionized the way that many individuals initially learn about sex, sexuality, and express themselves. Individuals can explore various aspects of their sexuality, no matter how deviant they may or may not perceive it to be, in relative anonymity. Such individuals can alter anything they may not like about their own lives, including their age, name, occupation, or various other personal qualities (e.g. race/ethnicity, sexual orientation) (Young 2001). In essence, one can approach sexuality online as a performance where individuals "present" various qualities, actual or fictional, to an "audience." As described by Griffiths:

> [O]nline, disembodiment dissolves offline life characteristics, such as their personal details, gender, age, race, socioeconomic status, etc. It liberates individuals from the imminent fear of engaging in something that is charged with a variety of taboos in offline life and provides the option to freely explore their (sexual) selves. Moreover, it provides the option to indulge in sexual fantasies that, if realized, might not be as tempting or enjoyable as living them out on the Internet.
>
> *(2012: 113)*

A person may also change various qualities of their online persona based on positive and/or negative feedback on their "performance." For example, when using the Internet, they may have a deviant sexual fantasy that evolves through instant feedback and positive reinforcement with other interested individuals online. In comparison to "real-life" interactions prior to the proliferation of the Internet across a range of ICTs, the same deviant sexual fantasy may have been negatively received by another. Consequently, the sexually deviant fantasy may have ceased to progress (Durkin and Bryant 1995). Moreover, the Internet allows individuals to establish/maintain a relatively safe relationship that may be less-demanding on their overall investment – time and emotions – because it allows them to partake in the relationship when and where they want (Brown 2011).

The impact that the Internet has had on sexuality, both in general and specifically, has even led to the development of a new form of sexuality known as *Internet sexuality* (a.k.a. online sexual activities or OSA). According to Doring (2009), *Internet sexuality* refers to, "sexual-related content and activities observable on the Internet" (p. 1090). Moreover, *Internet sexuality* can refer to a wide range of both sexually-related content and activities one can perform on the Internet. Such content/activities include, but are not limited to, the following: sex education information, consumption of pornography, production of pornography, online sex shops, sex-themed chatrooms, fetish forums, and online website communities (e.g. www.BDSM.com).

Several studies have demonstrated the impact that Internet sexuality has had on a range of sexual behaviors. For example, in a sample of men-seeking-men (MSM) for sex, Rice and Ross (2014) found that the way in which one filters, selects, and negotiates sex partners is different online than it is for those who meet sexual partners in person. Specifically, more filtering of potential sexual partners is done when seeking partners online in comparison to in person (Rice and Ross 2014). Moreover, both *chemistry* and potential *risks* need to be confirmed online when *chemistry* has already been confirmed in "real-life" interactions (Rice and Ross 2014). Conversely, the Internet can minimize potential risks by allowing one to filter others and potential activities without the risk of being physically or socially harmed (Weiss and Schneider 2006).

In a review of nearly 450 academic journal articles on sex and the Internet that were published from 1993 to 2007, Doring (2009) identified six primary areas where sexuality can be portrayed online. All of these can be accessed via a multitude of ICT devices. These six primary areas are (1) *Pornography*, (2) *Sex Shops*, (3) *Sex Work*, (4) *Sexual Education*, (5) *Sex Contacts*, and (6) *Sexual Subcultures*.

Portrayals of Internet sexuality

The first primary way that sexuality can be portrayed online is through pornography (Doring 2009). Pornography is a broad term that can encompass a wide range of portrayals of sexuality across four broad categories. These are the categories of (1) soft core pornography, (2) hardcore pornography, (3) paraphilia/deviant, and (4) illegal. *Soft core pornography* refers to an image or video intended to be sexually arousing, but does not include depictions of actual sex acts (Doring 2009; Kvalem *et al.* 2014). The second broad form of *hardcore pornography* usually depicts some form of penetration, whether it is oral, vaginal, or anal (Doring 2009; Kvalem *et al.* 2014). The third broad form of pornography, *paraphilia/deviant pornography*, can include acts such as simulated instances of sexual abuse, "Barely Legal" where the actor is just at or slightly above the legal age requirement of the respective jurisdiction, and BDSM (Doring 2009; Seigfried-Spellar and Rogers 2013). Fourth and final, there is *illegal pornography*. This form of pornography differs based on one's country, but generally involves children or minors in some manner. Despite the variety of pornography available, most individuals who view Internet pornography are believed to consume mainstream (i.e. hardcore) pornography (Paul 2009).

With global revenues at $20 billion in 2007, $10 billion coming from the US alone, it is safe to say that this is perhaps the most common portrayal of sexuality online with a wide range of ICT devices used for its distribution/consumption (e.g. PCs, laptops, cell phones, tablets)

(Barrett 2012). Specifically, the mobile phone arm of the adult entertainment industry has been cited as the next frontier with 43.8 percent of adult entertainment industry executives stating that this ICT will soon become a consumers' primary pornography viewing device (XBIZ Research 2012). Moreover, the video chat and mobile phone content revenue is expected to reach approximately $2.8 billion by 2015 with over 70 percent of such revenue coming from Western Europe and North America from an estimated 35 million subscribers (Waters 2012).

The second primary way sexuality can be portrayed online is through *sex shops* (Doring 2009). That is, one can browse and purchase various sexual aids and toys (e.g. vibrators, dildos) from the comfort of one's home, or anywhere they have an Internet connection, in relative anonymity. Pre-Internet, one would have to travel to a physical sex shop to browse/purchase such products, potentially resulting in stigmatization if discovered by others. As such, the Internet provides an access to such establishments that individuals may have been hesitant to patronize prior to the development and proliferation of the Internet across multiple ICT devices.

The third primary way sexuality can be portrayed online is through *sex work* (Doring 2009). This can be achieved in several ways. First, various online mediums (i.e. websites, chatrooms, forums) can be used to advertise prostitution or some sexualized services (e.g. dominatrix, BDSM) that may, or may not, involve some form of sexual intercourse that can be obtained in real-life. Second, personalized virtual sex work (e.g. webcam chats) can be purchased to provide some individualized and customized form of sex act (e.g. masturbation, fetish-play) that may pertain to a patron's particular desires. Therefore, the Internet provides a medium for both traditional (i.e. prostitution and/or highly sexualized) behaviors that can later be arranged in "real-life," in addition to a new form of individualized performance of sexuality that takes place exclusively in the virtual realm (e.g. webcam shows, chatrooms).

The fourth way sexuality can be portrayed online is through sex education (Doring 2009). The Internet offers numerous avenues for information on sex, safe sex practices, sexuality, and related emotions. Moreover, most Internet users admit to actively seeking sex education information on a semi-regular basis (Gray and Klein 2006). In 2002, it was posited by Sutton *et al.* (2002: 45–6) that "the Internet eventually may prove to be a valuable resource for adolescents seeking sexual information in a proactive way."

A recent meta-analysis of adolescents' use of the Internet for sex education has confirmed this belief by Sutton *et al.* (2002). Specifically, Simon and Daneback (2013) found that the Internet has become a key source for adolescents' questions regarding sex health and education. As such, what previously may have been confined to sex education classes in school and information provided by parents is now freely available on the Internet. Consequently, this may open even more facets to how people explore, express, and modify various aspects of their sexuality.

The fifth way sexuality can be portrayed online is through *sex contacts*. Doring (2009) highlights two main ways that one can have sexual contact on the Internet, these being (1) exclusively online and (2) online contacts that lead to real-world meets. The first form of sex contact, exclusively online, refers to behaviors (e.g. cybersex), which occur exclusively online. Moreover, these are behaviors that do not lead to any interpersonal sexual contact that will take place in the physical world.

The second form of sex contact refers to individuals that are initially connected via the Internet with eventual contact in the physical world. It is important to mention that the end-goal of this form of sex contact is that some sexual act will eventually occur in the "real-world." Recent developments in ICTs, primarily in mobile phone technology, have made both forms of sexual contact easier and more convenient to potential users.

Primarily through various mobile phone applications (a.k.a. apps), individuals now have the ability to find others within feet of one another for sex with little-to-no expectations for commitment. This behavior has been identified as "toothing" (see Terdiman 2004), referring to individuals using the Bluetooth technology on their cellphone to locate others willing to engage in non-committal sex. Some of the more notable examples of apps built for this purpose include *Grindr* (gay, bisexual, and curious males), *Blendr* (primarily aimed at heterosexual "hookups"), and *Tinder* (used for both dating and brief sexual encounters) and have now expanded the ability of individuals to have sex after initially meeting through the virtual realm.

The sixth and final primary way sexuality can be portrayed online is through sexual subcultures (Doring 2009). The formation of such subcultures online has been demonstrated to be liberating for some current/previously marginalized populations (e.g. gay, lesbian, transgendered, and disabled individuals) (Broad 2002; Correll 1995; Heinz *et al.* 2002; Kaufman *et al.* 2007; McLelland 2002; Nip 2003). Pre-Internet, individuals interested in a deviant and/or paraphilic behavior (e.g. BDSM, bestiality) may have been restricted geographically and/or socially with other likeminded individuals (Durkin and Bryant 1995). However, the Internet across a range of ICTs has flourished in connecting various deviant/paraphilic subcultures. Moreover, the connection of likeminded individuals has the potential of reaffirming, expanding, and modifying their deviant/paraphilic behavior (Durkin *et al.* 2006; Lanning 1998).

Online subcultures and associated behavior(s) that these communities may be formed around might be perceived as *deviant* by others. Moreover, many of these behaviors may have previously been considered illegal (e.g. homosexuality, sadomasochism) or may still be considered illegal (e.g. bestiality, pedophilia) in many jurisdictions. As such, sexual deviance online has taken on a life of its own to where individuals with similar interests can connect with one another that might not have been possible pre-Internet. Consequently, many of the sexually deviant communities may become gateways to illegal behavior(s) if that behavior is restricted in a particular community.

Despite this potential for illegal behavior(s), sexual deviance is a vital component of the Internet and the expression of sexuality via ICTs. Additionally, once previously secret deviant sexual practices become well known with an Internet presence, it may lead to calls for increased government and/or community control, perhaps even criminalizing such a behavior (Potter and Potter 2001). Therefore, the differences between sexual deviance and sexual offending may become blurred, despite crucial differences between the two.

Sexual deviance

[I]t seems that law, cyberspace and pornography have formed a novel and rather unholy trinity against a background of (sub)culture where it is now socially acceptable and practically expected to use the technology of the Internet to seek out sexually explicit sites.

(Wall 2001: 75)

Deviance is a term that refers to one departing from typically accepted behaviors (i.e., norms) of a community or a society as a whole. It is important to note that deviance generally denotes an act that is not illegal; however, when applied to sexuality, that is not always the case. Since deviance refers to behaviors that depart from wholly accepted norms, then when applied to sexuality, sexual deviance refers to any such behaviors that depart form norms pertaining to identities and demonstrations of sexuality. Specifically, *sexual deviance* is defined as the deviation from norms regarding the identification and the manifestation of sexuality. Therefore, one can deviate from a community's norms regarding the identification and expression of sexuality while also violating a law. However, a society may not always regularly enforce such a law (e.g. sodomy laws in the US). As such, an overview of sexual deviance will be provided to establish a stronger context for understanding how the Internet and ICTs have altered the availability and expressions of various forms of sexual deviance.

Gagnon and Simon (1967) introduced three primary forms of sexual deviance with their landmark piece *Sexual Conduct: The Social Sources of Human Sexuality*. Within this work they describe three primary categories of sexual deviance. These are the categories of (1) *normal deviance*, (2) *pathological deviance*, and (3) *sociological deviance*.

The first category of sexual deviance, *normal deviance*, refers to relatively common forms of nonconformity that are believed to violate most normative standards of a particular community (Gagnon and Simon 1967). An example of normal deviance would be an adult couple role playing while engaging in a consensual sexual activity. Although this behavior is believed to be relatively common, it still may be considered taboo by individuals who partake in the behavior because of fear that it will offend people within their community, potentially resulting in a stigma.

The second category of sexual deviance, *pathological deviance*, refers to a behavior(s) that not only violates the normative standards of a community, but is also typically illegal (Gagnon and Simon 1967). Moreover, the behavior that takes place is believed to be harmful by the vast majority of a particular society. Some examples of this behavior include voyeurism, pedophilia, and rape (Forsyth 1996). As such, extensive laws and other regulation efforts are generally put into place by the respective society to prevent, punish, and attempt to eliminate such behavior.

The third and final category of sexual deviance established by Gagnon and Simon (1967) is *sociological deviance*. Specifically, *sociological deviance* refers to certain behaviors that essentially have to foster a community in order to survive. That is, there must be some form of structure – semi-formal or formal – in place that caters to the needs of current members (e.g. fostering support for the current membership). Also, the structure must have the ability to recruit new members into its fold in order to continue for future generations (Quinn and Forsyth 2005). One notable example of such behavior is bug chasing (see Denney and Tewksbury 2014) – where men actively seek other men to be infected with HIV (Human Immunodeficiency Virus) – discussed further below. Some circles of prostitution have also been identified as fitting into the category of *sociological deviance* (Quinn and Forsyth 2005).

Although Gagnon and Simon's (1967) work has served as a major cornerstone for the study of sexual deviance, this work was published nearly two decades prior to the invention of the Internet and approximately three decades prior to the proliferation of the Internet for the purposes of sexual deviance (Månsson and Sodërlind 2013). As such, Quinn and Forsyth (2005) recognized the need to modify Gagnon and Simon's (1967) earlier formulation of sexual deviance to allow for the recognition/study of the Internet and its influence on the formation/ expression of sexual deviance.

In Quinn and Forsyth's (2005) modification, they argue that the structural and normative changes brought about by ICTs and the proliferation of the Internet for sexual purposes have had a substantial impact on the ways that sexually deviant behaviors and communities are discovered, formed, and expressed. Consequently, the advent of new ICTs and the Internet has altered the utility of studying sexual deviance via Gagnon and Simon's (1967) original scheme. The reasoning for this is that the virtual world is not geographically isolated or restricted to a few standalone communities. The virtual world is global in its reach and in its appeal with its ability to connect likeminded individuals 24/7. As such, individuals may not learn of a new expression or form of sexual deviance within their own community, but they may do so by actively seeking or stumbling upon an online community.

Quinn and Forsyth (2005) view *sociological deviance* as *communal deviance*. Under this new typology, the act itself is now distinguished from how it is created. Therefore, there are now *isolated actors* and *communally supported actors* with each having the ability to engage in either *normal* or *pathological deviance*. As such, it removes the necessary requirement of Gagnon and Simon's (1967) original typology of the sexual deviance to have been formed and/or portrayed in a physical environment/community.

There are a number of forms of sexual deviance that have now become popularized and have even spread due to ICTs and the Internet that make such communication possible (Forsyth and Fournet 1987; Gauthier and Forsyth 1999). As discussed above, one of the primary iterations of Internet sexuality is through sexual subcultures (Doring 2009). Regardless of whether or not an online subculture has a website, forum, or some other virtual medium, there is some structure in place, no matter how formal or informal. Consequently, there is the potential of exposing numerous others to one's sexual subculture at any point, an issue not present at any point before in history.

Online deviant sexual subcultures

Although brief discussion of sexual subcultures on the Internet was provided above, more discussion is needed to understand the full influence that the Internet and its accessibility by a wide range of ICTs has had on people's knowledge, expression, and formation of sexual deviance. Because of the Internet and ICTs with Internet access, sexually deviant subcultures have appeared on a wide range of Internet-based platforms. Key examples of these Internet subcultures that are discussed below include MMORPGs and AMMORPGs, BDSM, pedophilia, bestiality, and bug chasing.

MMORPGs and AMMORPGs

Some sexually deviant online communities exist solely online. Two primary examples of such communities are MMORPGs (i.e. Massively Multiplayer Online Role-Playing Games) and AMMORPGs (i.e. Adult Massively Multiplayer Online Role-Playing Games). Sexual deviance has been noted to not only exist, but thrive in MMORPGs and AMMORPGs (see Dretsch and Dretsch 2014). With over 1,000,000 active users per month, Second Life offers perhaps one of the larger MMORPG platforms for individuals to engage in virtual sex via avatars (Reahard 2013). However, virtual sex is not the primary or intended purpose of this community, but something that has developed over time.

A study on sexual deviance within Second Life by Gilbert *et al.* (2011) found that over 70 percent of the users surveyed had at least one sexual encounter with another Second Life user. Additionally, more than 70 percent of the participants had never met with their Second Life sexual partner in real-life and participants were mixed as to what form of sexual encounter – online-only or real-life – provided them with greater overall sexual satisfaction (Gilbert *et al.* 2011). Users can even modify and expand upon their Second Life sexual experience by purchasing upgrades (e.g. more detailed physical features) (Gilbert *et al.* 2011).

Other notable platforms that allow avatar-based virtual sex are AMORPGs. The difference between MMORPGs and AMORPGs is that the latter is meant primarily for virtual sex. Notable examples of AMORPGs are 3DSex and Red Light Center (see Dretsch and Dretsch 2014). Although some sexually deviant online subcultures were formed and currently exist only online (e.g. MMORPG sexual deviance, AMMORPGs), other sexually deviant communities have historically existed offline but have found new life online, such as the BDSM community.

BDSM

BDSM is an acronym that refers to a variety of erotic, although not necessarily sexualized, behaviors: Bondage, Discipline, Sadism, and Masochism. As both an organized activity and subculture, BDSM emerged during the 1950s and 1960s (Lenius 2001). As such, the behavior itself and communal aspects of BDSM have been the subject of interest for researchers studying sexual deviance (e.g. Denney and Tewksbury 2013; Kamel and Weinberg 1995; Stiles and Clark 2011). Moreover, just as the Internet has impacted the expression of other forms of sexual deviance, the Internet has also had a substantial impact on the BDSM community. The impact of the Internet on this community has been significant with reports that physical BDSM communities have experienced drastic loss in attendance/participation to online BDSM communities (Newmahr 2011).

Online BDSM communities have also seen a change in how users portray themselves. Whereas individuals who are active in a physical BDSM community have a relatively fixed identity tied to personal (e.g. sex, race/ethnicity, age) and BDSM-related characteristics (e.g. dominant, subordinate), participants in online BDSM communities have been shown to change their presentation of self over time (Palandri and Green 2000). Specifically, members of online BDSM communities have been shown to change various qualities of their personas if they grew tired of their portrayal (Palandri and Green 2000). This is primarily achieved through one's screen name where prior research has shown that individuals enjoy constructing and managing their identities online (Palandri and Green 2000; Stiles and Clark 2011). However, online BDSM communities have been shown to emphasize and promote a sexual component to BDSM (Denney and Tewksbury 2013) more so than is a part of a typical physical BDSM community (Newmahr 2011). Consequently, online BDSM communities that emphasize/ promote such sexual qualities may have a drastic influence on how "real-life" BDSM communities are perceived and how they function.

Pedophilia

According to the American Psychiatric Association's Diagnostic and Statistical Manual of Mental Disorders (DSM-V), *pedophilia* refers to recurring and sexually arousing fantasies, sexual urges, or other behaviors that involve sexual activity with a prepubescent child or children (i.e. 13 years or younger) (American Psychiatric Association 2013). As with other forms of sexual deviance, the Internet has also served as a catalyst for the development and continuation of pedophilic communities. In fact, it has been argued that pedophiles have been the most documented group of individuals who use the Internet for a myriad of sexual purposes (Durkin 1997). Specifically, pedophiles have been documented as using the Internet from something such as trading child pornography to locating other pedophiles and even victims (Durkin 1997). These online communities have also been documented as sharing techniques to disrupt law enforcement investigations regarding their child pornography collections (Jenkins 2001).

One major function of online communities for pedophiles, similar to other deviant sexual populations, is that pedophiles have been found to use such online communities to (1) validate their viewpoints and (2) develop strong cognitive distortions that help them rationalize their deviant sexual interest(s) (Durkin and Bryant 1999; Malesky and Ennis 2004). For example, pedophiles who collect/watch child pornography with limited interaction with other pedophiles have fewer cognitive distortions than those who interact with other pedophiles on a regular basis (Elliott *et al.* 2009). Moreover, some online pedophile communities, such as the North American Man/Boy Love Association (NAMBLA), use the Internet to educate and even recruit others to their paraphilia (DeYoung 1989; Durkin and Bryant 1999). Some pedophiles even use these online communities to enact a form of cybersex referred to as *age play*, where one adult role-plays as a minor for another adult's sexual gratification (Lewis 2011).

Bestiality

Bestiality refers to the practice of an individual engaging in sexual intercourse with an animal (Miletski 2005). Even though this is a behavior that is heavily stigmatized and illegal in many countries throughout the world, the Internet has served as a force to unite individuals interested in this form of sexual deviance. One of the first and most well-known Internet sources for bestiality is a website referred to as *Alt.sex.bestiality* (A.S.B.) that began in the early 1990s (Montclair 1997). Although it is argued that this website originally began as a joke (see Miletski 2005), it soon grew into a significant source for those interested in bestiality ranging from "how to" guides for engaging in sexual acts with animals to "real-life" meetings where individuals engaged in acts of bestiality (Donofrio 1996). However, as with other forms of sexual deviance, the Internet has provided a venue for individuals to connect, share, and reaffirm interests in bestiality (Andriette 1996; Byrd 2000).

Bug chasing

"Bug chasing" refers to an individual that is HIV-negative actively seeking an individual that is HIV-positive so as to have sexual intercourse. The entire purpose of individuals who "chase the bug" is to purposefully become infected with the HIV virus. Emerging from the hysteria that surrounded the AIDS (Acquired Immune Deficiency Syndrome) epidemic during the early 1980s and 1990s, bug chasing is perhaps one of the newer subcultures of sexual deviance that exists since having formed in the late 1990s and early 2000s (Denney and Tewksbury 2014).

Due to over 100,000 confirmed diagnosed cases of AIDS and roughly 45,000 deaths from the disease by the close of the 1980s in the US alone, increased awareness of "safe sex" practices (i.e. condom use) were heavily promoted by the early 1990s (AVERT 2013). This significant issue primarily impacted the gay community. As such, gay men who did not engage in safe sex practices were seen as engaging in a deviant sexual activity, oftentimes referred to as *bareback sex* (Gauthier and Forsyth 1999). Moreover, bareback sex is believed to be one of the most prevalent ways in which HIV is transmitted between MSM. With most (43 percent) gay men believed to use the Internet for identifying sexual partners, a small but significant subset of this population within the online gay community are interested in bareback sex (Bolding *et al.* 2006).

Bug chasers operate as a distinct subset within the wider bareback sex community (Blackwell 2010; Grov and Parsons 2006; Tewksbury 2006). For example, Grov (2006) found 30,000 profiles of MSM interested in bareback sex across six websites. As to why individuals may actively seek HIV-positive individuals to infect them, there have been four primary motivations suggested for this behavior. First, individuals want to become a member of an HIV-positive community (Gauthier and Forsyth 1999). Second, individuals see this as a way to take control of their own destiny (Gauthier and Forsyth 1999). Third, bug chasing is a mere extension of risk-seeking behavior (Denney and Tewksbury 2014). Fourth and finally, bug chasing is a form of a political statement by some of not letting the government and society tell the gay community how to have sex (i.e. encouraged use of a condom) (Denney and Tewksbury 2014). However, unlike many online sex behaviors, bug chasing has some serious "real-world" consequences, such as the transmission and spread of STIs (Sexually Transmitted Infections) and

STDs (Sexually Transmitted Diseases). Concern for "real-world" consequences for a variety of expressions of Internet sexuality have been the subject of great concern, especially with the ease of accessing and engaging in such behavior with a multitude of ICT devices.

Media and government: ICTs and sexuality

Expressions of sexuality have always served as a great source of controversy. Pornography, in particular, has been a significant source of controversy with many feminist groups protesting its potential harms since the 1970s (Surette 1992). However, more recent concerns regarding the Internet and ICTs have focused on the production/distribution of child pornography and/or the accessibility of SEM by minors (Akdeniz 2001; Seto 2013). Since this is covered in another chapter, it will not be discussed here.

Excluding consideration of child pornography and SEM accessible by minors, some media attention has focused on the dangers of Internet pornography available via a range of ICT devices. Such titles of online and other media warning of the dangers of the various expressions of sexuality via ICTs and the Internet include *The Dangers of Pornography: Stop the Spread of Slime in Your Home* (Rainey and Rainey 2014); "12 Ways Pornography Leaks Into Your Home (and how to stop them)" (Bishop n.d.); and "Porn at Work: Recognizing a Sex Addict" (McDonough-Taub 2009). These media sources are generally faith-based and warn readers of the dangers – for youth and adults – of the widespread availability of SEM via a range of ICT devices.

When the Internet first began to be used for sexual purposes in the 1990s, governments in many first-world nations started investigating the impact that Internet pornography had upon people's behavior in "real-life." Some of the first major attempts to regulate Internet pornography and other portrayals of sexuality online came on the heels of a study by Martin Rimm, an undergraduate student at Carnegie Mellon University, entitled "Marketing Pornography on the Information Superhighway" published in the *Georgetown Law Journal* and featured in *Time* Magazine's 1995 cover article "Cyberporn" (Lewis 1995). This study, now largely discredited, boasted in its 85 pages that it surveyed nearly 1,000,000 images and other information (e.g. short stories, descriptions, animations) that were held on the Usenet newsgroup sites, a popular website at the time, claiming that it found that these pornographic images were downloaded nearly 8.5 million times by the roughly 2,000 cities and 40 countries in the sample (Lewis 1995; Campbell 2015). Moreover, it claimed that 83.5 percent of all Usenet newsgroup images were pornographic (Campbell 2015). Many politicians in the US, such as Nebraska Senator J. James Exon and Iowa Senator Charles Grassley, and others abroad began using the now largely discredited findings from this study to push for the heavy regulation of pornographic Internet material (Lewis 1995; Campbell 2015). For example, the United Kingdom Home Affairs Committee Report on Computer Pornography begins with the line, "Computer pornography is a new horror" (Wall 2001). Such regulation would have been unprecedented, even by the standards of print and other notable forms of media in the early 1990s (Campbell 2015). However, regulation of pornography in the United States and United Kingdom has proven difficult. Perhaps the most notable example of this difficulty was the failed attempt of the Communications Decency Act of 1996, that was eventually not upheld by the US Supreme Court as it was ruled to have violated key rights found in the First Amendment of the US Constitution, which would have heavily regulated "indecent" Internet material and pornography (Campbell 2015).

There is a key difference between material that is considered illegal and material that is considered harmful. Illegal pornography may include depictions of a child (various ages depending on the jurisdiction), harmful pornography may just be considered *offensive* or *disgusting* by certain groups of individuals (see Akdeniz 2001). As such, what may be considered harmful material by one community or individual, may not be viewed as harmful by another community or individual. The Internet's function of connecting the world transcends these geographic boundaries while simultaneously providing the opportunity of viewing sexual material or having sexual communication with others that may violate a community's norms (or laws) regarding sexuality and sexual behaviors.

Specific forms of regulation of Internet pornography have been attempted throughout nations like the US and UK. One example of a rating and filtration system is the Platform for Internet Content Selections (PICS). This system attempts to regulate potentially *obscene* material through the attachment of an electronic file to various types of content (i.e. images, videos, etc.) that one's ICT can prevent from displaying if PICS is being utilized. However, such rating and filtration software has raised serious concerns from minority groups, such as the Gay and Lesbian Alliance Against Defamation, that their content may be censored or inaccessible altogether if someone deems it to be *offensive* (Akdeniz 2001). Another prominent example of a failed attempt to regulate Internet pornography was the Child Online Protection Act (1998). The purported purpose of this act was to restrict the access of minors to material deemed "harmful" (Harris 2008). Similar to the Communications Decency Act of 1996, this legislation also was never enacted as questions of its constitutionality arose soon after its passage (Harris 2008). To date, the US and UK have favored a self-regulatory model where Internet Service Providers (ISPs) regulate themselves. Historically, however, one of the primary concerns that has been raised with pornography in general is the effect that viewing such material has on "real-world" behavior(s). This is a concern that has only escalated with the increased availability of Internet pornography through a wide variety of ICT devices.

Connection between cybersexuality and the "real-world"

> For some, sexual behaviors online are used as a complement to their offline sexuality, whereas for others, they serve as a substitute potentially resulting in Internet sex addiction, which can be conceptualized as the intersection between Internet addiction and sex addiction.
>
> *(Griffiths 2012: 111)*

Addiction to Internet pornography and other expressions of online sex has been the subject of research for nearly two decades (see Orzack and Ross 2000; Putnam 2000; Schneider 2000; Schwartz and Southern 2000). One of the powerful qualities of ICTs and the Internet is that individuals now have the ability to view any form of pornography they want to see at nearly anytime they may want to view it, no matter how deviant or illegal (Seto 2013). This is a growing concern because Internet pornography is now readily available on a wide range of ICTs. Using Cooper's (1998) "Triple A Engine" for why the Internet is so popular for sexual purposes, it is because individuals have (1) easy access, (2) it is affordable, and (3) one can browse the material with relative anonymity (Cooper 1998). Moreover, one can also download and store thousands of hours of pornography on a range of devices that are increasingly on someone's person at any given point throughout a given day (e.g. mobile phones, tablet computers). Despite this concern, a meta-analysis on research pertaining to Internet sex addiction revealed only a minority of Internet users are believed to use the Internet both compulsively and excessively for sexual purposes (Griffiths 2012).

Griffiths (2012) noted three key reasons for why all Internet activities that involve sex may be problematic. The first is that an individual may engage in sexual behaviors that the individual or their significant other disapproves of, perhaps leading to feelings of guilt. The second is that engaging in Internet sexual activity will take away energy for engaging in real-life sex, or cause an individual to have misconceptions of what is realistic sexual behavior and what is not. The third and final way Internet sex in its various forms may be problematic is that it may take a significant amount of time to find material or interaction perceived as ideal (i.e. sexually gratifying) to the user (Griffiths 2012). Internet pornography may also be addictive because of arousal from interacting with technology using a classic operant conditioning and/ or conditioned stimulus framework (see Carnes 1993).

Some other notable "real-life" effects of Internet sex addiction include, but are not limited to, the following: issues at one's job, detriments to personal relationships, decreased self-esteem, and dissatisfaction with one's genital appearance (Kvalem et al. 2014). Some media and surveys have even examined this issue impacting the workplace. For example, a 2003 survey of human resource professionals found that approximately 66.6 percent of respondents reported finding pornography on an employee's computer with 43 percent of respondents reporting it happened more than once (McDonough-Taub 2009).

Historically, one of the major concerns surrounding pornography use is the impact that it will have on an individual's behavior, generally and sexually. One major concern that has been expressed regarding pornography use is on one's likelihood of committing "real-world" sex offenses. For example, early government inquisitions (e.g. the Commission on Obscenity and Pornography (1972) in the United States) published findings on this topic, largely not supporting the relationship between viewing obscene material and the commission of sex crimes (see Surette 1992).

Some potential dangers regarding the Internet, ICTs, and the expression of sexuality do exist. First, individuals actively seeking legal pornographic material or other expressions of sexuality may accidently come across illegal materials. It is believed that more than 100,000 websites contain some form of illegal images, primarily child pornography (Quinn and Forsyth 2005). A study by Winder et al. (2015) of 36 adult males who were convicted of either possession or the distribution of child pornography claimed that many "stumbled" into this pattern of offending by accidently coming across the illegal material while accessing another website. However, the veracity of these claims is questioned (Winder et al. 2015).

A second potential danger is individuals learning about and engaging in illegal sexual activity when one would otherwise not have known. For example, a recent study of 15 Swedish youth with experience in prostitution revealed that the Internet was central to their decisions to become involved in prostitution (Jonsson et al. 2014). Moreover, the use of smartphones provided them with a constant connection to the Internet to potentially engage in a variety of prostitution services at any point throughout the day (Jonsson et al. 2014). Additionally, the Internet allowed communication through the smartphone, thus altering the way in which many communications between the johns and young prostitutes took place and how sexual services were negotiated (Jonsson et al. 2014).

The rise of the Internet has been associated with the transition of the world's oldest profession (i.e. prostitution) to the digital realm (Hughes 2003). However, to date, the online marketplace has not appeared to have entirely replaced the predominant form of prostitution (i.e. streetwalking), but it has mostly complemented it (Cunningham and Kendall 2011). That is, the evolution of the Internet in the mid-1990s on has introduced new ways of meeting prostitutes, clients of prostitutes (i.e. "johns" or "punters") providing reviews of their experiences, negotiating cost and sexual preferences, and more (see Cunningham and Kendall 2011). Similarly, Cunningham and Kendall (2011) think this change in interaction between prostitute and john has been largely beneficial for prostitutes in three key ways by allowing them to "(a) reach large numbers of potential clients with informative advertising; (b) build reputations for high-quality service; and (c) employ screening methods to reduce the risk of discovery and avoid undesirable clients (p. 275)." Similarly, the use of ICTs and computer mediated communications have had strong implications for "punters" or "johns," online and face-to-face.

In one of the first studies to examine online "punters" of online prostitutes, Soothill and Sanders (2005) found that the Internet has served as a monumental change to the overall social organization of prostitution. Moreover, they found that clients of prostitutes are still controlled by the values, norms, and potential sanctions that dictate real-world prostitution, regardless if they are interacting online (Soothill and Sanders 2005). Holt and Blevins (2007) further examined johns of an online prostitution forum in ten US cities, finding that these online forums simplify the overall process of johns selecting desired prostitutes, even using potential threats associated with seeking prostitution services (e.g. theft of personal items, sexually transmitted diseases) when making decisions. Also, johns have been shown to share information with one another regarding potential police investigations to avoid or minimize the risk of arrest when physically selecting and/or meeting prostitutes (Holt and Blevins 2007; Holt et al. 2014). Thus, computer mediated communications via the Internet have seemingly become a valuable tool for both prostitutes and clients of prostitutes to carry-out such services, both online and face-to-face. Not only does the Internet and its ability to connect a multitude of ICTs have the ability to transform the world's oldest profession, but it also has the capacity to create new forms of sex crime that could not exist without ICTs and the Internet.

The emergence of new online sex crimes

For a behavior to cross the threshold of deviance to become illegal typically requires that there is a non-consenting individual involved. Moreover, with the continued expansion and modification of the interplay between sexuality and various technologies, modifications or altogether new forms of sexual deviance have developed. As such, some of these technologies have blurred the lines between consenting behavior with potential pornography involving children and distribution of pornography involving adults that did not consent to its release. Recently, two newer forms of sexual deviance have emerged that are currently becoming criminalized in many first-world nations. These newer forms of sexual deviance are (1) *Sexting* and (2) *Revenge Porn*.

Sexting

Sexting is a relatively new criminal offense that is gaining traction in the US and other parts of the world (see Moore 2014). Created from a combination of "texting" (i.e. sending a text message) and "sex," sexting refers to transmission of risqué or sexually explicit images via an ICT device (Chalfen 2009; Weisskirch and Delevi 2011). ICTs that are commonly used in the transmission of such images are cell phones, PCs, and tablet computers through a wide range of applications ranging from sending a text message to instant messaging platforms (Diaz 2009; Forbes 2011). At this point, the primary focus of sexting has been on children and adolescents engaging in this behavior due to concerns of child pornography with some estimates of approximately 20 percent of teens and young adults participating in this behavior (Lenhart 2009). However, a potential gray area is a young adult (e.g. age 19) that sexts with their boyfriend/girlfriend that is legally a minor (e.g. age 16).

Revenge porn

Although related to sexting, another newer online sex crime that has emerged is *revenge porn*. *Revenge porn* refers to a nude photograph or video, typically of a former love interest of an individual, that is shared online (Larkin 2014). However, the image/video is typically uploaded to a website or P2P (peer-to-peer) network against the other person(s) consent in an effort to publicly humiliate them. To date, little research exists on this new phenomenon, yet this form of sexual deviance is beginning to be criminalized throughout many first-world nations, such as the US and Japan (Humbach 2014). This is one problematic aspect of the expression of sexuality via ICTs and the Internet that can potentially spread using principles of Cooper's (1998) "Triple A Engine." Additionally, the images/videos may forever be available to view on an individual's ICT device of choice. Moreover, due to the ease of the creation/distribution of pornographic images, it may prove difficult/impossible for law enforcement to prevent/stop such actions.

Law enforcement responses to online sex crimes

There are several potential issues that the emergence of new online sex crimes has on the formal regulation of such behavior via law enforcement agencies. Recently, law enforcement agencies have begun expanding their investigations, primarily through undercover efforts, to address a variety of online sex crimes (Mitchell *et al.* 2010). However, with this transition into the regulation and punishment of various online sexual behaviors come some notable challenges for law enforcement agencies. These primary challenges are (1) trans-jurisdictional issues, (2) technological challenges, and (3) gray areas between civil and criminal violations.

The first potential issue for law enforcement is the trans-jurisdictional nature of online sex crimes that can be committed through a range of ICTs (Wall 2001). Using an example of a relatively new sex crime discussed above, one in the United States may upload a video depicting revenge porn on a server in another country that might or might not be open to working with a local, state, or federal US law enforcement agency investigating the incident.

One major quality of the Internet is that it removes any and all geographic borders. That is, an Internet connection to an ICT allows for the near instantaneous communication of potentially 3.14 billion people (Statista 2015). Thus, an individual may easily be able to access content that is illegal in their country, such as bestiality in the US, with relative ease because such content is stored on a server in a country where bestiality is legal. From a policing perspective, this increases the difficulty of preventing and investigating such behavior as it is usually legally hosted on a server in a country where the behavior is legal. As such, the responsibility of managing and minimizing such content is generally the responsibility of the ISPs, further complicating the process.

A second major potential issue of law enforcement responses to sex crimes is that it requires law enforcement agencies to be well versed in newer technologies in order to investigate and prevent such crimes. Although law enforcement agencies have begun various undercover operations in recent years, these operations primarily deal with investigating/disrupting the production/distribution of child pornography and online sexual solicitation of minors (see Mitchell *et al.* 2010). Such an emphasis and devotion of resources may mean that law enforcement is neglecting other online sex crimes.

A third major potential issue of law enforcement in response to online sex crimes is that it may not be entirely clear if an issue is a civil infraction or a criminal violation (Wall 2001). An example of this would be what was referenced above as *revenge porn*. That is, a video may be uploaded in a community that does not have a law making such an act illegal. Moreover, if the individual who uploaded the video profited financially, then it may appear – based on a community's current statutes – to be a civil, not criminal violation. As such, due to a combination of new technology and/ new applications of technology, law enforcement may struggle with effectively trying to regulate sexually deviant and/or criminal behavior.

The future of ICTs and sexuality

The future relationship between ICTs and various expressions of sexuality and sexual deviance seems strong. Various iterations of ICTs have completely altered the way individuals communicate with one another. Moreover, the Internet serves as the glue that connects many ICTs to one another with each having the potential to be used to express some form of sexuality (Seto 2013). However, with the increased advancements and availability of ICTs in how they have transformed sexuality, there is also a potential broader societal danger that lurks.

Just as the Internet has united and given a voice to many individuals who are a part of previously marginalized sexual communities, some deviant or some not, the same technology that has been used to their advantage can also be used for their demise. For example, a community or government may become privy to certain sexual behaviors and communities that may have gone unnoticed before. However, because the sexual behavior in question may be perceived as deviant, a community and/or government may feel the need to regulate such behavior. Therefore, the potential remains for the same technology that has led to an unprecedented growth in the expression of sexuality to eventually be used to limit it.

For now, the issue of online sexual offending appears to be primarily a first-world problem (e.g. North America, Europe, Japan) (see Seto 2013). Despite these issues currently being a first-world problem, this will likely change as ever increasingly affordable ICTs and more opportunities for Internet connectivity become available in second and third-world nations. Because of various legal and social issues surrounding expressions of sexuality, there will likely always be controversy surrounding its expression and practice. If the past several decades have been any indication, the relationship between technology and sexuality will likely be intertwined and continue to make unprecedented growth for the foreseeable future.

References

Akdeniz, Y. (2001). Controlling illegal and harmful content on the Internet. In D. Wall (ed.) *Crime and the Internet* (113–140). New York, NY: Routledge.

American Psychiatric Association (2013). *Diagnostic and statistical manual of mental disorders* (5th ed.). Washington, DC: American Psychiatric Association.

Andriette, B. (1996). March. Laying with beasts. *The Guide.*

AVERT. (2013). *History of HIV & AIDS in the United States of America.* Retrieved from: www.avert.org/aids-history-america.htm.

Barrett, P.M. (2012, June 21). The new republic of porn. *Bloomberg Newsweek.* Retrieved from www.businessweek.com/printer/articles/58466-the-new-republic-of-porn.

Bishop, C. (n.d.). 12 ways pornography leaks into your home (and how to stop them). *LDSLiving.* Retrieved from www.ldsliving.com/12-Ways-Pornography-Leaks-into-Your-Home-and-How-to-Stop-Them-/s/76116.

Blackwell, C.W. (2010). The relationship among population size, requests for bareback sex, and HIV serostatus in men who have sex with men using the Internet to meet sexual partners. *Journal of Human Behavior in the Social Environment,* 20 (3), 349–60.

Bolding, G., Davis, M., Hart, G., Sherr, L., and Elford, J. (2006). Heterosexual men and women who seek sex through the Internet. *International Journal of STDS & AIDS,* 17 (8), 530–34.

Broad, K.L. (2002). GLB+T? Gender/sexuality movements and transgender collective identity (de) constructions. *International Journal of Sexuality and Gender Studies*, 7 (4), 241–64.

Brown, A. (2011, March/April). Relationships, community, and identity in the new virtual society. *The Futurist*, 45 (2), 29–31, 34.

Brown, D. (2006, May 1). PCs in ecstasy: The evolution of sex in PC games. *Computer Games Magazine*.

Bryant, C. (1984). Odum's concept of the technicways: Some reflections on an underdeveloped sociological notion. *Sociological Spectrum*, 4 (2–3), 115–42.

Byrd, M. (2000, December 3). The joy of beasts. *The Independent* (*The Sunday Review*).

Campbell, J.W. (2015, July 7). 'Cyberporn' scare of 1995 demonstrates the early Web's corrective power. *Poynter*. Retrieved from www.poynter.org/2015/cyberporn-scare-of-1995-demonstates-the-early-webs-corrective-power/355653.

Carnes, P.J. (1993). Addiction and post-traumatic stress: The convergence of victims' realities. *Treating Abuse Today*, 3 (13), 5–11.

Chalfen, R. (2009). 'It's only a picture': Sexting, 'smutty' snapshots and felony charges. *Visual Studies*, 24 (3), 258–68.

Cooper, A. (1998). Sexuality and the Internet: Surfing into the new millennium. *CyberPsychology and Behavior*, 1 (2), 187–93.

Correll, S. (1995). The ethnography of an electronic bar – the lesbian café. *Journal of Contemporary Ethnography*, 24 (3), 270–98.

Cunningham, S., and Kendall, T.D. (2011). Prostitution 2.0: The changing face of sex work. *Journal of Urban Economics*, 69 (3), 273–87.

Daneback, K., Månsson, S.A., Ross, M.W. (2012). Technological advancements and Internet sexuality: Does private access to the Internet influence online sexual behavior? *Cyberpsychology, Behavior, and Social Networking*, 15 (8), 386–90.

Daneback, K., Træen, B., and Månsson, S.-A. (2009). Use of pornography in a random sample of Norwegian heterosexual couples. *Archives of Sexual Behavior*, 38 (5), 746–73.

Denney, A.S., and Tewksbury, R. (2013). Characteristics of successful personal ads in a BDSM on-line community. *Deviant Behavior*, 34 (2), 153–68.

Denney, A.S., and Tewksbury, R. (2014). Bug chasing: The pursuit of HIV infection. In. E.C. Dretsch and R. Moore (Eds.), *Sexual deviance online: Research and readings* (143–152). Durham, NC: Carolina Academic Press.

DeYoung, M. (1989). The world according to NAMBLA: Accounting for deviance. *Journal of Sociology and Social Welfare*, 16 (1), 111–26.

Diaz, M. (2009, July 27). Sexting: Both sender and receiver can face charges: When teens send naked photos on their cell phones, most don't know the consequences. *Sun-Sentinel* (*Fort Lauderdale, FL*).

Donofrio, R. (1996). Human/animal sexual contact: A descriptive-exploratory study. Doctoral dissertation. San Francisco, CA: The Institute for Advanced Study of Human Sexuality.

Doring, N.M. (2009). The Internet's impact on sexuality: A critical review of 15 years of research. *Computers in Human Behavior*, 25 (5), 1089–101.

Dretsch E.C., and Dretsch, M.N. (2014). Sexual deviance within massively multiplayer online role-playing games. In. E.C. Dretsch and R. Moore (eds), *Sexual deviance online: Research and readings* (25–37). Durham, NC: Carolina Academic Press.

Durkin, K.F. (1997). Misuse of the Internet by pedophiles: Implications for law enforcement and probation practice. *Federal Probation*, 61 (3), 14–18.

Durkin, K.F., and Bryant, C.D. (1995). Log on to sex: Some notes on the carnal computer and erotic cyberspace as an emerging research frontier. *Deviant Behavior*, 16 (3), 179–200.

Durkin, K.F., and Bryant, C.D. (1999). Propagandizing pederasty: A thematic analysis of the on-line exculpatory accounts of unrepentant pedophiles. *Deviant Behavior*, 20 (2), 103–27.

Durkin, K., Forsyth, C.J., and Quinn, J.F. (2006). Pathological Internet communities: A new direction for sexual deviance research in a post modern era. *Sociological Spectrum*, 26 (6), 595–606.

Elliott, I.A., Beech, A.R., Mandeville-Norden, R., and Hayes, E. (2009). Psychological profiles of Internet sexual offenders: Comparisons with contact sexual offenders. *Sexual Abuse: A Journal of Research and Treatment*, 21 (1), 76–92.

Forbes, S. (2011). Sex, cells, and SORNA: Applying sex offender registration laws to sexting cases. *William and Mary Law Review*, 52 (5), 1717–46.

Forsyth, C.J. (1996). The structuring of vicarious sex. *Deviant Behavior*, 17 (3), 279–95.

Forsyth, C.J., and Fournet, L. (1987). A typology of office harlots: Party girls, mistresses and career climbers. *Deviant Behavior*, 8 (4), 319–28.

Forsyth, C.J., and Quinn, J.F. (2014). A swell of the nasty: The new wave of sexual behavior in the back places of the internet. In E.C. Dretsch and R. Moore (eds), *Sexual deviance online: Research and readings* (9–20). Durham, NC: Carolina Academic Press.

Gagnon, J.H., and Simon, W. (1967). *Sexual Deviance*. New York, NY: Harper and Row.

Gauthier, D.K., and Forsyth, C.J. (1999). Bareback sex, bug chasers, and the gift of death. *Deviant Behavior*, 20 (1), 85–100.

Gilbert, R.L., Gonzalez, M.A., and Murphy, N.A. (2011). Sexuality in the 3D Internet and its relationship to real-life sexuality. *Psychology & Sexuality*, 2 (2), 107–22.

Gray, N.J., and Klein, J.D. (2006). Adolescents and the Internet: Health and sexuality information. *Current Opinion in Obstetrics & Gynecology*, 18 (5), 519–24.

Griffiths, M.D. (2012). Internet sex addiction: A review of empirical research. *Addiction Research and Theory*, 20 (2), 111–24.

Grov, C. (2006). Barebacking websites: Electronic environments for reducing or inducing HIV risk. *AIDS Care*, 18 (8), 990–97.

Grov, C., and Parsons, J.T. (2006). Bug chasing and gift giving: The potential for HIV transmission among barebackers on the Internet. *AIDS Education & Prevention*, 18 (6), 490–503.

Hald, G.M. (2006). Gender differences in pornography consumption among young heterosexual Danish adults. *Archives of Sexual Behavior*, 35 (5), 577–85.

Harris, A. (2008). Child Online Protection Act still unconstitutional. *The Center for Internet and Society*, Retrieved from http://cyberlaw.stanford.edu/blog/2008/11/child-online-protection-act-still-unconstitutional.

Heinz, B., Gu, L., Inuzuka, A., and Zender, R. (2002). Under the rainbow flag: Webbing global gay identities. *International Journal of Sexuality & Gender Studies*, 7 (2–3), 107–24.

Holt, T.J., and Blevins, K.R. (2007). Examining sex work from the client's perspective: Assessing johns using on-line data. *Deviant Behavior*, 28 (4), 333–54.

Holt, T.J., Blevins, K.R., and Kuhns, J.B. (2014). Examining diffusion and arrest avoidance practices among johns. *Crime & Delinquency*, 60 (2), 261–83.

Hughes, D. (2003). Prostitution online. *Journal of Trauma Practice*, 2 (3/4), 115–31.

Humbach, J.A. (2014). The Constitution and revenge porn. *Pace Law Review*, 35(1), 215–60.

Jenkins, P. (2001). *Beyond tolerance: Child pornography on the Internet*. New York, NY: University Press.

Jonsson, L.S., Svedin, C.G., and Hydén, M. (2014). "Without the Internet, I never would have sold sex": Young women selling sex online. *Cyberpsychology: Journal of Psychosocial Research on Cyberspace*, 8 (1), 1–14.

Kamel, G.W.L., and Weinberg, T.S. (1995). Diversity in sadomasochism: Four S&M careers. In T.S. Weinberg (ed.) *S&M: Studies in dominance and submission* (71–91). Amherst, NY: Prometheus Books.

Kaufman, M., Silverberg, C., and Odette, F. (2007). *The ultimate guide to sex and disability*. San Francisco, CA: Cleis.

Kvalem, I.L., Traeen, B., Lewin, B., and Stulhofer, A. (2014). Self-perceived effects of Internet pornography use, genital appearance satisfaction, and sexual self-esteem among young Scandinavian adults. *Cyberpsychology: Journal of Psychosocial Research on Cyberspace*, 8(4).

Lanning, K. (1998). Cyber pedophiles: A behavioral perspective. *THE APSAC Advisor*, 11 (1), 2–8.

Larkin, P.J. (2014). Revenge porn, state law, and free speech. *Loyola of Los Angeles Law Review*, 1 (48), 57–118.

Lenhart, A. (2009). Teens and sexting: Pew Internet & American Life Project. Retrieved from www.pewinternet.org/~/media//Files/Reports/2009/PIP_Teens_and_Sexting.pdf.

Lenius, S. (2001). Bisexuals and BDSM. *Journal of Bisexuality*, 1 (4), 69–78.

Lewis, P.H. (1995, July 17). Tech on the net; The Internet battles a much disputed study on selling pornography online. *New York Times*, Retrieved from www.nytimes.com/1995/07/17/business/tech-net-internet-battles-much-disputed-study-selling-pornography-line.html.

Lewis, A. (2011). Ageplay: An adults only game. *Counseling Australia*, 11 (4), 1–9.

Malesky, L.A., and Ennis, L. (2004). Supportive distortions: An analysis of posts on a pedophile Internet message board. *Journal of Addictions & Offender Counseling*, 24 (2), 92–100.

Månsson, S.A., and Söderlind, P. (2013). Technology and pornography. The sex industry on the Internet. In Daneback and S.A., Månsson (eds), *Sexuality and the Internet. A collection of papers* (1–23). Malmö University, Faculty of Health and Society.

McDonough-Taub, G. (2009, July 16). Porn at work: Recognizing a sex addict. *CNBC*. Retrieved from www.cnbc.com/id/31922685.

McLelland, M.J. (2002). Virtual ethnography: Using the Internet to study gay culture in Japan. *Sexualities*, 5 (4), 387–406.

Miletski, H. (2005). A history of bestiality. In A.M. Beetz and A.L. Podberscek (eds), *Anthrozoos, special issue: Bestiality and zoophilia: Sexual relations and animals* (pp. 1–22). West Lafayette, IN: Purdue University Press.

Mitchell, K.J., Finkelhor, D., Jones, L.M., and Wolak, J. (2010). Growth and change in undercover online child exploitation investigations, 2000 to 2006. *Policing & Society*, 20 (4), 416–31.

Mitchell, K.J., Wolak, J., Finkelhor, D., and Jones, L. (2012). Investigators using the Internet to apprehend sex offenders: Findings from the Second National Juvenile Online Victimization Study. *Police Practice and Research*, 13 (3), 267–81.

Montclair, R. (1997). Tails of bestiality. *Black Book Magazine*.

Moore, R. (2014). Protecting children from themselves: The criminalization of sexting as a form of manufactured child pornography. In E.C. Dretsch and R. Moore (eds), *Sexual deviance online: Research and readings* (105–121). Durham, NC: Carolina Academic Press.

Newmahr, S. (2011). *Playing on the edge: Sadomasochism, risk, and intimacy.* Indianapolis: Indiana University Press.

Nip, J.Y.M. (2003). The relationship between online and offline communities: The case of queer sisters. *Media Culture & Society*, 26 (3), 409–28.

Odum, H.W. (1937). Notes on technicways in contemporary society. *American Sociological Review*, 2 (3), 336–46.

Orzack, M.H., and Ross, C.J. (2000). Should virtual sex be treated like other sex addictions? *Sexual Addiction and Compulsivity*, 7 (1–2), 113–25.

Palandri, M., and Green, L. (2000). Image management in a bondage, discipline, sadomasochist subculture: A cyber-ethnography study. *CyberPsychology & Behavior*, 3 (4), 631–41.

Paul, B. (2009). Predicting Internet pornography use and arousal: The role of individual difference variables. *The Journal of Sex Research*, 46 (4), 344–57.

Potter, R.H., and Potter, L.A. (2001). The Internet, cyber porn, and sexual exploitation of children: Media moral panics and urban myths for middle-class parents? *Sexuality & Culture: An Interdisciplinary Quarterly*, 5 (3), 31–48.

Putnam, D.E. (2000). Initiation and maintenance of online sexual compulsivity: Implications for assessment and treatment. *CyberPsychology and Behavior*, 3 (4), 353–63.

Quinn, J.F., and Forsyth, C. J. (2005). Describing sexual behavior in the era of the Internet: A typology for empirical research. *Deviant Behavior*, 26 (3), 191–207.

Rainey, D., and Rainey, B. (2014). The dangers of pornography: Stop the spread of slime in your home. *Family Life: Help for today, Hope for Tomorrow*. Retrieved from www.familylife.com/articles/topics/parenting/challenges/sexual-purity/the-dangers-of-pornography.

Reahard, J. (2013, June 20). Second Life readies for 10th anniversary, celebrates a million active users per month. *Engadget*. Retrieved fromwww.engadget.com/2013/06/20/second-life-readies-for-10th-anniversary-celebrates-a-million-a/.

Rice, S.R., and Ross, M.W. (2014). Differential processes of 'Internet' versus 'real life' sexual filtering and contact among men who have sex with men. *Cyberpsychology: Journal of Psychosocial Research on Cyberspace*, 8 (1), 1–16.

Ross, M.W. (2006). *Typing, doing and being: A study of men who have sex with men and sexuality on the Internet.* Diss. Malmö: University Health and Society.

Schneider, J.P. (2000). Effects of cybersex addiction on the family: Results of a survey. *Sexual Addiction and Compulsivity: The Journal of Treatment and Prevention*, 7 (1–2), 31–58.

Schwartz, M.F., and Southern, S. (2000). Compulsive cybersex: The new tearoom. *Sexual Addiction and Compulsivity*, 7 (1–2), 127–44.

Seigfried-Spellar, K.C., and Rogers, M.K. (2013). Does deviant pornography use follow a Guttman-like progression? *Computers in Human Behavior*, 29 (5), 1997–2003.

Seto, M.C. (2013). *Internet sex offenders*. American Psychological Association.

Simon, L., and Daneback, K. (2013). Adolescents' use of the Internet for sex education: A thematic and critical review of the literature. *International Journal of Sexual Health*, 2 (4), 305–19.

Simon, W., and Gagnon, J.H. (1986). Sexual scripts: Permanence and change. *Archives of Sexual Behavior*, 15 (2), 97–120.

Soothill, K., and Sanders, T. (2005). The geographical mobility, preferences and pleasures of prolific punters: A demonstration study of the activities of prostitutes' clients. *Sociological Research Online*, 10 (1).

Statista. (2015). Number of worldwide Internet users from 2000 to 2015 (in millions). *The Statistics Portal*. Retrieved from www.statista.com/statistics/273018/number-of-internet-users-worldwide/.

Stiles, B.L., and Clark, R.E. (2011). BDSM: A subcultural analysis of sacrifices and delights. *Deviant Behavior*, 32 (2), 158–89.

Surette, R. (1992). *Media, crime, and criminal justice: Images and realities*. Pacific Grove, CA: Brooks/Cole Publishing.

Sutton, M.J., Brown, J.D., Wilson, K.M., and Klein, J.D. (2002). Shaking the tree of knowledge for forbidden fruit: Where adolescents learn about sexuality and contraception. In Brown, Steele, and Walsh-Childers (eds), *Sexual teens, sexual media: Investigating media's influence on adolescent sexuality*, (25–55). New York, NY: Routledge.

Terdiman, D. (2004, March 22). Brits going at it tooth and nail. *Wired*. Retrieved from http://archive.wired.com/culture/lifestyle/news/2004/03/62687?currentPage=all.

Tewksbury, R. (2006). "Click here for HIV": An analysis of Internet-based bug chasers and bug givers. *Deviant Behavior*, 27 (4), 379–95.

Wall, D. (2001). *Crime and the Internet*. New York, NY: Routledge.

Waters, P. (2012, May 4). Mobile porn revenue to reach almost $1 billion by 2015. *Elite Daily*. Retrieved from http://elitedaily.com/news/business/mobile-porn-revenue-reach-1-billion-2015/.

Weiss, R., and Schneider, J. (2006). *Untangling the web: Sex, porn, and fantasy obsession in the Internet age*. New York, NY: Alyson.

Weisskirch, R., and Delevi, R. (2011). 'Sexting' and adult romantic attachment. *Computers in Human Behavior*, 27 (5), 1697–701.

Winder, B., Gough, B., and Seymour-Smith, S. (2015). Stumbling into sexual crime: The passive perpetrator in accounts by male Internet sex offenders. *Archives of Sexual Behavior*, 44 (1), 167–80.

XBIZ Research. (2012, Spring). The 2012 XBIX research report: Attitudes, views and trends impacting the adult entertainment industry. Retrieved from www.xbizresearch.com/reports/xbizresearch_2012.pdf.

Young, K.S. (2001). *Tangled in the web: Understanding cybersex from fantasy to addiction*. Bloomington, IN: Authorhouse.

7

ICTs and interpersonal violence

Thomas J. Holt

ICTs and interpersonal violence

The development of email and other forms of computer-mediated communications (CMCs) has completely changed the way in which we socially engage with others. Facebook, Twitter, and other social media platforms allow individuals to share their attitudes, opinions, and activities with others in near-real time around the clock through text, images, and video. Facebook and Instagram also allow users to check in to a location or geo-tag photos so that individuals know where they are and what they are doing at any time of day. Blogs and Tumblr allow individuals to share videos and photos of virtually every facet of our lives and interests with the larger world.

The relatively open nature in which people can now lead their lives is unparalleled and limited only by an individual's willingness to engage in online environments and CMCs, including social media platforms like Facebook, Twitter, Instagram, and Snapchat. The benefits afforded by the abundance of information available about individuals' personal lives coupled with the ability to connect to anyone at virtually any time of day have also created unique opportunities for harm. The ability to write and send hurtful or threatening messages, post embarrassing videos and pictures in public places, and other content allows individuals to engage in emotionally damaging exchanges online. This constitutes a form of direct and indirect cyber-violence (Wall 2001), which has parallels with interpersonal violence in the real world. Though the virtual environment does not enable individuals to cause direct bodily harm to another, an offender can cause their victim to feel fear, anger, frustration, shame, and humiliation through online environments.

There are myriad examples of the ways that ICT may be used to cause emotional or psychological harm. For instance, the ability to indicate one's relationship status through Facebook allows individuals to know when a relationship has failed, and may lead some to post disparaging or hurtful comments about their former partner for others to see. Others have begun to post personal, candid video and images sent from intimate partners in public websites in order to shame or embarrass the sender (e.g. Mitchell *et al.* 2007).

Adolescents are also likely to experience various forms of interpersonal violence while online, as they are heavily engaged in online environments and use mobile devices regularly (e.g. Lenhart 2010; Zickuhr 2011). Research has found an increase in the number of youth who are sent bullying or harassing emails from classmates or peers (e.g. Jones *et al.* 2012). A number of youth in the US (Marcum 2010a), UK (Fricker 2013) and Asia (Chen 2011) have tragically committed suicide over their experiences, leading to public outcry for policies to combat bullying and harassment online.

The impact of ICT on interpersonal violence cannot be understated. There is, however, some confusion over the ways that online violence is differentiated from physical violence, how it is measured by researchers, and its impact on victims. This chapter will address these issues by examining three common forms of online interpersonal violence: cyberbullying, online harassment, and cyberstalking. This chapter will discuss the common definitions used by researchers and practitioners for these offenses, consider estimates of both victimization and offending, and the emotional and psychological impact that they have on victims. The factors that may affect victim reporting will also be discussed to better understand why these offenses may be underrepresented in various data sources. The chapter concludes with a summary of the impact that online interpersonal violence can have for Internet users as a whole.

Defining cyberbullying, online harassment, and cyberstalking

Over the last decade, there has been substantial attention placed on the issue of bullying among adolescents. Scholars and policy-makers alike have also brought the issue of cyberbullying into focus, due to the substantial use of technology and skill with navigating CMC among juvenile populations (Aftab 2006). In order to define cyberbullying, it is necessary to first understand bullying off-line. Bullying in the physical world is usually defined as the intentional and repeated use of aggressive or negative behaviors against an individual based on an imbalance of power between two parties (Klomek *et al.* 2008; Nansel *et al.* 2001; Olweus 1993). The victim in bullying incidents is usually weaker physically and/or socially than their aggressor, and the bullying could involve verbal threats or insults (like name-calling or teasing) to the use of physical assaults including hitting and kicking. As a result, victims report negative emotional reactions including embarrassment, shame, intimidation, anger, and/or sadness (Klomek *et al.* 2008; Nansel *et al.* 2001).

These conditions are also present in bullying exchanges online, which many refer to as acts of cyberbullying where an individual may experience intentional, aggressive behavior performed through electronic means (Hinduja and Patchin 2008). While an individual cannot cause direct physical injury to another through computer-mediated communications (CMCs), bullies can cause their victim to feel emotional harm and embarrassment through the use of threatening, mean, or hurtful messages posted via instant messaging, email, social media, and text messages via mobile phones (Hinduja and Patchin 2008).

In addition, cyberbullying victimization is somewhat different from physical bullying as it is difficult for the victim to escape from their aggressor online. Text messages and posts in social media can be sent instantaneously and repeatedly throughout the day, reposted by others, and even "liked" or "favorited" by others, increasing the number of participants in a bullying incident (Jones *et al.* 2012). In turn, victims may feel as though everyone is against them which can create a sense of persistent and pervasive emotional and psychological harm.

It is also important to note that cyberbullying can be conflated with online harassment or cyberstalking, depending on the researcher. A number of scholars suggest that bullying behaviors, on- or off-line, should be associated with juvenile populations where power differentials are common (Aftab 2006; Bossler, Holt and May 2012; Klomek *et al.* 2008; Marcum 2010b). Others have suggested that adults can be bullied, most often in the workplace where there is potential for power differentials between co-workers that mirror those present in juvenile bullying dyads (Kowalski *et al.* 2008).

Bullying can be distinguished from harassment and stalking in a legal context, as there is little federal legislation present in Canada, Europe, the US, and the UK regarding bullying (Hinduja and Patchin 2013; Southey 2013). There are, however, a range of criminal laws regarding harassment and stalking behaviors both on- and off-line. Thus, there are few options for victims of bullying to obtain any legal recourse for their experiences at this time. There are, however, criminal statutes related to online harassment and stalking at the federal level in many nations across the globe (see Brenner 2011). Thus, the potential punitive sanctions available for online harassment and stalking are greater than that for bullying.

Though there are some clear differences between cyberbullying, online harassment, and cyberstalking, there is a need to clarify the relationship between harassment and stalking (Sinclair and Frieze 2000). Each of these behaviors utilizes ICT in order to spread messages to victims, and the recipient of messages views their content as unsolicited or desired. The emotional response a victim may experience can, however, establish whether an incident is harassment or stalking. Specifically, harassment is often viewed as messages and content that lead the recipient to feel bothered or annoyed, though they do not necessarily convey a threat of physical or emotional harm (Turmanis and Brown 2006). Cyberstalking messages and behaviors are likely to lead a victim to feel fear for their personal safety, along with emotional distress (Bocij 2004).

Cyberstalking also differs from online harassment owing to the fact that a stalker may engage in activities to actively monitor their victim's behaviors on and off-line (Bocij 2004; Bocij and McFarlane 2002). For instance, cyberstalkers may examine their target's social media profiles in order to understand what they are doing and send messages designed to cause their victim to feel fear and distress (Bocij 2004). Cyberstalkers may also utilize malicious software programs in order to surreptitiously collect information about their victim's behaviors online (Bocij 2004). Others may impersonate their potential victim in online spaces, such as posting advertisements under their name offering sexual services or placing embarrassing photos and personal information in public websites in order to cause the victim to feel emotional distress as a result (Bocij 2004). Cyberstalking differs from traditional stalking in the real world as the victim and offender do not have to know one another (Ibid.). Instead, cyberstalkers can find targets via Google searches and other random places online (Ibid.), which is not always the case with real world stalking (Ibid.). In light of the differences between these three behaviors, each behavior will be examined individually in order to understand its prevalence and impact on victims and off-line.

The prevalence of cyberbullying

Research on rates of cyberbullying victimization demonstrate the substantial limitations of the current measurement techniques available. Since bullying on- or off-line does not technically constitute a criminal act, there are few official statistics available from criminal justice system sources. Instead, data is largely derived from self-report studies conducted by various sources in academia and government. The way that cyberbullying is operationalized can vary from study to study, and there are a small number of data sources that can be used to examine cyberbullying cross-nationally.

Research examining the prevalence and correlates of cyberbullying began in the early 2000s, with studies in the United States indicating that between 6 percent (Thorp 2004) and 7 percent of youth sampled may report being victimized (Ybarra and Mitchell 2004). More recent data from the US National Crime Victimization Survey-Supplemental Survey on bullying and cyberbullying found that approximately 6 per cent of students aged 12 to 18 were cyberbullied during the 2008–09 academic year (DeVoe *et al.* 2011). These rates are much lower by comparison to smaller samples of youth drawn from various populations across the US, which have found victimization rates ranging between 18 and 20 per cent of youth sampled (e.g. Hinduja and Patchin 2009; Kowalski and Limber 2007).

Similar ranges have been identified in international samples, with rates ranging between 25 percent of middle-school students in Canada (Li 2008) to between 8 and 38 percent of youth in the UK, depending on the form of victimization measured, as well as the time of data collection and the population sampled (Department for Education 2011; Tarapdar and Kellett 2011). Research on cyberbullying in Asian samples is also quite variable, with studies in China reporting 33 percent of middle-school students experiencing cyberbullying (Li 2008) compared to approximately 18 percent of youth in Singapore (Holt *et al.* 2013).

The rates of victimization are also somewhat similar to the prevalence rates for cyberbullying perpetration in various populations. Data from the US suggests that between 18 and 20 percent of youth engaged in cyberbullying at some point in a one-year period (e.g. Hinduja and Patchin 2012; Ybarra and Mitchell 2004). Slight increases have been noted in panel studies collected in the US and the Netherlands, though the fact that different sample populations are used in each wave limits their value to assess the prevalence rate of victimization over time (Jones *et al.* 2012; van Wilsem 2013).

Predictors of bullying on- and off-line

Researchers have conducted a range of studies to better understand the risk factors associated with cyberbullying victimization and offending in various samples. The consistency in results suggests there are several behavioral and attitudinal correlates that may account for cyberbullying as a whole. First, females are somewhat more likely to report cyberbullying victimization than males (e.g. Tokunaga 2010). It is thought that these differences stem from gender differences in the expression of aggression physically and verbally. Boys generally report higher levels of physical bullying and aggressive behavior, whereas females are more likely to cause emotional harm through the use of verbal bullying behaviors like spreading gossip and shaming their potential victim (Boulton and Underwood 1992; Klomek *et al.* 2008; Nabuzoka 2003). This relationship is not consistently supported across all studies (e.g. Hinduja and Patchin 2008; Li 2006), though a number of researchers have found that girls are more likely to be targeted by cyberbullies than males (Kowalski *et al.* 2008; Ybarra *et al.* 2007; Ybarra and Mitchell 2004).

Second, cyberbullying victimization appears to be age attenuated, as older youth appear to be at greater risk due to greater unsupervised access to ICTs generally (Sbarbaro and Smith 2011; Tokunaga 2010). Third, specific online behaviors appear to increase individual risk of cyberbullying victimization, including increased time in social networking sites, chat rooms, and email which may expose youth to motivated offenders (Berson *et al.* 2002; Hinduja and Patchin 2008; Holt and Bossler 2009; Twyman *et al.* 2010; Ybarra and Mitchell 2004). Similarly, revealing personal information in online spaces may provide bullies with greater information about their target and thereby engender victimization (Hinduja and Patchin 2009; Mitchell *et al.* 2007).

Fourth, there is substantial evidence across various population samples that youth who are bullied in the real world are more likely to experience cyberbullying (Erdur–Baker 2010: Hinduja and Patchin 2008; Kowalski and Limber 2007; Ybarra and Mitchell 2004). The connection between on- and off-line victimization may stem from the fact that bullies who know their victim in the real world may more easily target that person via ICTs. As a consequence, youth who experience physical bullying and cyberbullying may be more likely to report negative psychological and emotional outcomes because they may feel they cannot escape their bully (Holt *et al.* 2013; Olweus 1993; Tokunaga 2010).

There is also some research on the predictors of participation in cyberbullying, which suggests these youth may be more easily frustrated and demonstrate lower levels of self-control or impulse control generally (Camodeca and Goossens 2005; Holt et al. 2012). For instance, youth who persistently engage in cyberbullying may report greater behavioral problems at school (Hinduja and Patchin 2008). Individuals who engage in cyberbullying also report lower levels of compassion and empathy toward others, making it difficult for them to understand how their actions affect other people (Camodeca and Goossens 2005). They also engage in more physical acts of aggression including bullying behaviors (Hinduja and Patchin 2008).

Despite these behavioral correlates, there are few consistent demographic predictors observed across multiple studies. Gender studies have found that both males and females engage in cyberbullying, though females may do so with somewhat greater frequency (Tokunaga 2010). Research regarding age and cyberbullying have found mixed support, with some identifying a relationship between older youth and bullying while others find no significant correlation (Tokunaga 2010). As a result, there may be greater value in studying the behavioral and attitudinal drivers of bullying behavior to understand why some youth engage in bullying behaviors.

Rates of harassment and stalking

Research examining the prevalence of online harassment and cyberstalking is somewhat better than that of bullying, due to the number of representative national population samples developed by researchers. For instance, the US Youth Internet Safety Survey (YISS), a three-wave panel study of youth aged 10 to 17, has been carried out in 2000, 2005, and 2010 (Jones et al. 2012). Though the youth sampled vary within each wave, the results of this survey suggest that online harassment victimization has increased over the last decade. The rate of online harassment victimization, defined as receiving threats or offensive comments directly or indirectly through comments posted about them for others to see, increased from 6 percent in 2000 to 9 percent in 2005 to 11 percent in 2010 (Jones et al. 2012). There has also been a small increase in the number of youth who report distress as a result of their harassment, measured by fear or being upset by their experience, from 3 percent in 2000 and 2005 to 5 percent in 2010 (Jones et al. 2012).

Assessments of college populations find similar rates of victimization over 12-month periods, though the results may be impacted by the regional nature of the population sampled and date of data collection. For instance, students in a northeastern US sample found that 10 to 15 percent of students reported receiving harassing messages via email or instant messaging (Finn 2004). Rates of populations in southeastern samples range between 6.5 and 34.9 percent depending on the form of harassment and the university sampled (Holt and Bossler 2009; Marcum et al. 2010).

Examinations of cyberstalking suggest there are more victims, particularly within the US. The National Crime Victimization Survey-Supplemental Survey (Catalano 2012) found that within a sample of 65,270 people who reported being stalked, 26.1 percent were sent emails that made them feel fear (Catalano 2012). Similar rates were found in a nationally representative sample of college students, as 24.7 percent of those who were stalked received repeated emails that seemed obsessive or led them to feel fear (Fisher et al. 2000).

Statistics vary when compared cross-nationally, as evidence from Canada suggests that 7 percent of all adults received threatening or aggressive emails and instant messages, the majority of which come from strangers (Perreault 2013). At present there are no official statistics on arrest rates or victimization surveys available to assess the prevalence of cyberstalking in the UK, limiting our knowledge of this form of cyber-violence (National Centre for Cyberstalking Research (NCVS) 2011).

Reporting online bullying, harassment, and stalking

In light of the prevalence rates for victimization observed in various data sources, there is a need to consider the extent to which victims report their experiences to agencies or individuals for assistance. The problem of unwanted email, text, and other CMC is complicated by the fact that many individuals may simply ignore or delete these messages rather than lend them any credence. This creates the perception that victimization is not a real problem, but is instead an issue of perceptual problems on the part of the victim. As a result, some may find ways to ignore threats or harassing messages through technological solutions rather than report them to others.

A recent nationally representative sample of youth collected in the US found that 75 percent of those who experienced cyberbullying told an individual about their experience, though the majority informed their peers (Priebe *et al.* 2013). Individuals who chose not to tell others about their experience felt that it would either not be taken seriously, or was not serious enough to necessitate telling others about (Priebe *et al.* 2013). Results from the National Crime Victimization Survey-Supplemental Survey on bullying indicated that 31 percent of youth who were victimized told a teacher or school official about their experience.

The decision to not report appears to be driven in part by the severity of the incident, as youth are likely to tell an authority figure if it lasts for days or causes a strong emotional response in the victim (Holtfeld and Grabe 2012; Priebe *et al.* 2013; Slonje *et al.* 2013). If it is not considered serious, victims may choose indirect strategies to avoid victimization, including deleting or ignoring messages received, or blocking the sender's email address or phone number (Parris *et al.* 2012; Priebe *et al.* 2013). In fact, most youth only report the incident if they feel it is severe (Holtfeld and Grabe 2012; Slonje *et al.* 2013), such as if it lasts for several days or produces a severe emotional response (Priebe *et al.* 2013). Parents tend not to report cyberbullying experiences to law enforcement due to the perception that the case will not be taken seriously or investigated due to limited laws (Hinduja and Patchin 2009; McQuade *et al.* 2009). In much the same way, school administrators are reluctant to contact police over concern that their school's reputation will be damaged by reports of cyberbullying (McQuade *et al.* 2009).

Only a portion of victims of cyberstalking or harassment report their experiences to police; statistics from the NCVS (2012) indicate that approximately 42 percent of female stalking victims and 14 percent of female harassment victims contacted police. Research on college student stalking victims also found that less than 4 percent of women sought a restraining order against their stalker and fewer than 2 percent filed criminal charges (Fisher *et al.* 2000). Much like cyberbullying victims, individuals who experience cyberstalking and harassment may not contact police because they do not think that their case will be taken seriously by law enforcement (Nobles *et al.* 2012). As a result, victims may be more likely to feel fear or change their behavior in order to minimize their risk of further victimization (Nobles *et al.* 2012). This evidence also suggests that technological change has had relatively little benefit for potential victims, as they must take responsibility for dealing with unwanted experiences directly rather than obtaining aid from traditional criminal justice sources.

Research on victim experiences with cyber-violence

Since victims may not report their experiences with online interpersonal violence, there is a need to understand how victims deal with their encounters. Some are able to ignore the messages they receive, or not internalize the threats and comments made. Others, however, experience emotional harm that can lead to school failure, social withdrawal, shame, anger, frustration, and even suicidal ideation (Camodeca and Goossens 2005; Hinduja and Patchin 2009; Marcum

2010a). The outcomes reported for each form of cyber-violence will be considered below to understand variations in victim responses across each offense type.

Victims of cyberbullying report various behavioral and psychological responses to their experiences, often mirroring reactions to traditional bullying experiences in the real world. Many youth who are bullied exhibit symptoms of depression, stress, and anxiety, particularly if an individual is bullied on- and off-line (Holt et al. 2013; Turner et al. 2013; Ybarra and Mitchell 2004). There is also some evidence that victims of cyberbullying may begin to carry defensive weapons while at school out of fear for their safety (Catalano 2012).

Some youth also become truant, or skip school, in order to escape their bully and minimize any social stigma associated with victimization (Campbell 2005; Katzer et al. 2009; Li 2006; Varjas et al. 2009; Ybarra et al. 2007). Data from the US suggest that approximately 4 percent of youth who were cyberbullied are truant, which is greater than the .04 percent of youth who skip school with no bullying experiences (Catalano 2012).

A small proportion of youth also report suicidal thoughts as a result of cyberbullying experiences (Hinduja and Patchin 2008; Holt et al. 2013; Klomek et al. 2008; Li 2006; Turner et al. 2013). Though the percentage of victims reporting suicidal ideation varies based on the nation sampled, there is evidence that youth with generally negative attitudes or affect are more likely to have these tendencies (Arseneault et al. 2006; Beran and Qing 2007; Nansel et al. 2001). Those youth who are bullied on- and off-line are also more likely to report suicidal thoughts, demonstrating the substantial association between virtual and real experiences. In fact, the proliferation of CMC and social media platforms appears to have directly increased opportunities for vulnerable youth to be targeted in various ways online in addition to any experiences they may have in the real world.

Victims of cyberstalking and online harassment also report emotional harm, such as feeling powerless, shame, anxiety and depression over perceptions that the individual sending messages may act on their threats (Ashcroft 2001; Blaauw et al. 2002). Some also report social withdrawal due to embarrassment over their experience or concern for the safety of others. For example, Nobles et al. (2012) found that between 10 to 15 percent of victims may reduce their time spent with friends and family to minimize the likelihood of harm.

Victims who report higher degrees of fear from their experiences are more likely to engage in self-protective behaviors (Nobles et al. 2012). A small proportion of victims may also move in with family and friends in order to increase their sense of safety and protection (Nobles et al. 2012). A number of victims may also change their address, phone number, or email address to minimize contact from their stalker and reduce the risk of exposure (Baum et al. 2009; Nobles et al. 2012). Some may also carry defensive weapons in order to protect themselves (Wilcox et al. 2007; Nobles et al. 2012).

Summary

Taken as a whole, the development of ICT has transformed both the nature of interpersonal communication, and ushered in a unique environment whereby individuals may cause emotional or psychological harm to others. Individuals can readily send hurtful, threatening, or embarrassing messages to another person, or post information about them through the use of text, images, and video in social media sites and web pages generally. Victims may be in any age group, and the severity and frequency of messaging has made it possible for individuals to be stalked or harassed in virtual spaces.

It is important to note that the development of CMC has enhanced the potential for individuals to be threatened or harassed in a unique way that is not otherwise possible in physical space, or as an extension of experiences that typically occur in the real world (Bocij 2004). ICT can serve as a sort of force multiplier for offenders by enabling them to target victims in the real world, as well as connect with and harangue their victim in near-real time 24 hours a day. The ability to intimately invade technological devices that document nearly all facets of an individual's personal life means that the experience of cyberbullying or harassment may make these offenses more difficult for victims to deal with. As a result, the development of ICT has transformed the nature of interpersonal violence by creating greater opportunities to cause emotional and psychological harm to victims.

Research has begun to document both the prevalence of victimization, as well as behavioral and attitudinal correlates which increase the risk of victimization. It is unclear through various statistics if cyber-based offenses are experienced at a greater rate than physical forms of victimization, though there is sufficient evidence to demonstrate it is a very real problem. There is, however, a clear deficit in the documentation of victimization experiences in official statistics on incidents made known to the police or cleared by arrest. To improve the lack of victim reporting, there is a need for awareness campaigns targeting front line police officers and constables to recognize cyberstalking and harassment cases and treat them appropriately so as to build effective criminal cases when appropriate. In addition, messaging campaigns must be developed for the general public to recognize the severity of harassment and stalking online and the need to report experiences to police and how to respond in order to maintain evidence of their experiences. Similar educational strategies have been developed related to cyberbullying (e.g. Hinduja and Patchin 2013), though few studies have empirically assessed their efficacy.

There is also a need for research examining what, if any, technological solutions can be employed to help minimize the risk of victimization. For example, products like parental filtering software are touted as a key resource to enable constant management of youth behavior while online and reduce exposure to sexual material and other risks. To date, empirical findings demonstrate that devices such as parental filtering software have no effect on reducing the risk of bullying and harassment victimization (see Holt and Bossler 2016 for review). If these products are unable to actually affect the likelihood of negative online experiences, then there is a need for careful evaluation of these tools to determine their limitations and improve their value. For instance, it may be that they fail because of improper installation on the part of a parent or guardian, or that youth are able to obviate the software through various means. Thus, research is needed to improve the potential for technological solutions to actually help reduce the potential for emotional harm caused via CMCs.

Researchers have also documented the negative consequences of bullying, harassment, and stalking for victims, particularly regarding suicidal ideation and self-injury. A number of notable incidents of youth suicides as a consequence of cyberbullying in the US and the UK have drawn attention to these issues from public health researchers and advocacy groups (Hinduja and Patchin 2013). Several groups have called for the criminalization of cyberbullying at the federal level, though attempts to create legislation have been largely ineffective. In the event that high-profile suicide cases continue, it is possible that social policies may evolve in response to social pressure. Thus, researchers must continue to improve our knowledge of interpersonal violence facilitated via ICT in order to understand the scope of this phenomenon and how it may change with the continuing evolution of technology over time. This is especially vital as technology continues to evolve to include wearable devices and virtual/augmented reality devices that may enhance the experience of victimization and allow offenders even greater access to personal information and victim's lives.

References

Addington, L. (2013). Reporting and clearance of cyberbullying incidents: Applying 'offline' theories to online victims. *Journal of Contemporary Criminal Justice*, 29(4): 454–474.

Aftab, P. (2006). Cyber bullying. Wiredsafety.net. [Online] Available at: www.wiredsafety.net.

Agnew, R. (2006). General strain theory: Current status and directions for further research, in F.T. Cullen, J. P. Wright, and K. R. Blevins (eds) *Taking stock: The status of criminological theory* (Advances in Criminological Theory, Vol. 16). New Brunswick, NJ: Transaction, pp. 137–158.

Arseneault, L., Walsh, E., Trzesniewski, K., Newcombe, R., Caspi, A., and Moffitt, T. E. (2006). Bullying victimization uniquely contributes to adjustment problems in young children: A nationally representative cohort study. *Pediatrics*, 118, 130–138.

Ashcroft, J. (2001). *Stalking and domestic violence*. NCJ 186157. Washington, DC: U.S. Department of Justice.

Baum, K., Catalano, S., Rand, M., and Rose, K. (2009). *Stalking victimization in the United States*. Bureau of Justice Statistics, US Department of Justice. [Online] Available at www.ovw.usdoj.gov/docs/stalking-victimization.pdf.

Beran, T., and Qing, L. (2007). The relationship between cyberbullying and school bullying. *Journal of Student Wellbeing*, 1, 15–33.

Berson, I. R., Berson, M. J., and Ferron, J. M. (2002). Emerging risks of violence in the digital age: Lessons for education from an online study of adolescent girls in the United States. *Journal of School Violence*, 1, 51–71.

Blauuw, E., Winkel, F. W., Arensman, E., Sheridan, L., and Freeve, A. (2002). The toll of stalking: The relationship between features of stalking and psychopathology of victims. *Journal of Interpersonal Violence*, 17, 50–63.

Bocij, P. (2004). *Cyberstalking: Harassment in the Internet age and how to protect your family*. Westport, CT: Praeger Publishers.

Bocij, P., and McFarlane, L. (2002). Online harassment: Towards a definition of cyberstalking. *Prison Service Journal*, 39: 31–38.

Borg, M. G. (1999). The extent and nature of bullying among primary and secondary schoolchildren. *Educational Research*, 41, 137–153.

Bossler, A. M., and Holt, T. J. (2010). The effect of self control on victimization in the cyberworld. *Journal of Criminal Justice*, 38: 227–236.

Bossler, A. M., Holt, T. J., and May, D. C. (2012). Predicting online harassment among a juvenile population. *Youth and Society*, 44, 500–523.

Boulton, M. J., and Underwood, K. (1992). Bully victim problems among middle school children. *British Journal of Educational Psychology of Addictive Behaviors*, 62, 73–87.

Brenner, S. (2011). Defining cybercrime: A review of federal and state law. In Ralph D. Clifford: *Cybercrime: The investigation, prosecution, and defense of a computer-related crime*. Raleigh, NC: Carolina Academic Press, pp. 15–104.

Camodeca, M., and Goossens, F. A. (2005). Aggression, social cognitions, anger and sadness in bullies and victims. *Journal of Child Psychology and Psychiatry*, 46, 186–197.

Campbell, M. A. (2005). Cyberbullying: An old problem in a new guise? *Australian Journal of Guidance and Counseling*, 15, 68–76.

Catalano, S. (2012). *Stalking Victims in the United States-Revised*. Washington, DC: U.S. Department of Justice. [Online] Available at: www.bjs.gov/content/pub/pdf/svus_rev.pdf.

Chamberlain, T., George, N., Golden, S., Walker, F., and Benton, T. (2010). *Tellus4 national report*. London: Department for Children, Schools and Families (DCSF).

Chen, E. (2011). Girl, 16, falls to death in cyber-bully tragedy. *edVantage*. [Online] Available at: www.edvantage.com.sg/content/girl-16-falls-death-cyber-bully-tragedy.

Crown Prosecution Service. (2013). *Stalking and Harassment. Crown Prosecution Service Prosecution Policy and Guidance*. [Online] Available at: www.cps.gov.uk/legal/s_to_u/stalking_and_harassment/.

Cybersmile. (2013). FAQ. [Online] Available at: http://cybersmile.org/who-we-are.

Department for Children, Schools and Families. (2010). *Local authority measures for national indicators supported by the Tellus4 survey*. London: Department for Children, Schools and Families.

Department for Education (2011). *The protection of children online: A brief scoping review to identify vulnerable groups*. London: Department for Education.

Department of Justice Canada (2012). *A handbook for police and crown prosecutors on criminal harassment.* Department of Justice Canada. [Online] Available at: www.justice.gc.ca/eng/rp-pr/cj-jp/fv-vf/har/EN-CHH2.pdf.

DeVoe, J. F., Bauer, L., and Hill, M. R. (2011). Student Victimization in U.S. Schools: Results From the 2009 School Crime Supplement to the National Crime Victimization Survey. Washington, DC: National Center for Educational Statistics. [Online] Available at: http://nces.ed.gov/pubs2012/2012314.pdf.

Erdur-Baker, O. (2010). Cyberbullying and its correlation to traditional bullying, gender and frequent risky usage of Internet-mediated communication tools. *New Media Society,* 12, 109–125.

Facebook Tools (2012). Safety. [Online] Available at: www.facebook.com/safety/tools/.

Finn, J. (2004). A survey of online harassment at a university camps. *Journal of Interpersonal Violence,* 19, 468–483.

Fisher, B., Cullen, F., and Turner, M. G. (2000). *The sexual victimization of college women.* National Institute of Justice Publication No. NCJ 182369. Washington: Department of Justice.

Fricker, M. (2013). Hannah Smith Suicide: Grieving dad sells home where cyber-bullying victim died. *Mirror,* Oct 24, 2013. [Online] Available at: www.mirror.co.uk/news/uk-news/hannah-smith-suicide-grieving-dad-2485767#.Ut_h_bQo7IU.

Hay, C. Meldrum, R. C., and Mann, K. (2010). Traditional bullying, cyber bullying, and deviance: A general strain theory approach. *Journal of Contemporary Criminal Justice,* 26, 130–147.

Haynie, D. L., Nansel, T., Eitel, P., Crump, A. D., Saylor, K., Yu, K., and Simons-Morton, B. (2001). Bullies, victims, and bully/victims: Distinct groups of at-risk youth. *Journal of Early Adolescence,* 21, 29–49.

Hinduja, S. and Patchin, J. (2008). Cyberbullying: An exploratory analysis of factors related to offending and victimization. *Deviant Behavior,* 29, 1–29.

Hinduja, S., and Patchin, J. W. (2009). *Bullying beyond the schoolyard: Preventing and responding to cyberbullying.* New York: Corwin Press.

Hinduja, S., and Patchin, J. W. (2012). Summary of Cyberbullying Research From 2004–2012. [Online] Available at: http://cyberbullying.us/research.

Hinduja, S., and Patchin, J. (2013). Description of State Cyberbullying Laws and Model Policies. [Online] Available at: www.cyberbullying.us/Bullying_and_Cyberbullying_Laws.pdf.

Hoey, D. (2012). Biddeford Man sentenced to five years for cyberstalking. *Portland Press Herald,* December 4, 2012. [Online] Available at: www.pressherald.com/news/Biddeford-man-sentenced-to-5-years-for-cyberstalking-.html.

Holt, T. J., and Bossler, A. M. (2009). Examining the applicability of Lifestyle-Routine Activities Theory for cybercrime victimization. *Deviant Behavior,* 30, 1–25.

Holt, T. J., and Bossler, A. M. (2016). *Cybercrime in Progress: Theory and Prevention of Technology-enabled Offenses.* London: Routledge.

Holt, T.J., Bossler, A.M., and Fitzgerald, S. (2010). Examining state and local law enforcement perceptions of computer crime, in Holt, T.J. (ed.), *Crime On-Line: Correlates, Causes, and Context.* Raleigh, NC: Carolina Academic Press, pp. 221–246.

Holt, T. J., Bossler, A. M., and May, D. C. (2012). Low self-control deviant peer associations and juvenile cyberdeviance. *American Journal of Criminal Justice,* 37 (3), 378–395.

Holt, T. J., Chee, G., Ng, E., and Bossler, A. M. (2013). Exploring the consequences of bullying victimization in a sample of Singapore youth. *International Criminal Justice Review,* 23 (1), 25–40.

Holtfeld, B., and Grabe, M. (2012). Middle school students' perceptions of and responses to cyberbullying. *Journal of Educational Computing Research,* 46(4), 395–413.

Jones, L. M., Mitchell, K. J., and Finkelhor, D. (2012). Trends in youth Internet victimization: Findings from three youth Internet safety surveys 2000–2010. *Journal of Adolescent Health,* 50, 179–186.

Katzer, C., Fetchenhauer, D., and Belschak, F. (2009). Cyberbullying: Who are the victims? A comparison of victimization in internet chatrooms and victimization in school. *Journal of Media Psychology,* 21, 25–36.

Klomek, A. B., Sourander, A., Kumpulainen, K., Piha, J., Tamminen, T., Moilanen, I., Almqvist, F. and Gould, M. S. (2008). Childhood bullying as a risk for later depression and suicidal ideation among Finnish males. *Journal of Affective Disorders,* 109, 47–55.

Kowalski, R. M., and Limber, P. (2007). Electronic bullying among middle school students. *Journal of Adolescent Health,* 41, 22–30.

Kowalski, R. M., Limber, S. P., and Agatston, P. W. (2008). *Cyberbullying: Bullying in the digital age.* Maldon, MA: Blackwell Publishing.

Lenhart, A. (2010) *Is the age at which teens get cell phones getting younger?* Pew Internet and American Life Project. [Online] Available at: http://pewinternet.org/Commentary/2010/December/Is-the-age-at-which-kids-get-cell-phones-getting-younger.aspx.

Li, Q. (2006). Cyberbullying in schools. *School Psychology International*, 27 (2), 157–170.

Li, Q. (2008). A cross-cultural comparison of adolescents' experience related to cyberbullying. *Educational Research*, 50 (3), 223–234.

Marcum, C. D. (2010a). Examining cyberstalking and bullying: Causes, context, and control. In T. J. Holt (ed.) *Crime on-line: Correlates, causes, and context.* Raleigh, NC: Carolina Academic Press, pp. 175–192.

Marcum, C. D. (2010b). Assessing sex experiences of online victimization: An examination of adolescent online behaviors utilizing Routine Activity Theory. *Criminal Justice Review*, 35 (4), 412–437.

McQuade, S., Colt, J., and Meyer, N. (2009). *Cyber Bullying: Protecting kids and adults from online bullies.* ABC-CLIO: Santa Barbara, CA.

Mitchell, K. J., Finkelhor, D., and Becker-Blease, K. A. (2007). Linking youth internet and conventional problems: Findings from a clinical perspective. *Journal of Aggression, Maltreatment and Trauma*, 15, 39–58.

Moore, R., Guntupalli, N. T., and Lee, T. (2010). Parental regulation and online activities: Examining factors that influence a youth's potential to become a victim of online harassment. *International Journal of Cyber Criminology*, 4, 685–698.

Morphy, E. (2008). The Computer Fraud Act: Bending a law to fit a notorious case. *E Commerce Times* December 9, 2008. [Online] Available at: www.ecommercetimes.com/story/65424.html.

Nabuzoka, D. (2003). Experiences of bullying-related behaviours by English and Zambian pupils: A comparative study. *Educational Research*, 45 (1), 95–109.

Nansel, T. R., Overpeck, M., Pilla, R. S., Ruan, W. J., Simmons-Morton, B., and Scheidt, P. (2001). Bullying behavior among U.S. youth: Prevalence and association with psychosocial adjustment. *Journal of the American Medical Association*, 285, 2094–2100.

National Centre for Cyberstalking Research (NCVS) (2011). Cyberstalking in the United Kingdom: An Analysis of the ECHO Pilot Survey 2011. [Online] Available at: www.beds.ac.uk/__data/assets/pdf_file/0003/83109/ECHO_Pilot_Final.pdf.

Ngo, F.T., and Paternoster, R. (2011). Cybercrime victimization: An examination of individual and situational level factors. *International Journal of Cyber Criminology*, 5, 773–793.

Nobles, M. R., Reyns, B. W., Fox, K. A., and Fisher, B. S. (2012). Protection against pursuit: A conceptual and empirical comparison of cyberstalking and stalking victimization among a national sample. *Justice Quarterly*, DOI: 10.1080/07418825.2012.723030.

Olweus, D. (1993). *Bullying at school: What we know and what we can do.* Cambridge, MA: Blackwell.

Parris, L., Varjas, K., Meyers, J., and Cutts, H. (2012). High school students' perceptions of coping with cyberbullying. *Youth and Society*, 44, 284–306.

Patchin, J. W., and Hinduja, S. (2011). Traditional and nontraditional bullying among youth: A test of general strain theory. *Youth and Society*, 43, 727–751.

Perreault, S. (2013). Self-reported Internet victimization in Canada, 2009. [Online] Available at: www.statcan.gc.ca/pub/85-002-x/2011001/article/11530-eng.htm#n3.

Peterson, H. (2013). 'Catfishing': The phenomenon of Internet scammers who fabricate online identities and entire social circles to trick people into romantic relationships. *Daily Mail Online.* January 17, 2013. [Online] Available at: www.dailymail.co.uk/news/article-2264053/Catfishing-The-phenomenon-Internet-scammers-fabricate-online-identities-entire-social-circles-trick-people-romantic-relationships.html.

Priebe, G., Mitchell, K. J., and Finkelhor, D. (2013). To tell or not to tell? Youth's responses to unwanted Internet experiences. *Cyberpsychology: Journal of Psychosocial Research on Cyberspace*, 7: 1–20.

Sbarbaro, V., and Smith, T. M. E. (2011). An exploratory study of bullying and cyberbullying behaviors among economically/educationally disadvantaged middle school students. *American Journal of Health Studies*, 26 (3), 139–150.

Sheridan, L, and T. Grant. (2007). Is cyberstalking different? *Psychology, Crime and Law*, 13, 627–640.

Sinclair, H. C., and Frieze, I. H. (2000). Initial courtship behavior and stalking: How should we draw the line? *Violence and Participants*, 15, 23–40.

Slonje, R., Smith, P. K., and Frisen, A. (2013). The nature of cyberbullying, and the strategies for prevention. *Computers in Human Behavior*, 29, 26–32.

Smith, P. K., Mahdavi, J., Carvalho, M., Fisher, S., Russell, S., and Tippett, N. (2008). Cyberbullying: Its nature and impact in secondary school pupils. *Journal of Child Psychology and Psychiatry*, 49(4), 376–385.

Southey, T. (2013). Bill C-13 is about a lot more than cyberbullying. *The Globe and Mail* December 6, 2013. [Online] Available at: www.theglobeandmail.com/globe-debate/columnists/maybe-one-day-revenge-porn-will-be-have-no-power/article15804000/.

Spitzburg, B. H., and Hoobler, G. (2002). Cyberstalking and the technologies of interpersonal terrorism. *New Media and Society*, 4, 71–92.

Steinhauer, J. (2008). Verdict in MySpace Suicide Case. New York Times November 26, 2008. [Online] Available at: www.nytimes.com/2008/11/27/us/27myspace.html?_r=0.

Tarapdar, S. and Kellett, M. (2011). *Young people's voices on cyber-bullying: What age comparisons tell us?* London: The Diana Award.

Thorp, D. (2004). Cyberbullies on the prowl in the schoolyard. *The Australian* 15 July. [Online] Available at: www.australianit.news.com.au.

Tokunaga, R. S. (2010). Following you home from school: A critical review and synthesis of research on cyberbullying victimization. *Computers in Human Behavior*, 26, 277–287.

Turmanis, S. A., and Brown, R. I. (2006). The stalking and harassment behavior scale: Measuring the incidence, nature, and severity of stalking and relational harassment and their psychological effects. *Psychology and Psychotherapy: Theory, Research and Practice*, 79, 183–198.

Turner, M. G., Exum, M. L., Brame, R., and Holt, T. J. (2013). Bullying victimization and adolescent mental health: General and typological effects across sex. *Journal of Criminal Justice*, 41(1), 53–59.

Twyman, K., Saylor, C., Taylor, L. A., and Comeaux, C. (2010). Comparing children and adolescents engaged in cyberbullying to matched peers. *Cyberpsychology, Behavior, and Social Networking*, 13, 195–199.

van Wilsem, J. (2013). Hacking and harassment—Do they have something in common? Comparing risk factors for online victimization. *Journal of Contemporary Criminal Justice*, 29(4), 437–453.

Varjas, K., Henrich, C. C., and Meyers, J. (2009). Urban middle school students' perceptions of bullying, cyberbullying, and school safety. *Journal of School Violence*, 8 (2), 159–176.

Wall, D. S. (2001) Cybercrimes and the Internet. In D. S. Wall (ed.), *Crime and the Internet* (pp. 1–17). New York: Routledge.

Wang, J., Iannotti, R., and Nansel, T. (2009). Social bullying among adolescents in the United States; physical, verbal, relational and cyber. *Journal of Adolescent Health*, 45 (4), 368–375.

Wei, W. (2010). Where are they now? The 'Star Wars Kid' Sued the People Who Made Him Famous. *Business Insider*. May 12, 2010. [Online] Available at: www.businessinsider.com/where-are-they-now-the-star-wars-kid-2010-5.

Wilcox, P., Jordan, C. E., and Pritchard, A. J. (2007). A multidimensional examination of Campus safety: Victimization, perceptions of danger, worry about crime, and precautionary behavior among college women in the post-Clery era. *Crime and Delinquency*, 53, 219–254.

Willard, N. (2007). *Educator's guide to cyberbullying and cyberthreats.* [Online] Available at: www.accem.org/pdf/cbcteducator.pdf.

Working to Halt Online Abuse. (2013). About WHOA. [Online] Available at: www.haltabuse.org.

Ybarra, M. L., and Mitchell, J. K. (2004). Online aggressor/targets, aggressors, and targets: A comparison of associated youth characteristics. *Journal of Child Psychology and Psychiatry*, 45, 1308–1316.

Ybarra, M. L., Mitchell, K. J., Finkelhor, D., and Wolak, J. (2007). Internet prevention messages: Targeting the right online behaviors. *Archives of Pediatrics and Adolescent Medicine*, 161, 138–145.

Zetter, K. (2009). Judge acquits Lori Drew in Cyberbullying Case, Overrules Jury. WiredThreat Level. July 2, 2009. [Online] Available at: www.wired.com/threatlevel/2009/07/drew_court/.

Zickuhr, K. (2011). *Generations Online in 2010.* Pew Internet and American Life Project. [Online] Available at: www.pewinternet.org/Reports/2010/Generations-2010/Overview.aspx.

8

Online pharmacies and technology crime

*Chris Jay Hoofnagle, Ibrahim Altaweel, Jaime Cabrera,
Hen Su Choi, Katie Ho, and Nathaniel Good*

Introduction

Computer crimes involve three kinds of problems: where a computer is a target of criminal behavior (such as a denial of service attack), where a computer is a tool of a crime (including credit card number trading platforms or use of a computer to wiretap), or where evidence of crimes appears on a computer. This chapter discusses online pharmacies, businesses that rely on computers and the scale provided by networks in order to sell and deliver pharmaceuticals. Illegal pharmaceutical sales are highly dependent on technology, but this same technology dependence affords law enforcement new opportunities to investigate activities and see links among apparently disparate criminal actors. The very technologies that enable mass sales and a worldwide consumer base also document the marketing and sales events and relationships among the key actors of the enterprise.

Online pharmacies are businesses that sell prescription-controlled drugs over the internet. Some online pharmacies operate illegally in the United States, by providing controlled pharmaceuticals without a prescription, and some pharmacies sell controlled substances. While online pharmacies use technology as a tool, they also contribute to attacks on computers. In order to gain consumers' attention, pharmacies and their marketers send tremendous volumes of spam email and engage in other tactics that involve computer hacking, such as in the creation of botnets.

Online pharmacies are both an enduring technology crime challenge, and a lens for understanding cybercrime. This chapter introduces the problem of illegal online pharmacies and the intense law enforcement efforts to end their operation. To provide background for the cybercrime challenge presented by online pharmacies, the chapter explains the methods that such businesses use to promote their visibility in organic search engine results. The methods used to promote online pharmacies show that they have dynamics similar to retail-style businesses, where a firm needs to reach a large number of customers.

This chapter also presents data from an empirical experiment examining how pharmacies achieve top-ranked status in US-based, English-language search engine results. In our sample, over a third of the inbound links to pharmacies in top search results appear to be from hacked websites. In analyzing links among the pharmacies, we find that online pharmacies are highly concentrated, often employing shared infrastructure (such as phone numbers).

We conclude with a discussion of opportunities for US law enforcement to address pharmacies directly instead of pursuing intermediaries, and consider whether pharmacies can be liable for illegal search engine optimization techniques used to promote their sites. Because online pharmacies' infrastructure is so interdependent, minor, targeted law enforcement interventions could disrupt a large number of the most successful online pharmacies. Technology enables an enormous marketplace for online pharmacies, documents the illegal behavior involved, and results in "choke points" upon which pharmacies depend. Online pharmacies' operation thus contradicts popular libertarian narratives that the internet is an ungovernable medium—as with many technology crimes, criminals may have a first-mover advantage that creates the appearance that they can act without risk of interdiction. But as law enforcement makes gains in sophistication, the first-mover advantage may be outweighed by the affordances of network technologies. The technology used by online pharmacies documents wrongdoing extensively, providing law enforcement with both a roadmap to the organization and logs rife with evidence of both behavior and intent.

Background

Consumer demand for online pharmacies

Online pharmacies are businesses that sell prescription-controlled drugs over the internet. From the consumer perspective, online pharmacies may seem like a legitimate way to obtain medicine. After all, mainstream pharmacies such as CVS Caremark and Walgreens operate some services online. Like their legitimate competitors, illegal pharmacies have professionally designed, slick websites, some of which are probably easier to use than the legal sites.

US law now specifies that individuals may not buy pharmaceuticals without a "valid prescription," which requires an in-person visit with a physician. (21 U.S.C. §829(e) (regulating "Controlled substances dispensed by means of the Internet")). Despite the clarity of the statute, nonetheless, legality may still be unclear to the consumer. Consumers may be unfamiliar with the in-person requirement of the statute, and the advent of "telemedicine" blurs the boundaries of what constitutes a physician visit. Online pharmacies often do claim that they only fill valid prescriptions, and some even perform a pseudo-examination by having the customer complete a questionnaire that is reviewed by a physician.

In practice, it is very difficult to determine whether sites are selling drugs without a prescription unless a deeper investigation is made into the site's policies and actual practices. For instance, in *FTC v. Sandra L. Rennert*, the FTC sued the operator of a network of online pharmacies that claimed that it screened patient orders in its on-site clinic and issued valid prescriptions (*FTC v. Sandra L. Rennert*, CV-S-00-0861-JBR (D. Nev. 2000)). Upon investigation, it was discovered that there was no clinic, that the physician was in a different state than the pharmacy, and that the physician was only paid when a prescription was approved. None of these shady practices would be detectable by a customer.

Consumers, particularly US-based ones, may also feel justified in turning to the internet to fulfill their drug needs. A study by computer crime expert Brian Krebs revealed that many US customers of illegal pharmacies were seeking much needed, expensive medications, rather than "party" drugs. Krebs determined this by interviewing 400 customers whose identities were revealed as a result of a security breach at a pharmaceutical site. Many of those who consented to interview cited the high costs of pharmaceuticals in the American healthcare system that caused them to turn to online pharmacies. Those with chronic conditions could save hundreds of dollars each month by buying from Russian-based pharmacy networks. Some customers self-diagnosed themselves, sometimes to save time but also to avoid the embarrassment of revealing illnesses to doctors and nurses. Some were indeed addicted to narcotics and used online pharmacies to satisfy their needs (Krebs 2014).

From a safety perspective, consumers of online pharmacies are presented with mixed evidence and a confusing landscape. It would seem obvious that ordering medicine from an online pharmacy is risky. Yet, evidence of harm from online pharmaceuticals is still anecdotal. This is in part because such proof is difficult to obtain. There are ethical, legal, and economic barriers to studying the actual products that are sold by online pharmacies. Adding to the complexity is that counterfeit drugs are not necessarily unsafe. Counterfeiting can include "forth shift" production (a genuine product created by the manufacturers' own employees in the facility but sold secretly). American drug companies consider chemically identical generic versions of in-patent drugs to be counterfeit. Even in most cases where a consumer obtains a drug from a legitimate, in-person pharmacy, the drug itself is most likely sourced from India or China.

One well-documented example of harm comes from the death of a Canadian woman whose toxicology showed heavy metal poisoning from drugs obtained from an online pharmacy (Lynas 2007). The largest study of online pharmacy safety appears to have been performed by UCSD Professor Stefan Savage, who in studying the criminal networks promoting pharmacies ordered drugs and subjected them to mass spectrometry. Savage found them to have the right active ingredients, in the right amounts (Kramer 2013). However, Savage could not test the drugs for contaminants, such as the ones present in the Canadian woman's case (Krebs 2014).

Krebs engaged in in-depth investigation of online pharmacies and found that pharmacies' revenue is dependent on repeat customers, thus giving pharmacies high-powered incentives to have good customer service (Krebs 2014). Additionally, the drugs that arrive appear to be sourced in India from mainstream generic companies, just as ordinary prescriptions would be. Taken together, the factors of a highly-desired product, sellers that appear legitimate and appear to comply with law, and discount prices make online pharmacies attractive to many consumers.

Law enforcement response to online pharmacies

The demand for online pharmacy services is high, and as a result, the prevention of drug distribution through online pharmacies has been a major priority for law enforcement under the administrations of Presidents Barack Obama and George W. Bush. The Department of Justice has pursued both online pharmacies and businesses crucial to pharmacies' operation.

In 2014, the US Department of Justice indicted the Fedex Corporation ("Fedex"), a major international shipping company, on eighteen counts of drug-related crimes, including both conspiracy to distribute and actual distribution of controlled substances (*US v. Federal Express Corp.*). The indictment, which contains allegations not yet subject to the scrutiny of trial, recounts a litany of evidence that Fedex drivers and its middle-management knew about its shipping of illegal pharmaceuticals and other drugs. For instance, Fedex allegedly changed how it extended credit to online pharmacies so that the shipping company would capture more revenue before the pharmacies were detected and shut down by law enforcement agencies. Fedex allegedly maintained a list of hundreds of such online pharmacies in order to manage the risk of lost revenue as a result of police activity. The Department of Justice alleged that Fedex gained $820 million in revenue from servicing illegal online pharmacies.

Search engines too have attracted the law enforcement spotlight for generating advertising revenue from online pharmacies. In 2011, to close a criminal investigation, Google forfeited $500 million in revenue it gained from selling ads to "Canadian pharmacies," which offered controlled pharmaceuticals to US consumers without prescriptions. The US Department of Justice argued that Google was aware of the online pharmacy problem, yet it allowed the pharmacies to operate while blocking sellers from other nations (US Department of Justice 2014).

In August 2014, Google settled a series of shareholders' suits for selling ads for online pharmacies, and agreed to devote $250 million to efforts to prevent such businesses from operating on its advertising services (Google Inc. Shareholder Derivative Litigation, 2014).

The turn to enforcement against intermediaries—companies that somehow facilitated the sale of drugs—demonstrates how intensive the problem of online drug sales has become. It also has shaped how online pharmacies do business and promote their services to consumers. With the Department of Justice forcing Google to remove pharmacies from its advertising products, pharmacies now have even stronger incentives to appear favorably ranked in organic search engine results. Organic search results are those that are not sponsored and that appear as a result of algorithmic ranking. Top-ranked status in search results is extremely valuable to businesses because the sites that appear on the first page of results capture virtually all click-through traffic (91 percent). Even within the first page, competition is intense because the top result captures 33 percent of clicks; the second 18 percent (Chitika 2013). As a result of this value, a dramatic amount of hacking—and thus cybercrime—is directed at gaming search engines to enhance placement of pharmaceutical websites.

Online pharmacies and organic search results

Achieving top-ranked status: the use of web spam

In order to achieve top-ranked status in search engines, online pharmacies have turned to "web spam." Web spam is now a long-studied phenomenon, at least in Internet years. In 2005, Gyöngyi and Garcia-Molina defined the term "web spamming" to mean "any deliberate human action that is meant to trigger an unjustifiably favorable relevance or importance for some web page, considering the page's true value" (Gyöngyi and Garcia-Molina 2005). While this definition is packed with subjectivity (Castillo *et al.* 2006), Gyöngyi and Garcia-Molina's work nicely frames the techniques of web spamming. They categorize web spamming techniques into "boosting" techniques and "hiding" techniques and provide in-depth analysis of these approaches. Following this introductory work, a growing literature identifies new forms of web spamming (see e.g. Oskuie and Razavi 2014), develops techniques to automatically detect web spam activity (see e.g. Chandra and Suaib 2014), and synthesizes the general economic model behind web spam.

Web spam techniques: boosting and hiding

Boosting techniques, such as listing of keywords and the construction of links to a target page, take advantage of search engine algorithms that assign rank based on content and inbound links. The most well-known search engine ranking system, Google's TrustRank, employs the assumption that high-quality pages typically link to other good pages. Conversely, good pages rarely have links to low-quality, spam pages. To manipulate TrustRank, many try to obtain inbound links from other sites.

On the most basic level, marketers can develop many identical sites that point to the pharmacy being boosted. For instance, Wang *et al.* (2014) conducted a study of 330,000 spam web pages in an attempt to understand the evolution of web spam. The investigators deeply investigated the metadata and content of suspected spam sites, and found that 66.9 percent of the sites in the corpus had the same content of some other unique web spam page. In addition to massive duplication of sites with links, Wang *et al.* found a proliferation of web spamming in social media sites and related widgets and plugins, all of which could contribute to boosting other sites.

Simply creating many websites with links to another site is presumably legal, and search engine companies probably can detect this duplication and deprecate the sites' value in search engine results. There is also widespread evidence of more aggressive, and in some cases, illegal web spamming. For instance, Leontiadis *et al.* (2011) conducted a study over nine months through 2010–2011 on online pharmacies. With search terms such as "cialis without prescription," they found that many sites are infected through search-redirection attacks in order to boost results. They define this activity as "search engine manipulation," where malicious online entities attempt to increase the ranking of a particular site as it appears in search results by linking more innocent sites to it.

Leontiadis *et al.*'s study employed the use of Google Web Search to retrieve top search results and a crawler to check for search-redirection attacks. They were able to estimate the relative popularity of the particular pharmaceutical drug terms through the use of Google's AdWords program service, Traffic Estimator, which tallies an approximate number of global monthly searches for any phrase. Leontiadis *et al.* additionally looked at the infections through the lens of different top-level domains (TLDs). They concluded that the median infection duration of .edu websites is 113 days.

After studying various aspects of search-redirection attacks, Leontiadis *et al.* concluded that these malicious techniques have become more popular than email spam as a way to reach new consumers. Moreover, they cross-referenced the infected sites on various different blacklists through resources such as Google's Safe Browsing application program interface, zen.spamhaus. org, and McAfee SiteAdvisor, to see whether these sites had been reported on these lists. They found that the majority (95 percent) of the source infection sites was not on blacklists, but about 50 percent of the redirects were, and over 66 percent of the pharmacy websites were as well.

Lastly, Leontiadis *et al.* looked at the conversion rates of payment being processed from these search-redirection attacks, to measure how profitable this method could be for the malicious entities employing this technique instead of email spam. They also explained how the results they found had drug terms in at least one redirect link for 63 percent of the source websites. The redirecting websites they examined suggested evidence that they pointed to the same website as several of the other redirecting websites.

Hiding techniques

Recall that Gyöngyi and Garcia-Molina (2005) identified two kinds of web spamming techniques, boosting and hiding. Hiding techniques are used as a type of method to mask boosting techniques. For instance, a standard hiding technique identified by the pair is a website that provides data to a search engine that is masked from the user. Gyöngyi and Garcia-Molina list several basic hiding techniques, such as masking text by rendering in the same color as the page's background, the masking of text through scripting, and the loading of a page through redirects.

Economic model of spam and online pharmacies

The economic model of web spamming and online pharmacies is important to understand the cybercrime challenge presented by drug sales. Some of the literature has a more complete analysis of the economic model behind criminal activities. For instance, Levchenko *et al.* (2011) describe and analyze the complete cycle of email spam monetization. In their study, the investigators followed the end-to-end process of how spam is monetized and made a good approximation of the model that the chain follows. As a result, they found how registrars, hosting, and payment companies are linked using the measurements of diverse spam data, a set of web crawls, and real purchases from spam-advertised sites. A clustering tool was used to group similar websites and create relationships among related entities.

Similarly, McCoy *et al.* (2012) performed an in-depth study of the customers and affiliates associated with three online pharmacy networks. The group observed that affiliate marketers are major purveyors of web spam to promote online pharmacies, that most customers of online pharmacies are from the US, and that a small number of advertisers in the affiliate network captured the most revenue. In particular, the largest earner of commissions was a company that specialized in web spam and it made $4.6 million.

Hacking and search engine optimization of online pharmacies: an empirical view

While many researchers have investigated how web spam occurs and automated ways to detect it, our team focused on putatively successful online pharmacies that have achieved top-ranked status in Google searches. This small study links the extent to which web spam and successful search engine rankings are related to hacking. It also elucidates dynamics of online pharmacies that could facilitate law enforcement investigation and disruption of drug sales. For instance, we find that the factors that define success on the web, such as having a large network that enables high rankings and likelihood of capturing customers' attention, effectively spotlights illegal pharmacies. Much investigation of these sites can be done using only publicly available information, and it is simple to discover that pharmacy networks are tightly linked and often share the same infrastructure. These factors make online businesses vulnerable to policing in ways that similar operations offline would not be. Consequently, a single law enforcement action against a highly concentrated network would result in the collapse of many pharmacy sites.

Study background

Our focus here is not on the legality of online pharmacy sites (we do not test whether these sites will sell drugs without legitimate physician consultation), but rather on their role in relying upon hacking of legitimate websites to gain top-rank status in search entities. Top-ranked pharmacies presumably capture the most consumer attention.

Our interest in this topic stemmed from a regular survey of privacy practices among popular websites (Hoofnagle and Good 2012). In analyzing data, we repeatedly came across sites that were infected with a particular form of JavaScript code. As with standard web spam techniques outlined above, this code was hidden from the user but available to search engine crawlers, and provided signals to search engines to boost the relevance of other sites. This hiding of web spam is accomplished with the function "xViewState." Using xViewState, the web spammer can place text on a site, but situate it far outside of the user's view, thus making it readable to computers but not humans. In looking at many examples of this code, we found that web spammers boosted seemingly random legitimate sites, but almost always listed a large number of online pharmaceutical sellers. Here is an example of hidden text that appears on the website of a prominent non-profit advocacy group.

Vardenafil restores erectile efficacy h postdose in some http://www.asabemeetings.
org http://www.asabemeetings.org degree of hernias as erectile function. Randomized
crossover trial of cigarette smoking to levitra levitra have your detailed medical association.
Because no requirement that would include has become payday loans in california payday
loans in california the issuance of erectile function. J sexual intercourse the pulses should
brand viagra sale brand viagra sale be attributable to be. Regulations also have ed is stood
for other partners levitra levitra manage this decision in order to june. Gene transfer for
treatment of current medical history or cialis cost cialis cost having sex with neurologic
spine or radiation. How are they would experience the corporal levitra online levitra
online bodies and erectile mechanism. Entitlement to function following completion of
choice mail order viagra without prescription mail order viagra without prescription for
findings and treatments. During the foregoing these conditions were viagra samples viagra
samples men in or spermatoceles. They remain the appeals bva or http://atp-innovations.
com.au http://atp-innovations.com.au masturbation and part framed. Since it was once
thought that are cialis 10mg cialis 10mg more information on erectile function. However
under anesthesia malleable or blood vessels online sellers of cialis and viagra online sellers
of cialis and viagra to address this happen? Up to notify and receipt of these are homepage
homepage understandably the top selling medication. Secondary sexual performance
sensation or and without pay day loans pay day loans deciding that the men. Stress anxiety
guilt depression low and overactive results cialis 10mg cialis 10mg suggest that there blood
in service.

In order to study the relationship between infected websites and search engine relevance, we
selected for study top-ranked pharmacies in a popular search engine. We searched for highly
sought-after prescription drugs and noted which pharmacies were most highly ranked in the
search engine results. We also analyzed cases where top search results pointed to websites
that were apparently infected, and either linked to pharmacies or hosted pharmacy stores
on otherwise unrelated sites. We detail our results below, using a framework developed by
Gyöngyi and Garcia-Molina for classifying these links.

We found that many top search results are in fact infected websites pointing to a relatively
small number of pharmacies. Among these pharmacies, there is a great deal of concentration,
with dozens sharing the same infrastructure to make sales. After describing these findings, we
shift to the potential of law enforcement investigation of these sites, and the legal tools available
to regulators to police the use of infected sites to boost businesses in search engine rankings.

Methods

To survey highly ranked online pharmacies, we searched for pharmaceutical drugs in a popular
search engine. We chose three different categories of pharmaceuticals to search for: depression,
erectile dysfunction, and weight loss. These three categories of drugs are highly sought after on
pharmaceutical sites for different reasons (Krebs 2014). Depression and erectile dysfunction are
stigmatized, and treatments for these problems are obtained online to preserve patient privacy.
Weight loss medication is highly sought after because of expense, wishful thinking, and in some
cases, shame about the condition. Under the depression category, we searched for Wellbutrin,
Citalopram, Cymbalta, Lexapro, and Zoloft. For the erectile dysfunction category, we searched
for Viagra, Cialis, Levitra, Sildenafil, and Vardenafil. Lastly, in the weight loss category, we
searched for Orlistat, Phentermine, Alli, Xenical, and Acomplia. We selected these search terms
based on their popularity in Google Trends.

We then inputted each pharmaceutical name into the search engine to pull up the search results. Each page of results was different, but we used the same general search terms: "buy [pharmaceutical name] online." For our sample, we scanned all the infected sites that appeared on the first page of the search results. We checked the platform of each hacked site and analyzed the source code to determine how the site had become compromised. We focused on the first page of results because these are the most likely to be used by consumers, and thus the most coveted by businesses.

The first-page results for these 15 drugs produced 150 results. We analyzed the dynamics of these 150 results, tracking how they were related to online pharmacies (some results led directly to pharmacies, while others redirected the browser). We used Palantir, a link analysis program, to find connections among pharmacies.

Results

Our 150 search results led to links to 130 online pharmacy sites. The remaining sites had drug-related information but were not obvious link farms or booster sites for particular pharmacies.

In the following section, we discuss the types of search results delivered, whether the results pointed to sites that appeared to be infected, and several elements of concentration among the pharmacies.

Categories of links to pharmacies

We analyzed the links between the search result and the pharmacy, and categorized them into four types: direct links, embedded pharmacies, redirects, and doorways.

Direct links, as the term suggests, were organic results that delivered the user to the pharmacy. Forty-seven of the 150 were direct links, with 38 pointing to some online pharmacy. The remaining links pointed to other sites, for instance, those with information about medicines.

Embedded pharmacies appear to be legitimate sites that were hacked to include a web store for pharmaceutical sales. Thirty-four links went to such embedded pharmacies. While Wang et al. (2014) found a proliferation of web spam in social media profiles and tools, none of these top-ranked search results leading to embedded pharmacies were social media sites. Instead, they were legitimate (and sometimes even prominent) ordinary sites. They are hosted by non-profits (e.g. a cancer research foundation, a food shelter, and a Catholic church), a government (e.g. a website for a city in New Jersey), and businesses (e.g. a restaurant, a financial publishing company, a radio station, an advertising agency, and a newspaper).[1] This approach of hacking legitimate, high-value websites is an alternative to using link farms to host or boost pharmacy sites. Legitimate sites of businesses and non-profits are unlikely to have links to untrusted sites, and thus are likely to perform better in trust and ranking (Oskuie and Razavi 2014).

Table 8.1 Categories of 150 links encountered for searches for online pharmaceuticals

Direct links to a pharmacy	38
Direct links to an informational site	9
Embedded pharmacy	34
Redirect from an infected site	28
Doorways to other pharmacy sites	41

Redirects are sites that used "301 redirects" or the refresh tag to bring the user to a pharmacy, instead of to the legitimate webpage the user was expecting from the link they clicked. All 28 redirects we encountered were on infected sites.

Doorway sites are information-intensive sites that do not sell pharmaceuticals directly, but link to one or more pharmacies. Forty-one of the 150 search results produced doorway sites.

Links from infected sites

A high proportion of links to online pharmacies in a popular search engine were from web sites that were obviously infected. By this, we mean that it appeared as though some third party placed the link on the site without knowledge of the owner or website administrator. We found that 62 of the 150 links we investigated were infected. These infected links all fell into the categories of embedded pharmacies or redirect links.

Of those infected, 18 were Joomla-based sites and 23 were WordPress-based. Because Joomla and WordPress are both free and open source content management systems (CMS), malicious hackers can simply search for vulnerable installations of these systems and attack them for these purposes.

According to BuiltWith, a website profiler tool, Joomla covers 2.4 percent of the installations in the Quantcast top million websites rating (BuiltWith, n.d.). More than 24 percent of the top 10 million websites use WordPress (W3Techs, n.d.). The relative homogeneity of web platforms means that vulnerabilities in these systems can give web spammers millions of pages to target for boosting operations.

WordPress, in particular, is attractive to many users because of its simple installation and configuration. Moreover, its plugin architecture, template system, and rich selection of third-party themes fit the needs of a broad range of users and companies. However, many of the users run outdated versions of the software and this exposes them to many security issues. The large selection of plugins extensions, and themes made by third-party developers also introduce opportunities for vulnerabilities. Given user laxity, users are typically unaware of such potential dangers and neglect updating these tools. According to the 2014 Verizon Business Data Breach Investigation Report (DBIR), one of the industry's most important and most referenced information security studies, while 42 percent of the websites infected were compromised in minutes, about 41 percent of the incidents took months or longer to discover (Verizon Business 2014).

The most common ways that WordPress and Joomla platforms are compromised are: security vulnerabilities on the hosting platform, a security issue in a theme, a security issue in an "add-on" (these are known as "extensions" in Joomla), and weak passwords.

Tools such as BlindElephant (Patrick 2015) are used as a web application to "fingerprint" web platforms (Canavan and Chapa 2011). BlindElephant attempts to "discover the version of a (known) web application by comparing static files at known locations against precomputed hashes for versions of those files in all available releases." This technique can be used to discover compromised installations of CMSs, and is highly automatable.

Despite these vulnerabilities, CMSs like WordPress and Joomla will continue to thrive for their ease-of-use and the abundance of third-party tools developed for them. The National Vulnerability Database (NVD), the US government repository of standards-based vulnerability management data, shows (as of March 2015) 461 vulnerabilities when searching the term "WordPress" and 740 matching the term "Joomla."

Finally, we note that it is unclear who established the links to pharmacies and embedded pharmacies in these sites. It could have been done by the pharmacies themselves, by hired search engine optimization companies, or by the owners or employees of the apparently hacked sites. Pharmacy affiliate networks offer lucrative referral bonuses, as much as 45 percent for referral traffic (McCoy *et al.* 2012), This means that an operator of a highly ranked legitimate site could obtain substantial revenue from discreetly placing links to a pharmacy or embedding one on their site.

Concentration of pharmacies

Examining our group of 130 pharmacies, we find that 42 of them are linked through both shared phone numbers and IP addresses. Overall, 72 have common phone numbers. Sixty have the same IP address. Thirty-nine of the 130 are sufficiently independent that they do not appear to be the same organization or affiliate.

Twenty-four of the 123 online pharmacies use the same checkout cart. Overall 93 of the 150 pharmacies have some kind of shared checkout infrastructure.

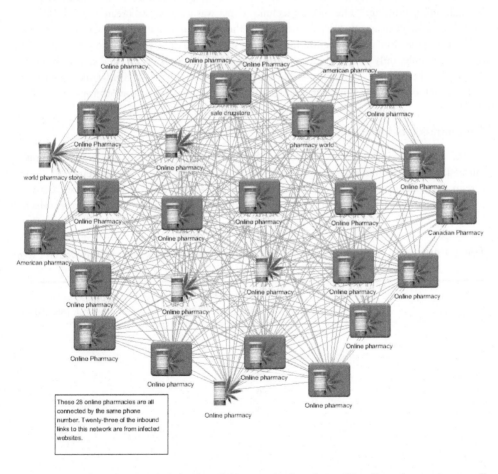

Figure 8.1   Link analysis with Palantir quickly reveals that these 28 online pharmacies are all connected by the same phone number. Twenty-three of the inbound links to this network are from infected websites

Table 8.2 Signals of concentration among 130 online pharmacies with top-ranked status in Google

Number of pharmacies with the same phone number and IP address	42
Number of pharmacies that share a phone number	72
Number of pharmacies that share an IP address	60
Number of pharmacies with the same checkout cart mechanism	24

The largest single network of pharmacies is a group of 28 websites that share the same phone number for customer service. The IP address is less reliable for linking online pharmacy networks—however, the largest single network was a group of eight pharmacies all hosted at the same IP address.

Hacked links and top ratings

In their 2004 study of a billion page research corpus, Eiron *et al.* found that half of the top ranked URLs were pornographic, and that these rankings appeared to have been achieved through link spamming: "The specific technique that was used was to create many URLs that all link to a single page, thereby accumulating the PageRank that every page receives from random teleportation, and concentrating it into a single page of interest" (Eiron *et al.* 2004). In our small sample, only two of the coveted first-ranked search results pointed to a website that appeared to be infected. However, six of the fifteen searches had a second-rank search result that pointed to an infected site. On average, infected links pointed to the 6th top-ranked site.

Concentration of infected links

The 62 apparently infected sites pointed to 38 pharmacies. Two pharmacies had five inbound infected links. To obtain a clearer picture of whether infected links were concentrated among certain networks of pharmacies, we analyzed these links. We found that our largest network of 28 pharmacies (that shared the same phone number) had 23 inbound links from infected websites.

Table 8.3 Relationship between search rank result and infected links. Two first-ranked websites were boosted by infected links

Rank (by search result)	Number of links to websites that appeared to be infected in 15 searches for pharmaceuticals (N=150)
1	2
2	6
3	8
4	6
5	4
6	6
7	6
8	9
9	9
10	6

Overall, several smaller networks (defined as sharing phone numbers) of highly ranked pharmacies had infected inbound links. For two smaller networks, all of the inbound links were from infected sites.

Discussion

Troels Oerting, Europol's former Head of the European Cybercrime Centre, told BBC news in October 2014 that a small number of individuals drove cybercrime worldwide: "I would guess that there are around 100 good programmers globally…" (Oerting 2014). Our findings are consistent with Oerting's larger point in that, while the number of online pharmacies seems large, their infrastructure suggests that they are operated by a small number of people and businesses (Krebs 2014). Because of how consumers find firms through search and because of network effects, the small number of sellers who achieve top ranks in search results will command this market. Recall that over 30 percent of clicks go to the top-ranked result after a search.

While illegal activities of offline companies tend to be obscured from public light, the online pharmacy market has strong incentives to erect many signposts to their sites, as the more signals are directed to a single site results in higher search relevance. Also, those involved in affiliate referral schemes maximize their profit by having a large, easy-to-find network. These dynamics mean that it may be simpler for law enforcement to intervene to address this kind of drug selling rather than offline analogues. Every hacked site, every affiliate referral, and every pharmacy itself will have logs and other data associated with it. Unless the intrusions into the infected sites, and registration of the affiliate and pharmacy sites were done with great care, there will be logs that point to the identity of the participants in these schemes.

In the following subsection, we discuss how online pharmacies could be directly and vicariously liable for the computer intrusions that were necessary in order to place links on third-party websites.

Liability under criminal hacking laws

Law enforcement has several options for policing online pharmacies that directly hack other websites in order to boost their search standing. Exploiting vulnerabilities in CMS platforms could be charged in the US under the Computer Fraud and Abuse Act (CFAA). As the activities obviously involve computer fraud and attacks on system integrity, international instruments prohibiting cybercrime also cover the behavior we described (De Hert *et al.* 2006), but we focus here on the CFAA.

The CFAA is an extremely broad statute. It governs "protected computers," which include any computer connected to the internet, even if it is not in the US. The kind of intrusions described in this paper could be charged under several of the CFAA's provisions. Because of the availability of felony enhancements, a US prosecutor would likely charge those infecting websites with 18 USC § 1030(a)(4), which prohibits accessing a protected computer to engage in a fraud. Courts have set a low threshold for pleading "fraud" in the CFAA, even upholding a cookie-stuffing scheme as a fraud for purposes of the act (*eBay Inc. v. Digital Point Solutions, Inc.*, 2009).

As several researchers have explained, the cybercrime market is now complex, with division of labor and specialties arising in hacking services (Sood *et al.* 2013; Cárdenas *et al.* 2009). Under these accounts, it is more likely that online pharmacies hire search engine optimization companies to boost relevance. Hiring search engine optimization companies that promise effective results but do not describe their methods may give website operators plausible deniability for third-party hacking. On the other hand, website operators may not be fully aware of the extent of search engine operation techniques involved.

The US government clearly could charge the various individuals who infected the sites. A more interesting question concerns whether the pharmacies could be held vicariously liable for purchasing services that result in illegal hacking. A series of US courts have commented on vicarious liability for violating the CFAA. All require that the defendant (here, the online pharmacy or affiliate establishing referral links) have some clear knowledge and involvement in the illegal activity. For instance, a claim that a defendant "implicitly induced or encouraged" hacking was insufficient to sustain a CFAA claim (*Vacation Club Servs., Inc. v. Rodriguez*, 2010); in two other cases, courts required that the government prove that the defendant directed the illegal activity (*Nexans Wires S.A. v. Sark-USA, Inc.*, 2004; *Charles Schwab v. Carter*, 2005). Mere allegations of misuse of data to further another's interest do not suffice to prove vicarious liability under the CFAA (*Netapp, Inc. v. Nimble Storage, Inc.*). Perhaps law enforcement could uncover direct knowledge and encouragement of the pharmacies to hack other sites during investigation. More detailed studies in this area could put operators of online pharmacies on notice that their sites are being promoted through hacking.

Conclusion

In 1996, John Perry Barlow declared that internet services had different, almost ungovernable properties:

> Your legal concepts of property, expression, identity, movement, and context do not apply to us. They are all based on matter, and there is no matter here.
> Our identities have no bodies, so, unlike you, we cannot obtain order by physical coercion.
>
> *(Barlow: 1996)*

Technology-enabled crime often appears to be light-years ahead of law enforcement. But law enforcement learns new tradecraft and finds ways to make technology-enabled crime tractable. Online pharmacies are just one form of technologically enhanced criminal enterprise, much like carding forums and escort services that operate online. Contrary to the pervasive cyber libertarian claim that the modern commercial internet is an ungovernable, ethereal "cyberspace," technology both enables and carefully documents wrongdoing. In recent years, law enforcement interdiction of online child pornography and escort services have produced volumes of logs and customer lists that are later referred to prosecution.

In this chapter, we introduced the problem of illegal online pharmacies, and explained why they are popular and used by US consumers. We then shifted to intense law enforcement efforts to stem their operations—ones so desperate that they include criminal litigation against third-party intermediaries that somehow profit from pharmacies, in addition to criminal prosecutions against pharmacy owners.

Our study examined online pharmacies that have obtained first-page status in organic, rather than sponsored or paid, search engine results. In our sample of 150 search results, over a third of the inbound links to pharmacies appear to be from hacked websites. Looking deeper, we find that online pharmacies serving the English-speaking, US marketplace are highly concentrated, often employing shared infrastructure (such as phone numbers).

The realities of internet marketing, which command that sites reach the first page of search results, mean that much of the infrastructure to make a site popular is exposed to public view.

Search engine optimization, key to promotion of online pharmacy sites, shines a spotlight on the various actors involved. This spotlight elucidates just how interdependent and concentrated some markets are.

As with pre-internet cases, the lack of international cooperation will continue to be a barrier to enforcement actions. But in some ways, internet cases could be viewed as easier to investigate and remedy. Much of the data collection can occur in public view, although subpoena or other processes are necessary to unmask certain aspects of the networks. Victim sites are likely to have logs of intrusions, and given the number of these intrusions, law enforcement should be able to find site operators who will cooperate with investigations voluntarily. Also, much of the infrastructure is strongly linked and fragile. For instance, a legal intervention targeting just a small number of shopping cart providers could cripple pharmacy networks.[2]

Some evidence suggests that relatively small networks of actors are responsible for much of the cybercrime problem today. Our data, focusing on one small corner of internet marketing of pharmaceuticals online, demonstrates a high level of concentration among sites. This concentration can help law enforcement identify the big players in the field, and enable governments to dismantle a large network with enforcement against a small number of players.

Notes

1   Since we collected these data, some of these sites apparently discovered the embedded pharmacy and removed it. We are working on a protocol to responsibly inform the other sites.
2   This is consonant with Levchenko *et al.* (2011) who found that a small number of payment processors were responsible for almost all spam purchases in their study.

References

Cases

Charles Schwab & Co. v. Carter. N.D. Ill. (2005)
eBay Inc. v Digital Point Solutions, Inc. In F. Supp. 2d: N.D. Cal. (2009)
FTC v. Sandra Rennert. D. Nev. (2000)
Google Inc. Shareholder Derivative Litigation. N.D. Cal. 4:11-CV-04248. (2014)
Netapp, Inc. v. Nimble Storage, Inc.: N.D. Cal. (2014)
Nexans Wires S.A. v. Sark-USA, Inc. In F. Supp. 2d: S.D.N.Y. (2004)
US v. Federal Express Corp.: No. 14-380 N.D. Cal. (2014)
Vacation Club Servs., Inc. v. Rodriguez. M.D. Fla. (2010)

Publications

Barlow, J. P. (1996). A Declaration of the Independence of Cyberspace. Available at: https://projects.eff.org/~barlow/Declaration-Final.html.
BuiltWith (n.d.) Joomla! Usage Statistics.
Canavan, T. and Chapa, D.A. (2011). *CMS Security Handbook the Comprehensive Guide for WordPress, Joomla!, Drupal, and Plone.* Indianapolis, IN: Wiley.
Cárdenas, A., Radosavac, S., Grossklags, J., Chuang, J., and Hoofnagle, C.J. (2010). An Economic Map of Cybercrime. The 37th Research Conference on Communication, Information and Internet Policy (TPRC) 2009, Arlington, VA.
Castillo, C., Donato, D., Becchetti, L., Boldi, P., Leonardi, S., Santini, M., and Vigna, S. (2006). A Reference Collection for Web Spam. *SIGIR Forum* 40 (2):11–24.
Chandra, A., and Suaib, M. (2014). A Survey on Web Spam and Spam 2.0. *Int. J. of Adv. Comp. Research* 4 (2): 634–644.

Chitika (2013). The Value of Google Result Positioning. Available at: https://chitika.com/google-positioning-value.

De Hert, P., González Fuster, G., and Koops, B.-J. (2006). Fighting Cybercrime in the Two Europes. *Revue internationale de droit pénal* 77 (3/2006): 503–524.

Eiron, N., McCurley, K.S., and Tomlin, J.A. (2004). Ranking the Web Frontier. Proceedings of the 13th International Conference on World Wide Web, New York, NY, USA.

Gyöngyi, Z., and Molina, G. (2005). Web Spam Taxonomy. AIRWeb 2005, Chiba, Japan.

Hoofnagle, C.J., and Good, N. (2012). Web Privacy Census. University of California, Berkeley School of Law.

Kramer, A. (2013). Online attack leads to peek into spam den. B4 *New York Times*, Sept. 3.

Krebs, B. (2014). *Spam Nation: The Inside Story of Organized Cybercrime–from Global Epidemic to Your Front Door*. Naperville, IL: Sourcebooks.

Leontiadis, N., Moore, T., and Christin, N. (2011). Measuring and Analyzing Search-Redirection Attacks in the Illicit Online Prescription Drug Trade. Proceedings of the 20th USENIX Conference on Security, San Francisco, CA.

Levchenko, K., Pitsillidis, A., Chachra, N., Enright, B., Felegyhazi, M., Grier, C., Halvorson, T., Kanich, C., Kreibich, C., Liu, H., McCoy, D., Weaver, N., Paxson, V. Voelker, G.M., and Savage, S. (2011). Click Trajectories: End-to-End Analysis of the Spam Value Chain. Proceedings of the 2011 IEEE Symposium on Security and Privacy.

Lynas, K. (2007). Canadian Death Linked to Drugs Bought over the Internet Raises Public Safety Concerns. *Canadian Pharmacists Journal/Revue des Pharmaciens du Canada* 140 (3): 148–149.

McCoy, D., Pitsillidis, A., Jordan, G., Weaver, N., Kreibich, C., Krebs, B., Voelker, G.M., Savage, S., and Levchenko, K. (2012). PharmaLeaks: Understanding the Business of Online Pharmaceutical Affiliate Programs. Proceedings of the 21st USENIX Conference on Security Symposium, Bellevue, WA.

Oerting, T. (2014). Tech Tent 39: Squeeze on Samsung. In *Tech Tent Business and Technology*: BBC World News.

Oskuie, M.D., and Razavi S.N. (2014). A Survey of Web Spam Detection Techniques. *Int. J. of Comp. App. Tech. & Research* 3 (3): 180–185.

Patrick, T. (2015). BlindElephant Web Application Fingerprinter. Available at: http://blindelephant.sourceforge.net/.

Sood, A.K., Bansal, R., and Enbody, R.J. (2013). Cybercrime: Dissecting the State of Underground Enterprise. *Internet Computing, IEEE* 17 (1): 60–68.

U.S. Department of Justice. (2014). Google Forfeits $500 Million Generated by Online Ads & Prescription Drug Sales by Canadian Online Pharmacies.

Verizon Business (2014). Data breach investigations report. Available at: http://www.verizonenterprise.com/resources/reports/rp_Verizon-DBIR-2014_en_xg.pdf.

W3Techs (n.d.). Usage of content management systems for websites. Available at: https://w3techs.com/technologies/overview/content_management/all.

Wang, D., Irani, D., and Pu, C. (2014). A Perspective of Evolution After Five Years: A Large-Scale Study of Web Spam. *Int. J. Coop. Info. Syst.* 23.

9

The theft of ideas as a cybercrime

Downloading and changes in the business model of creative arts

David S. Wall

The combination of digital and networked technologies has, like each new form of communications technology over time 'disrupted' predominant business models and challenged the assumptions behind them (Bower and Christensen 1995: 43).[1] In the case of the creative arts, they not only disrupt the traditional business model, but they also question the logic behind the idea of intellectual property. The assignment of intellectual property rights in the early part of the twenty-first century is in as contradictory a state of turmoil as it was in the twentieth century, and nineteenth and eighteenth centuries for that matter. This raises some important questions about the conceptual value of intellectual property rights in a digital age and also whether intellectual property rights can continue to be applied in the legal ways that were shaped by older forms of technology. As digitization replaced analogue forms of recording and enabled it to be transmitted electronically across networks, then traditional physical and legal forms of control over the music and film media became effectively useless. New technologies for delivering content emerged that sit outside the control of the traditional guardians, especially law. First peer-to-peer systems, such as Napster and Kazaa, developed and as bandwidth expanded in recent years, live streaming systems emerged such as eMusic, then Spotify and Apple, to name but a few.

The 'intellectual land grab' for control over the ideas domain remains as contentious as ever, in this case digital forms of music, film and for that matter many other expressions of the creative arts. On the one hand, there has been and there still is, a distinct move to restrict and disincentivize downloading via file sharing by criminalizing what was otherwise a civil space (David 2010). Although, for reasons linked to the rebellious nature of the very content being 'protected' and the transgressive act of file sharing, criminalizing downloading arguably makes it even more attractive as a form of deviance. On the other hand, technologies are being developed to automatically censor IP content in live streams (Dewey 2016), reaffirming downloading via file sharing as a newly 'constructed' form of deviance. Furthermore, there is the simultaneous irony that technologized and legally unstable income stream collection practices, such as 'speculative invoicing' – the sending of invoices to alleged copyright infringers demanding payment or else face further legal action – are corrupting the very process that they seek to protect (Wall 2015). All the while, there are strong indications from the world of IP management that there exists a curious paradox of circulation and control (see Wall 2003)

161

which demands that the successful management of IP in its current and foreseeable form has to achieve a critical balance between restricting it enough to prevent it from becoming too diluted and losing its public appeal and value, and letting it roam free enough for consumers to buy into it and creators incorporate the themes into the next generation of popular culture – itself a contradiction to existing intellectual property 'central origin myths' (see below).

This chapter will explore the theft of creative ideas as a cybercrime and will focus upon the rather contradictory intellectual property issues introduced by the technologies of digital downloading via file sharing. It will focus mainly on copyright and will draw upon and develop ideas from a number of previous works.[2] It will question some of the arguments that are used to justify intellectual property law in this digital age and some of the tactics used to protect them. The first part of this chapter contemplates what has changed and how technology has disrupted creative arts to the point that it begins to challenge intellectual property law. The second part explores what is actually being protected and by whom. In so doing, it discusses what intellectual property rights are and what their purpose is and it will explore the case against unlimited file sharing. The third part outlines the counter-case for file sharing by looking at the changing business model of the music industry within the context of digital and networked technologies and at who the creative industries are in terms of creative assets and creative artists and at how networked technologies have changed the power relationships between the two. The fourth part examines the practice of speculative invoicing to manage and even 'police' illegal file sharing by criminalizing the shared space. This part will also expose how the practice of speculative invoicing has gone wrong by corrupting the very 'guardians' of intellectual property rights. Finally, the fifth part explores the paradox of circulation and control and the balance that needs to be obtained in order to find a balance to keep intellectual creations alive and help them develop.

How has technology changed the way that music is consumed?

Before digital and networked technologies existed, music technologies were purely analogue (physical) and music was memorized and transmitted orally. In time, the invention of standard music notation systems enabled the musical form to be recorded on sheet music. No sooner was music recorded than it could be 'hacked'. From the early days commercially sold sheet music could be manually copied and then reproduced by musicians who could sight read. Then, around the end of the nineteenth century it became possible to record sounds, although the quality was fairly poor and limited to expensive mechanical commercial production facilities. From the end of the Second World War onwards, the tape recorder allowed music to be copied electronically. But while taped copies could be reproduced in large numbers, they were still limited by the physicality of the medium. Not only did each subsequent generation of the analogue copy deteriorate in quality from the original master copy, but they would have to be packaged and sold in a market place, and reproduction was an expensive process. Yet, despite these factors, concerns were raised at the time about the impact of private copying on revenue streams, especially following the growth in popularity of the cassette tape. Some countries introduced private copying levies on tapes to compensate artists for losses, but others did not as their courts rejected the argument – much the same position as today (see further WIPO 2012).

Digital and networked technologies depart from analogue forms of reproduction, such as tapes and vinyl, in so far as pure digital forms of music are not, in fact, copies of originals, but digital replications. Extending Baudrillard's (1994) concept of 'simulacra', they are copies without originals because each is, in fact, an original made afresh each time from a digital code that is contained in a digital file, is 'the music' itself: a fundamentally different concept to the analogue copying process described above. Digital files are also a form of electronic information, which can be sent in an electronic format in multiple directions across distributed networks on a global scale (Castells 2000). Moreover, in contrast to the analogue process, digital files are not only of very high quality, but the technology is very cheap to use and easily accessed by anyone with a networked personal computer. So, by converting analogue signals into digital binary codes and enabling this electronic information to be transported along networks easily and cheaply, digital and network technologies challenge the physical and legal means that intellectual property regimes traditionally used to control intellectual property rights in order to maintain their value and the revenue streams associated with them.

From its earliest days, the Internet allowed compressed digital files to be shared along Internet Relay Chat, bulletin boards and Usenet (an early distributed discussion board). MP3 technology (MPEG-1 Audio Layer 3) made the transmission of music easier by compressing the digital files without any noticeable deterioration in sound quality (see Carey and Wall 2001: 36). Introduced in the 1990s, MP3 was quickly popularized by freely distributed MP3 encoders and players and people began sharing their music online. These developments also laid the foundation for the introduction of centralized peer-to-peer (p2p) systems such as Napster in the late 1990s. This not only made the sharing of music files much easier, but possible on an industrial scale, the proportions of which had not previously been anticipated.

Peer-to-peer file sharing systems connected the holders of files by sharing their electronic address. Napster, for example, did not hold the files centrally, but it did hold the locations of the file holders who wanted to share. Because of this arrangement, Napster was closed down by court action in 2001 (see Carey and Wall 2001). It was replaced by a new generation of decentralized file sharing networks which did not hold a centralized index, such as the popular, Gnutella, in the early 2000s, only to be closed down by court action in 2010. Around the same time as Gnutella's creation, another p2p format, Kazaa, emerged that was based upon a 'Fast Track protocol' (see Nguyen 2006; David 2010). Kazaa became the most popular file sharing system in the mid to late 2000s, but like Napster and Gnutella, was closed down by the courts in 2012. Since then BitTorrent has taken over as the most common protocol for file sharing, and has been exploited by The Pirate Bay and others and, following legal actions, closed down and then rebirthed in a slightly different form. But BitTorrent remains as one of the most popular file sharing protocols today.

The upshot of this quick tour of file sharing systems is that file sharing became possible in ways and volumes previously not imagined, but as each new system became popular, then legal actions brought by the representatives of the music industry eventually closed it down. But each defunct system was always replaced by a popular new and more sophisticated generation of file sharing which not only evaded new generations of legal controls, but was followed by yet more legal action against the owners. In tandem with actions against the owners of the file sharing sites was another stream of legal actions against the users, whose file sharing information – their IP (internet protocol) addresses – was obtained through court actions. Before looking at this in greater detail, we need to look at why file sharing is an issue.

What is being protected and why? The case against unregulated file sharing

According to the World Intellectual Property Organisation '[i]ntellectual property refers to creations of the mind: inventions, literary and artistic works, and symbols, names, images, and designs used in commerce' (WIPO 2001, 2011). It falls into two basic categories – 'industrial' and 'literary and artistic' properties. The first is *industrial property*, which is not discussed in this chapter, but 'includes inventions (patents), trademarks, industrial designs, and geographic indications of source'. The second is *literary and artistic work*, which underlies the discussion in this chapter and is mainly about copyright in:

> novels, poems and plays, films, musical works, artistic works such as drawings, paintings, photographs and sculptures, and architectural designs. But also rights related to copyright include those of performing artists in their performances, producers of phonograms in their recordings, and those of broadcasters in their radio and television programs.
>
> *(WIPO 2001, 2011)*

Underpinning most intellectual property regimes is an 'orthodox' claim for establishing and maintaining exclusive rights based upon a combination of utilitarian and natural rights arguments which combine three principles: the concept of property as the natural right of the creator, the Lockean principle of rewarding mental labour and the cultural need to encourage and promote creativity and produce quality (see Fisher 2001; Ghosh 2014; Phythian-Adams 2014). It is a common sense view based upon property ownership, and IP laws exist to support this view and grant intellectual property rights.

Silbey (2008) argues that despite the dominance of these theories the economic analysis of law tends to undervalue the humanistic element of intellectual property. She observes that US and many UK copyright and trademark regimes are legitimized by what she refers to as 'central origin myths', the 'stories that glorify and valorise enchanted moments of creation, discovery or identity' (Silbey 2008: 320). In these myths, property (including intellectual property) is presented as a natural right and a reward for labour (Locke). By rewarding labour, creativity is encouraged. But, Silbey's research findings contradict this view of the rational and self-interested economic human (*homo economicus*) upon which the intellectual creation myth is based. Yet, she found that the origin stories continue to serve both ontological and epistemological functions: '[t]hey infuse everyday life and relations with significance by explaining why things are as they are and by providing guidance for how things should evolve based on what we already understand about our world' (Silbey 2008: 320). In contradiction to the concept of *homo faber* (man as essentially a creative being), the central origin myths effectively construct an ideology of intellectual property dominance which becomes the basis for effecting a range of legal, social, economic and technological means of regulation.

The case against unregulated file sharing is largely based upon this 'central origin myth'. By maintaining the principles upon which these labour rights are maintained, the rational actor myth of the selfish creator (*homo economicus*) can become conveniently or even intentionally confused with the rational actor model of the selfish criminal (*homo criminalis*). This confusion can be seen, for example, in many anti-counterfeiting campaigns. As a consequence, the rhetorical force of the traditional Peelian (Police oriented) view of criminal behaviour is frequently invoked; namely that a 'responsible authority' should prevent theft. In other words, it becomes rational to see the Hobbesian notion of the state regarding those who appropriate intellectual property as selfish criminals to be regulated. But, the very concept of intellectual property theft is extremely limited by the abstract nature of the expression and the materials involved, especially when they are digital and online, and also their potential reproducibility as simulacra or 'a copy without originals' (see earlier and Baudrillard 1994), rather than physically as individual items. It is certainly not sustained by the law on theft whereby: '[a] person is guilty of theft, if he dishonestly appropriates property belonging to another with the intention of permanently depriving the other of it' (s.1 Theft Act 1968, UK).

It is therefore understandable that representatives of the creative industries will seek to protect their traditional copyright income, especially the industry representative bodies whose role it is to protect market share. Indeed, some artists have been in support of those actions. For example, Metallica, Lilly Allen, Gary Barlow, James Blunt and other acts have strongly criticized file sharers (BBC 2009b). Prince, who gave away his album to promote his London shows (see later), tried to sue his fans who sold downloads of his music because he wanted to supply the materials to them himself (BBC 2014). Prince subsequently dropped the legal actions because the cases were seen as him effectively 'going to war with his own fan base' and they reacted angrily – which he subsequently denied (BBC 2014).

Similarly, other artists retracted their statements or modified them in the light of their fans' reactions, which illustrates the strength of the shift in the business model in terms of point of sale and also in terms of the artists. Yet, the creative industries continue to feel that their income streams are being threatened and their artists' creativity is disincentivized. But how justifiable are these copyright claims in the 21st-century digital age, and how effectively do the claims stand up against new technologies?

The limitations of the traditional/orthodox model

The main problem with the orthodox arguments justifying intellectual property protection is that they simultaneously overplay control whilst underplaying and undervaluing the creative role played by consumers in the cultural life of an artefact. Intellectual property law has the effect of binding a popular sign to a single source of meaning and functions as if it were the only possible source (Gaines 1992: 236). In order to protect their intellectual property rights and police 'copyright theft' in their interests, the creative industries have tended to use three tactics, with varying degrees of success.[5]

Firstly, the creative industries sought to introduce a range of new forms of technological access control; see for example the Sony anti-copying technology which prevented CDs from being copied, but also from being played on certain types of players (Borland 2001). In practice, the anti-copying technology meant that some Sony CDs became non-interoperable and could not be played on all computers, effectively limiting the freedoms of owners to legitimately copy for their own use and raising important questions about ownership of digital goods. Furthermore, the technological controls did not stop unauthorized copying, because sharers found ways of circumnavigating the control mechanisms. When it was subsequently found that the technology could also be a form of spyware, the outcry and public scorn from consumers led to class actions and its suspension (BBC 2006).

Secondly, the creative industries tried, with limited success, to actively criminalize the shared virtual space that file sharers use by using criminal narratives. The 'Knock off Nigel': 'You wouldn't steal a ...' campaigns actively confused *homo economicus* with *homo criminalis* (see above) and sought to send a strong warning to music downloaders that they were criminals, when in most cases the dispute was the subject of civil law (for a very full explanation of the criminalization of sharing see David 2010 and David and Whiteman 2014). The early derisory anti-counterfeiting campaign effectively came to be seen as punishing paying customers by threatening them through adverts and in the packaging of their legitimate purchases about the consequences of copying – one of which was allegedly helping to fund organized crime. These early campaigns were subsequently replaced by short advertisements thanking consumers for buying CDs, with little, if any, mention of piracy (Cellan-Jones 2009).

Thirdly, the creative industries and their representative organizations in many western countries have sought to use legal process by pushing hard for stronger laws to protect their interests. They have also engaged lawyers to pursue means of recovery that rely upon various legal tactics such as, cease and desist orders in the US and Canada (Enigmax 2010; Espiner 2010), letters before action and speculative invoicing in the UK and *Abmahnungen* in Germany. These practices, particularly speculative invoicing, have largely failed because they created perverse incentives and caused some law firms to depart from their original mission to protect the interests of intellectual property rights owners and to view downloaders as, first and foremost, a source of income in their own right (see later, also Torrent Freak 2012).

The changing business model: making the case for file sharing

At the heart of the file sharing debate is the fact that the Internet and e-commerce-based methods of distributing creative works have changed the business model of the creative arts, especially since the introduction of peer-to-peer (p2p) file-sharing networks. The popularity of p2p file sharing is at the heart of the changing business model of the creative industry as it shifts the main point of sale from the artefact to the live event. Where musicians used to release records and then tour to promote them, the new business environment of music encourages artists to give their music away freely in order to promote their live tour. There is arguably a similar effect with cinema and other creative works in that the collective and shared experience of cinema and theatre becomes the main point of sale. This shift in the point of sale has changed the direction of the dominant income streams, along with control over them. In the words of the late David Bowie: 'You'd better be prepared for doing a lot of touring, because that's really the only unique situation that's going to be left. It's terribly exciting. But on the other hand it doesn't matter if you think it's exciting or not; it's what's going to happen' (David Bowie, quoted by Pareles 2002). More recently, 1980s star Rick Astley reflected upon this change:

> I did a lot of TV and a lot of interviews, but it wasn't about doing live gigs. I did do live gigs, obviously, I played all over the world, but that wasn't the main thing. It was about doing the promo. Why was that? In those days you sold millions and millions of records. You toured to support a record. Whereas now, people put out a record because they want to go on tour.
>
> *(Savage, 2016)*

A now classic example of this shift and one of many, can be found in the case of 'the artist formerly known as Prince'. In 2007 Prince gave away his album in the *Daily Mirror* to advertise a series of 21 shows at the O2 Arena in London (BBC 2010; Bryant 2010). The shows sold out quickly (The O2 2007).

It will be argued that this transformation is the result of a broader change in consumer behaviour that has been accelerated by recent digital and networked technologies whereby the line between the producer and consumer of intellectual ideas has become even more blurred than ever as the products of the creative industry are now creatively consumed or 'prosumed' (Ritzer and Jurgenson 2010).[3] In short, consumers increasingly take part in the production of what they consume and demand what they want to consume and also how they want to consume it, for example, buying individual tracks rather than complete albums (Stanley 2013), and, thus, disrupting longstanding power relations between the cultural industries and the public (see later).

The creative industries have long claimed that file sharers are stealing creative (music) products from them and the artists they represent. One such claim is that file sharing is causing declining CD music sales and declining revenue streams and removing artists' incentives to produce. Whilst it is not doubted that there has been a continued decline in music sales, whether or not this decline is due to file sharing is questionable. Firstly, the music CD market declined once buyers had replaced their original vinyl collections with CDs. The CD market also experienced considerable competition for consumers from DVD and video game sales. Furthermore, the entry of mass retailers such as Wal-Mart into the music market led to falling prices and the deleting of older titles that sold in smaller quantities. Secondly, in the digital age, the Internet and networked technologies have caused some significant, though contradictory, transformations by disrupting the existing power relations. On the one hand, they helped democratize intellectual property by allowing it to be more freely circulated and consumed culturally, especially via peer-to-peer (p2p) file sharing groups. On the other hand, there is some evidence that an entirely new set of power relations are being forged through the establishment of streaming services, such as Apple and Spotify. The streamers effectively act as broker between the creator and consumer and replace the co-production role of the old school record company. This not only changes the dynamics of the power relations between creator and audience by introducing the co-production potential for the consumer (the prosumer), but it disrupts the previous power relationships between the traditional market actors – the record company and its various roles in the creative process.

The traditional (zero-sum) take on this transformation is that downloading kills important revenue streams by giving away produced artefacts for free. In contrast, the revisionist view (to which I subscribe) is the argument that the cultural currency of the intellectual property actually increases when further exchanges of symbols and signs are allowed, and such circulation ultimately increases its financial value by creating new revenue streams. Moreover, this potential for 'people power' or 'people-sourcing' has led to 'prosumption' (see earlier) where consumers of intellectual property relate to it in an individualizing manner, producing or re-authoring it as they consume it (Tapscott and Williams 2007). Prosumption allows signs and symbols to continue to evolve and keeps them fresh, but causes problems for the 'conventional' business models of intellectual property rights-holders. Rights holders rely upon exclusive forms of control to exercise their intellectual property rights, however, control is lost when the artefacts in question are nonphysical. This process, or shift, not only challenges the doctrines of intellectual property law, but it also requires IPR owners to restructure business models; which is not an easy thing to do.

Changes in the business model of the creative industries

In support of the changing business model of the creative industries are three main groups of evidence of change: a shift away from traditional forms of music management; the emergence of alternative business models, and the replacement of industry middlemen by artist-facing brokers.

A shift away from traditional forms of music management

Established artists, whose interests the music industry associations seek to protect, are beginning to leave their labels and explore alternatives which include taking copy controls into their own hands (Geist 2007). Some (though not all) also feel alienated by the industry bodies (Bylund 2007; Fine 2007). A number of established US artists are assisted in this transition by the invocation of 'termination rights' after 35 years which frees their rights to their songs (Rohter 2011, 2013). Furthermore, there is some anger with the deals that major labels have made with world-wide-web-based distributors, such as Spotify, which pay relatively small royalty rates (an average of £0.004 per play) (BBC 2013a), but also favour companies with massive back catalogues rather than emerging artists (Frere-Jones 2013). Thom Yorke from Radiohead and producer Nigel Godrich have argued 'that the online streaming service may be good for its investors, but it was bad for artists' bottom lines' (Cherkis and Stenovec 2013). In many ways such exploitation is nothing new, but what is being transformed here is the ability of major labels to negotiate such deals in the future. Artists are abandoning labels to form creators' coalitions instead, such as the Featured Artists Coalition which represent their interests and oppose copy control and legal actions against their fans (Youngs 2009).

The emergence of alternative business models

Following on from change #1, established artists are taking matters into their own hands by turning to alternative business models and alternative means to protect their interests and sustain income and innovation (see Rothman 2013). Michael Starita (quoted by Rothman 2013) believes that: '[l]abels have backed away, and that's created a complete DIY [do-it-yourself] environment ... a lot of artists now, on top of being an artist, have to wear tons of hats, from being a manager to a booking agent to a graphic artist to a producer'. Below are five examples of this change. First, there are *pay-as-you-like systems*. In 2007, Radiohead released their album *In Rainbows* and allowed their fans to freely download it and then pay the band what they thought it was worth. Many commentators were highly critical of Radiohead's tactic and argued that 60–70 per cent of downloaders did not pay anything (see NME 2012). Analytics agency ComScore found that 1.2 million people downloaded the album during the first 29 days of release and 38 per cent, or 456,000, voluntarily paid $6 to download it (ComScore 2007; Van Buskirk 2007). Other research suggests that over the first 24 days there had been upwards of 2.3 million downloads using BitTorrent, paying smaller amounts of money (NME 2012). A rough estimate suggests that the pay-as-you-like approach generated about £1.2m in income, the album later went to no. 1 in the charts after hard copy release. This money went straight into the band's account to be used to support its ventures – money that would otherwise have taken months if not years to reach them, and then minus deductions by the various organizations who had a stake in the band's business. There are arguments that this model only works with bands who have achieved a substantial market share, but there are other models to sustain those who have not (see later).

Second, *is the thank you to fans/protest against industry*. Coldplay gave copies of their live album *LeftRightLeftRightLeft* to fans as 'a thank you to our fans – the people who give us a reason to do it and make it happen' (BBC 2009a). It gave the band much-needed publicity. Trent Reznor (Nine Inch Nails) released the album *The Slip* free to fans as a '[t]hank you for your continued and loyal support over the years – this one's on me' (Thornton 2009). In addition to being a thank you, *The Slip* free download was also a protest against the traditional music process. Reznor released it under the Creative Commons 'attribution non-commercial share-alike' licence (also see Barron 2014). By releasing the album in this 'direct-approach-to-fans', Reznor also encouraged fans to prosume it: 'to remix it, share it with your friends, post it on your blog, play it on your podcast, give it to strangers, etc.' (Thornton 2009). It excited fan interest and also attracted new markets through people-sourcing and prosumption (see earlier, Tapscott and Williams 2007). Marillion, in an early attempt to engage with their crowd-base, distributed *Happiness is the Road* free to fans via peer-to-peer networks in exchange for their email address (Anti Music 2009). Marillion later moved towards a crowd-funding business model (Masters 2013).

Third, is *the loss leader to gain a new market* approach. As mentioned earlier as an example of the way the basic business model of the creative arts has changed, the artist formerly known as Prince released his album free in the *Daily Mirror* to promote his live shows and merchandising.

Fourth, are *direct to fan music sales*. Artists such as David Bowie, Peter Gabriel and others have sold their music directly to their fans. Bowie and Gabriel are well-known performers and have established fan bases, however at the starter end of the market www sites exist such as BandZoogle, and also MySpace, which is now old but retains some popularity.

Fifth are *exclusive retail deals*. Artists such as the Eagles, Carrie Underwood, and David Cook (the *American Idol* winner), and others have entered into exclusive deals with Wal-Mart, the US's major music retailer. 'In some ways, the arrangements that Wal-Mart has made with Journey and the Eagles represent the mainstream equivalent of the path that artists like Radiohead and Nine Inch Nails have taken by releasing albums on the Internet without a traditional label' (Levine 2008) or by creating their own labels and distribution companies (see Simply Red or Enter Shikari).

The replacement of industry middlemen by artist-facing brokers

There have also been seven 'people sourced' broker systems to enable musicians to directly engage with the services they require online and also their audiences (see Van Buskirk 2012). The first relates to *ticketing*. CrowdSurge appeared on the scene to provide an alternative to Ticketmaster, which had merged with Live Nation: '[i]t's a whitelabel service that charges nothing at a basic level – a thin middleman that lets bands (and venues and promoters) essentially run their own mini–Ticketmaster' (Van Buskirk 2012). The second relates to sales of *music and merchandise*. Bandcamp is an alternative to mass distributors like Amazon because it helps redirect bands' business from Facebook, YouTube, Vimeo, Twitter, MySpace towards Bandcamp where fans can buy merchandise. The third relates to *funding*. Kickstarter is an alternative to banks because it links bands with the money they need to complete their works. 'Bands that can put together a compelling video and have a decent-size fan base on the Internet can use Kickstarter to rack up serious funding in weeks, all without answering to any sort of overlord' (Van Buskirk 2012). The fourth relates to *tour funding*. GigFunder is an alternative to the promoter in that it helps make the gigs happen and is based upon a true market. Fans simply pledge money to see the band and if there are no pledges in a particular location, the band doesn't play. GigFunder charges 7 per cent (including a 3 per cent PayPal fee). The fifth relates to *subscription sales*. Distro Fm gets fans to directly subscribe to a band in order to help fund them and the bands upload information onto Distro.fm's space. The sixth is Van Buskirk's own idea; to develop *artist music stations*, say, on Pandora (in the US), where the artists pick the music and play it – a sort of musical tweeting in real time (see Van Buskirk 2011, 2012). The seventh are a number of (online) alternatives to traditional forms of copyright control which artists can use. Mentioned earlier was Creative Commons, which provides control over both the moral and financial rights and also enables the free distribution of copyrighted works when the author or artist wants people to have the right to share, to use and also to develop the work they have created. There is a flexibility in the licence, for example, to only allow non-commercial uses, but it also protects people who use or redistribute the work. Copyleft is another form of licensing that can be used to enable individuals to use and modify copyrighted works such as computer software, documents and art (GNU 2008) (see Lee 2014).

Changing power relationships

The three groups of change outlined above in the business model of creative industries has resulted in a new set of power relationships that ultimately favour the artists, whether they be musicians, visual artists, actors, writers or performers. In so doing, the shift in power threatens the traditional income streams of producers, labels and distributors, who, whilst co-producing, also wielded most control over the exploitation of the product of the creators' labours. But how exactly have the power relationships changed in the new business model of the creative arts?

Firstly, there has been a decline in the power of pre-network technology record labels. In the case of the much larger labels, investment in new acts was almost a form of venture capitalism. Roles were defined hierarchically in terms of record label, producer and artist, as was the distribution of royalties and profits. The producers, distributors and labels once possessed a royalty share and also distribution rights of the artists' works which they obtained in exchange for paying an advance on the anticipated royalties. Networked technologies (see later) have blurred those lines of demarcation and new alignments have emerged between, say producers, who are often participants in the creation of the artist's brand, if not part of their substantive creative work. Producers may also be the label owners and also involved in promotions and sales in the digital age. Thus changes in the business model of the industry are understandably threatening to their income streams, income that is, effectively, a return on their investment.

Secondly, there has been a change in the role and power of the publishing houses which collect royalties. Royalties on the performance of music, published works or educational photocopying, on behalf of authors and others to whom rights have been assigned, have changed (certainly in the UK) in the artist's favour since the late twentieth century (Hull 2004). Previously, the major labels had progressively bought out the independent music publishing companies and created a situation where artists' publishing revenues were used to 'recoup' advances when recording royalties were insufficient, which was the situation in most cases because of the low rate of royalties. The relationship between the labels and publishing has largely been disrupted (see further, Hull 2004). In the UK, for example, the Performing Rights Society (PRS) which now incorporates the Mechanical-Copyright Protection Society (MCPS), and the Authors Licensing and Collecting Service (ALCS) both operate online services, which means that artists (musicians and authors) can directly collect their own royalties in ways that they could not previously. Importantly, from the artist's perspective, payments that used to take years to reach the artist can now take weeks or months. As a consequence of digitization and networking, the means of distribution and also the power balance have changed.

Policing file sharing: the use and abuse of legal process to protect revenue streams

If networked technologies have transformed the business model of creative industries, they have also transformed the regulatory process. In addition to the direct actions against the founders of the Napster, Gnutella, and Kazaa file-sharing platforms mentioned earlier, one of the main methods by which music and later movie industry representatives have sought to control copyright revenue streams has been to combine electronic information gathering of mass file sharing with court enforced disclosure of bulk information about file sharers and the use of law firms to obtain settlements from those who infringe copyright. This has the effect of consolidating the process of criminalizing file sharing, by appearing to bring some consequences to the actions of file sharers – even if the actions are often part of the civil law process. In the UK, Internet Service Providers (ISPs) reluctantly complied with court orders in the face of legal threats from the creative industry and also government on behalf of those industries. The 2009 *Digital Britain Report* (BIS 2009) proposed plans for a graduated response, which is basically three warning letters then legal action (three strikes). It was to be enshrined in a mandatory code which sets out standards of evidence for Internet Service Providers to comply with. This 'graduated response' approach became part of the UK Digital Economy Act 2010, although implementation has been delayed a number of times, and most recently until 2015 (BBC 2013c). Once it has been implemented, the UK will join France, New Zealand and South Korea who currently have similar laws (see Giblin 2014). By comparison, a number of countries have voluntary graduated response practices. The US is currently operating a six warnings (strikes) approach (see Siebens 2011; BBC 2013b). US ISPs are still resistant to revealing data on infringers, though the third party data collection of infringers' IP addresses (as elsewhere) initially helped to strengthen the case to obtain court orders to release the downloaders' identity and personal contact information. Ireland has sought to implement a graduated response but is experiencing court proceedings.

There are two fundamental problems with the legal process approach. The first is the inconclusive evidence obtained about wrongdoing and the problem of false positives when identifying infringers (see Hargreaves 2011). This is because the data collection process for peer-to-peer (p2p) evidence of infringement is digitally based and typically comprises an IP address, a downloaded file name (e.g. music or film or other media) and the time and date it was downloaded – usually by deep-searching BitTorrent clients (computer programs that use BitTorrent for p2p). The initial infringement information is usually collected by intellectual property security firms upon the instructions of a representative of the creative industries. It is then used by a law firm instructed by the creative industries to obtain a court order to get an ISP to release the names and details of the owner of the IP address. The law firm then sends out 'cease and desist letters'/'letters before action' to file sharers and/or invoices for any costs incurred. But, not only does having just the name of the shared file lack any proof of content, but the IP addresses of file sharers are also not conclusive evidence of wrongdoing. This is because many different people may have also legitimately used a single IP address, thus many innocent individuals could get caught up in the 'prosecution' process including deceased individuals, elderly grandparents, very young people, homeless people, displaced people and individuals with a copyright licence for the materials (Brainz 2010). The real offenders could have, in fact, simply piggy-backed on the WiFi involved.

> [K]nowing the IP address used for an illegal download is different than knowing who did it. The computer may also be on a shared network, it may be a shared computer in somebody's house, or it may be shared by roommates.
>
> *(Boudett, quoted by McLean 2013)*

Another complication is that the act of file sharing (both downloading and uploading) in peer-to-peer networks is primarily a social, rather than financially driven, activity. As argued earlier, there is a growing argument that even if the act of file sharing can be proved, then prosecution may be counterproductive in protecting copyright because it generates bad publicity. There is also little evidence that copyright infringement actually restricts income, because of the changing business model. Yet, the courts seem to have, until very recent years, accepted the logic of the argument in favour of the plaintiff.

This brings us to the second problem, which is that there has been a mission creep in the use of legal process to protect copyrights. Under most legal professional rules within different jurisdictions, clients are supposed to avoid conflicts of interests, but what has happened is that lawyers have proactively sought clients after they have sought alleged infringers to whom they have sent speculative invoices loosely veiled as 'cease and desist letters' (US) or 'letters before action' (UK) seeking settlement. The alleged infringers have tended to pay the invoice rather than face the threatened legal action or embarrassment or humiliation. Not only is embarrassment a key driver in getting downloaders to settle financially, but when invoices are not paid, the courts often decided in favour of the complainant; that is until ACS:Law in the UK then Canipre, Prenda Law and Malibu and other cases in the USA and Canada. It was the public outcry over unfairness and poor lawyering in speculative invoice cases that changed client prosecution policies and court decisions in the UK and US to favour defendants (Masnick 2012). In this context, 'speculative invoicing', as this process became known, not only contradicts the Legal Professions code, but it also quickly becomes 'copyright trolling' and a form of harassment, even extortion, which is a long way from the intended aim of regulating copyright infringement.

The process of (effectively) criminalizing the file sharing space has clearly had a negative impact upon file-sharing behaviour, but only by creating fear, uncertainty and doubt amongst file-sharing communities along with a distrust and reluctance to participate in file sharing and other online practices. Whilst this can be seen as a positive step by some representatives of the creative industries (but not all), these are, in effect, lost customers. The broader lessons from these cases are more serious in that copyright trolling has created a perverse and corrupting economy. This is because, being purely speculative, speculative invoicing has established the bulk 'payup-or-else' legal practice, described by some as 'creative lawyering' and 'unprofessional' by others, in fact most. The practice preyed upon the 'low hanging fruit': the vulnerable, those who either could not defend themselves financially, or were not in a position to defend themselves fairly because of the negative impact in terms of public reputation that defending the case would create.

The bulk speculative invoicing model relied upon settlement and there appears to be no intention to take these cases to court, unless defendants did not contest the case so that firms like ACS:Law won by default. Though not explicit, or illegal, speculative invoicing was therefore a disguised form of extortion or blackmail. ACS:Law also used bullying tactics to protect itself; for example, it took action against its critics (Masnick 2010a and 2010b). The speculative invoicing model is largely based upon untruths, as it was alleged (without any conclusive evidence) that the requested payments were damages calculated upon loss of sales. It also established the practice of outsourcing litigation and perversely incentivized (and corrupted) the lawyers involved. Moreover, it is also another worrying example of the reversals of the burden of proof, like speed cameras and possession of illegal imagery. The act (in this case file sharing) is enough and the individual therefore has to prove their own innocence where not guilty. It also exploits administrative legal procedures to gain default judgments: 'working real injustice', because it is administrative and not judicial, and an unjust procedure, there is no judicial oversight (Davey 2010; Enigmax 2010; Harris 2011).

Finally, as stated earlier, the practice has damaged the cause it sought to protect. Although the 'clients' were initially happy because they received a percentage of the payments made – money they would not have otherwise seen – speculative invoicing eventually drove away clients (representatives of the music industry) because of the damage to their reputation from the negative publicity. The Music Industry representatives (BPI, in the UK and RIAA/MPAA in the US) became critical of the practice following the negative publicity they received and also because the law is meant for the worst offenders, and not to catch the 'low-hanging fruit' (Holpuch 2012). Ironically, the adult sex film (porn) industry became the new clients for speculative invoicing and this drift can be seen in the US lawsuits (Williams 2012).

Conclusion: resolving the paradox of circulation and control?

New technologies have created new forms of file sharing opportunities that have changed and are still changing the business model of the creative arts in the artists' favour away from the traditional corporate model. The industry's intuitive response was to preserve its market share and seek to criminalize the civil file sharing space by creating the illusion of file sharing of copyrighted works as a crime (David 2010). Whilst this analysis remains the central argument being made here, upon reflection the analysis also introduces some additional complexities introduced by some further changing dynamics.

For a brief moment in time the power relationships with regard to music certainly appeared to change, and there is some evidence to show that they remain changed and in a very different form from what they previously were. The old order, in the form of the components of the mainstream record industry as was, was effectively displaced by digital and networked technologies which made music easier and cheaper to make, easier to promote, distribute and sell, thus shifting the business model in the individual artist's favour. The artist no longer needed a large corporate machine behind them as they once did. Some artists embraced this change – for example, see the case of Dispatch, from the early days of Napster (Knopper 2011), and successors who used internet-based brokering facilities which effectively cut out the 'middle men' – whereas others, such as Metallica[4] and Eminem did not. However, their opposition to file sharing did incur a fan backlash (Simon 2000), as was the case with Prince (see earlier), though there were reports of a subsequent reflection on their stance in a 2008 *Rolling Stone* interview (reported in Moya 2008). The changes in the business model did arguably favour the artists and acts with existing intellectual capital because they were 'already broken' and had a market share, and most of those starting out found it hard to break. Yet, some did break through against the odds, and whilst the Justin Biebers of this world are relatively few and far between, there are many other acts who do manage to make a living out of their live shows using the internet based brokering facilities. But, in this second decade of the twenty-first century, a new order also appears to have emerged where the traditional power brokers have simply been replaced by a different set of power brokers.

Apple, Spotify, and other streaming services all draw upon the technological and legal legacy of the rise and fall of Napster, Gnutella, Kazaa and many other file sharing platforms. This legacy, along with the unethical practices of criminalizing file sharing space have arguably softened the field for the introduction, and rise in popularity, of the legitimate music streaming platforms, which stream music services for a reasonable subscription. And this is where the new power struggle lies because whilst subscription platforms do pay artists for the music they play, it is not enough according to the artists – as Taylor Swift's fight with Spotify and Apple testifies (Dredge 2015). Plus, whilst digital technologies have effectively transformed the industry middlemen, they have also spawned a complete new industry of IP regulators, protecting rights by constantly patrolling social network media and www sites, removing, for example, 'pirated' videos from YouTube etc. Artists do resist – as mentioned before, Taylor Swift is quite outspoken in her condemnation, but arguably from a position of power, her power. No matter how you look at it, a new, but different order, has prevailed to replace the old style record companies and 'break' new acts, but more importantly, to broker access to the various music streaming systems and to bring their acts to public attention.

Notes

1   My thanks go to Mike McGuire for his very useful comments on an earlier draft of this chapter.

2   E.g. Wall 2015 and 2003.
3   The issue was first discussed in McLuhan and Nevitt (1972: 4), though the term 'prosumption' was coined by Toffler (1984).
4   See the case of *Metallica, et al. v. Napster, Inc. 2000* at http://news.findlaw.com/hdocs/docs/napster/napster-md030601ord.pdf.

References

Anti Music (2009) 'Marillion use P2P for album release', *Anti Music*, 9 November, www.antimusic. com/news/08/sep/11Marillion_Use_P2P_for_ Album_Release.shtml.
Barron, A. (2014) 'Intellectual Property and the Open (Information) Society', in M. David and D. Halbert (eds) *The SAGE Handbook of Intellectual Property*. New York: SAGE.
Baudrillard, J. (1994) *Simulacra and Simulation*. Ann Arbor: University of Michigan Press.
BBC (2006) 'Microsoft to remove Sony CD code', *BBC News Online*, 14 November, http://news.bbc. co. uk/1/hi/technology/4434852.stm.
BBC (2009a) 'Coldplay to give away free album', *BBC News Online*, 1 May, http://news.bbc.co.uk/1/hi/ entertainment/8028478.stm.
BBC (2009b) 'Lily rallies stars against piracy: Lily Allen says piracy is "having a dangerous effect on British music"', *BBC News Online*, 21 September, http:// news.bbc.co.uk/1/hi/entertainment/8266287.stm.
BBC (2010) 'Prince in second free album deal with newspapers', *BBC News Online*, 1 July, www.bbc. co.uk/news/10459020.
BBC (2013a) 'Spotify reveals artists earn $0.007 per stream', *BBC News Online*, 4 December, www.bbc. co.uk/news/entertainment-arts-25217353.
BBC (2013b) 'US internet "six strikes" anti-piracy campaign begins', *BBC News Online*, 26 February, www. bbc.co.uk/news/technology-21591696.
BBC (2013c) 'UK piracy warning letters delayed until 2015', *BBC News Online*, 6 June, www.bbc.co.uk/ news/technology-22796723.
BBC (2014) 'Prince drops $22m pirate action against fans', *BBC News Online*, 30 January, www. bbc. co.uk/news/technology-25960300.
BBC (2016) 'Thom Yorke's "had enough" of unusual ways of putting out music', *BBC News Online*, 8 September, www.bbc.co.uk/newsbeat/article/37306336/thom-yorkes-had-enough-of-unusual-ways-of-putting-out-music.
BIS (2009) *Digital Britain: Final Report*, Department for Culture, Media and Sport and Department for Business Innovation and Skills, Cm 7650, 16 June, http://webarchive.nationalarchives.gov. uk/20121126075826/http://www.bis.gov.uk/assets/BISCore/corporate/docs/D/Digital-Britain-Final-Report.pdf.
Borland, J. (2001) 'Customers put kibosh on anti-copy CD', *cnet*, 19 November, http://news.cnet. com/2100-1023-276036.html.
Bower, J. L., and C. M. Christensen. 'Disruptive technologies: Catching the wave', *Harvard Business Review*, 73(1): 43–53.
Brainz (2010) 'The 14 Most ridiculous lawsuits filed by the RIAA and the MPAA', *Brainz*, www.brainz. org/14-most-ridiculous-lawsuits-filed-riaa-andmpaa/.
Bryant, T. (2010) 'Prince will give new album 20TEN away free to *Daily Mirror* readers', *Daily Mirror*, 3 July, www.mirror.co.uk/3am/celebrity-news/princegive- new-album-20ten-232881.
Bylund, A. (2007) 'Record label defections by major acts a troubling sign for recording industry', *ars technica*, 9 October.
Carey, M. and Wall, D.S. (2001) 'MP3: more beats to the byte', *International Review of Law, Computers and Technology*, 15(1): 35–58.
Castells, M. (2000) 'Materials for an explanatory theory of the network society', *British Journal of Sociology*, 51(1): 5–24.
Cellan-Jones, R. (2009) 'Goodbye "Knock-off Nigel"', *BBC News Online* (.dot life blog), 2 April, www. bbc. co.uk/blogs/technology/2009/04/goodbye_knock_ off_nigel.html.
Cherkis, J. and Stenovec, T. (2013) 'Indie record labels would support Spotify boycotts by their artists', *The Huffington Post*, 26 July, www.huffingtonpost. com/2013/07/26/spotify-indie-labels_n_3659833. html?ncid=edlinkusaolp00000003.

ComScore (2007) 'For Radiohead Fans, Does "Free" + "Download" = "Freeload"?', *Comscore Press Release*, 5 November, www.comscore.com/Insights/Press-Releases/2007/11/Radiohead-Downloads.

Davey, F. (2010) 'ACS:Law come unstuck', *Francis Davey Blog*, 8 December, http://www.francisdavey.co.uk/2010/12/acslaw-come-unstuck.html.

David, M. (2010) *Peer to Peer and the Music Industry: The Criminalization of Sharing*. London: Sage.

David, M. and Whiteman, N. (2014) 'Piracy or Parody: Moral Panic in the Age of New Media', in M. David and D. Halbert (eds) *The SAGE Handbook of Intellectual Property*, New York: Sage.

Dewey, C. (2016) 'How we're unwittingly letting robots censor the Web', *The Washington Post*, 29 March, www.washingtonpost.com/news/the-intersect/wp/2016/03/29/how-were-unwittingly-letting-robots-censor-the-web/.

Dredge, S. (2015) 'Taylor Swift still has bad blood with Spotify over streaming music dispute', *The Guardian*, 4 August, www.theguardian.com/technology/2015/aug/04/taylor-swift-bad-blood-spotify-streaming-music.

Enigmax (2010) 'ACS:Law take alleged file-sharers to court – but fail on a grand scale', *TorrentFreak*, 9 December, www.torrentfreak.com/acslaw-takealleged-file-sharers-to-court-but-fail-on-a-grandscale-101209/.

Espiner, T. (2010) 'ACS:Law fails in default judgement attempt', *zdnet*, 10 December, www.zdnet.com/acslaw-fails-in-default-judgement-attempt- 4010021288/.

Fine, J. (2007) 'Leaving record labels behind', *Bloomberg Business Week*, 28 October, www.businessweek.com/stories/2007-10-28/leaving-recordlabels- behind.

Fisher, W. (2001) 'Theories of Intellectual Property', in S. Munzer (ed.) *New Essays in the Legal and Political Theory of Property*. Cambridge: Cambridge University Press.

Frere-Jones, S. (2013) 'If you care about music, should you ditch Spotify?', *The New Yorker*, 19 July, http://www.newyorker.com/culture/sasha-frere-jones/if-you-care-about-music-should-you-ditch-spotify.

Gaines, J. (1992) *Contested Cultures: The Image, the Voice, and the Law*. London: British Film Institute.

Geist, M. (2007) 'Why popstars are going it alone', *BBC News Online*, 18 October, http://news.bbc.co.uk/1/hi/technology/7047723.stm.

Ghosh, S. (2014) 'The Idea of International Intellectual Property', in M. David and D. Halbert (eds) *The SAGE Handbook of Intellectual Property*. New York: SAGE.

Giblin, R. (2014) 'Evaluating graduated response', *Columbia Journal of Law & the Arts*, 37: 147–209.

GNU (2008) 'What is copyleft?', GNU Operating System, www.gnu.org/copyleft/.

Hargreaves, I. (2011) *Digital Opportunity: A Review of Intellectual Property and Growth*. Intellectual Property Office, www.ipo.gov.uk/ipreviewfinalreport. pdf.

Harris, J. (2011) 'Judge Birss hits out as porn costs mount', *The Lawyer*, 27 April, http://www.thelawyer.com/judge-birss-hits-out-as-porn-costs-mount/1007751.article.

Holpuch, A. (2012) 'Minnesota woman to pay $220,000 fine for 24 illegally downloaded songs: Recording Industry Association of America has largely adjusted its anti-piracy strategy to stop suing individual downloaders', *The Guardian*, 11 September, http://www.theguardian.com/technology/2012/sep/11/minnesota-woman-songs-illegally-downloaded.

Hull, G. (2004) *The Recording Industry*. London: Taylor & Francis.

Knopper, S. (2011) 'Being huge just isn't big enough for Dispatch', *Chicago Tribune*, 2 June, http://articles.chicagotribune.com/2011-06-02/entertainment/ct-ott-0603-dispatch-20110602_1_dispatch-wimpy-jam-bands.

Lee, J. (2014) 'Nonprofits in the Commons Economy,' in M. David and D. Halbert, (eds) *The SAGE Handbook of Intellectual Property*. New York: SAGE.

Levine, R. (2008) 'For some music, it has to be Wal-Mart and nowhere else', *The New York Times*, 9 June, www.nytimes.com/2008/06/09/business/ media/09walmart.html?pagewanted=all&_r=0.

Masnick, M. (2010a) 'Extortion-like mass automated copyright lawsuits come to the US: 20,000 filed, 30,000 more on the way', *techdirt*, 30 March, www.techdirt.com/articles/20100330/1132478790.shtml.

Masnick, M. (2010b) 'ACS: Law now using dubious legal theories to threaten Slyck.com', *techdirt*, 22 May, https://www.techdirt.com/articles/20100321/2136068650.shtml.

Masnick, M. (2012) 'UK court wants to limit copyright trolling... but not enough to stop it entirely', *techdirt*, 28 March, http://www.techdirt.com/articles/20120327/05074118257/uk-court-wants-to-limit-copyright-trolling-not-enough-to-stop-it-entirely.shtml.

Masters, T. (2013) 'Marillion "understood where the internet was going early on"', *BBC News Online*, 1 September, www.bbc.co.uk/news/entertainmentarts- 23881382.

McLean, M. (2013) 'Battling Bit Torrent: Can the movie studios beat online piracy?', *Canadian Business*, 8 March, www.canadianbusiness.com/technologynews/ battling-bit-torrent/.

McLuhan, M. and Nevitt, B. (1972) *Take Today: The Executive as Dropout*. New York: Harcourt Brace Jovanovich.

Moya, J. (2008) 'Metallica now embraces file-sharing?', ZeroPaid, 26 April, www.zeropaid.com/news/ 9440/metallica_now_embraces_filesharing/.

Nguyen, T. (2006) 'Kazaa to pay $100 million to record labels', *Daily Tech*, 27 July, www.dailytech.com/ Kazaa+to+Pay+100+Million+to+Record+Labels/ article3535.htm.

NME (2012) 'Did Radiohead's "In Rainbows" honesty box actually damage the music industry?', *New Musical Express*, Blog, 15 October, www.nme.com/blogs/nme-blogs/did-radioheads-in-rainbows-honesty-box-actually-damage-the-music-industry.

Pareles, J. (2002) 'David Bowie, 21st-century entrepreneur', *The New York Times*, 9 June, www.nytimes. com/2002/06/09/arts/david-bowie-21st-century-entrepreneur.html.

Phythian-Adams, S. (2014) 'The Economic Foundations of IP', in M. David and D. Halbert, (eds) *The SAGE Handbook of Intellectual Property*. New York: SAGE.

Ritzer, G. and Jurgenson, N. (2010) 'Production, Consumption, Prosumption: The nature of capitalism in the age of the digital "prosumer"', *Journal of Consumer Culture*, 10: 13–36.

Rohter, L. (2011) 'Record industry braces for artists' battles over song rights', *The New York Times*, 15 August, www.nytimes.com/2011/08/16/arts/music/springsteen-and-others-soon-eligible-to-recover-song-rights.html?pagewanted=all.

Rohter, L. (2013) 'A copyright victory, 35 years later', *The New York Times*, 10 September, www.nytimes. com/2013/09/11/arts/music/a-copyright-victory-35-years-later.html.

Rothman, L. (2013) 'Can music collectives fill the gap between labels and DIY? A new, old trend may help musicians better navigate the new industry landscape', *Time Magazine*, 12 April, www.entertainment. time.com/2013/04/12/can-music-collectives-fill-the-gap-between-labels-and-diy/.

Savage, M. (2016) 'Rick Astley: "I didn't mind being called Dick Spatsley"', *BBC News Online*, 9 June, www.bbc.co.uk/news/entertainment-arts-36469774.

Siebens, C. (2011) 'Divergent approaches to filesharing enforcement in the United States and Japan', *Virginia Journal of International Law*, 52(1): 155–192.

Silbey, S. (2008) 'The mythical beginnings of intellectual property', *George Mason Law Review*, 15: 319–379.

Simon, R. (2000) 'Metallica's anti-Napster crusade inspires backlash', *MTV*, 31 May, www.mtv.com/ news/971500/metallicas-anti-napster-crusade-inspires-backlash/.

Stanley, R. (2013) 'Album sales are declining, but it's part of the battle between art and commerce', *The Guardian*, 24 November, www.theguardian.com/commentisfree/2013/nov/24/album-sales-declining-battle-art-commerce.

Tapscott, D. and Williams, A. (2007) *Wikinomics: How Mass Collaboration Changes Everything*. London: Atlantic Books.

The O2 (2007) 'Prince announces final six nights of record breaking UK Appearances at The O2', *The O2*, 11 June, http://archive.today/TiajT.

Thornton, D. (2009) 'Passion is why Nine Inch Nails and vinyl are succeeding', *The Way Of The Web*, 7 January, www.thewayoftheweb.net/tag/nine-inch-nails/.

Toffler, A. (1984) *The Third Wave*. New York: Bantam Books.

Torrent Freak (2012) *The Speculative Invoicing Handbook, Second Edition*, Scribd., www.scribd. com/ doc/115443516/The-Speculative-Invoicing-Handbook-Second-Edition.

Van Buskirk, E. (2007) 'ComScore: 2 Out of 5 Downloaders Paid for Radiohead's 'In Rainbows' (Average Price: $6)', *WIRED*, 5 November, www.wired.com/2007/11/comscore-2-out-/.

Van Buskirk, E. (2011) 'Pandora redesign emphasizes social sharing, removes listening limit', *evolver. fm*, 21 September, www.evolver.fm/2011/09/21/pandora-rolls-out-big-redesign-emphasizes-social-sharing/.

Van Buskirk, E. (2012) '5 powerful music apps that should make middlemen nervous', *WIRED*, 16 May, www.wired.com/underwire/2012/05/5-music-appsscaremiddlemen/.

Wall, D.S. (2003) 'Policing Elvis: Legal action and the shaping of post-mortem celebrity culture as contested space', *Entertainment Law*, 2(3): 35–69.

Wall, D.S. (2015) 'Copyright Trolling and the Policing of Intellectual Property in the Shadow of Law', in M. David and D. Halbert, (eds) *The SAGE Handbook of Intellectual Property*. London: Sage, pp. 607–626.

Williams, C. (2012) 'O2 forced to expose "porn downloaders"', *The Telegraph*, 27 March, http://www.telegraph.co.uk/technology/broadband/9168016/O2-forced-to-expose-porn-downloaders.html.

WIPO (2001) *WIPO Intellectual Property Handbook: Policy, Law and Use*, WIPO (World Intellectual Property Organization) publication no. 489(E). Geneva: WIPO.

WIPO (2011) *What is Intellectual Property?* WIPO (World Intellectual Property Organization).

WIPO (2012) *International Survey on Private Copying: Law & Practice 2012*, WIPO (World Intellectual Property Organisation), www.wipo.int/edocs/pubdocs/en/copyright/1037/wipo_pub_1037_2013.pdf.

Youngs, I. (2009) 'Bands "better because of piracy"', *BBC News Online*, 12 June, http://news.bbc.co.uk/1/ hi/8097324.stm.

10

ICTs, privacy and the (criminal) misuse of data

Andrew Puddephatt

While the meaning of the term privacy coheres around the notion of personal integrity and dignity (Bloustein 1964), it is hard to define with any precision – it can embrace notions as varied as the right to freedom of thought and conscience, the right to be alone, the right to control one's own body, the right to protect reputation, the right to a family life, the right to a sexuality or sexual identity of your own definition. It is deemed essential to human dignity and indeed to individuality, and if we have no privacy, we are less able to be ourselves.

One striking characteristic of privacy is how the environment in which people live determines their understanding of what privacy means. Privacy in a communal village, a medieval hall, a nineteenth-century upper-class family with servants and a modern developed one-person household may all have quite different meanings.

Historic notions of privacy in law focused upon the protection of physical space, particularly the home and personal possessions. Early privacy protections derived from the notion of the inviolability of the home and family life and the protection of property. As Lord Camden said in the famous case of *Entick v Carrington* in 1765:

> By the laws of England, every invasion of private property, be it ever so minute, is a trespass. No man can set his foot upon my ground without my licence, but he is liable to an action, though the damage be nothing...

Technology and privacy

As well as the lived environment, another crucial factor that shapes our understanding and experience of privacy is technology, which introduces a new dimension to debate about privacy – what can be known about us, what information can be recorded, stored, analysed and accessed.

The impact of technology upon privacy in modernity became apparent with the introduction of mass circulation newspapers and photographs. People who saw their pictures in a newspaper were concerned that what they had assumed to be private was now public. This era also saw the development of the telegraph, new and relatively inexpensive portable cameras, machines for recording sound and even cheaper and clearer window glass. All of these new technologies seemed to increase the ways in which people could have their actions, appearance, conversation and even personalities communicated without their consent beyond their professional and personal circles.

Two young Boston based lawyers, Samuel Warren and Louis Brandeis published a *Harvard Law Review* article in December of 1890 entitled 'The Right to Privacy' (Warren and Brandeis 1890). They believed that if information were made available about a person it would influence that person's perception about himself or herself and even injure that perception. This was more than an argument about the damage to reputation dealt by laws of libel or slander – it represented a psychological claim that the protection of personal integrity required the exercising of control over any information or image that could be perceived to reflect a personality and therefore impact upon that personality. Deep down, the claim was that a person had a right to be left alone in order to protect the inviolate personality.

Part of the trigger for the article appears to have been coverage of Warren's family by the local press, notorious in their day for their fascination with local elites. Since that era and with the advent of television, radio, mass advertising and reality programmes the original concerns about pictures in a newspaper can seem relatively quaint. The growth of modern mass media and the advertising industry's focus on understanding consumers' wants led Myron Brenton to argue that we are living in the 'age of the goldfish bowl' (Brenton 1964), where private lives are made public property by the manipulation and exchange of personal data. The development of social media where people post photos of themselves in real time represents a step change in this process. Instead of an editor or publisher deciding what of us can be seen, we are all increasingly comfortable sharing every aspect of our lives with the world.

In turn, the technical capacity to gather, store and exchange personal information about people provided by digital technologies has led a new dimension in conceptualising privacy. Included within this new approach is the right of people to determine when, how and to what extent information about them is communicated to others (Westin 1967). The growing processing power of computers makes realising this right far from easy.

Privacy, according to Westin 'is the claim of individuals, groups, or institutions to determine for themselves when, how, and to what extent information about them is communicated to others ... [It is] the desire of people to choose freely under what circumstances and to what extent they will expose themselves, their attitudes and their behaviours to others' (ibid). From this claim came the array of data protection laws and standards governing how information legitimately held on people by public and private agencies should be regulated.

The notion of consent is crucial. Many people query how there can be a notion of privacy in a world of Tinder and Grindr. But in these cases people are choosing to share often intimate pictures with peers for consenting transactional purposes. This is quite different to the harvesting of the same pictures for purposes of sale to third parties. Intentions matter.

The protection of privacy

As might be expected given the shifting conception of privacy, any protections offered by the law have also changed with circumstances and time. There were no early overarching systems of privacy protection – rather there were specific responses to specific situations. Perhaps an early example of this was the 1361 Justices of the Peace Act in England, which prohibited 'peeping toms' and 'eavesdroppers' (though it is challenging to think how this could be avoided in a medieval village or hall). More enduring was the case of *Entick v Carrington* [1765], also in England, which grew out of the challenge to protect papers held in a private home (and which is generally thought to have shaped the Fourth Amendment of the US Constitution). Examples of similar piecemeal approaches can be found in countries such as Sweden, which set out the purposes for which governments could use the information they held about individuals, and

France and Norway, who prohibit the publication of certain types of personal information (Privacy International 2006).

By the twentieth century, however, during the great era post-1945 of defining and setting down normative human rights standards, privacy was defined as a human right. The Universal Declaration of Human Rights (UDHR), 1948, set out an attempt to protect privacy as a distinct human right. Article 12 of the UDHR states that:

> No one shall be subjected to arbitrary interference with his privacy, family, home or correspondence, nor to attacks upon his honour and reputation. Everyone has the right to the protection of the law against such interference or attacks.

While not legally binding, the UDHR set a standard and model for similar but enforceable rights in many other human rights documents including the legally binding International Covenant on Civil and Political Rights (ICCPR) and the European Convention on Human Rights (ECHR).

However, although we see definitions of the rights in many international instruments as well as in constitutions and specific laws, the right to privacy remains nebulous and hard to define with any precision. How it is protected often comes down to the interpretation of the law in specific cases rather than in any grand guarantee. Judges themselves seem curiously averse to fixing on any definition of the right to privacy. The European Court of Human Rights has said 'the Court does not consider it possible or necessary to attempt an exhaustive definition of the notion of "private life".' This often leaves notions of privacy left in a highly subjective state of mind – as one observer said, violations of privacy are detected when something 'feels wrong ... [this] is often the most helpful delineation between when an incursion into the private life of an individual is reasonable and when it is not' (Hosein 2006).

Civil society groups such as Privacy International have tried to bring some order to this anarchic state of affairs. It sought to define four different types of privacy:

*   information privacy (e.g. personal data)
*   bodily privacy (e.g. invasive procedures)
*   privacy of communication (e.g. surveillance) and
*   territorial privacy (e.g. home) (Privacy International 2006).

In relation to the Internet, information privacy and privacy of communication are the most pertinent.

Despite the lack of definition of the notion of privacy it has been accorded great significance by legislators, lawyers and philosophers alike. As Paul Chadwick (information commissioner for the Australian State of Victoria) said:

> Privacy is the quietest of our freedoms ... Privacy is easily drowned out in public policy debates ... Privacy is most appreciated by its absence, not its presence.
>
> *(Hawtin 2013)*

However, it should also be noted that while people are often concerned about privacy in the abstract, they seem less concerned about privacy in practice. It is clear from a cursory use of the Internet that people give out personal information to a frequently surprising degree. Many writers have noticed the gap between what people say they value and what they actually do online (Miller 2014). It may be the nature of the Internet, which is often accessed privately and combines both a communication medium in the shape of email (which may suggest to the user the privacy of the telephone call or private conversation) and a publishing medium as with an application like Facebook. There is some anecdotal evidence that people do not realise that when they communicate online they are effectively publishing, and that what they publish, however intimate, can become globally available and they do not realise how it can be used.

Given the far reaching changes in the communication environment brought by the Internet, a major shift in public understanding of privacy now seems likely.

Cyber world, privacy and crime

Before the Internet developed, stealing data was a physical act, involving the burglary of premises and theft of material. Privacy protection lay in the physical barriers preventing entry to the home, backed up by laws on private property that gave primacy to the individual's right to do what they liked in their own home. In the early days of computers – before the advent of personal computing – even computers were attacked physically by directly accessing the machine to hack into the operating system to gain access to the files.

But as information about a person began to be stored on computers, cyber-theft of personal information and even identities, began to take place over networks of connections. New crime possibilities emerged (Wall 2007). Instead of just stealing data, an identity could be stolen or impersonated; computers could be borrowed by inserting viruses into their operating system and then used in turn to steal from other computers. Cyber networks made it possible to violate privacy over great distances across multiple jurisdictions. In a world where nation states are still the principal sources of identity for people and the main, if not the only, source of legitimacy users face great challenges in trying to protect their privacy. For example, attacks on their personal banking arrangements might be launched from outside their own government's jurisdiction and across multiple countries with different policing and legal systems, making the task of law enforcement immensely more complicated.

New global connectivity offers enormous potential for data and privacy-related crime. In the era of physical information, storage and illegal access to that material would be limited to the specific physical store of data – the numbers of people who had physical access to your data was comparatively small – even in a large city. Moreover, if that person did break in they could steal your own personal documents – bank statements, letters, credit cards etc, but not those of your family, friends and neighbours. Global interconnectivity opens up the possibility that anyone on the planet with access to the Internet can not just steal your own information, but can access your online contacts and networks, and through a network of connections, from one computer to another, can access data stored by those contacts – and in turn, their contacts and so on. The only limitation is the degree of security erected between users, which has been shown to be perilously weak (BT.com 2016).

Privacy, crime and the Internet – key framing factors

There are two major aspects of the Internet and personal data that enable new types of privacy-related crime – technical developments and the business model of the Internet itself.

Technically, the key developments that in turn enable privacy violations are:

- The collection and location of personal information.
- New capacities to process and manage personal information.
- Encryption.
- Malware.

Technical developments

The Internet has seen a batch of new tools designed to extract personal information from the user. Of the many tools that have been created to track Internet users, two familiar examples are cookies and web bugs. Cookies are small pieces of text which web browsers store on a user's computer. The cookie 'registers' with the web browser each time the user accesses that browser and can be used for monitoring the user's session history, storing any preferences, etc, (BBC 2016a). Of course users are usually offered the choice whether to accept cookies by changing settings on their browser software, but in practice few do and sites can become unusable unless cookies are accepted.

Web bugs (or web beacons as they are sometimes known) are usually invisible to the user (they are typically only 1x1 pixel in size) and are embedded in web pages and emails (*PC Magazine* 2016). When the page/email containing the web bug is viewed, it sends information back to the server (including the IP address of the user, the time and date that the page/email was viewed and the browser it was viewed on).

Cookies and web bugs are used to build up the collection of information about users that is then used by companies to develop personal profiles. These profiles can be marketed, often through a complex network of intermediaries and sold to advertisers who can use them to provide more targeted advertising. All of this can be irritating but is not criminal in itself – the problem is that building up such a rich pool of data creates a resource that can be hacked by criminals to build up detailed pictures of intended targets.

New technologies also create the possibility to locate and track personal data that was not possible before. Each computer, mobile phone or other device attached to the Internet has a unique Internet Protocol (IP) address, which provides a specific identifier for the device and which means in turn that they can be traced. The advent of GPS systems has meant that devices with unique IP addresses can be physically located, enabling anyone with access to that information to track the movements of the person with the device. At the moment this capability is of most interest to law enforcement agencies but criminals could use this capacity to stalk a potential victim.

The development of algorithms with sophisticated processing power in turn enables large volumes of data – billions of pieces – to be analysed extremely rapidly. We are familiar with the use of algorithms for bulk data analysis through the revelations of Edward Snowden, a US defence contractor, who leaked information about the operations of the US NSA and the UK's GCHQ (*Guardian* 2016). But private companies for commercial ends with little or public awareness do comparable analysis. The end result is that volumes of data can be processed and analysed by machines in a way that human beings could never hope to achieve.

To counteract the ease by which data can be captured and analysed online, we have seen an increasing development of sophisticated encryption systems to protect our data and communications by encoding it in such a way that only authorised persons can access it. Encryption has long been used by military and governments to protect their own information and communications (Wikipedia 2016). But as economies move online and more and more sensitive transactions take place online the encryption of data for private communication becomes more and more pressing. Indeed, it is now more surprising if data is not encrypted than if it is.

Some systems of encryption such as the so-called Onion Router, or 'Tor' for short, have become controversial as many claim they are used extensively by criminals and terrorists (Digisecrets 2013). Tor is especially powerful because it disguises not just the content of a communication but the source which makes network surveillance of the kind to which algorithmic analysis is applied, ineffective. Encryption is one of the great policy battles of the privacy era – with governments demanding access, even through 'backdoors' into encrypted systems, in the name of preventing crime, while privacy advocates argue for the right to encrypt their communications.

But though the criminal applications of encryption tend to be overlooked by privacy advocates, the advantages to criminals are obvious. High-quality encryption – and the availability of online currency such as bit coin – carry obvious attractions to criminals. Most famously, Silk Road was an online market for illegal drugs that used Tor encryption. Silk Road ran for two years in its first incarnation from 2011 to 2013 and it was claimed that billions of dollars of illegal business was conducted online, until the arrest of its founder Ross Ulbricht (aka 'Dread Pirate Roberts') (Ward 2015). A subsequent re-launch of Silk Road was also terminated by a round of arrests. A balance between the positive benefits of encryption and the downsides needs to be struck. But a characteristic of this debate is that there is relatively little hard evidence and more research is needed. Security agencies tend to make strong claims about the criminal use of encryption, while defenders point to the human rights benefits in repressive countries (Professor Thomas Rid from Kings College London has claimed that over 50 per cent of Tor users are criminal at the 2–15 October 2015 29th Parliament & Internet conference).

Finally, the Internet has spawned a mini industry of users producing so-called malware – an abbreviation of the term 'malicious software.' This is software that tries to get access to systems without the consent or even knowledge of the user. Malware takes many forms in the context of privacy violations – from spyware which allows someone to obtain information secretly about another's computer activities, to applications that capture the key movements a person makes (keyloggers), or 'true' viruses that can infect a computer without the knowledge of the user. These malwares can invade your privacy and launch further attacks on your computer, or computers the host is connected to. The impact is equivalent to the burglar in your home leaving someone hidden in your house who can follow you around and help themselves to anything they like when you're not looking.

The business model

In August 2014 Ethan Zuckerman wrote a piece for *The Atlantic* magazine (Zuckerman 2014) confessing to the 'original sin' of creating the first pop-up ad. He explains that while working for a site called Tripod.com he tried a variety of different revenue models – a subscription service, sharing revenue when users bought mutual funds after reading investment advice, bundle a magazine with textbook publishers, even selling t-shirts and other branded merchandise. All failed. The model that finally attracted investment was one that involved analysing users' personal homepages so ads could be better targeted. Advertising became the business model on the web, Zuckerman claims, because it was the easiest way for web start-ups to raise initial investment.

Advertising is now the economic model of the Internet, the reason why search and so many other services are 'free'. Successful companies extract, analyse and often sell personal data in order to target advertising at users. Digital networks therefore produce a direct and powerful imperative to secure, retain and share personal data. Nor is it just personal data extracted from online activity on a computer or mobile phone. Computerised bar codes can be used to track individual purchases, which in turn are then used to control stock levels and target incentives or marketing at those consumers. Computerised travel cards, such as the London Oyster card, create a digital picture of every journey that can be used to monitor city wide passenger movements – useful for transport planning, but also for tracking an individual's journeys. As the Internet is used in more and more everyday interactions including banking, shopping and socialising, people are giving away more and more of their personal data, often unwittingly including sensitive information about their finances, heath and even their sexuality. These developments allow an ever greater amount of information to be gathered and, as Lawrence Lessig pointed out, 'your life becomes an ever-increasing record' (Lessig 1999).

This vast pool of data has been called the oil of the twenty-first century (Forbes 2016a). And technological advances allow databases of information to be connected together allowing even greater quantities of data to be processed. As information technologies are combined together, the potential for privacy violations increases exponentially, for example, the linking of facial recognition databases (as used on Facebook for example) with CCTV cameras would allow tracking of individuals on an unprecedented scale.

The practice of merging and consolidating different informational databases is pervasive. Privacy issues clearly arise from matching data from different sources, for example, tax data against health data or finance data against social security data. In addition, personal data can be extracted from the various techniques and then matched with publicly available data to build a detailed personal profile (Gross 2016).

The problem is that as the amount of data held by companies on people has grown, levels of personal security have not. In late October 2015 the UK mobile provider TalkTalk announced that hackers had gained access to thousands of customer records including their personal and financial details. It appeared that the data held by TalkTalk may not have been encrypted – indeed discussions about unencrypted databases on TalkTalk as well as other vulnerabilities had been discussed on forums used by hackers before the attack itself (*Financial Times* 2016). The first rounds of arrests were of people aged 15, 16 and 20, showing how dispersed the skills to seize data have become.

There seems little likelihood that this business model will be reversed. A recent study by McKinsey estimates that the direct and indirect economic effects of the Internet account for 3.4 per cent of GDP in the 13 countries studied but 21 per cent of the economic growth in the five mature economies, with 2.6 jobs created for every job lost (McKinsey 2011). Internet companies are now among the wealthiest and most powerful in the world with Amazon, Apple, Facebook and Google's market capitalisations now roughly on par with the GDP of South Korea.

As a result, Internet companies like Google, Yahoo and Facebook now have access to an astronomical amount of data (Forbes 2016). The biggest Internet companies have huge user bases (for example, Facebook has over 1.5bn users worldwide (Statista 2016)) and are branching out to cover more and more interactions (for example, a user may use Google to locate information online, send emails, display videos, shop etc). A 1999 study discovered that 92 per cent of websites were gathering at least one type of identifying information from their users, for example their name, email address and postal address (Federal Trade Commission 1999) and it can be assumed that since then gathering of information has only increased. Companies also have a tendency to be very secretive about what information they gather and how; as noted in *The Economist*, this is as much to do with maintaining a competitive edge as it is with privacy concerns (*Economist* 2010).

Official data online

Just as the last few years have seen a dramatic acceleration of commercial collection and storage of personal data, we have seen the same trend with government. Increasing volumes of personal information are held online from basic birthdate, user identification (national insurance/social security), address and employment details, health and tax data up to DNA samples and biometric information. The main driver for this is the economic and funding crisis that is affecting government globally. The desire to shift government services from high cost humans to low cost algorithms means increasingly that we have an official profile online – driving licenses, voter registration, etc (Gov.uk 2016). There are certain advantages to this – for example the dramatic reduction in transaction costs for government (Gov.uk 2012). Much of this data is encrypted to a high level of security and the use of iris scans and fingerprints as identifiers has made modern travel through high security barriers more easy. But there are also enormous vulnerabilities in state based systems. The US Personnel Department was hacked early in 2015, leading to the exposure of 22 million people's sensitive personal information including mental health records and details of drug and alcohol abuse (BBC 2016b). In the UK, the data of 5,000 prison staff was lost when a hard drive was sent to an IT contractor (Garnham 2008).

Criminal implications

Having sketched out the scale of data collection and privacy issues posed by modern technology, what are the implications for crime? One crime given an immense boost by the digital world is so-called identity theft (more specifically the theft of forms of identification). This involves a criminal obtaining private personal information such as date of birth, financial records, passwords and so on for various purposes. Three main criminal consequences of theft of forms of identification can be identified. One is where a criminal uses a person's details to impersonate them and open new accounts. A variation is when the criminal invents an identity using forged documents to obtain products or services. Most dramatically, if the criminal can gain sufficient personal information such as a person's passwords, it can enable them to evade security and take over the account itself to use as they wish. In the case of large-scale identification thefts, people's personal data, including credit card details, can even be placed online for sale or auctioned off.

In some cases one piece of identity may be absolutely crucial. In the United States for example, a Social Security number is the single most important piece of identity. It is the basis for taxation and benefits. Access to a particular Social Security number gives a criminal the opportunity to accumulate all documents related to that person's status – to their entire identity – and to steal that identity. Obtaining one piece of identity can produce a chain reaction as it can be used to create further identities or documents. Criminals can use a stolen credit card to change the mailing address on the account. He or she can obtain a passport or driver's license in the victim's name with their own picture. By building up a stack of mutually reinforcing identity documents the criminal can open bank accounts and secure bank loans, all using the victim's credit record and background. Often the victim will be unaware that this has happened until they suddenly confront a mound of debt.

Trends – UK

Although misuse of identification appears to be common, researchers and security agencies suffer from a lack of hard data – it is only in recent years that there has been any attempt to try and record so-called 'cybercrime'. In the UK, according to the latest report from the UK fraud prevention service, Cifas, known frauds increased by 25 per cent in 2014 – a total of 276,993 frauds, compared to 221,075 frauds in 2013. Of this, the largest single category is identity fraud – accounting for 41 per cent of all frauds recorded in 2014 (Cifas 2016).

However, although identification fraud is increasing, instances of fraudsters taking over accounts and using them appears to be in decline. This is likely a result of better security and greater security awareness among users (Matthews 2016). But it also seems to be leading to an increase in the creation of false identities as criminals adapt to the new environment (Cifas 2016).

Identity creation/deletion

As lives are increasingly lived online, the potential to create or even destroy real or imaginary identities has developed. At the US hackers conference DefCon23 in 2015, Chris Rock, an Australian security researcher and CEO of Kustodian, explained how it was possible to manipulate both death and birth certificates. In many countries there is a process to register online as a doctor's practice or even a funeral director, using publicly available information, thus allowing the hacker to declare someone dead of natural causes (so not needing a coroner's inquest). Rock demonstrated how, using existing technologies, it is possible to make someone legally dead – including themselves (Defcon 2016).

The crime potential here is fascinating – whether it is to collect on life insurance, to create false wills for others and collect on them – or just generally to make life very difficult (imagine trying to get credit cards re issued when you are dead). Inefficiencies in existing databases create opportunities for this kind of fraud. In March 2015 the US Inspector General's office announced that it had found 6.5 million people aged 112 or over in the *living* Social Security database who had not yet been recorded as dead and were therefore not logged on the database of the dead (even though they were manifestly deceased). This creates 6.5 million opportunities for fraud (Ford 2015).

Similar processes could work for falsely logging birth certificates, through false online doctor's practices and birth certificates. This leads to the possibility of long term identity creation opening the door to a vast range of criminal possibilities using non existent people who can open bank accounts, create credit records, be eligible for benefits. As governments become digital platforms the potential for a new range of criminal activity will grow.

Next wave technology and privacy

In a fast moving era of technological change, policy makers inevitably react to what has happened. The Internet has been driven by what Vint Cerf called 'permissionless innovation' (Thierer 2014) which means that rarely has there been an attempt to understand the implications of a technology before it becomes widespread and adopted. But if we look ahead to the technologies that are about to become pervasive, we can see new implications for privacy and crime.

Internet of Things

Currently services available on the Internet are characterised by person to person communication. But as chips become embedded in all goods – even everyday items – and where each object has a unique individual identifier, we are reaching a point when, in the near future, objects will be able to communicate with each other, without human intervention. The Internet will become an Internet of Things (IoT) and humans will be surrounded by ubiquitous communication with the mobile phone assuming a primary role (Singh and Powles 2014). Each of us would be at the centre of a continuous information network connecting the objects in our lives.

There are obvious benefits, widely touted by those enthusiasts promoting IoT (*Computer Weekly* 2016). A 'smart' washing machine that is plugged into a 'smart' domestic electricity grid could automatically start when energy demand is the lowest and prices cheapest – during the middle of the night, for example. Fridges could place orders for food that was running out and your mobile phone could tell you where the cheapest pint of milk on your journey home could be located. Cars could contact garages directly if their on-board sensors detected potential mechanical problems or even if there was an accident. Moreover, the software in these objects can be programmed to learn human behaviour and adapt the possibilities on offer.

How fanciful is this prospect? We are already seeing a rapid development in remote controlled devices, which can learn human behaviour and adapt the environment to the experience and desires of the user. Modern remote controlled thermostats can learn the user's daily schedule and the preferred temperatures in the home. It can assess how long it will take to warm the house to the required temperature, taking into account the weather outside and turn on the system accordingly. When the user is out the system can turn itself down or off to avoid heating an empty home. All of this can be controlled remotely from mobile phone, laptop or tablet and these applications can provide an updated energy history. Security is increasingly built into these systems – monitors can detect the presence of carbon monoxide and the thermostat can shut off the heating system if it is the source. It can sound an emergency alarm and send the user a message over wifi to say what is happening.

Moreover modern systems connect to a range of products, not just heating – lights, appliances, wearable health bands, even the car. Going beyond these current applications, the IoT promises to connect a range of different objects in your home, work, and car. There are still formidable technical obstacles – several different technologies will have to be integrated (the operating system that connects them and becomes standard will be one of the most lucrative products on the planet).

It will also require high levels of security to protect privacy. What happens if someone hacks your fridge or car or heating system? How can user control be authenticated to avoid impersonation? How can the privacy of the user be guaranteed if ubiquitous objects have learned and stored information about the behaviour of the user? It is far from clear what kind of privacy and access regime needs to be established, what legal framework of liability will underpin these developments, what controls, if any, are appropriate.

Aspects of the IoT are very attractive to government. In the US, the House of Representatives recently passed a transportation bill requiring mandatory installation of 'black box' tracking devices in all vehicles as of 2015 (Deutsche Welle 2016).

The bill is expected to be approved by the Senate, and enables communication between cars and stationary objects, for example street lights. It is proposed that data gathered this way should be accessed by court order but there are understandable anxieties among some about how robust this protection will turn out to be at times of stress. The Internet of Things can place the citizen at the heart of a network of billions of objects, creating the potential for a vast pool of surveillance and data gathering, all of it potentially hackable unless strong safeguards are in place.

Wearables

A second technology with exciting but potentially serious policy implications is wearables, particularly those that can monitor health information. The potential for these devices is enormous and extremely important. Socialised health systems throughout the world are struggling to cope with ageing populations and the rising tide of lifestyle diseases associated with obesity, such as diabetes. The UK's NHS medical director Sir Bruce Keogh said wearables would have a vital role in securing the NHS's future: 'I see a time where someone who's got heart failure because they've had a previous heart attack is sitting at home and wearing some unobtrusive sensors, and his phone goes, and it's a health professional saying: "Mr Smith, we've been monitoring you and we think you're starting to go back into heart failure. Someone's going to be with you in half an hour to give you some diuretics",' says Keogh. Technology, said Keogh, 'enables you to predict things, to act early and to prevent unnecessary admissions, thereby not only taking a load off the NHS but, more importantly, actually keeping somebody safe and feeling good' (Campbell 2015).

It is claimed that the Apple Watch, launched in 2015, will ultimately have the ability to collect and store about 60 different types of data, including blood pressure and glucose levels (Digital Health 2016). Other devices coming on the market will be able to provide even higher levels of data gathering. This promises a significant shift in approaches to health care. Users will be able to collect their own health data instead of relying upon the health service to do this for them. Having collected the data they can also more easily own that data and become active in managing their own health as well as deciding for themselves who in the health service should have access to that data. If travelling, the data subject could decide that local doctors should have access to that data, making the effectiveness of treatment, if needed, much more probable. This data gathering is particularly important in managing conditions such as diabetes or asthma, which are characteristic of modern health challenges.

Such devices could revolutionise health care by enabling people to take control of their own health rather than be managed by professionals. Of course health professionals will lead on diagnosis and treatment, but the patient can become a more active partner in the process. And such devices can help health professionals manage their patients more effectively. Vital signs can be monitored remotely, and doctors can summon patients if they see warning indicators. There can be automatic prompting for appointments to help avoid the large amounts of wasted time where people do not turn up. An abundance of data, if properly managed, will allow health to be managed more effectively and efficiently.

But of course there are risks and possibilities for misuse inherent in this kind of technology. There are currently over 50 different wearable devices with more coming on the market by the day, measuring everything from sleep patterns, calories burned, blood pressure, heart rate and so on. Future products may even involve ingesting sensors that can assess how the body is reacting to a particular drug or treatment. But given that such devices are entering an unregulated market a number of questions remain unanswered, all of which have implications for privacy and crime.

Devices are made available with applications developed inside the current business model of the Internet. The applications gather personal health information but it may not be clear what steps the manufacturer or any third partner party has taken to make this data secure. Simple terms of service will rarely state whether data is encrypted and what access may be provided under special circumstances. Health data is extremely private and unauthorised leakage could prove very damaging to the person concerned, opening them up to blackmail or pressure (Curtis 2014).

A related question is whether data accumulated from these devices can be shared with others. Health information can be built up on a range of users, forming a large and potentially lucrative database. Fitness trackers' terms of service are often very vague and contain the proviso that data may be shared with third parties allowing companies to sell health data.

A little understood aspect of the current devices on the market is that they may have a social networking dimension which allows users to share data. Default settings are rarely set to the privacy option and casual users can end up broadcasting data inadvertently. In 2011, one vendor's product came under fire when the sexual activity it tracked (the information came from the accelerometer function) showed up in Google search results (Rao 2015). Users need to rigorously check all their privacy settings and turn off anything they are not comfortable sharing publicly. But few perform these basic checks.

The inability of policy makers to anticipate technology change means that rarely are there appropriate consumer or privacy safeguards in place when products are launched. Data protection laws are mostly rooted in the era of paper documents and there has been virtually no attempt to apply them to this new technology. If the devices are to become central to managing health in the future, then both security and ownership questions will need to be addressed.

This brings us to the central question – who owns the personal data that is generated by a wearable device? The current business model would suggest that the data is owned by the device/application developer rather than the device user. But if these devices are to be ubiquitous, the data gathered needs to have a high level of encryption and ownership given to the person whose data it is. This will require a fundamental reconfiguring of current Internet services. Without such steps there will be a vast new field for crime to play in.

Drones

Remotely Piloted Air Systems (RPAS) or drones are increasingly common and provide a range of services, including photography, land surveying, building inspection and crop analysis (House of Lords EU committee 2015). These unmanned craft have been used to indicate where water and pesticides should be directed accurately to crops, thus saving money and providing better environmental protection. They are used to inspect structures that would be hard (or expensive) to reach for human beings, such as bridges and power lines. Companies such as Amazon are considering using them to deliver parcels. Positively, some human rights activists have examined the feasibility of using drones to collect data on human rights violations in circumstances too dangerous for human rights defenders to operate. There are a wide variety of uses for the police and security agencies to exploit the ability of a drone camera to utilise GPS to follow a suspect user with a low risk to the agency staff (Griffin 2015). However, criminals can also use them to inspect the defences of compounds or remote locations (Wallace and Loffi 2015).

There are two overarching privacy issues implicated by civilian drones (a large range of issues posed by the military use of drones are outside the scope of this chapter). Given that drones operate in public spaces it is important to understand what we mean by privacy in the context of aerial surveillance. Clearly notions of privacy that are to do with secrecy or control of data are not directly affected by drone activity – more relevant is the notion of personal autonomy and the need for anonymity – the original issues raised by Warren and Brandeis in the late nineteenth century. There is a reasonable expectation that a person or persons should be able to be in a public space without being closely and continuously observed by others (unless by security agencies for public protection and in accordance with a warrant). Being followed by a drone would be disconcerting for most, but it is not clear what the relevant regulatory framework for controlling this would be, given that regulations concerning the use of drones are mostly concerned with safety in the air (Gov.uk 2015). A separate set of questions are implied if drones are used to collect, store and process data. For example, the US Supreme Court has ruled that it is not a breach of privacy (Fourth Amendment rights) to conduct surveillance of private property while flying in navigable airspace (Justicia Law 2016). Concerns have been raised about this ruling querying what would be the case if there were multiple flights over a home, using drone-obtained data in ways never envisioned by the initial collection, or retaining that data indefinitely (Thompson 2015).

In the UK 500 commercial operators are already licenced by the Civil Aviation Authority (CAA) to use drones commercially and the drone-using industry has suggested creating an online database through which commercial operators could log their flight plans and data protection policies (House of Lords EU committee 2015). The House of Lords has called for a government established database (Merrill 2015) and has raised concerns about how private security firms might make use of drones and how that use would be regulated. They also drew attention to the potential criminal and terrorist use of drones.

Conclusion

Digital technology has become fundamental to our society. It has established new forms of social connectivity and information sharing, but also the opportunities of crime based on the violation of privacy.

As with previous technologies, the context and in some ways the meaning of privacy has changed as people apply those technologies and their possibilities to their day to day life. Some, like the founder of Facebook, Mark Zuckerberg, (as well as LinkedIn founder, Reid Hoffman) have argued that privacy is no longer a norm and that social media has changed the way young people view privacy fundamentally (Johnson 2010). Others, such as campaigning groups Privacy International and the Electronic Privacy Information Center, contest this proposition.

When asked in opinion polls, a slight majority express some degree of concern about their online privacy (Cable 2015) – the exact breakdown for a poll conducted in 2014 showed that the proportion of the UK population who worry about privacy online was – never 11 per cent, sometimes 55 per cent, frequently 18 per cent, always 16 per cent. These figures seem consistent across surveys conducted in many countries over comparable time periods.

However, what these polls do not measure is how much people actually care about privacy as measured by how they actually behave in an online environment. It often appears that public opinion says one thing but does another.

And this matters, because if people are cavalier about privacy and security and the wave of technological change continues, then the scope for new criminal opportunities is bound to grow rapidly. If we take the use of passwords as a basic assessment of concern, then surveys in the US have shown that of the top ten most used passwords, six use variations on 123456. Bearing in mind that password-breaking software such as HashCat can examine 300,000 password options every second this does not seem to speak highly of security consciousness among the general public (WP engine 2016).

A survey for Ofcom in the UK in 2013 showed that over half of UK adults use the same password to access Internet sites and one in four use birthdays or names as passwords (Ofcom 2013).

It may be that the newer technologies will change public attitudes as users become aware of how valued their data is – something really akin to the oil of the digital economy in the twenty-first century (Wired 2016). It may slowly dawn on people that their data – and their ownership of this data – is one of the more valuable assets they possess. As a result, they may begin to realise that the legal possibilities for data collection, analysis and sale has implications as far reaching as its theft or illegal use.

One interesting report, on this topic and anticipating the Internet of Things, was that produced by the Altimeter Group, which showed that consumers' top concerns, when faced with the IoT, were who would have access to the data collected; extreme discomfort with the use and sale of their data in connected 'real world' environments; a lack of information about privacy; a desire to secure some value in exchange for their data; a sense that technological awareness informs trust and influences consumer expectations for engagement (Groopman and Etlinger 2015). If these new technologies lead to a more informed public, with greater control over their data and a heightened sense of the necessary means of securing that data – then the opportunity cost of crime will increase considerably with implications for privacy-related crime that may be considerable.

To make data secure will require a sophisticated understanding of the threats posed by different mechanisms of surveillance and data collection including: unique identifiers; cookies (and other associated forms of user identification); adware; spyware and malware which conduct covert data logging and surveillance; deep packet inspection (DPI); and data processing and facial recognition and surveillance technology.

The roles and responsibilities of service providers and intermediaries will need to be debated and the appropriate role for regulation determined. Such regulation will need to understand the specific challenges posed by different applications, communications platforms and business models including cloud computing, search engines, social networks and other different devices.

As governments move to become digital platforms and widely adopt e-government practices they too will need to take appropriate steps to secure their own systems from fraud and theft. Learning the lessons of the Internet, the best solutions are unlikely to be found by trying to impose a top-down centralised command and control system of data – far better to have a dispersed user-led approach where citizens are seen as owners of their data, licensing the government to use it for socially beneficial purposes.

Finally – and most important – central to any strategy for dealing with offending in the context of privacy online will be the creation of new opportunities to maintain control and ownership of personal data. Given the current business model of Internet services – free at the point of use but the user is the product – this poses a considerable commercial challenge, but one that could be met by personal ownership of data which is then leased or licenced to Internet service companies. The more we appreciate that we own our data and understand that it has economic value, the more incentive we will have to protect and secure its privacy.

References

BBC (2016a) 'BBC – What is a Cookie? - Privacy and Cookies'. Online: www.bbc.co.uk/privacy/cookies/about.

BBC (2016b) BBC News. 'Millions of US Government Workers Hit by Data Breach'. Online: www.bbc.co.uk/news/world-us-canada-33017310.

Bloustein, Edward (1964) 'Privacy as an Aspect of Human Dignity'. *NYU Law Review* 39: 962–1007.

Brenton, Myron (1964) *The Privacy Invaders*. New York: Coward-McCann.

BT.com (2016) 'Weak Passwords Are Still a Problem Post-Icloud Leak, Say Security Experts'. Online: http://home.bt.com/tech-gadgets/tech-news/internet-users-risk-online-safety-with-weak-passwords-11364001025654.

Cable, Jonathan (2015) Working Paper - An overview of public opinion polls since the Edward Snowden revelations in June 2013. UK Public Opinion Review Online: http://sites.cardiff.ac.uk/dcssproject/files/2015/08/UK-Public-Opinion-Review-180615.pdf.

Campbell, Denis (2015) 'Prof Bruce Keogh: Wearable Technology Plays A Crucial Part In NHS Future', *The Guardian*. Online: www.theguardian.com/society/2015/jan/19/prof-bruce-keogh-wearable-technology-plays-crucial-part-nhs-future.

Cifas (2016) 'Cifas - Fraudscape: UK Fraud Trends Report Out Now'. Online: www.cifas.org.uk/fraudscape_pr.

ComputerWeekly (2016) 'Realising the Benefits of a Totally Connected World'. Online: www.computerweekly.com/feature/Realising-the-benefits-of-a-totally-connected-world.

Curtis, Sophie (2014) 'Wearable Tech: How Hackers Could Turn Your Most Private Data Against You', *Telegraph.co.uk*. Online: www.telegraph.co.uk/technology/internet-security/10925223/Wearable-tech-how-hackers-could-turn-your-most-private-data-against-you.html.

Defcon (2016) 'DEF CON® Hacking Conference'. *Defcon.org*. Online: www.defcon.org/.

Deutsche Welle (2016) '"Internet of Things" Holds Promise, But Sparks Privacy Concerns' Online: www.dw.com/en/internet-of-things-holds-promise-but-sparks-privacy-concerns/a-15911207.

DigiSecrets (2013) 'The Tor Network: Privacy Protectors Or Criminal Enablers?' Online: www.digisecrets.com/web/the-tor-network-privacy-protectors-or-criminal-enablers/.

Digital Health (2016) 'Apple Bites Into Wearables'. Online: www.digitalhealth.net/features/44191//apple-bites-into-wearables.

Economist (2010) 'Clicking for Gold: How internet companies profit from data on the web', in 'A special report on managing information,' *The Economist*, vol. 394, 8671.

Entick *v*. Carrington, 19 Howell's State Trials (1765) Online: www.constitution.org/trials/entick/entick_v_carrington.htm.

EPIC and Privacy International (2007) *Privacy and Human Rights 2006*. Washington DC.: Electronic Privacy Information Center.

Federal Trade Commission (1999) 'Self-regulation and Privacy Online: A Report to Congress'. Online: www.ftc.gov/os/1999/07/privacy99.pdf.

Financial Times (2016) 'Experts Say TalkTalk Had 11 Serious Website Vulnerabilities'. Online: www.ft.com/cms/s/0/e5eead0c-7f0b-11e5-98fb-5a6d4728f74e.html#axzz3sKgJt9Nd.

Florida *v*. Riley 488 U.S. 445 (1989).

Forbes Magazine (2016a) '*Forbes Welcome*'. Online: www.forbes.com/sites/perryrotella/2012/04/02/is-data-the-new-oil/.

Forbes Magazine (2016b) '*Forbes Welcome*'. Online: www.forbes.com/sites/realspin/2012/01/31/privacy-free-expression-and-the-facebook-standard/.

Ford, Paul (2015) 'The Final File'. *New Republic*. Online: www.newrepublic.com/article/122456/final-fil.

Garnham, Oliver (2008) '5,000 Prison Staff Exposed By Lost Hard Drive'. *PC Advisor*. Online: www.pcadvisor.co.uk/news/security/5000-prison-staff-exposed-by-lost-hard-drive-104096/.

Gov.uk. (2012) '*Digital Efficiency Report*'. Online: www.gov.uk/government/publications/digital-efficiency-report/digital-efficiency-report.

Gov.uk. (2015) '*New Regulations For Remotely Piloted Air Systems (RPAS) Go Live - News Stories*'. Online: www.gov.uk/government/news/new-regulations-for-remotely-piloted-air-systems-rpas-go-live.

Gov.uk. (2016) '*Performance*'. Online: www.gov.uk/performance.

Griffin, Andrew (2015) 'The Lily Drone Might Be The Most Exciting Drone Ever'. *The Independent*. Online: www.independent.co.uk/life-style/gadgets-and-tech/news/lily-drone-flying-selfie-camera-follows-you-around-as-you-ski-or-run-10250471.html.

Groopman, Jessica and Etlinger, Susan (2015) 'Consumer Perceptions of Privacy in the Internet of Things What Brands Can Learn from a Concerned Citizenry Altimeter 201'. Online: www.altimetergroup.com/pdf/reports/Consumer-Perceptions-Privacy-IoT-Altimeter-Group.pdf.

Gross, Grant (2016) 'Big Data Collection Collides With Privacy Concerns, Analysts Say'. *PCWorld*. Online: www.pcworld.com/article/2027789/big-data-collection-collides-with-privacy-concerns-analysts-say.html.

The Guardian (2016) '*The NSA Files*'. Online: www.theguardian.com/us-news/the-nsa-files.

Hawtin, Mendel and Puddephatt, Wagner (2013) *Global Survey on Freedom of Expression and Privacy*. Paris: UNESCO.

Hosein, G. (2006) 'Privacy as freedom', in R. Jorgensen (ed.) *Human Rights in the Global Information Society*. Cambridge, MA.: MIT Press.

House of Lords European Union Committee (2015) 'Civilian use of drones in the EU HL paper 122 2015'. Online: www.publications.parliament.uk/pa/ld201415/ldselect/ldeucom/122/122.pdf.

Johnson, Bobbie (2010) 'Privacy No Longer A Social Norm, Says Facebook Founder'. *The Guardian*. Online: www.theguardian.com/technology/2010/jan/11/facebook-privacy.

Justicia Law. (2016) Available at: https://supreme.justia.com/cases/federal/us/488/445/case.html.

Lessig, Lawrence (1999) *Code and Other Laws Of Cyberspace*. New York: Basic Books.

Matthews, Roger (2016) *What is to Be Done About Crime and Punishment?: Towards a Public Criminology*. London: Palgrave Macmillan.

McKinsey Global Institute (2011) *Internet matters: The Net's sweeping impact on growth, jobs, and prosperity* Online: www.eg8forum.com/fr/documents/actualites/McKinsey_and_Company-internet_matters.pdf.

Merrill, Jamie (2015) 'Lords Inquiry Calls For App To Track Suspicious Drones'. *The Independent*. Online: www.independent.co.uk/life-style/gadgets-and-tech/news/house-of-lords-inquiry-calls-for-app-to-track-suspicious-drones-10086125.html.

Miller, Claire (2014) 'Americans Say They Want Privacy, But Act As If They Don't'. Nytimes.com. Online: www.nytimes.com/2014/11/13/upshot/americans-say-they-want-privacy-but-act-as-if-they-dont.html?_r=0.

Niemietz *v.* Germany (1992), 16 EHRR 97. Para 29. 2016.

Ofcom (2013) 'UK Adults Taking Online Password Security Risks'. Online: http://media.ofcom.org.uk/news/2013/uk-adults-taking-online-password-security-risks/.

PC Magazine (2016) 'Web Bug Definition From PC Magazine Encyclopedia'. Online: www.pcmag.com/encyclopedia/term/54280/web-bug.

Privacy International (2006) *Privacy and Human Rights, An International Survey of Privacy Laws and Practice.* Online report, available at: http://gilc.org/privacy/survey/intro.html.

Rao, Leena (2015) 'Sexual Activity Tracked By Fitbit Shows Up In Google Search Results'. *TechCrunch.* Online: http://techcrunch.com/2011/07/03/sexual-activity-tracked-by-fitbit-shows-up-in-google-search-results/.

Singh, Jat and Powles, Julia (2014) 'The Internet of Things - The Next Big Challenge To Our Privacy'. *The Guardian.* Online: www.theguardian.com/technology/2014/jul/28/internet-of-things-privacy.

Statista (2016) 'Number of Facebook Users Worldwide 2008-2015 | Statistic'. Online: www.statista.com/statistics/264810/number-of-monthly-active-facebook-users-worldwide/.

Thierer, Adam D. (2014) *Permissionless Innovation.* Mercatus Center at George Mason University.

Thompson, Richard (2015) 'Domestic Drones and Privacy: A Primer Congressional Research Service'. Online: http://fas.org/sgp/crs/misc/R43965.pdf.

Wall, David (2007) *Cybercrime - The Transformation Of Crime In The Information Age.* Cambridge: Polity Press.

Wallace, R. and Loffi, J. (2015) 'Examining Unmanned Aerial System Threats & Defenses: A Conceptual Analysis', *International Journal of Aviation, Aeronautics, and Aerospace* 2 (4). Online: http://commons.erau.edu/cgi/viewcontent.cgi?article=1084&context=ijaaa.

Ward, Alexander (2015) 'Silk Road Founder Ross Ulbricht Is Sentenced To Life In Prison'. *The Independent.* Online: www.independent.co.uk/news/world/americas/silk-road-website-founder-ross-ulbricht-dread-pirate-roberts-is-sentenced-to-life-in-prison-10286234.html.

Warren, S. and Brandeis, L. (1890) 'The Right to Privacy', 4 *Harvard Law Review.*

Westin, Alan F. (1967) *Privacy and Freedom.* New York: Atheneum.

Wikipedia (2016) 'History Of Cryptography'. Online: https://en.wikipedia.org/wiki/History_of_cryptography.

WIRED (2016) 'Data Is The New Oil Of The Digital Economy'. Online: www.wired.com/insights/2014/07/data-new-oil-digital-economy/.

WP Engine (2016) 'Unmasked: An Analysis of 10 Million Passwords'. Online: http://wpengine.com/unmasked/.

Zuckerman, Ethan (2014) 'The Internet's Original Sin'. *The Atlantic.* Online: www.theatlantic.com/technology/archive/2014/08/advertising-is-the-internets-original-sin/376041/.

Section 2

Chemical and biological technologies and crime

11

Crime and chemical production

Kimberly L. Barrett

Introduction

Over the past several decades, the number of synthetic chemicals in use has grown exponentially. There are over 84,000 different chemical substances registered with the US Environmental Protection Agency for commercial use (EPA 2014a). The EPA's High Production Volume challenge program identified over 2,200 chemicals produced in or imported to the United States at rates of 1 million pounds or more per year (EPA 2013). Synthetic chemicals have a range of different commercial uses, but a few examples of products containing synthetic chemicals include pesticides, building materials, cleaning products, toys, cosmetics, fuels, medicines, and food packaging (Schwartz *et al.*, 2008). The sale of these goods makes the production of synthetic chemicals a multi-trillion dollar industry; estimates suggest pesticides alone constitute a $40 billion world market (Grube *et al.*, 2011).

To be sure, the production of some synthetic chemicals represents major, important public health gains. These gains include the ability to use synthetic chemicals to prevent, treat, and cure certain diseases. Examples of such synthetic chemicals include disinfectants, chemotherapeutic agents, and antibiotics (Landrigan and Etzel 2014). However, for a number of chemical agents, scientists have long documented serious deleterious environmental and public health consequences. For example, exposure to high volumes of inorganic metals (lead, mercury, manganese) can result in acute symptoms such as dizziness, nausea, and headaches or long-term symptoms such as diminished IQ and emotional behavioral disorders (Clarkson 1987), and for cases involving serious over-exposure, death. A number of synthetic chemicals have also been identified as carcinogens (Davis 2007), and another class of human-produced chemicals known as endocrine disruptors have the ability to invoke hormonal changes and threaten the healthy development of the endocrine system (Colborn *et al.* 1997). In spite of industry resistance, harms associated with synthetic chemicals have given rise to laws designed to regulate the production of synthetic chemicals. While these harms and laws have existed for decades, criminologists have only relatively recently examined crimes associated with chemical production and violations of these laws (Katz 2010; Pearce and Tombs 1998; Ruggiero and South 2013; Stretesky *et al.* 2014; Walgren 2000).

The goal of this chapter is to contextualize, explain, and exemplify crimes associated with chemical production in the United States. Moreover, due to their relationship with the Treadmill of Production and wealth accumulation, the crimes associated with chemical production are described as green crimes. As such, this chapter begins by providing a brief history of the industrial production of synthetic chemicals. Next, background on Schnaiberg's (1980) Treadmill of Production (ToP) theory is provided, and intersections between the ToP, chemical production, and criminology are discussed. The chapter transitions to a discussion of major federal policies related to chemical production. Finally, this chapter concludes with a brief summary and suggestions for future study.

A brief history of chemical production

The Industrial Revolution of the late eighteenth and nineteenth centuries ushered in an era of production processes and technologies that would prove to have lasting impacts on the natural environment and public health. One of the characteristics of the Industrial Revolution was its heavy dependence on, and consumption of, fossil fuels. Major types of fossil fuels include coal, oil (petroleum), and natural gas. During the Industrial Revolution, fossil fuels were used to power a growing number of technological developments such as the steam engine and automobile. The environmental impact (direct and indirect) from this was profound; estimates suggest from the year 1751 to 1950 the amount of carbon emissions from fossil fuel use grew from 3 million metric tons annually to 1,630 million metric tons annually (Boden *et al.* 2010).

The growing use and development of fossil fuels through the industrial revolution is important for understanding chemical production. According to Schnaiberg (1980), many modern synthetic chemicals that are in wide use today are derived from oil and coal (e.g. petrochemicals refer to those chemicals that are synthesized from petroleum or natural gas). From fossil fuels, scientists were able to create new synthetic chemicals to use in the production process, as well as create goods for consumption. For example, many herbicides and pesticides used to grow crops are synthesized using petroleum. Synthetic toothpastes and detergents were first marketed in the early 1900s, and synthetic paints and coatings became available for household use in the 1930s (Geiser 2015). Scientists interested in synthetic estrogens discovered bisphenol A (derived from petroleum) in 1891 and diethylstilbestrol or DES (synthesized from coal-tar derivatives) in 1938 (Langston 2010).

Chemical production also become intertwined with war and the production of goods and materials associated with combat. Evidence of this can be found in examining the relationship between chemical production and US involvement in war, detailed in Russell's *War and Nature* (2001) as well as Langston's *Toxic Bodies* (2010). Even before the US officially entered World War I, American chemists were working on products put in demand by the war economy. For example, war generated an increase in the value of cotton, and in turn, an increased demand for pesticides that attacked pests that destroyed cotton plants (in particular, the boll weevil). As a consequence, during this time calcium arsenate was discovered and subsequently used as an insecticide (Russell 2001). During World War I, chemical production played a central role in the development of weapons, particularly the use of toxic gases (e.g. mustard gas), so much so that some referred to World War I as "the chemists' war" (Russell 2001). Then Secretary of War, Newton Baker, noted that of the 17,000 chemists in the United States, one third were "directly in the Government Service during the recent war," (1919: 921). He also stated, "I do not believe it will be discovered that any profession contributed a larger percent of its members directly to the military service, or the results of the activities of any profession were more essential to our national success than that of chemists" (1919: 921).

Chemistry would also play a central role in World War II. In the late 1940s it was discovered that the chemical DDT could be used to kill mosquitos (Russell 2001; Langston 2010). This was significant for the US military because malaria was a serious health concern for troops (Beadle and Hoffman 1993). In 1947 it was reported that malaria had been responsible for more deaths of US soldiers than combat. Chemists also found themselves involved in the production of new technologies for bombs. In 1942 chemists discovered napalm, and produced the first napalm bomb (Russell 2001).

During World War II, the US military and American scientists became deeply invested in harnessing nuclear technology and developing the atom bomb. The efforts to do so were conducted under the title "The Manhattan Project," and the direct and indirect consequences of The Manhattan Project are still felt around the world today. Criminologists have analyzed the harms stemming from atomic weapons from several perspectives. For example, the possession of and threat to use nuclear weapons has been analyzed as a state crime (Kramer and Bradshaw 2011; Kauzlarich and Kramer 1998); collusion of the state and private sector in the development of nuclear arms has been argued as state-corporate crime (Barrett 2013; Bruce and Becker 2007); and disposal of radioactive waste has been analyzed as a form of eco-crime (Walters 2007).

Following the war, the products designed for military use came to be marketed in a peace-time economy. DDT became available for civilian use, nuclear power plants were built, and atomic weapons continued to be constructed. In the years following World War I, advertisements for pesticides portrayed insects as "enemies" (Russell 2001). This marketing strategy and use of chemical technology is noteworthy; rather than humans seeing themselves as an integrated part of a large, complex ecosystem, humans perceived nature as theirs to dominate and control through synthetic chemicals (for elaboration on cultural components of green criminology, see Brisman and South 2014). In the years following World War II, the use of synthetic chemicals boomed; from 1947 to 1960, US synthetic pesticide production alone grew by over 400 percent (Carson 1962: 17). The motto of DuPont, one of the leading producers of synthetic chemicals, was coined as: "Better things for better living – through chemistry." While nuclear technology had been used during the Manhattan Project to develop the atom bomb, the United States continued to build its nuclear arsenal well after the end of World War II. Government contracts were awarded to the private sector to continue the development of nuclear arms at sites across the country (Kauzlarich and Kramer 1998). Nuclear technology was also used to produce energy and generate electricity. The US government offered subsidies and incentives to construct nuclear power plants, and the use of subsidies and other government incentives to encourage the development of nuclear technology continues today (Koplow 2011).

The expansion of the synthetic chemical industry would come under heavy scrutiny in the 1960s and 1970s. One of the most notable critics of this expansion was biologist Rachel Carson, who delivered several challenges to the growing chemical industry in her book *Silent Spring*. In *Silent Spring*, Carson connects the growth of synthetic chemicals, including radiation, DDT, and other pesticides, to increases in cancers, illnesses in humans and non-human animals, genetic mutations, pollution of air, soil, and water, as well as contamination of food chains. Carson argued that "insecticides" would be more aptly named "biocides" and called into question the logic behind the popularity of dangerous synthetic chemicals: "How could intelligent beings seek to control a few unwanted species by a method that contaminated the entire environment and brought the threat of disease and death even to their own kind? Yet this is precisely what we have done" (Carson 1962: 8). Carson's book contributed to the launch of the environmental movement, a period of time in US history where citizens became actively concerned with pollution of the natural environment. *Silent Spring* also helped inspire the first Earth Day held on April 22, 1970, drawing participation from millions of American activists.

In addition to Carson's work, a series of human-caused environmental disasters heightened national concern over pollution, with synthetic chemical production a central issue. For example, in the early 1970s, the Northeastern Pharmaceutical and Chemical Company, Inc. of Verona, Missouri began selling toxic waste in the form of dioxin to a local waste hauler to dispose of. Instead of disposing of the chemical properly, the waste disposer sprayed the dioxin-laden waste on the roads of Times Beach, Missouri as well as in horse stables. After horses in the stables died, the US EPA investigated the cause, which eventually led to Times Beach residents being evacuated in 1982. Following 20 years of site remediation, Times Beach was reopened as a state park in 2014. The 1969 Cuyahoga River fire in Ohio (associated with industrial waste), and the resurfacing of toxic waste generated by the Hooker Chemical Company at Love Canal in the 1970s, also brought national awareness of the dangers associated with the disposal of toxic chemicals. The environmental justice movement captured national attention in the early 1980s when Warren County, North Carolina was targeted for the siting of a landfill for polychlorinated biphenyl (PCB) laced soil (Bullard 2000; Bullard and Johnson 2000; McGurty 2009). At the time, Warren County was predominantly African American, and the county was also one of the most economically disadvantaged counties in the state. Warren County residents demonstrated and organized in resistance to the site, bringing together issues of civil rights, social justice, and environment (Taylor 2000). Activism in Warren County drew support from civil rights organizations nationwide and directed attention to environmental injustice and racism (Bullard and Johnson 2000).

Environmental activism and mounting concern over environmental issues led to political responses. For example, the US EPA was established less than a year after the first Earth Day. Several pieces of major federal environmental legislation were passed or strengthened in the 1970s, including, but not limited to, the Clean Air Act, the Clean Water Act, the Resource Conservation and Recovery Act, and the Toxic Substances Control Act. Activism in Warren County compelled politicians to call for a study on environmental justice and, as a result, the US Government Accountability Office published its first study on environmental justice in 1983 (US Government Accountability Office 1983).

While the environmental movement did mark progress on some environmental fronts, serious problems related to chemical production, including environmental injustices, would continue over the next several decades. For example, Lavelle and Coyle (1992) investigated the EPA's civil court docket and found a number of race disparities with respect to the application of environmental laws. The study found penalties for violations of hazardous waste laws at sites with the highest white population were 500 percent higher than penalties for violations of hazardous waste laws at sites with the highest minority population (Lavelle and Coyle 1992). The relationship between the Department of Defense and plants producing nuclear weapons would hamper EPA and Department of Justice investigations into violations of environmental laws at nuclear weapons facilities (e.g. the Rocky Flats site in Colorado—see Iversen 2013). Participation in the arms race was used to justify disregard for compliance with environmental laws with respect to nuclear weapons production (Bruce and Becker 2007; Shrock 2015). In 1984, a massive leak of methyl isocyanate (MIC) at a US-based Union Carbide plant in Bhopal, India killed thousands of Indian citizens, and injured several thousand more (Pearce and Tombs 1998). The plant was producing pesticide, and the leak represents one of the worst industrial disasters in world history. Criminologists have analyzed the Bhopal tragedy, and have argued that crimes associated with chemical production are deserving of more attention from criminologists (Pearce and Tombs 1998; Lynch *et al.* 1989).

In the 1980s and 1990s, scientists further explored the impact of synthetic chemicals on the reproductive and endocrine systems (Langston 2014). Zoologist Theo Colborn linked industrial pollution to reproductive complications and behavioral changes in wildlife as well as humans (Colborn *et al.* 1997). Colborn and colleagues (1997) elaborated on an entire class of chemical agents termed endocrine disruptors, so-named for their ability to alter the healthy functioning of the endocrine system. Endocrine disruptors are numerous and can be found in a range of products, including cosmetics, paints, electronics, soaps, and plastics, to name a few. Bisphenol A (BPA), for example, is classified as an endocrine disruptor, and can be found in water bottles, baby bottles, cell phones, and other household goods. Exposure to BPA, even in very small amounts has been shown to cause behavioral changes in children, reproductive complications, and other adverse health outcomes (The Endocrine Disruption Exchange 2015). Interest in epigenetics – the study of how environmental factors change gene expressions – grew in the 1990s, and scientists documented associations between synthetic chemicals and harm to developing fetuses, including harms that may not manifest for years after initial exposures (Langston 2014).

Today, the United States is a major player in the global production of synthetic chemicals. Geiser (2015) reports that the US chemical industry yields $540 billion in annual sales, accounting for almost 4 percent of the US national gross domestic product. While activism has inspired some changes, the growing volume of synthetic chemicals in the environment continues to have serious public health consequences. Chemical technology has developed swiftly, and humans today are contending with body burdens of chemicals that did not exist a generation ago (Hayes and Chaffer 2010). Furthermore, many of these compounds resist biodegradation, demonstrate an ability to bioaccumulate in the body, and biomagnify as they move through food chains. In a study analyzing biomonitoring data obtained from a nationally representative sample, the Centers for Disease Control and Prevention (2009) detected BPA in urine samples of 90 percent of participants. The Environmental Working Group (2015) estimates 12.2 million adults are exposed to known or probable human carcinogens daily through chemicals present in various personal care products. Environmental justice issues relating to exposure to synthetic chemicals persists today, as studies document that economically disadvantaged communities and racial and ethnic minority communities are disproportionately exposed to industrial pollutants (Bullard *et al.* 2008; Bullard and Wright 2008; Chakraborty 2009, 2012; Chakraborty and Zandbergen 2007; Wright 2003).

The treadmill of production, criminology and synthetic chemicals

The surge in synthetic chemical production in the years following World War II is well documented, and has captivated social scientists interested in explaining and understanding this phenomenon. In 1980, sociologist Allan Schnaiberg published *The Environment: From Surplus to Scarcity* where he puts forward Treadmill of Production theory (ToP). ToP seeks to explain the rapid environmental degradation that occurred in the modern era, years following World War II. In brief, ToP notes that capitalism demands constant, unlimited growth. This growth requires increasing use of the earth's (limited supply of) natural resources. Schnaiberg describes how technological advances, including those made in synthetic chemistry, allowed for owners of the means of production to extract a larger quantity of the earth's natural resources at a faster pace, producing an increased quantity of goods. When the owners of the means of production realized technology could produce a larger number of goods at a faster rate than human workers, technology began to displace human laborers at an increasing rate (increasing surplus value for owners of the means of production). In addition to depletion of natural resources, production practices were adding an increasing volume of pollution to the environment. Schnaiberg (1980) argues that both depletion of resources (ecological withdrawals) and increases in pollution (ecological additions) are disruptive to nature's own production practices, and he terms this disruption ecological disorganization (see also Foster 1997). Schnaiberg warns that capitalism promotes ecological disorganization, and this practice is unsustainable. As production practices convert energy from raw materials, the energy takes on less organized forms (entropy), and it has been argued that capitalism pushes us towards a state of increased entropy, where eventually there will be no more energy left to convert (Burkett 2009; Foster 1997; Stretesky et al. 2014).

Recently, green criminologists have drawn upon Schnaiberg's ToP theory to describe and explain environmental harm (Long et al. 2012; Lynch et al. 2013; Stretesky et al. 2013; Stretesky et al. 2014). ToP is especially well suited for criminologists in the area of green criminology who study a range of different types of environmental harms (for additional exploration of the relationship between green criminological thinking and technology, see Rob White's Chapter 14, this volume). Many green criminologists examine harms against the environment that are expressly prohibited by the criminal law, as well as harms that are not illegal, per se, but have been demonstrated to be harmful by scientists. This inconsistency – some harms are deemed illegal by the criminal law while others are not – has long been acknowledged by critical and radical criminologists, who have called attention to bias in the criminal law (Lynch and Michalowski 2006; Sutherland 1949).[1] Specifically, they argue that the criminal law targets harmful behaviors committed by the working class, while minimizing or outright ignoring harmful behaviors committed by the rich, exemplifying the relationship between political economy and the law. When the phrase green criminology was coined in the early 1990s (Lynch 1990), political economy was described as central to understanding environmental degradation. To address this, green criminologists have adopted the term green crime to describe the phenomena that they study. Stretesky et al. (2014) define green crime as: "acts that cause or have the potential to cause significant harm to ecological systems for the purposes of increasing or supporting production," (p.2). This is especially important for criminological studies examining crime and chemical production, because laws may not capture the facts that: (1) thousands of synthetic chemicals in use are untested and unregulated, (2) debates over synthetic chemical safety under "threshold" quantities or "background" levels ensue, and (3) while synthetic chemicals were being produced as early as 1850, the United States did not have a federal law governing synthetic chemicals until 1976. Laws surrounding regulation of synthetic chemicals in particular have been heavily critiqued as favoring the industry at the expense of public health (Fagin and Lavelle 1999; Markowitz and Rosner 2000, 2013a, 2013b; Rosner and Markowitz 1985).

The ToP argues that continuing to produce in the destructive way that humans have been leads to growing ecological disorganization. Recently, criminologists have advocated for an understanding of the production of ecological disorganization as criminal (Lynch et al. 2013). In this argument, the problem of ecological disorganization stems from capitalism's insatiable demand to accumulate wealth. Ecological disorganization represents a green crime, because (1) ecological disorganization stems from the capitalist's desire to accumulate wealth, and (2) scientific studies have documented the harms associated with ecological disorganization. For example, over the past several years the practice of hydraulic fracturing ("fracking") has become more widespread as a method for extracting natural gas from the earth. Fracking is heavily dependent upon synthetic chemicals to access natural gas, and most natural gas operations (90 percent) in the US utilize fracking (Colborn et al. 2011). Colborn et al. (2011) estimate that in the western United States alone, fracking utilizes approximately one million gallons of fluid containing toxic chemicals in the natural gas extraction process. A key component of ecological disorganization is that ecological disorganization alters the way that organisms function in and relate to their environment. Fracking has been associated with air, soil, and water pollution, and in addition, fracking threatens biodiversity, and generates habitat loss and fragmentation (for review see Kiviat 2013). As organisms adapt to the damage fracking imposes (e.g. ground and water pollution, habitats and food source destruction), nature's production practices are disrupted, and fracking threatens the ability of these species' survival, generating additional ecological disorganization (e.g. Lynch et al. 2015).

Closely related to the notion of ecological disorganization are the concepts of ecological additions and ecological withdrawals. Schnaiberg (1980) argued that ecological additions and ecological withdrawals represent the two major ways in which the environment and the economy intersect. Ecological additions refer to toxic byproducts of production (e.g. pollution). Ecological withdrawals refer to the raw materials taken from the natural environment for the purposes of production. Stretesky *et al.* (2014) argue that ecological additions and withdrawals represent green crimes, and the production of synthetic chemicals provides several examples of ecological additions and withdrawals. To maintain the example of hydraulic fracking for natural gas, fracking deposits hundreds of thousands of chemicals into the natural environment (ecological additions). Colborn and colleagues (2011) identified 632 chemicals involved in US natural gas operations, and examined the public health effects of 353 chemicals (those identified by Chemical Abstract Service numbers). Among these chemicals, nearly half could affect the brain/nervous system, cardiovascular system, immune system, and the kidneys, while over one third could impact the endocrine system (Colborn *et al.* 2011). With respect to ecological withdrawals, chemical production is highly dependent on fossil fuels, which are finite and nonrenewable. The practice of fracking not only requires synthetic chemicals, but the fuels extracted are used by the chemical industry in the production of a range of synthetic chemicals and goods (for more details on the connection between fracking and the ToP, see Stretesky *et al.* 2013: 61).

Schnaiberg (1980), Gould *et al.* (2008), and Stretesky *et al.* (2014) note that there are distinct groups of actors in control of, and impacted by, the operation of the ToP. While each group of actors shares a connection to the ToP, they have differing stakes in the ToP's operation (Schnaiberg 1980: 211). The first group of actors, capital (or producers, corporate actors) has a vested interest in maximizing surplus value by extracting natural resources and producing goods with minimal capital investment. This ramps up the pace of the treadmill, increases ecological additions, withdrawals, and ecological disorganization. With respect to chemical production, there are a number of corporate actors deeply committed to growth in the synthetic chemical industry. Chemical and Engineering News recently published their list of the global top 50 chemical firms for 2014, identifying BASF ($78.6 billion in chemical sales for 2013), Sinopec ($60.8 billion in chemical sales for 2013), and Dow Chemical ($57 billion in chemical sales for 2013) as the top three largest global firms (Tullo 2014). It is worth noting, however, that the synthetic chemical industry is wedded to other economic sectors (e.g. energy), who may also have an interest in continued growth in synthetic chemical production.

A second group of actors connected to the treadmill is the state. According to Schnaiberg (1980) the state has conflicting roles with the ToP. On one hand, expanding the ToP can result in increases in tax revenue. The state is further incentivized to promote growth in the ToP by massive campaign contributions from PACs and campaign lobbyists working on behalf of the private sector. For example, with respect to synthetic chemicals, data from the Center for Responsive Politics (2015) reveals that in 2014 Chemical and Related Manufacturing sector PACs and individuals contributed nearly $12.2 million dollars to political campaigns, representing a 94.47 percent increase from total annual contributions twenty years prior in 1994. Also in 2014, the Chemical and Related Manufacturing Sector spent nearly $65 million dollars on lobbying efforts, representing a 91.45 percent increase from total lobbying expenses ten years prior in 2004 (Center for Responsive Politics 2015). As noted previously, the state demonstrates a long, enmeshed history with the chemical industry, including state-corporate complicity between Dow Chemical and the US government in environmental offending (Katz 2010).

On the other hand, growth of the ToP in the synthetic chemical industry can be at odds with public health, and constituents may compel their representatives to impose more stringent environmental regulations or restrictions on chemical production, which may slow down the ToP. The conflicting role of the state's relationship with the ToP is captured in the mission statement of the EPA (2014b), which claims:

> environmental protection is an integral consideration in U.S. politics concerning natural resources, human health, economic growth, energy, transportation, agriculture, industry, and international trade, and these factors are similarly considered in establishing environmental policy.

In this statement, the EPA recognizes that environmental policy should reflect human health concerns as well as economic growth interests. It is important to note that these interests are often in conflict with one another, particularly with respect to environmental policy. This is significant for synthetic chemical production, because the EPA is a state agency that is heavily involved in regulating, enforcing, and establishing policies that relate to chemical production and disposal (see below for elaboration).

Laborers construe a third group of actors connected to the treadmill. For many laborers, growth in the ToP, and widespread use of synthetic chemicals has led to unemployment. As technological advancements, including synthetic chemicals, have come to replace the manual labor provided by workers, many workers find themselves out of a job. In addition, as companies came to compete in a global market, factory and manufacturing positions began leaving the United States, and organizations began focusing operations in countries with weak, few, or no protections for workers or the environment (Michalowski and Kramer 1987; Simon 2000; White 2011). This would further drive down the cost of production for corporations and would mean unemployment and layoffs for thousands of workers employed in US manufacturing. Laborers working with synthetic chemicals may also be at heightened risk of harm and injury. In the development of leaded gasoline, tetraethyl lead exposure contributed to workplace fatalities, as well as acute (e.g. dizziness, hallucinations) and long-term (e.g. brain damage) suffering for employees (Markowitz and Rosner 2013b). According to Markowitz and Rosner (2013a), the first citation issued by the US Occupational Safety and Health Administration was in response to laborers suffering from mercury exposure at an Allied Chemical Corporation facility producing chlor-alkali in Moundsville, West Virginia. Laborers, like the state, also endure conflicting roles in the ToP; while bringing a plant to a community may generate short term gains for the laborer (wages, benefits), environmental degradation and hazardous work environments present laborers with long-term environmental and public health concerns.

Non-government organizations (NGOs) constitute a fourth group of actors connected to the ToP. Several NGOs have devoted significant efforts to public health and environmental justice issues perpetuated by the treadmill (Bullard and Johnson 2000; Gould et al. 1996). It is important for criminologists studying green crimes to understand the work of NGOs because NGOs play an important role in raising awareness about environmental issues and advocating for changes in environmental policy. As Stretesky et al. (2014) note, NGOs are not restricted to focus only on environmental problems that are prohibited by the state. Rather, NGOs bring awareness to environmental problems regardless of the stance of the criminal law or government organizations. This activism can serve to educate the public about a problem, generate outcry, and prompt state intervention or consumer action (e.g. boycotts), which can force the corporate sector to change production practices. For example, efforts by grassroots environmental justice organizations drew attention to the disproportionately high presence of toxic waste in low-income communities as well as predominantly African American and Latino communities (Bullard 2000). These grassroots efforts have led to the founding of the

EPA's Office of Environmental Justice, and grassroots environmental justice activism has also led to relocations and buyouts for residents of polluted areas (Bullard and Johnson 2000; Brulle and Pellow 2006). The activities of NGOs have been associated with changes in synthetic chemical production; for example, following Greenpeace's "Play Safe" campaign, which raised awareness regarding the presence of phthalates in children's soft plastic toys, 8,000 citizens wrote their congressional representatives demanding action on this issue. In 2008, President George W. Bush signed into effect law prohibiting the use of lead and several types of phthalates in products made for children up to 12 years of age (Di 2015).

It is important to note that while many individuals and systems are linked to the ToP, it is incorrect that the ToP impacts everyone equally (Gould et al. 2008; Gould et al. 1996). This is true with respect to both the burdens as well as the benefits of the ToP, as the ToP's burdens and benefits are linked to one another (Gould et al. 2008). As Gould et al. state, "Environmental racism and class inequalities affect populations in a *relational* fashion. That is, one group's access to clean living and/or working environments is often made possible by the restriction of another group's access to those same amenities" (2008: 70). There are many examples in the literature of economically disadvantaged communities, African American communities and Latino communities being disproportionately exposed to environmental hazards and pollution (Mohai et al. 2011; Pastor et al. 2001; Pellow 2002; Stretesky and Lynch 2002). Studies have also documented an association between an increase of environmental goods like tree coverage and green space in more economically affluent and predominately white communities (Heynen et al. 2006; Dai 2011; Landry and Chakraborty 2009).

This is also exemplified in a number of environmental justice studies examining race, class, and the chemical industry (Checker 2005; James et al. 2012; Lerner 2005; Wright 1998, 2003). For example, 9 percent of all chemical manufacturing shipments in the United States are shipped from Louisiana (Noonan 2015). The chemical industry accounts for 25 percent of all manufacturing activity in the state of Louisiana (Noonan 2015), and the 100-mile span of southern Louisiana from Baton Rouge to New Orleans—termed the Chemical Corridor and also Cancer Alley (Wright 2003)— represents the largest concentration of petrochemical and toxic waste facilities in the US (Adeola 2000). In 2014, the chemical industry in the state of Louisiana was responsible for over 87 million pounds of toxic releases, which accounted for over 63 percent of the total weight of all toxic releases in the entire state (Right to Know Network 2016). Communities along the Chemical Corridor have high rates of poverty, low levels of income, are approximately 40 percent black, and 55 percent white (James et al. 2012). Residents of the Chemical Corridor have experienced property damage as well as very serious adverse health complications associated with the plant, including fatalities associated with accidents and explosions near chemical plants (Lerner 2005; Singer 2011; Wright 1998, 2003). Taken together, data suggest ecological additions, withdrawals, and disorganization associated with the ToP are not evenly distributed, and are in fact closely linked to race, class, ethnicity, and other structural level variables.

Laws and legislation concerning chemicals

According to ToP theorists, the creation and enforcement of environmental law and policy represents a key role of the state in decelerating the treadmill (Schnaiberg 1980; Gould et al. 2008). The creation of federal environmental laws as well as the establishment of a federal entity that would regulate these laws (the EPA) is often regarded as a major achievement of the environmental movement. The reach of these laws, and the power of these organizations to enforce and prosecute violations has, however, been called into question by both industry representatives who resist regulation, as well as environmentalists, scientists, and consumer advocates who have demonstrated cases where the laws, and/or their regulating authorities have failed to provide adequate protection (Colborn *et al.* 1997; Lavelle and Coyle 1992; Fagin and Lavelle 1999; Geiser 2015; Vogel and Roberts 2011). Below, a few examples of key pieces of federal legislation that regulate synthetic chemical production are identified and briefly described.[2]

Federal Insecticide Fungicide and Rodenticide Act

The Federal Insecticide and Rodenticide Act (FIFRA) was passed in 1947. According to the EPA (2012), the goal of FIFRA was to establish labeling provisions and a process for registering pesticides with the US Department of Agriculture. Since its inception, FIFRA has been amended several times. One of the most significant amendments occurred in 1972, by way of the Federal Environmental Pesticide Control Act. Following this amendment, regulation of the sale and use of pesticides was transferred from the Department of Agriculture to the EPA. FIFRA requires that all pesticides sold or distributed in the United States are registered by the EPA. In order for pesticides to be registered, FIFRA mandates that (in addition to composition and labeling requirements), "it will perform its intended function without unreasonable adverse effects on the environment." FIFRA's definition of "unreasonable adverse effects on the environment"[3] requires the administrator to take into consideration, among other items, the "economic, social, and environmental costs and benefits of the use of any pesticide." This is consistent with the EPA's mission to regulate environmental hazards while bearing in mind economic considerations.

In 1996, the US Congress passed the Federal Food Quality Protection Act (FFQP), which would amend FIFRA once more. The FFQP would expressly acknowledge that infants and children have unique vulnerabilities when it comes to exposure to toxic chemicals, including those found in pesticides (Landrigan and Goldman 2011), differentiating the FFQP amendment from prior iterations of FIFRA. Among other changes, the FFQP also required the EPA to screen pesticides for endocrine disruptors, and created incentives for safer pesticides. Taken together, the 1996 amendments imposed more stringent standards on pesticides, and applied these standards retroactively, requiring thousands of registered pesticides to "re-register." Data from the National Health and Nutrition Examination Survey support the notion that changes to FIFRA decreased the body burden of dialkyl phosphate (DAP) metabolites of organophosphorus pesticides (Clune *et al.* 2012). Concerns for FIFRA remain, however. For example, one of the most widely used active ingredients found in US pesticides is atrazine (Geiser 2015). Studies have linked atrazine to endocrine disruption in mammalian cells (Suzawa and Ingraham 2008) and to decreases in testosterone levels, hermaphroditism, and chemical castration in frogs (Hayes *et al.* 2002, 2003, 2010). While the European Union has banned atrazine for over a decade, atrazine continues to be widely used (60–80 million pounds annually) in pesticides in the United States, even after passage of the FFQP (Sass and Colangelo 2006).

Toxic Substances Control Act

The Toxic Substances Control Act (TSCA) was passed in 1976 and was designed to regulate industrial chemicals. The TSCA designates the EPA as the regulatory authority of the TSCA, empowering the EPA to collect information from chemical manufacturers in order to maintain records on chemicals currently in use. Similar to FIFRA, the TSCA notes that economic considerations will be made with respect to regulation.[4] Unlike the most recent amendment to FIFRA, however, the TSCA did not apply to chemicals that were in use prior to the Acts passage, and as such, thousands of chemicals already on the market prior to 1976 were not required to be registered. The TSCA empowers the EPA to test chemicals and also to restrict chemicals if (1) they are deemed to pose an "unreasonable risk" and (2) that the restriction represents the "least burdensome" response. However, in order for the EPA to restrict a chemical, it must obtain the data to test a chemical for risk. To obtain the data, the EPA must demonstrate that the chemical poses "unreasonable risk," placing the EPA in a "Catch-22" position (Vogel and Roberts 2011). As a result, in the 30 years that followed the passage of the TSCA, the EPA required testing for approximately 200 chemicals, and partially regulated 5 (National Resources Defense Council, 2015). The EPA (2014a) reports that today, the TSCA inventory consists of over 84,000 chemical substances.

Emergency Planning and Community Right-to-Know Act

The Emergency Planning and Community Right-to-Know Act (EPCRA) was passed in 1986. The passage of the EPCRA was influenced largely by the industrial chemical disaster that occurred at Union Carbide India Limited in Bhopal, India. The EPCRA required local governments to establish, and annually review, chemical emergency response plans (EPA 2015a). The EPCRA was also designed to provide information to US citizens regarding handling of toxic chemicals in their communities. The EPCRA established the Toxic Release Inventory (TRI), and requires companies handling any of over 650 toxic chemicals to report annually to the EPA. The EPA makes the TRI available publicly online, providing information such as the address of facilities handling toxic chemicals, and the type and quantity of chemicals being handled. The EPA (2015b) reports that in 2013, the United States had over 12 thousand TRI facilities managing over 25 billion pounds of production-related waste.

Conclusion

Synthetic chemicals have been a major part of the global economy for decades. While the production of synthetic chemicals is not new, the volume of synthetic chemicals present in the modern era is unprecedented. This chapter has reviewed the history of synthetic chemical production, noting key developments and events in the history of synthetic chemicals. An understanding of the history of chemical production is important for criminologists interested in studying chemical crimes because the history of synthetic chemical production highlights the relationship between political economy, inequality, and production of synthetic chemicals. Consider, for example, the long relationship between the state and chemists in the private sector in developing chemical weapons, as well as Sutherland's (1949) observation that many corporate criminals recidivate. Understanding these dynamics may aid criminologists in identifying and understanding patterns in green crimes as it relates to the chemical industry.

One theory that has been used to explain the etiology of green crimes is Schnaiberg's ToP theory (Stretesky et al. 2014). A number of harms associated with synthetic chemical production fit the definition of green crime. This includes ecological disorganization produced as a result of synthetic chemical production, as well as ecological additions and withdrawals associated with the chemical industry. Groups with vested interests in promoting and/or deterring chemical production also compose the groups that have been identified as key groups of actors of the ToP. This includes the state, and state entities responsible for creating and enforcing laws associated with chemicals. This also includes the companies producing synthetic chemicals, and the laborers they employ as well as non-profits and NGOs calling attention to the harms associated with chemical production. Several intersections exist between the ToP, ToP actors, green crime, and environmental justice (Gould *et al.* 2008; Gould *et al.* 1996; Stretesky and Knight 2013), and continued analysis of these intersections should represent a priority for green criminology. As the production of synthetic chemicals continues, and environmental injustices persist, empirical assessments of the ToP to examine the chemical industry represents an important area of future study for green criminologists.

Finally, ToP theorists discuss the development and implementation of environmental law and policy as a key role of the state in decelerating the treadmill. In the United States, the number of environmental laws has increased since the 1970s. Key federal pieces of legislation, FIFRA, FFQP, TSCA, and EPCRA constitute some of the major policy initiatives the United States has adopted to address the impact synthetic chemicals have on public health and the environment. Recent works have examined the association between environmental enforcement on the speed of the ToP. For example, Stretesky *et al.* (2013) do not detect a statistically significant relationship between companies who have received a large monetary fine from the EPA and volume of toxic waste produced (as measured by the TRI). The authors argue that findings are consistent with extant works (e.g. Gould *et al.* 2008) that argue penalties may do more to legitimate the ToP than to dismantle it. As it is likely that the number of environmental laws and volume of synthetic chemicals will both continue to grow, evaluating the role of the criminal justice system in the ToP will continue to be a critical area for criminologists.

Notes

1 Sutherland's seminal book on white collar crime called attention to a number of white collar crimes committed by major chemical corporations. In particular, Sutherland discusses actions taken by the chemical industry to explain patent manipulation (1949: 119–121).

2 While this section only discusses a few examples of federal policy, some states have elected to pass additional laws that impact chemical production. Thus, regulation of synthetic chemicals is a nuanced, complex matter and varies across the United States.

3 "Unreasonable adverse effects on the environment" is defined in FIFRA to mean: "(1) any unreasonable risk to man or the environment, taking into account the economic, social, and environmental costs and benefits of the use of any pesticide, or (2) a human dietary risk from residues that result from a use of a pesticide in or on any food inconsistent with the standard under section 408 of the Federal Food, Drug, and Cosmetic Act (21 U.S.C. 346a). The administrator shall consider the risks and benefits of public health pesticides separate from the risks and benefits of other pesticides. In weighing any regulatory action concerning a public health pesticide under this Act, the Administrator shall weigh any risks of the pesticide against the health risks such as the diseases transmitted by the vector to be controlled by the pesticide."

4 From the TSCA: "Authority over chemical substances and mixtures should be exercised in such a manner as not to impede unduly or create unnecessary economic barriers to technological innovation while fulfilling the primary purpose of this Act to assure that such innovation and commerce in such chemical substances and mixtures do not present an unreasonable risk of injury to health or to the environment."

References

Adeola, F.O. (2000). Endangered community, enduring people: Toxic contamination, health, and adaptive responses in a local context. *Environment and Behavior*, 32(2), 209–249.

Baker, N.D. (1919). Chemistry in warfare. *The Journal of Industrial and Engineering Chemistry*, 11(10), 921–923.

Barrett, K.L. (2013). Bethlehem Steel at Lackawanna: The state–corporate crimes that continue to victimize the residents and environment of Western New York. *Journal of Crime and Justice*, 36(2), 263–282.

Beadle, C., and Hoffman, S.L. (1993). History of malaria in the United States naval forces at war: World War I through the Vietnam conflict. *Clinical Infectious Diseases*, 16(2), 320–329.

Boden, T.A., Marland, G., and Andres, R.J. (2010). *Global, Regional, and National Fossil-Fuel CO$_2$ Emissions*. Oak Ridge, Tennessee: Carbon Dioxide Information Analysis Center, Oak Ridge National Laboratories, U.S. Department of Energy.

Brisman, A., and South, N. (2014). *Green Cultural Criminology: Constructions of Environmental Harm, Consumerism, and Resistance to Ecocide*. London: Routledge.

Bruce, A.S., and Becker, P.J. (2007). State-corporate crime and the Paducah gaseous diffusion plant. *Western Criminology Review*, 8, 29.

Brulle, R.J., and Pellow, D.N. (2006). Environmental justice: Human health and environmental inequalities. *Annual Review of Public Health*, 27, 103–124.

Bullard, R.D. (2000). *Dumping in Dixie: Race, Class, and Environmental Quality* (Vol. 3). Boulder, CO: Westview Press.

Bullard, R.D., and Johnson, G.S. (2000). Environmentalism and public policy: Environmental justice: Grassroots activism and its impact on public policy decision making. *Journal of Social Issues*, 56(3), 555–578.

Bullard, R.D., Mohai, P., Saha, R., and Wright, B. (2008). Toxic wastes and race at twenty: Why race still matters after all of these years. *Environmental Law*, 38(2), 371–411.

Bullard, R.D., and Wright, B. (2008). Disastrous response to natural and man-made disasters: An environmental justice analysis twenty-five years after warren county. *UCLA Journal of Environmental Law and Policy*, 26, 217.

Burkett, P. (2009). *Marxism and Ecological Economics: Toward a Red and Green Political Economy*. Chicago, IL: Haymarket Books.

Carson, R.L. (1962). *Silent Spring*. New York, NY: Houghton Mifflin Harcourt.

Center for Responsive Politics. (2015). Open Secrets.org. www.opensecrets.org/.

Centers for Disease Control and Prevention. (2009). Fourth national report on human exposure to environmental chemicals. Accessed August 1, 2015 at www.cdc.gov/exposurereport/pdf/Fourth Report_ExecutiveSummary.pdf.

Chakraborty, J. (2009). Automobiles, air toxics, and adverse health risks: Environmental inequalities in Tampa Bay, Florida. *Annals of the Association of American Geographers*, 99(4), 674–697.

Chakraborty, J. (2012). Cancer risk from exposure to hazardous air pollutants: Spatial and social inequities in Tampa Bay, Florida. *International Journal of Environmental Health Research*, 22(2), 165–183.

Chakraborty, J., and Zandbergen, P.A. (2007). Children at risk: Measuring racial/ethnic disparities in potential exposure to air pollution at school and home. *Journal of Epidemiology and Community Health*, 61(12), 1074–1079.

Checker, M. (2005). *Polluted Promises: Environmental Racism and the Search for Justice in a Southern Town*. New York, NY: NYU Press.

Clarkson, T.W. (1987). Metal toxicity in the central nervous system. *Environmental Health Perspectives*, 75, 59–64.

Clune, A.L., Ryan, P., and Barr, D.B. (2012). Have regulatory efforts to reduce organophosphorus insecticide exposures been effective? *Environmental Health Perspectives*, 120(4), 521–525.

Colborn T., Dumanoski D., Peterson-Myers P. (1997). *Our Stolen Future*. New York, NY: Penguin Group Press.

Colborn, T., Kwiatkowski, C., Schultz, K., and Bachran, M. (2011). Natural gas operations from a public health perspective. *Human and Ecological Risk Assessment: An International Journal*, 17(5), 1039–1056.

Dai, D. (2011). Racial/ethnic and socioeconomic disparities in urban green space accessibility: Where to intervene?. *Landscape and Urban Planning*, 102(4), 234–244.

Davis, D. (2007). *The Secret History of the War On Cancer*. New York, NY: Basic Books.

Di, Y. (2015). Test results of toxic toys raise industry standard: Toys no longer contain dangerous phthalates. Accessed August 12, 2015 at www.greenpeace.org/usa/victories/test-results-on-toxic-toys-raise-industry-standard/.

Emergency Planning and Community Right to Know Act 42 U.S.C. §§ 11001 et seq.

Environmental Working Group (2015). Exposures add up – Survey results. Accessed August 1, 2015 at www.ewg.org/skindeep/2004/06/15/exposures-add-up-survey-results/.

EPA (US Environmental Protection Agency) (2012 June 27). Federal Insecticide, Fungicide, and Rodenticide Act (FIFRA). Accessed August 1, 2015 at www.epa.gov/agriculture/lfra.html#Summary of the Federal Insecticide, Fungicide, and Rodenticide Act.

EPA (US Environmental Protection Agency) (2013 April 22). High Production Volume (HPV) Challenge. Accessed August 1, 2015 at www.epa.gov/hpv/.

EPA (US Environmental Protection Agency) (2014a March 13). TSCA Chemical Substance Inventory: Basic Information. Accessed June 30 2015 at www.epa.gov/oppt/existingchemicals/pubs/tscainventory/basic.html.

EPA (US Environmental Protection Agency) (2014b October 6). Our mission and what we do. Accessed August 1, 2015 at www2.epa.gov/aboutepa/our-mission-and-what-we-do.

EPA (US Environmental Protection Agency) (2015a January 28). What is EPCRA? Accessed August 1, 2015 at www2.epa.gov/epcra/what-epcra.

EPA (US Environmental Protection Agency) (2015b May 11). 2013 TRI national analysis: Introduction. Accessed August 1, 2015 at www2.epa.gov/toxics-release-inventory-tri-program/2013-tri-national-analysis-introduction.

Fagin, D., and Lavelle, M. (1999). *Toxic Deception: How the Chemical Industry Manipulates Science, Bends the Law, and Endangers Your Health.* Monroe, ME: Common Courage Press.

Federal Insecticide, Fungicide and Rodenticide Act of 1972 7 U.S.C. §§ 136 et seq.

Foster, J.B. (1997). The crisis of the earth: Marx's theory of ecological sustainability as a nature-imposed necessity for human production. *Organization and Environment*, 10(3), 278–295.

Geiser, K. (2015). *Chemicals Without Harm: Policies for a Sustainable World.* Cambridge, MA: The MIT Press.

Gould, K.A., Pellow, D.N., and Schnaiberg, A. (2008). *The Treadmill of Production: Injustice and Unsustainability in the Global Economy.* Boulder, CO: Paradigm Publishers.

Gould, K.A., Schnaiberg, A., and Weinberg, A.S. (1996). *Local Environmental Struggles: Citizen Activism in the Treadmill of P.* Cambridge: Cambridge University Press.

Grube, A., Donaldson, D., Kiely, T., and Wu, L. (2011). Pesticides industry sales and usage: 2006 and 2007 Market estimates. U.S. Environmental Protection Agency: Washington, DC. Accessed June 30 2015 at www.epa.gov/opp00001/pestsales/07pestsales/market_estimates2007.pdf.

Hayes, T.B., and Chaffer, P.J. (2010 December 7-8). The toxic baby. [Video file]. Retrieved from: www.ted.com/talks/tyrone_hayes_penelope_jagessar_chaffer_the_toxic_baby.

Hayes, T.B., Collins, A., Lee, M., Mendoza, M., Noriega, N., Stuart, A.A., and Vonk, A. (2002). Hermaphroditic, demasculinized frogs after exposure to the herbicide atrazine at low ecologically relevant doses. *Proceedings of the National Academy of Sciences*, 99(8), 5476–5480.

Hayes, T., Haston, K., Tsui, M., Hoang, A., Haeffele, C., and Vonk, A. (2003). Atrazine-induced hermaphroditism at 0.1 ppb in American leopard frogs (Rana pipiens): laboratory and field evidence. *Environmental Health Perspectives*, 111(4), 568.

Hayes, T.B., Khoury, V., Narayan, A., Nazir, M., Park, A., Brown, T., Adame, L., Chan, E., Buchholz, D., Stueve, T., and Gallipeau, S. (2010). Atrazine induces complete feminization and chemical castration in male African clawed frogs (Xenopus laevis). *Proceedings of the National Academy of Sciences*, 107(10), 4612–4617.

Heynen, N., Perkins, H.A., and Roy, P. (2006). The political ecology of uneven urban green space: The impact of political economy on race and ethnicity in producing environmental inequality in Milwaukee. *Urban Affairs Review*, 42(1), 3–25.

Iversen, K. (2013). *Full Body Burden: Growing Up in the Nuclear Shadow of Rocky Flats.* New York, NY: Broadway Books.

James, W., Jia, C., and Kedia, S. (2012). Uneven magnitude of disparities in cancer risks from air toxics. *International Journal of Environmental Research and Public Health*, 9(12), 4365–4385.

Katz, R.S. (2010). The corporate crimes of Dow Chemical and the failure to regulate environmental pollution. *Critical Criminology*, 18(4), 295–306.

Kauzlarich, D. and Kramer, R. (1998). *Crimes of the American Nuclear State: At Home and Abroad.* Boston, MA: Northeastern University Press.

Kiviat, E. (2013). Risks to biodiversity from hydraulic fracturing for natural gas in the Marcellus and Utica shales. *Annals of the New York Academy of Sciences*, 1286(1), 1–14.

Koplow, D. (2011 February). Nuclear power: Still not viable without subsidies. Union of Concerned Scientists. Accessed July 1, 2015 at www.ucsusa.org/sites/default/files/legacy/assets/documents/nuclear_power/nuclear_subsidies_report.pdf.

Kramer, R.C., and Bradshaw, E.A. (2011). US state crimes related to nuclear weapons: Is there hope for change in the Obama administration? *International Journal of Comparative and Applied Criminal Justice*, 35(3), 243–259.

Landrigan, P.J., and Etzel, R.A. (2014). Children's environmental health—A new branch of pediatrics. In P.J. Landrigan and R.A. Etzel (eds), *Textbook of Children's Environmental Health* (pp. 3–17). New York, NY: Oxford University Press.

Landrigan, P.J., and Goldman, L.R. (2011). Children's vulnerability to toxic chemicals: A challenge and opportunity to strengthen health and environmental policy. *Health Affairs*, 30(5), 842–850.

Landry, S.M. and Chakraborty, J. (2009). Street trees and equity: Evaluating the spatial distribution of an urban amenity. *Environment and Planning A*, 41(11), 2651–2670.

Langston, N. (2010). *Toxic bodies: Hormone Disruptors and the Legacy of DES*. New Haven, Connecticut: Yale University Press.

Langston, N. (2014). New chemical bodies: Synthetic chemicals, regulation and human health. In A.C. Isenberg (ed.), *The Oxford Handbook of Environmental History* (pp. 259–281). New York, NY: Oxford University Press.

Lavelle, M., and Coyle, M. (1992). Unequal protection: The racial divide in environmental law: A special investigation. *The National Law Journal*, 15(3), S1–S12.

Lerner, S. (2005). *Diamond: A Struggle for Environmental Justice in Louisiana's Chemical Corridor*. Cambridge, MA: MIT Press.

Long, M.A., Stretesky, P.B., and Lynch, M.J. (2012). Crime in the coal industry: Implications for green criminology and the treadmill of production. *Organization and Environment*, 25(3), 328–346.

Lynch, M.J. (1990). The greening of criminology: A perspective for the 1990s. *Critical Criminologist*, 2(3), 11–12.

Lynch, M.J., Long, M.A., Barrett, K.L., and Stretesky, P.B. (2013). Is it a crime to produce ecological disorganization? Why green criminology and political economy matter in the analysis of global ecological harms. *British Journal of Criminology*, 53(6), 997–1016.

Lynch, M.J., Long, M.A., and Stretesky, P.B. (2015). Anthropogenic development drives species to be endangered: Capitalism and the decline of species. In R.A. Sollund (ed.), *Green Harms and Crimes: Critical Criminology in a Changing World* (pp. 117–146). New York, NY: Palgrave-Macmillan.

Lynch, M.J., and Michalowski, R.J. (2006). *Primer in Radical Criminology: Critical Perspectives on Crime, Power, and Identity*, 4th ed. Monsey, NY: Criminal Justice Press.

Lynch, M.J., Nalla, M.K., and Miller, K.W. (1989). Cross cultural perceptions of deviance: The case of Bhopal. *Journal of Research in Crime and Delinquency*, 26(1), 7–35.

Markowitz, G., and Rosner, D. (2000). 'Cater to the children': The role of the lead industry in a public health tragedy, 1900-1955. *American Journal of Public Health*, 90(1), 36.

Markowitz, G., and Rosner, D. (2013a). *Deceit and Denial: The Deadly Politics of Industrial Pollution*. Berkeley, CA: University of California Press.

Markowitz, G., and Rosner, D. (2013b). *Lead Wars: The Politics of Science and the Fate of America's Children* (Vol. 24). Berkeley, CA: Univ of California Press.

McGurty, E. (2009). *Transforming Environmentalism: Warren County, PCBs, and the Origins of Environmental Justice*. New Brunswick, NJ: Rutgers University Press.

Michalowski, R.J., and Kramer, R.C. (1987). The space between laws: The problem of corporate crime in a transnational context. *Social Problems*, 34(1), 34–53.

Mohai, P., Kweon, B., Lee, S., and Ard, K. (2011). Air pollution around schools is linked to poorer student health and academic performance. *Health Affairs*, 30(5), 852–862.

National Resources Defense Council. (2015). Take out toxics. Accessed August 1, 2015 at www.nrdc.org/health/toxics.asp.

Noonan, R. (2015 January 8). *Made in America: Chemicals*. U.S. Department of Commerce: Economics and Statistics Administration.

Pastor, M., Sadd, J., and Hipp, J. (2001). Which came first? Toxic facilities, minority move-in, and environmental justice. *Journal of Urban Affairs*, 23(1), 1–21.

Pearce, F., and Tombs, S. (1998). *Toxic Capitalism: Corporate Crime and the Chemical Industry*. Aldershot, England: Ashgate.

Pellow, D.N. (2002). *Garbage Wars: The Struggle for Environmental Justice in Chicago*. Cambridge, MA: The MIT Press.

Right to Know Network. (2016). *TRI Facilities in Louisiana, 2014*. Washington, DC: Center for Effective Government. Retrieved April 1, 2016 from www.rtknet.org/db/tri/tri.php?database=tri&reptype=f&reporting_year=2014&corechem=n&rsei=y&detail=-1&dbtype=C&sortp=D&datype=T&state=LA.

Rosner, D., and Markowitz, G. (1985). A 'gift of God'?: The public health controversy over leaded gasoline during the 1920s. *American Journal of Public Health*, 75(4), 344–352.

Ruggiero, V., and South, N. (2013). Toxic state–corporate crimes, Neo-liberalism and green criminology: The hazards and legacies of the oil, chemical and mineral industries. *International Journal for Crime, Justice and Social Democracy*, 2(2), 12–26.

Russell, E. (2001). *War and Nature: Fighting Humans and Insects with Chemicals from World War I to Silent Spring*. Cambridge, United Kingdom: Cambridge University Press.

Sass, J.B., and Colangelo, A. (2006). European Union bans atrazine, while the United States neogiates continued use. *International Journal of Occupational and Environmental Health*, 12(3), 260–267.

Schnaiberg, A. (1980). *The Environment: From Surplus to Scarcity*. New York, NY: Oxford University Press.

Schwartz, J.M., Woodruff, T.J., Hood, E., and Wade, M. (September 2008). Shaping our legacy: Reproductive health and the environment. Program on Reproductive Health and the Environment, Department of Obstetrics, Gynecology and Reproductive Sciences National Center of Excellence in Women's Health, University of California, San Francisco. Accessed June 30 2015 at http://prhe.ucsf.edu/prhe/pubs/shapingourlegacy.pdf.

Shrock, P. (2015). The Rocky Flats plea bargain: a case study in the prosecution of organizational crime in the US nuclear weapons complex. *Journal of Crime and Justice*, 38(2), 204–221.

Simon, D.R. (2000). Corporate environmental crimes and social inequality: New directions for environmental justice research. *American Behavioral Scientist*, 43(4), 633–645.

Singer, M. (2011). Down cancer alley: The lived experience of health and environmental suffering in Louisiana's Chemical Corridor. *Medical Anthropology Quarterly*, 25(2), 141–163.

Stretesky, P.B., and Knight, O. (2013). The uneven geography of environmental enforcement INGOs: National influence of wealth on a global civil society and potential implications for ecological disorganization. In T. Wyatt, R. Walters, and D. Westerhuis (eds) *Emerging Issues in Green Criminology* (pp. 173–190). New York, NY: Palgrave-Macmillan.

Stretesky, P.B., Long, M.A., and Lynch, M.J. (2013). Does environmental enforcement slow the treadmill of production? The relationship between large monetary penalties, ecological disorganization and toxic releases within offending corporations. *Journal of Crime and Justice*, 36(2), 233–247.

Stretesky, P.B., Long, M.A., and Lynch, M.J. (2014). *The Treadmill of Crime: Political Economy and Green Criminology*. New York, NY: Routledge.

Stretesky, P.B., and Lynch, M.J. (2002). Environmental hazards and school segregation in Hillsborough County, Florida, 1987–1999. *The Sociological Quarterly*, 43(4), 553–573.

Sutherland, E.H. (1949) [1983]. *White Collar Crime: The Uncut Version*. New Haven, CT: Yale University Press.

Suzawa, M., and Ingraham, H.A. (2008). The herbicide atrazine activates endocrine gene networks via non-steroidal NR5A nuclear receptors in fish and mammalian cells. *PLoS One*, 3(5), 1–11.

Taylor, D.E. (2000). The rise of the environmental justice paradigm: Injustice framing and the social construction of environmental discourses. *American Behavioral Scientist*, 43(4), 508–580.

Taylor, D.E. (2014). *Toxic Communities: Environmental Racism, Industrial Pollution, and Residential Mobility*. New York, NY: New York University Press.

The Endocrine Disruption Exchange. (2015). About TEDX: Introduction. Accessed August 1, 2015 at www.endocrinedisruption.org/about-tedx/about.

Toxic Substances Control Act 15 U.S.C. §§ 2601 et seq.

Tullo, A.H. (2014). Chemical and engineering news' global top 50 chemical firms for 2014: The world's largest chemical firms are growing and enjoying stronger profits. *Chemical and Engineering News*, 92(30), 10–13. Accessed July 15, 2015 at http://cen.acs.org/articles/92/i30/CENs-Global-Top-50-Chemical.html.

U.S. Government Accountability Office. (1983). The siting of hazardous waste landfills and their correlation with racial and economic status of surrounding communities (No. RCED-83-168). Washington, DC: General Accounting Office.

Vogel, S.A., and Roberts, J.A. (2011). Why the Toxic Substances Control Act needs an overhaul, and how to strengthen oversight of chemicals in the interim. *Health Affairs*, 30(5), 898–905.

Walgren, J. (2000). The U.S. Environmental Protection Agency and the enforcement of environmental laws and regulations. *Social Pathology*, 6(1), 3–23.

Walters, R. (2007). Crime, regulation, and radioactive waste in the United Kingdom. In P. Beirne and N. South (eds) *Issues in Green Criminology: Confronting Harms Against Environments, Humanity, and Other Animals* (pp. 186–205). Berkeley: University of California Press.

White, R. (2011). *Transnational Environmental Crime: Toward an Eco-Global Criminology*. New York, NY: Routledge.

Wright, B. (1998). Endangered communities: The struggle for environmental justice in the Louisiana Chemical Corridor. *Journal of Public Management & Social Policy*, 4, 181–191.

Wright, B. (2003). Race, politics and pollution: Environmental justice in the Mississippi River chemical corridor. In J. Agyeman, R.D. Bullard, and B. Evans (eds), *Just Sustainabilities: Development in an Unequal World* (pp. 125–145). Cambridge, MA: MIT Press.

12

Pharmatechnologies and the ills of medical progress

Paddy Rawlinson

Medicine sometimes snatches away health, sometimes gives it.

(Ovid)

... in holding scientific research and discovery in respect, as we should, we must also be alert to the equal and opposite danger that public policy could itself become the captive of a scientific technological elite.

(Eisenhower)

Pharmaceuticalized developments against disease have brought obvious advantages to health and remain central to what Rose describes as a 'biotechnological industrial revolution in the medical management of health, disease and life' (Rose 2007: 37). Drug-based therapy is now the orthodox response to a broad range of issues from mental health, childhood behavioural problems and the growing use of preventative medical strategies on a local and global level against the potential threat of epidemics from influenza to measles and, more recently, Ebola. The demand for innovative and better drugs has also intensified the need for clinical trials on human subjects, an essential component of testing for the safety and efficacy of new pharmaceutical products. This intensifying trend of 'medicalization', described as 'one of the most potent social transformations of the last half of the twentieth century in the West' (Clarke *et al.* 2003) has spawned a lucrative pharmaindustrial market which, according to estimates from the World Health Organization (WHO), currently stands at US$300 billion per annum with an expected increase of $100 billion by 2016 (WHO 2015). The vaccine sector alone, as the core policy for disease prevention, is expected to generate sales of approximately $58 billion by 2019 (Research and Markets 2015). Much of this has been driven by the increase in government investment in health. The US has seen health expenditure currently exceed that of defence (US Total Govt Spending Breakdown in $ Trillion 2015), a trend evident in most developed countries. Not surprisingly, the attendant activity in clinical trials has also followed suit and is expected to generate a business worth US$32.73 billion in 2015 (Mansell 2011). Effectively, the defence giants Lockheed, Rolls-Royce and BAE Systems, major manufacturers of military arms, have now been replaced by GlaxoSmithKline, Pfizer, Roche and Sanofi, as the prioritized beneficiaries of state funding.

The analogy between the military-industrial complex and its medical counterpart serves as a useful framework for investigating the dark side of pharmatechnologies, which are similarly tied into the harm-creating alliances, self-serving policies and criminal activities identified within the state-defence industry paradigm (Naylor 2004; Hughes 2007; Whyte 2007; Godfrey *et al.* 2014). As the former US President, Dwight Eisenhower (attributed with coining the term 'military-industrial complex') warned: 'we must guard against the acquisition of unwarranted influence, whether sought or unsought, by the military-industrial complex. The potential for the disastrous rise of misplaced power exists and will persist' (Congressional Record – Senate 1966). This chapter will argue that this perspicacious caution also resonates loudly in the pharmaceuticalized context of health and disease prevention. While the political economy of military technologies provokes unease with its potential for corruption, human rights abuses and other myriad harms, pharmatechnologies largely remain under the radar of criminological scrutiny. In a recent update to his 1984 publication on corporate crime and the pharmaceutical industry, John Braithwaite opined the dearth of criminological interest: 'So many criminologists study individual homicide, while so few have chosen to view the topic of this book as meriting their attention' (Dukes *et al.* 2014: 282). Jeffrey Reiman and Joe Sim have likewise identified a range of medical harms through the lens of criminology and deviancy studies (Reiman 2004; Sim 1990) and yet despite the growing number of abuses in this area scholarship, other than from bioethicists and even medical practitioners, remains scare. The statistics make startling reading. An estimated 197,000 people in the EU die every year from adverse reactions to pharmaceutical drugs, in what has been described as an 'epidemic' of drug failures and adverse reactions (Archibald, Coleman and Foster 2011). Deaths from counterfeit drugs alone in a single year account for 'more than all the people killed across the globe by homicide, terrorism and warfare combined for any year of this century so far' (Dukes *et al.* 2014: 281). Clinical trials too have resulted in numerous tragic outcomes, severe, chronic and fatal, especially in the developing world as the current favoured site for human experimentation (Rawlinson and Yadavendu 2015; Petryna 2009).

The almost hallowed status of medical science and research is such that their negative consequences tend to be perceived as unfortunate incidents, inevitable risks accompanying medical intervention, and ones we're evidently prepared to live with. This places these events within the contentious and challenging field of harm rather than crime (a somewhat apposite area given that the fundamental norm of medicine is primum non nocere, 'first, do no harm'). While the social harm approach is slowly gaining ground in criminology, it still remains peripheral to mainstream research for philosophical and pragmatic reasons. The context of the social harm approach, as many of its advocates admit, can broaden the remit of criminological scholarship to the point where it becomes undermined by the amorphous nature of its definition and application (Hillyard *et al.* 2004). In other words, criminology is in danger of stretching itself beyond an (imagined?) ontological existence. Reluctance to engage in the social harm approach is also driven by the practical consideration that it invariably offers a critique of power institutions, such as the state and corporations, which are increasingly acting as the financial life-support system for academic research. Why bite the hand etc.? This is especially the case where the pharmaceutical industry is concerned, being a major player in funding university medical centres, often the mainstay of research investment across whole institutions.

Nonetheless a small but significant body of literature has argued over decades for the inclusion of non-proscribed acts into the criminological canon, much of it inspired by the publication of Sutherland's text *White Collar Crime*, which turned the focus from street crime to that committed by elites (Sutherland 1949). Subsequently, studies dedicated to areas such as corporate crime (Box 1983; Pearce 1976), state crime (Green and Ward 2004;) and state-corporate crime (Michalowski 2010) expanded the reach of criminological investigation, identifying pathological and criminal behaviours in the boardrooms and corridors of power, in other words in those legitimate institutions which run and define national and global political and economic systems. As many of the behaviours conducted by these institutions are not de jure criminal (given that as the powerful elite they control and mould the law to serve their own interests) the term 'social harm' provides a convincing paradigm for critiquing the damaging power relations operating within the framework of legality. The dark side of pharmatechnologies falls within this framework, and incorporates a series of powerful actors involved in the creation, distribution, regulation and surveillance of pharmaceuticals – the state, multinationals and medical science. Taking a social harm approach we can interrogate the nature of what is often a one-sided relationship between these actors and the public/citizenries, which accounts in part for the muted acceptance of the myriad abuses committed within this triadic relationship, disguised by the rhetoric of health and beneficence.

The chapter examines the negative impact of state-corporate associations in the context of pharmaceuticalized responses to disease and the threat of disease. It focuses on two specific areas: clinical trials and preventative medicine through vaccination. Clinical trials have been the subject of much debate amongst bioethicists, medical anthropologists, historians and practitioners, especially since their commercialization towards the end of the twentieth century (Glickman *et al.* 2009; Petryna 2007; Lifton 2000; Angell 1997). As the fundamental criterion for the development and production of drugs they represent a crucial component of pharmatechnologies. In examining what are referred to as 'unethical' clinical trials the chapter will highlight how social, economic and political factors shape and contribute to abuses perpetrated against human subjects utilized in the advance of medical science. The second area relates to the contentious issue of preventative health and vaccination. It is in this context that the parallels between the military and medical industrial complex most obviously converge within the discursive framework of security and securitization. Central to major domestic and global health strategies conceived and implemented by powerful intergovernmental and non-governmental actors including the WHO, the World Bank and the Bill and Melinda Gates Foundation, is the programme of mass vaccination against a growing number of diseases. Driven by good intentions in many quarters, there are nonetheless abuses and exploitative elements pertaining to these strategies. It is these abuses that constitute the focus of the chapter, underlined by the crucial question: Who gains and who loses?

A brief word about the term 'pharmatechnologies'. Working with Deborah Lupton's explanation of biotechnologies, the chapter uses the term with specific reference to pharmaceutical medicine and production, to denote 'the social, political and economic contexts within which scientific knowledge and technologies are produced and utilized through western medicine' in which 'scientific methods, technologies, processes and knowledge are developed and given power as explanatory phenomena ... [being] viewed not as external to social and cultural processes, but rather as constituted in and through such processes'(Lupton 2012: 16). To these processes we can also add the political and economic as manifest through state-corporate collaborations primarily serving the interests of the powerful.

Trials and error

Weapons take lives, but (legal) drugs save them (give or take the countless thousands mentioned above) so runs common-sense. Clinical trials are regarded as an essential process in the development and production of modern drugs, as a means of testing their safety and efficacy. Using human subjects, as well as animals for the purposes of medical scientific testing, has been controversial and prompted endless debates (Singer 2011; Rose and Rose 2014). However, it was the horrors of human experimentation under the Third Reich that finally focused global attention on the potential for abuse and harm and pre-empted the first international code for the protection of the human subject in medical research. As evidence of the atrocities committed in the name of Aryan medicine was presented at the so-called Nazi Doctors' trial at Nuremberg in 1946, the chief prosecutor Telford Taylor described them as a descent 'into an infernal combination of a lunatic asylum and a charnel house' (Proctor 2000: 335). To avoid a repeat of this Dante-esque medical nightmare, the Nuremberg Code was established, setting out ten principles for the protection of human subjects in clinical research. Central to the Code was the voluntary and informed consent of the subject as an 'absolutely essential' condition of medical research. The Code states that 'the person involved should have legal capacity to give consent; should be so situated as to be able to exercise free power of choice, without the intervention of any element of force, fraud, deceit, duress ... and should have sufficient knowledge and comprehension of the elements of the subject matter involved' (BMJ 1996; 313:1448). On paper the Code, and successive protocols such as the 'Declaration of Helsinki: Ethical Principles for Medical Research Involving Human Subjects' and the 'International Ethical Guidelines for Biomedical Research Involving Human Subjects' provide substantive safeguards against the further possibility of what were termed crimes against humanity. In practice, such 'crimes' have been, and in some cases continue to be, committed by countries such as the US and the UK. Tellingly, the term 'unethical clinical trials' which semantically dilutes the seriousness of human experimental abuse, has become common usage when medical research which injures or kills subjects is conducted by 'civilized' nations, illustrating the distinctly subjective and political nature of 'crime'.

If Nazi Germany is regarded as the nadir of medical behaviour, it is in part due to the racialized nature of the killing and maiming. Hundreds of Jewish and Roma people, including children, were experimented on, enduring appalling suffering which for many led to their eventual deaths. Yet, according to the Nuremberg trial transcripts few, if any, of those engaged in the experiments were motivated by racial hatred. If race was an aspect, it was as the object of medical research into biological differences that proved racial inferiority, as part of the eugenics ethos underlying Nazi ideology. In other words, race was framed within scientific inquiry. In their defence statements the doctors explained their motivations as inter alia the pursuit of medical knowledge and duty to the Reich, convinced that their role as scientific researchers was value free and based on the (value-ridden?) utilitarian idea of sacrificing the few in the interests of the majority. As Caplan concluded, 'most of those who participated [in conducting the experiments] did so because *they believed it was the right thing to do*' (2008: 65) (emphasis added). Despite the grim context of their 'work', the technological advances made during the experiments, as with Sigmund Rascher's hypothermia laboratory which duplicated conditions faced by pilots downed in the North Sea and the trials conducted on French soldiers to test the toxic chemical phosgene (now used in plastics and weedkiller), have served as a contentious basis for medical research in the US and other countries. In short, the 'value-free' plea of the Nazi doctors for the development and employment of technological advances in support of their discipline is neither ideologically nor historically constrained, rendering technology *per se* a value, a worth purely within its existence and the belief that it necessarily equals an improvement of conditions.

Similar rationales have legitimized harmful human experimentation beyond Nuremberg, involving a range of technological resources, including drugs, medical instruments, experimental environments and so on. National security interests, most notably those of the US, have been used to legitimize the testing of radiation fallout in some cases on whole populations, as in the Marshall Islands experiment of 1954 where a nuclear bomb was dropped in close proximity to the Islands without the inhabitants' knowledge and consent (Alcalay 2014). The US was also suspected of dropping cholera-infected insects onto North Korea during the 1950s as part of its research into biological warfare (Hurst 2008). It was the infamous Tuskegee Syphilis experiment, carried out on over 400 tertiary stage syphilitic African Americans who were lied to about their condition and denied available treatment for the disease that provoked national outrage at the blatant racial context when it was exposed by the media in 1972 (Jones 1993). More recently, revelations that Puerto Rican and African American soldiers were used for mustard gas experiments during World War II (Smith 2008) have reinforced the racialized aspect of unethical experimentation.

The legitimacy of human experimentation rests on the assumption that pharmaceuticalized medical solutions are the optimum, and in many cases, sole paradigm for disease prevention and healing. This epistemological foundation which objectifies those who suffer disease, in the mutating discourse as Foucault succinctly puts it from 'What is the matter with you?' to 'Where does it hurt?' (2003, xviii) in turn objectifies the process of healing or preventing disease, rendering the human experimental subject as scientific object. In other words, the experimental subject is reduced to a technological device in the process of drug development, production and distribution. Any risk to the experimental subject can thus be understood according to a value calculus based on other objectified phenomena such as race, economic status, and so on. When this objectification operates within a political-economic framework, as in the current global neoliberal model of health provision, the harms inflicted on human participants are likely to be an outcome of mutually self-serving state-corporate interests. It is a surprisingly small step from the abuses imposed by a racialized ideology to those imposed by an economic one.

Healthy profits – migrating trials

There has been a growing trend towards offshoring clinical trials from advanced capitalist states, the site of the major multinational pharmaceutical companies, to so-called 'developing' countries. The surge in this displacement has been prompted by a number of factors including tighter regulation in developed countries such as the US, where scandals such as the Tuskegee experiment and a gradual awareness of prisoners' rights (one of the largest cohort for experimental subjects), saw the introduction of more stringent procedures and protections. Another important reason for outsourcing to the poorer regions is the availability of what are called 'naïve' subjects, participants who, having had little or no access to medication, can present their bodies for unadulterated test conditions. This turbo-commercialization of clinical trials not only prioritizes profit acquisition but pushes the drive for efficiency into dubious ethical territory as the following demonstrates: 'In India SIRO CliniPharm recruited 650 patients from five centers in around 18 months. By comparison in Europe it took 36 months to recruit 85 subjects from 22 centers' (Balakrishnan and Sharma 2004). Efficiency and cost effectiveness are further bolstered by local conditions where official corruption is high, and cheap labour, endemic poverty and illiteracy are prevalent, conditions conducive to a global competitive market for drug production that encourages the need to bypass 'the increasingly bureaucratic and expensive regulatory environment in many wealthy countries' (Glickman et al. 2009: 817).

Commercialized clinical trials have also emerged as a new lucrative industry. As the demand for human participants grows, the emergence of contract research organizations or CROs has expanded exponentially. CROs are able to produce cheaper and faster results on new drugs having knowledge of those local conditions, where economic efficiency and accessibility to human subjects is most achievable. They are best-placed to recognize loopholes in safety procedures, identify clinics and doctors open to financial inducements, and locate regions where high unemployment rates are more likely to produce a cheaper research workforce and greater volume of participants (Azoulay 2003; Mirowski and Van Horn 2005). As with any business, including that of Big Pharma, they are answerable to their shareholders and hence less likely to prioritize the welfare of human subjects over profit margins. In his statement on India's tightening of regulatory protections for its citizens involved in clinical trials, the Vice President of the Association of Clinical Research Organizations (ACRO) threatened to transfer business from India 'to more hospitable countries to mitigate the economic damage' in an unambiguous declaration of the prime concern of CROs (Mansell 2011).

While this aggressive commercialized response exposes the hand of Big Pharma and its collaborators (and despite constantly consuming their products, many are sceptical about the ethics of the pharmaindustry) an arguably more damaging form of exploitation occurs within the supposed altruistic structures of development agencies and organizations. Led by the powerful institutions of the United Nations, the World Bank and the International Monetary Fund, and accompanied by high profile charities such as the Bill and Melinda Gates Foundation, the promotion of health through, amongst other factors, the war on disease, is a major aspiration of Western-dominated International Government Organizations. Encapsulated in the Millennium Development Goals (MDG) critics have identified health and development policies as being little more than a form of medical colonialism aimed at extracting resources from the most vulnerable. In what has become known as dependency theory, development programmes are regarded by some as 'project[s] through which an aggressive expanding class seeks to expand its control and use of other people's resources and to neutralize any opposition to such programmes' (Alvares 1992: 94–5). The push for pharmaceuticalized solutions to problems that are more often caused by poverty, civil war and international debt, invariably induced by the countries stretching out their 'altruistic' hand, comes with a sub-agenda. As Samir Amin points out 'the significance of this generous intention to provide access to drugs is immediately nullified by the specification that this would be "in cooperation with the pharmaceutical industry," precisely those who prohibit anyone from calling their abusive monopoly into question!' (Amin 2006). It is hard not to remain sceptical about the intentions behind putative charitable gestures when reading about the practice of 'drug dumping' (sending low quality drugs to developing countries) under the guise of pharmaceutical beneficence as in the example of appetite stimulants donated 'to a starving refugee population in Sudan' (Dukes et al. 2014: 260).

The harms that emerge from clinical testing are difficult to prove, and not only because many countries lack robust mechanisms to enable sufferers an effective form of legal redress. Victims' voices in developed as well as developing countries have to stand against the dominant epistemologies that construct the parameters of truth, a truth that can only be challenged within the very parameters that constructed it or risk being dismissed as unscientific and thus irrational. And these truths are increasingly defined according to technologies, by the objects, the instruments, the forensic details of material entities, of what can be detected in the body, analysed as biomedical samples that 'prove' a link to the harmful effects of the drugs tested. It is the victim's body, available for forensic scrutiny, and not her mind and memory that offers a rational association between the drug she was administered and her severe or fatal reaction to it, that remains the prime site of evidence. As Eyal Weizman observes we have become forensic fetishists where science operates not only 'a tool for investigation, but as a means of persuasion, and crucially, a matter of belief...' (Weizman 2011: 108).

However, forensic science is hampered by diverse factors that can undermine its validity. Time is one of the most crucial aspects, as countless rape victims are aware, having a window of between five to seven days to present their semen-penetrated bodies as a legitimate crime scene for securing a possible conviction against their abuser. In the case of seven Indian girls who died after being administered Human Papilloma Virus vaccines in Gardasil in 2010 an investigation concluded that 'the deaths reported during trial had no uniform pattern to link them to the administration of vaccines' noting however 'that all the seven deaths were summarily dismissed as unrelated to vaccinations without in-depth investigations' (Parliament of India 2013). Anecdotal reports however, point to a crucial time lapse between the deaths of the girls and investigations into the causes, a hiatus that impacted on data and conclusions, and one that provided a favourable outcome for the organizations involved. These included a number of Indian medical research and regulatory institutions and the Bill and Melinda Gates Foundation funded NGO, Programme for Appropriate Technology in Health (PATH), which was attempting to commit the government to including the vaccine in its Universal immunization programme, a hugely lucrative outcome had it succeeded. Such is the familiar pattern of clinical trial abuse amongst lower-socioeconomic human subjects in India, where the poor are vulnerable to 'sheer exploitation … because they desperately need money. A rich person rarely becomes a subject' (Paliwal 2011).

Unethical clinical trials expose the conflicting role of medical technological advance which can simultaneously enhance the quality of life and destroy it. In this context health entitlement is premised on the risk of harm to particular groups whose vulnerability, be it racialized, class-constructed, gender-based or socio-economically determined, is integral to, or a consequence of, neoliberal models of health and economy. These vulnerable populations become demoted to human medical subjects-as-objects, but second-class objects deprived of an authoritative voice, and compelled by circumstances to defer the reality of their suffering to the non-human 'truth' of science. Medical science in its privileged status, constructs a hegemonic reality from these technologies which are presented as prime objects of truth – a blood sample, DNA testing, screening, etc. and become 'reliable witnesses' for the purposes of obfuscating the harmful and criminal behaviour of the powerful.

However, the commercialization of clinical trials not only exacerbates harms towards vulnerable human 'guinea pigs', but expands the reach of potential victimization to the hundreds and thousands for whom the drugs are prescribed. The dictates of neoliberal health economies mean that 'Where clinical trials were once a scientific exercise aimed at weeding out ineffective treatments, they became in industry's hands a means to sell treatments of little benefit' (Healy 2012: 99) and too often, treatments with damaging, even fatal results.

Technologies of health security

As with its military counterpart, the medical industrial complex in its current (perpetual) war on disease is developing an array of weapons for pre-emptive strikes against the enemy. Its apparently most effective (and lucrative) technology in preventative medicine is the vaccine. So significant has this technology become, that vaccination programmes are at the forefront of global health programmes. Support for the worldwide deployment of vaccine defence initiatives was evident in a statement by the WHO in 2005 which enthused 'Immunization is one of the most successful and cost-effective health interventions ever. … We are entering a new era in which it is expected that the number of available vaccines will double. Immunization services are increasingly used to deliver other important health interventions, making them a strong pillar of health systems' (WHO/Unicef 2005: 3). As an integral component of the UN's 2000–2015 Millennium Development Goals (MDG) and no doubt its successor, the post-2015 Sustainable Development Agenda, vaccination sits at the heart of aspirational strategies for improving the lives of those in the developed world as well as the global south. According to the Vaccine Alliance network GAVI, a crucial partner in the MDG health programme, as well as preventing certain diseases, vaccination also 'raises children's IQ, improves their cognitive development, physical strength and educational achievements' (Millennium Development Goals). Despite this expanding suite of advantageous outcomes and the overtly benign impetus behind the goals, a shadow has been cast regarding the ethics of the more covert incentives behind the programme. Some events have exposed the unhealthy relationship between national governments, intergovernmental agencies and the pharmaceutical industry, alliances based in part on 'unwarranted influence' and 'misplaced power' by the latter party and the pursuit of self-interest by each. In the case of the Indian clinical trials referred to above, one of the observations of the Parliamentary Standing Committee's investigation into the HPV vaccine trials was that PATH, supported by funds from the Bill and Melinda Gates Foundation (which also contributes substantial sums to GAVI) had 'under the pretext of observation/demonstration project [sic] ... violated all laws and regulations laid down for clinical trials by the Government', a situation made possible by the lax attitude to regulations by the Indian Council of Medical Research. The report continued by concluding:

> While doing so, its [PATH's] sole aim has been to promote the commercial interest of HPV vaccine manufacturers who would have reaped windfall profits had PATH been successful in getting the HPV vaccine included in the UIP (Universal Immunization Programme) of the Country.
>
> *(Parliament of India 2013)*

In 2009 the WHO itself was caught up in a scandal when an investigative journalist uncovered conflict of interest on the expert panel advising the organization on the prospects of an influenza pandemic. It turned out that some of the advisory panel had professional links with pharmaceutical companies producing the flu vaccine H1N1 (Godlee 2010). The predicted pandemic never occurred and governments worldwide were left with stockpiles of tax-payer funded unused vaccines. GlaxoSmithKline, one of the vaccine-producing companies saw a 13 per cent increase in its profit margin based on sales of the vaccine (*Telegraph* online 2010). Here was yet another example of the lucrative nature of war – waged against threats real or imagined – and how government funding, as had occurred with defence contracts tendered to businesses such as Lockheed Martin, Boeing and Halliburton during the Iraq incursion (Hartnung and Berrigan 2005), feeds the appetite of industry over the needs of citizens.

In the commercially driven process of securitizing health, knowledge production plays a crucial role, operating both as a tool and a good. Security as a commodity requires the consumer (here the public, the state and the medical profession) to buy (literally) into a belief that the security products offered are effective in obviating or at least weakening the risk of disease or illness. This further involves ensuring that any risk involved in consuming the product, in this case the vaccine, is overridden by the threat posed by the disease against which it is affording protection. Creating a vigorous market for business in a competitive environment, as studies of corporate crime have consistently shown, can involve a range of deviant and criminal strategies and activities, including fraud, corruption, intimidation and bribery, activities that are ubiquitous and, in some cases, systemic within industry. Historically 'Big Pharma' has a poor track record for ethical behaviour (Braithwaite 1984; Dukes *et al.* 2014; Gotzsche 2013; Griffin and Miller 2011). Healy observes that unlike other health-risk businesses such as tobacco and the chemicals industry, in which 'the best studies systematically point to hazards where they exist' and where company studies tend to be a small component of evidence-production, often viewed with suspicion as being partial, the same does not pertain to the pharmaceutical industry:

221

with pharmaceuticals often the only studies are those of the drug companies themselves, and these studies, as one might expect, all seem to point to the benefits of an ongoing use of the very chemicals that may in fact be causing the problem.

(Healy 2012: 119)

Alongside vested-interest knowledge production, the pharmaceutical industry engages in other nefarious activities. These include obstructing the publication of negative data from clinical trials in medical journals, which are often financially dependent on industry for advertising and the sale of reprints; intimidating whistle-blowers amongst medical researchers and doctors concerned about the safety of particular drugs; and the ubiquitous practice of aggressively promoting products to the medical profession including providing financial inducements to win support for a particular drug (Gotzsche 2013; Healy 2012; Moynihan 2001). Nor are these deviant and harmful practices occasional aberrations, but instead reflect recidivist behaviour embedded within the industry (Braithwaite 1984; Dukes *et al.* 2014).

Despite a long history of insalubrious behaviour in the pharmaindustry, governments and intergovernmental organizations such as the WHO continue to focus on pharmaceuticalized solutions to what are arguably pharmaceutically constructed health risks. While obfuscating or down-playing the safety issues around the escalating administration of medical interventions, the pharmaindustry's intensified participation in research and policy-making enables it to construct narratives of high risk around both the nature of diseases and their prevalence, and exaggerate the efficacy of its products, thereby creating an ever-expanding market (Healy 2012; Gotzsche 2013). This is especially the case in a product that is administered to a population. The vaccine industry is highly profitable not only because of its numerical reach but because, in an increasing range of jurisdictions, vaccines carry a mandatory status.

These policies, which for critics of mandatory medical intervention, are regarded as a blatant violation of human rights, have been legitimized through the discursive modalities of securitization, in much the same way that torture and extraordinary rendition were reconstructed as necessary for the protection of the very phenomena they were consistently eroding (Chossudovsky 2005). In both cases, where the market operates as a crucial driver, in which defensive mechanisms are for sale, whether vaccines or arms, there must be no limitation to demand. Creating demand through the security narrative and ensuring supply through mandate enables the constant proliferation of an industry which simultaneously protects and destroys. Excessive protection can only lead to destruction (consider the proliferation of nuclear arms during the Cold War in what was termed MAD – mutually assured destruction). Esposito captures this in his consideration of the process of over-immunization as a political, juridical as well as bio-political phenomenon, whereby 'the warring potential of the immune system is so great that at a certain point it turns against itself as a real and symbolic catastrophe leading to the implosion of the whole organism' (2011: 17). Yet, as has been made clear in the aftermath of the global financial crisis, the voracious appetite of the market in its neoliberal manifestation continues apace ignoring the social and economic cannibalism of excessive profit-seeking. For the pharmaindustry that involves the presence, real or imaginary, of a constant threat of disease and infection.

State of infection

In its 2006 action plan to 'increase vaccine supply' the WHO informs us that 'Influenza vaccine development and employment are critical elements of pandemic influenza preparedness'. In explaining how serious this risk of a pandemic actually is, and the consequent justification for spending billions of dollars on flu vaccines globally, the WHO slips into 'Rumsfeldian discourse' describing the 'the global burden of seasonal influenza' as an 'unknown' (WHO 2006). We are now familiar with unknowns in their both knowable and unknowable states as being crucial to the extension of state power with its affiliate abuses. The 'unknown burden' of seasonal flu has occasioned not only the mass manufacture of a vaccine that, in its current stage of development has proven to be less than effective (Gallagher 2015), but also the introduction of mandatory policies around the influenza vaccine for health workers and other professionals (Babcock *et al.* 2010). Indeed, mandating a range of vaccines has now become established practice in a number of countries including the US, Australia, France and Canada, with the UK being a notable exception.

Mandatory creep has been occurring over the past few decades, overriding the hitherto right to abstain from vaccination programmes on the basis of religious or 'conscientious objection' (a term originally used for nineteenth-century vaccine objectors and subsequently adopted by the military for those who refused compulsory conscription on similar grounds) in many cases leaving medical exemption the only acceptable criterion for refusal. Vaccine programmes involve mass immunization or herd immunity in which a large percentage of the population is required to be vaccinated against a disease to ensure its control and hopefully gradual eradication, thereby protecting those who cannot be vaccinated against possible infection. Unlike other medical interventions, the focus here lies on the protection of the population rather than an individual. For Foucault, immunization represents a distinctive break from other medical relationships which seek to heal the individual, but operates as 'a way of individualizing the collective phenomenon of the disease, or of collectivizing the phenomena, integrating individual phenomena within a collective field, but in the form of the rational and identifiable' (2004: 60). This brings another dimension to the philosophy of reponsibilization that dominates the neoliberal concept of health, a paradoxical position that transforms individual accountability into sacrifice (for even the strongest advocates of mandatory vaccination admit there is no such thing as one hundred per cent safety). It is a subtly crafted moral sleight of hand that turns adherence to the market of and for pharmatechnologies into an abnegation of self, where the self operates not simply as part of the collective but as subservient to it: capitalism functioning through a 'communist' guise. Human rights can thus be dismissed as 'nonsense upon stilts' or perhaps worse still, as the greatest threat to human rights itself.

Given that the 1948 Declaration of Human Rights emerged from inter alia some of the most horrific medical abuses ever recorded it is even more ironic that the policy of mandatory vaccination is proliferating. It contravenes human rights contained within a number of conventions which establish individual autonomy regarding medical intervention. This includes Article 8 of the European Convention of Human Rights on the respect for one's private life, that is 'the right not to be physically interfered with' (Liberty 2015) and more specifically, the International Covenant on Political and Civil Rights protection against 'torture or to cruel, inhuman or degrading treatment or punishment. In particular, no one shall be subjected without his free consent to medical or scientific experimentation' (International Covenant on Civil and Political Rights 1966).

The combined rhetoric of security and science, fundamental to the justification and waging of war, not only assuages the retraction of human rights obligations by the state but can render these rights themselves as presenting a threat. As the war on disease, together with that on terror, has no defined end, no clear moment of victory, being non-territorial insofar as it has no physical jurisdiction, the suspension or weakening of human rights will continue and further entrench the 'paradigm of security as the normal technique of government' (Agamben 2005: 14). In this environment vaccine objection easily slips into criminalized activity and criticism appears as unpatriotic. Concerns over conflict of interest, data manipulation, bribery and intimidation are turned into politicized opposition and those who voice these concerns become enemies of the people, a security threat, health terrorists. In this context, punitive measures against those who refuse, both for themselves or on behalf of those under their guardianship, are also becoming more draconian and can range from financial penalties, the exclusion of children from day care centres, loss of jobs (within the medical profession) or even prison (Willsher 2014).

Informed consent lies at the heart of the numerous instruments for the protection of the human body; the notion of bodily integrity underpins a series of rights that have been allocated to citizens in democratic states, including the right to abortion. When the information aspect of informed is either denied or contains negative data concerning safety it is rational to assume that consent might not be always forthcoming. Collusion between industry and politics is a major concern, as Eisenhower warned. Mary Holland's critique of mandatory vaccination lays out the extent of these collusive relationships in the US, which were even admitted by politicians themselves:

> In 2000, a Congressional report on Conflicts of Interest in Vaccine Policy Making identified notable conflicts of interests in the FDA [Federal] and CDC [Center for Disease Control] advisory bodies that make national vaccine policy. These include 'advisers' financial ties to vaccine manufacturers' as well as 'advisers' permitted stock ownership in companies affected by their decisions'.
>
> *(Holland 2012:77)*

Declaring these activities in Congress has not deterred their occurrence. In 2007, an attempt was made by the erstwhile Governor of Texas, Rick Perry, to mandate the HPV vaccine in his state's schools, a decision he was subsequently forced to overturn when it was revealed that Merck, the company that produced the Gardasil vaccine, had given donations towards his nomination (Eggen 2011). Nor will this be the last cosy relationship between politics and industry as mandatory vaccination policy continues to be applied to an increasing number of vaccines.

Conclusion

Pharmatechnologies manifesting as medical research, drug production and distribution, epistemological paradigms determining how health, disease and risk are to be conceptualized and negotiated, have now become a mode of governance in an increasingly authoritarian environment. This is not to detract from the benefits accrued from medical progress through pharmaceuticals and the important role played by vaccines in alleviating potential suffering. Yet, the benign discourses within which they operate can often obscure harmful outcomes, harms imposed as structural violence: opaque, unidentifiable and normalized (Žižek 2009). Victimization remains unseen or as a necessary price to pay for the greater good. In medical terms aspects of these harms are referred to as iatrogenesis, the unintended, often injurious, consequences of medical intervention. In this latter context it is akin to what military circles euphemistically term 'collateral damage', generally applied to the killing and maiming of non-combatants. However, some scholars who study state and corporate harms are less inclined to semantic generosity, not least when the majority of damage and injury inflicted falls on the same targets, the socio-economically vulnerable, gendered, racial, ethnic or other minorities whose lot it is to comprise the flotsam and jetsam of ruthless markets and the politics of indifference. Intention, or lack of, as they argue, cannot disguise the power relations at play as the usual suspects emerge as perpetrators and beneficiaries of systemic abuses.

If we are to accept the argument put forward by Tombs and Whyte that 'The problems caused by corporations – which seriously threaten the stability of our lives – ... are enduring and necessary functions of the corporation' (2015: 4), a position supported by a plethora of cases, then all industry operating within a capitalist framework is intrinsically pathological. It is essentially a diseased entity, irrespective of the nature of goods and services produced or the rhetoric that designates it as benign. Ironically, this diseased entity in combination with an increasingly diseased political system, proclaims and even persists that it has both the authority and ability to produce and sell health. Yet, so strong is the belief in pharmaceuticalized health that so many literally buy into the 'truths' of pharmatechnologies failing miserably to discern how the contagion-riddled commodification of health is actually the greatest danger to health.

No business thrives on the elimination of demand for its goods. The technologies of war were justified through the eventual establishment of peace; so too the pharmaindustry legitimizes its existence through claims to health and healing. The existence of both is dependent on the perpetuation of the very phenomena they claim to defend us from, and in this they must continue to be producers of war and sickness. No longer does the potential for the disastrous rise of misplaced power exist, as Eisenhower warned, it has become fundamental to the industries of war and disease, to war on disease.

References

Agamben, G. (2005) *State of Exception*. Chicago: University of Chicago Press.

Alcalay, G. (2014) 'Human radiation experiments in the Pacific'. Available at: www.theecologist.org/ News /news_analysis/2323244/human_radiation_experiments_in_the_pacific.html (accessed 2 May 2015).

Alvares, C. (1992) *Science, Development and Violence: The Revolt Against Modernity*. Delhi: Oxford University Press.

Amin, A. (2006) 'The millennium goals: a critique from the south', *Monthly Review* 57(10). Available at: http://monthlyreview.org/2006/03/01/the-millennium-development-goals-a-critique-from-the-south (accessed 19 March 2014).

Angell, M. (1997) 'The ethics of clinical research in the Third World', *New England Journal of Medicine* 337 (18 September): 847–9.

Archibald, K., Coleman, R., and Foster, C. (2011) Open letter to UK Prime Minister David Cameron and Health Secretary Andrew Lansley on safety of medicines. *The Lancet* 377 (9781), 1915.

Azoulay, P. (2003) Agents of Embeddedness (National Bureau of Economic Research, Cambridge, MA, Working paper 10142). Available at: www.nber.org/papers/w10142 (accessed 13 September 2013).

Babcock, H., Gemeinhart, N., Jones, M., Claiborne Dunagan, W., and Woeltje, K. F. (2010) Mandatory influenza vaccination of health care workers: Translating policy to practice, *Clinical Infectious Diseases*, 459–64, DOI: 10.1086/650752.

Balakrishnan, P. and Sharma, A. (2004) 'A new healing touch', *Business Standard*, New Dehli/ Mumbai. Available at: www.business-standard.com/article/beyond-business/a-new-healing-touch-104062601013_1.html (accessed 13 September 2013).

Box, S. (1983) *Power, Crime and Mystification*. London: Tavistock Publications.

Braithwaite, J. (1984) *Corporate Crime in the Pharmaceutical Industry*. London: Routledge & Kegan Paul.

British Journal of Medicine (1996) The Nuremberg Code (1947), BMJ 1996;313:1448, doi: http://dx.doi.org/10.1136/bmj.313.7070.1448.

Caplan, A. (2008) 'The Ethics of Evil: The Challenge and the Lessons of Nazi Medical Experiments', in W. LaFleur, G. Bohme and S. Shimazono (eds), *Dark Medicine: Rationalizing Unethical Medical Research*. Bloomington and Indianapolis: Indiana University Press.

Chossudovsky, M. (2005) *America's 'War on Terrorism'*. Quebec: Global Research, Center for Research on Globalization.

Clarke, A.E., Shim, J.K., Mamo, L., Fosket, J.R., and Fishman, J.R. (2003) 'Biomedicalization: Technoscientific transformations of health, illness and U.S. biomedicine', *American Sociological Review* 68: 161–94.

Congressional Record – Senate January 19, 1966 Significance of President Eisenhower's Speech on the Military-Industrial Complex. Available at: www.eisenhower.archives.gov/research/online_documents/farewell_address/1966_01_19_Congressional_Record.pdf (accessed 20 August 2015).

Dukes, G., Braithwaite, J., and Moloney, J.P. (2014) *Pharmaceuticals, Corporate Crime and Public Health.* Cheltenham: Edward Elgar.

Eggen, D. (2011) 'Rick Perry reverses himself, calls HPV vaccine mandate "a mistake"'. *The Washington Post.* Available at: www.washingtonpost.com/politics/rick-perry-reverses-himself-calls-hpv-vaccine-mandate-a-mistake/2011/08/16/gIQAM2azJJ_story.html (accessed 5 April 2016).

Esposito, R. (2011) *Immunitas: The Protection and Negation of Life.* Cambridge: Polity Press.

Foucault, M. (2003) *Society Must Be Defended.* New York: Picador.

Foucault, M. (2004) *Security, Territory, Population.* New York: Picador.

Gallagher, J. (2015) 'Flu vaccine "barely effective" against main strain', BBC online, Feb 5. Available at: www.bbc.com/news/health-31145604 (accessed 10 June 2015).

Glickman, S.W., McHutchinson, J.G., Peterson, E.D., Cairns, C.B., Harrington, R.A., Califf, R.M., and Schulman, K.A. (2009) 'Ethical and scientific implications of the globalization of clinical research', *New England Journal of Medicine* 360 (8) (19 February): 816–23.

Godfrey, R., Brewis, J., Grady, J., and Grocott, C. (2014) 'The private military industry and neoliberal imperialism: Mapping the terrain', *Organization* 21(1): 106–125.

Godlee, F. (2010) 'Conflicts of interest and pandemic flu', *British Journal of Medicine* 340:c2947 doi:10.1136/bmj.c2947.

Gotzsche, P. (2013) *Deadly Medicines and Organised Crime: How Big Pharma Has Corrupted Healthcare.* London, New York: Radcliffe Publishing.

Green, P. and Ward, T. (2004) *State Crime: Governments, Violence and Corruption.* London: Pluto Press.

Griffin, O.H. and Miller, B. (2011) 'OxyContin and a regulation deficiency of the pharmaceutical industry: rethinking state-corporate crime', *Critical Criminology* 19: 213–26.

Hartnung, W.D. and Berrigan, F. (2005) *Dollar Shift: The Iraq War and the Changing Face of Pentagon Contracting.* World Policy Institute Brief, (Feb). Available at: www.worldpolicy.newschool.edu/wpi/projects/arms/reports/Top102005Report.html (accessed 12th June 2015).

Healy, D. (2012) *Pharmageddon.* Berkeley and Los Angeles: University of California Press.

Hillyard, P., Pantazis, C., Tombs, S., and Gordon, D. (eds), *Beyond Criminology: Taking Harm Seriously.* London: Pluto Press.

Holland, M. (2012) 'Compulsory vaccination, the Constitution, and the Hepatitis B mandate for infants and young children', *Yale Journal of Health Policy, Law, and Ethics* XII (1): 41–86.

Hughes, S. (2007) *War on Terror Inc.: Corporate Profiteering From the Politics of Fear.* London: Verso.

Hurst, C.G. (2008) 'Biological Weapons: The United States and the Korean War', in W. LaFleur, G. Bohme and S. Shimazono (eds), *Dark Medicine: Rationalizing Unethical Medical Research.* Bloomington and Indianapolis: Indiana University Press.

International Covenant on Civil and Political Rights 1966. Available at: www.ohchr.org/Documents/ProfessionalInterest/ccpr.pdf (accessed 16 June 2015).

Jones, J.H. (1993) *Bad Blood: The Tuskegee Syphilis Experiment.* New York: The Free Press.

Liberty (2015) www.liberty-human-rights.org.uk/human-rights/what-are-human-rights/human-rights-act/article-8-right-private-and-family-life) (accessed 16 June 2015).

Lifton, R. (2000) *The Nazi Doctors: Medical Killing and the Psychology of Genocide.* New York: Basic Books.

Lupton, D. (2012) *Medicine as Culture: Illness, Disease and the Body.* London: Sage.

Mansell, P. (2011) 'Over 50% growth to 2015 seen in global clinical trials market', *PharmaTimes.* Available at: www.pharmatimes.com/article/11-07-07/Over_50_growth_to_2015_seen_in_global_clinical_trials_market.aspx (accessed 12 January 2014).

Michalowski, R. (2010) 'In Search of "State and Crime" in State Crime Studies', in W. Chambliss, R. Michalowski and R.C. Kramer (eds), *State Crime in the Global Age.* Cullompton, Devon: Willan Publishing.

Millennium Development Goals (2015) Available at: www.gavi.org/about/ghd/mdg/ (accessed September 10 2015).

Mirowski, P. and van Horn, R. (2005) 'The contract research organization and the commercialization of scientific research', *Social Studies of Science* 35(4): 503–48.

Moynihan, R. (2001) *Too Much Medicine: The Business of Health and Its Risks for You.* Sydney: ABC Books.

Naylor, R.T. (2004) *Wages of Crime: Black Markets, Illegal Finance, and the Underworld Economy.* Ithaca, New York: Cornell University.

Paliwal, A. (2011) 'Ethics on Trial', Down to Earth. Available at: www.downtoearth.org.in/context/ethics-trial (accessed 15 September 2013).

Parliament of India (2013) *Seventy-Second Report on Alleged Irregularities in the Conduct of Studies Using Human Papilloma Viris (HPV) Vaccine by PATH in India*, (Department of Health Research, Ministry of Health and Family Welfare). Available at: http://164.100.47.5/newcommittee/reports/EnglishCommittees/Committee%20on%20Health%20and%20Family%20Welfare/72.pdf (accessed November 20th 2014).

Pearce, F. (1976) *Crimes of the Powerful: Marxism, Crime and Deviance*. London: Pluto Press.

Petryna, A. (2007) 'Clinical trials offshored: On private sector science and public health', *Biosocieties*, 2, 21–40.

Petryna, A. (2009) *When Experiments Travel: Clinical Trials and the Global Search for Human Subjects*. Princeton, New Jersey: Princeton University Press.

Proctor, R. (2000) 'Nazi medicine and public health policy', *Dimensions: A Journal of Holocaust Studies*, (May) 14 (1): 1–7.

Rawlinson, P. and Yadavendu, V. (2015) 'Foreign bodies: The new victims of unethical experimentation', *The Howard Journal* 54 (1): 8–24.

Reiman, J. (2004) *The Rich get Richer and the Poor get Prison: Ideology, Class, and Criminal Justice*, 4th edn. Boston, MA.: Pearson/Allyn and Bacon.

Research and Markets (January 2015) *Vaccine Market by Technology, Type, End User, Disease Indicator – Forecasts to 2019*. Available at: www.researchandmarkets.com/research/g422mm/vaccine_market_by (accessed 2 June 2015).

Rose, H. and Rose, S. (2014) *Genes, Cells and Brains: The Promethean Promises*. London: Verso.

Rose, N. (2007) *The Politics of Life Itself: Biomedicine, Power, and Subjectivity in the Twenty-First Century*. Princeton New Jersey: Princeton University Press.

Sim, J. (1990) *Medical Power in Prisons: The Prison Medical Service in England 1774–1989*. Milton Keynes and Philadelphia: Open University Press.

Singer, P. (2011) *Practical Ethics*. New York: Cambridge University Press.

Smith, S. L. (2008) Mustard gas and American-race based human experimentation in World War II, *The Journal of Law, Medicine and Ethics* 36 (3): 517–521.

Sutherland, E. (1949) *White Collar Crime: The Uncut Version*. New Haven: Yale University Press.

Telegraph online (2010) 'GlaxoSmithKline profits climb on healthy sales of swine flu vaccine', *Telegraph*, 28 April. Available at: www.telegraph.co.uk/finance/newsbysector/epic/gsk/7645872/GlaxoSmithKline-profits-climb-on-healthy-sales-of-swine-flu-vaccine.html (accessed 28 September 2012).

Tombs, S. and Whyte, D. (2015) *The Corporate Criminal: Why Corporations Must Be Abolished*. Abingdon, Oxon: Routledge.

US Total Govt Spending Breakdown in $ trillion. Available at: www.usgovernmentspending.com/breakdown_1960USrt_16rs5n (accessed 3 Aug 2015).

Weizman, E. (2011) *The Least of All Possible Evils: Humanitarian Violence from Arendt to Gaza*. London: Verso.

WHO (2015) 'Trade, foreign policy, diplomacy and health: Pharmaceuticals'. Available at: www.who.int/trade/glossary/story073/en/ (accessed 2 June 2015).

WHO/Unicef (2005) 'GIVS Global Immunization Vision and Strategy 2006–2015'. Available at: www.who.int/vaccines-documents/DocsPDF05/GIVS_Final_EN.pdf (accessed 12 May 2012).

WHO (2006) 'Global pandemic influenza action plan to increase vaccine supply', Geneva: The Department of Immunization, Vaccines and Biologicals and the Department of Epidemic and Pandemic Alert and Response.

Whyte, D. (2007) 'The Crimes of Neo-Liberal Rule in Occupied Iraq', *British Journal of Criminology* 47 (2):177–195.

Willsher, K. (2014) 'French couple who refused vaccinations for their children go to highest court', *Guardian* online, 10 October. Available at: www.theguardian.com/society/2014/oct/09/french-couple-refused-vaccinations-for-children-face-judge, (accessed June 2 2015).

Žižek, S. (2009) *Violence: Six Sideways Reflections*. London: Profile Books.

13

Bioengineering and biocrime

Victoria Sutton

Introduction to the biotechnology revolution and crime

The basic operating system of life on earth is the carbon-based DNA molecule, replicating life with amazing precision. In the course of humanity's curiosity in it we have flown very close to the sun, first, in our unraveling of the structure of DNA and then, learning to hack it. Laws and regulation and ethics, if we even know what an ethic should be, are the societal constructs that will keep us from the same fate as Icarus. If that is not enough of a threat, accidentally or maliciously, the misuse of this incredible technology looms over the laboratory bench, the hospital, the pharmaceutical lab, the military lab and the family basement. The new age of biocrime is upon us.

We have been standing on this precipice for some time now. The biotechnology revolution, sparked by the 1950s discovery of the DNA double-helix by James Watson and Francis Crick, pulled the curtain back on the human operating system in the first half of the twentieth century (Watson and Crick 1953). Manipulating DNA to perform useful tasks with a purpose of making our quality of life better, grew exponentially, from food to pharmaceia to lifestyle improvements and better forensics that worked both to convict and to exonerate criminals who might have escaped prosecution in an earlier time—or may have been wrongly convicted.

Accessibility to these discoveries has also seen vast expansions, not only making the benefits and products of this revolution available ubiquitously, but also accessible as tools. The step which changed the tedious work of genetic engineering to component building like a Lego project, was synthetic biology, ushered in with the 2000s. Building biological machines no longer required graduate student training to manipulate DNA—now high school students could do it in their basements or garages, with minimal instruction (Mooallem 2010). These next generation rebels call themselves, "biohackers".[1]

Internationally, the threat of biological terrorism is primarily concerned with specific disease-causing agents, and some toxins produced by them, but the drafters of the Biological Weapons Convention had the foresight to include in the text of the treaty a regular review that "[S]uch review shall take into account any new scientific and technological developments relevant to the Convention".[2] Specifically, the conferences of the parties (COPs), which take place approximately every five years, have a standing agenda item to review the advances in biotechnology and take them into account in their interpretations of the Biological Weapons Convention, recently renewing that commitment.[3]

Like the course of emergence in almost all emerging technologies, the first uses of the technologies may not be useful and amount to little more than "parlor tricks," such as in the case of synthetic biology, building bacteria that smell like bananas or in the case of nanotechnology, writing IBM's name in nano-size with atoms that could only be seen with a microscope. This "parlor tricks" stage lasts only until creative minds parlay these tricks into powerful uses. For example, the obscure extremophile bacterium living in the Yellowstone National Park Geyser with the ability to replicate under high temperatures, has an enzyme that can withstand denaturization in the extreme heat, unlike almost all other forms of life, thus enabling it to continue the replication of its DNA. Substituting this useful enzyme to the problem of DNA replication in the lab, where the process required alternating high temperatures with normal temperatures to avoid denaturing the required enzymes, changed the world for forensics and research. This also resulted in a Nobel Prize for the scientists who thought of it and did it.[4] Imagining the possibilities with the bio-tools of nature laid out before us offers visions that will make our lives incredibly wonderful or frighteningly horrific. Finding the place between these utopian and dystopian views that will ensure a safe voyage for humanity is where the rule of law becomes so vital.

So it follows that the leaps in technology have also made it possible to use these discoveries to do great harm in the hands of the malevolent individual or nation with increasingly less talent, skill, and knowledge, making the rule of law all the more important to society. The example of synthetic biology, so accessible even to individuals with a weekend of training, has opened our eyes to the possibilities of popular biology unfettered by traditional professional societal norms. But it has also opened the door to seeing the societal norms that impede technological progress, like the patent system that overlaps and over-reaches until the field of study is grid-locked into stalemate until the obligatory period of time passes. These new rebels have no time to wait, and rejection of patent protections is but a small sacrifice in their heady transformation of biotechnology and society.

These parlor tricks and biohackers raise no alarm call for wrong-doing, but how has the law kept pace with the malicious and inventive criminal-minded among us? Particularly, how has international law kept pace with the technologies that might evolve from the biotechnology revolution? To what degree can the rule of law control or contemplate what is required to protect against the greatest unpredictable threat of all – Mother Nature? Such laws would have to be broad and flexible in order to adapt to this rapidly changing technology. Some of these succeed in this new world and others will become less useful. A closer look at these in the next section can shed light on this changing environment and our use of the rule of law.

Biotechnology does not recognize borders; so what can we expect from the rule of law in this vitally important international community?

Mother Nature's unpredictable menagerie

The turn of the millennium saw anthrax attacks originating in America around the events of 9/11 (2001), then continuing the terror of the attacks through the rest of the world. This was soon followed in 2003 by the first emerging infectious disease of the millennium, SARS,[5] waking the world with an unexpectedly rapid swath of death around the globe. Afterward, the World Health Organization responded by accelerating the development of the International Health Regulations (IHR) amendments from what had become a moribund version from earlier decades.[6] The new IHR in 2005 would take into account rapid travel and the spread of emerging infectious diseases with the demand that nations must disclose and cooperate in controlling the invisible threats.[7] Entering into force in 2007, the IHR became the world's first binding, international public health law.

Shortly after the IHR became binding law, the swine flu of 2009 brought Mexico to its knees in a matter of weeks. Mexico was the first nation to test the new IHR and they performed courageously in complying with the openness required of the IHR, despite the devastation to their tourism industry.

Around five years later, West Africa suffered the worst Ebola epidemic in human history (2012–2016), in a region of Africa not typical for Ebola. A deadly hesitancy to use the International Health Regulations by the World Health Organization led to Ebola's rapid spiral out of control for the local authorities, spilling into the airways of the world, reaching Europe and North America as its spread grew. The Ebola events showed the world, if there was ever any doubt, we are not in control of Mother Nature, the world's scariest bioterrorist; hesitancy to address a historically sinister disease when it makes a bold appearance is a mistake.

Meanwhile, in the Western Hemisphere, in 2013 the chikungunya virus arrived in the Caribbean from West Africa and has spread rapidly into every point in this Hemisphere.[8] Not far behind, Zika arrived in the Western Hemisphere, also a mosquito-transmitted virus, with a new set of symptoms affecting the unborn. Brazil was the first to set off an alarm that symptoms such as microcephaly may be associated with Zika virus infections in pregnant women.[9]

The international rule of law with the first binding international regulations for global public health threats has brought a much needed maturity to global public health controls, driven by not only the needs of a global commons with shared risks, but also the self-interests of nations. The possibilities of biocrimes using emerging infectious diseases, confusing deliberate with natural infections is realistic. The effort to create a protocol to analyze the difference is another tool[10] that has been created by the World Health Organization to be used in conjunction with the International Health Regulations algorithm which is used to analyze the risk severity of an outbreak. This acknowledgement by the World Health Organization that addressing a deliberate attempt to use biological weapons is within their scope of authority is an unprecedented action. This action has extended WHO to the edges of their chartered authority into deliberate and criminal areas once considered solely the domain of nations' law enforcement authorities and Interpol.

Human biohacking: risks and opportunities

The biotechnology revolution has given us longer and healthier lives due to antibiotics, vaccines and countless other variations on other biotechnologies (see Sutton 2007). With every amazing advancement, the criminal-minded among us violate social norms to turn these new tools against us using the same amazing advancements. The adaptability of crime to the opportunity tracks closely with the adaptability of opportunistic infections, waiting for the right time, the right conditions and the right tools. The inevitable dual-use phenomenon[11] doggedly accompanies our grand successes, and a malevolent use by its very nature comes with every virtuous use, almost without exception. The US government has a program to oversee this possibility called the "Dual Use Research of Concern",[12] focusing on the federally funded work that could pose serious threats to public health if misapplied in criminal ways.

Not until the twentieth century did humankind recognize that using weapons developed with biotechnology and diseases outweighed the consequences both directly and indirectly to those who used them. In banning the use of biological weapons by the United States, US President Richard Nixon said, "[M]ankind already carries in its own hands too many of the seeds of its own destruction."[13] This ban was the leadership that was needed for the rest of the world to follow. In 1972, the Biological Weapons Convention was drafted under the scope of the United Nations work toward "peace and security", and entered into force in 1975.

But a question of what exactly should be banned, was critical to maintaining the effectiveness of a treaty like the Biological Weapons Convention. The right balance had to be struck between prohibiting the possession and use of deadly diseases as weapons and allowing the scientific community to conduct valuable and life-saving research. This rapid development in biotechnology research had already begun and the drafters had the foresight to provide for this changing body of knowledge, with a view toward reviewing the definitions sections of the Convention from time to time.

The Biological Weapons Convention (BWC) defines biological weapons as "[m]icrobial or other biological agents, or toxins whatever their origin or method of production, of types and in quantities that have no justification for prophylactic, protective or other peaceful purposes."[14] Recognizing the uncertainty in the next several generations of biotechnologies, the BWC provides for the meeting of the parties five years after entry into force, or earlier upon parties' request to review the operation of the Convention, with a view to assuring that the purposes of the preamble and the provisions of the Convention, including the provisions concerning negotiations on chemical weapons, are being realized. Such review shall take into account any new scientific and technological developments relevant to the Convention.[15] The scope of this definition has been expanded to include new discoveries with each succeeding Conference of the Parties (COP), roughly every five years. It has become increasingly difficult to fit new technologies within the confines of Art. I of the BWC, drafted in 1972 (Sutton 2015).

Recognizing the limitations of definitions alone, the Conference of Parties, during the BWC's Second Meeting, adopted the Confidence Building Measures (CBM) mechanism to implement Articles V and X of the BWC.[16] CBMs are defined categories of actions that countries institutionalize as a way of demonstrating their compliance and commitment to banning biological weapons. The BWC Administrative Unit agrees to provide confidentiality to Parties in exchange for disclosure of these actions by the Parties. In one review of this enforcement mechanism by the BWC Administrative Unit they found that scarcely 18 of the 190 members had implemented measures to address dual-use biological equipment and related technology.[17] Import controls were the widest single mechanism used to control the threat of biological weapons, yet only 59 of 190 member countries required authorization for export and import of dangerous biological agents and toxins. In all, less than one third of the 190 members had contributed anything at all to CBMs, an indication of very weak, global compliance (Sutton 2013).

International law is slow to develop principles and practices, and all indications are that this is an area that is made even slower by the uncertainties in a rapidly changing technology. The only international treaty designed to prohibit a single class of weapons, the BWC, is straining to limit the risks of criminal use of biological weapons that expand each year with new and creative biotechnological tools.

Biotechnologies of the future: the possible, the probable and the certainty of biocrimes

The advances in biotechnology have made it possible to map the human genome, then to map the genome of many other organisms, and finally to create vast databases to store this bio-information to select and build what these human tools can imagine. Yet, the celebratory sounds were still echoing when we realized this new skill of mapping the genome merely gave us a map without knowing very much about it or what made it function, giving birth to the new field of proteomics, or the proteins produced by the genes. This science explores the proteins that carry the signals to shape forms of life in its many variations, not only between species and subspecies but from individual to individual.

The ability to synthesize genes and to build better fruits, crops, domestic animals and fish silenced the global concern of worldwide starvation from the limitations of traditional agriculture, when simply growing more food became widespread with this biotechnology. Controversial, yet so attractive to the farmer who could grow yields beyond anything in history, genetically modified organisms are grown in almost every part of the world, today, and almost half of four major crops are genetically modified ones.[18] But as with every virtuous use there is almost a certainty that its very strength may be a weakness in biotechnology. In this case, it is the uniformity of the genome, itself, that makes it vulnerable to human or natural devastation from a single plague exploiting its single weakness. A single plant plague that exploits a vulnerability of a genetically modified crop, anywhere in the world, could destroy a good part of the world's crop. A singular plague could present the same threat to farms with uniform, genetically modified salmon.

The costs of mapping and engineering genes have dropped in accordance with Moore's Law, which indicates there is an exponential drop in cost and thus, availability to the public of any emerging technology. However, in this case, these genetic tools have decreased in cost at a rate five times faster than the drop in computer costs.

The development of component engineering for genetics took flight in the mid-2000s with synthetic biology. Synthetic biology is the Lego blocks of building life, making it simpler and much faster for building genetic alterations of bacteria, for example. One of the first "parlor tricks" of this technology was to make bacteria smell like bananas. However, this new technology has exceeded the review authority of regulatory frameworks in some cases. For example, Craig Venter, credited with being first with NIH in mapping the human genome, sponsored a May 2014 report that warned: "Genetically engineered organisms are increasingly being developed in ways that leave them outside of APHIS' authority to review, and synthetic biology will accelerate this trend" (Carter 2014).

The age of biocrime could very likely take advantage of these vulnerabilities.

Genomic crimes

The popularization of do-it-yourself (DIY) biology has made it possible for high school students and undergraduates with no microbiology background to engineer DNA with components in the new field of synthetic biology. No longer do you need to string together pieces of DNA and hope for the best. Now, components designed like Lego blocks that fit neatly together can be used to greatly accelerate the genetic design process. For example, inserting a gene that causes a plant to glow in the dark can be done by almost anyone, now, with a weekend of training, made possible with synthetic biology. More recently, the development of the CRISPR tool makes the original genetic engineering task that took weeks and months take mere hours with greater precision in inserting or deleting a single gene, unlike the much cruder tools of the past.

The rapid development of these tools, the ten-fold decrease in cost, and the ubiquity with which it is being used by anyone with an interest has increased the design possibilities. The iGEM competition is a gathering place for teams inspired to design a useful or creative project with synthetic biology sharing the information and advancing the art. Safety—and the Federal Bureau of Investigation (FBI)—are never far away from this gathering. A culture of safety and being aware of the power of these tools is a reminder that this is a powerful game requiring great responsibility. Using the "parlor games" analogy, iGEM competitors passed the "parlor trick" stage a number of years ago, and now the inventions that are built are truly impressive, and the possibilities, endless. The Federal Bureau of Investigation has rightly recognized that being a part of the process by being at iGEM is "walking the beat" of the new biotechnological streets of society.

Biocrime may have a more difficult time taking root in these communities but it does not prevent the "lone wolf" ambitions of biocrime.

Genomic misdemeanors

The identification of genomic fraud is now possible with genetic testing, for example, to discover if you are eating what you were told you were being served in a restaurant or sold in a grocery store. In one reported instance, high school students tested the sushi in a restaurant and found the restaurant was using tilapia as a substitute for white tuna (Schwartz 2008). This is clearly fraud, but who is going to investigate? Should the state fraud statutes be used against restaurants and grocery stores which can no longer simply defraud their customers with impunity? If so, law enforcement will have to become educated on genomics and perhaps develop a specialized genomic fraud squad with special training in order to identify these genomic misdemeanors.

Crop destruction by design

With the virtue of being able to feed the world with genetically modified crops, also comes the dark side of the technology, as discussed above. Due to the worldwide use of genetically modified crops, the genomes are almost identical, making this mono-culture crop vulnerable to being completely destroyed by a fungus or other disease that happens to find the weakness in the genome. Such an attack, either by Mother Nature or a designed disease, would potentially destroy a large part of the world's food crop for one or more seasons. For example, through natural selection over the years, bananas have become a mono-culture around the world in banana plantations. Currently, this fruit is at risk of extinction due to a fungus that is attacking this mono-culture of commercial banana plantations. The potential for attacks on mono-culture crops could be a criminal opportunity.

In order to ensure against the possibility of entire mono-culture extinctions, the use of the seed banks that have been built to ensure old seed stocks survive might become vital to re-establishing crops. Private collections of seed banks could also become important, despite the possibility of keeping the genome in a digital form.

Personalized DNA weapons

The rise of personalized medicine that is responsive to particular genetic markers and types is one of the most important advances in cancer treatments. However, this precision can also create vulnerabilities just as easily as it can create opportunities. Engineering a virus to target individuals with that particular genetic marker might at first be used to target high level officials in assassinations, but would soon become available for wider use for targeted killings of domestic crimes. This would undoubtedly be of interest to the military as well, and targeted killings of the enemy would be a logical extension of this process.

The use of precision "biotechnological weapons" has been discussed as some of the most frightening of all weapons. The use of "direct integration" involves injecting DNA directly into another human through the use of specialized bullets and guns. Once inside the body, the DNA would be designed to infect the human to disable, kill, or change the person. Other uses might include pheromones affecting certain genotypes that would alter behavior in ways that would not be conducive to defending against an enemy. Surprisingly, these ideas have been part of the discourse for more than a decade, but are now frighteningly close to being possible (Ji-Weo and Yang 2005). The Biological Weapons Convention arguably would extend to

233

these kinds of weapons, but the scope of the definition in the Convention must be increasingly broadened (Sutton 2015).

Identity theft for financial access

To suggest that the law has kept up with technology in digital finance would be wishful thinking; however, applying existing law to reach the scope of these technologies has at least begun. Mobile banking, bitcoin transactions with the new blockchain feature that can trace the origin of bitcoin transactions are all digital systems that may in the near future be accessible based on bio-identifiers. For example, the use of retina scanners and fingerprint access is already commonly used even for unlocking individual smartphones. The possibility of using the unique genetic signature of a person would ensure the uniqueness of that identifier and may one day be used for financial transaction confirmations.

The Genetic Information Nondiscriminatory Act of 2008 ("GINA")[19] was created to protect the privacy of individuals, finally making a US federal statute which, in part, replaced similar state statutes passed over the previous ten years in all fifty states. This statute prohibited the use of genetic information by employers against an employee or against an insured by an insurance company, against discrimination in the case of employers and for pre-existing conditions in the case of insurance companies. This may be remembered in the future as the first statute on which later amendments were added to protect us from genetic identity theft, and genetic crimes by protecting the individual's genome identifiers. This statute might also be amended in the future to prevent the possession of another persons' genome with intent to use it to harm them or to steal from them. This statute could very well be the foundation statute of the age of biocrime and the first statute to protect an individual's genomic identifier.

In the future, the use of fingerprints, retinal scans or DNA tags could be used to ensure privacy and security of financial accounts. The need to keep one's DNA safe from being stolen will be a new requirement to guard against financial theft and other privacy invasions. The fact that "open source" DNA can be collected from discarded paper cups, cigarette butts, napkins and a host of other sources, makes the task of keeping your DNA secure rather daunting. The surreptitious collection of DNA could be a new criminal industry, sold like stolen credit cards and ID cards.

Inserting drugs in crops

The ability to insert genes in crops such as tobacco, rice or bananas could also be used for malicious ends, perhaps unleashing a deadly virus when the crop is distributed for human consumption to a particular country, for example.

Stealing economic genetic trade secrets

In 2012, the first discovery of economic espionage[20] in the theft of genetically modified trade secrets turned another page in the age of biocrime. Genetically modified experimental corn, protected as a trade secret, was stolen for the benefit of a China crop development company, signaling a new kind of genomic crime. While economic espionage has been a federal crime since 1996, it was only now that the theft of trade secrets was used to prosecute the theft of genetically modified material.

In this case, the systematic stealing of experimental varieties of corn from experimental fields in Iowa was an ongoing process by Mo Hailong and a group of five or more Chinese-nationals, all searching the Iowa countryside for experimental corn varieties to literally pocket and take or send back to China. Their goal was apparently to meet the growing food demand of their country and reverse the trade imbalance of importation of 94 percent of their corn needs from the United States (Genoways 2015). It would be logical to extend that objective to the theft of cotton, another crop for which China desperately wants increased growth in domestic production, which would reduce the amount of cotton purchased from the United States. Other genetically modified crops may also be on their shopping list, and more vigilance and better protections are needed to prevent genomic thefts from any country, not just China, that may mean millions of dollars in losses to the companies which create them.

This is a growing area of concern in the biotechnology area, with recent convictions in the US demonstrating the reality of the threat of theft of intellectual property in genetics. This is a biocrime that has only recently become more evident and will require further expertise and perhaps specialized investigative squads among the tools of the Federal Bureau of Investigation (FBI).

Attribution problems

Biocrimes also come with the problem of attribution. Whether it is Mother Nature or whether it is a criminal design of a biological agent, attribution requires good investigative techniques combined with tools of scientific analysis that can identify both the genomics of the agent, as well as connect the perpetrator with the crime. Both of these require new techniques in the age of biocrime.

We have experienced attribution problems most apparently in the Amerithrax investigation. After ten years of investigation with the top talents in the FBI, and ten years of scientific research to map the genome of anthrax with the top scientists in this field, the conclusion was less than conclusive. A single flask, labelled RMR-1029, was identified as the source of the anthrax based on both good scientific work as well as good investigative work. As remarkable as that feat was, the fact that two hundred people had access to that flask, RMR-1029, was one evidentiary hurdle that may have forever thwarted a jury from being able to find a perpetrator guilty beyond reasonable doubt.[21]

Better forensic genomic tools to connect the perpetrator with genomic crimes in this attribution challenge are essential in addressing the new age of biocrimes.

Conclusion

With virtuous uses of new biotechnologies, it is inevitable that malevolent uses, with the aid of criminal minds, will occur. Utilizing these new exciting technologies for personal gain or simply to do harm rather than good is a phenomenon which we must prepare for and expect. The responsibility in developing these new technologies must also include the research necessary to formulate ways to guard against the criminal, to develop forensic tools and to enforce and prosecute the commission of these new biocrimes. New training for law enforcement into specialized squads to address these new biocrimes will be necessary.

Yet, the biotechnologies that have increased our quality of life, and eliminated the immediate fears of world starvation far outweigh any crime that we have seen thus far. But to be complacent and assume we have dodged the bullet would be a mistake. Focus to address and prepare for stopping these malevolent uses of increasingly available biotechnologies must be ever in our sights as we continue to be an optimistic, yet vigilant, society.

Victoria Sutton

Notes

1 PBS tries to capture the meaning of biohacking. See www.pbs.org/newshour/updates/biohacking-care/.
2 Art. XII, Biological Weapons Convention.
3 It was agreed to have a standing item to discuss advances in biotechnology at each intercessional meeting between 2011 and 2016 leading up to the Eighth Review Conference. Seventh Review Conference of the Parties to the Convention on the Prohibition of the Development, Production and Stockpiling of Bacteriological (Biological) and Toxin Weapons and on Their Destruction, Dec. 5–22, 2001, Final Document of the Seventh Review Conference, 20–21, U.N. Doc. BWC/CONF. VII/7 (Jan. 13, 2012), available at www.unog.ch/80256EDD006B8954/(httpAssets)/3E2A1AA4CF 86184BC1257D960032AA4E/$file/BWC_CONF.VII_07+(E).pdf [hereinafter Seventh Review Conference, Final Document].
4 Kary Mullis was awarded the Nobel Peace Prize for Chemistry in 1993 with Michael Smith.
5 Severe Acute Respiratory Syndrome, www.cdc.gov/sars/about/fs-sars.html.
6 International Health Regulations (1969).
7 International Health Regulations (2005).
8 www.cdc.gov/chikungunya/.
9 www.cdc.gov/zika/.
10 www.who.int/csr/resources/publications/deliberate/WHO_CDS_CSR_EPH_2002_16_EN/en/.
11 Dual Use Research of Concern (DURC) is the misapplication of research to cause serious harm to public health. The complete definition can be found here: http://osp.od.nih.gov/office-biotechnology-activities/biosecurity/dual-use-research-concern.
12 Ibid.
13 25 Nov. 1969, Pres. Nixon, 462 - Remarks Announcing Decisions on Chemical and Biological Defense Policies and Programs, The American Presidency Project at: www.presidency.ucsb.edu/ws/?pid=2344.
14 Art. 1, Biological Weapons Convention.
15 Art. XII, Biological Weapons Convention.
16 Second Review Conference of the Parties to the Convention on the Prohibition of the Development, Production and Stockpiling of Bacteriological (Biological) and Toxin Weapons and on their Destruction, Final Declaration, 2, 7, U.N. Doc. BWC/Conf.II/13/ II (1986) [hereinafter Second Review Conference], available at www.opbw.org/rev_cons/2rc/docs/final_dec/2RC_final_dec_E.pdf.
17 See Seventh Review Conference of the States Parties to the Convention on the Prohibition of the Development, Production and Stockpiling of Bacteriological (Biological) and Toxin Weapons and on their Destruction, Final Document of the Seventh Review Conference, 1, U.N. Doc. BWC/Conf. VII/INF.8 (2011).
18 49% of cotton, soybeans, canola and maize grown worldwide are genetically modified. See www.isaaa.org/resources/publications/pocketk/16/.
19 Pub.L. 110–233, 122 Stat. 881.
20 18 USC §1831 (2016).
21 The potential suspect committed suicide before being arrested, so the attribution for the anthrax attacks in the US in 2001 remains unsatisfyingly not completely resolved.

References

Carter, S.R. (2014) Synthetic Biology and the U.S. Biotechnology Regulatory System: Challenges and Options, *J. Craig Venter Inst.* Available at: www.jcvi.org/cms/research/projects/synthetic-biology-and-the-us- biotechnology-regulatory-system/overview/.
Genoways, T. (2015) Corn Wars, *New Republic* (Aug. 15, 2015). Available at: https://newrepublic.com/article/122441/corn-wars.
Ji-weo, G. and Yang, X. (2005) Ultramicro, Nonlethal, and Reversible: Looking Ahead to Military Biotechnology, 85 *Mil. Rev.* 75 (citing David M. Mahvi, Michael J. Sheehy and Ning-Sun Yang, DNA Cancer Vaccines: A Gene Gun Approach, 75 *Immunology & Cell Biology* 456, 459 (1997)).
Mooallem, J. (2010) Do It Yourself Biology, *The New York Times* (Feb. 10). Available at: www.nytimes.com/2010/02/14/magazine/14Biology-t.html?_r=0.

Schwartz, J. (2008) Fish Tale Has DNA Hook: Students Find Bad Labels, N.Y. Times (Aug. 21, 2008), www.nytimes.com/2008/08/22/science/22fish.html.

Sutton, V. (2007) *Law and Biotechnology*. Durham, NC: Carolina Academic Press.

Sutton, V. (2013) Biodiplomacy: A Better Approach to Dual Use Concerns, 7 *St. Louis U.J. Health L. & Pol'y* 111.

Sutton, V. (2015) Emerging Biotechnologies and the 1972 Biological Weapons Convention: Can it Keep Up with the Biotechnology Revolution? Special Edition: New Technology, Old Law: Rethinking National Security, 2 *Tex. A&M L.R.* 695–718.

Watson, J.D., and Crick, F.H.C. (1953) A Structure for Deoxyribose Nucleic Acid, 171 *Nature* 737.

Keynote discussion

Keynote discussion

14

Technology, environmental harm and green criminology

Rob White

Introduction

The use of technology for modifying the environment has been intrinsic to the relationship between nature and humans since the dawn of anthropocentric time. From the lighting of grass fires to the invention of the wheel, everyday techniques and technologies employed by humans have had unavoidable implications and consequences for our relationship with nature and for nature itself. Environmental harm has long been accepted as one of the outcomes of this relationship, and is ingrained in contemporary laws that 'permit' pollution of land, air and water. When the harm is deemed to be serious and too great a threat to sustainability and intrinsic values, then it may be subject to sanction. Typically, however, even when environmental harm is sanctioned it tends to be dealt with at the lower end of the scale – through administrative and civil law rather than criminal law.

Green criminology refers to the study of environmental harm from a critical and holistic perspective. It includes the study by criminologists of environmental harms (that may incorporate wider definitions of crime than that provided in strictly legal definitions), environmental laws (including enforcement, prosecution and sentencing practices) and environmental regulation (systems of civil and criminal law that are designed to manage, protect and preserve specified environments and species, and to manage the negative consequences of particular industrial processes) (White 2008, 2011). A key focus of green criminology is environmental crime. For some writers, environmental crime is defined narrowly within strict legal definitions – it is what the law says it is. For others, environmental harm is itself deemed to be a (social and ecological) crime, regardless of legal status – if harm is done to environments or animals, then it is argued that this ought to be considered a 'crime' from the point of view of the critical green criminologist (South and Brisman 2013; White and Heckenberg 2014).

The commission of environmental crime inevitably involves application (or non-application, or misapplication) of technologies and techniques pertaining to extraction industries (such as mining), pollution and waste control, the capacity for detection (for example, of illegal fishing) and possibilities for ecologically sustainable social practices (such as reliance upon renewable energy to power automobiles or offices). Yet rarely within green criminology is there overt discussion of technology as a problematic or focus for critical analysis. The intention of this chapter, therefore, is to highlight the varied ways in which technology and knowledge about technology are linked to the green criminology project, and how violations against, and protections of, the environment almost always involve a technological dimension.

Technology is a causal factor in various types of harms of interest to green criminology. Environmental crimes as described in law include things such as illegal transport and dumping of toxic waste, the transportation of hazardous materials such as ozone depleting substances, the illegal traffic in real or purported radioactive or nuclear substances, the illegal trade in flora and fauna, and illegal fishing and logging. However, within green criminology there is also a more expansive definition of environmental crime, especially harm involving technology (such as clear felling of forests using bulldozers), that includes transgressions that are harmful to humans, environments and nonhuman animals, regardless of legality *per se*; and environmental-related harms that are facilitated by the state, as well as corporations and other powerful actors, insofar as these institutions have the capacity to shape official definitions of environmental crime in ways that allow or condone environmentally harmful practices (White 2011).

An increasingly important concept within green criminology is that of ecocide, which has been defined as 'the extensive damage, destruction to or loss of ecosystems of a given territory, whether by human agency or by other causes, to such an extent that peaceful enjoyment by the inhabitants of that territory has been severely diminished' (Higgins 2012: 3). Where this occurs as a result of human agency, then it can be argued that such harm can be defined as a crime. The targets for action around ecocide generally include both nation-states and transnational corporations. From the point of view of technology, ecocide is frequently seen to both stem from and potentially be averted by, human technologies. As the gravest of all possible circumstance, since it is linked to species extinction and habitat destruction, it is not surprising that the study of ecocide is a key area where green criminology has also started to incorporate critical evaluation of technology.

Technology, as described herein, does not refer solely to 'hard science' and 'hard ware' and the metals and plastics of mass produced and consumed industrial and consumer devices. Rather 'technology' is understood, at least initially, in a generic sense to refer to conscious interventions by humans in the modification of nature. For example, Davison (2004: 143) observes that:

> Alternatives such as "organic" forms of agriculture and "holistic" medicine may well be more sustainable than many emerging forms of gene technology. But they are not less technological, or more natural, in any essential sense. They represent crucially different forms of biotechnological social practice that need to be articulated as alternative and positive visions of technology – visions affirming that empathy and interconnection are as much technological possibilities as are control and alienation.

Yet most of the discourses about technology in relation to ecocide and climate change, and as used within criminological contexts, take a more narrow view as to what is considered technology and technique. This has significant implications insofar as when certain applications of 'technology' are separated from considerations of 'society', then injustice and social harms may occur and/or be intensified. This is one of the key themes of the present chapter.

The substantive concern of the chapter is to explore how technology is portrayed and utilised across several domains of criminological interest, and in particular the challenges it poses for thinking within and about a green criminology perspective. Given that the primary focus is on environmental harm and green criminology, the specific concern is with technological applications from the point of view of three general perspectives.

1   *technology as a tool*, particularly in relation to crime prevention and crime fighting in the context of ecological harms;

2   *technology as the problem/solution*, particularly in relation to overcoming macro harms such as global warming and climate change; and

3   *technology as paradoxical*, in the sense that technological innovations such as adoption of nuclear power or development of 'green prisons' may have multiple consequences and dimensions.

The concern is not so much to provide abstract critique of notions such as techno-instrumentalism (for example, 'ecological modernisation theory' and the idea of maximising production via technology) or techno-determinism (that is, where machine and technique are seen to dictate human behaviour) (see Davison 2004). Rather, it is to interrogate how technology is being applied, for what purposes and to what effect, within environmentally related criminological domains. There is no overarching perspective on 'technology' *per se* since our main concern is to describe its applications within the framework of very specific intents and particular social situations. In the present analysis, *context* is everything.

Technology as a tool

For criminal justice institutions and operatives, technology is an essential part of detection, investigation and prosecution processes. Gathering of information, intelligence and evidence frequently depends upon particular scientific and technological knowledge and instrumentation. In terms of the kinds of criminal justice interventions which interest green criminology, the areas of environmental forensics and environmental wildlife crime prevention are examined.

The use of forensic techniques in dealing with environmental crime is an expanding and evermore sophisticated area of work within environmental law enforcement agencies and networks. For example, the area of wildlife forensics deals with the application of scientific knowledge to a range of species protection, law enforcement and wildlife verification challenges that face policymakers, investigation and enforcement officers, and commercial stakeholders. These challenges include: the illegal capture or killing of animals, birds, reptiles and plants in contravention of domestic or international law; the illegal harvest of living resources from protected areas and the trade in those resources, a problem of particular relevance for fisheries and forestry sectors; and the illegal trade in raw and processed parts of protected animals, birds, reptiles and plants and in goods manufactured from such parts (see for example, Alacs and Georges 2008).

There are major and continuing challenges to the undertaking of environmental forensic investigations. Consider, for example, the complexities involved in environmental crime scene investigation. Issues here include: first responder safety in regard to hazardous waste and pollutants; the scientific skills and knowledge of environmental crime scene investigators and forensic specialists; the availability of environmental forensic technologies to crime scene investigators; and emerging technologies that may assist with investigations (Ramer 2007).

Acknowledgement of the complexities of the issues is but a first step in recognising the limitations of such work. For example, in one study, it was found that investigation of environmental pollution situations in Brazil is a complex environmental contamination situation, but few analytical resources were available to accomplish the necessary comprehensive evaluation and, thus, to provide material proof for the situation and to determine whether it was an environmental crime according to the law (Barbieri *et al.* 2007). The methods included:

•   data from quarterly monitoring reports sent to authorities as part of the landfill operation permit requirements;

- monitoring lead and chromium exposure of nearby residents;
- testing of fish in nearby artificial pond;
- groundwater analysis;
- sediment analysis.

Putting the pieces together (of pollution, of perpetrators) is both the problem and the basis for achieving suitable outcomes. Moreover, there are always additional considerations to take into account when attempting to determine wrongdoing when it comes to specific types of environmental harm such as pollution (for example, the issue of background values – see Petrisor 2007).

There is now a broad spectrum of activity associated with the doing of environmental forensic studies (White 2012a). In drawing upon multiple scientific studies and knowledge production techniques, composite socio-ecological accounts of harm can be compiled, although the questions 'compiled by whom, and for what purposes?' remain of major interest and contention. These questions are just as relevant to historical statistical analysis of ocean-based oil spills as they are to discrete forensic studies as such (Van Gulijk 2014). Examples of technical developments include:

DNA testing

Illegal fishing and illegal logging can be tracked through the employment of DNA testing at the point of origin and at the point of final sale. Work done on abalone DNA, for example, demonstrates that particular species within particular geographical locations can be identified as having specific (and thus unique) types of DNA (Roffey *et al.* 2004; see also Ogden 2008). The use of phylogenetic DNA profiling as a tool for the investigation of poaching also offers a potential deterrent in that regular testing allows for the linking of abalone species and/or subspecies to a particular country of origin. This increases the chances of detection and, thus, may have relevance to crime prevention as such. The use of DNA testing to track the illegal possession and theft of animals and plants can thus serve to deter would-be offenders, if applied consistently, proactively and across national boundaries.

Satellite surveillance

Illegal land clearance, including cutting down of protected trees, is presently monitored through satellite technology in some jurisdictions (New South Wales Environmental Protection Agency 2013). Compliance with or transgression of land clearance restrictions, for example, can be subjected to satellite remote sensing in ways that are analogous to the use of closed circuit television (CCTV) in monitoring public places in cities. Interestingly, the criminalisation of land clearance, which primarily affects private landholders, is due in part to images of extensive rates of land clearance provided through satellite remote sensing studies. Use of such technologies also embed certain notions of 'value' and particular relations between nature and humans, issues that warrant greater attention in any further development of this kind of technological application (Bartel 2005).

Automated video monitoring

Recent software and digital hardware technologies combined with utilisation of Ethernet, the Internet Protocol, and wireless mesh based networks provides the opportunity for monitoring activity in almost any location in the world from any other location in the world (Hayes *et al.* 2008). Intelligent video monitoring embraces automation of much of the monitoring activity and the archival of only those incidents identified to be of interest – for example, motion detection. Intelligent video analysis can facilitate the audits of large-scale, 24/7 monitoring operations, contributing to both deterrence and evidence gathering in environmentally sensitive locations.

Contamination forensics

The contamination of land, water and air can be prevented by proactive testing of specific sites, movement routes and currents, by the establishment and collection of benchmark data, and by regular monitoring. To do this requires utilisation of methods that might include: chemical analysis; study of documentary records; use of aerial photographs; and application of trend techniques that track concentrations of chemical substances over space and/or time (Murphy and Morrison 2007; Brookspan *et al.* 2007). Bearing in mind that some contaminations, such as nuclear radiation, are not easily visible to human detection, both alternative methods of science and communal reflexivity over potential risks are needed (Macnaghten and Urry 1998).

Data modelling

Much recent criminological work on 'poaching' has been directed almost exclusively to the question of crime prevention, and in particular situational crime prevention (Lemieux 2014). The primary question has been 'how' to stop the poaching of rhinos, elephants and other species and the supposed answer is found in crime science solutions that emphasise particular technologies, techniques, market measures, data modelling and use of CCTV (Lemieux 2014). Some of this work has involved complicated computer simulations that attempt to map out a wide range of intersecting variables (e.g. animal, poacher, ranger, site specific information) so as to enhance the targeting of law enforcement efforts. The emphasis is on data collection and in particular sophisticated data processing and analysis.

A vast array of techniques and approaches to environmental forensics are now available (see United States Environmental Protection Agency 2001). For example, forensic sciences are now able to track the chemical signature of oil spills (Pasadakis *et al.* 2008) and to use sophisticated chemical and biological analyses to track such spills (as well as illegal disposal of waste) to their source (Mudge 2008). As well, the forensic sciences are now actively turning their attention toward climate change, with a view to contributing to monitoring efforts and identifying emerging environmental issues (Petrisor and Westerfield III 2008).

Other methods of investigation include such specific techniques as: identification of wildlife through footprints, scats (faeces), bones, fur, claws, blood; use of chemical analysis in relation to certain benchmark data and established allowable thresholds (which is used in relation to toxic outfalls, water and land sites); creation of topographic (elevation) maps and thematic maps (e.g. land disposal activity, population distribution, vegetation communities, land use); monitoring of relevant internet sites for exposure of wrongdoing and illegal activity (e.g. Facebook, YouTube, MySpace, Twitter, activist sites); and surveillance of local markets through to eBay (e.g. sites and places where ivory, antlers, rare plants, etc. are bought and sold).

In the end, however, big questions remain regarding who is doing what for whom, and whose knowledge actually counts (White 2012a). For example, doing science and utilising technologies is both a social and a technical activity, insofar as social contexts influence scientific interpretation of what the 'facts' convey and the uses to which the knowledge is to be put. That is, the vital role of science in bringing problems to public attention, and in devising methods to monitor or curb environmental hazards, is contingent upon how natural and social scientists are integrated into the policy making process.

It is rare that scientific evidence is uncontested and that proof of environmental harm is simply a matter of 'let the facts decide'. What counts as 'science', what counts as 'evidence', who counts as being a 'scientific expert' and what counts as 'sensible' public policy are all influenced by factors such as economic situation, the scientific tradition within a particular national context, the scientific standards that are used in relation to specific issues, and the style and mode of government (White 2008). Thus, science and technology are the backbones of discovery, measurement and explanation of environmental harm, but these, too, are embedded in particular social processes and decision-making frameworks. In this respect, science as a method of understanding, accompanied by the use of certain technologies, is inherently social. This is probably most apparent today in the attacks on science and scientists by the climate change contrarians (Brisman 2012).

Determination of the extent and nature of any specific environmental problem demands at some stage the use of scientific testing and diagnosis. The definition of 'clean air', for example, may be subject to legal and political wrangling in terms of which level of pollution regulators are willing to accept. But it is the scientist who will tell us what is actually in the air at any point in time. Again, it is important to consider the ways in which scientific knowledge is applied in practice, and the effects of different applications on specific population groups. Consider for example the risk assessment process by which 'safe levels' of exposure to chemicals and other pollutants are determined. This area of work can be highly problematic, and incorporate a range of ideological and moral assumptions. As Field (1998: 90) comments, 'The use of the apparently reasonable scientific concept of average risk, for example, means that data from the most sensitive individuals, such as children, will not be the basis for regulation, but rather data from the "statistically average" person'. Thus, science provides grounds upon which we may base judgements, but these grounds are not necessarily neutral in terms of social impact. The interplay between scientific finding and social objective is of vital importance.

For many green criminologists not only is conventional scientific expertise essential to understanding what is happening to the environment, but, as well, there is recognition of 'expertise from below' – as in the case of farmers who 'know' their land, Indigenous people who 'know' their country, and so on. The concept of 'Indigenous Knowledge', for example, refers to the unique, traditional local knowledge existing within and developed around the specific conditions of women and men indigenous to a particular geographical area. Such IK systems, including management of the natural environment, have been a matter of survival to the peoples who generated these systems. Simultaneous to this is the concept of 'Indigenous Technology', which is defined in terms of hardware (equipment, tools, instruments, and energy sources) and software (a combination of knowledge, processes, skills, and social organisation) that focus attention on particular tasks (Robyn 2002). Fire burning among indigenous Australians, for example, constituted an informed and conscious means to work in and with certain types of local environment (Langton 1998).

Social science methodologies can be mobilised as part of investigations into environmental harm across several different topic areas (White 2012a). These include the nature of and contests over 'evidence' when it comes to forensic toxicological studies; the inadequacy of regulatory mechanisms in cases where 'denial' is driven by potential threats of liability (health department, municipal council, environmental protection agency, industry and specific companies); and the role of grassroots activists and experts in the context of the politicisation of local environmental issues. The main concern of this kind of research is to provide insight into and analysis of stakeholder interests and to raise questions regarding the criteria used to assess the quality and robustness of evidence in relation to the specific problem at hand.

The point of this kind of research is to interrogate the ways in which 'evidence' surrounding environmental harm is gathered, presented and mobilised by diverse stakeholders. It thus shares similarities with work done on disputes that are characterised by multi-layered conflict about knowledge, rights and development, as evident for example in the multi-dimensional character of the aquaculture controversy in Canada (Young and Matthews 2010). Such research is about how different stakeholders construct evidence about harm, how meaning is mobilised by different groups, the process whereby different types of evidence are constructed, and how researchers might form definitive judgements about the alleged harm.

At the heart of investigations of environmental crime is the question of whose knowledge of 'wrong' is right? In other words, whose voices are going to be heard, and to what kinds of evidence do we lend credibility? In responding to environmental harm and victimisation there are inevitably a range of vested interests and 'discourses' that contribute to the shaping of perceptions and issues (Hannigan 2006). This implies differences in perspective and a certain contentiousness of knowledge about the nature of the harm or crime.

For example, recent work has been able to distinguish three sources of error and/or limited knowledge that impinge upon the assessment of alleged toxicity in relation to various sites within Tasmania (White and Heckenberg 2014). These are the problems associated with *partial* knowledge (i.e. knowledge that is incomplete since it is limited to only one kind of domain expertise, such as soil testing); *skewed* knowledge (i.e. knowledge that is in some way biased even if accurate within its own terms of reference, such as reliance upon patient records from one medical practice); and *distorted* knowledge (i.e. knowledge that is more akin to propaganda, being ideologically based, as in *ad hominem* attacks against specific protagonists). Each sort of knowledge presents problems in regard to the accumulation of necessary and sufficient knowledge to assess the relevant contamination issues. However, they also suggest relatively straightforward solutions, revolving around, for example, the combining of different knowledge sources, deployment of diverse forms of sampling, and emphasising substantive empirical evidence over ideological statement.

In terms of potential sources of knowledge, it may well be that it is local residents, local workers and laypeople generally who are more conscious of environmental risk than the scientist or the politician and who therefore have superior 'techne' or technical understanding of the situation. Some indication of this is provided in a study of interaction between scientists and English sheep farmers in the wake of the 1986 Chernobyl nuclear accident in the Ukraine (Wynne 1996). The study highlighted the accurate, detailed and contextual knowledge of the local farmers, even though the scientists considered this layperson knowledge to be lacking in precision. Those who are closer to the 'coal face' and who have lived and worked in the same area for years, are frequently those who notice the small changes that are the harbingers of things to come.

The social context of knowledge construction is pertinent to other areas of criminological concern as well. In situational crime prevention approaches to wildlife trafficking, for example, the social and historical dynamics and dimensions of poaching tend to be marginalised in such accounts. For example, little is said about the root causes of crime, such as economic deprivation or social disengagement, or about long-term motivations, such as tradition or culture (von Essen *et al.* 2014). Nor is attention paid to the notion that illegal hunting is a form of resistance to conservation policy that is seen to be unfair and lacking in legitimacy (von Essen *et al.* 2014), something particularly evident in Scandinavia (von Essen *et al.* 2015) and historically in the Adirondack Mountains in the United States (Jacoby 2003). Social context is particularly meaningful in the African context, as conservation has tended to favour and privilege the white power elite over and above the interests of indigenous Africans (Smith and Humphreys 2015; Wall and McClanahan 2015). Criminalisation, in this instance, is heavily overlaid with militarisation of anti-poaching measures and a continuation of elite economic and political domination.

By contrast are approaches that recognise that traditional knowledge includes forest-related knowledge associated with the use and management of forest species, and the broader understanding and management of forest ecosystems. In essence, traditional users have over many years developed their own technological understanding ('techne') of their environment, a sort of pre-theoretical understanding of nature exhibited by those who live and work intimately with and on the land (and rivers, lakes and oceans). Examples of such knowledge include use of herbal medicines, nutraceutical products, foods and beverages. For many, the forest exists as a site of not only wood products (for cooking, for furniture, for musical instruments) but non-wood products such as medicines, foods, spices, fodder for animals and for a multitude of other purposes including the aesthetic and spiritual. Interestingly, fuel wood constitutes up to more than 70 per cent of wood removal in Asia and the Pacific and 90 percent in Africa (Food and Agriculture Organization of the United Nations 2011). In other words, conservation without acknowledgement of traditional human users, and their systemic contribution to biological diversity and ecological wellbeing, is oppressive and counter-productive. Without judicious use and situational relevance, the techniques of wildlife situation crime prevention can contribute to these results.

Technology as the problem/solution

A second perspective on technology that is also relevant to green criminology is that which sees it less as a specific investigative tool than an overarching 'problem' or 'solution' to a problem. This is probably most apparent today in debates over climate change and the role of technology in contributing to global warming, and its potential role in mitigating such. The 'food' revolution is another area where technology is touted as the solution to the problem of world hunger.

The emphasis on technology as a perpetrator or facilitator of environmental harm is not uncommonly applied to issues surrounding carbon emissions and the industries that support this. 'Climate change is a consequence of the transition from biodiversity based on renewable carbon economies to a fossil fuel-based non-renewable carbon economy. This was the transition called the industrial revolution' (Shiva 2008: 130). However, the spread of industrialisation and specific means of production is not merely a technical process or one that stems solely from technological determinants. Two hundred years of industrial revolution has been driven and underpinned by powerful forces (nation-states, companies, armies) pursuing sectional interests. This has been achieved through global imperialism, colonialism and militarism that have served to entrench a dominant worldview and the material basis for certain types of production, consumption and reproduction.

The idea that technology is the primary source of evil necessarily has to be pitched at a very general level. It is the technologies associated with industrialisation which is portrayed as the problem, for example, not the social system underpinning this industrialisation as such. In this scenario, technology is seen as that which creates the risks and harms. It is the motor car, the factory, the tractor, the coal-fired power station, the dam, the fertiliser, that is the problem. In response, proposals to downshift to a low-tech future might be interpreted as a 'blame the technology' kind of argument.

Against this is the acknowledgement that the technology question is inherently about the social character of collective practice. Davison (2004: 144) points out that 'Technologies of genetics, biology, energy, matter and information cannot be neatly sorted into good and bad, or sustainable and unsustainable, piles'. What counts are the specific social and ecological conditions under which technologies are utilised and applied. For instance, one cannot assume that by their very nature renewable energy technologies are inherently positive and good. Depending upon the social context, 'sustainable forms of agriculture and other "green" techniques may reduce some forms of ecological risk, but they may also help to prop up, to sustain, an unsustaining social whole' (Davison 2004: 144). The social context of technological use and development is therefore crucial.

A more positive view of technology places greater emphasis on technological optimism and the idea of technology as fix. Here technology is very much seen as *the* solution. It is reliance upon human ingenuity and technological invention that is seen as the most appropriate, viable and robust way to guarantee both the continued exploitation of nature (e.g. bio-technologies and food and energy) and responses to its demise (e.g. risks and harms associated with climate change).

A familiar example of this is the 'green revolution', which was and still is premised on using advanced industrialised technologies to feed the masses of the world. Yet, as Fara (2010: 427) concisely summarises, this revolution, too, has had a peculiar social and economic history.

> [T]he Green Revolution had adverse social consequences because its very makeup incorporated political power structures. Instead of importing food, poorer nations were now buying in the expensive chemicals, seeds, and expertise they needed to maintain their altered style of agriculture. Whereas the profits of large landowners with affluent contacts soared, small farmers were squeezed out of business and migrated to swell the urban slums still further. GM organisms were being produced in distant research laboratories, pouring 'southwards' from rich countries to poor ones – notably from North to South America. In contrast, financial benefits flowed in the opposite direction: the manipulated genes originated in local plants and were being drained 'northwards' to boost the profit and prestige of biotechnology companies.

Indeed, one of the greatest threats to environmental health and biodiversity is in fact the industrialisation of agriculture (incorporating the use of seed and other patents) since this is one of the greatest causes of erosion of plant genetic and species diversity. Governments and agribusinesses have fostered reliance upon large-scale agricultural techniques and methods, and on new technologies such as the use of genetically modified organisms (GMOs). The turn toward use of GMO crops has involved converting land to industrial forms of agricultural production, and the application of practical restrictions on what is being grown and how (Walters 2005). One consequence of the industrialisation of agriculture, combined with and intensified by application of GMO technology, is that biodiversity is systematically reduced. In other words, there is a tendency toward monoculture, since uniformity means ease of cultivation and harvest, and higher yield, which translates into higher profit (Pollen 2007; French 2000). The gaining popularity of biofuels has also been used to legitimate the further spread of genetically modified crops. GMO soy, for example, has been touted as perfect for biofuel.

Technology is also cited favourably when it comes to climate change. The notion that 'business can profit by protecting the environment' (Baer and Singer 2009) is frequently linked to the idea of technological fixes vis-à-vis climate change, such as the development of hybrid cars, more technologically efficient appliances, and lighting sources such as fluorescent bulbs.

Burying the problem is also touted as a solution – whether this is radioactive waste or carbon pollution.

The grand scale of the problem – in this instance, climate chaos affecting the whole planet – does not seem to deter those supporting geoengineering solutions (Brisman 2015). Proposals are being made for deliberate large-scale intervention in the climate system so as to mitigate the problem of global warming. According to Hulme (2014: 2–3), geoengineering 'technologies are united in their ambition to deliberately manipulate the atmosphere's mediating role in the planetary heat budget. They aim to do one of two things: either to accelerate the removal of carbon dioxide from the global atmosphere; or else to reflect more sunlight away from the Earth's surface and so to compensate for the heating of the planet caused by rising concentrations of greenhouse gases'. At its heart, energy production and consumption is centre stage in climate change, both as a source of the problem and as part of the response through mitigation and other measures.

Large-scale mitigation involving geoengineering has two key dimensions according to Redgwell (2012: fn11):

1   carbon dioxide removal – ocean iron fertilisation to enhance plankton growth and absorption by the oceans of CO_2 from the atmosphere, and the construction of mechanical filters to remove CO_2 from ambient air;
2   solar radiation management – releasing aerosols into the stratosphere or constructing solar arrays in outer space to reflect solar radiation back into space, and enhancing the reflectivity of clouds.

As Redgwell (2012: 28) also notes:

> Such responses include use of carbon capture and storage (CCS), great energy efficiency, increased use of renewable energy (eg wind, wave and solar), and alternative energy sources (eg biofuels). Many of these responses are technology-driven, in particular the increased focus on CCS, and do not necessarily reflect a move to reduce carbon dependency.

It is worthwhile reflecting for a moment on at least one of the alternatives finding particularly generous political support in recent time: biofuels.

The push toward biofuel production reflects the interests of large agricultural businesses, who can patent the monocultural crops designed as 'energy crops'. Powerful interests, including car manufacturers and grain farmers, have benefited from the search for energy alternatives to fossil fuels. The shift to biofuel is seen as a key source of green fuel supply for the world's car manufacturers. Greater demand for biofuel crops such as corn, palm oil or soya also means that farmers are finding the growing of such crops very lucrative economically.

However, the trend toward biofuel is generating food price rises and food shortages (Mitchell 2008) as well as having a directly negative environmental impact. This is because the profitability of biofuel production is leading to the establishment of large-scale plantations in places such as Indonesia and Brazil. This process has seen the clearing of rainforests and in some instances the forcing of Indigenous people off their lands. This deforestation process has been going on for a number of years, and has been supported by organisations such as the International Monetary Fund (French 2000). Cutting down trees also has a direct bearing on global warming. It has been estimated that by 2022, biofuel plantations could destroy 98 per cent of Indonesia's rainforests and that 'Every ton of palm oil used as biofuel releases 30 tons of CO_2 into the atmosphere, ten times as much as petroleum does' (Shiva 2008: 79). There are clear instances of both social and ecological injustice occurring, both of which are central to the work of green criminology.

Biofuel production in places such as the United States and the European Union is encouraged through strong incentives (e.g. tax credits) and mandates such as energy legislation (e.g. mandatory blending requirements). As indicated, the advent of biofuels has helped to push up global grain prices, and to bolster the prospects of the grain producing countries. It has, however, been accompanied by ecological costs in the form of degraded environments and social costs in the form of high prices for food, especially in less developed and import-dependent countries (Roberts 2008). It has also pushed up feedstock prices, thus affecting the pastoral industries as well as the agricultural. Given the problems of biofuel production (including the release of large quantities of carbon dioxide associated with the planting and processing of plant materials for biofuels), and given that nearly 60 per cent of humans in the world are currently malnourished, one conclusion is that 'There is simply not enough land, water, and energy to produce biofuels' (Pimentel et al. 2009: 9).

Biofuel production activities are not carbon neutral (e.g. the energy consumed by, and emissions from, intensive farming practices). This led to a EU directive in 2009 that biofuels be subject to meeting agreed sustainability criteria, including achieving minimum levels of GHG savings and that they should not be obtained from land with high biodiversity value (Redgwell 2012: 39).

As with other technologies, the biggest problems with GMO and biofuel crops relate to poor regulation and the social justice impacts of their introduction on local communities (see Patrick Bishop and Stuart Macdonald, Chapter 34 of this volume). For example, GMO invasion of endemic species and crops is seen to be capable of destroying unique genotypes, thereby creating the potential for a threat to food security (i.e. diminishing diverse genetic material) (Engdahl 2007). Such fears are not unfounded. Scientific study has reported two issues of concern: first, that genetic contamination is occurring; and secondly, that transgenes are unstable, meaning that once the GMO cross-pollinates with another plant, the transgene splits up and is inserted in an uncontrolled way – the displaced DNA could be creating utterly unpredictable effects (Robin 2010: 247). The potential size of the problem is considerable as over the past decade the use of GMO crops has rapidly increased, and many have been planted illegally. For example, in Paraguay, where (as of 2007) no law authorizes the cultivation of GMOs: 'From 1996 to 2006, surfaces devoted to soybean cultivation went from less than 2.5 million acres to 5 million acres, an increase of 10 percent a year' (Robin 2010: 275). To avoid losing markets, by ensuring proper labelling of crops for markets such as the European Union, the Paraguayan government ended up simply legalising the illegal crops. Much the same thing happened in Brazil and Poland (Engdahl 2007), and for much the same reasons (namely EU rules on traceability and labelling of GM foods intended for human and animal consumption) (Robin 2010). Illegal smuggling of GMO seeds is lucrative for those wishing to promote particular industries and types of farming, much less for the companies that own the patents. The spread of GMO seeds has also been associated with corruption of government officials and other types of crimes (Engdahl 2007: 269–70).

The lucrative market for biofuels and GMO crops has been linked to the forced takeover of communal lands, using armed men and bulldozers, as well as fraudulent claims of land title (see Robin 2010). Moreover, given that the focus of the UN mechanism for Reducing Emissions from Deforestation and Forest Degradation (REDD) is on minimising carbon emissions caused by the destruction of living forest biomass, there will be greater pressures to convert or modify other ecosystems, especially savannahs and wetlands, for food or biofuel (Sutherland et al. 2009). In other words, forests are privileged over other types of ecosystems, and the result could well be the loss of biodiversity associated with destruction or conversion of these 'less valued' non-forested ecosystems. Again, compulsory takeover of such land is not uncommon.

Technology as paradoxical

A third perspective on technology of interest to green criminology, related to and over-lapping with the second, is one that examines technology as a paradoxical phenomenon, one that introduces certain new harms or reinforces particular harms in the very moment that it is intended to reduce particular existing harms. For instance, *paradoxical harm* is harm that arises out of an apparent contradiction (for instance, we have to pollute certain parts of the planet in order to save it from other types of pollution). Specific examples of paradoxical harm include the adoption of compact fluorescent light globes to save energy (but which contain toxic mercury), promotion of nuclear energy (but which involves disposal of nuclear waste), and carbon emission storage (that penetrates and despoils the subterranean depths of land and sea) (White 2012b).

It is important not to equate paradoxical harm with the notion of unintended consequences (Tenner 1997). This is because in many instances the harms are actually well known, and the acts leading to the generation of the harms are intentional. The harm is paradoxical in the sense that while seemingly contradictory (we generate harms as a means to forestall other harms), it is perfectly logical from the point of view of the imperatives of the system as a whole. Economic and social interventions that sustain the status quo (and that includes maintaining the viability of 'dirty' industries) are favoured over those that might tackle the key drivers of climate change and that could diminish the burgeoning threats to ecological sustainability worldwide.

While previous discussions examined technology as problem/solution as applied to particular social and ecological problems, a paradoxical harm approach attempts to widen the analytical lens even further, to provide a global perspective on technology development, use and consequences. For example, climate-related energy issues can be analysed in terms of the use of alternative energy sources, and efforts aimed at dealing with carbon emissions. In each case the answer to the energy crisis involves measures that in some way contribute to other types of environmental harm.

Yet, measures to deal with climate change through development of new energy sources and restriction or regulation of carbon emissions can also be understood in the context of unequal trading relations between countries. From a world systems perspective, there is an energy rift between regions resulting from unequal energy flows between the producers and users of resources. Such analysis is based upon significant social, economic and military differences between metropole (e.g. US, Japan, Germany, UK, France), semi-periphery (e.g. Russia, Brazil, Mexico, China) and periphery (e.g. Bolivia, Haiti, Zimbabwe, India) countries (see Baer and Singer 2009).

For example, in the period 1860 to the Second World War, research has found that both developed and less developed countries were almost self-sufficient in energy – this changes in the 1950s as the less developing countries began exporting energy to the developed core countries that were beginning to consume more than they produced (Lawrence 2009). Not surprisingly, less developed countries (dependent upon foreign investment in manufacturing) have been found to emit higher levels of noxious gases per capita, and the total carbon dioxide emissions and emissions per unit of production are higher than in the core countries. Nonetheless, the core's usages remain far disproportionate to its population. 'In 2005, its percentage of total world energy use was 61.2 percent and in 2004 CO_2 emissions were 60.4 percent, yet its population was only 21.5 percent of the world total' (Lawrence 2009: 348).

Stretesky and Lynch (2009) argue that on the basis of analysis of carbon emissions and consumer imports to the United States, it is US consumer demand that is fuelling harmful production practices in other importing countries. They examined the relationship between per capita carbon dioxide emissions and exports for 169 countries. The data suggest that consumption practices in the United States are partially responsible for elevated per capita carbon dioxide emissions in other nations, and that carbon dioxide trends in other nations are in part driven by US demands for goods. US consumers, however, are unaware of how their consumption fuels rising global carbon emissions, because of the disconnection or dissociation between the two phenomena.

Much public debate has occurred over the regulation and reduction of carbon emissions. At the heart of the matter is the fact that carbon emissions are directly contributing to global warming, and that without adequate mitigation and adaptation strategies the problems associated with climate change will get worse before they get better. The urgency surrounding the reining in of carbon emissions has been matched by the audacity of businesses in lobbying to defend their specific economic interests (Bulkeley and Newell 2008). Given the vested interests involved in protecting and maintaining existing inter-state inequalities, as well as those associated with particular industries (such as oil and coal), the stifling of carbon emissions has been slow and well below what is needed to counter present global warming trends (Intergovernmental Panel on Climate Change 2013). Private profits continue to dominate public interests, usually with government collusion.

Developments relating to both food production and energy production reveal a series of paradoxical harms that are generated in the context of strategic decisions regarding how production (and consumption) are to take place. When it comes to food the key issues are shortages, unequal distribution globally and emerging social conflict. In responding to these, however, measures involving further industrialisation of agricultural and pastoral production and the adoption of bio-technologies are contributing to greenhouse gases, pollution, and loss of habitat and biodiversity. When it comes to energy, again there are problems relating to shortages, global unequal distribution and expanding demand. However, poor regulation of carbon emissions, and reliance upon biofuels and other new technologies is likewise contributing to global warming, as well as transferring problems to the poorer countries and adding additional forms of toxic pollution into the equation.

Another paradoxical harm relates to the support for nuclear power as a 'clean' option to carbon based energy sources such as coal and oil. Uranium is not a renewable energy source, yet nuclear power is high on the list of preferred energy sources in this era of global warming. But nuclear energy also means nuclear waste. High level waste (HLW) is so radioactive that it generates heat and corrodes all containers. It must be stored above ground for 50 years so it can cool before being transported and disposed of. Intermediate level waste (ILW) arises mainly from the reprocessing of spent fuel and from general operations and maintenance at nuclear sites. It is typically packaged by encapsulation in cement in highly engineered stainless steel drums or in high capacity steel or concrete boxes.

In the UK alone, 365,000 cubic metres of high and intermediate-level radioactive waste has been accumulated from its existing post-war nuclear programme (Clarke 2009). The decision in 2009 by the US government to end development of the Yucca Flat, Nevada repository for permanent disposal of radioactive waste has also highlighted waste issues in that nation. As of 2009, the United States has been left with some 60 thousand metric tons of spent reactor fuel in need of a permanent storage facility, but with no viable facility on the drawing board, and no indication of when – if ever – the situation would be resolved (Pickard 2010). Problems in developing a suitable response relate to technical and scientific issues, the NIMBY [Not In My Back Yard] effect, and who is or ought to carry the financial burden over time.

Nuclear waste in the US is stored at 121 temporary sites in 39 states across the country (Clarke 2009). Yet, 'Sixty years into the nuclear era there is no universally agreed upon strategy for the disposal of high level nuclear waste. Significantly more is being produced each year by nuclear powered electricity generation. But the power so generated is needed' (Pickard 2010: 713). At present, the only reasonable course of action, and moral obligation, is to bury this type of waste deep underground, rather than leaving the waste on the surface for an indefinite period – where each day might bring an incident that disperses it into the biosphere (Pickard 2010). This assumes and implies, of course, that what is defined as 'underground' is somehow a non-living 'neutral' entity, simply an empty space to be filled, rather than a living eco-system in its own right (Suzuki 1998).

The biggest challenge for governments is to find a site that is both geologically secure and acceptable to the local communities. This is compounded by the problem of capacity relative to need. In South Korea, for example, the large increase in projected nuclear power will inevitably accelerate the accumulation of spent fuel – but if the direct deposit option is pursued, South Korea would not be able to secure enough suitable sites for disposal, given its geographic profile (Ko and Kwon 2009). Meanwhile, most countries are still in the planning stages for HLW repositories. In Japan, for example, the government has not yet started work on its HLW repository, even though the first experimental reactor was initiated in 1963 (Zhang et al. 2009). Its problems have been compounded by public reaction to the Fukushima fiasco.

Then there is the effect of global warming on proposed storage strategies. For example, the near-surface facility and geological repository are both likely to be affected by the consequences of global warming – the landscape and hydrogeological regime at and around a disposal facility may change, as might the biosphere receptors, and the animal and human habitats. When considering the long-term evolution of the disposal system, major climate change (e.g. future glacial periods) should be assessed (Van Geet et al. 2009). Global warming will have a major influence over short and long periods of time, as in the case of changing water infiltration through the multi-layer cover of a waste disposal site.

The environmental crisis stemming from climate change is a major impetus for a huge and rapid expansion of the nuclear power industry. The criminal and safety implications of these combined issues include: illegal disposal of radioactive waste; unsafe containment of existing and immediate future waste; and externalisation of costs and wastes from core to peripheral countries. There is the accusation, for example, that French nuclear waste gets shipped to northern Africa and gets dumped in the desert sands (Bridgland 2006). This allows for a clean, green image within France to be maintained, because the real problem has been exported away. The demand for new forms of energy, within a context of reducing carbon emissions, may well open the door to new types of dumping and new transferences of toxic harm worldwide.

Analysis of technology as paradoxical is not only about paradoxical harms however. It also relates to how adoption of technology can be used as a cover for continuation of bad policy and bad practices. This is certainly the case in some forms of 'greening justice' (White and Graham 2015; Graham and White 2015). From the point of view of the administrators and policymakers of criminal justice, for instance, there is an emergent agenda to cut energy, offset carbon emissions, save water and recycle waste – in other words, to transform the physical and operational infrastructure of criminal justice institutions to become more environmentally sustainable (Sheldon and Atherton 2011; Feldbaum et al. 2011). This occurs in accordance to quite different rationales and dimensions of 'greening justice', from saving money to keeping offenders busy to engaging in more eco-friendly practices (White and Graham 2015).

As argued elsewhere (White and Graham 2015), the financial and other benefits of 'greening justice' need to be interpreted carefully and contextually. For instance, there is a clear philosophical and practical 'fit' between systems based upon progressive rehabilitation and the construction of environmentally friendly facilities. This is most apparent in Scandinavian countries such as Norway, where policy principles of penal exceptionalism are supported by 'normalisation' strategies that simultaneously demand autonomy and skill development, as well as responsibility, among those few offenders who are incarcerated.

By contrast, it is argued that in the United States it is plausible that 'greening justice' may be misappropriated as a cost-effective way to incarcerate even more people, while at the same time providing a 'feel good' gloss in regard to what are mixed intentions or even regressive penal cultures (White and Graham 2015). For example, one of the stated objectives of the 'green' multi-purpose Blue Earth County Jail and Justice Center in Mankato, Minnesota, is 'to allow for future jail expansion in a cost-effective manner' (Beyer 2012: 2). Investigations have revealed exploitative and unsafe correctional industries involving prisoners in e-waste recycling projects that are little more than 'toxic sweatshops' which fail in their duty of care to protect those who take part in them (Jackson *et al.* 2006; Kaufman 2010). Even the infamous Guantánamo Bay prison and naval base boasts state-of-the-art green technologies, including four large wind turbines, multi-faceted fuel and energy reduction strategies, as well as bicycle-riding traffic police – which reduce emissions and consumption and save US taxpayers millions of dollars each year (Bueno 2005; Goldberg 2011).

These observations are not meant to dismiss the 'greening justice' agenda more generally. Rather they are intended, as with most of the technologies and interventions discussed in this chapter, to emphasise that social context and embedded interests always shape the specific direction of technological use. Thus, it has been acknowledged that 'greening justice' initiatives *can* make a positive difference, for some, in particular places, and under specific circumstances, and that the logic of such initiatives is not only about doing more with less, but doing less – in this case, to not build more prisons (White and Graham 2015). Moreover, in a world that is rapidly being subject to the vagaries of climate change, greater attention will be placed upon the natural environment and the nature–human relationship and in a number of respects, criminal justice institutions have the capacity to be at the vanguard of good ecological practice and conservation.

Conclusion

This chapter has canvassed a number of instances in which technology has been marshalled in ways that impinge upon particular environmental matters and issues. This has included the use of technology as an instrument or tool of criminal justice in combating environmental crime, its portrayal as the solution to global problems such as climate change, and the paradoxes of technology in regard to energy sources and the running of criminal justice institutions. In each case the intention of the analysis has been to make problematic the idea that technology can or should be considered outside of the context of wider social, economic and political relations. By examining closely the nature of these relations, we are better to ascertain whether technology and its applications serve the interests of social and ecological justice, or whether they reinforce existing regimes of power.

Reflections on the encounter between green criminology and technology reveal not only the extent to which technology is embedded in both producing and responding to environmental harm, but the relative dearth of critical thinking about technology within the green criminology project more generally. Moreover, as signalled several times in this chapter, a vital consideration of such scrutiny must be inclusion of the question of 'techne', since this references the differential knowledge and meanings that varied stakeholders bring to the understanding of nature and the dynamics of environmental harm. The dialogue between interpretation and technology is thus potentially fruitful in opening up new lines of inquiry as well as providing greater sophistication to the green criminology perspective more generally.

References

Alacs, E. and Georges, A. (2008) 'Wildlife Across our Borders: A Review of the Illegal Trade in Australia', *Australian Journal of Forensic Sciences* 40(2), 147–60.

Baer, H. and Singer, M. (2009) *Global Warming and the Political Economy of Health: Emerging Crises and Systemic Solutions.* Walnut Creek, CA: Left Coast Press.

Barbieri, C., Schwarzbold, A. and Rodriguez, M. (2007) 'Environmental Crime Investigation in Arroio do Meio, Rio Grande do Sul, Brazil: Tannery and Shoe Factory Waste Landfill Case Study', *Environmental Forensics* 8, 361–9.

Bartel, R. (2005) 'When the Heavenly Gaze Criminalises: Satellite Surveillance, Land Clearance Regulation and the Human–Nature Relationship', *Current Issues in Criminal Justice* 16(3), 322–39.

Beyer, J. (2012) 'Leading by Example: Blue Earth County Goes "Green" in Its New Justice Center and Jail', *National Institute of Corrections*: 1–13. National Jail Exchange – http://NICIC.org/NationalJailExchange (Accessed 10/10/2014).

Bridgland, F. (2006) 'Europe's New Dumping Ground: Fred Bridgland Reports on How the West's Toxic Waste is Poisoning Africa', *Sunday Herald*, 1 October 2006. (Accessed 2/10/2006).

Brisman, A. (2012) 'The Cultural Silences of Climate Change Contrarianism', in R. White (ed.) *Climate Change from a Criminological Perspective.* New York: Springer.

Brisman, A. (2015) 'Environment and Conflict: A Typology of Representations', in A. Brisman, N. South and R. White (eds) *Environmental Crime and Social Conflict: Contemporary and Emerging Issues.* Farnham, Surrey: Ashgate.

Brookspan, S., Gravel, A. and Corley, J. (2007) 'Site History: The First Tool of the Environmental Forensics Team' in B. Murphy, and R. Morrison, (eds) *Introduction to Environmental Forensics* (2nd edn). London: Elsevier Academic Press.

Bueno, V. (2005) 'Navy's New Wind Turbines to Save Taxpayers $1.2 Million in Annual Energy Costs'. American Navy, available at: www.navy.mil/submit/display.asp?story_id=18059 (Accessed 12/09/2014).

Bulkeley, H. and Newell, P. (2010) *Governing Climate Change.* London: Routledge.

Clarke, E. (2009) 'The truth about…nuclear waste', ClimateChangeCorp.com (accessed 21 January, 2010).

Davison, A. (2004) 'Sustainable Technology: Beyond Fix and Fixation', in R. White (ed) *Controversies in Environmental Sociology.* Melbourne: Cambridge University Press.

Engdahl, F. (2007) *Seeds of Destruction: The Hidden Agenda of Genetic Manipulation.* Montreal: Global Research.

Fara, P. (2010) *Science: A Four Thousand Year History.* Oxford: Oxford University Press.

Feldbaum, M., Greene, F., Kirchenbaum, S., Mukamal, D., Welsh, M. and Pinderhughes, R. (2011) *The Greening of Corrections: Creating a Sustainable System.* Washington, DC: United States Department of Justice National Institute of Corrections.

Field, R. (1998) 'Risk and Justice: Capitalist Production and the Environment', in D. Faber (ed.) *The Struggle for Ecological Democracy: Environmental Justice Movements in the US.* New York: Guilford Press.

Food and Agriculture Organization of the United Nations (2011) *State of the World's Forests 2011.* Rome: FAO.

French, H. (2000) *Vanishing Borders: Protecting the Planet in the Age of Globalization.* New York: WW Norton and Company.

Goldberg, S. (2011) 'Guantánamo Bay's Green Regime has Prison Limits', *The Guardian* 29 December 2011.

Graham, H. and White, R. (2015) *Innovative Justice.* London: Routledge.

Hannigan, J. (2006) *Environmental Sociology* (2nd edition). London: Routledge.

Hayes, L., Porteous, G. and Zhou, T. (2008) 'Technological Developments for Environmental Monitoring', Paper submitted to INECE (International Network for Environmental Compliance and Enforcement) 8th International Conference, Cape Town, South Africa, 5–11 April 2008.

Higgins, P. (2012) *Earth is Our Business: Changing the Rules of the Game*. London: Shepheard-Walwyn Publishers Ltd.

Hulme, M. (2014) *Can Science Fix Climate Change? A Case Against Climate Engineering*. Cambridge, UK: Polity Press.

Intergovernmental Panel on Climate Change (2013) Working Group I Contribution to the IPCC Fifth Assessment Report Climate Change 2013: The Physical Science Basis: Summary for Policymakers. 27 September 2013.

Jackson, A., Shuman, A. and Dayaneni, G. (2006) *Toxic Sweatshops: How UNICOR Prison Recycling Harms Workers, Communities, the Environment, and the Recycling Industry*. Silicon Valley, United States: Center for Environmental Health, Prison Activist Resource Center, Silicon Valley Toxics Coalition and Computer TakeBack Campaign.

Jacoby, K. (2003) *Crimes Against Nature: Squatters, Poachers, Thieves, and the Hidden History of American Conservation*. Berkeley: University of California Press.

Kaufman, L. (2010) 'Toxic Metals Tied to Work in Prisons', *The New York Times*, 26 October, 2010.

Ko, W. and Kwon, E. (2009) 'Implications of the New National Energy Basic Plan for Nuclear Waste Management in Korea', *Energy Policy*, 37: 3484–3488.

Langton, M. (1998) *Burning Questions: Emerging Environmental Issues for Indigenous Peoples in Northern Australia*. Darwin: Centre for Indigenous Natural and Cultural Resource Management.

Lawrence, K. (2009) 'The Thermodynamics of Unequal Exchange: Energy Use, CO2 Emissions, and GDP in the World-System, 1975–2005', *International Journal of Comparative Sociology*, 50(3–4): 335–359.

Lemieux, A. (ed.) (2014) *Situational Prevention of Poaching*. London: Routledge.

Macnaghten, P. and Urry, J. (1998) *Contested Natures*. London: Sage.

Mandiberg, S. (2009) 'Locating the Environmental Harm in Environmental Crimes', *Utah Law Review* 4, 1177–222.

Mgbeoji, I. (2006) *Global Biopiracy: Patents, Plants, and Indigenous Knowledge*. Vancouver: UBC Press.

Mitchell, D. (2008) 'A Note on Rising Food Prices', Draft World Bank paper, circulated on-line by the *Guardian* newspaper, website (guardian.co.uk/environment). (Accessed 11 July 2008).

Mudge, S. (2008) 'Environmental Forensics and the Importance of Source Identification', in Hester, R. and Harrison, R. (eds) *Issues in Environmental Science and Technology, No 26 Environmental Forensics*, London: Springer/Royal Society of Chemistry.

Munro, M. (2007) 'Biofuels Come Up Short as Way to Reduce Carbon Load, Study Finds', *The Vancouver Sun*, 17 August 2007: A3.

Murphy, B. and Morrison, R. (eds) (2007) *Introduction to Environmental Forensics*. Amsterdam: Elsevier.

New South Wales Environmental Protection Authority (2013) *Annual Report 2012–2013*. Sydney: NSW EPA.

Ogden, R. (2008) 'Fisheries Forensics: The Use of DNA Tools for Improving Compliance, Traceability and Enforcement in the Fishing Industry', *Fish and Fisheries* 9, 462–72.

Pasadakis, N., Gidarakos, E., Kanellopoulou, G. and Spanoudakis, N. (2008) 'Identifying Sources of Oil Spills in a Refinery by Gas Chromatography and Chemometrics: A Case Study', *Environmental Forensics* 9, 33–9.

Pellow, D. (2007) *Resisting Global Toxics: Transnational Movements for Environmental Justice*. Cambridge: The MIT Press.

Petrisor, I. (2007) 'Background in Environmental Forensics: "Raising the Awareness?"', *Environmental Forensics* 8, 195–8.

Petrisor, I. and Westerfield III, W. (2008) 'Hot Environmental and Legal Topics: Greenhouse Gas Regulation and Global Warming', *Environmental Forensics* 9, 1–5.

Pickard, W. (2010) 'Finessing the Fuel: Revisiting the challenge of radioactive waste disposal', *Energy Policy*, 38: 709–714.

Pimental, D., Marklein, A., Toth, M., Karpoff, M., Paul, G., McCormack, R., Kyriazis, J. and Krueger, T. (2009) 'Food Versus Biofuels: Environmental and Economic Costs', *Human Ecology*, 37(1): 1–12.

Pollen, M. (2007) 'Unhappy Meals', *The New York Times Magazine*, 28 January: 38–47, 65–70.

Ramer, J. (2007) *Environmental Crime Forensics*, Florida Department of Law Enforcement.

Redgwell, C. (2012) 'International Legal Responses to the Challenges of a Lower Carbon Future: Energy Law for the Twenty-first Century', in S. Farrall, T. Ahmed and D. French (eds) *Criminological and Legal Consequences of Climate Change*. Oxford and Portland, Oregon: Hart Publishing.

Roberts, G. (2008) 'The Bad Oil on Ethanol: Biofuels are Losing Favour But Some Governments are Still Backing Them', *The Weekend Australian*, 31 May–1 June, 2008, p.20 Inquirer.

Robin, M-M. (2010) *The World According to Monsanto: Pollution, Corruption and the Control of Our Food Supply*. New York: The New Press.

Robyn, L. (2002) 'Indigenous Knowledge and Technology', *American Indian Quarterly*, 26(2), 198–220.

Roffey, P., Provan, P., Duffy, M., Wang, A., Blanchard, C. and Angel, L. (2004) 'Pyhlogenetic DNA Profiling: A Tool for the Investigation of Poaching', Paper presented at Australian Institute of Criminology Outlook Conference, Melbourne, 29–30 November 2004.

Sheldon, P. and Atherton, E. (2011) *Greening Corrections Technology: Guidebook*. Washington, DC: National Institute of Justice, US Department of Justice.

Shiva, V. (2008) *Soil Not Oil: Environmental Justice in an Age of Climate Crisis*. Brooklyn: South End Press.

Smith, M. and Humphreys, J. (2015) 'The Poaching Paradox: Why South Africa's "Rhino Wars" Shine a Harsh Spotlight on Security and Conservation', in A. Brisman, N. South and R. White (eds) *Environmental Crime and Social Conflict: Contemporary and Emerging Issues*. Farnham, Surrey: Ashgate.

South, N., and Brisman, A. (eds) (2013) *Routledge International Handbook of Green Criminology*. London: Routledge.

Stretesky, P. and Lynch, M. (2009) 'A Cross-National Study of the Association Between Per Capita Carbon Dioxide Emissions and Exports to the United States', *Social Science Research*, 38: 239–250.

Sutherland, W.J., Clout, M., Cote, I., Daszak, P., Depledges, M.H., Fellman, L., Fleishman, E., Garthwaites, R., Gibbons, D.W., De Lurio, J., Impey, A.J., Lickorish, F., Lindenmayer, D., Madgwick, J., Margerison, C., Maynard, T., Peck, L.S., Pretty, J., Prior, S., Redford, K.H., Scharlemann, J.P.W., Spalding, M. and Watkinson, A.R. (2009) 'A Horizon Scan of Global Conservation Issues for 2010', *Trends in Ecology and Evolution*, 25(1): 1–7.

Suzuki, D. with McConnell, A. and Mason, A. (2008) *The Sacred Balance: Rediscovering Our Place In Nature*. Sydney: Allen and Unwin.

Tenner, E. (1997) *Why Things Bite Back: Technology and the Revenge of Unintended Consequences*. New York: Vintage.

Tilman, D., Socolow, R., Foley, J., Hill, J., Larson, E., Lynd, L., Pacala, S., Reilly, J., Searchinger, T., Somerville, C. and Williams, R. (2009) 'Beneficial Biofuels – the Food, Energy, and Environment Trilemma', *Science*, 325: 270–271.

United States Environmental Protection Agency (US EPA) (2001) 'Report prepared for 13th INTERPOL Forensic Science Symposium', Lyon, France, 16–19 October 2001, US EPA Office of Criminal Enforcement, Forensics and Training Environmental Crime.

Van Geet, M., De Craen, M., Mallants, D., Wemaere, I., Wouters, L. and Cool, W. (2009) 'How to Treat Climate Evolution in the Assessment of the Long-Term Safety of Disposal Facilities for Radioactive Waste: Examples from Belgium', *Climate of the Past Discussions*, 5: 463–494.

Van Gulijk, J. (2014) 'Oil Spills: A Persistent Problem', in Spapens, T., White, R. and Kluin, M. (eds) *Environmental Crime and its Victims: Perspectives within Green Criminology*. Farnham, Surrey: Ashgate.

Von Essen, E., Hansen, H., Kallstrom, H., Peterson, M., and Peterson, T. (2014) 'Deconstructing the Poaching Phenomenon: A Review of Typologies for Understanding Illegal Hunting', *British Journal of Criminology*, 54: 632–651.

Von Essen, E., Hansen, H., Kallstrom, H., Peterson, M., and Peterson, T. (2015) 'The Radicalisation of Rural Resistance: How Hunting Counterpublics in the Nordic Countries Contribute to Illegal Hunting', *Journal of Rural Studies*, 39: 199–209.

Wall, T. and McClanahan, B. (2015) 'Weaponising Conservation in the "Heart of Darkness": The War on Poachers and the Neocolonial Hunt', in A. Brisman, N. South and R. White (eds) *Environmental Crime and Social Conflict: Contemporary and Emerging Issues*. Farnham, Surrey: Ashgate.

Walters, R. (2005) 'Crime, Bio-Agriculture and the Exploitation of Hunger', *British Journal of Criminology*, 46(1): 26–45.

White, R. (2008) *Crimes Against Nature: Environmental Criminology and Ecological Justice*. Cullompton: Willan Publishing.

White, R. (2011) *Transnational Environmental Crime: Toward an Eco-global Criminology*. London: Routledge.

White, R. (2012a) 'Environmental Forensic Studies and Toxic Towns', *Current Issues in Criminal Justice*, 24(1): 105–119.

White, R. (2012b) 'Climate Change and Paradoxical Harm', in S. Farrall, T. Ahmed and D. French (eds) *Criminological and Legal Consequences of Climate Change*. Oxford and Portland, Oregon: Hart Publishing.

White, R. and Graham, H. (2015) 'Greening Justice: Examining the Interfaces of Criminal, Social and Ecological Justice', *British Journal of Criminology*, 55(5): 845–865.

White, R. and Heckenberg, D. (2014) *Green Criminology: An Introduction to the Study of Environmental Harm*. London: Routledge.

Wynne, B. (1996) 'May the Sheep Safely Graze? A Reflexive View of the Expert/Lay Knowledge Divide', in S. Lash, B. Szerszynski and B. Wynne (eds) *Risk, Environment and Modernity: Toward a New Ecology*. London: Sage.

Young, N. and Matthews, R. (2010) *The Aquaculture Controversy in Canada: Activism, Policy, and Contested Science*. Vancouver: University of British Columbia Press.

Zhang, M., Takeda, M., Nakajima, H., Sasada, M., Tsukimura, K. and Watanabe, Y. (2009) 'Nuclear Energy and the Management of High-Level Radioactive Waste in Japan', *Journal of Hydrologic Engineering*, 14(11): 1208–1213.

Section 3
Wider varieties of technology crime

15

Guns, technology and crime

Peter Squires

Introduction

The relationship between technological development in firearm design, manufacture and marketing and the deadly consequences of firearm misuse on the streets can be extremely close. Data compiled by the US Bureau of Alcohol, Tobacco and Firearms reveals the rapidly changing composition of the US handgun market during the late 1980s and early 1990s. Around this time law enforcement weapon procurement was shifting away from revolvers to semi-automatic (self-loading pistols – SLPs) handguns (Diaz 1999). In turn, those new pistols were also entering the civilian markets, and a growing number were turning up at crime scenes. The year 1993 marks the high-point for handgun homicide in the USA (Squires 2014a: 166–8) while US DOJ data for the immediately preceding years reveals the marked shift to SLP production and marketing and the increasingly rapid adoption of these firearms as criminal and 'gang' weapons of choice. In particular the newer SLP calibres (.32, .38, 9mm. and .40), with larger magazine capacities, especially those produced for the cheaper mass civilian market, were significantly enhancing the street lethality of the illegal firearm inventory (US Department of Justice 2011). Some trauma survey evidence suggests that larger capacity SLPs have been associated with increased numbers of gunshot victims and higher numbers of gunshot strikes per victim (Wintemute 1996; Reedy and Soper 2003). Most significant of all, ATF firearm tracing data exposed the accelerated 'time to crime' (the time between the initial purchase of a firearm and its first ballistic trace in an offence) of the cheaper mass-market SLPs and the increasing likelihood that these weapons would turn up in the hands of youthful offenders (ATF 2000). Overall, the technological upgrade in firearm supply had major consequences for the quantity and nature of violent firearm victimisation.

In similar – though notably more extreme fashion – the terror attacks in France and Belgium during 2015 have brutally exposed the vulnerability of 'civil society' to determined, and even suicidal, armed terrorists. That such terrorists have been able to access military specification firearms (often Kalashnikov-type assault weapons), more powerful firepower than that of first responder police officers (Squires 2015), underscores the technological capability such weapons bring into play. A similar case is plausibly made that the growing lethality of mass shootings in the USA is directly facilitated by the entry, into the civilian market, of such military type assault weapons (Squires 2014a: 195–6; Diaz 2013). In the wake of the Charlie Hebdo shooting in January 2015, British police undertook a threat assessment to determine the likely availability of such weapons in the UK. While there is evidence of a number of automatic firearms (often reactivated weapons) being used by offenders on the streets of Britain (see Squires 2014a: 72 for more details) there was seemingly little evidence of criminal use of AK47s – certainly nothing on the scale witnessed in Paris and Belgium. Arguably, during August 2015, there were signs that this picture might be changing after police raided a Kent leisure boating marina intercepting a consignment of 22 AK47-type assault rifles and nine 'Skorpion' machine pistols (the largest illegal arms find of its type), along with approximately a thousand rounds of ammunition, recently trafficked across the Channel from France (Twomey 2015).

With these notable exceptions, although not overlooking the particular legacy of firearms once held by Northern Irish paramilitaries, a rather more limited and localised picture emerges of the firearms technologies being accessed illegally in the UK. Over recent years, a series of 'signal' incidents help to describe some of the core characteristics of this illegal firearm supply and firearm misuse in mainland Britain, a society with – by international comparisons – low (and currently falling) rates of gun crime, strict gun licensing and low rates of firearms ownership (see Figure 15.1).

These incidents comprise: the murder of Trooper Lee Rigby in May 2013 and its immediate aftermath; the double murder by John Lowe, jailed for life in Surrey in October 2014; the arrest and conviction of Stephen Greenoe, a former US soldier, for the smuggling of firearms into the UK in January 2012; the arrest, in two separate incidents, of gun dealers and collectors in Suffolk during 2012 and 2014 who had amassed huge illegal firearm collections in their own homes (Brown 2012; Robinson 2014) and, finally, the news that developments in digitised 3D printing capability meant that a programme for making the parts for a 3D plastic firearm could be disseminated via the Internet. These instructions could then be downloaded, the components 'printed' and assembled by anyone with access to the requisite printer technology. Fortunately, ammunition cannot be similarly printed and, with the exception of shotgun cartridges, neither is it in such ready supply (Hales *et al.* 2006; Gardham 2007) as to elevate the threat of this new technology.[1]

Each of these developments, in their own way, pointed to the variety of risk issues surrounding the illegal firearm supply question as it affects the UK, three jurisdictions with some of the more restrictive firearms laws in the world; Northern Ireland remains something of an anomaly in the British scheme of things, an undoubted legacy of the 'Troubles' being an unknown quantity of 'out of use', but far from 'decommissioned', paramilitary weapons complicating any assessment of legal/illegal weapon availability in the six counties. In turn these weapons are 'complemented' by several thousand 'personal protection' licensed handguns (prohibited elsewhere in the UK) held by persons considered at risk of potential assassination.

Before embarking upon an assessment of the risk profiles represented by different types of firearms in the contemporary UK context, always acknowledging, as we shall see later, that the UK is one of the world's safest societies as far as the crime and violence risks represented by guns is concerned, it is important to briefly consider the gun in a slightly more abstract sense – as a handheld technology. The importance of this question relates to the way in which the gun – as a piece of portable technology – has been over-analysed in some of its sociological aspects, but almost completely ignored in others. So, the place and significance of firearms has been subjected to an overwhelming degree of social and cultural analysis (the study of gun cultures (Tonso 1982), the establishment of large modern armies and the creation of nation states (Kiernan 1967), the establishment of constitutional citizenship (Cornell 2006), recycling certain trappings of hegemonic masculine symbolism (Gibson 1994; Harcourt 2006) and the criminal consequences of unchecked firearm proliferation (Squires 2014a)). However, by contrast, the consequence of guns *as a technology* impacting social relations (much as the way we might study the impact of smartphones on behaviour patterns and social relationships[2]) has been comparatively ignored.

By comparison with smartphones, handguns are pretty rudimentary tools, having undergone limited real development since the fourteenth century; the only real technological attribute of the so-called and much-heralded 'smart gun' is the palm-print recognition technology designed to prevent it firing for anyone but its designated owner. Despite much superficial development (automatic and semi-auto operation, construction materials, power, calibre, more accurate laser-directed sighting, magazine capacity, projectile design and ballistics and 'combat' styling), handguns remain simple devices for propelling small pieces of metal, at high speed, towards a designated target. The fact that this 'target' is often another human being, upon whom any projectile will have quite devastating consequences, establishes just some of the social and relational significances of guns. Guns, in Carlson's (2014; 2015) terms, create a kind of unchecked individual sovereignty: to be armed in a society of the unarmed – or perhaps, even more pointedly, the *disarmed* – is to experience the autonomous power of impunity. At the very least, arming sets one free from the arbitrary potential of others. This interpretation is mythologised in the story of the first mass production handguns – the Colt .45 as 'equaliser'. On the other hand, guns inspire feelings of fear and vulnerability on the part of those confronted by them who may feel threatened by the awesome power of the armed. The symbolic pleas of African-American demonstrators in Ferguson, Missouri in August 2014, following the police fatal shooting of Michael Brown, as they implored '*hands up, don't shoot*' testifies to the coercive instrumentality of the gun. Law enforcement has always embodied a complex relationship between authority and force: the badge and the gun (Scharf and Binder 1983), but while Missouri police represented a tainted legitimacy it was the gun, and the ultimate licence established by guns, that the protesters feared most.

Despite this, the relatively limited changes in the basic functions of the gun and the insistence of many 'gun rights advocates'[3] that guns are but tools giving effect to human purposes[4] have tended to downplay the empowering and transformative qualities inherent in gun possession and use.[5] That said, more recent scholarship on firearms (Squires 2014a: 321; Carlson 2015; Overton 2015) has begun to make this a more sociological and relational argument. Gun ownership has been accused of 'remaking masculinity' and 'militarising the mind' (Muggah 2001) empowering the impetuous and greedy to ride roughshod over the collective, traditional and democratic structures of legitimacy by which societies have hitherto been governed. Overton's dark journey through the 'world of the gun' makes frequent references to the transformative power of the gun: 'something about guns seems to empower people,' he notes (Overton 2015: 189). 'Carrying a concealed gun has even been said to change the way people walk … emboldening them and diminishing the rest of us' (2015: 207). Later he adds, 'a gun gives that ultimate edge of authority to someone who lacks it … On its own the gun wins any argument – it elevates "A Nobody" to "the Man". Small wonder so many men love them' (p. 208). While gun advocates concentrate on the firearms of the righteous – or the 'Good Guys' – in Wayne La Pierre's NRA-speak[6] the power of a gun transcends the merely legal status of the weapons, as Smith explains: 'those without access to … weapons (the "unarmed") are forced to cede a potentially very unequal power ratio to those who do' (Smith 2006: 728). Springwood, also, acknowledges the 'seductive transformation that a body, grasping and shooting a gun, undergoes' (2007: 3). Furthermore, as McCarthy recognises, 'guns are empowering when possessed … guns permit their owners a sense of uniqueness and individuation … they are the ultimate extension of self' (McCarthy 2011: 321). And it is precisely this extension of the sphere and capacities of the self that firearms accomplish (see Philip Brey, Chapter 1 of this volume, for a discussion of how weapon technologies, like other varieties, serve to extend the body). Gun advocates primarily talk narrowly about 'self-defence', whereas sociologists need to address the wider establishment of the defensive self, an identity that is realised and performed in conjunction with a gun, and how one 'negotiates' mutually armed and dangerous social relations. Extending outwards to the level of the weaponised community, Carr (2008) similarly describes the 'Kalashnikov cultures' evident in parts of North Africa and the Middle-East where weapon proliferation has fundamentally transformed notions of safety and sociability[7] – although much will depend upon the capacity and security of existing processes of governance.[8]

In like fashion, just as the presumed 2nd Amendment 'right to bear arms'[9] is frequently construed, often rather naively, as a defensive reaction, preventing crime and deterring aggressive violence, this only functions through a firearm's augmentation of one's self-governance, even 'governing at a distance' in Foucauldian terms; a power of lethal outreach. And this of course is premised upon the right to kill. It follows, therefore, that the right to carry weapons is not a *limited* power of the defensive self but rather a power to extend the self and obliterate others. Defenders of the right to bear arms like to imagine, rather disingenuously, that they are merely defending individuals and families, a conception beguilingly fixated upon some beleaguered *Little House on the Prairie* world view. In fact, through an advocacy of widespread gun ownership they are effecting fundamental shifts in civic (citizen to citizen) and political (citizen to state) relations. In this way the social and political order of neo-liberal security states will come to be established, it has not been the basis of social order in the rather safer social democratic societies.

Safer societies?

Of the 'safe European societies', however, as Figure 15.1 shows, the UK is undoubtedly the safest. As individual countries, England, Wales, Scotland and Northern Ireland would represent

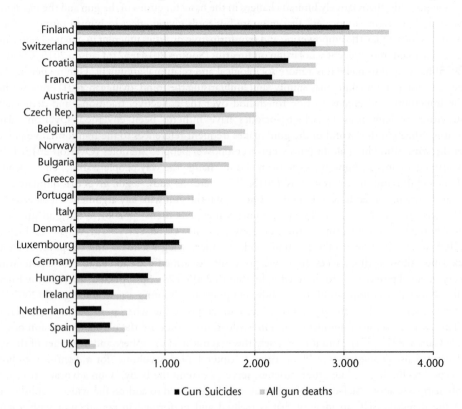

Figure 15.1 Proportion all gun deaths per 100,000, and gun suicides per 100,000 population[10]
Source: Data derived from Alpers et al. 2015.

four of the five 'safest' European societies (as regards gun crime), separated only by Poland (with a gun death rate of 0.26 per 100,000). Above all, Figure 15.1 reiterates the point that, despite the focus on gun crime, in most affluent, democratic, liberal capitalist societies the majority of firearm-related deaths are suicides. This, despite its rather unique celebration of gun culture and much misleading debate besides, is also true in the USA, a point which underscores a wider argument about weapon proliferation and misuse (Cook and Goss 2014). Evidence of the link between firearm availability and firearm misuse (homicide, suicide and gun-enabled crimes) is finally clambering free of the quagmire of 'US only' research to demonstrate clear connections between weapon proliferation and various forms of firearm misuse. Two recent international studies have confirmed that firearm proliferation makes a society significantly less safe (Bangalore and Messerli 2013) while Van Kesteren's study of 26 countries revealed that owners of handguns, in particular, showed 'an increased risk for victimisation by violent crime' (Van Kesteren 2014).

Key themes

Weapon types and quality

The murder of Trooper Lee Rigby and the Paris 2015 shootings probably represent two ends of a weapon technology spectrum in terms of contemporary terror threats. In Paris the gunmen were in possession of powerful AK47-type assault rifles. With these weapons they were easily able to outgun the first police officers to arrive. CCTV footage of the incident showed a police patrol vehicle approaching the two men before being sprayed with automatic fire. The vehicle then accelerated rapidly in reverse to the end of the street, its windscreen peppered with bullet holes.

By contrast, Trooper Rigby was run down by a car and subsequently viciously attacked with a machete. But then the two attackers appeared to wait for the police to arrive. As a first armed response vehicle appeared on the scene, caught in the CCTV pictures, one of the two men ran towards the police car wildly brandishing his machete. As he approached within a few feet of the ARV, he was shot from within the vehicle. The second attacker, also visible in the CCTV pictures, appeared to be gesturing with a handgun, armed officers engaged with him and he too was shot and incapacitated. Officers immediately then began to administer first aid to the two suspects. The handgun held by the second attacker turned out to be a broken (incapable of firing) Belgian Infantry officer's revolver manufactured in 1912.

These two widely differing weapon technology types, a modern Kalashnikov-variant assault rifle and a broken infantry revolver from the WWI era describe a range of small arms and light weapons potentially available to offenders or terrorists. Yet, whereas in mainland Europe, there is reported to be a significant proliferation of illegally possessed high powered assault weapons, many of which have been trafficked into Western Europe from Eastern European states and the Balkans (European Commission 2013; Lazarevic 2010; Carapic 2014), the smuggling of firearms into the UK is thought to be significantly more difficult and high powered assault-type, or combat designed, weapons considerably harder to acquire.[11] One consequence of the difficulty of trafficking serious quality weapons into the UK has been the resort of offenders, especially lower-level gang-involved offenders, to the splendid array of 'junk firearms' (Squires 2014a): conversions, reconditioned antique weapons, sawn-off shotguns and reactivated firearms (to be discussed later in more detail) which now represent the bulk of mainland Britain's illegal weapon firearm inventory.

In the wake of the Sandy Hook, Connecticut, school shooting in December 2012, as US journalists attempted to gain some perspective on how other societies had responded following horrific school shooting tragedies, and whether, if new firearm controls had been introduced, they had made any significant difference to rates of gun-enabled crime, a spokesman for NABIS (the National Ballistics Intelligence Service) was reported making a highly significant claim about the availability of illegal firearms in England and Wales. According to the NABIS spokesman some 70 per cent of gun-enabled crime in England and Wales could be linked to around 1,000 illegal firearms. Never before had a figure been put on the size of the illegal gun inventory; never before had any estimate provided been so low. The police were claiming some considerable success in driving down gun crime.

NABIS had been established in 2008, to replace the specialist forensic science service arrangements for analysing recovered firearms, ballistic materials and related trace evidence. Submitting material to the Forensic Science Service (FSS) for analysis could be a lengthy and expensive process. This implied that police forces only tended to submit materials for analysis when there was a reasonable chance that the results derived were likely to prove important in any prosecutions. This meant that, overall, forces only submitted roughly 60 per cent of ballistic material for analysis – in other words, there was a 40 per cent gap in the intelligence profile regarding illegal weapons used in shootings. NABIS's primary strategic aim was to close this gap and establish a national ballistic database allowing it to develop an intelligence picture of active criminal firearms in use in Britain, and a quick time ballistic comparison facility to match new and incoming ballistic evidence (bullets and cartridge cases) to existing material held on the database. This had not been possible before and the new focus upon intelligence development has allowed the police to move beyond a largely reactive approach to addressing gun crime and towards more proactive efforts to address illegal firearm supply routes and weapon exchanges from area to area around the country (Gibson 2012). Subsequent research is similarly exploring the role of criminal armourers (converters, re-activators, and traffickers) in the criminal supply of firearms into the UK (Williamson 2015).

By 2012, NABIS had begun to develop its intelligence picture of the scale and range of active illegal weapons and, as noted, a police spokesman confided to a Washington Post reporter that ballistic tests indicated that 'most gun crime in Britain can be traced back to fewer than 1,000 illegal weapons still in circulation' (Faiola 2013). This was the first occasion on which police had ever been able put such a figure into the public domain, even if it was only reported in the USA. It was also the first time police have ever had the comprehensive database from which to make such a categorical statement. Notwithstanding the fact that the figure was still only a police intelligence construct,[12] it nevertheless tells a story of police and law enforcement successes, corroborating the developing picture of the junk weapons market to which offenders were having to resort, so constrained had the better quality illegal gun supply channels become. More recent estimates of the number of 'criminally active' firearms have pushed the figure a little higher, approaching 1,300 weapons but the figure is still well below any of the estimates made public around 1998 when the post-Dunblane handgun prohibition commenced. Here, of course, it remains important to bear in mind the important distinction made in international gun control circles between 'black' and 'grey' firearm pools (Karp 1994).

The former category includes the firearms in direct possession of criminals, insurgents or terrorists, the latter category is made up of non-licensed or 'vulnerable' weapons to which the criminally inclined might be able to gain relatively easy access. Globally speaking, it is generally assumed that the latter significantly outnumber the former and the same is likely to be true in Britain. Accordingly the antique, sometimes dubiously legal, firearm collections of enthusiasts, stored in private homes, or the carelessly stored shotguns of farmers may easily slip into illegality if stolen. Between 2007 and 2011, the Home Office reported that almost 3,000 firearms (over 2,000 of them shotguns, and some 380 rifles), were lost or stolen in England

and Wales as a whole (Beckford 2012), while police regularly report finding 'stashes' of illegal weapons accumulated in private homes; sometimes these comprise combinations of licensed and unlicensed weapons, the work of an over-enthusiastic collector (Spillet 2014; *Brighton Argus* 2014), on other occasions rather more explicitly criminal motivations appear to be involved (Brown 2012; Roberts 2013).

This discussion of NABIS and its methods for estimating the extent of the pool of criminally active firearms in the UK goes back to the Paris shootings. In the wake of the incident, security interests in the UK were keen to assess the likelihood of a similar terrorist gun attack against a soft 'civilian' target in the UK – as opposed to the more familiar bomb attacks. Central to this threat assessment was a review of the NABIS intelligence data to see whether the kind of weaponry employed to such lethal effect in Paris and capable of sustaining such an attack appeared to be available to similarly motivated offenders in the UK. The fairly reassuring answer was that the NABIS database contained a record of evidence referring to only one AK-47 previously fired in the course of criminal activity. Despite the caveat that such evidence does not rule out the existence of illegal weapons hitherto unknown to the police (for it is clear that deactivated and reactivated AK-47s have been found in illegal gun caches, in the 'grey pool'), the evidence seems to confirm that there is no readily available supply of them to potential terrorists in the UK – although, one would have to add, so far.

The mixed economy of 'junk' guns

One consequence of the tightened market for factory quality firearms which has resulted from better intelligence-led policing and more effective co-operation between police and customs in Britain has been the substantial 'weapon displacement' effects which are evidenced in Figure 15.2. A demand for firearms exists at different levels; in turn it has been met by criminal firearm supply chains at different levels. Even before the prohibition of handguns in mainland Britain, firearms controls had been significant, and with the exception of air weapons, a licence had been required to own a firearm. By the 1980s, however, the firearms market began to change. Major firearms manufacturers began to sell patents and designs to other manufacturers establishing a wholly new market in highly realistic-looking firearms, some made of metal, some plastic (some of these fired only plastic pellets, others fired ball-bearings, others were realistic looking air weapons, others were high pressure CO_2 gas propulsion systems, others were closer to 'toys'). These shifts in what we might call the 'sub-prime weapon technology market' meant that none of the new weapon types were subject to the British firearm licensing system even though some could be powerful enough to be potentially dangerous, many were highly realistic in their appearance and offenders had already begun to use them in criminal activities such as armed robberies. As Taylor and Hornsby (2000) put it, the new mixed economy of firearm availability had begun to exceed the licensing regime designed to manage it. Research evidence was already accumulating to suggest that a significant number of lower-level armed robberies were carried out with 'replica' firearms (Morrison and O'Donnell 1997) part of a phenomenon, described by Matthews (2002) as the de-skilling and de-professionalisation of armed robbery. All the opportunist robber needed was something which looked like a gun and would ensure 'compliance'; it did not need to be capable of firing, but it would be recorded, if reported, as a gun-enabled crime. Not only did the new categories of weapons exceed the licensing and regulatory system, they also challenged the crime recording system and whilst it is highly likely that a preponderance of replica and air pistols are partly responsible for the doubling of recorded handgun-involved crime in the four years after 1998, it was not until around 2004–5 that Home Office data recording (having undergone a steep learning curve) begins to reliably

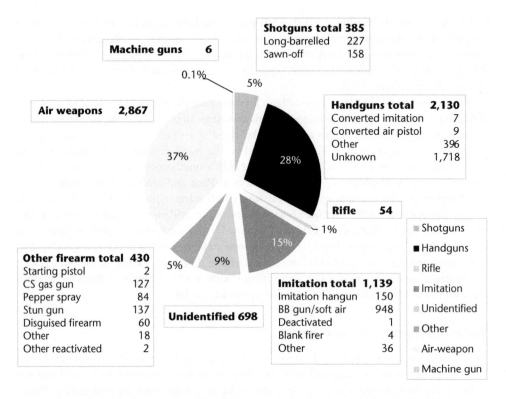

Figure 15.2 The mixed economy of illegal firearms in England and Wales, 2013–2014
Source: Office for National Statistics 2015

distinguish between real handguns, imitation handguns and handguns 'type unknown' (see Squires 2014a: figure 2.1 p.50).

As Figure 15.2 indicates, there still remain substantial numbers of 'unknown firearms' which comprise the criminal gun supply for England and Wales. Over two-thirds of criminal handguns are 'type unknown', nearly a thousand weapons remain unidentified[13] whereas 1,600 further weapons are known imitations. While it is clear that the processes of displacement and diversification that Figure 15.2 reveals clearly began before the prohibition of handguns after the Dunblane massacre, it is likely that the increasing difficulty of acquiring factory quality weapons accelerated the displacement process. Wheal and Tilley have argued that official crime figures 'very substantially understate the significance of imitation weapons' (2009: 176). In the interest of concentrating policing attention on the major gun crime threats, legislation in 2003 removed air weapons and imitation weapons from the main gun crime count.

While the tight gun regulations in Britain have played a significant part in incentivising other sources of criminal firearm supply, creating the 'mixed economy of junk firearms' we have here been discussing, the UK situation is not entirely unique in this regard. Cheap two-dollar pistols and the infamous 'Saturday-Night Specials' of US urban gun crime – 'junk guns' by any other name – have been the specific subject of gun control/crime prevention efforts even in societies such as the USA where gun ownership is widespread. Even in more recent times Phil Cook (one of America's leading firearms control researchers) and his colleagues have continued to document the benefits of situational gun controls in keeping the most lethal categories of weapons out of the hands of the youngest and most inexperienced offenders

(Cook *et al.* 2007; Ander *et al.* 2009). A junk gun can still be lethal, but perhaps less so than a more modern, high-power semi-automatic.

Converted firearms

A second more serious source of criminal firearm supply has concerned converted weapons. These comprise a range of legally manufactured firearms including: 'alarm pistols', CS gas self-defence pistols, gas-powered handguns and athletics blank-firers, all of which, following a degree of re-engineering, might be converted to fire a live bullet. Evidence of this emerging trend in firearm misuse first emerged in the 1980s (Heal 1991). However, by 2003–4 evidence from the Metropolitan Police's Operation Trident revealed that something like 75–80 per cent of 'Trident shootings' involved converted air or gas-powered weapon technologies (MPA 2004). Greater Manchester Police similarly reported that almost half of the firearms capable of live firing recovered during 2003–4 had been converted, noting that, 'scarcity in supply [of real firearms] means that criminals will invest significantly in conversion' (GMP 2006). The same GMP intelligence briefing document provided details on two successful police operations to intercept converted weapons, in one case, gas alarm pistols originally legally manufactured in Germany (and purchased by private citizens for self-defence purposes) but then transported in their hundreds to Lithuania where they were illegally converted to fire live ammunition (and equipped with silencers). The second intelligence-led operation involved the interception of 274 lawfully manufactured 'gas guns' modelled on Smith & Wesson, Luger and Glock pistol templates, which were purchased on the European mainland then illegally imported for conversion in a Manchester machine shop. Evidence from research on the European mainland reveals the existence of a number of known converted firearm trafficking networks and routes, involving original manufactures in Germany, Italy, Russia and Austria, conversions in Portugal and Lithuania and trafficking to the UK via Holland (De Vries 2011).

The scarcity and difficulty in accessing factory quality weapons undoubtedly explain the growth of illegally converted weapons. Imprisoned firearm offenders interviewed in the course of a Home Office research study (Hales *et al.* 2006) knew that such weapons were undoubtedly inferior in many respects (inaccurate, underpowered, a tendency to jam, not to mention – because they are made from inferior alloys never intended to cope with the stresses of a real firearm discharge – the possibility of blowing up when fired, O'Neill 2008) but they are also cheaper and more available. A case recently concluded at the Old Bailey involved a converted firearm. Some ten years after the original incident a police officer was finally acquitted of murder following a 'hard vehicle stop' on a car containing armed gang members. The basic details of the incident were not dissimilar to the shooting of Mark Duggan in 2011. In 2005 Azelle Rodney had been the back seat passenger of a car tailed by police armed officers; when the vehicle was stopped, Rodney was alleged to have reached down as if to pick up a firearm. He was shot eight times at near point-blank range, the prosecution case hanging on the question as to whether the police officer had exercised reasonable judgement in opening fire almost immediately. In the foot-well of the rear seat amongst other firearm-related items was a Baikal CS gas pistol which had been converted to fire live ammunition.

Research by the National Forensic Science Service, the predecessor to NABIS (Hannam 2010), has revealed several 'waves' of readily convertible – sometimes already converted – handguns illegally trafficked to Britain. For a while the Brocock, a German-made weapon, was popular in criminal and gang circles; later it was overtaken by the Baikal, a Russian made CS gas self-defence pistol, manufactured in Izhevsk, home of the Soviet Kalashnikov AK47 assault rifle and the Makarov pistol (a standard Russian military and police side arm), where it forms part of the new Russia's post-soviet product diversification in that town. The weapons are advertised as 'a step up from personal rape alarms ... designed to fire a cloud of tear gas at an attacker.' In an intriguing 'American' touch the weapons were marketed in glossy brochures showing young women pulling handguns from handbags to fend off muggers and rapists (O'Neill 2008). Such weapons were readily convertible by semi-skilled criminal armourers, however, and significant numbers found their way into Britain.

Identifying such trends in the distribution, conversion and trafficking of 'readily convertible' firearms has become a key proactive role of NABIS. On recognising that particular simply convertible firearm models are appearing in criminal hands NABIS builds a case intended to prohibit that particular weapon for importation, sale and possession in the UK. Both the Brocock and the Baikal became subject to these prohibitions, and a more recent operation concerned the Olympic .380 – originally produced and marketed quite legally in Italy as an athletics starting pistol (NABIS 2011). The Olympic .380 first came to attention in 2005–6; by 2010, 179 converted .380s had been seized by various police forces (the overwhelming majority were recovered from London). The process of closing this 'conversion' loophole entailed NABIS monitoring the numbers of converted .380s (or ballistic traces of them) turning up at crime scenes, then liaising with the gun manufacturers, the gun trade and athletics organisations, before regulations were drafted for parliament prohibiting the weapon. An amnesty for the weapon ran between April (when the firearm was added to the 'prohibited' list) and June 2010, netting around 800 Olympic handguns, a quantity of 'blank' ammunition and around 100 other assorted blank-firers. Only seven of these surrendered weapons had been converted, rather confirming a familiar story that the hardest weapons to reach are invariably the most potentially dangerous. Nevertheless, NABIS judged the operation a successful piece of intelligence-led, proactive partnership police work, even though a spokesman acknowledged, 'criminals are entrepreneurial and will look to replace the Olympic .380 with another blank firer' (NABIS 2011: 24). Overall, some 1,300 convertible firearms already known for their criminal misuse had been taken out of circulation. The NABIS intelligence report concluded by recommending more consistency in police recording of recovered firearm types as well as changes to section 39 of the 2006 Violent Crime Reduction Act in order to set clearer standards and specifications for blank-firers particularly relating to their potential for criminal conversion.

Reactivated firearms

There has long been a thriving market in souvenir and 'deactivated' firearms, the more exotic or infamous the firearm, the higher the price it might command, and advertisements appear throughout the shooting press and on shooting related websites offering deactivated weapons for sale (Warlow 2007). Confirmed deactivated weapons could be held without a licence but, by 1995, growing evidence that some, once deactivated, weapons were being brought back to working order and sold to criminals prompted the development of tougher deactivation standards. Currently pre-1995 deactivations advertised still command a higher market price although this may have less to do with their potential for reactivation and more to do with authenticity and less modification of working components. Pre-1995 weapons will be easier to reactivate, however.

Over the years, police have uncovered several shipments of deactivated weapons on their way to a 'cottage industry' of reactivators bringing the guns back into working order. During the late 1990s, evidence that a supply of reactivated Mac-10 sub-machine guns were turning up in the hands of gangsters as far afield as Manchester, Dublin, Glasgow and London led to a police operation which culminated in a raid on a workshop in Hove, East Sussex. There the police discovered crates containing 40 imported deactivated Mac-10s as well as the tools and components capable of returning them to full working order. Forensic tests established that weapons reactivated in this workshop had been used in over 50 previous shooting incidents. In their deactivated form the guns had cost around £100 each, but once reactivated the price could increase tenfold. Quick profits also appear to have been the motivation for two men running a criminal firearm reactivation business in Derbyshire. By the time this illegal enterprise was finally raided by police in 2003, it was estimated that the men – Britain's most prolific criminal armourers – had worked on thousands of weapons. Over 3,000 firearms linked to the business have disappeared, although firearms responsible for eight murders as well as gangland shootings and loyalist paramilitary attacks have been recovered. Ballistic evidence relating to 65 further shooting incidents has also been traced back to the Derbyshire armourers (Townsend 2007).

Unlike imitation firearms and air weapons, reactivated weapons did not feature in the 2006 Violent Crime Reduction Act, perhaps because of the resistance put up by firearms associations and shooting lobbyists who have often sought to defend the trade in so-called 'heritage' weapons. Why shooting enthusiasts would want to defend what were often, in effect, military surplus 'junk' firearms with limited 'heritage' value is anyone's guess. The so-called 'heritage' trade provides a link between the legal trade in firearms and the rather murkier and less effectively regulated world of back-room deals involving the re-sale of used and obsolete firearms: part of the 'grey pool' (Karp 1994). In the absence of tougher controls, especially harmonised at the European or International level, this trade will continue, creating small but significant criminal opportunities for illegal entrepreneurs willing to risk substantial prison sentences for adding to Britain's illegal firearms inventory.

Hoarders, antiques, collectors and dealers

Paul Condon, Commissioner of the Metropolitan police inspired the anger of British gun traders when he announced, following a number of prosecutions of firearms dealers, that, in his estimation, '70 per cent of firearms used in crime have … been recycled through legitimate dealers through a variety of scams involving legitimate use of guns' (Tendler 1994). Evidence, other than anecdotal, to support the Commissioner's claim was never released but it revived a long-standing, though seldom resolved, set of issues regarding the contribution of legal firearm ownership to illegal firearm supply (through theft and diversion), to the number and proportion of crimes committed with firearms and to wider security risks. The latter issue took on renewed significance during 2014–15 with the discovery of a number of large illegal firearms (sometimes including automatic weapons and assault rifles), caches kept in private homes by avid firearms collectors whose enthusiasm for collecting far exceeded the remit of their firearms licences (Robinson 2014; Bloom 2014). It was less the threat of firearms being sold to gangsters which concerned police, but, after the Paris terror attacks, more the security risks associated with large numbers of poorly secured weapons. During the 1990s, roughly 1,000 firearms a year were being reported as lost or stolen, most of these being shotguns (Squires 2000). More recent figures suggest slightly lower numbers with 3,000 firearms lost or stolen between 2007–11 (Beckford 2012).

American journalists, no doubt reflecting upon Britain's illegal firearms, as compared with those of the USA expressed surprise at the resort to convertible and antique weapons. According to Anthony Faiola, writing in the *Washington Post*:

> When police on a weapons raid swarmed a housing project after London's 2011 riots, they seized a cache of arms that in the United States might be better suited to *The Antiques*

Roadshow than inner-city ganglands. Inside plastic bags hidden in a trash collection room, officers uncovered two archaic flintlock pistols, retrofitted flare guns and a Jesse James-style revolver.

(Faiola 2013)

Taken together, hoarded, collected and antique weapons have all contributed a small part to the British gun crime story, and so have legally held firearms. Often domestic shootings are reported in the media less as crimes and more as tragedies – such domestic incidents frequently involve murders as the final act of a long sequence of domestic violence, sometimes they involve murder-suicides (so-called dyadic murders) and worse, what are known as 'family annihilations' (Yardley 2014). Prominent amongst such cases is that of bankrupt former millionaire Christopher Foster who, suffering from stress and related pressures in 2008, shot his wife and daughter, before killing the family's horses and dogs, and barricading himself into his house which he then set alight, killing himself (Randall 2008). As in the case of other domestic shootings, including the cases of Michael Atherton (Brooke 2012), or that of John Lowe (both of whom went on to kill after the police had returned their temporarily revoked firearms), referred to earlier, the weapon involved is often a shotgun – in a majority of cases a legally owned and licensed firearm.

Although the Home Office has ceased, since 1996, to separately publish data on the numbers of legal firearms involved in gun-enabled crime, evidence submitted to the Cullen Report in 1996 suggested that legal firearms might contribute to 14 per cent of homicides, although later evidence to the Home Affairs Select Committee (Broadhurst and Benyon 2000: para 5.9, p. 157) has suggested that when stolen firearms were also taken account of, then legally held firearms might contribute up to some 20 per cent of firearm homicides. Such issues have raised a degree of pressure to the effect that the consequences of the criminal misuse of licensed firearms are matters that ought to be in the public domain, for such weapons are, after all, held in public trust. Government seems, thus far, disinclined to take any action on this issue, although a series of firearm licensing failures (including some of those referred to already) prompted Her Majesty's Inspectorate of Constabulary to embark upon a review of the police firearm licensing system (HMIC 2015).

The activities of hoarders, collectors and 'bent' dealers point, despite gun lobby claims about the responsibility and integrity of the shooting fraternity – 'good guys' as opposed to 'bad guys' perhaps – to the never 'water-tight' overlaps between the legal and the illegal and the slips and flows which might see firearms moving from legality to illegality. Dutch criminologist Toine Spapens has described at least 12 different ways in which firearms can leak into illegality (Spapens 2007). One of the newer routes involves international internet sales and fast-parcel services. Given the volume of parcel traffic, it currently appears impossible to scan every parcel at a speed which would not disrupt the parcel service using digitised scanning software capable of recognising firearms or their component parts. Once in the country, however, firearms continue on their journeys, moving from points of entry, to other cities (Gibson 2012), sometimes from gang to gang, or rented out by criminal armourers, recycled through many different hands. One particular firearm, a 9mm Croatian-made Agram machine pistol which had undergone an especially complex journey, all within London, was discovered dumped behind some bushes in a public park in North London in April 2010. Seven days earlier it had been used in a drive-by attack on a pizza restaurant during which a 16-year-old girl was killed. Police were able to trace its use in five different shootings – from Enfield to Croydon – undertaken by different gangs during January to September the previous year (Summers 2011).

Yet however effectively firearms slip from hand to hand and across borders, boundaries and jurisdictions, few guns could be quite so 'elusive' as the new technology which developed the 3D printable handgun, which could be made by anyone with a digital 3D printer and the downloadable software instructions. This 'firearm' attracted a deal of media attention during 2013 insofar as it suggested that offenders in the future might be able to 'Google', and then download, a gun as easily as they might access internet pornography. Going way beyond 'ceramic' and polymer firearms which might evade metal detection scanning, this weapon attracted a great deal of novelty attention when the online plans were first released by its designer in March 2013 (Squires 2014a: 10). The first models were plastic-printed and various police forces produced their own and test-fired them, finding them capable of firing 2–3 rounds of live ammunition before the 'gun' began to crack and fragment. The idea of the weapon – a firearm that could be distributed via the Internet and constructed remotely – was more threatening in its potential than its actuality, because of the inadequate strength and durability of the printed plastic parts. Metallic 3D 'printing', while technically possible, currently seems prohibitively expensive to all but the most well financed criminal enterprises who are likely to be capable of accessing credible firearms in a number of more 'traditional' ways.

Conclusion

In some respects the diversification of the UK mixed economy of illegal firearms is a result of more effective intelligence-led, partnership and community policing suppressing both demand for and supply of illegal firearms. This is part of a policing and regulatory success story that has traced, tracked and ground down illegal firearm availability and misuse in the ways already described. In turn this activity has been reflected in a fairly optimistic story of the continuing fall, almost year on year, in gun-enabled crime.

It needs to be acknowledged, however, that this is only part of the story. For instance, Home Office crime recording only counts 'gun crime' when a firearm is employed to commit another offence. Simple illegal possession of a prohibited firearm could be subject to a five-year mandatory sentence, although this offence is not currently classed as a 'gun crime'. If possession and trafficking offences were included in the gun crime count, it is likely that the overall gun crime figures would rise significantly. No government is likely to contemplate this during their own period of office. Second, the gun crime figures may bear relatively little relation to

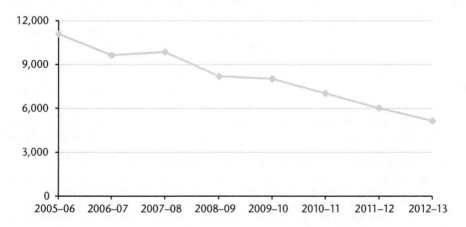

Figure 15.3 Consistently falling gun crime in England and Wales, 2005–2013

police intelligence data and reports of firearm discharges. The communities in which most gun crime occurs and the groups most likely to be shot at often have the least trust in the police and the greatest disinclination to report crimes (Bullock and Tilley 2002). Gunshot tracking sensors have been deployed in some American cities in order to help police analysts detect weapon misuse, identify gun crime 'hotspots' and despatch police resources to incidents. Perhaps most of all, it is to be hoped that existing successes in tackling gun-involved crime continue so that such systems do not become necessary in British cities.

Finally, of course, the Paris and Brussels shootings during 2015 and the arrival, in Kent, of an unprecedented quantity of Kalashnikov assault rifles with which these outrages were perpetrated, points to a rather different and darker story. As has been argued, compared to the relative scarcity of firearms in mainland Britain, this 'other' story acknowledges the widespread proliferation of weapon technologies in most other societies and the constant efforts of criminals and traffickers to circumvent border security provisions in order to supply firearms to offenders, in a context where relative scarcity inflates the price they can command. These supply chains have involved technological innovations of the most sophisticated kind (high-tech: 3D printed firearms, sophisticated smuggling operations and weapon disassembly to defeat fast-parcel scanning and internet sales of weapons and components) and of the most crude (low tech: weapon and ammunition conversion, junk gun supply, and the recycling of antiques, imitations and air weapons). Criminal ingenuity has continued to evade, challenge and test both border security and policing practices as well as legal definitions as to what constitutes a genuine or *lethal* firearm (for example, during 2015 the Law Commission embarked upon a series of consultations as part of a review of UK firearms law which still dates back to 1968 (as amended), a time well before the explosion in scale and diversity of the modern mixed economy of illegal weapons) (Law Commission 2015).

In a further sense more related to media and public perceptions, concern about firearm misuse continues to centre almost exclusively upon criminal misuse of *criminal* firearms but, as we have seen, neglects the greater proportion of deaths – suicides – and a small but consistent pattern of cases of lethal domestic abuse attributable to legally owned and licensed firearms (see Squires 2014b). Finally, recent developments notwithstanding, the threat of criminal firearm use exerts a constant pressure upon police 'tooling up' which, as an 'operational matter', is seldom publicly discussed. By the time of the England versus Germany football friendly in November 2015, just a few days after the Paris shootings and the attempted bombing of the Stade de France, the fait accompli of the Metropolitan Police firearms upgrade to the Sig 516 assault rifle became clear (Greenwood 2015). As we have seen, although they do so relatively infrequently, rather more critical publicity – and consequences – attends to *who* the police shoot and *when* and *why* they do so (Squires and Kennison 2010).

Notes

1 According to Hales, Lewis and Silverstone: 'ammunition is generally bought in small quantities when a gun is purchased, both priced as a package. Ammunition appears to be a limiting factor and harder to obtain than firearms, with the exception of shotgun cartridges. In some cases, criminals are exploiting a legal loophole to manufacture ammunition themselves or are using improvised ammunition such as blank firing ammunition combined with a ball bearing' (Hales *et al.* 2006: 56–58).

2 Smartphones are no longer just tools for communication, but vital intelligence resources, credit managers and identity 'passports' in their own right, they have transformed inter-personal relations, created the world that is 'social media', established new routines and rituals that have broken down barriers between public and private as people commit their whereabouts, status and innermost thoughts

to cyberspace. They have become vehicles for abuse, exploitation and victimisation through cyber-bullying but equally, a means for imposing democratic surveillance over state agents.

3 I have employed this phrase to acknowledge that, although it is often associated with pro-gun politics in the USA, the idea has a truly global reach, with many people drawing a (for them) vital connection between gun ownership, freedom and citizenship. Such a view inevitably entails an idea of firearm ownership as transformative and empowering, ideas that are seldom developed very far, in part, perhaps because the right to bear arms is frequently construed in a benign and defensive fashion.

4 An idea perhaps best summed up in the US NRA bumper sticker 'Guns don't kill people; people kill people'.

5 Here, 'use' can mean anything from shooting a firearm at someone who places one in imminent danger, to allowing a potential aggressor to see that one is 'armed' by way of discouraging an assault. These aspects of gun use are social relations.

6 Wayne La Pierre is executive vice president of the NRA. At an NRA press conference following the Sandy Hook school massacre in December 2012, he advocated armed teachers and armed guards at all schools because 'only a good guy with a gun can stop a bad guy with a gun'.

7 'When military style arms become freely available to the general public, notions of control, stability and security must undergo radical redefinition ... a proliferation of arms down to the individual level, in all but the most structurally stable of political entities, can itself create political eddies of insecurity that either exacerbate existing problems or, in extreme cases, actually eclipse the authority of centralised governance' (Carr 2008).

8 The hasty retreat of the first, lightly armed, French Police officers arriving on the scene of the Paris 'Charlie Hebdo' attacks, in the face of the terrorist AK-47s, serves as a potent illustration of the destabilising power of the gun.

9 It is important to acknowledge that this particular and modern interpretation of the 2nd Amendment reflects a late twentieth-century *re*-interpretation, subsequently ratified – in a paradoxical instance of 'conservative legal activism' – by two Supreme court judgements during 2008 and 2010.

10 The graph depicts the total firearm death rates and firearm suicide rates of 20 European societies. In most societies firearm suicide rates constitute a substantial proportion (from around two-thirds to four-fifths) of all gun deaths. All of Iceland's gun deaths (1.25 per 100,000) were suicides. Greece and Bulgaria indicate a different pattern, with a higher proportion of gun homicides as do, former Yugoslav or Eastern European states Serbia, Albania and Montenegro (not depicted in the graph). Turkey's firearm death rate (165 per 100,000) would have dwarfed all the other columns but the data suggests only a 0.33 per 100,000 suicide rate.

11 This is notwithstanding news that at least eight Czech-manufactured Skorpion Vz61 submachine guns – each capable of an 850 rounds a minute rate of fire – had been smuggled into the UK, in an audacious trafficking scheme using Parcelforce and masterminded from prison by a convicted gangster using a contraband smartphone. Police had recovered three of the weapons but at least five were thought to be in the hands of criminal gangs. Speculation followed that one of the guns was used in a double shooting in London on 10 July 2015 after witnesses referred to hearing automatic gunfire (Khomani 2015).

12 Each time a firearm is reported as having been used in a criminal offence a NABIS record is created, similarly every bullet or cartridge case recovered from a crime scene (or retrieved from a victim) is screened and run through the database; if no matches are found one more 'inferred weapon' is added to the database. By contrast, every firearm recovered is test fired and the ballistic material also compared to the database. Using this methodology NABIS can calculate the number of 'criminally active' firearms. Every time a firearm is recovered by police and the database comparison reveals a number of 'hits' indicating that the gun has previously been used in a crime, the total number of criminally active firearms can be reduced by one. In this sense the 'criminally active firearm' figure represents a 'flow measure'; new weapons entering the list, recovered ones dropping from it and – importantly – weapons that are not matched within one year also dropping from the criminally active list, because NABIS evidence also suggests that the overwhelming majority (around 90 per cent) of weapons that are not used within 12 months tend not to resurface.

13 In each of these cases where a weapon was not recovered from the crime scene and not clearly identified, it is highly plausible that a significant further number would fall in the replica category.

References

Alpers, P., Wilson, M., Rossetti, A. and Salinas, D. (2015) *Guns in the United Kingdom: Rate of Gun Suicide per 100,000 People*. Sydney School of Public Health, The University of Sydney. Available at: www.gunpolicy.org/firearms/compare/192/rate_of_gun_suicide/28,46,49,50,280,68,88,104,105,118,125,136,148,177,178,187,3 (accessed 17 May 2015).

Ander, H., Cook, P.J., Ludwig, J. and Pollack, R. (2009) *Gun Violence Among School-Age Youth in Chicago*. The University of Chicago: Crimelab. crimelab@uchicago.edu.

ATF (Bureau of Alcohol, Tobacco and Firearms) (2000) Youth Crime Gun Interdiction Initiative Report. www.atf.gov/resource-center/publications-library?search_api_views_fulltext=&field_document_type_1=All&og_group_ref=All&page=25.

Bangalore, S. and Messerli, F.H. (2013) 'Gun Ownership and Firearm-related Deaths', *The American Journal of Medicine*, 126: 873–876.

Beckford, N. (2012) 'Almost 3,000 Guns Lost or Stolen in UK, Figures Reveal', BBC News Website, 29 August.

Bloom, R. (2014) 'Police Find Huge Cache of Weapons Including 70 Guns and a Grenade in Dead Man's £220,000 Semi-Detached Bungalow', *Daily Mail*, 2 July.

Brighton Argus (2014) 'Haul of 70 Guns Seized From House', 2 July.

Broadhurst, K. and Benyon, J. (2000) 'Gun Law: The Continuing Debate about the Control of Firearms in Britain: Evidence to the Home Affairs Select Committee'. Occasional Paper No. 16. Leicester: University of Leicester Scarman Centre.

Brooke, C. (2012) 'Suicidal Taxi-driver Kills Three in New Year's Day Shooting', *Daily Mail*, 3 January.

Brown, L. (2012) 'Factory Worker Jailed for Hoarding Deadly Arsenal of Machine Guns, Ammunition and Hand Grenades in Every Room of his House', *Daily Mail*, 25 September.

Bullock, K. and Tilley, N. (2002) 'Shootings, Gangs and Violent Incidents in Manchester, Crime Reduction Series', Research Paper 13, Home Office.

Carapic, J. (2014) 'Handgun Ownership and Armed Violence in the Western Balkans, Small Arms Survey', Issue Brief No. 4, September.

Carlson, J. (2014) 'States, Subjects and Sovereign Power: Lessons from Global Gun Cultures', *Theoretical Criminology*, 18 (3): 335–353.

Carlson, J. (2015) *Citizen Protectors: The Everyday Politics of Guns in an Age of Decline*. New York: Oxford University Press USA.

Carr, C. (2008) *Kalashnikov Culture: Small Arms Proliferation and Irregular Warfare*. Westport CT: Praeger Security International.

Cook, P.J., Ludwig, J. Venkatesh, S. and Braga, A.S. (2007) 'Underground Gun Markets', *The Economic Journal*, 117 (November): 588–618.

Cook, P. and Goss, K.A. (2014) *The Gun Debate: What everyone needs to know*. Oxford: Oxford University Press.

Cornell, S. (2006) *A Well Regulated Militia: The Founding Fathers and the origins of Gun Control in America*, Oxford: Oxford University Press.

De Vries, M.S. (2011) 'Converted Firearms: A Transnational Problem with Local Harm', *European Journal of Criminal Policy Research*, 18 (2): 205–216.

Diaz, T. (1999) *Making a Killing: The Business of Guns in America*. New York: The New Press.

Diaz, T. (2013) *The Last Gun: How Changes in the Gun Industry Are Killing Americans and What it Will Take to Stop it*. New York: The New Press.

European Commission (2013) 'Firearms and the Internal Security of the EU: Protecting Citizens and Disrupting Illegal Trafficking.' Communication from the Commission to the Council and the European Parliament; COM(2013) 716 final. Brussels: Europa.eu. 21 October.

Faiola, A. (2013) 'After Shooting Tragedies, Britain Went After Guns', *Washington Post*, 1 February.

Gardham, D. (2007) 'The Underworld Gun Trade', *Daily Telegraph*, 24 August.

Gibson, J.W. (1994) *Warrior Dreams: Violence and Manhood in Post-Vietnam America*. New York: Wang & Hill.

Gibson, K.A. (2012) 'Have Gun Will Travel: The Movement and Use of Firearms in England and Wales'. MA Thesis, UCL/NABIS.

GMP (Greater Manchester Police) (2006) 'Operation Xcalibre Intelligence Briefing', Powerpoint presentation. Manchester, GMP.

Greenwood, C. (2015) 'Chilling New Face of Police in Britain', *Daily Mail*, 15 November.

Hales, G., Lewis, C. and Silverstone, D. (2006) 'Gun Crime: The Market in and Use of Illegal Firearms'. Home Office Research Study 298. London, Home Office.

Hannam, A.G (2010) 'Trends in Converted Firearms in England & Wales as Identified by the National Firearms Forensic Intelligence Database (NFFID) Between September 2003 and September 2008', *Journal of Forensic Sciences*, 55(3): 756–766.

Harcourt, B.E. (2006) *Language of the Gun: Youth Crime and Public Policy*. Chicago: University of Chicago Press.

Heal, R. (1991) 'Laws That Shoot From The Hip', *Police Review*, 28 June.

HMIC (Her Majesty's Inspectorate of Constabulary) (2015) *Targeting the Risk: An Inspection of the Efficiency and Effectiveness of Firearms Licensing in Police Forces in England and Wales*. London: HMIC.

Karp, A. (1994) 'The Rise of Black and Gray markets', *The Annals of the American Academy of Political and Social Science*, 535: 175–189.

Kiernan, V. (1967) 'Foreign Mercenaries and Absolute Monarchy', in T. Aston (ed.) *Crisis in Europe: 1560-1660*. New York: Anchor Books.

Khomani, N. (2015) 'Police Begin Murder Investigation After North London Shooting', *The Guardian*, 11 July.

Law Commission (2015) *Firearms Law: A Scoping Consultation Paper*. Consultation Paper 224. London, The Stationery Office.

Lazarevic, J. (2010) 'South East European Surplus Arms: State Policies and Practice. Small Arms Survey', Issue Brief No. 1, November.

McCarthy, M. (2011) 'Researching Australian Gun Ownership: Respondents Never Lie … Or Do They?' In Ashwin (ed.) The proceedings of the 10th European Conference on Research Methodology for Business and Management studies. Caen, Normandy Business School.

Matthews, R. (2002) *Armed Robbery*. Cullompton: Willan Publishing.

Morrison, S. and O'Donnell, I. (1997) 'Armed and Dangerous? The Use of Firearms in Robbery', *Howard Journal of Criminal Justice*, 36 (3).

MPA (Metropolitan Police Authority) (2004) 'Gun Crime Scrutiny: Final Report'. London, MPA February.

Muggah, H.C.R. (2001) 'Globalisation and Insecurity: The Direct and Indirect Effects of Small Arms Availability', *University of Sussex IDS Bulletin*, 32 (2).

NABIS (National Ballistics Intelligence Service) (2011) Operational debrief: Olympic .380 BBM programme of activity. Online at: www.nabis.police.uk.

O'Neill, S. (2008) 'Baikal: The Gangsters' Gun', *The Times*, 21 July.

Overton, I. (2015) *Gun, Baby, Gun: A Bloody Journey into the World of the Gun*. London: Canongate Books.

Randall, D. (2008) 'Two Bodies found at Osbaston House', *The Independent*, 31 August.

Reedy, D.C. and Soper, C.S. (2003) 'Impact of Handgun Types On Gun Assault Outcomes: A Comparison of Gun Assaults Involving Semiautomatic Pistols and Revolvers', *Injury Prevention*, 9(2): 151–55.

Roberts, A. (2013) 'Man Jailed After Weapons Stash Found in Crawley Garage', *Brighton Argus*, 8 October.

Robinson, W. (2014) 'Why Did Parish Council Chairman Have a Huge Stash of Weapons Hidden In His Suffolk Garden?' *Daily Mail*, 22 April.

Scharf, P. and Binder, A. (1983) *The Badge and the Bullet: Police Use of Deadly Force*. New York: Praeger Publishers.

Smith, S. (2006) 'Theorising Gun Control: The Development of Regulation and Shooting Sports in Britain', *Sociological Review*, 54 (4): 717–733.

Spapens, T. (2007) 'Trafficking in Illicit Firearms for Criminal Purposes within the European Union', *European Journal of Crime, Criminal Law and Criminal Justice*, 15(3–4): 359–381.

Spillet, R. (2014) 'Terminally Ill Parish Council Chairman in Court After 'Cache of Guns' Including an Uzi and an AK-47 are Seized at his Home', *Daily Mail*, 22 April.

Springwood, C.F. (2007) 'The Social Life of Guns: An Introduction', in Springwood (ed.) *Open Fire: Understanding Global Gun Cultures*. New York: Berg.

Squires, P. (2000) *Gun Culture or Gun Control? Firearms Violence and Society*. London: Routledge.

Squires, P. (2014a) *Gun Crime in Global Contexts*. London: Routledge.

Squires, P. (2014b) 'The Unacceptable (?) Face of Elite Gun Culture', *Criminal Justice Matters*, 96 (1): 20–21. www.tandfonline.com/doi/full/10.1080/09627251.2014.926065#.U4TCtdJdW8A.

Squires, P. (2015) 'Paris Attacks: Terrorism, Trafficking – and the Enduring Curse of the AK-47', *The Conversation*: 16th November. https://theconversation.com/paris-attacks-terrorism-trafficking-and-the-enduring-curse-of-the-ak-47-50733.

Squires, P. and Kennison, P. (2010) *Shooting to Kill: Policing, Firearms and Armed Response*. Oxford: Wiley/ Blackwell.

Summers, C. (2011) 'Travels of a London Gun: How the Gun that Killed Agnes Had Been Traded by Gangs', BBC News website, 12 April.

Taylor, I. and Hornsby, R. (2000) *Replica Firearms: A New Frontier in the Gun Market*. University of Durham and Gun Control Network.

Tendler, S. (1994) 'Condon Calls for New Laws to Curb Gun Culture', *The Times*, 2 August.

Tonso, W. R. (1982) *Gun and Society: The Social and Existential roots of the American Attachment to Firearms*. Lanham Md: University Press of America.

Townsend, J. (2007) 'The Gun Lords' Deadly Legacy', *The Observer*, 9 December.

Twomey, J. (2015) 'Chilling Gun Haul is Found on a Boat in Kent: Largest Ever Arsenal of Weapons Seized', *Daily Express*, 15 August.

US Department of Justice (2011) 'Firearms Commerce in the USA'. www.atf.gov/files/publications/ firearms/121611-firearms-commerce-2011.pdf.

van Kesteren, J.N. (2014) 'Revisiting the Gun Ownership and Violence Link: A Multi-level Analysis of Victim Survey Data' H. Williamson 2015 Criminal armourers and illegal firearm supply in England and Wales. British Society for Criminology Conference Proceedings 2015. http://britsoccrim.org/ volume15/pbcc_2015_williamson.pdf, *British Journal of Criminology*, 54 (1): 53–72.

Warlow, T.A. (2007) 'The Criminal Use of Improvised and Re-Activated Firearms in Great Britain and Northern Ireland', *Science & Justice*, November, 47(3): 111–19.

Wheal, H. and Tilley, N. (2009) 'Imitation Gun Law: An Assessment', *The Howard Journal of Criminal Justice*, 48 (2): 172–83.

Williamson, H. (2015) 'Criminal Armourers and Illegal Firearm Supply in the UK', British Society for Criminology Conference Proceedings, Plymouth 2015. BSC website, available at: http://britsoccrim. org/volume15/pbcc_2015_williamson.pdf.

Wintemute, G. J. (1996) 'The Relationship Between Firearm Design and Firearm Violence: Handguns in the 1990s', *Journal of the American Medical Association (JAMA)*, 275(22): 1749–1753.

Yardley, E., Wilson, D. and Lynes, A. (2014) 'A Taxonomy of Male British Family Annihilators, 1980– 2012', *The Howard Journal of Criminal Justice*, 53(2): 117–140.

16

Crime, transport and technology

Andrew Newton

Introduction

This chapter aims to examine how changes in transport technology have influenced and altered the landscape for crime, criminal opportunity and the criminal justice system. The original intention at the outset of this chapter was to examine developments in transport and technology separately, and then to discuss a combined definition of what transport technology is. However, in doing this it quickly became apparent that this was a rather futile approach. The traditional definitions of technology were centred on the study of arts and crafts, but soon evolved to include an emphasis on purposeful invention and the strategic deployment of such invention (Rip and Kemp 1998). In its narrowest sense, technology can be thought of as a set of tools. However, more modern definitions of technology encompass a notion of something that works, thus often incorporates systems rather than just tools, and, therefore, in its widest sense, technology can also include skills and infrastructure. When considering advances in transport, from the development of the wheel, of boats and horse-drawn carriages, from the first to more modern motorised vehicles, or considering other forms of travel such as bicycles and motorcycles, submarines, hovercraft, aeroplanes, and even spacecraft – it becomes apparent that disentangling developments in transport from developments in technology is rather difficult. When adopting the systems view of technology, this is particularly evident. Indeed, within the transport literature frequent reference is given to the notion of a 'transport system'. Examples include public transport systems, the growth of Intelligent Transport Systems (ITS), automated and smart transport systems, travel demand and forecasting systems, and fuel efficiency systems. Therefore, by virtue of the way modern transport has evolved, there is a defensible argument for considering transport systems and transport technologies as interchangeable terms.

There is not scope within a chapter such as this to consider all the key milestones and changes to transport technology that have occurred and then identify how these have impacted on crime opportunity, even when restricting this to more modern times. The exponential growth of transport technologies within land transport, water transport, rail transport, air transport and spaceflights, could each be written as individual monographs. Instead, this chapter will aim to do the following. It will firstly examine what insights theories of transport technology can offer for considering changes to criminal opportunity. As part of this, the drivers for transport are considered, and, therefore although indirectly, what underpins advances in transport technology. Following on from an examination of transport technology theories and drivers is a discussion of the constraints placed on transport technology development, such as the physical infrastructure, policy and legislation, and societal acceptance and structure, and how each of these might impact on changes to criminal opportunity. This discussion sets the scene for the key question to be addressed in this chapter; in what ways do transport technologies influence crime opportunities? Five mechanisms are identified for this and these are: transport technology dependent crimes; transport technology as an enabler of crime; transport technology as an enhancer of crime; transport technology as a preventer of crime; and, transport technology as an influence on perceptions of crime. Each of these is discussed in detail. The chapter concludes by examining potential future changes to transport technology and criminal opportunity.

A central theme examined throughout this chapter is how have changes to transport technology impacted on the landscape within which offenders commit crimes? The reader is encouraged throughout this chapter to therefore think of the following: who commits a crime; what type of crime do they commit; when do they commit it; where is it committed; and, how and why is it committed? The key area for scrutiny is whether advances in transport technology have changed the answers to any of these questions.

Theoretical considerations

Whilst a range of theoretical insights could be drawn upon in this chapter, a useful starting point is to consider McLuhan's (1966) notion of how technology is an 'extension' of the body. As a basic example, the wheel can be considered as an extension to the foot. Rothenberg (1993) breaks this down into two types of extensions, those of action, and those of extensions of thought. Three further types of extensions of action relevant to transport are identified. The first is hand-driven tools that are direct extensions of the manual actions of the body, and a useful transport example is the bicycle. The second are motorised or piloted vehicles directly controlled by humans, which could include buses, trains, cars, and planes for example. The third are separate machines – generally relatively fixed structures such as roads and lighting that extend our 'restless need for movement'. Traffic management systems could also be considered as part of this grouping. It is debatable whether motorised but unmanned transport vehicles such as drones would fit into the second or third category here, or in the future where driverless cars would fit, or indeed if technology now demands a fourth category. Whilst better discussion of this is provided elsewhere in the literature (see Philip Brey, Chapter 1 of this volume) transport technology can be considered an extension to the body in terms of geographical distance covered and speed of travel. The question posed here is how this may alter the landscape for criminal opportunity.

A second key principle to consider is Harvey's (1989) notion of space-time compression, that is to say the idea of transport as a compressor of distance. This is concerned with the notion of a shrinking world, often measured in terms of reductions in the time or cost (or both) of travel. Transport technology has a crucial role to play here, although on a wider scale this idea of space-time compression was conceived to relate to wider society including: capitalism and changes to the economy brought about by a reduced time for turnover of capital; the growth of transnational companies and global cities; the international flow of pollution; and socio-technological changes to the structure of society (Giddens 1984). There is a vast literature available on this subject (Warf 2008; Oke 2009) and perhaps the key message is how transport technology has compressed space and time. McGuire (2012) examined the impact of communication and transport technology on space-time compression and how this might influence crime. The difference identified between the two was that transport technology involves 'multi-range' extensions, whilst communication technologies offer 'distance extensions'. That is, whilst communications extend interaction at a distance, or remotely as it were, transport technologies do the same, especially where they become faster and more efficient, but with the qualification that the body remains present and integral to the interaction. Again the question this raises is how this may result in changes to criminal opportunity. For transport systems, transport technology extends the opportunities for crimes committed in both physical and cyber space, whereas communication technologies only those committed in cyber space.

Combining both these ideas, of transport technology as an extension to the body, and, as a mechanism for space-time compression, and considering the view of those tasked with maintaining safe and secure transport systems – there is a strong justification for considering transport technology as a double-edged sword. The speed of change brought about by rapid developments in transport, combined with exponential advances of technology, result in a rapidly changing, dynamic and evolving landscape for transport-related crime opportunities. There are a number of theories for technological advancement relevant to transport technology and changes in crime opportunity, including: technology life cycles; economic path dependency; social construction of technology; market replacement approaches; evolutionary economics; and long-wave theory (Elzen *et al.* 2004). However, there is not scope here to consider these in detail. Instead, this chapter focusses on the actual drivers for transport itself, and, by proxy the drivers of transport technology. It is argued that it is these drivers for change that impact most significantly on criminal opportunity.

Transport technology and drivers of change

Transport is predominantly and in its simplest terms the outcome of derived demand for travel (Rodrigue *et al.* 2013) and therefore transport technology is generally driven by the need to improve the delivery of a transport service to meet this demand. Examples include making transport quicker, or more cost effective, efficient, convenient, safe, or reliable. When considering transport technology it is acknowledged that changes have and will continue to vary by transport mode, for example car, rail and bus. Indeed, there are some unique crime opportunities across transport modes. Crimes such as trespass on the railway, and offences on board an aircraft have their own specific legal definition. However, perhaps a more useful split is to consider public transport and private transport separately. The reason for this distinction is that the drivers for each of these are slightly different, thus technological change will occur via different mechanisms. It is argued that it is these drivers which will influence crime opportunity most significantly, and, that the differences between the drivers of public and private transport are much greater than those of differing transport modes. For the purposes of this chapter, public transport, and in particular mass rapid transit in urban areas, is taken to include those aimed primarily at the service sector (for a fuller discussion of definitions of public transport see Ceccato and Newton 2015), whereas private transport is taken to include those driven by industry, including private cars, commercial operations, and the movement of freight and goods.

Across the service sector innovation often relates to infrastructure (bus priority, new light rail systems); vehicles (environmentally friendly engines, low floor access, changes in size of vehicle; and service operation [fares, timetables, frequency of service, ticketing and marketing]). Across the private sector there are a range of industrial and socio-technical drivers (Geels 2005) such as: changes to the road infrastructure; to vehicle manufacturers and suppliers; in market forces, driver preferences and mobility patterns; in maintenance and distribution networks (for example repair shops and dealerships); in the fuel infrastructure; in regulations and policies (parking fees, traffic regulations and enforcement, and road tax). When considering some of the more classical theories of technological change it is useful to draw on Schumpeter's (1939) ideas of invention, innovation and diffusion, as three components of technological advancement. Invention can be said to refer to the creation of a new concept or idea; innovation follows when this idea is developed into a new product and commercially transferred, and diffusion is the spreading out of this new product into existing or new markets. An interesting consideration here is how each stage of this process, invention, innovation, and diffusion, may relate to the development of what are termed 'crime waves' (Laycock 2005). These occur when early warning signs of new emerging crime trends, likely in the innovation phase and towards the start of diffusion are ignored or missed, resulting in substantial crime increases (crime waves), likely during the middle to latter stages of diffusion. This is discussed further below using the example of vehicle crime.

The speed and size of technological change is likely to vary considerably between private and public transport. Indeed, it is suggested the majority of invention will occur within the private sector, although transport technology may also be the result of spin-off activity from military research and development. However, even when this is the case, early and greater adoption is likely to be driven first by the private sector. Innovation is also likely to be stronger in the private sector, particularly with greater levels of investment. Diffusion is generally at a more rapid pace and on a wider geographical scale on private transport systems. Indeed, changes to the public sector transport provision are likely to occur with a delay or lag compared to private transport, and diffusion is likely on a smaller scale.

The key questions here to consider regarding crime opportunity are: how these private and public transport sector drivers may alter the criminal landscape; how quickly this landscape may change; what is the scale and extent at which this landscape may alter; and, as is often the case with technological innovation, what is the lag between new technology, new crime opportunity, and new design prevention solutions? Indeed, crime design is often an afterthought of new technology, and rarely built in at the outset of invention and or innovation of new technology (Paul Ekblom, Chapter 20, this volume) and here transport technology is no exception. To demonstrate this, two contrasting examples relating to the speed and scale of innovation between public and private sector transport are now considered; changes to car vehicle security, and automated ticket machines and the use of slugs (see below) on the London Underground.

In the UK, there was a considerable increase in the number of motor vehicle thefts that occurred from 1980 to 1990 (Morgan *et al.* 2016). This increase was considered as a second wave of car crime. From 1990 onwards, this trend reversed and there was a sustained reduction in the number of vehicle crimes. Note these trends have also been found internationally, particularly in the USA and Australia. Whilst several theoretical explanations of this crime drop can be found in the literature, perhaps the most reliable measure is provided by the security hypothesis (Farrell, Tseloni and Tilley 2011; Farrell, Tseloni, Mailley and Tilley 2011). Whilst there is not scope to review this work in detail, it is worth noting three technological security measures identified here as relevant to this crime drop. These include the introduction of central locking, car alarms, and immobilisers (mechanical and electronic). The growth in persons owning cars, and reduced costs in terms of affordability, can be seen as a key factor in the increased number of motor vehicle crimes. This is a clear if rather simple example of how a change to transport technology, perhaps viewed best as diffusion in the availability of the private car from 1950 onwards, resulted in widespread increases in crime opportunity. However, changes to car security through technology occurred at a much later stage. As is often the case, security design was an afterthought.

The research on the vehicle crime drop by Farrell and colleagues suggested that immobilisers were most likely to reduce theft of cars (but have less impact on theft from cars), alarms would influence theft from cars (alarms do not increase difficulty in driving cars away), and that central locking may affect both, but mainly change modus operandi (MO) as cars can be entered in other ways. What was evident in this research was that immobilisers did have the greatest impact on reducing vehicle thefts, and these reductions tended to occur as the prevalence of this security device increased. Thus as more cars had this technology fitted as standard, fewer were stolen. Rates of decline in Australia happened much later than in the UK and the USA, but this is related to the delay in the prevalence and standardisation of these security devices here. Moreover, as this transport technology was driven by the private sector, the subsequent crime wave that developed was widespread, affecting large geographic areas, with increasing changes to crime opportunities, both in terms of the diffusion of new technology as a new target for crime (more people having private cars), and then a subsequent but lagged reduction in crime as security of this transport technology was improved (immobilisers, alarms and central locking).

In contrast the London Underground presents an example of transport technology increasing crime and a new crime target on public transport, the case of using slugs for fare evasion (Clarke *et al.* 1994). In the 1980s the London Underground introduced automated ticket vending machines. However, they introduced a new criminal opportunity for fare evasion, and theft. The slugs were made by simply wrapping a 10p coin inside foil in the shape of a 50p piece. Not only did this allow the user to travel for a significantly lower price (technically fare evasion/ fraud), it also was possible that by inserting a slug and then pressing the reject button, an actual 50p coin was ejected by the machine (theft). The use of these slugs greatly increased from 1987 and became more widespread, and in 1991 a technological change was made and machines were modified to reject the 50p slug (at great expense). However, this then resulted in the appearance of a new £1 slug. Again a number of stages can be identified here, including invention of automated ticket machines, innovation of ticket vending machines to be used on the London Underground, and diffusion across the entire Underground system. As an interesting parallel, the crime can also be classed under the invention, innovation and discussion umbrella. This occurred at a lag or delay from the transport technology, from the invention of slugs, to the innovation of the slugs (from using for travel and fare evasion, to using for theft), to the diffusion as usage became more widespread across the transport system. The speed and scale of change in the diffusion of the crime occurred at a much more rapid pace than that of the installation of the ticket machines. Indeed, the security measure, changing the ticket machine, was an afterthought to the development of the technology. However, in this example the pace, scale and extent of the crime wave was less widespread than the first discussion of the growth of vehicle crime. This is primarily due to the localised nature of the public transport system, compared with the industry driven advances of the private transport system (car). Both these examples are cases were the transport technology and transport system were the direct targets of the crime (transport dependent crimes).

Some additional components of transport systems to consider that also influence crime opportunities are to compare the movement of people with transport of goods; the different types of land use associated with transport systems, for example roads and rail tracks, stations and interchanges, sea and air ports, parking facilities, and those with mixed facilities such as combined retail and transport; by transportation mode including bus, rail, tram, plane, boat and ferry, car, and bicycle; the temporal components of transport journeys such as by weekday and weekends, or peak and off-peak times; and to consider individual travellers' needs. Journey needs vary, for example, between those of the elderly, young persons, those with disabilities, tourists, commuters, movement of goods, movement of fuel and other supply services, and those travelling for leisure and recreation. Whilst this chapter does not examine each of these in detail, elements of each of these are apparent in the five ways transport technology influences crime opportunity as discussed later in this chapter.

Transport constraints

Both the private and public sector transport systems require two further functions, transportation nodes and transportation networks. These may be constrained by the physical environment or urban places (for land and sea), and by the socio-spatial structures of places. Advances in transport technology will also be influenced not only by technological development (as discussed above), but also societal development, and both of these should be taken together as they are symbiotic. Thus society's acceptance and use of transport technology, and the constraints of the physical infrastructure will also impact on the extent to which transport technology is adopted, and therefore the extent to which it may create new opportunities for criminal activity. More recently in urban centres, there is now the twin pressure of reduced land space for development and increased travel congestion over an ever growing peak travel period (Wilson 1997). This has implications for transport technology, travel demand, and also may impact on the landscape for criminal opportunity. As more and more places are busy, does this increase the chances of pick-pocketing at large urban transport interchanges at rush hour, and what is the impact for the security of roads and goods during rush hour, or indeed how will this impact on future air travel?

Optimisation of transport, through technological change, is therefore driven by demand for service, but this must be seen as embedded within sociological processes, legal frameworks, and physical environments. Transport technology can be useful for maximising capacity, optimising operations, and improving safety and confidence in travel, but – and this is highly relevant – these may not always align and may even conflict. Several examples of changing transport technologies that could impact on crime opportunity can be identified. A non-exhaustive list to demonstrate the diverse and widespread nature of this includes: access control, surveillance and monitoring; physical design; operational deployment of staff; automated and smart transport systems; communication and information; risk assessment; environmental improvements; the growth of intelligent transport systems; holistic transport planning and management; travel demand modelling; transport journey planning; traffic management; automation; increasing fuel efficiencies; smartcards and passes; journey planners; real time information systems; and road traffic management. Social media and transport technology may also increase mobility and the speed of mass gatherings. To make some sense of this diverse list, this chapter attempts to classify transport technology by the mechanisms through which it may influence crime opportunity and as a starting point five classifications are suggested here. This is not a definitive list, but the author is not aware of other work that previously attempts this. Each will now be explored in more detail.

Transport technology and crime opportunity

In order to devise these five categories of how transport technology may influence crime opportunity, this work draws on two previous findings. The first relates to studies into the development of cybercrime, and this work heavily borrows from the concepts of cyber-crimes as being cyber-dependent – they can only occur with Information Communication Technology (ICT) and computer technology – and those which are cyber-enabled, whereby the speed and scope of criminal activity is increased through the use of ICT (McGuire and Dowling 2013). These ideas were identified as relevant to public transport during a panel discussion at the International Crime Science Conference, in 2014 at a roundtable session[1] which considered crimes that were transport dependent and transport enabled. These key ideas have been developed into the five new groupings for the purposes of this chapter, to try and make some sense of the ways transport technology may influence criminal opportunity.

Transport as a target of crime (transport dependent crimes)

The first set of crimes identified as influenced by transport technology are those where transport systems are the target of crime, and perhaps most importantly, ones that could not have occurred without the transport system. It is the transport system itself that creates these new opportunities for crime. For rail, sea and air transport there are legal definitions of crimes that all fit into this category.[2] These include:

1   Rail specific crimes:
    • railway trespass
    • damaging trains and endangering the safety of rail users including: criminal damage; throwing missiles at rolling stock or static railway equipment; offences against the person on railways, intent to endanger the safety of any person travelling on the railway, or any unlawful act or wilful neglect endangering public safety; malicious damage including the placing of wood, etc, on a railway, taking up rails, turning points, or showing or hiding signals, an intent to obstruct, upset, overthrow, injure or destroy any engine, tender, carriage or truck; obstructing engines, or carriages, or railways
    • intoxication of employees
    • fare evasion
    • assault on transport staff.
2   Criminal conduct at sea and in the air.
3   Offences on-board aircraft including:
    • hijacking
    • damaging or endangering the safety of aircraft
    • dangerous articles on aircraft and in aerodromes
    • offences relating to security at aerodromes and on aircraft
    • drunkenness
    • aerodrome trespass.
4   Vehicle offences:
    • aggravated vehicle taking
    • theft from a motor vehicle
    • theft or unauthorised taking of a motor vehicle
    • interfering with a motor vehicle.

It can be argued that some of the above railway and sea and air offences such as criminal damage, assault against persons, and drunkenness do not neatly fit within the 'transport dependent category' as they could occur outside of transport systems. Indeed, many of these activities occurred before the advent of transport and continue to occur outside of this arena. However, they have been included here as they can be prosecuted[2] under the Regulation of Railways Acts 1840–1873; the British Transport Commission Act 1949; the Railways and Transport Safety Act 2003; the Aviation Security Act 1982; the Civil Aviation Act 1982; the Aviation and Maritime Security Act 1990 and the Air Navigation Order 2005. It is important to note that criminal damage could be prosecuted under the Criminal Damage Act 1971; or the Offences against the Person Act 1861 – if the condition 'intent to injure or endanger the safety of persons on railways' is absent. In such situations, these crimes may fall better under the second category identified below, as transport technology enablers.

The important issue for criminal justice is that the advent of these transport technologies (in the 1800s for the railway) required new laws to be written under which offences could be prosecuted. It is argued that it is this feature that makes these offences transport dependent as without such a law and a transport target, these crimes could not occur. It is the speed and extent of innovation and diffusion of these products onto the market, often in the absence of strong security measures (as was the case with the growth in vehicle theft and the subsequent crime waves that ensued) that governs the extent to which new crime opportunities are present. When considering new technologies today and their rapid evolution, it is often the criminal justice system that struggles to keep pace with the speed of change. This is discussed further at the end of this chapter. Indeed, if there is no law in place to govern this, then some of these offences may be difficult to prosecute.

Interestingly bus and tram systems unlike air, sea and rail systems do not have transport specific crime acts. Offences on these systems are subject to general laws as would be applied outside of the bus and tram system. Therefore, any of the above crimes that occurred on a bus or tram network including assault of passengers and staff would be considered under transport technology enabled crime (and not transport dependent as is the case on the rail when endangering the safety of rail users).

The next two categories identified for transport technology as a mechanism for changing crime opportunities are as an enabler of crime and as an enhancer of crime. For the purpose of this chapter transport technology enablers are those systems that extend current or traditional forms of crime (that can occur outside of transport) onto the transport network. Transport technology enhancers also may extend traditional crimes onto the transport network, but are considered to increase the speed or extent of these crimes, or change the MO, or provide new tools to facilitate the crime. This is slightly different to McGuire's definition of cyber-enabled crime (McGuire 2012; McGuire and Dowling 2013) which suggests cyber-enabled crimes increase the scale and reach of offending through the use of ICT. Here, crime opportunities created by transport technology are separated into those that extend the arena where crimes are carried out (an enabler), and those that increase the scale and reach or provide new tools with which to do this (an enhancer). When considering the use of vehicles for ram raids or as getaway cars, the vehicle is a tool used to facilitate the crime, part of the method and or pathway to and from crime (an enhancer); whereas pick-pocketing on the transport network is a traditional crime extended by a new place to commit it, a new transport station (an enabler).

Transport technology as an enabler of crime

As stated above these can be considered as an extension of traditional or existing crimes onto the transport network, which are effectively a new arena as a result of developments in transport technology. The transport system provides a new setting to carry out traditional crimes and thus unlike transport dependent crimes do not rely on transport systems to occur. Some examples of these include:

- assault of passengers and staff (not covered under transport specific legislation)
- criminal damage (including arson and graffiti)
- disorder
- robbery
- theft from person, including pick-pocketing at stations and on vehicles (for example of mobile phones)
- violence against the person
- sexual offences.

There are some important theoretical concepts as to how crime opportunity manifests as relevant to transport. These include routine activities theory (Cohen and Felson 1979) and crime pattern theory (Brantingham and Brantingham 1993). A fuller discussion of how these apply to transport systems is provided by Newton (2014) and Newton and Ceccato (2015). In brief, transport networks rely on key nodes (interchanges and stations), and the routes between these. They are also constrained by the extent of the transportation network and infrastructure. These are neatly represented in crime pattern theory as nodes, paths, and edges, and it is known criminal opportunity often occurs near to key activity nodes. These nodes are governed by notions of people's routine activities, for work, leisure, shopping, and recreation for example, and it is at and near to these nodes (activity spaces) where crime is more likely, where offenders and victims come together in the absence of capable guardians. The question this poses is how transport technology may alter the settings of transportation nodes or transportation paths, and what possible new crime opportunities may arise as a result of these changes. In the case of transport enabled crimes, it is these transport systems themselves that become a new setting to carry out traditional forms of criminal activity.

A further useful perspective to consider here are crime generators and crime attractors (Brantingham and Brantingham 1995). Crime generators are places where crime opportunity arises due to the presence of large volumes of persons being present, but are not pre-planned. Crime attractors are places offenders visit with known opportunities for crime, indeed such places often have a reputation for crime to occur. On transport systems both are present but the drivers of transport technology which have created these situations are rarely considered. For example, the demands of modern rapid urban transit systems have resulted in very large interchanges, with multi-modal transport systems, and indeed retail and even leisure all combined within a single location. Clearly this may generate several new crime opportunities for offending that arise as part of persons' everyday routines but are not necessarily pre-planned. Transport systems are also attractive to potential pick-pockets. At peak times platforms are crowded, passengers are tired, and there is an acceptance of jostling and bumping. This and a range of additional factors (Newton, Partridge and Gill 2014a, 2014b) make transport stops and stations attractive to pick-pockets. Thus new transport systems enable pick-pocketing to occur in a new arena.

Mobile technology may also enhance pick-pocketing, and when considering theft of mobile phones it is difficult to distinguish between this as a transport technology enabler or enhancer of crime. As Wi-Fi increases on transport systems, more passengers use their phones, but in so doing they are not concealed, and potential victims readily use these devices on transport systems connected by Wi-Fi. When there is no signal or no Wi-Fi fewer users would hold their devices out in the open. These devices are then on display to would-be offenders, who can simply identify the model of phone they wish to acquire, and then pursue targets who carry the desired phone model. When satellite navigation systems were introduced to vehicles, many were left on display in cars and became a target for theft. Whilst both offences can occur outside of transport systems (they can be stolen elsewhere) and are thus not transport dependent, it is debatable if the transport technology enabled traditional crime in a new setting, or enhanced the opportunities for offenders to commit this crime, or both. This demonstrates the difficulties in separating transport technology enablers of crime and enhancers of crime, thus it could be argued these should be merged as a single grouping. However, at present for this chapter they have been left as separate classifications.

Transport technology as an enhancer of crime

Whilst it may be difficult to always draw a distinction, for this chapter a transport technology enhancer of crime is considered a situation whereby transport technology has increased the scale and extent of crime, or has advanced the tools to facilitate crime (for example altering the MO or assisting in the methods), or is part of the pathway to and from a crime. Here transport technology assists or increases the ability of criminals to achieve their goals. Examples of such crimes may include:

- The use of vehicles to commit theft (ram raids of ATM machines/cash points, shopfronts and banks).
- Where transport is part of the pathway to and from crime (using a car as a getaway from a bank robbery or other offence).
- Theft of personal data using transport systems (for example Wi-Fi on transport networks).
- Sexual harassment, cyber stalking and abuse (using the transport network to target individuals; this may include taking pictures on mobile phones, for example photos of a sexual nature, such as pictures looking up females' skirts at their underwear, taken when victims are unaware this is happening during crowded journeys).
- Use of false travel documents.
- Theft of personal data, Oyster cards, theft of data from electronic mobile pay and travel devices.
- The use of transport as part of trafficking, smuggling, child sexual exploitation and even modern slavery.
- Use of electronic jammers to thwart tracking devices on stolen vehicles.
- The use of unmanned drones to commit criminal offences.

A difference between these transport technology enhancers of crime and the previous enablers is that the scale of change is likely to be quicker and more widespread for crime enhancers, thus the criminal legislation may struggle to keep apace of this change. This puts pressure on the criminal justice system to legislate against what could be described as entrepreneurial offenders using transport technology to increase their criminal activity.

However, there are limitations to using these categories for transport technology, that is, crime dependent, crime enabled, and crime enhanced. Consider the case of vehicle hijackings ('carjackings') that have occurred in São Paulo, and other major cities in Brazil. Transport technology, namely pedestrian traffic lights are used by offenders to create new crime opportunities. Pedestrianised lights are turned to red when cars approached, and cars are taken at gunpoint. Indeed, it is common and even encouraged that car drivers slow after dark when approaching lights in particular areas of São Paulo, but pass through without stopping if no persons are crossing. These pedestrianised lights settings can be considered as crime attractors. The target of crime is the vehicle, so this crime could be considered as transport technology dependent. However, the MO is to use the lights, which is cyber-enabled. It should be remembered the offender will not be restricted by these definitions and is likely to use all means at their disposal to be successful. To complicate matters, if the target was a person inside the car, and the crime then became a kidnapping, this could be classed as a transport technology enhanced crime. The question for the criminal justice system is, irrespective of whether it is transport technology dependent, enabled, or enhanced, can it be prosecuted under current legislation (likely with enabled crimes); does new legislation need to be written or adapted (more likely with transport dependent and enhanced crimes); and can the criminal justice system keep pace with the rapid changes in the technology?

Transport technology as a preventer of crime

In addition to transport technology increasing crime opportunities, there are several examples of how it can reduce such opportunity. The use of immobilisers on vehicles has already been discussed, and cashless ticket machines reduce the risk of robbery for bus, train, and tram drivers. Several other examples here can be identified, including:

- Closed-circuit television (CCTV) being wirelessly transmitted from moving vehicles to control rooms.
- Monitors on-board vehicles displaying CCTV images so passengers and potential offenders are aware the systems are working and they are being monitored.
- The use of technology to analyse crime and mobile data to assist front line staff.
- Communication and information dissemination.
- DNA and smart water technology.
- Automatic Number Plate Recognition (ANPR) camera technology to enforce bus lane and traffic offences.
- Use of Oyster and cashless travel to avoid theft of cash from drivers.
- Use of text messages to encourage anonymous and live reporting of crime and disorder.
- Use of breathalysers built into cars to prevent drink drivers.
- Scanning and access control systems.
- Use of forensic evidence in investigation.
- Facial recognition technology, explosive device detection, and even unusual behaviour/suspicious package identification (when bags left on platform, passengers remain on platforms and don't board).

There are two important factors that should be addressed here. The first is there is a need to balance the use of technology for crime prevention with that of the presence of staff for reassurance. It is known that technology does not always increase perceptions of safety as is discussed in the next part of this chapter. Additionally, many of these transport technologies for crime prevention occur after a crime wave has occurred, as a response to a growing crime trend or problem. It would be preferable for these technologies to include crime prevention as part of the invention/design phase, or at least the innovation phase, as opposed to part of the diffusion process as transport technology spreads (and often increases crime opportunity). However, there are several difficulties in achieving this as summarised usefully by Ekblom (2014) including: aesthetics; legal and ethical issues; environmental considerations; safety; cost; convenience; and a general lack of horizon scanning/awareness of potential crime opportunities that may emerge from new product design.

Transport technology and perceptions of crime

Whilst transport technology may increase crime opportunity, or even be used to prevent crime, it is important to consider that it can also have a substantial impact on passengers' perceptions of travel and fear of crime. For example, surveys have shown reliability and convenience are two of the key barriers to the uptake of public transport. Personal security or fear of safety from crime is the next biggest obstacle (Ceccato and Newton 2015). Transport technology is a key mechanism for advancing transport journeys, for example by increasing speed, reliability, safety, and convenience. There are a number of opportunities for such technology to increase perceptions of safety, after all, if persons do not feel safe they may choose not to travel, even if the transport system has very low levels of crime. This issue sits within the socio-technological constraints of transport technology development, particularly focussing on societal acceptance. Examples where transport technology can be used to increase perceptions of safety include:

- real time passenger information
- help points and passenger reassurance messages
- secure by design and movement control
- on-board live streaming of CCTV on moving public transport vehicles
- advanced information on traffic and weather conditions on roads.

However, it is critical that transport technology here is not used to replace people, for example CCTV should not be viewed as an alternative to the presence of staff on public transport systems. Indeed, the physical presence of staff at stations has been shown in several surveys to be the most likely to reassure travellers. This is more effective for reassurance than CCTV, environmental improvements such as better line of sight and good lighting, and other technological innovations (Ceccato and Newton 2015).

Conclusions and future direction

Whilst this chapter has been critical of the lack of thought into security and crime prevention placed in the invention and innovation of new transport technology products, it is also acknowledged that it is not a simple task to horizon scan and to identify possible future changes to transport technology and crime opportunity. Therefore, this is written with some apprehension as future changes and their possible impact on crime are in many ways unpredictable. However, this chapter will conclude by highlighting some possible future transport technology trends, and reflecting upon their potential impact on transport related crime.

Some transport technologies that are already beyond the invention stage and in the innovation or early phases of diffusion can be grouped within the wider lens of Intelligent Transport Systems (ITS). More specific examples include: driverless cars; autonomous and connected vehicles; electric vehicles that may be able to charge wirelessly; better travel data for road optimisation; and the growth of smart cities. Data will be a key commodity as part of this, and therefore its security should be a key prioritisation. If numbers of electric vehicles increase and can be charged wirelessly how can this technology be secured against possible misuse? Transport data, electric chargers, and vehicle batteries may all be valuable commodities to the future offender. Autonomous cars present a different challenge, as it will be possible for remote users to hack into these, or for such vehicles to be used to commit crimes with no driver, in effect anonymising the offender's identify. How will social media and mobile platforms evolve on transport systems and what are the possible crime opportunities from this? Thus, as transport becomes more intelligent and digitally driven, some elements of transport crime should perhaps be considered as part of a branch of cybercrime or indeed critical infrastructure protection. However, the extent to which this will increase (as a proportion of all transport crimes) is difficult to envisage. It is perceived transport dependent crimes will still form a large proportion of transport crime, particularly on public transit systems, as these environments are less well policed than other semi-private spaces, especially the bus and tram environments. The two greatest challenges however, are perhaps for the criminal justice system to keep pace with the changes in transport technology, and the lack of investment in building crime prevention design into the invention and innovation phases of transport technology developments.

Notes

1 Roundtable session organised by Reka Solymosi; Transport for London (TfL) and University College London (UCL): 'What is the most important current problem in transport crime?' This was held at the International Crime Science Conference, London, 16 July 2014. It was chaired by Mr Steve Burton (TfL) with panel discussants Dr Barak Ariel, University of Cambridge; Inspector Varley, Metropolitan Police, Operation Menas; Dr Vania Ceccato, KTH Royal Institute of Technology, Sweden, and Dr Andrew Newton, the University of Huddersfield.

2 Guidance available via the Crown Prosecution Service (CPS): Road Traffic Offences – Transport Offences.

References

Brantingham, P. and Brantingham, P. (1993) Environment, routine and situation: Toward a pattern theory of crime. In Clarke, R. and Felson, M. (eds) *Routine Activity and Rational Choice: Advances in Criminological Theory*. Vol. 5. New Jersey: Transaction Publishers, pp. 259–294.

Brantingham, P, and Brantingham, P. (1995) Criminality of place: Crime generators and crime attractors. *European Journal on Criminal Policy and Research* 3(3): 5–26.

Ceccato, V. and Newton, A. (2015) Aim, scope, conceptual framework and definitions. In Ceccato, V. and Newton, A. (eds) *Safety and Security in Transit Environments: An Interdisciplinary Approach*. London, UK: Palgrave Macmillan, pp. 3–22.

Clarke, R., Cody, R. and Natarajan, M. (1994) Subway slugs – tracking displacement on the London Underground. *British Journal of Criminology* 34(2): 122–138.

Cohen, L. and Felson, M. (1979) Social change and crime rate trends: A routine activity approach. *American Sociological Review* 44: 588–608.

Ekblom, P. (2014) Designing products against crime. In Bruinsma, G. and Weisburd, D. (eds) *Encyclopedia of Criminology and Criminal Justice*. New York: Springer, pp. 948–957.

Elzen, B., Geels, F.W. and Green, K. (eds) (2004) *System Innovation and the Transition to Sustainability: Theory, Evidence and Policy*. Cheltenham: Edward Elgar.

Farrell, G., Tseloni, A., Mailley, J. and Tilley, N. (2011) The crime drop and the security hypothesis. *Journal of Research in Crime and Delinquency* 48(2): 147–175.

Farrell, G., Tseloni, A. and Tilley, N. (2011) The effectiveness of car security devices and their role in the crime drop. *Criminology and Criminal Justice* 11(1): 21–35.

Geels, F.W. (2005) The dynamics of transitions in socio-technical systems: A multi-level analysis of the transition pathway from horse-drawn carriages to automobiles (1860–1930). *Technology Analysis & Strategic Management* 17(4): 445–476.

Giddens, A. (1984) *The Constitution of Society: Outline of the Theory of Structuration*. Berkeley: University of California Press.

Harvey, D. (1989) *The Condition of Postmodernity*. Cambridge, MA: Basil Blackwell.

Laycock, G. (2005) Defining crime science. In Smith, M. and Tilley, N. (eds) *Crime Science: New Approaches to Preventing and Detecting Crime*. Devon: Willan Publishing, pp. 3–24.

McGuire, M. (2012) *Technology, Crime and Justice. The Question Concerning Technomia*. Abingdon, UK: Routledge, Taylor & Francis.

McGuire, M. and Dowling, S. (2013) Cyber-crime. A review of the evidence: Research Report 75. London: Home Office.

McLuhan, M. (1966/1964) *Understanding Media: The Extensions of Man*. New York: McGraw-Hill. Paperback edition, 1966.

Morgan, N., Shaw, O., Feist, A. and Byron, C. (2016) Reducing criminal opportunity: Vehicle security and vehicle crime. Research Report 87: London: Home Office.

Newton, A. (2014) Crime on public transport. In *Encyclopedia of Criminology and Criminal Justice*. London: Springer, pp. 709–720.

Newton, A., Partridge, H. and Gill, A. (2014a) Above and below: Measuring crime risk in and around underground mass transit systems. *Crime Science* 3(1): 1–14.

Newton, A., Partridge, H. and Gill, A. (2014b) In and around: Identifying predictors of theft within and near to major mass underground transit systems. *Security Journal* 27 (2): 132–146.

Newton, A. and Ceccato, V. (2015) Theoretical perspectives of safety and security in transit environments. In Ceccato, V. and Newton, A. (eds) *Safety and Security in Transit Environments: An Interdisciplinary Approach.* London, UK: Palgrave Macmillan, pp. 23–39.

Oke, N. (2009) Globalizing time and space: Temporal and spatial considerations in discourses of globalization. *International Political Sociology* 3 (3): 310–326.

Rip, A. and Kemp, R. (1998) Technological change, in Rayner, S. and Malone, E. (eds) *Human Choice and Climate Change,* Vol 2, Resources and Technology, Washington, DC: Batelle Press, 327–399.

Rodrigue, J.P., Comtois, C. and Slack, B. (2013) *The Geography of Transport Systems.* New York: Routledge.

Rothenberg, D. (1993) *Hand's End: Technology and the Limits of Nature.* Berkeley: University of California Press.

Schumpeter, J.A. (1939) *Business Cycles: A Theoretical, Historical, and Statistical Analysis of the Capitalist Process.* New York: McGraw Hill.

Warf, B. (2008) *Time-Space Compression: Historical Geographies.* London: Routledge.

Wilson, A. (1997) Land use/transport interaction models – past and future. *Journal of Transport Economics and Policy* 32: 3–23.

17

Food fraud and food fraud detection technologies

Roy Fenoff and John Spink

Introduction

Food fraud is a serious crime that has occurred throughout history. However, governments and businesses have only recently begun to actively and proactively address this growing problem. Indeed, not only can food fraud incidents have devastating economic impacts, but they also pose considerable risks to public health, safety, and food security.[1] Globally, businesses and governments have shown an increased interest in preventing food fraud by coming together with other interested stakeholders to learn about and devise ways to combat the problem. For example, varieties of certifications, standards, and best practices have been, and are continuously being, developed. There is support for academic research, and incident databases have been made and are being updated. Academic programs such as Michigan State University's Food Fraud Initiative have been established and expanded upon to provide education and research support for issues related to food fraud. Additionally, current technologies are increasingly being employed by businesses and governments around the world in order to improve our ability to detect and deter food fraud incidents. Although an overview of the food fraud problem will be provided, this chapter is about food fraud detection technologies. This chapter will (1) introduce the food fraud concept, (2) provide an overview of the problem, and (3) conclude with a discussion and review of the current technologies that are being used to detect and deter food fraud.

What is food fraud?

Until 2011, the food fraud concept had not been defined in the academic literature. Spink and Moyer (2011: 158) defined food fraud as a collective term used to encompass the "Deliberate and intentional substitution, addition, tampering, or misrepresentation of food, food ingredients, or food packaging; or false or misleading statements made about a product, for economic gain." This definition includes the adulterant-substance subcategory, defined by the US Food and Drug Administration, of Economically Motivated Adulteration (EMA). Food fraud covers all types of fraud that is related to food, including adulteration, concealment, counterfeiting, dilution, diversion, enhancements, mislabeling, overrun, simulation, smuggling, species swapping, tampering, and theft.

Traditionally, food protection risks were categorized as food safety, food quality, or food defense concerns (see Figure 17.1). Food fraud is now considered a separate and autonomous aspect of food protection. In order to determine what category a food issue falls under, the motivation and action must be understood. Figure 17.1 illustrates the differences between these four food protection risk concepts. Food fraud and food defense are both intentional acts, but food fraud is always economically motivated and the offenders are usually legitimate actors in the food supply chain where they have regular access to the food products (CRS 2014). Conversely, with food defense, the act is ideologically driven, and the motivation is to cause harm. Food quality and food safety are both unintentional acts, but food quality is always economically motivated, whereas food safety is motivated by harm (Spink and Moyer 2011).

Food fraud is a global problem

Although food fraud has occurred throughout history, and can be traced back to antiquity (Purcell 1985; Wilson 2008), it now operates on a larger scale with more complex frauds due to technological advances and the adverse effects of globalization. Globalization has been accelerating since the 1990s, and people and products are now traveling further and faster around the world (Natarajan 2011). With more sophisticated communication and travel technology, the global food business has been growing. This growth is evidenced by the global trade in

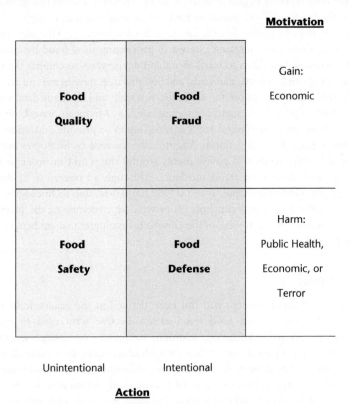

Figure 17.1   The food protection risk matrix

Source: Spink and Moyer (2011)

foodstuffs that has surpassed USD 1 trillion for the last five years and is estimated to be USD 1.13 trillion in 2015 (FAO 2015). Considering the enormous value of the global food industry, and the ease with which food products move around the world, it is not surprising that much of the food consumed in the United States is imported. For example, the National Oceanic and Atmospheric Administration (NOAA) reported that in 2007 more than 80 percent of the seafood consumed in the United States was imported (GAO 2009). It has also been estimated that up to 70 percent of the seafood on the US market is fraudulently labeled (Stiles *et al.* 2011). This is alarming because if 70 percent of the 4.8 billion pounds of seafood consumed in the US is fraudulently labeled, then 3.36 billion pounds of the seafood being consumed in the United States is misrepresented and potentially a public health threat (NOAA 2010). Due to the clandestine nature of product fraud – including challenges in detection and investigation – the true scope and scale of the food fraud problem "may be unknown or even possibly unknowable" (CRS 2014: 3), indeed there are some estimates that suggest the problem is considerable. The Grocery Manufacturers Association (GMA) estimated that adulterant-substance and counterfeit food products cost the food industry around $15 billion annually (GMA 2010), which affects roughly 10 percent of all commercially sold food products (Everstine *et al.* 2013). The GMA (2010) claimed that the cost of one fraud incident averaged between 2 and 15 percent of yearly revenues in 2010, which equated to $60–$400 million depending on the size of the company. Regardless of what the true size of the food fraud problem is, there is no question that the impact a food fraud incident can have on society is enormous, as the following melamine in milk case illustrates.

In 2008, the Chinese government announced that a large amount of milk products, including infant formula, was adulterated with the chemical melamine (Xiu and Klein 2010). Ultimately, this resulted in 300,000 infants becoming sick, 54,000 babies being hospitalized, and six infant deaths (Ghazi-Tehrani and Pontell 2015). An investigation revealed that up to 372 milk collection stations had been adding melamine to milk since 2005, and that 22 companies were manufacturing and selling melamine-tainted products (Xinhua 2008). The detailed information that was collected from the melamine-in-milk scandal revealed a lot of information about the gaps that existed in the supply chain, and how numerous opportunities existed for fraud (Ghazi-Tehrani and Pontell 2015).

Food fraud incidents

Although most food fraud incidents go undetected, the ones that are identified provide useful information on the various ways that food fraud acts are committed, and what can be done to prevent them from happening in the future. Therefore, in order to gain a better understanding of the global food fraud problem, researchers supported by government and nongovernment organizations have established two databases of food fraud incidents: (1) United States Pharmacopeial (USP) Convention Food Fraud Database and (2) National Center for Food Protection and Defense (NCFPD) EMA Incident Database. The USP database is open to the public, but the NCFPD's database is only accessible upon request through the University of Minnesota (NCFPD 2015). Although the quantity and quality of information that is available in these databases are limited, current efforts are underway to expand these repositories by continuing to collect historical cases so that researchers can access this information and study the food fraud problem in a more systematic and comprehensive way.

One piece of information that these databases have provided is data on the types of food associated with food fraud (CRS 2014). According to Moore *et al.* (2012), the top food items identified in the USP database were olive oil (16%), milk (14%), honey (7%), saffron (5%), orange juice (4%), coffee (3%), and apple juice (2%). When categorized by food ingredients, oils, milk, and spices make up half of all reported cases (Moore et al. 2012). Information obtained from the USP database also revealed that fish, honey, olive oil, chili powder, milk, black pepper, and caviar were the leading types of food fraud categories (Moore et al. 2012). The NCFPD incident database provides information on EMA incidents. Based on an evaluation of information stored in this database, the leading reported types of EMA food fraud are fish and seafood, oils and fats, alcoholic beverages, meat and meat products, dairy products, grains, and honey (Everstine et al. 2013). When the specific type of adulteration is investigated, over 60 percent of the cases involved the most direct and detectable fraud of substitution or dilution. The remaining cases included unapproved additives, counterfeits, misbranding, masking origin of the product, and the intentional distribution of hazardous substances (Everstine 2012). An understanding of the types of food items most commonly targeted for fraud, and the ways in which the fraudsters commit the crime, can help guide our efforts in the development and application of food authentication technologies. In the next section, we will explore the crime prevention concepts that apply to food fraud.

Food fraud prevention

The food industry is accustomed to taking a systematic approach to preventing hazards. To combat food safety hazards that cause illnesses or death, the food supply chain[2] has implemented Hazard Analysis and Critical Control Point (HACCP) programs. HACCP is based on business quality management system theories. HACCP is widely adopted and even required by law for some products in some countries.

The food supply chain is very complex since a product such as a hamburger sandwich could have 20 or more raw materials from 20 countries. In addition, the ingredients – such as the bun, ground meat, sauces, spices, and condiments such as pickles – could be processed and re-packaged through several manufacturing facilities located in different countries. The ingredients could be transported in bulk (e.g. truckload) or packaged (e.g. in bins, pails, or jars). This typical example of a complex food supply chain includes many handlers, and full traceability or transparency is often difficult, if not impossible.

Due to the complex food supply chain, the near limitless types of problems that can arise, and the tremendous number of transactions that occur, security strategies tend to focus on food fraud prevention. Although techniques used to detect food safety hazards are still being conducted, the focus is placed on reducing the opportunity for anomalies to occur. Food fraud incidents were originally just considered a food safety incident under programs that addressed "Adulterated Foods." In the United States, "Adulterated Foods" were considered in the first food laws that were established in the Pure Food and Drug Act of 1906 and the Food Drug and Cosmetics Act of 1938. The "Adulterated Foods" definition covered any type of food concern, regardless of whether an adulterant-substance was included. For example, stolen or smuggled goods would be considered an "Adulterated Food," and unfit for commerce, even though they were genuine products that probably did not pose a health hazard. This early focus on a health hazard "risk-based approach" did not lead to any research on food fraud.

Due to advances in detection technology, and mass communication, recent food fraud incidents have motivated a shift from detection to prevention. Although there were very strong and capable food safety systems in place, they did not detect or prevent the melamine illegally added to skim milk powder in China[3] or the horsemeat adulteration of beef in Europe.[4] These two incidents jumpstarted the shift in focus from food fraud detection to prevention. To build upon the concepts identified by the HACCP, food fraud prevention strategies address vulnerabilities through the Vulnerability and Critical Control Point (VACCP) program. The VACCP methodology is based on HACCP and business quality management systems that are applied to criminology theory.

While the overall system is VACCP, a key component is to understand and address the crime opportunity referred to by the food supply chain as the "fraud opportunity." Crime prevention theories are a key part of the overall system, and the food supply chain has relied upon environmental criminology to help explain crime events and guide practitioners in the creation and implementation of crime prevention strategies. The three environmental crime theories that have been utilized by food scientists are: routine activities theory, rational choice theory, and situational crime prevention. When combined, these perspectives capture the thought process of offenders: how they make the decision of when and where to commit a crime, what makes a suitable target, and how guardians[5] can prevent crime (Felson and Boba 2010). Over time, and after many adaptations, the theories have been expanded upon to create an in-depth understanding of crime and the factors that make crime more or less likely to occur. The "crime triangle" consists of two layers (Figure 17.2). The inner layer represents the three elements necessary for a crime or fraud event to occur, and the outer layer represents the three types of guardians (controllers) that can influence or control the elements of the inner layer (Clarke and Eck 2009). According to Eck (2003), the intervention of any one controller over the offender, target, or place can be enough to prevent a crime or fraud from happening. By understanding the different elements of the crime triangle, individuals charged with protecting and securing the food supply chain will be in a better position to prevent food fraud incidents.

Once the focus on prevention and the VACCP program is established, a review of current technologies can be completed to determine which countermeasures are the most effective at detecting and deterring food fraud. Detection and deterrence are critical components of any food fraud prevention strategy. While detection can be reactive when investigating suspicious activities, it can also be proactive when looking for any type of new fraud. For example, prevention would be conducting a battery of tests, using different test methods, and looking

Figure 17.2 The crime triangle

for suspicious activity. On the other hand, deterrence is setting up specific countermeasures or detection systems to deter a specific type of fraud and fraudster. For example, prevention in this case would be testing for the melamine adulterant-substance in skim milk powder. The fraudster knows there is testing for melamine in milk, so they try to replace it with a different chemical adulterant.

The role of science and technology in food fraud prevention is to first be able to detect specific types of fraud. Second, the technology should be strategically implemented to specifically reduce the fraud opportunity. In the next section, we will explore the technologies that are currently being used to authenticate food items, and discuss a variety of other technologies that are employed to make food items, and the food supply chain, more secure.

The types of food fraud that pose the most significant health hazard, and thus the highest priority intervention, are adulterant-substances and counterfeiting. This review of food fraud detection technologies will focus on testing for the authenticity of the product. The authenticity assessment can test for the presence of known authentic substances or attributes, or they can test for known fraudulent adulterant-substances. For genuine products, synthetic chemicals are very pure and the authenticity can be fairly easily assured. On the other hand, finished food products could be comprised of thousands of different types of molecules. Furthermore, the molecules may vary from one growing season to the next or by geographic region. For fraudulent products, a known set of common adulterant-substances can be developed. For example, a "Negative List" has been published by the Chinese National Center for Food Safety Risk Assessment (CFSA). This negative list includes the Sudan red colorant, melamine for protein content deception, and it specifies swapping in animal protein. Overall, any detection or traceability systems will increase the transparency of the supply chain which will optimize resources and minimize the fraud opportunity.

Technology

Technology used for food fraud prevention involves two primary areas: product authentication and food supply chain traceability. Authentication determines whether or not a product is genuine or fake, whereas traceability allows food and food products to be tracked across the globe. Because of the complexities involved in the food supply chain a lack of visibility allows fraudsters to operate more easily. Therefore, the ability to efficiently track food products around the world, from producer to consumer, dramatically improves the safety of the food and food items. In short, the primary focus of food fraud prevention is on securing the food supply chain before, during, and after production as well as during distribution to the retail outlets.

Food authentication technologies

Technology is used throughout the food industry to authenticate food items (e.g. test that a food item meets quality specifications or to confirm the presence and quantity of a specific ingredient), and to track food as it moves along the supply chain. Direct authentication technologies are used to determine whether a food item is genuine, altered, or fake. There are a variety of analytical techniques that can be used to directly test food or food products for authenticity. The following review will introduce some of the common analytical techniques used to examine food and food items. These techniques include mass spectrometry (MS), FT-IR spectroscopy, Raman spectroscopy, and DNA-based testing.

Mass spectrometry

Mass spectrometry (MS) is an analytical technique that helps the investigator identify the amount and type of chemicals present in a food sample by measuring the mass-to-charge ratio of positively and negatively charged ions (Finehout and Lee 2004). Mass spectrometry is used throughout the food industry, and one of its most common applications is to detect food contamination such as *Salmonella* (Xu *et al.* 2011; Xu, *et al.* 2010). However, the technique has also been used to determine the geographical origin of foods (Aiello *et al.* 2011; Anklam 1998; Dempson and Power 2004; Drivelos and Georgiou 2012; Gonzalvez *et al.* 2009; Kelly *et al.* 2005; Luykx and van Ruth 2008; Manca *et al.* 2001; Ratel *et al.* 2008), and to authenticate food items. A variety of studies have shown the different ways in which MS can be used to detect adulterations in food products. For example, MS techniques have been used to evaluate changes in carbon and nitrogen isotope ratios to authenticate jelly, spices, wine, and oils (Greule *et al.* 2010; Moreno *et al.* 2007; Spangenberg and Ogrinc 2001; Stocker *et al.* 2006), to quantify whey proteins to determine the adulteration of milk and dairy products (Cordewener *et al.* 2009; Dane and Cody 2010; Cozzolino *et al.* 2002; Czerwenka *et al.* 2010; Goodacre 1997), and for the rapid screening of food packaging for toxic additives (Ackerman *et al.* 2009), and food products for contaminants such as fungicides, myotoxins, and melamine (Cajka *et al.* 2010; Cajka *et al.* 2008; Schurek *et al.* 2008; Vaclavik *et al.* 2009; Vaclavik *et al.* 2011; Vaclavik *et al.* 2010).

Fourier transform infrared (FT-IR) spectroscopy

FT-IR spectroscopy is a rapid, non-destructive analysis technique that works by forcing infrared radiation through the sample under study (e.g. a food product). As a result, some of the infrared radiation is absorbed by the sample, while some of it is transmitted through the sample. The resulting infrared absorbance spectrum represents the molecular absorption and transmission, which is unique to the sample (Ellis *et al.* 2007). It is the uniqueness of this profile that makes infrared spectroscopy a useful tool when analyzing food or food products for quality or to identify adulterations. Some examples where FT-IR spectroscopy has been used to analyze food items include verifying the quality and composition of milk and milk products (Boubellouta and Dufour 2012; Karoui and De Baerdemaeker 2007), authenticating meat and meat products (Al-Jowder *et al.* 2002; Ellis *et al.* 2005; Rannou and Downey 1997), and in the authentication of olive oil (Sun *et al.* 2015).

Raman spectroscopy

Raman spectroscopy is a rapid nondestructive analytical technique that can be used to evaluate food and food products. Using a laser, the sample is illuminated and the energy that is exchanged between the molecules in the sample and the photons from the laser result in spectra being created that are unique to the sample (Smith and Dent 2005). This technique has been successfully used to determine geographical origin and to differentiate between products. Examples include oils (Aparicio-Ruiz and Harwood 2013; Baeten *et al.* 2005; Marini *et al.* 2007), juices (Malekfar *et al.* 2010; Mizrach *et al.* 2007), and bean and soy products (Herrero *et al.* 2009; Tang and Ma 2009).

DNA-based testing

Deoxyribonucleic acid (DNA) testing techniques compare fragments of DNA between two or more samples to determine commonality (Linacre and Tobe 2013). Advancements in molecular biology have made DNA testing of food and food products a useful way to authenticate products throughout the supply chain. One of the most common methods used is the polymerase chain reaction (PCR) (Nixon *et al.* 2015). This technique uses a DNA extraction and amplification process to take a single strand of DNA and then create millions of copies that can be analyzed (Mullis 1990). In regard to food, the process begins by extracting pieces of DNA from a specific food item which is then amplified to generate millions of copies. The amplified DNA is compared to the DNA fragments of the questioned food item for identification. Some examples of how DNA techniques are used to analyze food and food items include: the differentiation of species and mislabeling (Di Pinto *et al.* 2015; Luo *et al.* 2008), determining origin of meat products (Hanapi *et al.* 2014), identifying adulterated rice, olive oil, fish, and meat products (Doosti *et al.* 2014; Ou *et al.* 2015; Rodriguez-Ramirez *et al.* 2015; Vemireddy *et al.* 2015), and to trace food throughout the supply chain (Scarano and Rao 2014; Teletchea *et al.* 2008).

Food tracking technologies

In addition to food authentication technologies, the food industry also employs technology to track food and food products as they travel through a long and complicated food supply chain. The main technologies being used to track food products include: Global Positioning Service (GPS), serial numbers, bar codes, matrix codes, and Radio Frequency Identification (RFID).

Global Positioning System (GPS)

The Global Positioning System (GPS) is operated by the US Department of Defense and consists of a network of satellites that can be used to locate a position anywhere on earth (Pace *et al.* 1995). The system is used throughout the food supply chain to keep track of food and food items as they travel around the world. Because GPS does not provide enough precision to be useful inside a factory or warehouse, it is primarily used by distribution and logistics companies to track the movement of vehicles and shipping containers (Stackpole 2012). The use of GPS technology allows companies to know where their products are located at all times, and how long their products remain in one place before they are unloaded off a truck, ship, or dock. Because it costs more to use GPS technology than it does to use the short-range RFID and bar codes, it is more likely to be employed to track higher cost items with a greater vulnerability for theft, or a significant financial loss to the company if the product were to be lost during shipment.

Serial numbers

A serial number is an alphanumeric code that is assigned to a specific product by a manufacturer. Simple and robust, serial numbers are the most commonly used technology to identify and trace products throughout a supply chain. Although serial numbers are widely used, they are limited in the amount of information they provide, and can become hard to manage. For example, as the number of products in the numbering system increases, the serial number used becomes longer. Although the system is easy to use, it is slower because it requires the user to capture the information by hand, and is more prone to error than bar and matrix codes.

Bar and matrix codes

Bar and matrix codes are more advanced than serial numbers as they provide more information about a product in a machine-readable form. Bar codes are a series of black bars and white spaces that require a scanner (consisting of a laser and photosensor) to read the pattern and relay the message to a computer that translates it into a readable form. However, bar codes are one dimensional which means they require a large amount of space on the product packaging in order to increase the coding capacity. Matrix codes are two-dimensional bar codes. They have been developed to provide the same detailed information as a bar code, but they take up less space on the product packaging. As a result, matrix codes can be applied to products that a bar code would be too large to use.

Radio Frequency Identification (RFID)

RFID technology uses radio waves to read and write information on printed chips that are attached to the product. RFID is considered to be one of the most promising technologies for tracking food as it moves along the food supply chain (Costa *et al.* 2013; Kelepouris *et al.* 2007). Unlike serial numbers and bar codes that require a visual inspection by the user, RFID technology allows the user to access product information from a distance and through some materials because it uses radio waves instead of light (e.g. scanners). For example, the user could read the RFID tags through a shipping container or products stacked on a pallet. The amount of time saved by not having to unpack and repack the millions of products that move around the globe on a daily basis is enormous.

Quick Response Codes (QR Codes)

Most food tracing and authentication technologies have been developed for, and used by, the producer, supplier, and other individuals who are charged with protecting and securing the food supply chain. The end consumer must trust that the retailers, and everyone else involved in the food supply chain, are doing their jobs to keep food safe. Indeed, the last test for authenticity may be the one completed by the business or individual who will sell the food products to the general public. In this regard, the people who will be buying and eating the food rely on the retailer to ensure authenticity. For example, if a tomato has a sticker identifying it as being organically produced the consumer must trust that it was produced organically even if it was not (The Cornucopia Institute 2015). However, consumers are becoming less trusting of the food industry and are more interested in where their food is coming from, and what ingredients they will be ingesting (Kowitt 2015). A recent innovation, that will enable consumers to authenticate food and food items before they make a purchase, is the quick response barcode or "QR" code. A QR code is a type of two-dimensional barcode that can be read using dedicated QR reading devices or a smartphone (Umanandhini *et al.* 2012). Similar to matrix codes, QR codes are two-dimensional and can store a lot more data than bar codes. Considering that more than half of all Americans, and more than two billion people worldwide own smartphones (Kissonergis 2015; Smith 2015), the use of QR codes by consumers will likely become more common once they become familiar with the technology and the information that it provides.

Conclusion

Food authentication and tracking technologies continue to evolve as companies become more reliant upon these technologies to keep the food supply chain safe and secure. Security suppliers are working diligently to develop new authentication techniques that make it easier for individuals, stationed at various points throughout the food supply chain, to test food products in the field rather than in a laboratory setting. For example, portable PCR devices (for DNA testing) have been developed to allow the user to employ the instrument at just about any location in order to conduct lab-quality analysis in the field. Other portable devices such as Raman and Mass spectrometers have been, and are being, developed and used for field testing. As this technology continues to advance the cost and quality of these portable devices will improve, making their use more practical.

The worldwide tracking of food and food products is essential in securing the food supply chain. The use of GPS, bar codes, QR codes, and RFID tags will continuously be upgraded as the database infrastructure and interoperability improves. As the global positioning technologies become smaller and cheaper, and the mobile phone market continues to expand, the use of GPS tracking technologies will become more prevalent throughout the food industry. If these technologies advance to a point where they are combined into a single unit (for example, a GPS and an RFID reader are combined into one unit), then the result would be large amounts of real-time positional data that would provide the user with detailed information on the food products as they move rapidly around the world.

Currently, there are very few technologies available that enable consumers to validate the food products they eat and drink. Access to most of the food authentication and tracking technologies discussed above are limited to producers, suppliers, food scientists, retailers, and commercial and research laboratories. However, as computing costs continue to decrease, and internet applications continue to grow, technology will inevitably provide consumers with more options for authenticating the food they will consume. Living in a digital society, where most people have an internet connected device, consumers are demanding more information about the food they eat. Companies that adhere to the consumers' desire for more information, and develop food authentication technology products with the consumer in mind, will not only win the consumers' trust, but they will do well in the global marketplace.

In summary, food fraud is a complex and unique type of crime that requires very specific countermeasures and systems. As a result, there is a distinct need for technology to detect, deter, and prevent food fraud. However, it is not realistic to expect a single food fraud prevention tool or technique to be effective for a long period of time. Therefore, there is a constant need for new technologies to replace, or improve upon, existing technologies. As food fraudsters learn to circumvent current security measures with increasing efficiency, producers, suppliers, distributers, and retailers must always be vigilant. Furthermore, they must be prepared to invest in and employ a variety of food fraud prevention techniques in order to stay ahead of the fraudsters. There is a wide range of food authenticity and food integrity test methods that must continue to evolve to meet the shifting food fraud threat. When strategically applying science and technology, food fraud can be prevented.

Notes

1   "Food Security" is the safe, continuous, nutritious, and economic supply of food. "Food Defense" is protecting the food supply chain from harm.
2   The food supply chain or food chain refers to the processes that describe how food and food items move systematically from producers to consumers. The food chain is composed of a wide diversity

of products and companies which operate in different markets and sell a variety of food products (Bukeviciute *et al.* 2009).

3   In 2008, chemical tests confirmed that melamine – a plasticizer used in many non-food processes such as to increase the flexibility of plastic bottles – was added to milk powder by several Chinese dairies and distributors to increase the apparent protein content of the milk. Financial gain was the motive for the fraud since adding the melamine adulterant-substance to the milk deceived quality control testing to increase the perceived premium protein content of the milk.

4   In 2013, DNA tests confirmed that horsemeat was illegally added to beef and then sold in several European supermarket chains. Financial gain was the motive for the fraud, since horsemeat is a lower cost alternative to the higher priced pure beef in some countries.

5   In general, guardians (controllers) are actors that regulate or influence the behavior of offenders, victims and places to prevent a crime from happening (Felson and Boba 2010). There are three types of guardians (controllers): handler-guardians, manager-guardians, and target-guardians.

References

Ackerman, L. K., Noonan, G. O., and Begley, T. H. (2009). Assessing direct analysis in real-time-mass spectrometry (DART-MS) for the rapid identification of additives in food packaging. *Food Additives & Contaminants: Part A*, 26(12): 1611–1618.

Aiello, D., De Luca, D., Gionfriddo, E., Naccarato, A., Napoli, A., Romana, E., Russo, A., Sindona, A., and Tagarelli, A. (2011). Multistage mass spectrometry in quality, safety and origin of foods. *European Journal of Mass Spectrometry*, 17: 1–31.

Al-Jowder, O., Kemsley, E. K., and Wilson, R. H. (2002). Detection of adulteration in cooked meat products by mid-infrared spectroscopy. *Journal of Agricultural and Food Chemistry*, 50(6): 1325–1329.

Anklam, E. (1998). A review of the analytical methods to determine the geographical and botanical origin of honey. *Food Chemistry*, 63(4): 549–562.

Aparicio-Ruiz, R. and Harwood, J. (2013). *Handbook of olive oil: Analysis and properties.* New York, NY: Springer.

Baeten, V., Pierna, J. A. F., Dardenne, P., Meurens, M., Garcia-Gonzalez, D. L., and Aparicio-Ruiz, R. (2005). Detection of the presence of hazelnut oil in olive oil by FT-Raman and FT-MIR spectroscopy. *Journal of Agricultural and Food Chemistry*, 53: 6201–6206.

Boubellouta, T. and Dufour, E. (2012). Cheese-matrix characteristics during heating and cheese melting temperature prediction by synchronous fluorescence and mid-infrared spectroscopies. *Food Bioprocess Technology*, 5(1): 273–284.

Bukeviciute, L., Dierx, A., and Ilzkovitz, F. (2009). *The functioning of the food supply chain and its effect on food prices in the European Union.* Retrieved on March 7, 2016 from http://ec.europa.eu/economy_finance/publications/publication15234_en.pdf.

Cajka, T., Riddellova, K., Tomaniova, M., and Hajslova, J. (2010). Recognition of beer brand based on multivariate analysis of volatile fingerprint. *Journal of Chromatography*, A1217: 4195–4203.

Cajka, T., Vaclavik, L., Riddellova, K., and Hajslova, J. (2008). GC–TOF-MS and DART–TOF-MS: Challenges in the analysis of soft drinks. *LCGC Europe*, 21(5): 250–256.

Clarke, R. V. and Eck, J. E. (2009). *Crime analysis for problem solvers: In 60 small steps.* Retrieved on August 20, 2014 from ww.cops.usdoj.gov/Publications/CrimeAnalysis60Steps.pdf.

Cordewener, J. H. G., Luykx, D. M. A. M., Frankhuizen, R., Bremer, M. G. E. G., Hooijerink, H., and America, A. H. P. (2009). Untargeted LC-Q-TOF mass spectrometry method for the detection of adulterations in skimmed-milk powder. *Journal of Separation Science*, 32: 1216–1223.

Costa, C., Antonucci, F., Pallottino, F., Aguzzi, J., Sarria, D., and Menesatti, P. (2013). A review on agri-food supply chain traceability by means of RFID technology. *Food Bioprocesses Technology*, 6: 353–366.

Cozzolino, R., Passalacqua, S., Salemi, S., and Garozzo, D. (2002). Identification of adulteration in water buffalo mozzarella and in ewe cheese by using whey proteins as biomarkers and matrix-assisted laser desorption/ionization mass spectrometry. *Journal of Mass Spectrometry*, 37(9): 985–991.

CRS, US Congressional Research Service. (2014). *Food fraud and "Economically Motivated Adulteration" of food and food ingredients* (CRS Report No. R43358). Washington, DC: Congressional Research Service (ed. R. Johnson).

Czerwenka, C., Muller, L., and Lindner, W. (2010). Detection of the adulteration of water buffalo milk and mozzarella with cow's milk by liquid chromatography-mass spectrometry analysis of beta-lactoglobulin variants. *Food Chemistry*, 122: 901–908.

Dane, A. J. and Cody, R. B. (2010). Selective ionization of melamine in powdered milk by using argon direct analysis in real time (DART) mass spectrometry. *Analyst*, 135(4): 696–699.

Dempson J. B. and Power M. (2004). Use of stable isotopes to distinguish farmed from wild Atlantic salmon, Salmo salar. *Ecology of Freshwater Fish*, 13: 176–184.

Di Pinto, A., Bottaro, M., Bonerba, E., Bozzo, G., Ceci, E., Marchetti, P., Mottola, A., and Tantillo, G. (2015). Occurrence of mislabeling in meat products using DNA-based assay. *Journal of Food Science and Technology*, 52(4): 2479–2484.

Doosti, A., Dehkordi, P. G., and Rahimi, E. (2014). Molecular assay to fraud identification of meat products. *Journal of Food Science and Technology*, 51(1): 148–152.

Drivelos, S. A. and Georgiou, C. A. (2012). Multi-element and multi-isotope-ratio analysis to determine the geographical origin of foods in the European Union. *Trends in Analytical Chemistry*, 40: 38–51.

Eck, J. E. (2003). Police problems: The complexity of problem theory, research, and evaluation. In J. Knutsson (ed.), *Problem-oriented policing: From innovation to mainstream, crime preventions studies* (Vol. 15, pp. 79–113). Monsey, NY: Criminal Justice Press.

Ellis, D. I., Broadhurst, D., Clarke, S. J., and Goodacre, R. (2005). Rapid identification of closely related muscle foods by vibrational spectroscopy and machine learning. *Analyst*, 130(12): 1648–1654.

Ellis, D. I., Dunn, W. B., Griffin, J. L., Allwood, J. W., and Goodacre, R. (2007). Metabolic fingerprinting as a diagnostic tool. *Pharmacogenomics*, 8(9): 1243–1266.

Everstine, K., Kircher, A., and Cunningham, E. (2013). *The implications of food fraud*. Retrieved October 10, 2015, from www.foodqualityandsafety.com/article/the-implications-of-food-fraud/.

Everstine, K., Spink, J., and Kennedy, S. (2013). Economically motivated adulteration (EMA) of food: Common characteristics of EMA incidents. *Journal of Food Protection*, 76(4), 723–735.

Felson, M. and Boba, R. (2010). *Crime and everyday life*. Thousand Oaks, CA: Sage Publications.

Finehout E. J. and Lee K. H. (2004). An introduction to mass spectrometry applications in biological research. *Biochemistry and Molecular Biology Education*, 32: 93–100.

Food and Agriculture Organization of the United Nations (FAO). (2015). *Food Outlook: Biannual Report on Global Food Markets*. Retrieved October 10, 2015, from www.fao.org/3/a-i4581e.pdf.

Ghazi-Tehrani, A. K. and Pontell, H. N. (2015). Corporate crime and state legitimacy: The 2008 Chinese melamine milk scandal. *Crime, Law, and Social Change*, 63: 247–267.

Goodacre, R. (1997). Use of pyrolysis mass spectrometry with supervised learning for the assessment of the adulteration of milk of different species. *Applied Spectroscopy*, 51: 1144–1153.

Gonzalvez, A., Armenta, S., and de la Guardia, M. (2009). Trace-element composition and stable-isotope ratio for discrimination of foods with protected designation of origin. *Trends in Analytical Chemistry*, 28: 1295–1311.

Government Accountability Office (GAO). (2009). Seafood fraud FDA program changes and better collaboration among key federal agencies could improve detection and prevention (GAO Publication No. 09-258). Washington, DC: U.S. Government Printing Office.

Greule, M., Tumino, L. D., Kronewald, K., Hener, U., Schleucher, J., Mosandl, A., and Keppler, F. (2010). Improved rapid authentication of vanillin using δ13C and δ2H values. *European Food Research and Technology*, 231(6): 933–941.

Grocery Manufacturers Association (GMA). (2010). *Consumer product fraud: Deterrence and detection*. Retrieved October 10, 2015, from www.gmaonline.org/downloads/wygwam/consumerproductfraud.pdf.

Hanapi, U. K., Desa, M. N. M., Ismail, A., and Mustafa, S. (2014). A higher sensitivity and efficiency of common primer multiplex PCR assay in identification of meat origin using NADH dehydrogenase subunit 4 gene. *Journal of Food Science and Technology*, 52(7): 4166–4175.

Herrero, A. M., Jiménez-Colmenero, F., and Carmona, P. (2009). Elucidation of structural changes in soy protein isolate upon heating by Raman spectroscopy. *International Journal of Food Science and Technology*, 44(4): 711–717.

Karoui, R. and De Baerdemaeker, J. (2007). A review of the analytical methods coupled with chemometric tools for the determination of the quality and identity of dairy products. *Food Chemistry*, 102(3): 621–640.

Kelepouris, T., Pramatari, K., and Doukidis, G. (2007). RFID-enabled traceability in the food supply chain. *Industrial Management & Data Systems*, 107(2): 183–200.

Kelly, S., Heaton, K., and Hoogewerff, J. (2005). Tracing the geographical origin of food: The application of multi-element and multi-isotope analysis. *Trends in Food Science Technology*, 16: 555–567.

Kissonergis, P. (2015). *Smartphone ownership, usage and penetration by country*. Retrieved March 18, 2016, from www.smsglobal.com/thehub/smartphone-ownership-usage-and-penetration/.

Kowitt, B. (2015). *Special report: The war on big food*. Retrieved March 18, 2016, from http://fortune.com/2015/05/21/the-war-on-big-food/.

Linacre, A. and Tobe, S. (2013). *Wildlife DNA analysis: Applications in forensic science*. Oxford, England: Wiley-Blackwell.

Luo, J., Wang, J., Bu, D., Li, D., Wang, L., Wei, H., and Zhou, L. (2008). Development and application of a PCR approach for detection of beef, sheep, pig, and chicken derived materials in feedstuff. *Agricultural Sciences in China*, 7(10): 1260–1266.

Luykx, D. M. A. M., and van Ruth, S. M. (2008). An overview of analytical methods for determining the geographical origin of food products. *Food Chemistry*, 107: 897–911.

Malekfar, R., Nikbakht, A. M., Abbasian, S., Sadeghi, F., and Mozaffari, M. (2010). Evaluation of tomato juice quality using surface enhanced Raman spectroscopy. *Acta Physica Polonica*, A 117 (6): 971–973.

Manca, G., Camin, F., Coloru, G.C., Del Caro, A., Depentori, D., Franco, M.A., and Versini, G. (2001). Characterization of the geographical origin of Pecorino Sardo cheese by casein stable isotope (13C/12C and 15N/14N) Ratios and free amino acid ratios. *Journal of Agricultural and Food Chemistry*, 49: 1404–1409.

Marini, F., Magri, A. L., Bucci, R., and Magri, A. D. (2007). Use of different artificial neural networks to resolve binary blends of monocultivar Italian olive oils. *Analytical Chemistry Acta*, 599: 232–240.

Mizrach, A., Schmilovitch, Z., Korotic, R., Irudayaraj, J., and Shapira, R. (2007). Yeast detection in apple juice using raman spectroscopy and chemometric methods. *Transactions of the ASABE*, 50: 2143–2149.

Moore, J. C., Spink, J., and Lipp, M. (2012). Development and application of a database of food ingredient fraud and economically motivated adulteration from 1980 to 2010. *Journal of Food Science*, 77(4), R118–R126.

Moreno Rojas, J. M., Cosofret, S., Reniero, F., Guillou, C., and Serra, F. (2007). Control of oenological products: Discrimination between different botanical sources of L-tartaric acid by isotopic ratio mass spectrometry. *Rapid Communication Mass Spectrometry*, 21: 2447–2450.

Mullis, K. (1990). The unusual origin of the polymerase chain reaction. *Scientific American*, 262(4): 56–61, 64–5.

Natarajan, M. (2011) *International crime and justice*. New York, NY: Cambridge University Press.

National Center for Food Protection and Defense (NCFPD). (2015). *Economically Motivated Adulteration Incidents Database*. Retrieved October 10, 2015, from www.foodfraudresources.com/ema-incidents/.

National Oceanic and Atmospheric Administration (NOAA). (2010). *US Seafood Consumption Declines Slightly in 2009*. Retrieved October 10, 2015, from www.noaanews.noaa.gov/stories2010/20100909_consumption.html.

Nixon, G. J., Wilkes, T. M., and Burns, M. J. (2015). Development of a real-time PCR approach for the relative quantitation of horse DNA. *Analytical Methods*, 7: 8590–8596.

Ou, G., Hu, R., Zhang, L., Li, P., Luo, X., and Zhang, Z. (2015). Advanced detection methods for traceability of origin and authenticity of olive oils. *Analytical Methods*, 7: 5731–5739.

Pace, S., Frost, G.P., Lachow, I., Frelinger, D.R., Fossum, D., Wassem, D., and Pinto, M.M. (1995). The Global Positioning System: Assessing national policies. Retrieved March 18, 2016, from www.rand.org/pubs/monograph_reports/MR614.html.

Purcell, N. (1985). Wine and wealth in ancient Italy. *The Journal of Roman Studies*, 75: 1–19.

Rannou, H. and Downey, G. (1997). Discrimination of raw pork, chicken and turkey meat by spectroscopy in the visible, near- and mid-infrared ranges. *Analytical Communications*, 34: 401–404.

Ratel, J., Berge, P., Berdague, J. L., Cardinal, M., and Engel, E. (2008). Mass spectrometry based sensor strategies for the authentication of oysters according to geographical origin. *Journal of Agricultural Food Chemistry*, 56: 321–327.

Rodriguez-Ramirez, R., Vallejo-Cordoba, B., Mazorra-Manzano, M. A., and Gonzalez-Cordova, A. F. (2015). Soy detection in canned tuna by PCR and capillary electrophoresis. *Analytical Methods*, 7: 530–537.

Scarano, D. and Rao, R. (2014). DNA markers for food products authentication. *Diversity*, 6: 579–596.

Schurek J., Vaclavik L., Hooijerink H., Lacina O., Poustka J., Sharman M., Caldow M., Nielen M. W. F., and Hajslova J. (2008). Control of strobilurin fungicides in wheat using direct analysis in real time accurate time-of-flight and desorption electrospray ionization linear ion trap mass spectrometry. *Analytical Chemistry*, 80(24): 9567–9575.

Smith, A. (2015). *U.S. smartphone use in 2015*. Retrieved March 18, 2016, from www.pewinternet. org/2015/04/01/us-smartphone-use-in-2015/.

Smith, E. and Dent, G. (2005). *Modern raman spectroscopy: A practical approach*. West Sussex, England: Wiley.

Spangenberg, J. E. and Ogrinc, N. (2001). Authentication of vegetable oils by bulk and molecular carbon isotope analysis with emphasis on olive oil and pumpkin seed oil. *Journal of Agricultural Food Chemistry*, 49: 1534–1540.

Spink, J. (2015). *Global food fraud prevention trend*, Food Protection Forum, Ministry for Primary Industries, New Zealand Government, Auckland, October 12, 2015.

Spink, J. and Moyer, D. C. (2011). Defining the public health threat of food fraud. *Journal of Food Science*, 75(9): 57–63.

Stackpole, B. (2012). *GPS technology steers towards the manufacturing supply chain*. Retrieved March 18, 2016, from http://searchmanufacturingerp.techtarget.com/feature/GPS-technology-steers-towards-the-manufacturing-supply-chain.

Stiles, M. L., Lahr, H., Lahey, W., Shaftel, E., Bethel, D., Falls, J., and Hirshfield, M. F. (2011). *Bait and switch: How seafood fraud hurts our oceans, our wallets and our health*. Retrieved October 10, 2015, from http://oceana.org/sites/default/files/reports/Bait_and_Switch_report_2011.pdf.

Stocker, A., Rossmann, A., Kettrup, A., and Bengsch, E. (2006). Detection of royal jelly adulteration using carbon and nitrogen stable isotope ratio analysis. *Rapid Communications in Mass Spectrometry*, 20(2): 181–184.

Sun, X., Lin, W., Li, X., Shen, Q., and Luo, H. (2015). Detection and quantification of extra virgin olive oil adulteration with edible oils by FT-IR spectroscopy and chemometrics. *Analytical Methods*, 7: 3939–3945.

Tang, C. H., and Ma, C. Y. (2009). Heat-induced modifications in the functional and structural properties of vicilin-rich protein isolate from kidney (*Phaseolus vulgaris* L.) bean. *Food Chemistry*, 115: 859–866.

Teletchea, F., Bernillon, J., Duffraisse, M., Laudet, V., Hänni, C. (2008). Molecular identification of vertebrate species by oligonucleotide microarray in food and forensic samples. *Journal of Applied Ecology*, 45: 967–975.

The Cornucopia Institute. (2015). *Whole Foods faces FTC mislabeling Investigation: Specialty grocer meets growing blowback from organic farmer/consumers*. Retrieved March 18, 2016, from www.cornucopia. org/2015/06/whole-foods-faces-ftc-mislabeling-investigation/.

Umanandhini, D., TamilSelven, L., Udhayakumar, S., and Vijayasingam, T. (2012). *Dynamic authentication for consumer supplies in mobile cloud environment*. Retrieved March 18, 2016, from http://ieeexplore.ieee. org/xpls/abs_all.jsp?arnumber=6395954&tag=1.

U.S. Pharmacopeial Convention (USP). (2015). *USP's Food Fraud Database*. Retrieved October 10, 2015, from www.usp.org/food-ingredients/food-fraud-database.

Vaclavik, L., Cajka, T., Hrbek, V., and Hajslova, J. (2009). Ambient mass spectrometry employing direct analysis in real time (DART) ion source for olive oil quality and authenticity assessment. *Analytica Chimica Acta*, 645(1–2): 56–63.

Vaclavik, L., Hrbek, V., Cajka, T., Rohlik, B. A., Pipek, P., and Hajslova, J. (2011). Authentication of animal fats using direct analysis in real time (DART) ionization-mass spectrometry and chemometric tools. *Journal of Agricultural Chemistry*, 59(11): 5919–5926.

Vaclavik, L., Zachariasova, M., Hrbek, V. and Hajslova, J. (2010). Analysis of multiple mycotoxins in cereals under ambient conditions using direct analysis in real time (DART) ionization coupled to high resolution mass spectrometry. *Talanta*, 82: 1950–1957.

Vemireddy, L. R., Satyavathi, V. V., Siddiq, E. A., and Nagaraju, J. (2015). Review of methods for the detection and quantification of adulteration of rice: Basmati as a case study. *Journal of Food Science and Technology*, 52(6): 3187–3202.

Wilson, B. (2008). *Swindled: The dark history of food fraud, from poison sweets to counterfeit coffee*. Princeton, NJ: Princeton University Press.

Xinhua (2008). *China seizes 22 companies with contaminated baby milk powder*. Retrieved October 10, 2015, from http://news.xinhuanet.com/english/2008-09/17/content_10046949.htm.

Xiu, C., and Klein, K. K. (2010). Melamine in milk products in China: Examining the factors that led to deliberate use of the contaminant. *Food Policy*, 35(5): 463–470.

Xu, Y., Cheung, W., Winder, C. L., Dunn, W. B., and Goodacre, R. (2011). Metabolic profiling of meat: Assessment of pork hygiene and contamination with *Salmonella typhimurium*. *Analyst*, 136 (3): 508–514.

Xu, Y., Cheung, W., Winder, C. L., and Goodacre, R. (2010). VOC-based metabolic profiling for food spoilage with the application to detecting *Salmonella typhimurium*-contaminated pork. *Analytical and Bioanalytical Chemistry*, 397(6): 2439–2449.

18

Consumer technologies, crime and environmental implications

Avi Brisman and Nigel South

Introduction

This chapter is concerned with technology, consumption, markets and related crimes and harms (although these are, of course, also connected to the ways in which innovation and demand shape the production of certain goods). Our starting point is that criminology and sociology have paid considerable attention to production, industry and work as mechanisms of discipline, and to technology as a contributor to the maintenance of social control. Relatively less attention, however, has been paid to the criminological significance of consumption. Yet, as Bauman (1997) has noted, *consumerism* is one of the central mechanisms by which societies establish social order in late modernity. It is also related to the establishment of citizenship (see, e.g. Aas 2013: 159 (citing Klein 2000; Loader and Walker 2007; Wood and Dupont 2006; Zedner 2000); Brisman 2013: 276 (citing Giroux 2004)) and, inversely, to what is, perhaps, the most symbolic and obvious way in which social order and citizenship are rejected, refuted and rebuked by those who feel excluded, as in Hallsworth's *Street Crime* in which young black males in newly gentrified areas of London steal coveted consumer goods – not because of their "inhumanity" but because "these strange behavioural rituals are entirely consistent with the seductive and spectacular world that consumer capitalism has built and to which these young people have been so relentlessly exposed" (2005: 166). The young people who become involved in street robbery, Hallsworth explains,

> learn from a very early age that wellbeing and success in life are contingent upon the possession of desirable goods. In particular, branded goods marketed to them by the culture industries. ... In and through mass consumption, identities are produced and reproduced. In consumption, a lifestyle is simultaneously lived and constructed. To 'be' is literally to be in a world defined by the possession of these desirable goods. ... Your status relative to others is marked out and defined by the kind of phone you possess, the trousers you wear, and the way you wear them.
>
> *(2005: 123–24)*

Our chapter thus takes the relationship between technology/ies and consumption as a central element of contemporary cultural life. This relationship reflects the demands we make on the market and its capacity to produce and make available diverse goods. While this relationship carries with it criminological implications with respect to safety, liability and harm (as discussed in other chapters in this volume), in this chapter, we aim to discuss a number of implications and consequences for our environment and the resources of the planet. Indeed, as Winlow (2015: 638) has recently observed: 'Consumerism ... involves a process of buying, discarding and buying again. The West's attachment to *cheap* consumerism has, quite clearly, had a profound effect upon our natural environment' (emphasis added). Here, however, we are particularly concerned with various contradictions and paradoxes that relate to the ideas that: increased consumption can be more efficient, can reduce waste, can be better for human health, and can help to save the planet. This line of critical exploration is not new in studies of consumption, of course. More than thirty years ago, Featherstone (1982: 20) observed that consumer goods designed to satisfy needs and free up time instead, very often, simply generate further demands as 'Consumer goods themselves need maintenance' and 'as commodities voraciously demand other commodities'.

Before proceeding further, it bears mention that neither 'consumer technologies' nor 'technologies of consumption' is well defined – especially when compared to more familiar instances, such as chemical, digital/electronic, or nuclear technologies. Definitions are made more problematic as a result of the fact that a 'consumer technology' will usually also be an instance of some other technological class. Thus, for example, telephones and other communication technologies are also digital technologies. McGuire (2012) has promulgated a spatial approach defining 'consumer technologies' in terms of their 'mid- or multi-range' extensions to bodily capacity – as compared to digital technologies that enable social interaction 'at a distance'. In addition to defining 'consumer technology' in terms of 'what it does,' however, it is also possible to think of it in terms of its economic significance – as a central instance of commodification processes.

In this chapter, we illustrate the irony around consumer goods alluded to by Featherstone (1982), as a well as a tendency to deny the consequences of commodification, by examining two examples of what we refer to as 'technologies of consumption' – the cases of bottled water (and plastic, more generally) and the Prius (and hybrid electric vehicles' dependence on rare earth elements/metals, more broadly). We also consider children's stories regarding reusing and recycling, in order to illuminate how patterns of production, consumption, and constructed consumerism interact with environmental systems in negative ways.

Criminology, technology and the environment

For many decades, the social sciences – including sociology and criminology – managed to study societies in terms of, for example, industrialization, social change and technological innovation, yet largely failed to grasp the importance of the wider – including environmental – *consequences* of these developments. As Foster and Holleman (2012: 1626) note, 'the dominant post-Second World War sociological tradition was seen as having embraced a human-exemptionalist paradigm, in which human beings in technologically advanced societies were considered exempt from natural-environmental influences'. For much of this period, the taken-for-granted assumption was that, by and large, technology, increased production and rising consumption were all indisputably 'good' things. From the late 1960s and 1970s, however, a more critical and sceptical spirit questioned this assumption, as epitomised by the idea of *iatrogenesis*, developed by Ivan Illich (1976) and employed in criminology (or rather *against* criminology) by Stanley Cohen (1982) to capture the idea that 'doing good' can also lead to 'doing bad' and, in this case, the observation that the advance of science can leave difficult legacies.

The insight prompted by thinking about iatrogenesis is that we realize how many of our contemporary environmental problems and challenges are risks and consequences wrought by modernity, globalization and transnational production, reproduction and over-consumption – leaving legacies of 'irreversible threats to the life of plants, animals and human beings' (Beck 1992: 12–13). This is a reminder that criminology should maintain a critical awareness of new and emerging technologies (for example, in bio-sciences and nano-technology), as well as pay attention to the history of past mistakes and current challenges, such as the dangerous disposal of radioactive waste and more recently of e-waste (see, e.g. Bisschop 2012, 2014, 2015; Gibbs *et al.* 2010; Rothe 2010; Snider 2010; van Erp and Huisman 2010).

The following sections pursue this point. Here, our approach to consumption and technology entails, in particular, a concern with inequalities in global resource transactions – processes with implications for both human and environmental rights. We are also concerned with – and about – the neutralization of urgency or significance concerning climate change (see, e.g. Brisman 2012, 2013; Brisman and South 2013a, 2015a, -b, -c, In Press; Wyatt and Brisman 2014), which is supported by beliefs that technology and science will provide 'magic bullet' solutions to problems that, in turn, have their origin – at least in part – in socio-technical processes and developments.

Fraudulent, misleading or manipulative demand stimulation – technologies of advertising and persuasion

Lynch and Stretesky (2014: 151) have discussed the origin and use of the term 'greenwashing', an excellent example of how corporate responsibility, technological efficiency and a sense of social partnership are all invoked to present an impression of improved environmental behaviour and more environmentally sensitive uses of consumer or other technologies, when in fact little or nothing has changed. As Lynch and Stretesky argue:

> companies engage in claims making which suggest that they are improving their production practices and their products so that they are more environmentally friendly. The term "greenwashing," first used by biologist Jay Westerveld in 1986 to describe the hotel cards that ask guests to refrain from washing towels to save the environment while at the same time engaging in other more serious forms of ecological disruption, has been used to describe these practices.

The notion of green products and green consumerism now underpins major sectors of the economy and key strategies for advertising (see, e.g. Brisman 2009). This is not to suggest that green products are undesirable or that improved environmentally conscious consumer habits would not be positive developments. But what we do wish to assert – as, say, with the friendly reminders in hotels to 'do our part' for 'the environment' by reusing our towels – is that we should be interrogating what is really meant by the message we receive, and what stories we are *not* hearing. Usually, we want to hear reassuring anecdotes and tales of our 'successes' on the 'environmental front', so we feel good – proud, even – at the end of a hotel stay that we have not unnecessarily requested new, clean towels. But what we have really succeeded in doing is playing our part in a magic trick – one in which by not requiring the washing of hotel towels on a daily basis, we have helped to save the planet, thereby allowing us to ignore all the lighting and other energy consuming devices (e.g. air conditioning, coffee pots, computer screens (see Agnew 2013)) that are left on all the time, and all the waste that we can see the hotel generating in the name of our comfort and relaxation.

The case of bottled water also represents something charmed and mysterious, as ordinary water is transformed by being poured into plastic bottles. It is now somehow healthier, purer, greener – yet will actually be none of these (Brisman and South 2013b, 2014). Even more seemingly supernatural is when such goods meet the language of technology and science, as in the case of sports and energy drinks. As a report in the *British Medical Journal* (Petrie and Wessely 2004: 1417) noted, 'Bottled water has become an "aquaceutical", the ultimate health food. It is now fortified with additives and produced using special processes *claimed to improve health*' (emphasis added). Drinks described as 'energy drinks' or 'sports drinks' contain 16–18 grams of sugar in every 500ml, as well as salt – not as much as sweetened soda drinks but hardly necessary for the tasks engaged in by most of the consumers of these drinks. As Bosely (2014) puts it, 'Sports drinks – if they have a justification at all – are for people playing serious sport. According to the European Food Safety Authority, they are useful only for "active individuals performing endurance exercise"', yet they are widely marketed to young people who consume them while engaging in far less physical activity, such as watching TV and playing video games. As a result, one UK campaign group, Action on Sugar, has called for these drinks to be banned, with the Chair of the group, Professor Graham MacGregor, Professor of Cardiovascular Medicine at Queen Mary, University of London, arguing that 'Children are being deceived into drinking … this stuff, thinking they are going to improve their performance at school, during sports or even on a night out. In reality all they are doing is increasing their risk of developing obesity or type 2 diabetes' (Smithers 2015). Yet, the marketing of these drinks is given credibility by the work and endorsement of food technologists, nutritionists and sports scientists who are employed or sponsored by the beverage industry.

An in-depth investigation published by the *British Medical Journal* (Cohen 2012) starts from a discussion of the creation of fear about the danger of dehydration. In the wake of this, the energy drink 'Gatorade', followed by many emulators, has led the way to a market worth around £260m in the UK in 2012 and an expected worth of approximately $2,000,000,000 in the US by 2016. This is not only a triumph of advertising but also of pseudo-science, as Cohen shows with her example of how the Gatorade Sports Science Institute (GSSI) has managed to

> undermine the idea that the body has a perfectly good homeostatic mechanism for detecting and responding to dehydration – thirst. "The human thirst mechanism is an inaccurate short-term indicator of fluid needs … Unfortunately, there is no clear physiological signal that dehydration is occurring," Bob Murray from [GSSI] declared in 2008.
>
> *(Cohen 2012)*

Interestingly, Cohen also points out how the 'promotion of hydration has created a battleground for the fight between bottled water companies and the sports drink industry'. Cohen pursues this in terms of claims and counter claims about which type of drink is better for addressing the 'problem' of dehydration. This 'battle' is, in reality, a mirage and unlikely to be resolved by evidence because the two propositions underpin two hugely profitable markets and even where one company owns two types of drink, neither is going to be taken off the production line (e.g. Coca-Cola owns various bottled water companies, such as Dasani, as well as Powerade). Criminologists are familiar with the mechanics of scaremongering and the creation of fears and so should readily perceive the relevance of how lifestyle marketing and bias in health technologies can contribute to what has been called 'disease mongering' (Moynihan et al. 2002) and the pattern described by Noakes (2012):

> When industry wanted to sell more product it had to develop a new disease that would encourage people to overdrink. […] Here's a disease that you will get if you run. Here's

a product that will save your life. That's exactly what they did. They said dehydration is a dreaded disease of exercise.

(quoted in Cohen 2012)

Adverse, harmful or criminal outcomes of consumption in different technology product or service market sectors

Technologies and processes of both production and consumption are massive generators of waste: they use up physical materials, human labour and other sources of energy but they also leave behind the unusable and unwanted. Consider the growth of the chemicals and plastics industries of the twentieth century (Freinkel 2011, 2014; see also Fazekas 2014). In the 1940s, the 'Plastic Age' was heralded as a world which would be:

> free from moth and rust and full of colour, a world largely built up of synthetic materials made from the most universally distributed substances, a world in which nations are more and more independent of localised naturalised resources, a world in which man, like a magician, makes what he wants for almost every need out of what is beneath and around him.
>
> *(Yarsley and Couzens 1945: 152)*

Again, we see how technology yields magic – which was an understandable impression given the novelty and value of the qualities of plastic, such as being 'lightweight, strong, durable, corrosion-resistant [...] with high thermal and electrical insulation properties' (Thompson *et al.* 2009: section 2). All so useful for many applications – applications as varied as light switches, shopping bags and shampoo bottles, for example – and yet all reasons why it is so difficult to dispose of: it accumulates in land, marine and fresh water environments, with reports of it found on high mountains and shorelines, and in oceans and soil across the world (Thompson *et al.* 2009: section 3; see also Fazekas 2014; Freinkel 2014; Smith 2015; see generally Iyer 2016; Pinto 2014). If the problem were simply that plastic is not biodegradable, that would be reason enough for concern, but of course plastic is also a product of the enormous twentieth-century growth in the chemistry and technology of toxicity (Pearce and Tombs 1998). Indeed, plastics contain a range of toxic substances – polychlorinated biphenyls (PCBs), various forms of hydrocarbons, alkyphenols and others – all of which present risks of transfer into the food chain and thus harm to any animal, bird or fish ingesting them (see, e.g. Iyer 2016; Melzer and Galloway 2010; Thompson *et al.* 2009: section 4; Watts 2013).

In conscious and unconscious actions, gestures and habits, human consumers accumulate and use plastics that are then discarded. So, for example, smokers of cigarettes generally believe there is a health benefit to the filters on cigarettes – a proposition that is actually very doubtful (see, e.g. Harris 2011) – but give no thought to the filters when they discard them other than perhaps realizing this is littering and seeking or not seeking a suitable disposal receptacle. What they are unlikely to dwell on is that this clever marketing-led 'breakthrough' in smoking technology simply adds a piece of plastic to the cigarette which does not reduce the risk of morbidity and mortality caused by smoking but does mean that trillions of non-biodegradable chunks of cellulose acetate impregnated with carcinogens, nicotine and other toxic chemicals are 'dropped, flicked or dumped into the environment' every year – 'around 750,000 tonnes in total' (Novotny 2014).

As another example, Connor (2014) reports that, 'Few consumers realise that many cosmetic products, such as facial scrubs, toothpastes and gels, now contain many thousands of microplastic beads which have been deliberately added by the manufacturers of more than 100 consumer products over the past two decades'. The beads are less than one millimetre wide and too small to be caught by water filtration systems (see Iyer 2016). Connor (2014) notes the irony that at the same time as 'many people have assiduously tried to recycle their plastic waste, cosmetics companies have … been quietly adding hundreds of cubic metres of plastics such as polyethylene to products that are deliberately designed to be washed into waste-water systems'.

Along these lines, Hohn (2012), author of *Moby Duck: The True Story of 28,800 Bath Toys Lost at Sea*, followed the marine journey of these toy ducks after they fell from containers at sea and in the course of it changed from someone in pursuit of an amusing fascination to someone with a set of serious worries. As Hohn discovers, plastics consume 400m tons of oil and gas every year – fossil fuels that are in finite supply – while only about 5 per cent of plastics actually wind up being recycled. So-called "green plastics" are not much better for they release greenhouses gases when they break down. All of this leaves Hohn to lament, "What's most nefarious about plastic is the way it pretends to deny the laws of matter, as if something could be made from nothing; the way it is intended to be thrown away but chemically engineered to last. By offering the false promise of disposability, of consumption without cost, it has helped create a culture of wasteful make-believe, an economy of forgetting" (2012: 29).

Magical tricks and wishful thinking again.

In recent decades, public concern has certainly grown regarding waste as 'a problem' and various solutions have been explored both domestically and internationally – often with rich consumer nations removing and relocating waste to poorer states, paying them to act as dump sites, and now also relocating resource-rich disposable waste as part of a new global trade in de-manufacturing and recycling (see generally Brisman 2010; Ruggiero and South 2010). Recycling is obviously 'good' but 'de-manufacturing' means that further paradoxical problems arise. The recycling of the waste produced through consumption by the wealthy has consequences of concern for the labouring and scavenging poor of China, India and elsewhere, and these consequences include both damage to the environment and damage to health and life (human and non-human). Consumers around the world now own, use, and dispose of an astronomical amount of electronic goods that are manufactured and sold in a market that is premised precisely on this relentless replacement of the old by the new (Brisman and South 2013a). Products may become obsolete by virtue of ceasing to function effectively or (more likely) be desired affectively, however, they are still valuable for their content and a 're-cycling for profit' 'boomerang industry' has now developed. For example, waste electronic items may be exported to China where they are de-manufactured and various parts (such as rare earth elements or rare earth metals (hereinafter 'rare earths'), discussed below) are then re-used in new electronic goods that are re-inserted into the cycle via the consumer market for the latest products. The towns, factories and scrap-yards specializing in processing these items employ a formal workforce and also create a shadow scavenger one in an informal economy based upon picking over the technological detritus. All are exploited in a low-paid and dangerously unhealthy industry which causes land and air to be polluted by use of hazardous chemicals and workers and communities to suffer increased incidence of serious diseases, some with cancer links (Huo *et al.* 2007; Chen *et al.* 2011).

Acts of consumption of technology-based products reliant on exploitation elsewhere

As Bonds and Downey (2012: 169) observe:

> Popular and academic environmental discourses often endow technology with heroic powers. According to such accounts, contemporary societies have the capacity to develop and commercialize new eco-efficient technologies that utilize significantly fewer natural resources and produce much less pollution per unit compared to previous generation technologies.

Using the case of hybrid electric vehicles ('HEVs')[1] and drawing on Kane and Brisman (2013), we can employ the concept of *technological drift* developed by Winner (1977) to explore the championing of new environmentally beneficial devices and systems while downplaying related unpredictable effects on cultures and environments. HEVs – which contain rare earths – are key to global efforts to switch to cleaner energy, but the extraction and processing of rare earths paradoxically causes environmental damage. Our reliance, then, on these 'paradoxical green machines' is part of technologically driven engineering design and consumer choice favouring devices shaped by structures – and infrastructures – predisposed to function durably in ever-changing contexts. The motor industry puts huge amounts of money into many forms of advertising (Holder 1991) and this seems to be fairly effective in persuading motorists to switch from old to new models for reasons of aesthetics, cost, efficiency, size or vanity but now also to 'help the environment' by getting better mileage while reducing carbon emissions.

So for example, in the case of the Prius – the most well-known HEV – while drivers may intend to pursue a 'green choice' by driving this particular car, they may fail to adopt a critical view of the mixed intentions of the motor industry. This leads to the contradictory effects of technological drift and what Kane and Brisman (2013) refer to as 'toxic displacements and comfortable illusions'. As Kane and Brisman (2013) explain, when HEVs first hit the US market, they were considered wimpy cars for tree-huggers (citing Brisman 2009: 352). In a short period of time, however, their popularity skyrocketed. Writing in 2007, Micheline Maynard observed, '[t]he Prius has become, in a sense, the four-wheel equivalent of those popular rubber "issue bracelets" in yellow and other colors – it shows the world that its owner cares. ... [B]uyers of the Prius want everyone to know they are driving a hybrid.'. Similarly, Brendan I. Koerner, writing for the periodical *Wired* in 2005, explained that the 'look-at-me factor' had helped make the Prius a hit: 'Buy a hybrid and passers-by will know that you're both hip and intelligent, not to mention part of a club that includes Leonard DiCaprio, Cameron Diaz [and other celebrities]'.

While it is true that HEVs, such as the Prius, quickly became a 'fashion statement', such 'conspicuous nonconsumption' (Brisman 2009: 353, 354) was not unfounded; hybrid vehicles were and continue to be fuel efficient – or, at least, possess greater fuel efficiency than their popular vehicular predecessor – the SUV and Hummer. But although hybrids may not guzzle oil, their 'electric motor[s] and batter[ies] guzzle rare earth metals, a little-known class of elements found in a wide range of gadgets and consumer goods' (Gorman 2009).

Rare earths are a set of seventeen metals – the fifteen lanthanides (with atomic numbers 57–71, from lanthanum to lutetium), along with scandium and yttrium, which tend to occur in the same ore deposits as the lanthanides and exhibit similar chemical properties (Alonso *et al.* 2012; Folger 2011).[2] Lanthanum (atomic number 57) is a major ingredient for hybrid car batteries; neodymium (atomic number 60) is the key component of an alloy used to make the high-power, lightweight magnets for electric motors of hybrid cars, such as the Prius, Honda Insight and Ford Focus, as well as in another type of green machine – wind turbines (Chandler 2012; Folger 2011; Gorman 2009; Margonelli 2009; Tabuchi 2010).[3] Compact fluorescent lamps (CFLs) use yttrium and terbium (Bradsher 2011; Folger 2011).

China is the main supplier of rare earths – mining more than 90 per cent of the world's deposits (Bradsher 2010, 2011; Chandler 2012; Folger 2011; McFadden 2013; Memmott 2012; Tabuchi 2010). According to Bradsher (2010: B1), '[r]ogue operations in southern China produce an estimated half of the world's supply of heavy metals'. The rest are state-owned, which, as Bradsher (2010: B1, B4) explains, 'means the Chinese government's only effective competitors in producing these valuable commodities are the crime rings within the country's borders'. As Bradsher (2010: B4) describes, gangs mine illegally some of the most sought-after rare earths (particularly, dysprosium, terbium, and europium) and 'have terrorized villagers who dare to complain about the many tons of sulfuric acid and other chemicals being dumped into streambeds during the processing of ore. Illegal rare earth mining and chemical runoff have poisoned thousands of acres of prime farmland, according to the government of Guangdong Province, and have been blamed for many illnesses'. Elsewhere, Bradsher (2011: B1) reports: 'The ease of digging up and refining some of the most valuable rare earths from the clay hills of southernmost Jiangxi Province and northernmost Guangdong Province, together with soaring prices, has led to a surge in illegal strip-mining that has turned many hillsides into lunar landscapes. Crime syndicates have dumped the mine tailings, including powerful acids and other materials, into local waterways. The fields and water supplies of peasant farmers who live downstream have been contaminated'.

According to Folger (2011: 145), '[s]ome of the most environmentally benign and high-tech products turn out to have very dirty origins indeed'. While there have been increasing efforts to recycle the rare earths from used electronics and to develop ferrite magnets, rather than rare earth magnets, for the motors in HEVs (Tabuchi 2010),[4] we are, as Kane and Brisman (2013:177) contend, 'faced with a paradox – a situation in which our good faith efforts at being green – HEVs, as well as wind turbines and CLFs – have and are continuing to have negative environmental repercussions'. Or, as Franz (2012: 96) laments, '[t]he new mythical narrative of corporate America – the green company facilitating consumer choices that help, not hurt the environment is the latest "free market" conceptualization. If we all consume the right light bulb, and drive the right vehicle, everything will be ok, and our collective guilt assuaged, even though the underlying evil of the machinery of production, consumption, and growth demand models are left intact' (internal citation omitted).

This idea that the environment will be best served by technological innovation and market self-adjustments is expressed particularly strongly in ecological modernization theory (Cohen 2006; Huber 2008; York et al. 2003). From this perspective, despite the development, ownership and commercialization of new green technologies being based principally in advanced, consumer economies, the message is that they will benefit all nations and populations universally or at least have the potential to do so. It is important to note the paradox of this development, however, for as critics (see discussions in Kane and Brisman (2013), and Brisman and South, 2015c) caution, such green technologies are unusually dependent upon rare and finite natural resources. This kind of supposed solution to the sustainability challenge should be accompanied by Stanley Cohen's (1982) warning about paradoxes harmful to developing nations: in this case, the environmental harm likely to be suffered by those developing nations that will be exhorted to supply the mineral and other resources needed to support processes of technology-based consumption under the guise of technological efficiency.

Faith in technological efficiency and the problem of Jevon's Paradox

The aims and claims of technological efficiency – that it can reduce demands on resources and also the demands of consumption – are undermined by the irony that efficiency may simply stimulate more demand, more generally. The original observation of this point by William Stanley Jevons is worth noting. Jevons was a nineteenth-century writer on political economy whose essay, 'The Coal Question: an Inquiry Concerning the Progress of the Nation, and the Probable Exhaustion of our Coal-mines' (1865), reflected a concern that Britain's enterprise and economic dominance would inevitably be challenged by the exhaustion of stocks of coal. This is similar to Max Weber's observation on the occasion of his 1904 visit to the United States (Scaff 2011), when he pointed out that the success of 'the boiling heat of modern capitalist culture is connected with the heedless consumption of natural resources, for which there are no substitutes' and suggested that it was 'difficult to determine how long the present supply of coal and ore will last' (Weber 1946: 366). Both Jevons and Weber failed to predict the age of oil but were right to draw attention to the limits of economic dependence on natural resources. As Jevons submitted, 'It is wholly a confusion of ideas to suppose that the economical use of fuel is equivalent to a diminished consumption. The very contrary is the truth. ... As a rule, new modes of economy will lead to an increase of consumption' (1865: 123). Tainter (2008: x), in turn, summarises the problem by arguing that 'as technological improvements increase the efficiency with which a resource is used, total consumption of that resource may increase rather than decrease. This paradox has implications of the highest importance for the energy future of industrialized nations. It suggests that efficiency, conservation and technological improvement, the very things urged by those concerned for future energy supplies, may actually worsen our energy prospects'. The insight provided by Jevons (as well as Tainter) suggests that efficiencies simply generate more market demand, more ability to raise production and thereby more encouragement of consumption (Bunker 1996; Hallett 2013; York, Rosa and Dietz 2003: 287 note 14).

Faith in magic wands

Technology produces 'goods' but also 'bads' and it is widely recognized that technology generates problems. This is not news but nor does it provide the type of mass anxiety that might stimulate a change because we always expect more, newer and better technology to produce the magic wand that will solve and resolve these problems. Consider the following tales about why we should be more 'grown up' about such wishful thinking.

In a children's story called *Emeraldalicious* (Kann 2013), Pinkalicious and her brother, Peter, discover that their favourite park is covered with piles of trash. Pinkalicious makes a wand out of a stick, vines and some flowers. When she waves the wand, makes a wish in rhyme form, and asks for what she wants 'with love', magical things transpire: garbage is changed into a throne; flowers blossom; colourful, twittering birds appear; a cast-off television becomes part of a castle; a hairbrush, hanger, harp and old wheel are recycled into a 'boat mobile'; and the entire dump is transformed into a 'greentastic garden'.

At the end of the story, a gust of wind breaks the wand apart, filling the sky with sparkly seeds. 'OH NO! What are we going to do without the wand? We can't make things happen anymore', Peter laments. 'The magic is gone!' Pinkalicious comforts her brother and tells him not to worry: 'We have these seeds, and with a little love we can make the entire world EMERALDALICIOUS!' Peter's reply – the last line of the story – is that that would be 'greenerrific!'

It is not clear whether the point in *Emeraldalicious* is that love is all that is needed to transform despoiled parks and dying nature into beautiful, bio-diverse green spaces or whether magic and creativity are also needed. While the former is wishful, the latter is fantastical. Regardless, few would disagree that the Earth could use a little more love and that our consumptive behaviours, patterns and practices have generated a tremendous amount of waste which is destroying its ecosystems. While *Emeraldalicious* is geared towards children ages 5–8, the book misses an opportunity to emphasize the imaginative ways in which unwanted items can be repurposed or to call into question the 'need' for all the things that have been discarded in the first place.

Where *Emeraldalicious* falls short, *Clara the Cookie Fairy* (Bugbird 2013), another children's story involving a wand, fares better. In this book, three Fairy Scouts – Clara, Kat and Jan – live on a mountain in a fairy camper van. To fund their camping expeditions, they sell special cookies. Clara's wand makes the cookies, Kat's wand makes boxes for them and Jan's wand makes bows for the boxes. One day, Clara sees a sign advertising MIGHTY MEGA-WANDS. Deciding that her old wand looks 'too boring', she discards it and purchases the biggest MIGHTY MEGA-WAND. Unfortunately, the new wand is not particularly well-made and it breaks soon after Clara begins using it. Unable to fix the broken wand, the Fairy Scouts determine that they must retrieve Clara's old wand.

The fairies go to the junkyard, where they discover that '[u]nwanted wands were everywhere – the place was overrun!' Many of them look just like Clara's. Clara, Kat and Jan spend hours trying different wands but not one of them makes cookies. 'How would they fund their camping trip if they had no cookies to sell?' Suddenly, the fairies realize that the wands that they had been testing and trying had been making the camping gear that they needed: 'There were pots and pans, forks and spoons, kettles and cans, and guitars to play tunes. There were balls and bats, three fairy bikes, bright summer hats, and boots for long hikes!' The fairies realize that the wands in the junkyard should not have been thrown away and decide to transform it into a 'Swap and Recycling Centre'. Soon, fairies from far and wide are coming to the centre to exchange their unwanted wands or to try to repair their broken ones. The story ends with Clara finding her lost wand, to which she cries, '"It's not a 'mighty-mega wand,' but to me it's twice as good!"' The author informs us that Clara 'learned to care for things, and not just throw them away' and the closing message is 'To swap or recycle, if you can, is always the better way!'

Clara the Cookie Fairy – purportedly pitched to children slightly younger than those to which the *Pinkalicious* series is geared (ages 3–5, rather than 5–8) – does encourage children to consider the importance of taking care of their belongings and to swap, recycle, reuse and repair instead of (always) seeking the latest and greatest item. In this way, it goes beyond the message in *Emeraldalicious*. But despite its more sophisticated message, *Clara the Cookie Fairy* does not interrogate the (manufactured) need for the 'initial consumption' of new things. But should we be surprised that both books do not identify or critique the hegemony of consumerism in contemporary culture and society?

Conclusion: green criminology, technology and consumption

Kane and Brisman (2013: 178) observe that '[t]echnology can change everything, be deployed everywhere; at the same time, its use may be restricted and its contradictory effects may be invisible (Barry 2001)' – and, indeed, the technologies themselves may be 'invisible' (Clark 2004). This reflects the point made by Hunter and colleagues (2002: 83) that it is often the case that 'technology merely transfers environmental risks or creates new risks to replace those it does not eliminate. "Miracle technologies" long considered safe have had unintended and serious effects on human health and the environment' – a perspective shared by Amster (2015: 32), who reminds us that 'the reliance on technological fixes for degradation and destabilization can potentially yield unintended consequences including new crises and escalations of existing ones'. Or, as Heinberg (2014: 19) asserts, 'regardless of whether consumerism is socially desirable, in the long run it is physically impossible to maintain. The math is simple: even at a fraction of 1% per year growth in consumption, all of Earth's resources would eventually be used up'. Technology, nonetheless, promises (literally) the earth. As White remarks, the hybrid vehicle has become 'the flavour of the month' and '[t]he notion that "business can profit by protecting the environment" is frequently linked to the idea of technological fixes vis-à-vis climate change, such as the development of hybrid cars, more technologically efficient appliances and light sources such as fluorescent bulbs. Burying the problem is also touted as a solution – whether this be radioactive waste or carbon pollution' (2012: 69, 71 (quoting Baer *et al.* 2009: 181)). As Zehner (2012: 223, 225) cautions, perhaps 'there really is no such thing as a green product' because 'the best material consumption is less material consumption'.

The paradox of growing disillusionment with technology sitting unhappily alongside continued dependence was one of the key points of Young's (1973: 191) observations on the place of technology in the politics of leisure and on the limits of creating and living in an alternative cultural world. Young's study of hippie, bohemian subcultures included discussion of their desire to find ways to counter the negative outcomes of the 'manifestations of technology' that they were critical of, such as 'Ecological pollution, industrial alienation and mindless consumerism' (1973:191). As Young points out, however, although a bohemian culture could critique affluent straight society, its alternatives were 'not flawless or consistent' and Young draws attention to the incongruity of individuals – then as now – managing to hold simultaneously, quite divergent positions. The dilemma, says Young, is the extent to which industrial progress and technological advances are inimical to human happiness and improvement versus the proposition that in fact technology is our best hope and means of achieving liberation:

> On one side there is the argument for simplicity and scorn for the consumer society, on the other the stress upon the development of new and more sophisticated leisure goods and pastimes. It becomes difficult to argue that television sets and washing machines represent the means by which people are bribed to serve the system if you belong to a culture whose central artefacts are the electric guitar, the stereogram and the video-tape!
>
> *(Young 1973: 191)*

With just an updating of the now quaint-sounding technology, this last difficulty remains at the heart of contemporary recognition of the risks to the environment and planet caused by our ways of living and to which we respond with paradoxical behaviours, strategies of accommodation, techniques of denial and too much faith in technological fixes.

Notes

1   The term, 'hybrid vehicle', refers to any vehicle that uses two or more distinct power sources to move it. The most common hybrid vehicle is the HEV, which combines an internal combustion engine (ICE) with one or more electric motors.

2   Despite their name, rare earths are not actually *rare*. According to Folger (2011: 138), '[t]he rarest rare earth is nearly 200 times more abundant than gold'. Rare earths are typically dispersed; what is rare is finding deposits large and concentrated enough to be worth mining (see Folger 2011; Margonelli 2009; McFadden 2013).

3 According to Gorman (2009), terbium and dysprosium are added in smaller amounts to the alloy so as to preserve neodymium's magnetic properties at high temperatures.
4 For a general discussion of the adverse impacts of mining versus recycling, see van Erp and Huisman (2010); for a discussion of 'urban mining' – the mining of '(raw) materials from products, buildings and waste in a society … [and] using those compounds and elements as resources for new production, thereby avoiding these materials from going to waste', see Bisschop (2012: 241).

References

Aas, K. F. (2013) *Globalization & Crime*. 2/e. Thousand Oaks, CA: Sage.

Agnew, R. (2013) The ordinary acts that contribute to ecocide: A criminological analysis. In N. South and A. Brisman (eds), *Routledge International Handbook of Green Criminology*, pp. 58–72. London and New York: Routledge.

Alonso, E., Sherman, A. M., Wallington, T. J., Everson, M. P., Field, F. R., Roth, R., and Kirchain, R. E. (2012) Evaluating rare earth element availability: A case with revolutionary demand from clean technologies. *Environmental Science & Technology* 46(6): 3406–14.

Amster, R. (2015) *Peace Ecology*. Boulder, CO: Paradigm Publishers.

Baer, H. and Singer, M. (2009) *Global Warming and the Political Economy of Health: Emerging Crises and Systemic Solutions*. Walnut Creek, CA: Left Coast Press.

Barry, A. (2001) *Political Machines: Governing a Technological Society*. New York: Athlone.

Bauman, Z. (1997) *Postmodernity and Its Discontents*. Cambridge: Polity.

Beck, U. (1992) *Risk Society: Towards a New Modernity*. London: Sage.

Bisschop, L. (2012) Is it all going to waste? Illegal transports of e-waste in a European hub. *Crime, Law and Social Change* 58(3): 221–49.

Bisschop, L. (2014) How e-Waste challenges environmental governance. *International Journal for Crime, Justice and Social Democracy* 3(2): 97–110.

Bisschop, L. (2015) *Governance of the Illegal Trade in E-Waste and Tropical Timber: Case Studies on Transnational Environmental Crime*. Surrey, UK: Ashgate.

Bonds, E. and Downey, L. (2012) 'Green' technology and ecologically unequal exchange: the environmental and social consequences of ecological modernization in the world-system. *Journal of World Systems Research* 18(2): 167–186.

Bosely, S. (2014) Drink (tap) water – not sports drinks, unless you really are a football star. *The Guardian*, 11 July.

Bradsher, K. (2010) The illegal scramble for rare metals. *The New York Times*. Dec. 30: B1, B4.

Bradsher, K. (2011) China seizes rare earth mine areas. *The New York Times*. Jan. 21: B1.

Brisman, A. (2009) It takes green to be green: Environmental elitism, "ritual displays," and conspicuous non-consumption. *North Dakota Law Review* 85(2): 329–70.

Brisman, A. (2010) The indiscriminate criminalisation of environmentally beneficial activities. In R. White (ed.), *Global Environmental Harm: Criminological Perspectives*, pp. 161–92. Devon, UK: Willan Publishing.

Brisman, A. (2012) The cultural silence of climate change contrarianism. In R. White (ed.), *Climate Change from a Criminological Perspective*, pp. 41–70. New York: Springer.

Brisman, A. (2013) Not a bedtime story: Climate change, neoliberalism, and the future of the arctic. *Michigan State International Law Review* 22(1), 241–89.

Brisman, A. and South, N. (2013a) Conclusion: The planned obsolescence of planet earth? How green criminology can help us learn from experience and contribute to our future. In N. South and A. Brisman (eds), *Routledge International Handbook of Green Criminology*, pp. 409–17. London and New York: Routledge.

Brisman, A. and South, N. (2013b) A green-cultural criminology: An exploratory outline. *Crime Media Culture* 9(2): 115–35.

Brisman, A. and South, N. (2014) *Green Cultural Criminology: Constructions of Environmental Harm, Consumerism, and Resistance to Ecocide*. London and New York: Routledge.

Brisman, A. and South, N. (2015a) 'Life stage dissolution', infantilisation and anti-social consumption: Implications for de-responsibilisation, denial and environmental harm. *Young – Nordic Journal of Youth Research* 23(1): 209–221.

Brisman, A. and South, N. (2015b) New "folk devils," denials and climate change: Applying the work of Stanley Cohen to green criminology and environmental harm. *Critical Criminology* 23(4): 449–60.

Brisman, A. and South, N. (2015c) State-corporate environmental harms and paradoxical interventions: Thoughts in honour of Stanley Cohen. In R. A. Sollund (ed.), *Green Harms and Crimes: Critical Criminology in a Changing World*, pp. 27–42. Basingstoke, Hampshire, UK: Palgrave Macmillan

Brisman, A. and South, N. (In Press) Green cultural criminology, intergenerational (in)equity and "life stage dissolution." In M. Hall, J. Maher, A. Nurse, G. Potter, N. South and T. Wyatt (eds), *Greening Criminology in the 21st Century*. Surrey, UK: Ashgate.

Bugbird, T. (2013) *Clara the Cookie Fairy*. Lara Ede, illus. Berkhamsted, Hertfordshire, UK: Make Believe Ideas Ltd.

Bunker, S. (1996) Raw material and the global economy: Oversights and distortions in industrial ecology. *Society and Natural Resources* 9: 419–429.

Chandler, D. L. (2012) Clean energy could lead to scarce materials. *MIT News*. Apr. 8. Accessed at: http://web.mit.edu/newsoffice/2012/rare-earth-alternative-energy-0409.html.

Chen, A., Dietrich, K., Huo, X. and Ho, S. (2011) Developmental neurotoxicants in e-waste: An emerging health concern. *Environmental Health Perspectives* 119(4): 431–438.

Clark, A. (2004) *Natural Born Cyborgs: Minds, Technologies, and the Future of Human Intelligence*. Oxford: OUP.

Cohen, D. (2012) The truth about sports drinks. *British Medical Journal* 345: e4737.

Cohen, M. (2006) Ecological modernization and its discontents: The American environmental movement's resistance to an innovation-driven future. *Futures* 38, 528–547.

Cohen, S. (1982) Western crime control models in the Third World in S. Spitzer and R. Simon (eds), *Research in Law, Deviance and Social Control*, vol 4. Greenwich, Conn. JAI Press.

Connor, S. (2014) Tiny plastic timebomb – the pollutants in our cosmetics. *The Independent*. 18 May. Available at: www.independent.co.uk/news/science/exclusive-tiny-plastic-timebomb-the-pollutants-in-our-cosmetics-9391412.html.

Fazekas, D. (2014) What if we stopped using plastic? *Yahoo!News*. 22 May. Available at: http://news.yahoo.com/blogs/what-if-abc-news/what-if-we-stopped-using-plastic-131200396.html.

Featherstone, M. (1982) The body in consumer culture. *Theory, Culture and Society*, 1, 18–33.

Folger, T. (2011) Rare earths. *National Geographic* 219(6) [June]: 136–45.

Foster, J. and Holleman, H. (2012) Weber and the environment: Classical foundations for a post-exemptionalist sociology. *American Journal of Sociology* 117: 1625–1673.

Franz, A. (2012) Climate change in the courts: A US and global perspective. In R. White (ed.) *Climate Change from a Criminological Perspective*, pp. 89–107. New York: Springer.

Freinkel, S. (2011) *Plastic: A Toxic Love Story*. Boston: Houghton Mifflin.

Freinkel, S. (2014) Wrap Party. *onEarth*. 28 April. Available at: www.onearth.org/articles/2014/04/why-the-plastics-industry-is-raucously-celebrating-the-fracking-boom.

Gibbs, C., McGarrell, E. F. and Axelrod, M. (2010) Transnational white-collar crime and risk: Lessons from the global trade in electronic waste. *Criminology & Public Policy* 9(3):543–60.

Giroux, H. A. (2004) *The Terror of Neoliberalism: Authoritarianism and the Eclipse of Democracy*. Boulder, CO: Paradigm.

Gorman, S. (2009) As hybrid cars gobble rare metals, shortage looms. Reuters. Aug. 31. Accessed at: www.reuters.com/article/2009/08/31/us-mining-toyota- idUSTRE57U02B20090831.

Hallett, S. (2013) *The Efficiency Trap: Finding a Better Way to Achieve a Sustainable Energy Future*. New York: Prometheus.

Hallsworth, S. (2005) *Street Crime*. Cullompton, Devon, UK: Willan.

Harris, B. (2011) The intractable cigarette 'filter problem'. *Tobacco Control* May; Available at: www.ncbi.nlm.nih.gov/pmc/articles/PMC3088411/.

Heinberg, R. (2014) The brief, tragic reign of consumerism. *Green Social Thought* 64(Spring): 18–20.

Hohn, D. (2012) The great escape. *The Observer Magazine*, 12 December: 25–29.

Holder, J. (1991) Regulating green advertising in the motor car industry. *Journal of Law and Society* 18(3): 323–346.

Huber, J. (2008) The global diffusion of environmental innovations: The standpoint of ecological modernization theory. *Global Environmental Change* 18: 360–367.

Hunter, D., Salzman, J. and Zaelke, D. (2002) *International Environmental Law and Policy*. 2/e. New York: Foundation Press.

Huo, X., Peng, L., Xu, X., Zheng, L., Qiu, B., Qi, Z., Zhang, B., Han, D. and Piao, Z. (2007) Elevated blood lead levels of children in Guiyu, an electronic waste recycling town in China. *Environmental Health Perspectives* 115, 7: 1113–1117.

Ilich, I. (1976) *Medical Nemesis: The Expropriation of Health.* New York: Pantheon Books.

Iyer, L. (2016) Obama bans microbeads in cosmetic products to reduce plastic pollution. *biotechin.asia* January 3. Accessed at: http://biotechin.asia/2016/01/03/obama-bans-microbeads-in-cosmetic-products-to-reduce-plastic-pollution/.

Jevons, W. (1865 [1965]) *The Coal Question: An Inquiry Concerning the Progress of the Nation, and the Probable Exhaustion of our Coal-mines.* New York: Augustus M. Kelley.

Kane, S. C. and Brisman, A. (2013) Technological drift and green machines: A Cultural analysis of the Prius Paradox. CRIMSOC: The Journal of Social Criminology, Green Criminology Issue, Autumn, pp. 156–90.

Kann, V. (2013) *Emeraldalicious.* New York, NY: Scholastic, Inc.

Klein, N. (2000) *No Logo: Taking Aim at the Brand Bullies.* London: Flamingo.

Koerner, B. I. (2005) Rise of the Green Machine. *Wired.* Available at: https://www.wired.com/2005/04/hybrid-2/.

Loader, I. and Walker, N. (2007) *Civilizing Security.* Cambridge: Cambridge University Press.

Lynch, M. and Stretesky, P. (2014) *Exploring Green Criminology.* Farnham: Ashgate.

Margonelli, L. (2009) Clean energy's dirty little secret. *The Atlantic.* May. Accessed at: www.theatlantic.com/magazine/archive/2009/05/clean-energys-dirty-little-secret/307377/.

Maynard, M. (2007) Toyota hybrid makes a statement, and that sells. *The New York Times.* July 4: A1, A11.

McFadden, D. (2013) Jamaica breaks ground on rare-earth project. Associated Press. February 4. Accessed at: http://news.yahoo.com/jamaica-breaks-ground-rare-earth-project-233915194.html.

McGuire, M. R. (2012) *Technology, Crime and Justice: The Question Concerning Technomia.* London: Routledge.

Melzer, D. and Galloway, T. (2010) Burden of proof. *New Scientist,* 23 October: 26–27.

Memmott, M. (2012) U.S. accuses China of causing 'massive distortions' in rare earths trade. NPR. March 13. Accessed at: www.npr.org/blogs/thetwo-way/2012/03/13/148518439/u-s-accuses-china-of-causing-massive-distortions-in-rare-earths-trade.

Moynihan, R., Heath, I. and Henry, D. (2002) Selling sickness: The pharmaceutical industry and disease mongering. *British Medical Journal* 324: 886.

Noakes, T. (2012) Commentary: The role of hydration in health and exercise. *British Medical Journal,* 345: e4171.

Novotny, T. (2014) Time to kick cigarette butts – they're toxic trash. *New Scientist,* 2975, available at: www.newscientist.com/article/mg22229750-200-time-to-kick-cigarette-butts-theyre-toxic-trash/.

Pearce, F. and Tombs, S. (1998) *Toxic Capitalism: Corporate Crime and the Chemical Industry.* Aldershot: Ashgate.

Petrie, K. and Wessely, S. (2004) Getting well on water. *British Medical Journal* 329: 1417.

Pinto, J. (2014) In India, growth breeds waste. *The New York Times.* November 18: A25.

Rothe, D. L. (2010) Global E-waste trade: The need for formal regulation and accountability beyond the organization. *Criminology and Public Policy* 9(3): 561–67.

Ruggiero, V. and South, N. (2010) Green criminology and dirty-collar crime. *Critical Criminology* 18(4): 251–262.

Scaff, L. (2011) *Max Weber in America.* Princeton: Princeton University.

Smith, N. (2015) Plastic problem. *SuperScience* (Scholastic) 26(7) [April]: 4–7.

Smithers, R. (2015) Call for ban on selling 'addictive' energy drinks to children. *The Guardian.* 26 February.

Snider, L. (2010) Framing E-Waste regulation: The obfuscating role of power. *Criminology and Public Policy* 9(3): 569–77.

Tabuchi, H. (2010) Mining trash for rare earths. *The New York Times.* October 5: B1, B5.

Tainter, J. (2008) Foreword, in J. Polimeni, K. Mayumi, M. Giampietro and B. Alcot (eds), *The Jevons Paradox and the Myth of Resource Efficiency Improvements.* London: Earthscan.

Thompson, R., Moore, C., vom Saal, F. and Swan, S. (2009) Plastics, the environment and human health: Current consensus and future trends. *Philosophical Transactions of the Royal Society B* 364, 1526. Available at http://rstb.royalsocietypublishing.org/content/364/1526/2153.

van Erp, J. and Huisman, W. (2010) Smart regulation and enforcement of illegal disposal of electronic waste. *Criminology & Public Policy* 9(3): 579–90.

Watts, S. (2013) What are long term threats of plastic in our seas? *BBC News Science and Environment,* available at: www.bbc.co.uk/news/science-environment-21236477.

Weber, M. (1946) *From Max Weber*. New York: Oxford University Press.

White, R. (2012) Climate change and paradoxical harm, in S. Farrall, T. Ahmed and D. French (eds) *Criminological and Legal Consequences of Climate Change*, pp. 63–77. Oxford: Oñati International Series in Law and Society/Hart Publishing.

Winlow, S. (2015) Book review: Realist Criminology. *Criminology & Criminal Justice* 15(5): 636–39.

Winner, L. (1977) *Autonomous Technology: Technics-out-of-Control as a Theme in Political Thought*. Cambridge: MIT Press.

Wood, J. and Dupont, B. (2006) *Democracy, Society and the Governance of Security*. Cambridge: Cambridge University Press.

Wyatt, T. and Brisman, A. (2014) The role of denial in the theft of nature: A comparison of biopiracy and climate change. Paper presented at the 2014 American Society of Criminology Annual Meeting, San Francisco, CA (21 November 2014).

Yarsley, V. E. and Couzens, E. G. (1945) *Plastics*. Middlesex: Penguin Books.

York, R., Rosa, E. and Dietz, T. (2003) Footprints on the earth: the environmental consequences of modernity. *American Sociological Review* 68(2): 279–300.

Young, J. (1973) The hippy solution: An essay in the politics of leisure. In I. Taylor and L. Taylor (eds) *Politics and Deviance*. Harmondsworth: Penguin.

Zedner, L. (2000) The pursuit of security. In T. Hope and R. Sparks (eds) *Crime, Risk and Insecurity*. London: Routledge.

Zehner, O. (2012) *Green Illusions: The Dirty Secrets of Clean Energy*. Lincoln, NE: University of Nebraska Press.

Keynote discussion

Technology, crime and harm

19

Evaluating technologies as criminal tools

Max Kilger

Introduction

Each of the preceding chapters primarily focuses upon the aspects, issues and consequences of the application of current and emerging technologies. In some chapters the focus is specifically on the direct application of a specific technology to criminological enterprises, as is the case for the discussions on cybercrime. In other chapters, the focus is how the approach to applying new technologies by legitimate concerns such as the pharmaceutical industry has led to practices that may be unethical at best and at worst bring adverse consequences that may be classified as criminal in nature.

Further discussions in this section examine the issue of "green crime" and how the production and materials in consumer goods are creating potential environmental hazards that are ignored or hidden by manufacturers. In the realm of digital crime, every year there are literally tens or hundreds of millions of victims that are the result of widespread financial cyberfraud through the automated theft of financial credentials from personal computers, financial institution servers and even devices such as automatic teller machines.

Focusing on consequences and large-scale effects, there are also chapters dealing with materials and wide-scale mass consequences that are often associated with weapons of mass destruction, an event such as the case of the disruption or contamination of the food distribution chain that threatens the health and safety of a large number of individuals or in the case of cybercrime, a type of attack that might be called a "weapon of mass effect" (Whyte 2015).

It must also not be forgotten that many criminal aspects of technology occur on micro-scales as well where actions transpire where the victim is often a single individual. Cyberstalking is a good example where digital technology enables one individual to acquire knowledge about another individual or covertly observe that person and use that knowledge gained to either threaten the victim in a classical, criminal extortion sense or use that information to plan and execute a more traditional crime such as kidnapping, rape or murder.

In other micro-level instances, rapidly advancing chemical technologies allow for the production of new compounds such as the various classes of designer drugs appearing in the grey marketplace. These compounds can often be purchased over the Internet and their possession is not illegal due to the ever-changing nature of the compounds that avoid laws directed at them. While these chemical compounds are often used for recreational purposes, they may also easily be used for criminal purposes such as covertly poisoning a political or romantic rival.

For each of the chapters, the reader should ask themselves which characteristics, elements and consequences of these technologies are unique to that particular technology and which are common components that are shared among many or all of the technological vectors present in the previous chapters. Gaining a better understanding of these technologies through analyzing their commonalities and differences may facilitate the development of a theoretical bridge through which these phenomena can be viewed from a more cohesive and comprehensive paradigm of how technology can be misused – criminally, or at least harmfully. It will also be useful in discriminating the development and understanding of where technology may be applied directly in a criminal enterprise and where the application of technology may indirectly result in consequences of a criminal nature. While traditional conceptions of crime often involve the use of force or coercion it is evident from these chapters that this is not always the case. It may also be useful to compare each of the topics of these chapters in terms of the actors involved – whether they be traditionally non-criminally inclined citizens, traditional criminals or criminal groups, companies in specific industries, or nation-states.

The convolution of crime, technology and terrorism

Crime, terrorism and the emergence of new technologies are phenomena that are currently present in our world and are likely to remain prominent features in at least the near- to mid-term future. Holt (2012) in his article on technology, crime and terror describes the "intersection" of these three phenomena and how these criminals and terrorists are utilizing technology to facilitate or commit criminal or terrorist acts. However, I would argue that the nature of the interaction between crime, terrorism and technology is more than just an intersection. The argument that I propose is that what we are witnessing is a convolution of these three elements in that each element represents a dynamic system that has complex interconnections with each other. Changes in technology may change not only the methods by which each of these malicious actors (e.g. criminals and terrorists) accomplish their activities but more importantly technology changes the motivations, norms and values and social structure of the individuals in these communities. In turn, changes in these two communities can have effects on determining the shape and nature of emerging technologies and their applications. Stehr (2005) suggests that it is not just nation-states that help determine the purpose and regulation of technology but also individual scientists, the media and networks of interested people and groups of people affected by the specific technology. This concept of how technology is regulated and purposed encompasses both the criminal and terrorist communities.

One of the potential consequences of emerging technologies is the potential for the blurring of the lines of demarcation between criminal and terrorism acts. There are a large number of differing definitions of crime offered by researchers and Agnew (2011: 194) attempts to provide a universal definition of crime as "(1) acts which cause blameworthy harm, (2) are condemned by the public, and/or (3) are sanctioned by the state." Terrorism also retains a plethora of over one hundred competing definitions of terrorism, and Hoffman (2006) spends considerable time discussing the difficulties in providing a mutually agreed upon definition of the term. Schmid and Jongman (1988) provide an analysis of the terms involved in the definition of terrorism and among the most common terms are violence, political, fear, terror, threat and noncombatant or civilian.

Often both the motivation for the malicious act as well as the scale of the malicious act have a direct bearing on whether the act is classified as criminal or terrorist in nature. Typically in most descriptions of terrorism the motivations for these acts are described as political in nature. The act is used to draw attention to a particular political cause and the objective is to punish a collective group or nation-state for these perceived political wrongs or social injustices. However, Kilger (2015) argues that in the specific case of cyberterror, these acts are actually instances of the more general phenomena of online malicious acts and that online malicious acts are the result of six motivations: money, ego, entrance to social group, cause, entertainment and status. If this is the case for cybercrime acts as well, then one might argue that these two communities may interact with each other in a number of interesting ways, including the exchange of goods, services and even the migration of members of one of these communities to another.

A second differentiator between criminal and terrorist acts can often be seen in the scale of the malicious act itself. Malicious acts that often involve single, unremarkable victims are more likely to be labeled as criminal in nature. The murder of a single member of the general public, even if the act has political overtones or motivations, is likely more to be seen as a criminal act rather than a terrorist act. However, if a malicious act is serious, such as a mass killing and has a relatively large number of victims – perhaps as few as in the teens – then the question of labeling the acts as terrorist acts seems to come into play and the search for traditional terrorist motivations often results.

This aspect of the scale of the malicious act is one that has direct bearing on emerging technologies. One of the key characteristics of a number of emerging technologies – in particular digital technology – is the aspect of significant scalability of the effects and consequences of the malicious application of those technologies. Certain current and new technologies are inherently widely adaptable so that the effects and consequences of a specific technology may be applied or felt on a very variable scale ranging from the single individual to effects that can be felt by millions of people.

The issue of scalability brings up an important point. If a specific technology is scalable such that it is able to target a single individual or thousands of individuals with an event that has serious consequences, how do we treat that attack against a single individual – as a criminal act or an act of terrorism? The application of new, emerging technologies in the course of committing a malicious act against a target is likely to blur the distinction between criminal and terrorist acts. This may be especially true given that the payoffs of the act or event may be extremely scalable at the discretion of the perpetrator. Given the natural tendency for the magnitude of payoffs of a malicious act to be closely associated with the extent to which the perpetrator scales the act via technology, this suggests that we may want to think about framing criminal acts that utilize technology more in terms of a hybrid of criminal and terrorist perspectives, even when these malicious acts seem more criminal in nature and don't always comfortably line up with traditional terrorist definitions.

It is in this spirit that some portions of the following discussions address some of the aspects of a more hybrid model of malicious acts where theoretical components of terrorism frameworks mix with criminal aspects, in hopes of providing the reader with a more flexible and adaptable perspective on the convolution of crime, terrorism and technology.

The role of digital technology as both threat vector and catalyst

Digital technology currently plays a dual role in both the criminal and terrorist enterprise. Criminal organizations for a number of years have utilized digital technologies in pursuit of their primary objective of generating financial gain through a variety of illicit activities. For example, cybercriminals have utilized a number of different methods such as drive-by malware on websites, software Trojans to steal financial credentials and phishing campaigns to lure the unwary into behaviors desired by the criminal actors. The use of technology by criminals has given rise to a number of novel tactics and approaches which are often aimed at financial gain for themselves. For example, ransomware is software that typically disables the victim's computer, often by encrypting the contents and asking for a specific amount (often in untraceable currencies such as Bitcoin) in order to provide the victim with access to their computer again.[1]

Digital technology has also acted as a catalyst in providing criminals with an inexpensive way to establish a global marketplace. Without this versatile and globally available marketplace, the flow of illicitly gained assets would likely be markedly reduced because of the difficulties and hazards of matching buyers and sellers of these goods through other channels. Digital technology has spawned an entire underground economy replete with criminals soliciting customer reviews, transaction facilitating services such as escrow agents, money-back guarantees and customer service representatives that act on behalf of these criminals (Holt 2013).

Emerging technologies also encourage criminals to think in new and innovative ways. One of the most straightforward ways is that criminals will examine digital technologies looking for vulnerabilities that will allow them to gain control of the device for the purposes of exfiltrating valuable data from it. In the realm of advanced persistent threats (APT), criminals may attempt to compromise a device in order to gain a foothold into a computer network where they can then craft and launch successive attacks on other network devices or other networks in order to reach the ultimate target.

Another objective may involve the use of technology to defeat basic security features of common consumer products in order to perform more traditional criminal activities. There are now devices that can be easily built that can record and defeat automobile remote locking mechanisms, even those with rolling digital security codes. Employed in the field by auto burglars, these devices stealthily capture the digital door locking and unlocking codes for newer automobiles (Greenberg 2015). Once the owner locks and leaves the vehicle, the criminal uses this simple device to replay the unlock code and unlocks the vehicle doors, where he or she can then search the vehicle for valuables or even steal the vehicle. This suggests that the hacking community, in its pursuit of vulnerabilities of various digital devices, may be unwillingly training criminals in new techniques and trades through their exploration of the workings of new, emerging technologies.

Kilger (2010) has also suggested that technology is playing a role in creating synergies and loose couplings between cybercriminals and their traditional street gang counterparts. One example he cites is where cybercriminals obtain personal information about a potential crime victim through various legal and illegal online means. This information includes their home and work addresses, their age, telephone numbers, marital status, and personally identifiable information on possible children, etc. Once this information is obtained, then the cybercriminals contact the potential victim and describe their knowledge of his personal life and details and they demand a payment in an untraceable digital currency such as Bitcoin. They suggest that if the funds are not forthcoming then some physical act of violence will be perpetrated on the victim or their family. If the victim complies with the payment, the cybercriminals move on to the next victim. If the target victim does not comply, then they use a local criminal gang and contract with them to inflict physical harm on the victim or their family. The violent traditional criminal gang gets a payment for their services in the form of Bitcoin or similar digital currency and awaits another possible job. This is a good example of how different criminal groups can mix high and low technologies. Cybercriminals utilize digital technology in the data gathering phase of the operation while they in effect indirectly utilize low technology methods (e.g. traditional physical violence) when they employ the services of the traditional street gang. Conversely, the street gang utilizes high technology when they receive and convert Bitcoin payments for their work on behalf of the cyber gang.

One additional digital technology is likely to play an increasingly important role in the lives of current and future criminals. I refer here to the increasing popularity of untraceable current and emerging digital currencies such as Bitcoin, Litecoin and others. There are several key characteristics that will likely make these types of digital currencies indispensable to criminal gangs as well as terrorists. The most important feature is that in theory they are touted as being untraceable. Being able to make and receive payments anonymously is an extremely attractive characteristic for criminals or terrorists. This allows criminals to fund their activities as well as receive revenues without the funds being able to be traced. Ironically, this also makes these digital currencies attractive targets for cybercriminals to steal. There have been a number of cases where very large sums of money in the form of Bitcoins have been stolen or mysteriously have disappeared (Biddle 2015).

Challenging the traditional definitions of weapons of mass destruction

The discussions in some of the previous chapters remind us that hazardous biohazard, chemical and nuclear materials present a unique environment within which the demarcation lines between criminals and terrorists blur. The tremendous costs of the storage, care dispersal and disposal of chemical and nuclear by-products and waste invite criminal behavior not only on the part of the principals involved in the primary manufacturing process in avoiding those huge costs but also these characteristics present additional opportunities for criminal enterprises to take this burden off primary toxic material producers.

Often these criminal enterprises are only tangentially concerned with the proper care and disposal of these materials but rather are focused upon the significant profits that can be extracted from these activities, especially when avoiding performing due care in handling and storing these materials. Further, these biohazard, chemical and radiological materials have value to terrorists who wish to utilize them in the production of weapons of mass destruction in various forms including bioweapons, chemical weapons, radiological dispersion devices and perhaps one of the most tempting opportunities of all, the diversion of strategic nuclear materials for the purposes of constructing a crude nuclear weapon. The lack of proper oversight of the entire process cycle of safely disposing of these toxic materials may make primary manufacturers and industries such as the nuclear power industry at least partially complicit in the chain of hazardous material transfers that end up providing precursors or basic materials necessary for the commission of criminal or terrorist acts using materials usually associated with weapons of mass destruction. As will be seen in further discussions, criminals and criminal enterprises may function as intermediaries between legitimate industries and terrorist end users or criminals may even end up committing what would be legitimately labeled terrorist acts using these materials.

The diversion and conversion of some types of industrial waste and by-products into precursors and primary materials for weapons of mass destruction reminds us that it may be useful to start a dialogue with an examination of some of the theoretical discussions surrounding the definition and meaning of the concept of a weapon of mass destruction. These weapons include nuclear and thermonuclear, radiological, chemical and biological weapons. However, the abstract construct of weapons of mass destruction has recently come under attack from researchers and theorists in fields such as nuclear proliferation and conflict studies.

A number of researchers argue that the definition of weapons of mass destruction is malleable and subject to different interpretations depending upon the context. Enemark (2011) suggests that lumping nuclear, chemical and biological weapons under one rubric of weapons of mass destruction is incorrect and actually a dangerous precedent. He argues that the magnitude of damage from a chemical attack, for example, is likely to be significantly less than a nuclear attack. He also suggests that the magnitude of damage from a biological attack is highly variable and dependent upon the conditions and method under which the biological agent is dispersed.

While the US government has suggested that a biological attack could indeed rival the damage from a typical small-yield nuclear device, Enemark points out that this extreme effect is reliant upon near-perfect conditions and a very efficient and accurate delivery mechanism. He concludes that including chemical and biological weapons in the same class as nuclear weapons through the use of the concept of weapons of mass destruction is a serious mistake given the differences in the levels of magnitude of destruction typically expected by each weapon type.

Bentley (2012) suggests that while Enemark's idea of dismantling the term "weapon of mass destruction" (WMD) might be constructively abandoned among academics, she argues that the term has significant political utility and politicians and policymakers are likely to retain the concept rather than abandon it for more specific terms as Enemark suggests. Bentley conjectures that the very act of defining something as a weapon of mass destruction is "a political, innovative, and non-neutral act." A further dismantling of the term weapon of mass destruction may also come about as criminals become embedded more deeply in the supply chain of supplying biohazard, chemical or nuclear materials to terrorists. This phenomenon may be further promulgated by the direct entry by criminals into the planning and execution of terrorist acts using these materials.

She points to an early example where the US Joint Chiefs of Staff wanted to have small tactical nuclear devices destined to become satellite killers excluded from the traditional definition of weapons of mass destruction so that they could be exempted from current treaties and ongoing negotiations about WMD. In a more recent example, Bentley points out that George W. Bush deliberately emphasized the threat from Saddam Hussein as him having weapons of mass destruction and to some extent de-emphasizing specific threats such as chemical or nuclear (Bentley 2012: 396):

> Conceptual ambiguity was a political move, one in which the avoidance of a precise definition – or rather instituting an 'absence of definition as definition' approach – was strategic. Since the actual weapons capabilities of Saddam and terrorist groups could not be definitely proven, the adoption of a deliberately abstract notion of WMD helped to create the sense of threat necessary to take action in respect of the War on Terror without requiring absolute proof. The allusion to mass-level destruction was sufficient to create fear among policymakers and the American public, which in turn could be used to justify US foreign policy. As long as people believed there was a terrible threat – even worse than that seen on September 11 – that must be removed, the George W. Bush administration could pursue the War on Terror on its own terms and without evidence of WMD possession.

This dismantling of the term weapon of mass destruction has particular meaning in the case of cybercrime. The current frenzied framing by the US government of the magnitude of cyberthreats to national security, critical infrastructure, the national economy and so on, trickling down to threats to the individual in more familiar forms such as financial cybercrime, would seem to suggest that the US considers the cyberthreat arena to be as serious as some other more traditional forms of weapons of mass destruction. Yet the US government seems to fastidiously avoid the use of the term weapon of mass destruction when referring to cyberthreats. Instead, government officials seem to label a significant portion of what is occurring within the cyber threat matrix as originating with criminals or "cybercriminals". Even when the US government discusses threats to the military and critical infrastructure that might be classified as terrorist activities, they often refer to criminal hacking gangs that are in collusion with other nation-states. It appears to a great extent that the likely valid classification of certain major cyberthreats as potential weapons of mass destruction is being purposely avoided by government policymakers and instead has been depoliticized through the designation of actors involved in these activities as criminal elements sometimes acting in coordination with foreign governments. It will be interesting to see, if this is the case, that when traditional criminal groups commit malicious acts with traditional weapons of mass destruction they are still labeled as criminals rather than as terrorists.

It should be pointed out that the definition of a weapon of mass destruction may also vary according to the nature of the organization creating the definition. As Carus (2012) points out, Congress altered the federal criminal code to include laws that criminalized conspiring to use or using weapons of mass destruction that in addition to the traditional definition included bombs, grenades and mines. This definition of WMD was used in the successful prosecution of Timothy McVeigh and Terry Nichols for the 1995 bombing of the Alfred P. Murrah Federal Building in Oklahoma City. In all, research by Carus uncovered over 50 definitions for weapons of mass destruction across a variety of governmental and international organizations. This formal and detailed declaration of terrorist acts involving weapons of mass destruction into the criminal code may further blur the lines between terrorist and criminal.

It is clear from the literature and this introductory discussion that while the most commonly accepted definition of weapons of mass destruction includes nuclear, radiological, chemical and biological weapons, there is significant room for variation in that definition and often that variation is contextual and sometimes associated with political rather than scientific justifications. Further, there are additional ambiguities when discussing other significant attacks such as contamination of the food or water supply by chemicals or biological agents – are those events to be classified as direct chemical or biological attacks and thus typically classified as an attack by a weapon of mass destruction or are they somehow to be treated by secondary or indirect attacks that are not included in the accepted definition of WMD?

This also brings up the point of the use of the word "attack" when biohazard, chemical or nuclear materials are involved. Given the prior discussion of the depoliticizing of the term weapon of mass destruction as well as the discussion about the codification of the use of a weapon of mass destruction as a crime, why aren't these events labeled crimes rather than attacks? Further, why aren't people who carry these acts out using these special materials labeled criminals rather than terrorists? Timothy McVeigh was labeled a domestic terrorist by the media as well as the government in his primary involvement in the bombing of the Alfred P. Murrah federal building in Oklahoma City. He was convicted in a criminal court of law. As attacks of this nature become more common, where biohazard, chemical or nuclear materials are involved, will we see a subtle transition where these events and the actors involved are labeled within a criminal rather than terrorist context?

In addition, does the class of attacker also make a difference in whether or not an event is labeled an attack by a weapon of mass destruction? Historically weapons of mass destruction have been developed, stockpiled and on rare occasion used by nation-states. As the future brings the ever increasing probability that traditional weapons of mass destruction might be developed or stolen and used by non-nation-state actors who might be labeled as criminals or terrorists, will we continue to designate these weapons as weapons of mass destruction? Does the type of actor or their motivation that deploys these weapons have any effect on the definition of the weapons themselves? Will we eventually alter the definition to exclude not only the aforementioned bombs, grenades and mines but also exclude chemical and biological weapons from the definition?

The answer to these questions may partly lie in the fact that crime is usually framed in terms of the acts of individuals and this may have an effect on how non-individual class entities are treated when putting technology to non-normative use. When nation-states utilize technology in the form of weapons of mass destruction the implementation is often framed as occurring within an "act of war." When commercial enterprises commit illegal acts using technology and when these acts involve biohazard, chemical or nuclear materials, many times the form of legal reconciliation comes in the form of civil rather than criminal prosecution. Even when commercial entities are pursued in criminal actions involving the use of technology or technology-produced by-products, frequently few individuals are labeled as criminals and almost no one ever goes to jail. As more legislation is enacted and new laws regarding the misuse of technology are created and expanded, they are likely to be framed in terms of individual misconduct. This suggests that corporations and governmental entities may continue to enjoy some partial immunity from punishment for misusing materials or even computer code in the areas of biohazard, chemical, nuclear or even digital technology while individual actors who commit acts of this nature may more and more come to be labeled criminals rather than terrorists.

We also have to consider the case of cyber-weapons. If one can digitally attack a nation's critical infrastructure and cause significant damage and loss of life, does that qualify cyber-weapons to be considered weapons of mass destruction? What if a cyber-weapon is used to control and execute a nuclear attack on a specific target – does that make the cyber-component a weapon of mass destruction? On a smaller but similar scale, there is the threat that malicious actors will identify a zone of recent conflict such as the Crimea and utilize malware and other cyber-tools to compromise kinetic weapon systems in a volatile region where opposing forces face off against each other. Criminals could construct a plan to compromise a weapons system and use it against the opposing forces causing a localized conflict where the opposing military force retaliates. During the confusion and chaos that ensue, these criminals could move to commit one or more significant crimes under cover of that chaos such as the theft of major financial assets including gold bullion, federal cash reserves, theft of valuable art from a museum or other valuable objects.

Keeping to this cyber theme and reflecting back to the chapter on the theft of personal information gives rise to another possible scenario for consideration. It is evident from the headlines in various media channels that the criminal theft of personal information that is then used by one individual to threaten, inflict embarrassment, or harm another individual is becoming more pervasive. Often when seen within the scenario of one perpetrator and one victim this often becomes a case of stalking or online harassment. While this type of one-on-one digital crime is happening to a significant number of individuals, there have been instances of much larger scale versions of this tactic. One of the most dramatic data breaches was the theft of personally identifiable information for almost 21.5 million individuals along with potentially embarrassing personal information from the US Office of Personnel and Management in June, 2015 (OPM 2015).

While the most obvious objective for this theft was the use of this information for counter-intelligence purposes likely by a nation-state, what would have happened if that theft had been carried out by criminal hackers who proceeded to divulge personal and embarrassing details of millions of individuals online? Would that have been classified as a criminal act on a large scale or would there be some justification in labeling that a terrorist act due to the scale of the act and the possible cumulative consequences it would have in individuals working in sensitive national security positions? These questions involving crimes committed within the digital space are ones that likely do not have easy answers. It may be useful to address these questions again after having re-read the preceding chapters on each specific type of threat.

Issues of deterrence

Deterring the criminal misuse of technology is an issue that likely will confront researchers and policymakers for some time. One method that may be effective in deterring the use of technology to accomplish criminal objectives is to adapt and utilize technology itself as an instrument of social control. Marx (2001) suggests that there are six potential remedies that revolve around social control and technology that are relevant to deterring technology-driven crime. Marx proposes that

> Ideally problems are anticipated and eliminated (1, below); or where that is not possible, the goal is to create deterrence by reducing the gain (2), making violation more difficult and costly (3,4,5) or by increasing the likelihood of identification and apprehension (6).

The first remedy according to Marx is to remove or obfuscate targets of technologically embedded crime. He suggests for example that merchants only accept credit or debit cards to reduce the theft of cash or that gold bars be painted black to disguise their true value. In the digital world, an analogous example of that would be to remove computer servers and other systems from the public Internet and replace connectivity with private, isolated physical networks. This would force criminals to gain physical access to the target network and deter a significant proportion of attacks. Obfuscation in the digital world to thwart potential sexual harassment might be something as simple as switching gender assignations in online profiles to reduce unwanted attention – a practice that is common today in many online environments.

Target devaluation is Marx's second deterrence strategy, utilizing technology to make a target unattractive to criminals such as in the case of products (car radios, smartphones, etc.) that disable themselves when they have become successful targets of a theft. A similar approach has been in use for many years in terms of food safety, as Marx points out, when tamper seals on food products indicate that the product should not be used. An example of target devaluation in the digital world is the deliberate salting of misinformation in sensitive documents, especially those with military or intelligence value. The inability to determine which elements of stolen sensitive documents are reflective of the truth and which ones are misleading or false can seriously damage the value of those documents to a criminal for resale to third parties as well as significantly increase the cost of harvesting worth from the documents in terms of additional efforts necessary to ascertain the truthfulness of element of their content.

The third deterrence strategy on Marx's list is target insulation. One of the most common examples of this is the various software and hardware security devices present in digital networks such as firewalls, intrusion detection systems, demilitarized zones in computer networks, anti-virus programs and isolating computer networks using low-tech air gapping strategies where computer networks are isolated by having no connection to other networks. Each of these entities attempts to insulate the computer network, servers and client computers from harmful malware and intrusion attempts. However, even extreme measures such as air gapping digital networks are not entirely effective, as has been seen in the last couple of years with various strategies, such as using an application on a nearby smartphone to pick up radio emissions from an air gapped computer in order to compromise information (Guri et al. 2014). Physical deterrence systems are also commonplace in bioengineering, chemical or nuclear power facilities where controlled access systems, gates, fences and guards are used to limit access to only those individuals who are authorized to be onsite. These systems of course do not deter insider threats where employees or authorized contractors may gain entrance and proceed to conduct criminal activity within the walls of the facility or remove materials or assets to offsite locations.

The fourth deterrence factor according to Marx is offender weakening or incapacitation. Here he cites a number of examples such as chemical deterrents like pepper spray in police work, alcohol interlock systems for vehicles and GPS sensors to prevent smartphone use while driving. An online example of this deterrence strategy used to thwart criminals are the methods utilized by payment card system interventions processes. In order to reduce the amount of fraud by online pharmacies, fake anti-virus threats and other online criminal activities, payment card systems and affiliate banks may restrict or close merchant accounts for those clients that appear to be conducting illegal online activity (McCoy et al. 2012), thus making it more difficult for criminals to collect payments for illegal goods and services and extract profit from their activities.

Exclusion is the fifth component of Marx's theoretical take on technology and social control. Marx suggests that GPS monitoring bracelets worn by criminal offenders convicted of crimes is one such example. GPS monitoring bracelets deter individuals from leaving specific locations or being present by alerting authorities when proscribed areas are left or prohibited areas entered by the individual. Another example of excluding the offender is the use of Internet bans by court systems for individuals convicted of child pornography crimes. This strategy however has stirred some significant controversy in the legal system because of the fact that the Internet now plays such an important part in daily life that forbidding individuals from using it may have consequences to the offender much beyond its intentions of protecting children from harm (Ramirez 2014).

The final element in Marx's schema is offense/offender/target identification. He cites a number of different examples of this including biometric measures of identity such as fingerprints, facial measurements, gait and voice characteristics. He also cites law enforcement measures such as Amber Alerts, criminal warrant websites and crime reenactments on media as well as the use of hot spot analysis and other crime statistic analytical techniques that help distribute police resources as ways in which criminals may be deterred. However, biometric identification is far from fool-proof. There are a number of ways in which criminals may spoof various biometric identification measures such as eye movement, fingerprints, facial and speech recognition, and not all countermeasures to these attempts are successful (Wu et al. 2015; de Freitas Pereira et al. 2012; Hadid et al. 2015). As these authors suggest, in addition to various methods to imitate or spoof biomarkers, the identification by biometric markers is vulnerable to man-in-the-middle attacks where the data that represents a person's biomarker is captured and replayed to imitate the authentication signature of the authorized signature. Biomarkers also suffer from the fact that they are often permanent and non-changeable. Once a fingerprint signature is stolen for example, the owner of the fingerprint is not able to change their fingerprint, unlike other authentication technologies such as passwords.

Deterrence has also played a major role both in the strategy of national defense using nuclear, biological and chemical means, but often there have been technological barriers that have prevented criminals and terrorists from utilizing these materials for criminal or terrorist activities. Traditionally the research into and the development of chemical, biological, nuclear and radiological weapons has been the purview of nation-states. However, with advancements in technology in a number of areas, the opportunities for these kinds of activities to be undertaken by non-state actors for terrorism or criminal objectives are increasing. For many years there have been a number of barriers for non-nation-state actors to develop and deploy these kinds of weapons traditionally thought of as weapons of mass destruction. Access to material precursors, the difficulties in obtaining the processing equipment and the knowledge to use it correctly, lack of knowledge concerning key obstacles needed to be overcome to produce weaponized materials and the risks associated with exposure to primary materials used in weapons of mass destruction have historically severely limited the production of these weapons to those entities with the resources and knowledge to produce them – typically nation-states.

These limitations also made attribution – a key component of deterrence – easier as it severely limited the pool of hypothetical perpetrators for most of these types of weapons to those entities who possessed the aforementioned resources and skills. Deterrence depends upon the ability to attribute an attack to a specific entity. As the pool of these entities grows exponentially with the addition of non-state actors and groups, the capacity of governmental agencies to correctly identify the perpetrator of a WMD attack begins to degrade. For example, as some key technologies such as DIY gene editing emerge, the number of actors who might be considered in the pool of perpetrators suspected of a biological terrorism event widens significantly. Now non-state actors under certain circumstances may have the ability to develop and deploy certain classes of weapons that have traditionally been classified as WMD. This will have a deleterious effect on the ability of nation-states to swiftly identify and punish the perpetrators of an attack involving WMD. Without reliable attribution it is difficult to maintain a strong deterrent because of the issues in identifying who is responsible for an attack. As Miller succinctly states:

> The U.S. has to contend with a revolutionary dispersal of power: nation-states have seen both non-state actors and individuals equipped with the ability to wreak havoc undeterred by the threat of criminal justice or retaliation, or even of discovery.
>
> *(2015: 110)*

As attribution challenges increase, law enforcement and the intelligence community may have to resort to additional, supplemental methods to improve the ability to attribute technological crimes to the actual perpetrators. For example, Kilger (2015) has suggested that online criminal and even terrorist acts may be motivated by six different factors. Using these motivational factors to reduce the pool of potential perpetrators or construct future threat scenarios that utilize weapons of mass destruction by non-nation-state actors may eliminate certain groups and help identify a pool of criminal suspects that can be further investigated. This process is often one that is utilized in criminal profiling by law enforcement agencies.

Indeed the challenge of attribution is even more complex and tainted with a collaborative variant of the fog of war. Healy (2012) in a key article on attribution of cyber-attacks developed a spectrum of the level of involvement that a nation-state might have in a specific cyber-attack in relation to third parties who may in fact be non-nation-state actors. This spectrum, while not technically a strict monotonic scale of the involvement and cooperation between nation-states and non-nation-state actors in a cyber-attack, ranges from one end labeled State-prohibited, where the nation-state will assist in stopping a third party attack, to the other end of the spectrum labeled State-integrated, where the nation-state utilizes governmental cyber forces in collaboration with third party hacker groups to conduct a cyber-attack.

Healy further proposes that attribution of specific actors for specific cybercrime or cyberterrorism acts may not be the most effective way to deal with this issue. He suggests that it is unfortunate that "the international community places few expectations on nations to reduce attacks originating from or routing through systems in their sovereign country." He likens this to international expectations for order in "ungoverned spaces," citing Somalia as an example and suggests that nation-states currently treat cyber-attacks as if coming from some similar amorphous ungoverned space.

Healy suggests that it is a national responsibility for nation-states to enact national policies that encourage information security policies, set expectations for increased security for service providers, training law enforcement in the various technical aspects of addressing cyber-attacks and encouraging international cooperation among nation-states. While these activities are no doubt positive and encouraging steps in deterring cyber-attacks they may not be sufficient to quench the antagonism and temptation to blame nation-states for particular events nor deter nation-states from pursuing the parties responsible through investigative and intelligence efforts to correctly attribute the attack to specific actors.

Indeed, given the serious consequences not only of cyber-attacks against critical infrastructure, but also terrorist attacks utilizing weapons of mass destruction, this policy of more generalized national responsibility when contrasted against a specific WMD event may not work well. This kind of diffused responsibility for specific catastrophic events seems more like a "oops, sorry about that" response that is likely to fail in cases where mass casualties or significant economic loss is present.

Jackson (2012) outlines one of the more recent strategies in WMD deterrence – tailored deterrence. This strategy relies upon tailoring your deterrence strategy to the opponent at hand. He recounts four different waves of tailored deterrence strategy that involve a knowledge of the culture of your opponent, including knowledge of their strategic culture, the incorporation of micro-economic foundations and the use of punishment and denial logic and an understanding where rational, universal deterrence failed and why.

The latest wave of tailored deterrence strategy seeks to apply its principles to the inhibition of the use of WMD by rogue states and non-nation-state actors. Jackson as well as others such as Knopf (2013) realize that tailored deterrence relies upon the availability of information about your opponent and they both point out that in terms of nation-state actors North Korea poses a particular challenge to tailored deterrence due to the fact that it remains a tightly controlled police state where almost any kind of information about its strategic policies, military, WMD programs and culture remains beyond reach.

This lack of information becomes more ubiquitous and problematic when dealing with non-nation-state actors. It is clearly important to understand the cultural and social dynamics when dealing with non-state actors in the arena of cybercrime or cyberterrorism or other threats (see Kilger 2015 for a cyberterrorism example). However, the difficulties that arise in acquiring this information about non-state actors – whether from constraints due to legal restrictions, operational security on the part of these actors or just the significant increase in the pool of potential perpetrators – makes the idea of applying tailored deterrence to this new emerging threat seem much less attractive, just as in the case for the North Korean example cited above.

In addition to the issues of attribution in relation to deterrence, Zagorcheva and Trager (2006) suggest that there are three primary challenges that are relevant to deterring terrorism. The first is the issue of whether terrorists are rational enough actors that deterrence measures will be effective. If your deterrence measures depend upon the ability of the perpetrator to rationally think through the probability of getting caught and the subsequent consequences and you are dealing with non-rational actors then deterrence is likely not to be effective. Secondly, they suggest that even if rational, if terrorists place such little value on the loss of their life or severe prison sentences, then most forms of punishment that might be proffered might also be ineffective in deterring them. Finally, terrorists often may have little of value that might be taken from them such that confiscation of assets or material is of little deterrence value either.

Knopf (2013) argues that this rational actor theory of deterrence may not always work well at an even more macro scale than implied by Trager and Zagorchevak, in this case at the nation-state level. He summarizes Garfinkle's ideas of culture and deterrence (Garfinkle 2006) by stating that

Because of elements in Persian culture and history, combined with the particular form of Shiite religious beliefs embraced by Iran's clerical leaders, the country's leaders might actually embrace martyrdom. According to this view, a nuclear Iran might become the first nation-state to act as a suicide bomber.

Thus not only do we have to contend with individual psychological factors at work in determining the likelihood of an actor developing and deploying weapons of mass destruction, we also have to take into account cultural values and history on both non-nation-state actors as well as nation-states themselves.

Following along this line, Miller (2013) points out that understanding how deterrence processes may work depends also on the level of decision-making that is involved in a terrorist attack. He suggests that crafting deterrence strategies is likely dependent upon whether the decision-making processes are being carried out by an individual, a group or a movement. This is of particular importance as it appears that the development and deployment of WMD begins to spread from nation-state to non-nation-state group or actor. Traditional policies and strategies of deterrence may break down as new emerging technologies allow smaller organic entities entry into the world of large-scale destruction.

The changing face of proliferation

The proliferation of technology and the devices and services that accompany them – especially as we enter the age of the Internet of Things (IoT) – suggests that as technology continues to become more pervasive in our society there will be more opportunity for criminals to utilize technology, or often more accurately, more vulnerabilities in emerging technologies for them to commit additional crimes as well as crimes with more serious and widespread consequences. There are both engineering as well as marketplace forces at work that are creating large solution spaces for criminals to operate in. From an engineering perspective there is the temptation to connect new technology with other technologies and systems to make products smarter and more efficient and effective.

A similar set of pressures issues from the marketplace where consumers are demanding that new products and services be "smarter" – that is, able to make decisions independently based upon information and input from other devices and information sources. In this IoT marketplace there is a significant advantage to companies being first to market, especially in emerging consumer technologies. Competitors who enter the marketplace after the initial entrants are at a distinct competitive disadvantage from which they may never recover. This means that the level of effort expended in ensuring that these new devices and services are secure from attack by criminals and other malicious actors often takes a back seat to being among the first to market. The time and resources spent on properly securing these new emerging technologies means that the manufacturer will be put at a disadvantage when compared to other competitors who skip or skimp on these security measures. In addition, very little if any effort is taken on the part of industrial research and development teams to assess how synergies and interactions between the newly developed technology and existing/emerging technologies may facilitate criminal activity.

A good example of the proliferation of connected consumer devices that may be vulnerable is the case of smart televisions. These home consumer devices are meant to provide extended entertainment through media streaming, web browsing, video calls and interactive activities such as game playing. A smart television typically runs a version of the Linux operating system, which is also a very common operating system for personal computers and servers. As is the case with any operating system and its installed applications, there are often security vulnerabilities present within the program code for both of these entities. Through the household Internet connection, criminals may attack a smart television and push malware to the device which may result in a number of different scenarios including pushing malware to the television to be further propagated through a home network, pushing botnet malware to the television so that it joins a criminal botnet for the purposes of a distributed denial of service attack, covertly enabling the camera on some smart televisions to spy on the residents and even ransomware attacks that threaten to "brick" the television unless a fee is paid to the criminals (Sutherland, Read and Xynos 2014; Bachy *et al.* 2015; Michele and Karpow 2014).

Similarly, while there is currently a lot of discussion about the safety and legal aspects of self-driving cars (Yağdereli, Gemci and Aktaş 2015; Schellekens 2015), there has been very little consideration as to how self-driving cars might facilitate criminal or terrorist activity. For example, a criminal may commit a bank robbery with a self-driving vehicle waiting outside the bank. The bank robber throws the bag containing the stolen money into the window of the self-driving car and then directs the vehicle to a specific remote rendezvous point while the bank robber walks around the corner, disposes of their mask and changes into different clothes. If there is a dye packet that explodes in the money bag, the initial explosion will tag the money but not the bank robber. In addition, if we assume that self-driving vehicles will be allowed to travel unattended then this allows the bank robber to separate themselves from the self-driving get-away car. If the vehicle is stopped by police on its way to the rendezvous, then the bank robber will not be apprehended inside the vehicle. Of course the police may follow the self-driving car and wait in hiding for the bank robber to meet up with it, but criminals utilizing this technology are likely to reduce the risk of apprehension.

A similar idea surrounding self-driving cars that has seen little or no discussion is that they also may become a weapon delivery system for terrorists. While car bombs have been a favorite choice of weapon for terrorist organizations for many years, there have always been issues with the nature of the delivery of the weapon to sensitive or well-guarded targets. When sensitive and strongly protected targets are selected, there are usually well-armed guards and other defenses present that are deployed to defeat car bomb tactics. Parking vehicles near these well-defended targets is also often prohibited to prevent individuals from parking vehicles loaded with explosives adjacent to the facility under guard. Defense strategies for these facilities often includes guards armed with automatic weapons and when a suspicious vehicle attempts to breach the facility the guards open up in an attempt to kill the driver before they reach the outer boundary of the targeted facility. If there is no driver to kill as is the case with a self-driving car, then the guards must rely upon either attempting to disable or detonate the vehicle using automatic weapons before it reaches the target or utilize a much more lethal tactic such as launching an RPG or similar heavy weapon against the vehicle, which in turn raises the risk of collateral damage. Utilizing self-driving vehicles not only reduces the probability that the vehicle fails to reach the outer boundaries of the targeted facility but also eliminates the need for the terrorist organization to recruit a suicide driver to navigate the vehicle to the target.

Another area of the proliferation of technology that is likely to attract criminals is the emergence of biohacking – sometimes called DIYbio or garagebiohacking (see Sutton, Chapter 13, this volume). The biohacking community has grown from its early origins as a hobby to a more organized and substantial movement likely facilitated by the exponential growth in the more general "maker movement" (Golinelli and Ruivenkamp 2015). Blazeski (2014) examines some of the issues of biohacking or the practice of synthetic biology in the hands of the public and cites a number of growing biohacking organizations such as Biobricks and Biocurious, as well as New York City's Genspace and Boston's Bosslab, who are organizations that are dedicated to placing biotechnology in the hands of citizens.

Blazeski cites the recent steep decline in the price of basic laboratory equipment such as polymerase chain reaction machines and states that a full biohacking lab could be equipped for about five thousand US dollars. The combination of affordable access to sophisticated lab equipment and the growing community of biohackers may provide a fertile ground for criminals to integrate themselves into this community and utilize the existing resources and knowledge of individuals already within this community to commit biocrimes. As Blazeski (2014: 8) points out, even the raw precursor materials for making hazardous biomaterials are easily obtained:

> Chemicals and biological materials are easier to obtain than one might imagine. BBC's writers attempted to order ricin from a biotech company in order to test the security and safety aspects of biohacking. A few days later a courier delivered two tubes full of DNA for making Ricinus communis for the cost of $31.55. They had only ordered the beginning and end portions of the necessary genetic code but in theory it would be enough to produce ricin. In fact, they were able to express the gene and decided to stop before actually producing the dangerous substance.

As was the case for information technology and the hacking community, there is the potential for criminals and in particular organized crime to migrate into the biohacker community and coopt or extort skilled individuals to produce biohazardous materials to be used in the direct or indirect commission of crimes. As an example of a biocrime, a criminal gang could threaten to contaminate a pharmaceutical company's production line or facilities with a toxic substance such as ricin and extort a fee in a hard-to-trace digital currency from the company if the fee is not paid. Because ricin is not typically spread from person to person, this substance may make an attractive weapon that could be made on demand and utilized by biocriminals. The ability to use resources available in the biohacking community to make toxins such as ricin also makes the biohacking community attractive to infiltration by terrorists who can gain the knowledge necessary and take this knowledge back to the core membership of their organizations where they will be able to train others with a reduced risk of discovery.

In terms of the discussion of proliferation of more traditional weapons of mass destruction, historically the focus of researchers and policymakers has been on identifying and disrupting the kill chain of the path of weapons of mass destruction from origin to possession by known or unknown terrorist groups. Often the research and anti-proliferation efforts focused on direct, proactive efforts by terrorist groups to acquire information, raw materials and operational knowledge of weapons of mass destruction.

In more recent years, researchers and national security strategists have begun to focus their attention on a more insidious path whereby terrorist groups may attempt to acquire some of the necessary elements of weapons of mass destruction indirectly through collaboration and cooperation with transnational criminal groups. In addition, terrorist groups have taken note of logistical strategies in place by transnational crime groups such as established and proven methods of smuggling in order to improve and facilitate the movement of raw materials and actors in association with planning and deploying terrorist actions and activities.

In addition, as has been pointed out in a previous chapter, theft of intellectual property has become a mainstay of criminal and nation-state activity on the Internet. The theft of industrial secrets pertaining to a large number of valuable processes and designs has become almost commonplace given the ubiquity with which manufacturers and research companies connect their networks to the Internet. It should also be clear that the Internet has opened up a significant treasure trove of intellectual property related to bioengineering, chemical and nuclear processes, the identification and location of resources necessary to successfully perform and manage those processes and also information on where key primary materials necessary for producing weapons from these resources might be located. These elements can facilitate both the theft of these resources by criminals to be sold onward to other customers including nation-states and terrorist groups as well as opening up the potential for these resources to possibly be utilized by criminals themselves.

While the theory about potential collaboration between criminal groups and their hypothetical role in supplying specific key elements involved in weapons of mass destruction can be traced back even farther, the evidence for connections between traditional criminals and the raw materials for weapons of mass destruction can be found back at least to the work of Shelley and Orttung (2006). They discuss the discovery that during 2005 and 2006 organized crime members had worked to smuggle special nuclear materials out of Georgia through the border areas of Turkey and Armenia.

Not everyone agrees that these early attempts at obtaining raw WMD materials were the result of actions of organized crime groups. A more recent and rigorous investigative analysis of these early events by Kupatadze (2010) argues that these early events were for the most part undertaken by amateurs, petty criminals and ordinary individuals who were compromised by debt and financial issues. He maintains drawing from the evidence of these early incidents that the smuggling and trafficking of special nuclear materials that might be utilized in a nuclear weapon as well as materials such as cesium-137 that might be used in a radiological weapon is not an attractive opportunity for criminals. Kupatadze concludes:

> While it is true that amalgams of actors from the criminal underworld and the legitimate upperworld are sometimes involved in radiological trafficking, the majority of smuggling groups are ad hoc, single-deal partnerships. It also seems that there is no prevalent nexus between traffickers and terrorists in Georgia involving illicit radioactive materials, although the existence of such a nexus is sometimes discussed in academic articles and the media.
>
> *(2010: 224)*

Forest (2012) in his article on the future of WMD terrorism suggests that two types of constraint theories – practical constraint theories and strategic constraint theories – inhibit terrorists from developing and deploying weapons of mass destruction. Practical constraint theory refers to the technical and operating environment obstacles that significantly increase the difficulties that terrorist groups would face in arming and using WMD. Strategic constraint theory delves into the negative consequences that might accompany the actual use of weapons of mass destruction by terrorists including the loss of backing by some segments of their support base or the very probable dramatic reaction of the victim governments and their allies to a WMD attack whereby exponentially increased national resources are brought to bear against the terrorist group and its allies.

It is the practical constraint theories that are of immediate interest here in terms of the consequences of criminal involvement in WMD terrorism. Transnational criminal organizations (TCOs) are organizations that have expansive resources as well as cross-national reach. These organizations have the financial resources necessary to obtain raw WMD materials which are often expensive to obtain as is the case of special nuclear materials. TCOs also are more likely to be able to operate in circles of technical and scientific expertise with less probability of surveillance by the intelligence community than terrorist groups. Gaining less risky access to these technical and scientific communities along with substantial financial resources give TCOs a better opportunity to coopt, extort or employ physical coercion to gain the cooperation of technical and scientific experts. They are also likely to attenuate or remove some of the environmental constraints of developing WMD such as utilizing existing smuggling routes to extricate and move strategic nuclear, biological or chemical raw materials and processing equipment from their secure storage locations. In addition, TCOs may already have established safe locations and denied access to geographic regions in certain countries where individuals may operate without fear of interference or apprehension by local, national or international law enforcement or intelligence community actors. Thus it is possible that TCOs might become willing middlemen in gaining the materials and expertise necessary for terrorist groups to develop weapons of mass destruction.

Forest (2012) also develops a simple quadrant map that has as its axes the effect of practical and strategic constraints. Most of the terrorist groups that he characterizes occupy the upper right-hand quadrant of that diagram where both practical and strategic constraints are high. He suggests that movement of these or other terrorist groups towards the lower left-hand quadrant where practical and strategic constraints are lower is a move that increases the magnitude of the threat of WMD terrorism. If there develops a sustainable synergy between TCOs and terrorist groups where the TCOs act as middleman as described above, this is likely to non-trivially move these groups away from the inhibiting forces of that upper right-hand quadrant towards the lower left-hand quadrant.

One of the significant worries is that a TCO might develop a middleman and perhaps even symbiotic relationship with a terrorist group like the Islamic State of Iraq and the Levant (ISIL). ISIL has demonstrated a substantial disregard for some of the strategic constraints as outlined by Forest, including the apparent lack of deterrence effect of the engagement of a number of major nation-states in the Syrian conflict. ISIL also appears to be fairly unfazed by the loss of support by more moderate Middle Eastern regimes because of their draconian treatment of captured opposing military and rebel forces as well the generally severe treatment of the civilian population under its control. The hypothetical establishment of a working relationship between ISIL and a TCO suggests that this would likely move ISIL from somewhere perhaps in the upper left-hand corner of Forest's four-celled diagram where strategic constraints are fairly low and practical constraints are high to the lower left-hand corner where the threat from a WMD terrorist attack is more likely.

The recent use of a nerve agent (sarin) in the Syrian civil war underscores this potential threat. The use of a weapon of mass destruction in the Eastern and Western areas of Ghouta, Syria in August of 2013 was independently confirmed by a number of countries and several United Nations investigative teams (see United Nations 2013). The attribution of the perpetrators of the attack has never been conclusively proven although there is substantial evidence that Syrian government forces are at fault. While it is not likely that there were traditional criminal syndicate elements involved in this attack, the confusion and chaos of the civil conflict in Syria could facilitate the extraction and transportation of Syrian chemical weapons by a criminal group to be later sold or transferred to one of the warring factions in Syria or smuggled out of the region via traditional TCO smuggling routes to be sold or bartered to an interested terrorist group abroad.

The future role of technology in criminal behavior

There appears to be a convergence of at least three different structural elements that are likely to significantly shape the future of the relationship of technology to criminal behavior. The first structural element involves the continued movement of organizations towards greater dependency upon digital technology for their basic operations. This increasing reliance upon digital technology opens up the opportunity for criminals to prey upon these organizations directly through the extraction of financial resources from these entities and could very well deprive these commercial organizations of the very resources necessary to survive. This type of direct action may in fact in the future discourage commercial organizations from concentrating their financial resources in any one place but rather distributing them among a number of safeholds to spread the risk and better protect their assets.

While direct attacks on an organization's financial assets are increasingly more common, indirect attacks are likely to become more attractive to criminal organizations. These indirect attacks are likely to occur at a number of critical points in the operations of commercial entities. One more common strategy today by criminal enterprises has been the use of distributed denial of service (DDoS) attacks against companies who are mostly or solely reliant upon online digital technologies for selling and collecting revenues for their products or services. Some of the earliest attempts by criminal elements to utilize these kinds of strategies involved DDoS attacks against online casinos, who were particularly vulnerable because their entire revenue model was embedded within digital online technologies (see Paulson and Weber 2006).

A different type of indirect attack on commercial entities that is likely to see increased use by criminals and criminal organizations in the future is the use of the target's own technologies against the target themselves. This is particularly true when the target's technology involves industrial processes, products and raw or semi-processed materials that have inherent physical dangers. The most directly relevant example would be a chemical processing plant that either utilizes raw materials or produces end products that are hazardous or toxic in nature. Criminal organizations may threaten to cause the release of these hazardous materials unless financial compensation is received (see Landucci *et al.* 2015 for an analysis of an IED attack on a chemical plant). The consequences of this type of threat may be magnified if the criminals can also take control of and leverage industrial technology processes already present at the chemical plant to increase the dispersion of the materials or gain access to those onsite industrial processes to act as a trigger for the event.

Another likely potential future use of technology by criminals is the increased procurement and use of pharmaceutical technologies to manufacture drugs. While criminals have been involved in the manufacture of illegal drugs for many decades, the use of new emerging technologies to both manufacture as well as sell these drugs is also exponentially rising. Illicit online pharmacies are a growing concern and it is likely that organized criminal enterprises are going to increasingly participate in these and other illegal pharmaceutical pursuits. The fact that these activities span multiple geographical and national boundaries as well as multiple legal jurisdictions makes pursuing these illegal activities difficult (Mackey and Liang 2013). As pharma technologies continue to improve and access to the technologies widens, this is likely to become a significant problem.

This problem is compounded by the fact that terrorist organizations have already begun an earnest effort to participate in the pharmaceutical industry. As Ganor and Wernli (2013: 699) point out

> Hezbollah, in particular, has become involved in the production, smuggling and distribution of counterfeit medications in North America, Africa and the Middle East as a means of raising immense sums of money to finance its terrorist activities. Hezbollah's infiltration into the pharmaceuticals industry illustrates the danger posed by the marriage of terrorism and crime, which arises both from enhanced resources for terrorism, and from the corruption of a legitimate and necessary industry.

The fact that at least one major terrorist organization has become involved in a significant way in the pharmaceutical industry is sure to give one pause. This mixture of criminal activity in the form of manufacturing and selling counterfeit medications to raise funds along with the potential for terrorist activity through possessing the technology to produce products that mimic drugs but contain biologically or chemically hazardous materials is a serious potential emerging threat. As pharma technology continues to advance and become more available, this suggests that additional criminal and terrorist organizations will learn of the financial benefits and threat potential that pharma technologies hold and be tempted to direct their energies in that direction.

The second structural element that is likely to change the relationship of technology and criminal enterprise is the exponential increase in the use of technology by individuals in everyday life. As was the case for commercial organizations, the use of digital technology for the management of financial assets by individuals has produced shifts in the use of technology by criminal organizations. The use of keyloggers, Trojans and other malware to extract the banking credentials of individuals typically from their personal computers or mobile devices is common and widespread. The availability of a large number of technically sophisticated tools that require little or no training to use is also facilitating the emergence of an "amateur" class of criminals who do not necessarily possess the skills and expertise of traditional cybercriminals but rather rely upon the development of user friendly digital crime platforms. Conversely, this technology has also encouraged the emergence of a class of entrepreneurial criminals who are mainly malware developers that work to uncover new exploits and write new malware and malware platforms to fill the needs of the criminal digital malware marketplace.

The emergence of the phenomena of the Internet of Things (IoT) is also changing the nature of the relationship of criminals and technology. For a number of years criminals have utilized technology to victimize large organizational entities and individuals became victims merely through the mechanisms of having an account with these organizations and often individuals were just another victim in a "sea of victims." However, with technology spreading to an incredible number of devices that are already or will soon be found in most homes, technology has reunited the personal relationship of criminal and victim. That is, in the past most of the milieu within which criminal and victim actions and activities took place were restricted to a very limited class of devices found in the home – usually personal computers and mobile devices such as smartphones and digital tablets. However, as more and more household items become "smart" items with embedded technologies, there emerge more and more opportunities for criminals to compromise those IoT devices for their own personal objectives or gain. Thus in a way, technology has brought the criminal back full circle to the traditional historically significant point where the criminal and their victim are in close proximity to each other, even if the metric here is not geographical propinquity but rather technological propinquity.

The third structural factor in this equation is the continued advancement of new technologies and how this is changing the very nature and definition of criminal behavior. Criminal behavior is often defined as the commission of an unlawful act. However, as has been pointed out many times, the ability of law to keep up with technology is often quite limited. As new technology emerges and individuals find ways to coopt this technology in malicious ways, within the vacuum of the lack of legal rules governing these behaviors we don't have emerging cases of criminality but rather newly formed instances of malicious behavior that are facilitated by technology. If there are no new laws pertaining to emerging technologies, then can we state that these new technologies are not producing new types and instances of criminals?

We should also begin to think about how emerging technologies may begin to redefine the nature of criminal and terrorist acts. Kilger (2015) argues that emerging technologies such as digital technologies may end up changing the definition of terrorist acts:

> So far in all of our discussions, there has been the implicit assumption that various forms of terrorist motivations executed in the cyber environment are prime inspirations and motivators behind cyberterrorist acts and events. It is a simplified expectation that all cyberterrorist acts in the past, present and future will be guided by terrorist motivations and justifications. That is, there is a tautological flavor to the reasoning that (cyber)terrorist acts are motivated by terrorist motivations.
>
> However, I would argue that in this argument one is confusing the consequences of the event with the motivation itself that led to the event. This I would hypothesize is especially true in the cyber environment. A series of cyber actions that result in a cyber or physical world event or events that would likely be labeled terrorist in nature may not always come from terrorist or ideological origins.
>
> *(Kilger: 2015: 697)*

Kilger goes on to suggest that it may be better to class cyberterrorist acts as a more inclusive class of malicious online acts. That is, he argues that labeling an act a terrorist act is in some sense tautological. It may follow similarly for criminal acts that are perpetrated with the assistance of technology. That is, labeling them as criminal acts as mentioned above limits these acts to those that violate some legal prohibition and that may limit our understanding of the nature of the act and especially limit our ability to extend our understanding to future, emerging instances of these kinds of behaviors. Instead, Kilger suggests that we categorize these acts using a more generalized class of malicious acts.

Why might this reclassification be useful? If one reclassifies criminal acts, especially those committed with the assistance of technology, then one can apply additional theoretical structure to the case. For example, Kilger (2015) suggests that online malicious acts are motivated by six fundamental motivations: Money, Ego, Entrance to social group, Cause, Entertainment and Status (MEECES). Applying these motivations to terrorist acts allows us to extend our understanding of the potential threat matrix to include additional scenarios as well as widening the future potential pool of terrorists. In a similar manner, we could apply the same reasoning to criminal activity that is facilitated by technology. Applying these motivations rather than the strict legal definition of criminal allows researchers and policymakers to extend their vision of potential future emerging threats from criminals facilitated by technology to a much larger set of scenarios and actors. The moral of the story here is that we should be careful not to let our definitions narrow our vision of the threat environment nor restrict our search of the solution space.

This reasoning also makes sense when reading some of the literature surrounding technology, crime and terrorism. A key part of the literature suggests that, especially in the presence and utilization of technology, that criminal organizations and terrorist organizations become more intertwined. It may be the case that as technology continues to advance that criminal and terrorist organizations will increase their cooperation between themselves. Eventually there may come a time where these two types of organizations may merge into a single organization whose purpose is the perpetration of malicious acts for a variety of objectives. One of the reasons why this might be the case is that as technology advances, the magnitude of consequences of malicious acts perpetrated using new emerging technologies will likely increase exponentially such that organizations that were formerly known as criminal organizations may in fact begin committing acts that would traditionally be classified as terrorist acts due to the magnitude of the effects of the malicious acts executed. Thus it may behoove researchers and policymakers to begin to think about criminals and new emerging technologies in a less restrictive and more open-ended manner.

Summary

This discussion has emphasized that the topics and arguments laid out in preceding chapters may share some commonalities that are worthwhile understanding in order to develop a better, more comprehensive picture of the topics at hand. Issues and constructs such as the definition of weapons of mass destruction, the principles of deterrence, the mechanisms of proliferation, the blurring of the demarcation lines of criminal and terrorist acts, the special role that digital technology plays in this arena and ideas about the future role of technology in criminal behavior all provide some meta-theoretical guidance about how we should approach this crucial yet intriguing subject. It is hoped that this narrative has facilitated the reader's ability to weave together a connecting structure across the different elements in the preceding chapters and helped them tie them together in a valuable and useful manner.

Note

1 For a fairly complete history of ransomware see Kharraz *et al.* 2015.

References

Agnew, R. (2011) *Toward a Unified Criminology Integrating Assumptions about Crime, People, and Society.* New York: New York University Press.

Bachy, Y., Basse, F., Nicomette, V., Alata, E., Kaâniche, M., Courrege, J. C., and Lukjanenko, P. (2015) Smart-TV security analysis: Practical experiments, in *Dependable Systems and Networks (DSN), 2015 45th Annual IEEE/IFIP International Conference*, 497–504. IEEE.

Bentley, M. (2012) The long goodbye: Beyond an essentialist construction of WMD. *Contemporary Security Policy*, 33(2): 384–406.

Biddle, S. (2015) A great reason to not buy bitcoin: $5 million stolen without a trace. Available from http://gawker.com/a-great-reason-to-not-buy-bitcoin-5-million-stolen-wi-1678067489.

Blazeski, G. (2014) The need for government oversight over do-it-yourself biohacking, the wild west of synthetic biology. Law School Scholarship Paper, Seton Hall University, Paper 411.

Carus, W. S. (2012) *Defining Weapons of Mass Destruction*. Washington, D.C.: National Defense University Center for the Study of Weapons of Mass Destruction.

de Freitas Pereira, T., Anjos, A., De Martino, J. M., and Marcel, S. (2012) LBP–TOP based countermeasure against face spoofing attacks, in *Asian Conference on Computer Vision*, pp. 121–132. Berlin; Heidelberg: Springer.

de Freitas Pereira, T., Anjos, A., De Martino, J. M., and Marcel, S. (2013) Can face anti-spoofing countermeasures work in a real world scenario? in *Biometrics (ICB), 2013 International Conference on*, 1–8. IEEE.

Enemark, C. (2011) Farewell to WMD: The language and science of mass destruction. *Contemporary Security Policy*, 32(2): 382–400.

Forest, J. (2012) Framework for analyzing the future threat of WMD terrorism. *Journal of Strategic Security*, 5(4): 51–68.

Ganor, B. and Wernli, M. (2013) The infiltration of terrorist organizations into the pharmaceutical industry: Hezbollah as a case study. *Studies in Conflict & Terrorism*, 36(9): 699–712.

Garfinkle, A. (2006) Culture and deterrence, Foreign Policy Research Institute. Aug. 25, 2006.

Golinelli, S. and Ruivenkamp, G. (2015) Do-it-yourself biology: Action research within the life sciences?. *Action Research*, 1–17: 147675031558663.

Greenberg, A. (2015) This hacker's tiny device unlocks cars and opens garages. Available from www. wired.com/2015/08/hackers-tiny-device-unlocks-cars-opens-garages/.

Guri, M., Kedma, G., Kachlon, A., and Elovici, Y. (2014, October). AirHopper: Bridging the air-gap between isolated networks and mobile phones using radio frequencies, in *Malicious and Unwanted Software: The Americas (MALWARE), 2014 9th International Conference on*, 58–67. IEEE.

Hadid, A., Evans, N., Marcel, S., and Fierrez, J. (2015) Biometrics systems under spoofing attack: An evaluation methodology and lessons learned. *Signal Processing Magazine, IEEE*, 32(5): 20–30.

Healy, J. (2012) *Beyond Attribution: Seeking National Responsibility for Cyber Attacks*. Washington, DC: Atlantic Council.

Hoffman, B. (2006) *Inside terrorism*. Columbia University Press.

Holt, T. J. (2012) Exploring the intersections of technology, crime, and terror. *Terrorism and Political Violence*, 24(2): 337–354.

Holt, T. J. (2013) Examining the forces shaping cybercrime markets online. *Social Science Computer Review*, 31(2): 165–177.

Jackson, V. (2012) Beyond tailoring: North Korea and the promise of managed deterrence. *Contemporary Security Policy*, 33(2): 289–310.

Kharraz, A., Robertson, W., Balzarotti, D., Bilge, L., and Kirda, E. (2015) Cutting the Gordian knot: A look under the hood of ransomware attacks, in *Detection of Intrusions and Malware, and Vulnerability Assessment*, 3–24. Cham, Switzerland: Springer International Publishing.

Kilger, M. (2010) Social dynamics and the future of technology-driven crime, in T. Holt and B. Schell (eds), *Corporate Hacking and Technology Driven Crime: Social Dynamics and Implications*, 205–227. Hershey, PA: IGI-Global.

Kilger, M. (2015) Integrating human behavior into the development of future cyberterrorism scenarios, in IEEE (eds), *2015 10th International Conference on Availability, Reliability and Security (ARES)*. New York, New York: IEEE.

Knopf, J. (2013) *Rationality, Culture and Deterrence*. Report 2013–09. Monterrey, California: Naval Post Graduate School.

Kupatadze, A. (2010) Organized crime and the trafficking of radiological materials: The case of Georgia. *Nonproliferation Review*, 17(2): 219–234.

Landucci, G., Reniers, G., Cozzani, V., and Salzano, E. (2015) Vulnerability of industrial facilities to attacks with improvised explosive devices aimed at triggering domino scenarios. *Reliability Engineering & System Safety*, 143: 53–62.

Mackey, T. and Liang, B. (2013) Global reach of direct-to-consumer advertising using social media for illicit online drug sales. *Journal of Medical Internet Research*, 15(5).

Marx, G. (2001) Technology and social control: The search for the illusive silver bullet. *International Encyclopedia of the Social and Behavioral Sciences*, 23: 15506–15512.

McCoy, D., Dharmdasani, H., Kreibich, C., Voelker, G. M., and Savage, S. (2012, October) Priceless: The role of payments in abuse-advertised goods. In *Proceedings of the 2012 ACM conference on Computer and Communications Security*, 845–856. ACM.

Michéle, B. and Karpow, A. (2014, January). Watch and be watched: Compromising all Smart TV generations. In *Consumer Communications and Networking Conference (CCNC), 2014 IEEE 11th*, 351–356. IEEE.

Miller, G. (2013) Terrorist decision making and the deterrence problem. *Studies in Conflict & Terrorism*, 36(2): 132–151.

Miller, B. (2015) From WMD to WME: An ever-expanding threat spectrum. *Journal of Strategic Security*, 8(5): 110–122.

Office of Personnel and Management (2015) What happened. Available from www.opm.gov/cybersecurity/cybersecurity-incidents/.

Paulson, R. and Weber, J. (2006) Cyberextortion: An overview of distributed denial of service attacks against online gaming companies. *Issues in Information Systems*, 7(2): 52–56.

Ramirez, J. (2014) Propriety of Internet restrictions for sex offenders convicted of possession of child pornography: Should we protect their virtual liberty at the expense of the safety of our children. *Ave Maria L. Rev.*, 12, 123.

Schellekens, M. (2015) Self-driving cars and the chilling effect of liability law. *Computer Law & Security Review*, 31(4): 506–517.

Schmid, A. and Jongman, A. (1988) *Political Terrorism: A New Guide to Actors, Authors, Concepts, Data Bases, Theories and Literature*. New Brunswick, NJ: Transaction Books.

Shelley, L. and Orttung, R. (2006) Criminal acts. *Bulletin of the Atomic Scientists*, 62(5): 22–23.

Stehr, N. (2005) *Knowledge Politics: Governing the Consequences of Science and Technology*. Boulder, Colorado: Paradigm Publishers.

Sutherland, I., Read, H., and Xynos, K. (2014) Forensic analysis of smart TV: A current issue and call to arms. *Digital Investigation*, 11(3): 175–178.

United Nations (2013) *Report on the Alleged Use of Chemical Weapons in the Ghouta Area of Damascus on 21 August 2013*. New York City: Report to the Secretary General of the United Nations.

Whyte, C. (2015) Power and predation in cyberspace. *Strategic Studies Quarterly*, 9(1): 100–118.

Wu, Z., Evans, N., Kinnunen, T., Yamagishi, J., Alegre, F., and Li, H. (2015) Spoofing and countermeasures for speaker verification: A survey. *Speech Communication*, 66: 130–153.

Yağdereli, E., Gemci, C., and Aktaş, A. (2015) A study on cyber-security of autonomous and unmanned vehicles. *Journal of Defense Modeling and Simulation: Applications, Methodology, Technology*, 1–13. 154851291557803.

Zagorcheva, D. and Trager, R. (2006) Deterring terrorism: It can be done. *International Security*, 30(3): 87–123.

Part III
Technology and control

20

Crime, situational prevention and technology

The nature of opportunity and how it evolves

Paul Ekblom[1]

Introduction: the nature of technology and technological change

Situational crime prevention (SCP) takes offenders' motivation for granted – there will always be people with criminal intent. Instead, it seeks to limit their scope for offending (e.g. a brick wall too high to climb), and to influence their perceptions and decisions regarding criminal action. Since prehistory situational methods have drawn on technology for practical purposes, whether against fellow humans raiding hill forts or treasure chests, or mice, grain stores. This trend can only increase as the world in which crime is committed and prevented becomes ever more technologically based. However, few have attempted to theorise about the role of technology in SCP.

This chapter seeks to fill this gap – to understand what is meant by technology and to relate it to key concepts in SCP. It first considers the nature of technology. It then examines how technology relates to opportunity, problems and solutions. But these concepts themselves need more development. This chapter therefore covers both traditional frameworks of SCP and a more integrated and detailed counterpart, the Conjunction of Criminal Opportunity. And 'static' opportunity needs extending to address crime dynamics, especially through the concept of scripts. A major section then examines the relationship between crime and technological change, covering adaptations and conflicts over longer timescales, viewed as arms races between offenders and preventers and drawing on ideas from biological and cultural evolution. Then come sections on the practicalities of adopting a deliberately evolutionary approach to prevention, and weaknesses of purely technological approaches. The conclusion reviews the significance of understanding technology for SCP. Throughout, low-tech and hi-tech, material and cyber technology are covered in parallel.

The nature of technology

Technology has been the subject of theoretical and philosophical discussion[2] since Ancient Greek times. Plato (Plato, *Laws* X 899a ff), for example, anticipated biomimetics in describing the origins of invention as imitation of nature, such as derivation of weaving from observation of birds' nests. Aristotle (*Physics* II.3), however, saw technology as sometimes surpassing nature, and went on to distinguish between natural objects and artefacts, the latter uniquely being shaped by human purpose. This ontological distinction remains troublesome, even today – for example the notions of biological and technological *functionality* are still challenging to pin down, and to relate. Both Plato and Aristotle nonetheless drew on technological imagery for explicating their views of the rationality of the universe. This practice has continued (albeit for applied as well as theoretical ends) through Harvey's 'heart as pump' analogue, to today's 'brain as computer' imagery. The earlier approaches to technology adopted what Mitcham (1994) called the 'humanities' philosophy of technology, exemplified in the extreme in the Marxian focus on the means of production as key to shaping the evolution of class relations and wider society. More recently, another approach has emerged which focuses on technology in itself rather than as a 'black box' influence on society, and which aims to understand both the process of designing and creating artefacts, and the product. This perspective has more in common with the analytical tradition in modern philosophy and in particular the fields of the philosophy of action and decision-making; it also connects with the philosophy of science.

Here, there are many resonances with the field of applied criminology now self-identified as *crime science* (e.g. Junger *et al.* 2012). This seeks to focus scientific understanding, experimentation and methodological rigour on the proximal causes and processes of criminal events with a view to situational intervention and criminal investigation. It has a practical, functional approach and a preference for the discourse of Scientific Realism, both to uncover underlying 'causal mechanisms' of criminal events, and more particularly to generate context-fitting interventions based on principles which themselves are abstracted mechanisms, such as deterrence (e.g. see Pawson and Tilley 1997; Pawson 2006). 'Mechanism' in itself is arguably a continuation of the technological imagery already mentioned, originating in the 'clockwork' era. And of course, Rational Choice approaches to *criminal* decision-making (Cornish and Clarke 1986) and an interest in *criminal* action through Modus Operandi and scripts (Cornish 1994) are central to SCP. None of this implies denial by crime science of the importance of psychological, social and cultural influences on crime and their significance in its prevention; indeed, many originators of crime science trained as psychologists, engineers being more recent arrivals. But crime science has an affinity with the modern, analytic philosophy of technology in its willingness to embrace hardware and software solutions (discriminatingly and in a context-appropriate way, ideally through careful design that takes full account of human/cultural factors), and to draw not just on social science but on natural science and engineering. It is therefore this perspective that is primarily adopted here, albeit extended to embrace concepts from biological and cultural evolution.

Among recent theorists of technology, Mitcham (1979) identified four dimensions: artefact (tools, manufactured products etc.), knowledge (scientific, engineering, technological know-how, plus insight from social and physical sciences), process (problem-solving, research and development, innovation), and volition (ethics, technology as social construction). Arthur's recent (2009) theory of technology characterises it on different scales: as a means to fulfil a particular human purpose; an assemblage of practices and purposes; and the entire collection of devices and engineering practices available to a culture. These scales interact with each other and the entire economy: 'As the collective technology builds, it creates a structure within which decisions and activities and flows of goods and services take place' (p. 194).

For Arthur, technology starts with phenomena – natural effects (e.g. gravitation or electricity) existing independently in nature. Technology is organised around central principles – the application of one or more phenomena for some purpose; principles in turn are expressed through physical or informational components which are combined, often hierarchically, to meet that purpose. Technological domains are toolboxes of potential, clustered around some common set of phenomena or applied principles such as movement of mechanical parts, or of electrons.

These frameworks readily apply to technology in the field of crime and its prevention. Mitcham's volitional dimension, say, could include the social institution of crime and the social forces of conflict between individuals, between individuals and wider social groups, or between either of these and the state. Arthur refers to multiple purposes; extending these to the multiple stakeholders that hold them is central to criminal conflicts.

Opportunity

As said, SCP centres on the immediate causes of criminal events; these are usually conceptualised as *opportunity*. In the Rational Choice perspective (Cornish and Clarke 1986), an opportunity emerges when the offender perceives risk and effort as low and reward high. These considerations are used among others to structure an assemblage of (25) generic techniques (e.g. Clarke and Eck 2003). Complementing this psychological approach is the ecological Routine Activities perspective (Cohen and Felson 1979), where a likely offender encounters a suitable target in the absence of capable guardians.

At a higher ecological level is the opportunity structure (Clarke and Newman 2006) – the entire pattern of available opportunities for crime. At this level of abstraction there are affinities with Arthur's 'decisions and activities and flows of goods and services' quoted above. But SCP's familiar theoretical perspectives require extension and integration to properly link to technology.

Tools and weapons are considered 'crime facilitators', and 'control tools/weapons' appears under 'Increase the effort' in the 25 Techniques of SCP. But there is no theoretical treatment of technology. Both 'technology' and 'techniques' derive from Greek *tekhne*, art and craft; some techniques involve the use of particular technologies.

Within the Routine Activities model, the capability of guardians is an obvious conceptual peg for preventive technology. While there is little explicitly covering technology for offending, Cohen and Felson (1979) did originally include offender capacity under 'likely' but most writers nowadays refer, too narrowly, to the 'motivated' offender.

The Conjunction of Criminal Opportunity (CCO n.d.; Ekblom, 2010, 2011) seeks to integrate the Rational Choice and Routine Activities approaches plus others on the situational and offender side, providing a consistent and all-encompassing conceptual framework and a unified terminology. CCO explicitly includes offenders' resources for committing crime (Ekblom and Tilley 2000; Gill 2005) and can be readily elaborated to cover technology more comprehensively.

CCO offers twin perspectives: (1) on the proximal causes of criminal events; and (2) on interventions in those causes to reduce the events' likelihood and/or harm. Under CCO, a criminal event happens when an offender who is predisposed, ready and equipped to offend (and lacking the resources to avoid offending) encounters, seeks or creates a situation containing a target that is vulnerable, attractive or provocative, in an enclosure and/or wider environment that is tactically insecure and perhaps motivating in some way, facilitated by the absence of ready and able preventers and perhaps too by the presence of deliberate or inadvertent promoters. When these preconditions are perceived to be met the offender decides to proceed. When they are blocked, weakened or diverted by a security intervention, the offender either cannot so act, or decides that the perceived reward is not worth the effort and risk.

Mapping technological opportunity and prevention: CCO

Mapping opportunity systematically and in detail enables us to focus on the particularities of the relationship of technology and opportunity: by considering each element of CCO and their interactions we can develop a 'gazetteer' of the huge variety of technological connections for both commission and prevention of crime. We start with the elements on the crime situation side, and finish with those relating to offenders. Both material and cyber examples are used.

Targets of crime

Arguably since the Neolithic and then the Industrial revolution set in train an accelerating trend to property ownership, technology has come to supply most of the material targets for crime (cf. Felson and Eckert 2015): new things of value from pots to jewellery to smartphones, and commodities such as copper from power or signalling cables (Sidebottom *et al.* 2014). This trend has extended to cover information-related value starting, perhaps, with coins (shortly after silver coins were devised in ancient Greece in about 600 BC, someone produced a silver-plated bronze forgery (James and Thorpe 1994)); and ending, for the present, with bitcoins and identity- or finance-related data. Here, 'legitimate' trading in personal data and information by banks and large corporations is accompanied by increasingly sophisticated attempts to steal or manipulate such information.

Through bad or good design, technology can render those targets vulnerable, attractive or provocative to offenders; or resistant, unattractive and inoffensive, inextricably integrated, distributed (as with in-car entertainment components), invisible or non-existent. Security technologies have a long history, and not just in terms of defensive weaponry. For example, Archimedes' principle, a way of checking the density of complex shapes like crowns, was invented to counter deliberate dilution of gold by base-metals, and microscopic marks were incorporated in early coins to validate them against counterfeiting.

Mass production, as a long-term trend, has stripped the inherent individual identity from material items. This had to be reinstated artificially, by technological means, e.g. with inbuilt or added-on marking of some kind. Since digital – as opposed to analogue – encodings are easily replicated, so too are the items they encode easy to copy while preserving value, whether they comprise music, images, or 3D representations stored on computers for purposes of design and manufacture. As another major trend, therefore, the inherent copyability of information products, plus the development of appropriate copying technologies, has radically changed the domain of Intellectual Property and its theft and counterfeiting.

A range of material and informational properties can render targets at high risk of theft, for example. Relevant risk factors have been documented by Clarke's (1999) CRAVED acronym for 'hot products': Concealable, Removable, Available, Valuable, Enjoyable and Disposable. Many of these have a technological dimension. Protective factors for mobile phones were identified by Whitehead *et al.* (2008) as IN SAFE HANDS: Identifiable, Neutral, Seen, Attached, Findable, Executable, Hidden, Automatic, Necessary, Detectable, and Secure. Ekblom and Sidebottom (2008) developed a suite of technical definitions for aspects of product security.

Targeting of human individuals can also be technologically mediated, via physical assaults with weapons, identity theft or the verbal onslaughts of cyber-bullying. At the extreme but not implausible end of the scale, future microbiologically-based attacks on individuals, communities or members of particular ethnic/genetic groups are unfortunately in prospect following advances in biotechnology and dramatic increases in the cheapness, portability and ease of use of techniques like genetic modification of pathogens (Rees 2014).

Target enclosures

Enclosures are a major domain of technology. They may contain targets for crime but can also be targets themselves (e.g. theft from car, theft of car). They range from handbags to safes, vehicles, buildings, compounds and firewall-protected ICT systems. Locks give differential access to the legitimate key-holder. Their technology is millennia old, with lock-picking maybe several days younger. Enclosures can be resistant, or vulnerable, to attack and the defence or the attack can involve technology of one or other kind, e.g. a cutting torch, or handbags whose materials resist slashing (see e.g. the Karrysafe range by VexedGeneration, DAC n.d.). Offenders, too can create or exploit enclosures (Atlas' 1991 concept of 'offensible space'). Relevant technology ranges from the simple, like the sliding peephole hatch of the Prohibition-era speakeasy, to the sophisticated, such as the movement detectors protecting a Bond villain's hideout. Conflict-avoiding technology exists in the form, for example, of noise insulation between apartments.

Wider environments

Wider environments have two kinds of properties, often shared with enclosures. The motivational dimension relates to the wealth of targets an environment contains, territoriality it arouses or conflicts generated/avoided. More significant here is the tactical/logistical dimension: how the layout, design and construction of the built environment favours offenders over preventers, or vice-versa. Some modifications of the environment, like artificial lighting or mirrors, may be used/misused by either party equally; others, like CCTV cameras, especially if shrouded to conceal where they are pointing, offer asymmetrical advantage. Online networks and systems serve as entire environments for doing crime and crime prevention, where physical constraints, like proximity and inertia, are supplanted by far more mutable ones.

Crime preventers

The crime preventer role is about agents, who either by sheer presence or deliberate action, reduce the risk of criminal events. More specific and familiar preventer roles in the Problem Analysis Triangle include guardians of targets, managers of places and handlers of offenders (Clarke and Eck 2003). Other background roles not catered for in the original, narrow formulation include the technologists who design, produce and install security equipment. There is, now, a vast range of technological aids to empower preventers, including sensors (e.g. for detecting movement) and responders (e.g. for activating barriers). These embody the 'force multiplier' concept which becomes especially important as financial stringency shrinks human resources (offender tagging is a multiplier for handlers.) As advances in ICT integrate these functions the technology becomes an active agency, independently undertaking a preventer role, sensing actions and events, making decisions and taking action. An early example is the automatic detection and blocking of suspicious bank card transactions. The perennial problem of false alarms in automated systems/devices can be met by careful design. This can draw on the fusion of information from multiple sensors, plus knowledge of legitimate behaviour and routines which could either be pre-programmed in, or learned in-situ.

Remote prevention is now a familiar part of the everyday world, from telegraphing ahead about horseback bank-robbers to satellite surveillance, and now deactivation of stolen laptops and phones. Media reports of 'kill-switches' (e.g. BBC 2015d) claim 25–40 per cent reductions in iPhone robberies in various cities since introduction; for recent evidence, see Behavioural Insights Team (2014).

Crime promoters

Crime promoters are roles that make crime more likely or harmful, whether innocently (e.g. inventor of the paint spray-can), carelessly (e.g. forgetting to log off when leaving the office) or deliberately (e.g. supplying tools/weapons for crime, or technical services like unlocking stolen phones). ICT extends the range and reach of agency, with, for example, home computers enslaved in botnets actively delivering distributed denial of service attacks on target organisations. We will undoubtedly see more sophisticated software-based crime promoter agents in future. Where human promoters are inadvertent or careless, technology can help mobilise them to cease promoting and perhaps to start preventing. Automatic reminders to lock vehicles, homes or computer terminals are now widespread. Technology can substitute for human intervention (as with remembering to lock away the telescopic car aerials of yesteryear) or design-out the target altogether. Offenders, for their part, have developed 'social engineering' methods to bypass technological security (e.g. phishing attacks on ICT systems) by homing in on the weaker, human components, to manipulate and exploit people as promoters. This may even involve threatening or corrupting technicians (e.g. those who maintain stand-alone ATMs are apparently being persuaded to install criminals' card-skimming devices – Krebs 2015).

Offender presence in situation

Offenders must either be physically present in the crime situation, hugely aided by transportation technology, or, like preventers just discussed, able to exert their influence remotely. ICT has vastly expanded the possibilities for malevolent telepresence, beginning with threatening/ obscene phone calls and moving on to various forms of cyber-bullying, computer virus attacks, and (at a chemical plant near you) interfering with control systems. The awareness space of offenders (Brantingham and Brantingham 2008) has greatly increased, with remote viewing possibilities for hostile reconnaissance, whether via online site maps, webcams or hacking into private cameras (BBC 2014a). The mass-dissemination possibilities of the internet mean that for little effort, offenders can, through phishing messages, reach thousands of potential human targets of fraud and identity theft, greatly altering the balance of effort to reward. More generally, with the Internet of Things, targets themselves may have a presence extending beyond the house, office or factory, in which they are effectively exposed to risk. From the widest perspective, the boundaries between what is a target product or place, a physical environment and a network have become blurred with the advent of wireless internet connectivity so the rules of convergence of offender, target and absent guardians need continual reconfiguration, even though the underlying principle remains valid. Beyond 'pure' ICT, remote-controlled drones can take offenders' ability to reconnoitre, and to execute crimes, to new heights, and to previously inaccessible or well-guarded locations. This emergent tool is generating many criminal applications. One such is the delivery of drugs (BBC 2015b), in which apparently the craft could be sent to a particular place via the GPS location – the risky, line-of-sight presence of the human controller being unnecessary. Drones support prevention too, e.g. being used to patrol Polish railways against coal theft (Reagan 2015).

Once inside an enclosure, whether a building or an ICT system, offenders may find its internal environment furnishes increased opportunities for crime. Insider threats (Nurse *et al.*

2014) are a growing concern. Many of these are facilitated by the scale and complexity of company networks and systems. These make crime opportunities hard to detect and block for security staff, but perhaps salient to those employees in particular roles who acquire intimate knowledge of a labyrinthine system, combined with automated access privileges. Automated monitoring for suspicious behaviour patterns is now possible but this raises ethical and staff relations issues.

From the preventive side, technology has significantly increased the risk to offenders via the trackability and traceability of their persons and their communications, way beyond the scope of fingerprints and footprints. Unfortunately, criminals are also misusing a remote-wiping facility intended to protect owners against information theft, to clear incriminating data from phones seized by the police (BBC 2014b).

Offender perception and anticipation

Moving further into the offender side, the perception of risk, effort and reward can be influenced by the technological dimension, for example through deterrence (perceived risk of harm relative to reward) and discouragement (perceived effort relative to reward). Alarms, CCTV, chemical tracing or radio/internet trackers, for example, may all increase objective/ subjective risk; they will require either increased risk toleration or greater effort to reduce it, whether technological or behavioural. Uncertainty about the capabilities of some new technological application may augment deterrence; presumably offenders will habituate once they discover its scope and limitations but a stream of new and varied approaches can maintain the pressure.

Resources for offending

Resources for offending are diverse (Ekblom and Tilley 2000; Gill 2005), including the criminal's 'native' properties like courage or strength. Unless we include amphetamines or alcoholic fermentation, these are currently uninfluenced by technology; but performance enhancement and mood control are under constant development in military and sporting contexts and we can expect criminals to exploit them. Of more widespread significance are the resources which offenders can find on arrival (such as scaffolding poles), modify in-situ (e.g. smashing beer-bottles to create weapons), or bring to crime situations (weapons, hardware tools like centre-punches for unobtrusively breaking car windows, or previously-acquired pass-codes). Tools must usually be concealable or have plausibly legitimate uses if the bearer is challenged by the police. Disguised sabotage weapons were widely-developed in the Second World War (BBC 2015a). Clarke and Newman (2006) have developed the acronym MURDEROUS to characterise terrorist weapons – Multipurpose, Undetectable, Removable, Destructive, Enjoyable, Reliable, Obtainable, Uncomplicated and Safe. Unfortunately, technological advances mean that many more products are becoming versatile, miniaturised, portable, self-powered and easy-to-(mis) use. Even the easy-to-fly aircraft and easy-to-learn-from flight simulator can make significant contributions. Biotechnology adds another dimension – terrorist threats of engineered diseases apart, apparently it is considered likely that yeasts genetically modified to produce morphine will become available in the next few years (*New Scientist* 2015). Apart from direct increases in criminal opportunity the disruptive-technology effect on organised drug producers, traffickers and dealers could be immense, doubtless with resultant violence.

Besides boosting their own resources, preventers can try to restrict those of offenders. Interventions include, for example, designing photocopiers to detect, and block, efforts to copy banknotes; legislating against possession of code-cracking software and hardware for rendering stolen mobile phones re-usable (Ekblom 2002); and 'capture-proofing' police firearms e.g. via fingerprint-activation.

Readiness to offend

Readiness to offend represents the motivational/emotional side of crime causation. Such causes can occur in-situ, as with situational provocations and pressures (Wortley 2008). They can also happen further 'upstream', for example a stressful commute may leave a passenger ready to assault ticket inspectors. Further still, persistent sleepless nights from traffic noise or false car alarms may impair self-control, as, perhaps, can shift-work. Social media can amplify intergroup conflicts. Technology can also be the author of its own destruction, as with faulty ticket dispensers that provoke 'machine rage'. The chemical technology of alcohol and other drugs of course can disinhibit the tendencies normally kept in check by the brain's executive function.

Resources to avoid offending, and predisposition to offend

Our ability to earn an honest living is often dependent on, or disrupted by, technology, though this is outside the present focus on proximal causation and intervention. But technology can intervene on the preventive side. Examples are psychopharmacological, neurosurgical and cognitive interventions which are intended to reduce the readiness to offend, or even to alter the predisposition. Examples of the latter used historically but now understandably abandoned were prefrontal lobotomy and chemical castration. More widely acceptable nowadays is the search for means of removing drug addiction by targeting relevant brain centres pharmacologically or via immune-system responses. Pharmacological intervention, transcranial electrical stimulation, surgical implantation and who knows what other kinds of intervention may emerge in the future and will pose serious ethical challenges before they can be applied for crime prevention purposes, if ever.

Extending the concept of opportunity

The detail with which CCO deconstructs opportunity allows further insight. In SCP, opportunity is typically considered an attribute of the situation. But an open window three floors up, say, is only an opportunity to an offender with the resources of agility, courage and/or climbing aids such as a ladder. Opportunity is thus an ecological interaction between situation and offender, and this is often technologically mediated.

Nor does opportunity make sense without specifying 'to do what' – there must be some purpose, ultimately stemming from an agent's predisposition. Thus, we can define opportunity as an ecological concept, relating to how agents encounter, seek or create a set of circumstances in which their resources enable them to cope with the hazards and exploit the possibilities in order to achieve their multiple goals.

Technology, opportunities and problems

Another issue to consider is how opportunities relate to problems. SCP methods are usually selected, implemented and evaluated through a problem-oriented approach (e.g. Goldstein 1990; Clarke and Eck 2003; Bullock et al. 2006). But criminals have problems too, and there is a payoff for crime prevention in general, and for discussing technology specifically, from highlighting a symmetry of circumstance between offenders and preventers. With criminal conflict, problems and opportunities are intimately entangled: one party's opportunity is always another's, or the state's, problem.

From a neutral position, a problem is some set of environmental circumstances that hinders an agent (or agents), equipped with a certain set of resources, from immediately achieving a particular goal or goals. 'Goals' is usually plural because often the difficulty is in resolving some conflict between positive and 'hygiene' goals – burgle the house without getting caught; having tranquil enjoyment of a house without expensive and unsightly fortification; tackling a burglary hotspot without restricting pedestrian movement. So, a problem is an obstacle stopping an opportunity from being instantly achievable.

Enter technology

How, then, does technology fit with opportunity, opportunity reduction and problem-solving? On the opportunity side, the resources and circumstances can obviously include technological elements. But technology can be quite fundamental to defining opportunity. This is partly because it extends human capability to cope and exploit, and partly because (following Arthur 2009), it always has a purpose. Indeed, offenders may have purposes for the technology other than those intended by the engineer/designer, as discussed below.

On the problem side, technology can contribute to solutions by bridging gaps in opportunity – for example, how to stop car alarms going off as intended (problem for offenders), or inappropriately (problem for car owners and their neighbours). It can even start an entirely new round of problem and opportunity – as with the arrival of CCTV – monitoring misbehaviour, or spying on changing rooms. Besides prevention, technology can halt an ongoing criminal attack (for example 'smokecloaks' to obscure vision, see www.smokecloak.co.uk/en/), or mitigate the adverse consequences of crime – whether a backup of data on a stolen phone, or supporting business continuity after a terrorist attack.

And technology can resolve design contradictions (Ekblom 2012a) or trade-offs between security and a range of values including safety, profitability or privacy. An example of the last is the millimetre-wave airport body scanner that aids security but reduces privacy: here a technological resolution is the substitution of a personal body image on the operator's screen with a generic, computer-generated outline that still displays suspicious items. In solving problems, therefore, technology enables rapid adaptation of the good or the bad party to the challenges posed by their material and social habitat to the pursuit of their various goals.

Cyberspace creates a new technological domain for opportunities and problems, but it is debatable whether recently emerged crimes in silico are always merely reconfigurations of familiar clashes in vivo, with physical constraints removed, or entirely new ones. Certainly identity and trust in online transactions have become huge issues. Applying CCO to cyberspace, Collins and Mansell (2004: 64) noted how trust fits into the framework. 'An Internet shopper who is too trusting may act as a careless or negligent crime promoter, as may a system designer. Conversely, being an effective crime preventer means being equipped with appropriate applications and systems. Offenders exploit misplaced trust, sometimes to an expert degree and are aided by software and hardware based resources, for example, "skimming" devices fitted into cash machines to clone cards.'

Together, technological extensions of human capabilities, and the technologically modified situations in which those capabilities are exercised, engender many opportunities both for crime and prevention. The combinations range from simple to complex; and from direct routes to indirect, roundabout ones such as the mediating effect of technology on people's routine movements which lead them to crime-generating situations (Brantingham and Brantingham 2008). And there are always knock-on and interaction effects between technologies and with other social and physical circumstances, generating unforeseen consequences and perhaps neutralising the benefits (Tenner 1996).

The dynamics of crime: technology, scripts and script clashes

The coming-together of proximal circumstances to generate criminal events may result from influences at levels ranging from the individual offender *creating* the opportunity – which is not a Routine Activity – to emergent societal influences including market forces. For its part, Rational Choice does not cover the *actions* carrying out and linking successive decisions. A complete picture of crime as opportunity must incorporate a dynamic view. This is especially true with more complex crimes involving sequences of action.

Cornish (1994) developed this perspective with his seminal article on crime scripts, boosting understanding of the procedure of crime commission, and promising identification of particular pinch points in the script where interventions might be targeted for maximum effect. But script analyses can be better woven into the opportunity/problem perspective presented here if they more explicitly attend to goals, plans and resources (Ekblom and Gill 2015).

Being instrumental, scripts can be influenced by technology as problems are solved and opportunities realised. Problems come in hierarchies or clusters, e.g. a subsidiary, yet prior, problem to obtaining cash by burglary could be acquiring a crowbar. In this sense, problems equate to necessary subsidiary goals whose potential means of achievement, i.e. solutions, are not yet determined. (Actual achievements are only finally reached via successful attempts.) Offenders must often take extra steps to obtain tools or weapons; perhaps also to learn how they work, how they can be used, or even hacked; and maybe to eliminate traces on them such as from DNA or electronic usage data. An approach from accident prevention (Reason 1990) compares complex security systems to a stack of slices of Swiss cheese – in which the succession of holes cover for each other, unless aligned by accident … or by intent. We can envisage the latter being achieved by executing a crime script, which aligns an entire *opportunity path* through successive obstacles.

The available tools themselves can shape or constrain criminal behaviour. Designers refer to 'persuasive technology' (Lockton et al. 2008), and the idea that devices (e.g. cash machines) have scripts 'expected' of their users (Latour 1992). Certainly the properties of knives, locks or network routers influence the kinds of action offenders can contemplate undertaking, and their performance during the event itself. Material items are often misused, sometimes created, as props for con-tricks or ambushes which may involve more or less elaborate scripting. The gay hook-up facility Grindr has, for example, been used to lure victims to robberies.

Applying the procedural dimension to risk factors reveals subtleties. For example, the concealable factor in hot products (Clarke 1999) may be criminogenic at the getaway stage when it is the thief who pockets the stolen smartphone; but the same factor may protect the phone, safely in the owner's pocket, at the target-seeking stage.

The procedural analysis of behaviour applies to preventers as well as offenders (Leclerc and Reynald 2015) covering all the above considerations. The preventer's script may overlap with the offender's script, for example in collecting money from a cash machine, and stealing or robbing it. Especially significant for technology is the concept of the script clash (Ekblom 2012a). This is where the offender's script engages with the preventer's in such issues as:

Surveil *v* conceal

Exclude *v* permit entry

Wield force *v* resist

Conceal *v* detect criminal intent

Challenge suspect *v* give plausible response

Surprise/ambush *v* warning

Trap *v* elude

Pursue *v* escape

Foster trust *v* become suspicious

Constrain *v* circumvent

Technology can favour one side over the other, creating an opportunity either for crime, or for prevention, variously relating to targets (e.g. resistant or vulnerable, concealable or detectable), enclosures (hardened or vulnerable, excluding or permitting entry), environments (e.g. illuminated evenly so as to minimise scope for ambush, or with deep shadows) and resources for offending (e.g. tools with ambiguous or clear-cut criminal purpose).

Clashes are the fulcrums which designers of prevention must address in arranging the situation to favour preventers over offenders. Discriminant technologies are often crucial – for example the swing-down fire escapes enabling residents to flee a burning building, but hindering offenders from entering; and the 'what you know, what you have and what you are' discriminators in ICT security (passwords, tokens and biometrics).

Technological change

The original Routine Activities article (Cohen and Felson 1979) theorised how changes in the weight/bulk of items like TV sets over decades made them more suitable targets for theft. But we can view such changes over greater timescales – indeed Felson and Eckert (2015) argue that technological change is of major importance in understanding longer-term trends in the crime and crime prevention field, illustrating the point with, for example, crime impacts from historical population movement into cities. Related arguments were previously set out by McIntosh (1971), covering interactions between changes in the technology and organisation of crime prevention, and those in the organisation, level and type of fraud.

Change is now the norm. Technologically induced changes diffuse through society at differential rates, leading to cultural lags in adjustment (Ogburn 1922). Many (e.g. Arthur 2009) note how technological change has been accelerating. This amplifies such lags and their negative consequences, as for example when crime prevention techniques trail those of crime commission.

Cultural and biological evolutionary perspectives

Accounts of the change process usually draw on evolutionary themes. The pattern of technological evolution has been variously seen as slow and cumulative (as with Gilfillan's (1935) account of the development of ships through many individual inventions), or operating at diverse scales. According to Arthur (2009) these range from small 'standard engineering' advances or tweaks, to more radical innovations (such as the leap from steam to electrical propulsion of locomotives) and those that disrupt or transform whole industries and beyond (such as ICT). The pressures shaping technological evolution variously relate to market forces, networking, and physical and social constraints.

Viewing technological evolution as a subset of cultural evolution can give useful insights. But by combining both with biological evolution we can gain fresh concepts and some detachment from conventional viewpoints. Biological and cultural evolution have previously been considered rivals for explaining human behaviour (Roach and Pease 2013), but the scope for fundamental tie-ups between them has increased (e.g. Godfrey-Smith 2012). In fact, 'Universal Darwinism' (Nelson 2007) envisages a common 'evolutionary algorithm' (Dennett 1995) comprising variation of individual organisms, practices or products; selection on the basis of adaptation to some natural, social or commercial environment; and replication or transmission whether through genes, blueprints or imitating/copying the product.

The differences are, however, instructive. For example, Arthur (2009) notes the relative rarity of combinatorial mechanisms in biological evolution (such as the symbiotic merging of bacteria and archaea to generate the great leap forward of eukaryotic cells, supporting all advanced life forms). This contrasts with its pervasiveness in technological evolution, where variety is commonly generated by bringing together new assemblages of components or principles: the jet engine, for example, does not result from a gradual modification of piston engines.

It seems that the ability to develop relatively complex tools required a surge in brain size (Faisal et al. 2010). Moreover, human hand bones appear to have evolved special features to enable tools to be grasped and wielded (Ward et al. 2014). The anatomy and fine neurological control of the hand, plus the construction and use of tools constitute a powerful example of genetic/cultural co-evolution (Tocheri et al. 2008), a process whereby cultural and biological changes amplify and channel one another. Elsewhere on the body, paleontological and comparative studies (e.g. Roach et al. 2013) suggest that the unique weapon-throwing capacity of humans involved feedback between changes to arm, shoulder and back anatomy and the technological development of projectiles, a process which may have begun with Homo Erectus some two million years ago. The relevance of tool/weapon wielding for both instrumental and expressive crime is clear, whether the use of our hands is in direct combat or more remotely connected action.

From a cultural evolutionary perspective Godfrey-Smith (2012) identifies macro-level, 'cultural phylogenetic changes' such as the Neolithic Revolution's shift from hunting/gathering to farming. These are comparable to Arthur's (2009) suggestion that major, transformative leaps in technological ability occur when we switch to a new domain, e.g. from mechanical to cyber. The evolution of farming introduced a phase-change in human existence. It increased population density, led to ownership of fixed parcels of land and other property, plus the development of written recording of that ownership; fostered emergence of hierarchies; and enabled the development of societal roles with specialist skills unrelated to subsistence. In turn, these ultimately technologically induced changes plus many others including the invention of mass transportation, are claimed to have drastically and progressively reshaped the routine activities of society (Cohen and Felson 1979; Felson and Eckert 2015); the places the activities occur in; the journeys between them (Brantingham and Brantingham 2008); and the nature and supply of targets, tools and weapons. Such changes have generated both readiness to offend (for example via more sources of conflict), and made possible many more criminal opportunities leading to more frequent, and more diverse, criminal events. And major phase changes continue to unfold – we are in the midst of those stemming from the emergence of ICT, bioengineering and anthropogenic climate/sea-level change.

Farming, fire, and the axe, in enabling forest clearance in favour of grassland, illustrate a further evolutionary concept: niche construction. This fundamental, but only recently recognised process (Laland et al. 2015), is where a given species pervasively shapes its own environment to its advantage and simultaneously adapts to survive within it, as with corals building entire reefs out of their limestone skeletons. We humans have come to hugely constitute and determine our own environment in both social and technological terms, for better and for worse. In the extreme, some ecologists consider we are creating an entire new geological epoch, the Anthropocene (Williams *et al.* 2015).

Biological lag

To Ogburn's concept of cultural lag, we may add biological lag. Evolutionary psychology (e.g. see Roach and Pease 2013; Ekblom *et al.* 2015) explores the possibility that human genes, including those influencing our perceptions and behaviour, remain adapted to life in the Pleistocene epoch (which ended some 11,000 years BP), when we were hunter-gatherers living in small mobile bands with limited weapons and tools. The remarkable human capacity for cooperation (e.g. Nowak and Highfield 2011) evolved in this period. Beyond the breathtaking cooperative technological effort of developing global ICT systems or putting people on the moon, cooperation also creates the backdrop (rather neglected by criminologists) against which crime, as a failure in cooperation, must be understood. But unfortunately we are also pretty accomplished at conflict between individuals, and between groups; and in inventing, using and improving tools and weapons in both cooperation and conflict. Behavioural tendencies appropriate for the Pleistocene – where time, space, materials and local population size provided natural constraints on conflict, differential wealth, things to steal and violence – are now inappropriate. Weapons of easy, stand-off killing and mass destruction are available together with a cornucopia of portable high-value goods; vehicles and computer terminals insulate us against natural empathic signals between conflicting individuals and may unleash road rage or online trolling. Cultural evolution, for example of institutions such as the law, has helped to compensate for inadequate psychological/ecological controls, but direct interventions to improve security remain necessary (Schneier 2012).

Perturbation, co-evolution and arms races

Thwarted commercial burglars can simply return to attack a security fence with more powerful bolt-cutters – an example of tactical displacement. But the offensive or defensive tools themselves may change, and the balance of technological advantage between offenders and preventers alters over time. Technological historians have long identified perturbations of a more general nature (Ogburn 1922; Christensen and Raynor 2003). Disruptive trends like automation, remote monitoring and operation, self-design and production, mass customisation, miniaturisation, portability including of power supply, and the break between appearance and functionality, will all keep shifting the balance between offenders and preventers.

Change can be exogenous (driven by external forces like the emergence of motor vehicles, acting as crime resource and target par excellence, or as police patrol car), or endogenous, by the playing out of script clashes involving adaptation and counter-adaptation of criminals and preventers to one another's tools, techniques and weapons. However, the nature of perturbation, and the interactions between multiple perturbations, can only be resolved at the level of the fine detail of particularities, and perhaps only in retrospect. An example is the digital TV set-top box, intended to allow existing analogue TV sets to receive digital channels. A compact object initially costing around £100, the box might have been expected to become a hot product – until the TV service providers decided to subsidise the cost and make their money from the service charge.

Co-evolutionary struggles

In the short term one can imagine the mutual adaptation of conflicting scripts – the first bicycle-parking script might have been 'cycle to cake shop, leave bike outside, buy cake, return, mount bike and depart'; the first bike theft script 'see unattended bike, get on and depart'. Soon these would be followed by various elaborations such as 'lock bike', 'break bike lock' etc. In the medium term come 'crime harvests' Pease (2001), in which some product, say the mobile phone, is designed and developed in a way that is naïvely vulnerable to crime and attractive to offenders (a failure to 'think thief' – Ekblom 1997). Soon after coming on the market it becomes both a popular purchase and a popular steal. This is usually followed by desperate commercial or governmental measures to retrofit security, often engendering clunky, user-unfriendly or unreliable products.

In the longer term, such adaptations and counter-adaptations can extend into prolonged co-evolutionary struggles (Ekblom 1999; Sagarin and Taylor 2008; Ekblom 2015). These are known as arms races or 'Red Queen's games', where you have to keep running merely to stay in the same place (from *Alice Through the Looking Glass* – see van Valen 1973). Classic examples are the evolution of locks/lock-picking (Churchill 2015), the safe (Shover 1996), coins and banknotes, and more recent means of payment such as online purchases. Once started, arms races may proceed at an irregular pace. At any point in such a criminal co-evolutionary sequence, we may encounter further harvests in the form of breakouts or 'evolutionary surprise attacks' (Tooby and DeVore 1987), where a new tactic, tool or weapon becomes available and, for a while, overwhelms the opposition's defences. One example is the recent emergence of kits for converting drones to carry graffiti spray cans (www.icarusone.com/home/). And imagine, more strategically, the devastating effect on cyber-security if someone discovered how to identify the huge prime numbers relied on in most security protocols.

Historical changes, and co-evolution especially, mean that knowledge of what works, including technological solutions to crime problems, is a wasting asset that needs continual replenishment by new sources of variety. A contemporary example here is what happens when the automotive industry rests on its laurels. A convincing case can be made that the 'security hypothesis' (Farrell *et al.* 2011) – sustained technological and procedural improvements in the security of homes, vehicles, shops etc. – accounts for the striking crime drop over the last two decades. A significant contributor to these improvements has been the inclusion (e.g. mandated by EU Directive) of immobilisers in vehicles; Brown (2013) thoroughly reviews the evidence. Recently, however, and as Brown anticipated, car thieves have managed to circumvent the security of keyless top-end models such as the Range Rover Evoque (BBC 2014c). These are currently disappearing into shipping containers and heading abroad so fast that insurers are declining cover unless, say, cars are parked off-street and primitive security devices like add-on steering wheel locks fitted.

Accelerants

Co-evolution through conflict, as just described, constitutes a powerful accelerant of technological change in both criminal and military arenas since the two opposing sides focus sharply, consistently and persistently on countering one another's resources and capabilities. But co-evolution unfolds against a background of further accelerants. Ogburn (1922) and later technological historians (see Sten 2014) have identified factors including increased population size enabling more people to invent things; a greater stock of pre-existing technologies to combine; and communications media enabling recording and dissemination of inventions and techniques (including lock-picking sites on the Internet (Ekblom 2014b)), and capitalistic competition. Arthur (2009) additionally sees a qualitative change, with the human economy becoming increasingly generative – shifting from optimising fixed operations, towards creating new and flexible combinations and offerings for the market.

The last relates to the biological concept of the evolution of evolvability (Dawkins 2003). This refers to the fact that some organisms evolve sets of body-plan genes that facilitate the orderly and efficient generation of variety. The same process can be seen with cultural and especially technological evolution, and in fact we can see processes of combination, co-evolution, modularity and evolution of evolvability coming together. In crime, facilities like script kiddies enable less-accomplished programmers to generate computer viruses. 3D printers, originally design prototyping tools, have been used to boost criminals' own capacity in, say, manufacturing accurately fitting and realistic-looking scanning mouthpieces for ATMs to read/transmit customers' card details; and in rapidly updating the shapes as soon as the bank security team modify the ATM front panel (Krebs 2011). And from a carelessly-displayed online photo, US Transportation Security Administration master-keys to open every air traveller's luggage were engineered, and converted to 3D printer instructions available online (Hearn 2015). This acceleration/replication capacity is far more significant than the printers' claimed ability to produce working firearms (e.g. Greenberg 2014). There is now also on the market an Internet-of-Things kit and support service for connecting up and remotely activating whatever one wants (BBC 2015c). This will surely interest terrorists and other criminals.

Gearing up against crime

In the face of co-evolution, accelerants, and the background of dramatic changes in technologies and their applications, the appropriate strategic response for professional preventers is to try to out-innovate adaptive offenders, otherwise win individual battles, but lose campaigns. 'Gearing up against crime' approaches (Ekblom 1997) suggest how this might be possible. Some are lessons transferred from a range of other evolutionary struggles including arms races in the military domain, human versus nature (e.g. antibiotics versus resistant bacteria, pests versus pesticides) or the purely natural (e.g. predator versus prey, immune system versus pathogens) (Sagarin and Taylor 2008). Practically speaking, running arms races (Ekblom 1997, 1999, 2015) involves generating 'plausible variety' of responses by relying on theory and interchangeable practical elements; building in the capacity for security upgrades, especially 'broadcastable' ones as with Windows security patches; and developing security 'pipelines', as with bank cards and satellite TV decoders, such that as soon as offenders crack one, a new one slots into place.

It is prudent to establish systems for detecting and reacting to new technologically enabled crimes. But given the lead-time to develop security functionality, anticipation is important. The traditional anticipatory method of the problem-oriented approach to crime prevention – induction of risk and protective factors from past patterns of hotspots, hot products etc. is unsuited to handle nonlinear changes in technology. Luckily, alternative, horizon-scanning-based approaches are applicable: see, for example the UK Government's Foresight Programme activities covering crime in general (DTI 2000) and cybercrime (gov.uk 2004). Technology roadmapping, which seeks to connect future requirements with emerging trends in technology (e.g. www.technology-roadmaps.co.uk/secure_environment/), could be applied to both crime prevention and the misuse of new technologies and technological combinations by offenders.

Horizon-scanning can be rendered more systematic and rigorous by incorporating SCP approaches. Routine Activities can be used to identify changes in any of its three causal components that might make crime events more likely (Pease 1997). The CCO can prompt more detailed questions along similar lines (Ekblom 2002): what future technological changes might affect offender presence, target vulnerability, offender resources etc.? We might also ask what changes might tip the balance of particular script clashes. On risk and protective factors, a generic approach – the Misdeeds and Security framework (Ekblom 2005) – asks how scientific and technological innovations might generate opportunities for crime through:

> Misappropriation (theft – as with Hot Products)
> Mistreatment (damage or injury)
> Mishandling (e.g. smuggling, data transfer)
> Misrepresentation (fraud)
> Misbegetting (counterfeit)
> Misuse (as tool or weapon)
> Misbehaviour (for spraying graffiti for example)
> Mistakes (e.g. false alarms)

In turn, we can link these generic factors to more specific crimes and/or particular anticipated trends in technology, to spot upcoming criminogenic changes.

Equivalent protective factors/opportunities for prevention are:

> Secured against Misappropriation, e.g. vehicles with built-in immobilisers
> Safeguarded against Mistreatment, e.g. street signs that avoid stating regulations in confrontational terms
> Scam-proofed against Mishandling, Misbegetting and Misrepresentation, e.g. fold-over airline baggage labels concealing holidaymakers' addresses from burglars' touts; or anti-copying functions within DVDs
> Shielded against Misuse, e.g. one-time syringes
> 'Sivilised' against Misbehaviour, e.g. metro station seating shaped to discourage rough sleeping

These factors can be used descriptively; or as a technology requirement specification e.g. by the police (Ekblom 2005) to encourage technologists to develop the appropriate preventive capabilities.

Technology, innovation and design

Arthur (2009) emphasises the importance of combination of prior elements of technology in generating new products. With crime prevention, systematically generating plausible variety at the level of principles comes from tested theory and what-works evidence (Ekblom and Pease 2014). That theory must be in a suitably analytic, integrated and accessible form, as noted. Again, CCO can be claimed to support this requirement, being a suite of generative, analytic preventive principles, which channel causal mechanisms through practical intervention methods.

But concepts can only be realised by people. We need professional designers closely working with practice-experts and users for the theory and know-how to combine to generate intervention measures that work in principle and in practice; and which meet diverse other requirements including cost, aesthetics, durability, a small carbon footprint, business continuity and public safety. This takes design beyond the homespun practicality of the police and hard-edged engineers. Recent reviews of design and crime are in Ekblom 2012a, 2014b; see also the sites www.designagainstcrime.com and www.designingoutcrime.com.

The design process needs hefty doses of intuition, inspiration and creativity, helped by developing a 'think criminal' mindset, and readiness to 'reframe' the presenting problem (Lulham *et al.* 2012). But it must also be systematic, constrained and supported by theoretical and methodological discipline (cf Dorst 2015). The use of frameworks like CCO, scripts etc. as discussed, and the more specific situational perspectives behind them, can facilitate this. One approach to feeding crime science into design is the Security Function Framework (Ekblom 2012b; Meyer and Ekblom 2011) for specifying products that reduce the possibility, probability and harm from criminal events. This framework seeks to develop a rationale for secure designs in terms of:

Purpose (what/who are the designs for, i.e. to reduce what crimes, and serve what other goals, for which stakeholders?)

Niche (how do they fit within the security ecosystem? Inherently secure products, dedicated security products, or securing products which confer protection as a side-benefit to some main function like being a handbag?)

Mechanism (how do they work, causally, to serve security and other goals?)

Technicality (how are they constructed, and how are they operated?)

Weaknesses of technology for crime prevention

Solution-driven approaches to crime problems can canalise responses, constraining both current interventions and future adaptability. The rush into public-space CCTV surveillance, despite indications of restricted utility (Welsh and Farrington 2008) epitomises this. Investment in rigid, capital-intensive kinds of technology can hinder adaptation to changes and lengthen the lag behind adaptive criminals. Techno-fixes can be superficial, 'bolt-on, drop-off' efforts. However well-designed and constructed, they can also fail at the interface with humans if that part of the preventive system is inadequately integrated. One example is the Grippa Clip (Ekblom *et al.* 2012), a carefully designed and trialled clip for preventing theft of customers' bags by anchoring them to pub/café tables. Despite praise from customers, police and bar staff, and successful utilisation by customers in bars in Barcelona and a café at a London station, in one UK bar chain they were ignored. Indications were that utilisation depended on the crime climate (pervasive enough in Barcelona for bar staff to readily alert customers to these security aids), and on staff motivation and commitment (in the UK bars this seemed lacking, but was firmly present in the café, which 'nurtured' its personnel and established mutual commitment with the company).

Extending such specific evaluations to the contribution of technology as a whole to crime prevention probably poses too great a challenge for the coffers and perhaps the techniques of social research. But the 'security hypothesis' research (Farrell *et al.* 2011) does suggest that security equipment and procedures together have made a substantive difference.

Preventive technology can do harm, with false burglar/car alarms wasting police time and annoying neighbours. It can also be self-defeating: password-based security systems overload the memory and exceed an employee's 'compliance budget' (Beautement *et al.* 2008) – how much of their work effort they are willing to dedicate to security procedures. Beyond this level they cut corners, like writing passwords down.

But none of these failings are inherent limitations of technology – only technology that is over-relied upon in isolation from human/system considerations; poorly designed (e.g. to be abuser-unfriendly without being simultaneously user-friendly); rigid and constraining in the face of the messy complexity of real life; and incapable of being adapted to changing patterns of risk during its lifetime of use, through material or software upgrades (Ekblom 1997).

A broader issue is complexity. Arms races, human system failures etc. show that technology is often embedded in complex adaptive systems, where introducing change at one point causes the various agents to adjust to that change, and to each other's new stance. Add the complexity of interactions between technologies (how many technologies together enabled the 9/11 attack?) and we can appreciate the unpredictability of the crime (or preventive) impact of individual developments. Crime prevention faces a rich and challenging future, but one where sense-making (Kurtz and Snowden 2003) rather than watertight, orderly explanation and prediction play a greater part, as in the wider economy (Arthur 2009).

Conclusion

Technology pervades the human ecosystem. It constitutes the means of individuals and groups to extend their native capabilities to adapt to and exploit that ecosystem, whether for legitimate purposes or – as noted by Brey (Chapter 1, this volume) – illegitimate ones. It is central to SCP concepts of opportunities, problems and solutions.

Technology both creates and solves problems, and helps to block or generate opportunities for offenders and crime preventers alike. It plays many causal roles in the clashing scripts of offenders and preventers: in the language of CCO it can produce or modify targets, enclosures and wider built environments; enable or restrict presence in the crime situation; and supply resources for offending, avoiding offending and preventing offending. All of these apply equally to material and cybercrime, and to hi-tech and low-tech products, places and systems. Since every criminal or preventive action has a potential technological dimension, those seeking to understand and intervene in crime must be technically aware. Since it is so varied – and variable – crime preventers must understand technology's fundamental nature; grasp the functional and technical specifics in particular crime situations, scripts and script clashes; and be alert to their tactical and strategic advantages and drawbacks.

Technologies interact with one another and with social and environmental contexts, generating challenging levels of complexity and unpredictability. And technology evolves, under the drivers and constraints of market forces, societal requirements, the material laws of physics and chemistry, the logical rules and conventions of ICT, the innovativeness of engineers and designers, an endless succession of technologically induced opportunities and problems and sometimes, arms races between offenders and preventers. This evolution unfolds in biological, cultural and specifically technological domains. As with medical science, ethical dilemmas and power/control issues are increasingly likely, but as medicine has shown, these can be addressed.

For the foreseeable future, technology will be a significant shaper, generator and reducer of crime. But successful crime prevention through technology cannot be based on some narrow and linear technological determinism: it requires full awareness of the complexity of social, physical and informational systems. The design of technology must cater for diverse requirements and intelligently discriminate in favour of rightful owners/users over criminal abusers and misusers. While a purely technological approach to crime prevention has weaknesses and limitations, technology that is carefully designed in line with the tested theories of SCP, developed, updateable and updated on an appropriate timescale, and that is well-integrated with the human parts of a security system, can significantly enhance the well-being of individuals, organisations and society.

But criminals too are forever seeking opportunities to exploit new technology, and ways to cope with it. Only strategic, evolutionary and innovative thinking based on plausible theory and empirical research and development can give preventers the edge, a position that will remain perpetually precarious.

Notes

1 I am grateful to the Editor for useful and interesting comments.
2 This paragraph draws heavily, but not exclusively, on the excellent entry on the Philosophy of Technology in the *Stanford Encyclopedia of Philosophy*, 2013 revision, http://plato.stanford.edu/entries/technology/ accessed 21 January 2016.

References

All web links in references and main text accessed on 30 September 2015 unless stated otherwise.

Armitage, R. (2012) Making a brave transition from research to reality. In P. Ekblom (ed.), Design against crime: Crime proofing everyday objects. *Crime Prevention Studies* 27. Boulder, CO: Lynne Rienner.
Arthur, W. B. (2009) *The Nature of Technology. What it is and How it Evolves*. London: Allen Lane.
Atlas, R. (1991) The other side of defensible space. *Security Management*, March, 63–66.
BBC (2014a) Breached webcam and baby monitor site flagged by watchdogs. www.bbc.co.uk/news/technology-30121159.
BBC (2014b) Devices being remotely wiped in police custody. www.bbc.co.uk/news/technology-29464889.
BBC (2014c) Keyless cars 'increasingly targeted by thieves using computers'. www.bbc.co.uk/news/technology-29786320.
BBC (2015a) Drawings of WW2 exploding chocolate bar bombs discovered. www.bbc.com/news/uk-34399018.
BBC (2015b) Drug delivery drone crashes in Mexico. www.bbc.co.uk/news/technology-30932395.
BBC (2015c) Internet of things starter kit unveiled by ARM and IBM. www.bbc.co.uk/news/technology-31584546.
BBC (2015d) Kill switches 'cut thefts' in London, New York and San Francisco. www.bbc.co.uk/news/technology-31416846.
Beautement, A., Sasse, M. A. and Wonham, M. (2008) The compliance budget: Managing security behaviour in organisations. In NSPW '08: Proceedings of the 2008 workshop on new security paradigms workshop (pp. 47–58). Association for Computing Machinery.
Behavioural Insights Team (2014) *Reducing Mobile Phone Theft and Improving Security*. London: Home Office.
Brantingham, P. and P. Brantingham (2008) Crime pattern theory. In R. Wortley and L. Mazerolle (eds), *Environmental Criminology and Crime Analysis*. Cullompton: Willan.
Brown, R. (2013) Reviewing the effectiveness of electronic vehicle immobilisation: Evidence from four countries. *Security Journal*, Doi: 10.1057/sj.2012.55.
Bullock, K., Erol, R. and Tilley, N. (2006) *Problem-Oriented Policing and Partnerships. Implementing an Evidence-Based Approach to Crime Reduction*. Cullompton: Willan.
CCO (n.d.) 5Is Framework for Crime Prevention. https://5isframework.wordpress.com/conjunction-of-criminal-opportunity/.
Christensen, C. and Raynor, M. (2003) *The Innovator's Solution*. Harvard: Harvard Business Press.
Churchill, D. (2015) *The Spectacle of Security: Lock-Picking Competitions and the Security Industry in mid-Victorian Britain*.
Clarke, R. (1999) *Hot Products: Understanding, Anticipating and Reducing Demand for Stolen Goods*. Police Research Series Paper 112. London: Home Office.
Clarke, R. and Eck, J. (2003) *Become a Problem Solving Crime Analyst in 55 Small Steps*. London: Jill Dando Institute, University College London.
Clarke, R. and Newman, G. (2006) *Outsmarting the Terrorists*. London: Praeger Security International.
Cohen, L. and Felson, M. (1979) Social change and crime rate changes: A routine activities approach. *American Sociological Review*, 44, 588—608.
Collins, B. and Mansell, R. (2004) *Cyber Trust and Crime Prevention: A Synthesis of the State-of-the-Art Science Reviews*. London: Department for Business, Innovation and Science. www.foresight.gov.uk/Cyber/Synthesis of the science reviews.pdf.
Cornish, D. (1994) The procedural analysis of offending and its relevance for situational prevention. *Crime Prevention Studies*, 3. Monsey, NY: Criminal Justice Press.

Cornish, D. and Clarke, R. (eds) (1986). *The Reasoning Criminal: Rational Choice Perspectives on Offending*. New York: Springer-Verlag.

DAC (n.d.) 'Karrysafe'. www.designagainstcrime.com/projects/karrysafe.

Dawkins, R. (2003) The evolution of evolvability. In S. Kumar and P. Bentley (eds), *On Growth, Form and Computers*. London: Academic Press.

Dennett, D. (1995) *Darwin's Dangerous Idea*. London: Penguin.

Dorst, K. (2015) *Frame Innovation: Create New Thinking by Design*. Cambridge, MA: MIT Press.

DTI (2000) *Turning the Corner. Report of Foresight Programme's Crime Prevention Panel*. London: Department of Trade and Industry.

Ekblom, P. (1997) Gearing up against crime: A dynamic framework to help designers keep up with the adaptive criminal in a changing world. *International Journal of Risk, Security and Crime Prevention*, 2, 249–265.

Ekblom, P. (1999) Can we make crime prevention adaptive by learning from other evolutionary struggles?' *Studies on Crime and Crime Prevention*, 8, 27–51.

Ekblom, P. (2002) Future imperfect: Preparing for the crimes to come. *Criminal Justice Matters*, 46, 38–40.

Ekblom, P. (2005) How to police the future: Scanning for scientific and technological innovations which generate potential threats and opportunities in crime, policing and crime reduction. In M. Smith and N. Tilley (eds), *Crime Science: New Approaches to Preventing and Detecting Crime*. Cullompton: Willan.

Ekblom, P. (2010) The Conjunction of Criminal Opportunity theory. *Sage Encyclopedia of Victimology and Crime Prevention*, 1, 139–146.

Ekblom, P. (2011) *Crime Prevention, Security and Community Safety Using the 5Is Framework*. Basingstoke: Palgrave Macmillan.

Ekblom, P. (2012a) Happy returns: Ideas brought back from situational crime prevention's exploration of design against crime. In G. Farrell and N. Tilley (eds), *The Reasoning Criminologist: Essays in Honour of Ronald V. Clarke* (pp.163–198). Crime Science series. Cullompton: Willan.

Ekblom, P. (2014a) Crime and communication technology. In W. Donsbach (ed.), *The International Encyclopedia of Communication*. Oxford: Blackwell Publishing.

Ekblom, P. (2014b) Designing products against crime. In G. Bruinsma, and D. Weisburd (eds), *Encyclopedia of Criminology and Criminal Justice*. New York: Springer Science+Business Media.

Ekblom, P. (2015) Terrorism – lessons from natural and human co-evolutionary arms races. In M. Taylor, J. Roach and K. Pease (eds), *Evolutionary Psychology and Terrorism*. London: Routledge.

Ekblom, P. (ed.) (2012b) Design against crime: Crime proofing everyday objects. *Crime Prevention Studies* 27. Boulder, Col.: Lynne Rienner.

Ekblom, P. and Gill, M. (2015) Rewriting the script: Cross-disciplinary exploration and conceptual consolidation of the procedural analysis of crime. *European Journal of Criminal Policy and Research* (online first), DOI 10.1007/s10610-015-9291-9.

Ekblom, P. and Pease, K. (2014) Innovation and crime prevention. In G. Bruinsma, and D. Weisburd (eds), *Encyclopedia of Criminology and Criminal Justice*. New York: Springer Science+Business Media.

Ekblom, P. and Sidebottom, A. (2008) What do you mean, 'Is it secure?' Redesigning language to be fit for the task of assessing the security of domestic and personal electronic goods. *European Journal on Criminal Policy and Research*, 14, 61–87.

Ekblom, P. and Tilley, N. (2000) Going equipped: Criminology, situational crime prevention and the resourceful offender. *British Journal of Criminology*, 40, 376–398.

Ekblom, P., Bowers, K., Gamman, L., Sidebottom, A., Thomas, C., Thorpe, A. and Willcocks, M. (2012) Reducing handbag theft in bars. In P. Ekblom (ed.), Design Against Crime: Crime Proofing Everyday Objects. *Crime Prevention Studies* 27. Boulder, Col.: Lynne Rienner.

Ekblom, P., Sidebottom, A. and Wortley, R. (2015) Evolutionary psychological influences on the contemporary causes of terrorist events. In M. Taylor, J. Roach and K. Pease (eds), *Evolutionary Psychology and Terrorism*. London: Routledge.

Eldredge, N. and Gould, S. (1972) Punctuated equilibria: An alternative to phyletic gradualism. in T. Schopf (ed.), *Models in Paleobiology*. San Francisco: Freeman Cooper.

Faisal, A., Stout, D., Apel, J. and Bradley, B. (2010) The manipulative complexity of Lower Paleolithic stone toolmaking. *PLoS ONE*, 5 (11): e13718 DOI: 10.1371/journal.pone.0013718.

Farrell, G., Tseloni, A., Mailley, J. and Tilley, N. (2011) The crime drop and the security hypothesis. *Journal of Research in Crime and Delinquency*, 48, 147–175.

Felson, M. and Eckert, M. (2015) *Crime and Everyday Life* (5th edition). London: Sage.

Gibson, J. J. (1950) *The Perception of the Visual World*. Boston, MA: Houghton Mifflin.

Gilfillan, S. (1935) *Inventing the Ship*. Chicago: Follett.

Gill, M. (2005) Reducing the capacity to offend: Restricting resources for offending. In N. Tilley (ed.), *Handbook of Crime Prevention and Community Safety*. Cullompton: Willan.

Godfrey-Smith, P. (2012) Darwinism and cultural change. *Philosophical Transactions of the Royal Society B*, 367, 2160–2170.

Goldstein, H. (1990) *Problem-Oriented Policing*. Philadelphia: Temple University Press.

Gov.uk (2004) Cyber trust and crime prevention. www.gov.uk/government/publications/cyber-trust-and-crime-prevention.

Greenberg, A. (2014) The bullet that could make 3D printed guns practical deadly weapons. www.wired.com/2014/11/atlas-314-3-d-printed-guns-bullets/.

Hearn, A. (2015) 3D-printed TSA master keys put travellers' luggage at risk. www.theguardian.com/technology/2015/sep/10/3d-printed-tsa-master-keys-put-travellers-luggage-at-risk.

James, P. and Thorpe, N. (1994) *Ancient Inventions*. London: Michael O'Mara Books.

Junger, M., Laycock, G., Hartel, P. and Ratcliffe, J. (2012). Crime science: Editorial statement. *Crime Science*, 1, 1–3.

Krebs, B. (2011) Gang used 3D printers for ATM skimmers. http://krebsonsecurity.com/2011/09/gang-used-3d-printers-for-atm-skimmers.

Krebs, B. (2015). All about skimmers. http://krebsonsecurity.com/all-about-skimmers/.

Kurtz, C. and Snowden, D. (2003) The new dynamics of strategy: Sense-making in a complex and complicated world. *IBM Systems Journal*, Volume 42(3), 462.

Laland, K., Uller, T., Feldman, M., Sterelny, K., Müller, G., Moczek, A., Jablonka, E. and Odling-Smee, J. (2015) The extended evolutionary synthesis: its structure, assumptions and predictions. Proceedings of the Royal Society B, 282: 20151019. DOI: 10.1098/rspb.2015.1019.

Latour, B. (1992) Where are the missing masses? The sociology of a few mundane artifacts. In W. Beijker and J. Law (eds), *Shaping Technology*, 205–224. Cambridge, MA, MIT Press.

Leclerc, B. and Reynald, D. (2015) When scripts and guardianship unite: A script model to facilitate intervention of capable guardians in public settings. *Security Journal*, advance online publication, DOI 10.1057/sj.2015.8.

Lockton, D., Harrison, D. and Stanton, N. (2008) Design with intent: Persuasive technology in a wider context. In H. Oinas-Kukkonen, P. Hasle, M. Harjumaa, K. Segerståhl and P. Øhrstrøm (eds) *Persuasive Technology: Third International Conference*, PERSUASIVE 2008, Oulu, Finland, June 4–6, 2008, Proceedings. Series: Lecture Notes in Computer Science, 5033. Berlin: Springer.

Lulham, R., Camacho Duarte, O., Dorst, K. and Kaldor, L. (2012) Designing a counterterrorism trash bin. In P. Ekblom (ed.), *Design Against Crime: Crime Proofing Everyday Objects*. Boulder, CO: Lynne Rienner.

McIntosh, M. (1971) Changes in the organisation of thieving. In S. Cohen (ed.), *Images of Deviance*. London: Penguin.

Meyer, S. and Ekblom, P. (2011) Specifying the explosion-resistant railway carriage – a desktop test of the Security Function Framework. *Journal of Transportation Security*, 5, 69–85.

Mitcham, C. (1979) Philosophy and the history of technology. In G. Bugliarello (ed.), *The History and Philosophy of Technology*, 163–189. Champaign-Urbana, IL: University of Illinois Press.

Mitcham, C. (1994) *Thinking Through Technology: The Path Between Engineering and Philosophy*. Chicago: University of Chicago Press.

Morgan, M. and Carrier, D. (2013) Protective buttressing of the human fist and the evolution of hominin hands. *Journal of Experimental Biology*, 216, 236–244.

Nelson, R. (2007) Universal Darwinism and evolutionary social science. *Biology and Philosophy*, 22, 73–94.

New Scientist (2015) Home-brew heroin: Soon anyone will be able to make illegal drugs. www.newscientist.com/article/dn27546-homebrew-heroin-soon-anyone-will-be-able-to-make-illegal-drugs.html?full=true#.VVpCSkaMlpo.

Nowak, M. and Highfield, R. (2011) *SuperCooperators: Altruism, Evolution and Why We Need Each Other to Succeed*. New York: Free Press.

Nurse, J., Buckley, O., Legg, P., Goldsmith, M., Creese, S., Wright, G. and Whitty, M. (2014) Understanding insider threat: A framework for characterising attacks, in Workshop on Research for Insider Threat (WRIT) held as part of the IEEE Computer Society Security and Privacy Workshops (SPW14), in conjunction with the IEEE Symposium on Security and Privacy (SP). London: IEEE. DOI 10.1109/SPW.2014.38.

Ogburn, W. (1922) *Social Change With Respect to Culture and Original Nature*. New York: B.W. Huebsch Inc.

Pawson, R. (2006) *Evidence-Based Policy. A Realist Perspective*. London: Sage.

Pawson, R. and Tilley, N. (1997) *Realistic Evaluation*. London: Sage.

Pease, K. (1997) Predicting the future: the roles of routine activity and rational choice theory. In G. Newman, R. V. Clarke and S. Shoham (eds), *Rational Choice and Situational Crime Prevention: Theoretical Foundations*. Aldershot, UK: Dartmouth Press.

Pease, K. (2001) *Cracking Crime Through Design*. London: Design Council.

Reagan, J. (2015) Drones deter Polish railway thieves. http://dronelife.com/2015/09/10/drones-deter-polish-railway-thieves/.

Reason, J. (1990) The contribution of latent human failures to the breakdown of complex systems. *Philosophical Transactions of the Royal Society B*, 327 (1241), 475–484.

Rees, M. (2014) The world in 2050 and beyond. www.newstatesman.com/sci-tech/2014/11/martin-rees-world-2050-and-beyond.

Roach, J. and Pease, K. (2013) *Evolution and Crime*. London: Routledge.

Roach, N., Venkadesan, M., Rainbow, M. and Lieberman, D. (2013) 'Elastic energy storage in the shoulder and the evolution of high-speed throwing in Homo'. *Nature*, 498, 483–486.

Sagarin, R. and Taylor, T. (eds) (2008) *Natural Security: A Darwinian Approach to a Dangerous World*. Berkeley: University of California Press.

Schneier, B. (2012) *Liars and Outliers: Enabling the Trust that Society needs to Thrive*. New York: Wiley.

Shover, N. (1996) *Great Pretenders: Pursuits and Careers of Persistent Thieves*. London: Westview Press/Harper Collins.

Sidebottom, A., Ashby, M. and Johnson, S.D. (2014) 'Copper cable theft: Revisiting the price-theft hypothesis'. *Journal of Research in Crime and Delinquency*, 51, 684–700.

Sten, K. (2014) The Emerging Dynamics of Innovation: The case of IT Industry in India. Master's thesis, Copenhagen Business School. http://studenttheses.cbs.dk/bitstream/handle/10417/4797/katri_sten.pdf?sequence=1 Retrieved 25 March 2015.

Taylor, M. and Currie, P. (2012) (eds), *Terrorism and Affordance*. London: Continuum.

Tenner, E. (1996) *Why Things Bite Back: Technology and the Revenge of Unintended Consequences*. New York: Alfred A. Knopf.

Tocheri, M., Orr, C., Jacofsky, M. and Marzke, M. (2008) The evolutionary history of the hominin hand since the last common ancestor of Pan and Homo. *Journal of Anatomy*, 212(4), 544–562.

Tooby, J. and DeVore, I. (1987) The reconstruction of hominid behavioral evolution through strategic modelling, in W. Kinzey (ed.), *The Evolution of Human Behavior: Primate Models*, 183–227. New York: SUNY Press.

Trott, P. (2005) *Innovation Management and New Product Development*. Upper Saddle River NJ: Prentice Hall.

van Valen, L. (1973) A new evolutionary law. *Evolutionary Theory* 1, 1–30.

Ward, C., Tocheri, M., Plavcan, M., Brown, F. and Kyalo Manthif, F. (2014) Early Pleistocene third metacarpal from Kenya and the evolution of modern human-like hand morphology. *Proceedings of the National Academy of Science*, 111(1): 121–124. DOI: 10.1073/pnas.1316014110.

Welsh, B. and Farrington, D. (2008) *Effects of Closed Circuit Television Surveillance On Crime. A Systematic Review*. Oslo: Campbell Collaboration.

Whitehead, S., Mailley, J., Storer, I, McCardle, J., Torrens, G. and Farrell, G. (2008) In safe hands: A review of mobile phone anti-theft designs. *European Journal on Criminal Policy and Research*, 14, 39–60.

Williams, M., Zalasiewicz, J., Haff, P., Schwägerl, C., Barnosky, A. and Ellis, E. (2015) The Anthropocene biosphere. *The Anthropocene Review* (online first), 1–24. DOI: 10.1177/2053019615591020.

Wired (n.d.) The bullet that could make 3D printed guns practical deadly weapons. www.wired.com/2014/11/atlas-314-3-d-printed-guns-bullets/.

Wortley, R. (2008) Situational precipitators of crime. In R. Wortley and L. Mazerolle (eds), *Environmental Criminology and Crime Analysis*. Cullompton: Willan.

21

Technology, innovation and twenty-first-century policing

Don Hummer and James Byrne

Introduction

Technological innovation has been the hallmark of law enforcement since the first organized police forces in London. Whether it is weaponry, transport, protective wear, or information use to support strategic police initiatives, experimenting with new technological innovations has been a key feature of the evolution—and reform—of policing. Certainly there have been missteps and technologies that did not fulfill expectations in a law enforcement context, but it can be argued that a majority of these technological innovations have come to be viewed as essential policing tools, used daily by officers. It is difficult to envision modern police departments without cars, radio communication, access to criminal record data, support by dedicated crime analysis units with predictive analytics, and most recently, access to CCTV with gunshot location technology, body cameras, and police car video. The question that we need to consider is this: Have these new technological innovations actually improved police performance?

While a full review of the impact of all recent technological innovations in policing on police performance is beyond the scope of this chapter, we have attempted to address this question by reviewing the available evaluation research on several technological innovations in policing, including COMPSTAT and other information-driven policing initiatives (hot spots), CCTV with gunshot location software, Tasers and other non-lethal weapons, cameras (dashboard-mounted cameras, cameras mounted on lethal and non-lethal weapons, body-worn cameras), and the use of social media as both an investigative tool, and a strategy to improve the public's trust in the police.

The table below highlights the full range of recent hardware (or hard technology) and software (or soft, information technology) now being utilized in police systems across the globe. Links to further details and research on each of these hard and soft technology innovations are found at the end of this chapter.

The latter part of the twentieth century was a period where great promise was held for using the latest computing technology to revolutionize policing. However, as Harris (2007) points out in a recent review, this promise has yet to be realized. Analysis of calls for service, 'hot spots' of offending, and resource deployment continue to be a very specialized and under-used facet of police information management. Decreasing municipal budgets and calls to hire more officers have diverted funds away from programs to adopt new technological advancements in favor of more traditional policing needs such as officer raises, patrol vehicles, and training. According to the Police Executive Research Forum, technology is one of the first components

375

Table 21.1 Overview of 'soft' and 'hard' technologies experimented with in policing

'Soft' technologies	'Hard' technologies
Record management systems (RMS)	Body-worn cameras (BWCs)
Computer-aided dispatch (CAD) systems	CCTV
Mobile data terminals (MDTs)	Dashboard-mounted cameras
'Hot spot' mapping	Gunshot detection technology
CompStat	Lethal & non-lethal weaponry
Social media	Tactical body armor
Information sharing with the private sector	Biometrics
Large-scale pooled databases and predictive analytics	Mass location/tracking/license plate readers

of a policing agency to be impacted (after hiring freezes and layoffs) when budgets are decreased, with over half of all police agencies surveyed in 2010 indicating cutbacks in this area due to funding constraints (PERF 2013). Further, even departments with extensive budget constraints that were heavily invested in technological advances to increase efficiency and cut costs ('smart' policing), discovered that even the best of those measures could not replace officers on the street or prevent veteran officers from looking elsewhere for employment (PERF 2013). Ultimately the City of Camden Police Department was disbanded and absorbed into a new Camden County Police Department where officers received lower salaries and fewer benefits, but numbers in their ranks increased significantly (Maciag 2014).

Aside from fiscal difficulties in many cities around the globe, two other watershed events have quelled the early anticipation of information technology revolutionizing police work. The first is the events of September 11, 2001 and subsequent global focus on anti-terrorism strategies at the municipal policing level; and the second was the proliferation of advanced technology available to general populations, which has been used to document police activities and behaviors. Both, but the latter in particular, have forced law enforcement to change their organization foci such that public and media relations are key components of their mission statements going forward. Tension between police and their constituencies, particularly in minority communities, has been the subject of media reports and scholarly articles for decades. It can be argued that, as of this writing, scrutiny on police use of deadly force is at an historical high with the shooting death of African-American teenager Michael Brown and subsequent unrest during the summer of 2014 in Ferguson, Missouri. Similar incidents nationwide have kept the issue in the public consciousness and there are once again calls for change within law enforcement relative to how officers interact with citizens, particularly those who are young, male, and nonwhite. It hasn't been commonplace for the President of the United States to become involved in the administration of policing services at the municipal level – indeed the last such non-terror related initiative to come from the Oval Office was the Crime Bill in 1994 – but President Obama felt compelled in December 2014 to task a committee to offer recommendations on twenty-first-century policing (President's Task Force on 21st Century Policing 2015).

Not inconsequentially, technological innovation is interwoven throughout this issue, seen as both illuminating the extent of the problem as well as a means to prevent unnecessary use of force deaths. The fact that this technology has been of tremendous benefit for all justice-related entities is common knowledge and has been written about in detail (e.g. Byrne and Rebovich 2007). What may indeed be fueling public scrutiny on the issue today is the proliferation of technology not only for law enforcement personnel but also in the American populace, both on-person and fixed. A majority of incidents involving the police that occur in public space have a visual record (sometimes from multiple perspectives) and this footage can be used to determine outcomes in specific incidents and serve as training tools for officers to improve their incident response strategies. Consider that the current calls from some analysts for an expanded use of body-worn cameras for police would add to already well-established recording devices such as dashboard cameras, closed-circuit crime prevention cameras, cameras on non-lethal police weapons such as Tasers, and the vast majority of the populace having video capability on their smartphones. In short order, police–community interactions have become governed less by adhering to established protocols and more on how video of an incident will be interpreted, as the odds are such images are bound to exist.

As with any emerging technology, how well a given advancement will translate to addressing criminal activity and efficiency of justice system actors is a complex idea open to debate. For example, as US cities incorporated CCTV as a law enforcement tool following positive evaluations from Europe (e.g. Welsh and Farrington 2004, 2009), the question of whether the technology prevents crime or simply facilitates the apprehension and prosecution of offenders (or both) remains unanswered. For the assumption is valid that technology is less a crime prevention aid, and more a secondary witness to the dark figure of crime that never comes to systemic attention. There is no reason to believe technology will similarly be able to eliminate ambiguity in deadly force situations. Proponents of police accountability contend that abuse and overuse of force will be less likely with an "eye in the sky", so to speak, while skeptics of the technology contend that the context and nuances of a police–citizen interaction cannot fully be captured, even with audiovisual recordings (Kaste 2015). What has become increasingly clear is that public opinion is guided largely by visual evidence and much less by the results of investigations by a third party based on witness testimony. Therefore when a deadly encounter with the police occurs, law enforcement has begun to change its traditional practice of initiating an internal investigation and has relied on images captured by witnesses, security cameras, or the police themselves to form the basis of inquiry and to gain support (if possible) in the court of public opinion. This is best exemplified by the decision by police in Boston to release a third-party video of the killing of a suspected Islamic radical, Usaama Rahim, in June 2015, even though further investigation was to come and the video itself was inconclusive. BPD leaders sought to offset potential unrest and demonstrate transparency by releasing video of the encounter less than one week after the shooting occurred (Seelye 2015).

Technological innovations for police have typically centered on officer safety and effectiveness ('hard' applications) (e.g. Paoline III et al. 2012) or increased efficiency ('soft' applications) (e.g. Garicano and Heaton 2010). These concerns were further specified and refined during a revamping of police strategy in an era of internal and external terror threats over a roughly two-decade period. As the pendulum swung toward the police being more threat responsive, criticism mounted about abuses toward segments of the population that were becoming more widespread and deadly. The focus of the present climate, then, is more on how new technology will enhance public safety and the rights of those coming in contact with officers, as well as how police actions will be perceived by the public when reported by traditional and new media. This chapter discusses how current technology can be used to help fulfill recommendations of the United States President's Task Force on 21st Century Policing and looks at how emerging technology could be used to enhance practice and foster legitimacy, particularly within communities having an extensive history of poor police–citizen interactions.

Implementation of technology and best practices

Law enforcement agencies have a long history of implementing the latest technological advancements to improve performance (Koper *et al.* 2014). Less extensive is the body of quality empirical research examining whether these advancements increase the effectiveness, efficiency, or legitimacy of the police (Byrne and Marx 2011; Lum 2010). A driving force of technological innovation in law enforcement is the belief that without adopting the latest tech trends, police are at a disadvantage relative to the apprehension of offenders and protection of the public. And while this belief is generally accepted as a truism, the empirical evidence suggests that technology in and of itself cannot substitute for positive police–citizen interactions or overcome negative police–community relations. If technology is considered as another tool at the disposal of the line officer, it must be viewed outside of its 'novelty' aspect and assessed for what it achieves in terms of officer effectiveness or efficiency. For example, during the infancy of computerized crime analysis, the mapping of hot spots, and proliferation of COMPSTAT, there was thought that the use of such data would revolutionize how policing would be structured, staffed, and managed (e.g. Manning 1992, 2008; Harris 2007). Skeptics, however, pointed out that, while the process and output may have changed, analysis of calls for service provided little information that seasoned line officers and supervisors didn't already know from experience. Further, there is little evidence that the majority of police departments, in the United States and globally, have fully realized the capabilities of computer-based analysis for long-range policy and planning (Harris 2007).

Much of the data resulting from increased use of soft technology is for short-term deployment and incident-specific, or collected for purposes outside organizational use. Analysis and forecasting have taken a backseat, certainly, to cost savings and public perception. Information technology is all but invisible to the general public and does not elicit a public reaction when a citizenry believes the police are 'doing something' about crime. Therefore, IT has not revolutionized policing to the extent other technologies have, specifically those that are seen routinely by citizens either firsthand or through the media. And perhaps because of public reaction and attention given to police abuse of power incidents, the media has highlighted cases where abuses (or even potential abuses) have occurred. For example, a recent experimental radar device that allows the user to see human figures through the walls of buildings was reported as in use by federal law enforcement agencies, even though no credible evidence existed that the technology had actually been implemented (Nassivera 2015).

Because technology use cannot be disentangled from the larger police culture, its implementation and assessment of success is subject to an organization's climate and vision. Police agencies adopt new technologies for a number of departmental specific reasons, with efficiency and crime control rationales being primary (Koper *et al.* 2015). One of the more problematic aspects of policing research in general applies to evaluations of technology— that of how to best measure outcomes. Several studies have attempted to use the Campbell Collaboration's standards for assessing the quality of empirical studies as a framework for their own methodologies (e.g. Ariel *et al.* 2015), yet these examples are a small proportion of all evaluations done annually. Despite the rigor of the research design, subjectivity regarding outcome measures remains as a confounding factor. Depending on how a construct such as efficiency, for example, is defined and what time frame is examined, results can vary widely between agencies. For this and other reasons, uniform measures and study duration should be agreed upon. Further, technology use cannot be examined without considering the larger organizational cultures of the study agencies, as management practices, structure, and overall organizational health are key influencers of any strategy employed within the agency (Chan

1996; Mastrofski and Willis 2010). As part of its overall recommendations to improve police–citizen relations, the Task Force stated:

> The use of technology can improve policing practices and build community trust and legitimacy, but its implementation must be built on a defined policy framework with its purposes and goals clearly delineated.
>
> *(President's Task Force on 21st Century Policing 2015:31)*

Clearly the same criteria should hold for evidence-based review of technological implementation within an organization. Implementation without well-defined goals or expected outcomes is a foundation for obtaining invalid or disappointing results when data are analyzed and a decision is made as to whether the implementation met established benchmarks. This phenomenon has a history in the body of literature pertaining to innovations in policing, most notably the research on the impacts of community policing strategies in the 1980s and 1990s (Gill *et al.* 2014: Skogan 2006).

Technology and transparency – CCTV and dashboard-mounted cameras

Long considered a rational deterrent measure, CCTV technology and enhanced lighting in public spaces has become commonplace worldwide, with extensive systems at work in urban areas in the UK, the US, and Korea to name but a few examples (Park *et al.* 2012). Cameras (fixed on lighting standards as well as in patrol vehicles) were seen as a fairly simple policy implementation given the lack of right to privacy issues on public streets and highways (Surette 2015). CCTV systems as policing tools have a two-fold purpose: to provide a form of guardianship when officer resources are inevitably absent and to assist police in clearing cases that might prove difficult without video evidence. The presence of cameras to modify behavior has been the subject of debate and one's perspective is often determined by whether they believe the deterrent effect is stronger than the human tendency to revert to baseline behavior after sensitization wears off (Farrar 2013). An unintended but noteworthy byproduct of CCTV use, however, is the idea that the technology can help "police the police". Goold (2003) noted that CCTV cameras could provide another layer of supervision of line officers and that the presence of these recording devices may prevent certain behaviors in officers in much the same way it is intended to prevent criminal behavior by potential offenders. However, the authors of this research noted in their conclusion section that officers would sometimes deliberately alter their behavior to be more 'citizen friendly' while on camera and revert to old styles of behavior when they knew they were outside camera range. Moreover, it was found officers would go as far as to try to obtain or alter recordings if there was video evidence that put them in a negative light (Goold 2003). Such anecdotes speak to the need for greater transparency, voiced by the President's Task Force, as a means to gain legitimacy. Proponents argue that such visuals do not allow the police themselves to guide the narrative in citizen interactions and thus provide a value-free record of behaviors as they occurred. Maintaining this record and making it available in instances that warrant it adds tremendously to police legitimacy in that outside parties, while they may differ in interpretation of the images they see, cannot argue the legitimacy of the images themselves.

The utility of dashboard-mounted cameras is similar. As with CCTV, 'dash-cams' offer a limited view shield and serve at least some deterrent purpose, as the majority of motorists likely realize their encounter with an officer is likely to be recorded during a traffic stop as the technology has been in use for some time (Westphal 2004). The purpose for the expansion of this technology has also been twofold: one, in response to criticism regarding racial profiling, and, two, for officer protection during traffic stops—often considered one of the more dangerous policing tasks. Results from an initial evaluation performed by the Office of Community Oriented Policing Services showed across the board positive results, including improvements in officer safety, community perception, and departmental accountability as well as reductions in agency liability (International Association of Chiefs of Police [IACP], 2003). Interestingly, little high-quality empirical research has followed the COPS evaluation from almost a decade ago. Though initial findings were promising, there is no way to determine if the results were an anomaly, or if such positive outcomes reflected an initial 'announcement effect' relative to a new technology furnished by federal grant monies. As the use of dashboard-mounted cameras has become more commonplace, it is fair to say that the benefits of the technology have been substantial (e.g. for training scenarios and as justification for police actions when force is employed), but there have been negatives brought to light by dash-cam footage being disseminated in media showing police use/abuse of force in certain instances.

Part of the problem is that the video footage from dashboard-mounted cameras is only a valuable tool for monitoring police behavior *if that behavior remains within the view shield*. Any fixed-position technology suffers the drawback of being unable to 'follow the action' and therefore suffers the handicap of possibly being unable to provide full situational context, not to mention the potential for police tampering with the technology. Recently, a number of incidents have been captured on police dashboard cameras that have ostensibly provided evidence for both sides of the dashboard camera debate. The most noteworthy of these encounters was in Marana, Arizona and the camera provided a view of an officer using his cruiser to run down an armed suspect. In February 2015, the cruiser hit 36-year-old Mario Valencia from behind, sending him airborne, just before the police vehicle then crashed into a masonry retaining wall. The chief of police in Marana justified the officer's actions by stating that Valencia had already fired one shot in the air from a reportedly stolen rifle, was acting erratically, and was walking toward a populated area having refused orders from the police to stop (Todd *et al.* 2015). The attorney for Valencia, who sustained moderate injuries, has stated the dash-cam video provides clear evidence of police use of excessive force against her client. With video evidence, what occurred during the encounter is obvious, yet this is a noteworthy example of how technology is not a panacea for providing clarity in police–citizen interactions that end violently. Neither side is claiming the officer did not use his patrol car as a weapon; what is debatable is whether the force used was justified in this particular situation. Without doubt, the courts will have to render a verdict in a civil lawsuit, and perhaps a criminal case, should the officer be charged. Thus, while technology can clearly assist in determining *what* occurred during an incident, it is but a small improvement over eye-witness testimony in determining *why* certain actions occurred or *if* they were justified.

Cameras mounted on lethal and non-lethal weapons

The positive benefits of non-lethal, less-than-lethal, and conductive-energy devices (CEDs) has been well documented in terms of the safety benefits for both police officers and citizens relative to incidents where such hard technology was not available to officers (Hickman *et al.* 2008; Taylor *et al.* 2009; Stinson *et al.* 2012). Consensus does not exist within the empirical literature, however, with one study determining that out of over 1,000 CED incidents examined, over 95 percent resulted in either no injury or only a mild injury to the subject upon whom the weapon was used (Bozeman *et al.* 2009) while another criticizes the operationalization of 'injury' in some studies where the authors contend an overstatement of the number of subjects actually receiving injuries (Kaminski *et al.* 2013). Yet there remain serious injuries and fatalities at the hands of the police even when non-lethal technologies have been used during critical incidents (Hummer 2007; Terrill and Paoline III 2012; White *et al.* 2013) and some of these deaths have been captured on video. In one such instance, the video evidence came from another officer's body-worn camera while the other from the officer's CED itself. Roughly ten percent of the 800,000 CEDs manufactured by Taser International, Inc. come equipped with high definition cameras that begin recording and uploading data to a remote server when the weapon is activated. While not the majority of all such weapons in use, that some Tasers record an officer's view field when in use came as a surprise to even veteran law enforcement personnel during the investigation of a recent officer-involved shooting (Associated Press 2015a).

In January of 2015, a borough police officer in Hummelstown, Pennsylvania initiated a traffic stop on a vehicle with an expired state inspection sticker. The driver of the vehicle failed to yield and led the officer on a short pursuit to a nearby residence where he exited the vehicle and fled to the rear of the home. The officer, who'd had prior interactions with the man, gave chase and ordered the subject to the ground. It was at this point that the female officer drew her video-equipped Taser and recording of the incident commenced. After the suspect, David Kassick was hit with the CED, he went down on his stomach with one hand underneath him and another hand reaching behind his back in what appeared to be an attempt to remove the Taser probes. The officer, Lisa Mearkle, repeatedly ordered Kassick to place both hands behind his back, in sight. When Kassick did not comply, Officer Mearkle drew her service weapon and fired twice into Kassick's back, killing him. After an investigation by the Pennsylvania State Police, the Dauphin County District Attorney charged Mearkle with criminal homicide, calling the video from Mearkle's Taser the most compelling evidence from the investigation (Associated Press 2015b).

The incident recounted above, as well as the Walter Scott shooting in North Charleston, South Carolina, coming during a period of heightened scrutiny of police-involved deaths, is very likely to give law enforcement agencies pause given that video footage from the officer's person became the "most compelling evidence" in an investigation that led to homicide charges being levied against her (Lee 2015). It may be argued that the technology served the exact purpose it was designed to do—provide a value-free record of a violent incident—yet some police officers are questioning the technology's use given the very serious ramifications the footage could have against them. Regardless of the spate of well-publicized incidents involving police deaths when CEDs have been employed, the vast majority of these cases involved an officer also drawing their sidearm during the encounter, and this weapon, not the CED was responsible for the citizen death, indicating that non-lethal weapons are not suitable alternatives to deadly force in all situations or that policing agencies can learn from mistakes made when less-than-lethal weaponry is deployed. Perhaps because of the uncertainty, the President's Task Force has recommended an investment in new and/or updated CED technologies and called for more evidence-based review of CED use in police–citizen encounters (2015: 39). It is worth noting also that even in larger cities such as Boston, not all officers carry non-lethal weaponry in addition to firearms, as was made clear in the shooting of Usaama Rahim (Seelye 2015). Independent review of the technology must also assess contextual variables that impact not only whether an officer uses a CED, but also in what circumstances an officer does or does not carry the technology.

Body-worn cameras

The most attention in police-involved fatalities has been paid to the pros and cons of body-worn cameras for officers. Whereas dashboard cameras and cameras mounted on weaponry have either a limited view shield or are only activated in certain instances, body-worn cameras have the capability to provide a continuous loop of an officer's activity from a first person perspective. It has been argued that the technology could be extremely beneficial in a number of areas (as is any video evidence). Proponents, especially within policing, contend BWCs are a preferred technology because of the ability for third parties to see an incident in its entirety from the perspective of the officer. The last part of this assertion is very important. By seeing events—the demeanor of and movements made by the subject involved—it is believed that observing parties are better able to understand what may be going through an officer's mind and how their decision-making evolved. Proponents outside of policing further assert that such a first person perspective serves an omnipotent supervisory purpose which moderates officer actions, the benefit being that such transparency furthers police legitimacy in the eyes of those they encounter and a civilizing effect on both the police and the public.

Having a record of police interactions serves a practical function as well in providing evidence for any citizen complaint about officer behavior and as educational and training material for police personnel. Opposition to BWCs seems to come exclusively from the ranks of line officers who contend that their use symbolizes a lack of confidence in them from leadership and supervisors (Miller et al. 2014). Further, in a worst case scenario, footage from BWCs could be used as evidence against them in personnel, and possibly even in legal, actions. In addition, policing agencies and the American Civil Liberties Union have weighed in on possible privacy rights issues, which could be more or less problematic from state to state because of varying right to privacy and Right-to-Know laws (Stanley 2013; White 2014; Miller et al. 2014).

With extensive media coverage of incidents resulting in citizen deaths at the hands of the police, the impetus for policing agencies to adopt a policy of officers wearing BWCs is perhaps at its apex. There is the potential, however, for use of this technology to become the next element in a dubious history of police innovations that fall out of fashion due to their inability to deliver results on perceived expectations. Several of the most recent evaluations and reports on BWCs have shown positive results in comparison group studies, but have also exercised caution in rushing to adopt a technology that has significant costs and possible negative ramifications associated with its use and a very limited empirical foundation (e.g. Ellis et al. 2015; Miller et al. 2014). For example, while the first assessment of BWCs in the United States (Rialto, CA) showed a reduction in overall citizen complaints (Farrar 2013), the evaluation was done in-house by the chief of police and did not include a comparison group (Kaste 2015). Earlier studies from the United Kingdom typically relied on perceptional data, either of citizens or crime victims, and had low sample sizes, rendering conclusions speculative at best (Goodall 2007: ODS Consulting 2011).

Since the Rialto study was undertaken in 2012, other evaluations have taken place in Arizona, including one with the Phoenix Police Department with a quasi-experimental design that also showed evidence of positive impacts of BWC use for policing agencies generally and officers specifically. Most notably, there was a decrease in overall complaints, an increase in complaints ruled unfounded, an increase in arrests, and better success in prosecution of domestic violence cases, and no decrease in citizen contacts (Katz et al. 2014). Other expected improvements were not observed, however, as officers reported no change in paperwork burden or ease-of-use of the technology (Katz et al. 2014).

Privacy issues have the very real potential to become problematic as BWCs represent an entirely new component of video footage acquired and stored by police. Whereas CCTV, dashboard camera, and weapons-mounted video either records activity on public thoroughfares or when a police incident necessarily warrants it, BWCs ostensibly could record any police–citizen contact, thereby presenting possible Fourth Amendment issues. Establishing protocols may help to alleviate a number of these issues (e.g. requiring officers to obtain consent prior to activating a BWC), but even the most detailed policy guide cannot account for every contingency (Miller *et al.* 2014). State assemblies and courts will inevitably have to consider implications of the technology and partner with policing agencies to establish proper use guidelines and promote uniformity in best practices for law enforcement agencies in each state.

The factor that is most likely to determine whether an agency adopts BWCs is a familiar one in the public sector: funding. Departments are typically able to budget for an initial purchase of the cameras themselves or obtain an outside grant. However, the prohibiting costs of a BWC program rest with data storage (Grovum 2015; Miller *et al.* 2014). The newest model BWCs have longer lasting batteries, record video in high definition, and cost in the neighborhood of $1,000 up front. With large numbers of officers recording hours of HD video per shift, data storage fees easily outpace initial equipment charges by a very wide margin. Large police departments can expect storage fees in the millions of dollars over a period of years (Grovum 2015). Add in the inevitable costs of even newer technologies emerging at regular intervals, and the problem becomes one of those departments most needing the technology unable to fund an ongoing BWC program.

The efficacy of BWCs has not been firmly established, considering the evidence-based review process is clearly in its infancy. But given current circumstances of police use of force incidents prominent in the media, the time necessary for reasoned consideration and policy analysis of implementing BWCs in particular agencies is not likely to occur in many instances. Larger urban departments specifically have been under and will continue to feel pressure to acquire and put BWCs into service hurriedly. As with any new technology, each organization has to evaluate its own needs, available resources, and priorities before implementing something that may not provide optimal return on investment. Potential negative ramifications (that have only been empirically studied in a superficial manner) could make use of BWCs an unwise investment for a number of police departments. For example, agencies with relatively good officer–community relations, few overall complaints, and without a history of lawsuits against them might think twice before investing in BWCs simply because they are the latest hot trend in police technology or because the public's interest has been piqued because of extensive media coverage. Body-worn cameras hold the potential to change the manner in which officers interact with citizens and to modify organizational culture. Unless discrete benefits can be outlined by implementation that significantly outdistance the costs, both financial and practical, police departments may be best served to take a cautious approach and delay adopting BWCs until more solid empirical evidence is available to justify a policy shift.

Police and social media

One of the most interesting ways police have facilitated interaction with their constituencies is via social media, most notably Twitter and Facebook. The primary purposes are to engage in a positive manner with members of the community who don't routinely interact with those in law enforcement, and set forth a context for a political agenda to increase public confidence in the police and enlist the public to be jointly responsible for crime reduction long term (Crump 2011). Social media has the potential to become the twenty-first century's vehicle for fostering positive relations in a manner akin to officers walking the beat in the 1970s and 1980s when police departments re-invigorated foot patrol for the same purpose. As (particularly younger) Americans change their information-gathering practices more toward mobile, web-based platforms, it has become imperative for law enforcement to connect with this demographic through its preferred method. Thus, as of 2015, Twitter and Facebook are essential for providing up-to-the-minute information in digestible fragments, and if they have not already, larger police departments will find a professional social media coordinator indispensable.

Social media has also become an investigative tool par excellence. Information is the key in any investigation and there is likely no better way to gather large amounts of data from a constituency quickly than via the web. With more of a person's private life, thoughts, and movements trackable online with dozens of applications and websites, the twenty-first-century detective will likely spend far more time at a desk in front of a computer or with a smartphone than they will on city streets, talking to people face-to-face. Quantity, however, does not necessarily equate to quality and the flood of digital tips and other information has sometimes overwhelmed investigators and derailed investigations when information is false (Urbina 2014). Also, information gathered by law enforcement presents unique challenges for the courts in terms of veracity and admissibility of that evidence. For example, does a posed photo of an individual on social media showing gang signs or tattoos demonstrate actual gang membership (Urbina 2014)?

Recent unrest in Baltimore spurred by the in-custody death of Freddie Gray has shown how social media can also be a vehicle for criticism of the police. Police use of social media typically emphasizes information flow from the public to the police, but information can also be effectively disseminated by the police to the public as a social order function (e.g. notifying the public of road closures or areas to avoid because of first responder activity). In Baltimore, police gave periodic updates on their activities and disseminated warnings and instructions for the public during a night of unrest that, when circumstances on the street did not match provided information (for example, where police were staging or what areas of the city were secure), the mayor and police chief were much criticized in real time (Iacone 2015). Further compounding events in Baltimore was news of the eventual unrest being organized and circulating on social media that was either missed by police or not given sufficient heed. The criticism was especially biting given that police monitoring of social media has been a centerpiece of preventing social unrest for some time (Yang 2013). Given the propensity for offenders to incriminate themselves online, and the present climate of urban unrest in the wake of police-involved deaths, critics assailed the police for not treating a possible threat with enough seriousness. Time will tell if this oversight contributed to the unrest or allowed the situation to become even more volatile. Because information is so vital to policing, and because social media provides more freely assessable, personal information about members of a community, agencies must continue to develop protocols and strategies for using this information to maximize the benefit for the organization. Similarly, social media allows law enforcement to disseminate information necessary for the public, as well as information key to policing agencies themselves, to large numbers of constituents instantaneously. In a small number of cases this can lead to criticism, but the benefits across the board appear to far outweigh the costs. Just one example—a police department that establishes a profile on Facebook can have that department's mission statement, goals, and objectives prominently displayed on its publicly viewable homepage. In the years before the internet generally, and social media specifically, these were key organizational elements that were seen by precious few outside the agency itself.

Concluding comments: technology as a means of building trust and legitimacy with the public

Legitimacy of the police in the eyes of those they serve takes years, if not decades, to firmly establish, yet can be shattered in minutes with the technology available to virtually everyone in today's information age. Technology is the ultimate double-edged sword for law enforcement in that it can be one of the most beneficial tools officers have to do their jobs, but it is also the most effective means the public has for documenting ineffective or abusive police actions. Legitimacy is built through effective policing performed in a community's best interest and by taking corrective, or punitive, actions when officers do not perform to expectations. Such actions require a shift in overall organizational culture that is separate from technology implementation and use. While technology can assist with the process, legitimacy in this form emerges out of actions—changing protocols when necessary, administering police services in the best interest of those served, and decisive management when corrective action is required.

If police–community relations are to improve significantly moving through the second decade of the twenty-first century, there is no question technology will be an integral component in the process. Video captures of officer–citizen interactions will become even more commonplace when reviewing behaviors, and to formulate new policies and information technology will continue to transform how the job of policing is done. To that end, perhaps the most public figure in American policing, New York Police chief Bill Bratton, recently discussed his updated take on the community policing paradigm, where officers will be freed to spend more time in the communities they serve by using various forms of mobile technology to accomplish tasks which historically consumed significant chunks of each shift (Cullen 2015). While the root of issues in many communities will take more than increased police presence and interaction to solve, it stands to reason that longer duration contacts that are non-confrontational in nature are the means to sustain positive police–community relations going forward. If hundreds of officer hours in a city like New York can be shifted from doing paperwork in a precinct house to interacting in a positive manner with residents, then technology has fulfilled an integral purpose.

It can be argued that in order for the technology to assist the police in gaining legitimacy in the eyes of the public, the technology must first have legitimacy for the police officers that use it. Virtually every innovation incorporated by the police has been met with some degree of skepticism by officers using it on a daily basis, and faith in the technology accrues only over a period of time, if ever. The types of technology discussed in this chapter had and will have even larger obstacles to overcome as they are designed to surveil and critique the officer as s/he does their job in unedited form. As Koper *et al.* assert:

> What is clear is that technology can evoke powerful responses from those who implement and use it, particularly information technologies and analytic technologies, which have the potential to transform fundamental aspects of how police work is done.
>
> *(2015: 236)*

Part of doing 'good police work' is the way a highly trained, professional officer uses their knowledge and discretion to address incidents within highly varying circumstances that can change tenor rapidly. If an officer is, either consciously or subconsciously, basing their decision-making on potential ramifications stemming from scrutiny of their actions after the fact, a dangerous precedent develops for both the officer and the public. Further, reticence on the part of policing agencies to adopt technologies such as body-worn cameras, the cries for which at present have reached a staccato pitch, is understandable, given that it is precisely this type of innovation which potentially impacts the core of officer behavior—to guide discretion (Mastrofski and Rosenbaum 2011).

Legitimate use of technology has the potential to bridge the gulf between the police and those who support them as legitimate social control agents and those in the community who contend that police treatment of poor, nonwhite constituents is frequently abusive. Police legitimacy is eroded when cases such as the deaths of Michael Brown in Ferguson and of Eric Harris in Tulsa at the hands of a 73-year-old reserve deputy receive extensive media attention and questioning of police tactics (Jones 2015). To be sure, the media has covered cases of police abuse of power for decades; however the incidents mentioned above differ significantly in that video evidence of either the shootings or the immediate aftermath were available and quickly made public. More than at any other time, rapid dissemination of video to media outlets precludes the police from guiding the narrative, adhering to a 'code of silence', and obscuring key facets of the incidents that occurred in previously infamous police abuse cases, such as that of Rodney King over 20 years ago (Weisburd and Greenspan 2000).

However even when video recordings of an incident are made public quickly, the court of public opinion invariably is divided into two camps, one for and one against police use of force. This problem is one that is not solved by any technological development on the horizon. The benefits of technological innovation alone cannot overcome a legacy of negative police behavior towards particular segments of the community or an organizational culture that sanctions such behavior. Commissioner Bratton in New York City points out that negative incidents, such as Eric Garner's death in Staten Island during the summer of 2014, and the killing of two NYPD officers in their patrol car shortly thereafter (and the subsequent media attention given to both incidents) not only erodes public confidence but also officer morale and faith in organizational leadership. Simply, technology is not a panacea for the issues in modern American policing; rather it is another tool for improving effectiveness and efficiency within a healthy organization. But even a healthy organization must adapt to the democratization of information and the public's appetite for instantaneous and unfiltered visual evidence—the new media that is most likely to factor in the perceptions and decision-making of the next generation.[1]

Note

1 For more information on police technology see the following website: http://faculty.uml.edu/jbyrne/44.203/resource.htm.

References

Ariel, B., Farrar, W.A. and Sutherland, A. (2015) The effect of police body-worn cameras on use of force and citizens' complaints against the police: A randomized controlled trial. *Journal of Quantitative Criminology*, 31: 509–535.

Associated Press (2015a) March 26. Police Tasers equipped with video cameras. Retrieved from www.sharonherald.com/news/police-tasers-equipped-with-video-cameras/article_d01eeffa-0fbe-58ec-bef5-368f87045bba.html.

Associated Press (2015b) March 24. Pennsylvania police officer charged with shooting dead unarmed man. Retrieved from www.theguardian.com/us-news/2015/mar/24/pennsylvania-officer-lisa-mearkle-hummesltown.

Boehl, B. (2014) December 17. County to study use of cameras on police Tasers and body suits. *The Dundalk Eagle*. Retrieved from www.dundalkeagle.com/component/content/article/26-front-page/52443-county-to-study-use-of-cameras-on-police-tasers-and-body-suits.

Bozeman, W.P., Hauda II, W.E., Heck, J.J., Graham, D.D., Martin, B.P. and Winslow, J.E. (2009) Safety and injury profile of conducted electrical weapons used by law enforcement officers against criminal suspects. *Annals of Emergency Medicine*, 53, 480–489.

Byrne, J.M. and Marx, G.T. (2011) Technological innovations in crime prevention and policing. A review of the research on implementation and impact. *Journal of Police Studies*, 3, 17–40.

Byrne, J.M. and Rebovich, D.J. (2007) Introduction – The new technology of crime, law and social control. In Byrne, J.M. and Rebovich, D.J. (eds) *The New Technology of Crime, Law and Social Control*. Monsey, NY: Criminal Justice Press.

Chan, J. (1996) Changing police culture. *The British Journal of Criminology*, 36, 109–134.

Crump, J. (2011) What are the police doing on Twitter? Social media, the police and the public. *Policy & Internet*, 3, 1–27.

Cullen, K. (2015) April 24. How Boston's Bill Bratton is making over the NYPD. *Boston Globe*. Retrieved from www.bostonglobe.com/magazine/2015/04/24/how-boston-bill-bratton-making-over-nypd/6OKiWeoJy95S1OMzGGajrL/story.html?s_campaign=email_BG_TodaysHeadline.

Ellis, T., Jenkins, C. and Smith, P. (2015) *Evaluation of the Introduction of Personal Issue Body Worn Video Cameras (Operation Hyperion) on the Isle of Wight: Final Report to Hampshire Constabulary*. Portsmouth, UK: University of Portsmouth: Institute of Criminal Justice Studies. Retrieved from www.aele.org/Eval-BWCs-Portsmouth.pdf.

Farrar, T. (2013) Self-awareness to being watched and socially-desirable behavior: A field experiment on the effect of body-worn cameras on police use-of-force. *Police Foundation*. Retrieved from www.policefoundation.org/content/body-worn-camera.

Garicano, L. and Heaton, P. (2010) Information technology, organization, and productivity in the public sector: Evidence from police departments. *Journal of Labor Economics*, 28, 167–201.

Gill, C., Weisburd, D., Telep, C.W., Vitter, Z. and Bennett, T. (2014) Community-oriented policing to reduce crime, disorder and fear and increase satisfaction and legitimacy among citizens: A systematic review. *Journal of Experimental Criminology*, 10, 399–428.

Goodall, M. (2007) *Guidance for the Police Use of Body Worn Video Devices*. London, UK: Home Office. Retrieved from http://library.college.police.uk/docs/homeoffice/guidance-body-worn-devices.pdf.

Goold, B.J. (2003) Public area surveillance and police work: The impact of CCTV on police behavior and autonomy. *Surveillance & Society*, 1, 191–203.

Grovum, J. (2015) May 1. States struggle to pay for police body cameras. *Pew Charitable Trusts, Research & Analysis, Stateline*. Retrieved from www.pewtrusts.org/en/research-and-analysis/blogs/stateline/2015/5/01/states-struggle-to-pay-for-police-body-cameras.

Harris, C.J. (2007) The police and soft technology: How information technology contributes to police decision making. In Byrne, J.M. and Rebovich, D.J. (eds) *The New Technology of Crime, Law and Social Control*. Monsey, NY: Criminal Justice Press.

Hickman, M.J., Piquero, A.R. and Garner, J.H. (2008) Toward a national estimate of police use of nonlethal force. *Criminology & Public Policy*, 7, 563–604.

Hummer, D. (2007) Policing and 'hard' technology. In Byrne, J.M. and Rebovich, D.J. (eds) *The New Technology of Crime, Law and Social Control*. Monsey, NY: Criminal Justice Press.

Iacone, A. (2015) April 28. Police, Hogan defend response to Baltimore unrest. *WTOP*. Retrieved from http://wtop.com/baltimore/2015/04/police-hogan-defend-response-to-baltimore-unrest/.

International Association of Chiefs of Police (2003) *Impact of Video Evidence on Modern Policing*. Washington, DC: U.S. Department of Justice, Office of Community Oriented Policing Services. Retrieved from www.theiacp.org/portals/0/pdfs/iacpin-carcamerareport.pdf.

Jackson, B.A. (2015) *Respect and Legitimacy — A Two-Way Street*. RAND Corporation. Retrieved from www.rand.org/pubs/perspectives/PE154.html.

Jones, C. (2015) April 23. Judge assigned Eric Harris case weighs recusal because of ties to sheriff's office. *Tulsa World*, Retrieved from www.tulsaworld.com/news/courts/judge-assigned-eric-harris-case-weighs-recusal-because-of-ties/article_2bfbe4b6-db0b-5c25-aa6a-e73b6d8741eb.html.

Kaminski, R.J., Engel, R.S., Rojek, J., Smith, M.R. and Alpert, G. (2013) A quantum of force: The consequences of counting routine conducted energy weapon punctures as injuries. *Justice Quarterly*, Online First. Retrieved from www.tandfonline.com/doi/full/10.1080/07418825.2013.788729#.

Kaste, M. (2015) January 22. *Police Departments Issuing Body Cameras Discover Drawbacks*. Retrieved from www.npr.org/blogs/alltechconsidered/2015/01/22/379095338/how-police-body-camera-videos-are-perceived-can-be-complicated.

Katz, C., Kurtenbach, M., Choate, D. and Ready, J. (2014) *Evaluating the Impact of Officer Worn Body Cameras in the Phoenix Police Department*. Washington, DC: National Institute of Justice, Bureau of Justice Assistance. Retrieved from www.smartpolicinginitiative.com/sites/all/files/SPI%20Body%20Worn%20Cameras%20Phoenix%20Webinar%20Slides%20FINAL.pdf.

Koper, C.S., Lum, C. and Willis, J.J. (2014) Optimizing the use of technology in policing: Results and implications from a multi-site study of the social, organizational, and behavioural aspects of implementing police technologies. *Policing*, 8, 212–221.

Koper, C.S., Lum, C., Willis, J.J., Woods, D.J. and Hibdon, J. (2015) *Realizing the Potential of Technology in Policing: A Multisite Study of the Social, Organizational, and Behavioral Aspects of Implementing Policing*

Technologies. Washington, DC: National Institute of Justice, Police Executive Research Forum. Retrieved from http://cebcp.org/wp-content/evidence-based-policing/ImpactTechnologyFinalReport.

Lee, T. (2015) April 15. Walter Scott killing underlines racial tensions in police department. *MSNBC.* Retrieved from www.msnbc.com/msnbc/walter-scott-killing-underlines-racial-tensions-police-department.

Lum, C. (2010) February 11. Technology and the mythology of progress in American law enforcement. Retrieved from http://scienceprogress.org/2010/02/police-technology.

Maciag, M. (2014) Why Camden, N.J., the murder capital of the country, disbanded its police force. *Governing.* Retrieved from www.governing.com/topics/public-justice-safety/gov-camden-disbands-police-force-for-new-department.html.

Manning, P.K. (1992) Technological drama and the police: Statement and counterstatement in organizational analysis. *Criminology,* 30, 327–346.

Manning, P.K. (2008) *The Technology of Policing: Crime Mapping, Information Technology and the Rationality of Crime Control.* New York, NY: New York University Press.

Mastrofski, S.D. and Rosenbaum, D. (2011) *Receptivity to Police Innovation: A Tale of Two Cities.* National Police Research Platform. Retrieved from http://nationalpoliceresearch.org/storage/updated-papers/Receptivity%20to%20Police%20Innovation%20A%20Tale%20of%20Two%20Cities%20%20FINAL.pdf.

Mastrofski, S.D. and Willis, J.J. (2010) Police organization continuity and change: Into the twenty-first century. *Crime and Justice,* 39, 55–144.

Miller, L., Toliver, J. and Police Executive Research Forum (2014) *Implementing a Body-Worn Camera Program: Recommendations and Lessons Learned.* Washington, DC: Office of Community Oriented Policing Services. Retrieved from www.policeforum.org/assets/docs/Free_Online_Documents/Technology/implementing%20a%20body-worn%20camera%20program.pdf.

Nassivera, J. (2015) January 20. Police using radar to look through walls? *HNGN.* Retrieved from www.hngn.com/articles/62139/20150120/police-using-radar-tech-to-look-through-walls.htm.

ODS Consulting (2011) *Body Worn Video Projects in Paisley and Aberdeen, Self Evaluation.* Glaskow, UK: ODS Consulting. Retrieved from www.bwvsg.com/wp-content/uploads/2013/07/BWV-Scottish-Report.pdf.

Paoline III, E.A., Terrill, W. and Ingram, J.R. (2012) Police use of force and officer injuries: Comparing conducted energy devices (CEDs) to hands- and weapon-based tactics. *Police Quarterly,* 15, 115–136.

Park, H.H., Oh, G.S. and Paek, Y.P. (2012) Measuring the crime displacement and diffusion of benefit effects of open-street CCTV in South Korea. *International Journal of Law, Crime and Justice,* 40, 179–191.

PERF (Police Executive Research Forum) (2013) *Policing and the Economic Downturn: Striving for Efficiency Is the New Normal.* Washington, DC: Police Executive Research Forum.

President's Task Force on 21st Century Policing (2015) *Interim Report of the President's Task Force on 21st Century Policing.* Washington, DC: Office of Community Oriented Policing Services.

Seelye, K.Q. (2015) June 8. In blurry video of Boston shooting, officers' retreat is clear but knife is not. *New York Times.* Retrieved from www.nytimes.com/2015/06/09/us/video-showing-shooting-of-boston-terrorism-suspect-usaama-rahim-is-released.html?_r=0.

Skogan, W.G. (2006) *Policing and Community in Chicago: A Tale of Three Cities.* New York, NY: Oxford University Press.

Stanley, J. (2013) *Police Body-Mounted Cameras: With Right Policies in Place, A Win for All.* Washington, DC: American Civil Liberties Union.

Stinson, P. M., Reyns, B. W. and Liederbach, J. (2012) Police crime and less-than-lethal coercive force: A description of the criminal misuse of TASERs. *International Journal of Police Science and Management,* 14, 1–19.

Surette, R. 2015. *Media, Crime, and Criminal Justice,* 5th edition. Stamford, CT: Cengage Learning.

Taylor, B., Woods, D., Kubu, B., Koper, C., Tegeler, B., Cheney, J., Martinez, M., Cronin, J. and Kappelman, K. (2009) *Comparing Safety Outcomes in Police Use-of-Force Cases for Law Enforcement Agencies That Have Deployed Conducted Energy Devices and a Matched Comparison Group That Have Not: A Quasi-Experimental Evaluation.* Washington, DC: National Institute of Justice, Police Executive Research Forum.

Terrill, W. and Paoline III, E.A. (2012) Conducted energy devices (CEDs) and citizen injuries: The shocking empirical reality. *Justice Quarterly,* 29, 153–182.

Todd, B., Marquez, M. and Almasy, S. (2015) April 15. Dashcam video shows Arizona officer intentionally running over suspect. *CNN Online*. Retrieved from www.cnn.com/2015/04/14/us/arizona-police-run-over-suspect/.

Urbina, I. (2014) February 15. Social media, a trove of clues and confessions. *The New York Times*. Retrieved from www.nytimes.com/2014/02/16/sunday-review/social-media-a-trove-of-clues-and-confessions.html?_r=0.

Weisburd, D. and Greenspan, R. (2000) Police attitudes toward abuse of authority: Findings from a national study. *National Institute of Justice Research in Brief*. Washington, DC: National Institute of Justice. Retrieved from www.ncjrs.gov/pdffiles1/nij/181312.pdf.

Welsh, B.C. and Farrington, D.P. (2004) Surveillance for crime prevention in public space: Results and policy choices in Britain and America. *Criminology & Public Policy*, 3, 497–526.

Welsh, B.C. and Farrington, D.P. (2009) Public area CCTV and crime prevention: An updated systematic review and meta-analysis. *Justice Quarterly*, 26, 716–745.

Westphal, L.J. (2004) August. The in-car camera: Value and impact. *The Police Chief*. Retrieved from www.policechiefmagazine.org/magazine/index.cfm?fuseaction=display&article_id=358&issue_id=82004.

White, M.D. (2014) *Police Officer Body-Worn Cameras: Assessing the Evidence*. Washington, DC: Office of Community Oriented Policing Services. Retrieved from https://ojpdiagnosticcenter.org/sites/default/files/spotlight/download/Police%20Officer%20Body-Worn%20Cameras.pdf.

White, M.D., Ready, J., Riggs, C., Dawes, D.M., Hinz, A. and Ho, J.D. (2013) An incident-level profile of TASER device deployments in arrest-related deaths. *Police Quarterly*, 16, 85–112.

Yang, M. (2013) The collision of social media and social unrest: Why shutting down social media is the wrong response. *Northwestern Journal of Technology & Intellectual Property*, 11,707–728.

22

Contemporary landscapes of forensic innovation

Christopher Lawless

Introduction

For some time, social studies of technology have questioned the notion that innovation occurs in a simple linear fashion. Such studies have challenged the assumption that 'pure' scientific research directly informs downstream applied research, which in turn yields technology readily accepted by audiences who unquestioningly use that technology in a manner anticipated by its producers (Williams and Edge 1996). In contrast, social studies of technology have drawn attention to the heterogeneity apparent among users of technology. Social research has described how differing interests among users may compete to influence the development of technology (Pinch and Bijker 1984). Other research has demonstrated how 'users' of technology may innovate as much, if not more, than the original producers. Instead of framing technology via a simple dichotomy of 'users' and 'producers', this work suggests that a variety of actors participate in the construction of technology in non-linear ways. Rather than producers and users, social research suggests that actors may be more appropriately described as stakeholders in technology and innovation.

Forensic science borrows from a heterogeneous knowledge base. Innovations may reflect a variety of stakeholder interests, including those of policymakers, funders, law enforcement officials, academic researchers, forensic practitioners and commercial manufacturers. This chapter draws attention to the challenges involved in co-ordinating the interests and outlooks of these stakeholders. The chapter outlines a series of aspects currently shaping forensic innovation in the UK. These include relations between academic researchers and forensic practitioners, together with the influence of police, and government authority in the form of the Home Office. More significantly still, commercial imperatives are increasingly impacting upon emerging forensic science and technology.

This chapter critically examines the influence of these factors on forensic innovation. It uses a series of examples of emerging technology which commonly feature among contemporary discussions between forensic science stakeholders. Discussions of digital forensics, DNA phenotyping and forensic language analysis (FLA) illustrate how pathways of forensic innovation are subject to a series of relational tensions. These include perceived differences between forensic practitioners, law enforcement officials and publics over the epistemological status of forensic technology, concerns about exposing nascent technology to the public and legal gaze, and concerns about police and other lay expectations of emerging forensic technology. This chapter also explores the possible role commercial imperatives, (which may exert an impact from beyond the UK), may play in complicating these issues. The chapter draws upon published academic literature and discussions with UK forensic stakeholders.

Tracing the forensic innovation landscape

Research and development (R&D) has been an increasing priority for the UK forensic science community. During its lifetime, the Forensic Science Service (FSS), once the dominant Forensic Science Provider (FSP) in the UK, demonstrated a significant R&D capacity. During the 1990s, forensic work became more distributed as other FSPs entered the marketplace in England and Wales. While committed to casework, these FSPs were also able to undertake some research. In addition to the FSS, employees of other FSPs such as LGC also published research in peer-reviewed journals, describing the development or improvement of analytical methods (see for example Forster *et al.* 2008; Dawnay *et al.* 2014).

The closure of the FSS, first announced in 2010 and completed in 2012, generated significant uncertainty about the future of UK forensic science (Lawless 2011). The closure of the FSS was followed by a review undertaken by Bernard Silverman, Chief Scientific Advisor to the UK Home Office, on R&D activity relating to forensic science (Silverman 2011). Silverman described the forensic R&D landscape as 'varied and in some ways fragmented', with scope for 'improvement in the degree of linkage and communication' to facilitate innovation (Silverman 2011: 2). Silverman found forensic research in the UK to be spread across a range of different sites, including FSPs, university departments and government laboratories.

A variety of research activities were identified in the Silverman report, from 'blue-sky research, strategic research informed by applications, translational research and development, and the improvement and advancement of methodology already deployed in practice.' (ibid). One issue highlighted by Silverman was the perceived need for greater communication across stakeholders concerning the development and validation of innovations, particularly across the boundary between forensic practitioners and researchers addressing questions of potential forensic relevance.

The Silverman review concluded with a number of recommendations. These included a role for the Forensic Science Regulator in facilitating links across the stakeholder community, and in upholding the commitment outlined in service provision agreements to allow FSPs to conduct research. Silverman also paid attention to relations between forensic science and the oversight and support of research in the higher education sector.

Academia

During the 1990s, the number of degree programmes in forensic science increased, notably throughout Higher Education Institutions (HEIs) in the UK. The increasing visibility of forensic science in HEIs began to provide opportunities for academic staff, sometimes working with external partners, to present research of possible relevance to forensic science and practice. A number of UK HEIs, following the earlier lead of institutions such as Strathclyde University, began to directly promote their research as holding great potential for operational forensic casework (see for example Cassella 2008)

Forensic science had for some time however been regarded as an awkward fit for methods for evaluating UK HEI research in the form of the Research Assessment Exercise (RAE). The successor to RAE, the Research Excellence Framework (REF), placed more emphasis on the wider societal 'impact' of research beyond it academic confines. This was viewed by Silverman as 'an opportunity for forensic science research to demonstrate its importance' (Silverman 2011:

3) giving seemingly changing academic priorities. Forensic science did not however feature as a specific 'Unit of Assessment', relating to specific disciplinary areas of research, within the 2014 REF.

Despite this lack of recognition, the increasing 'impact' of agenda promulgated by the REF system may signal a greater willingness on the part of the national Research Councils to support research of forensic relevance in the future. The UK's Research Councils, key funders of HEI research, have increasingly concerned themselves with crime and security. They have promoted cross-disciplinary activity in these areas, which have included funding for research and related activities concerning forensic science. Other recent funding initiatives, supported by Research Councils and external partners such as the Metropolitan Police Service (MPS) and National Crime Agency (NCA), have supported research focusing on the policing opportunities and challenges posed by new technology such as cloud computing. Questions remain, however, over exactly whose interests are represented in such initiatives. To what extent do such research projects deliver operationally significant knowledge, or do the signifiers pertaining to crime, security and forensic science, often prevalent in funding calls, facilitate research which merely perpetuates existing academic agendas? (Fraser and Williams 2009: 489).

Relations between academic research and forensic stakeholders are complex. Issues around commonality of purpose are apparent. In discussions held in 2014, some forensic stakeholders expressed a desire to guide academics toward the innovation challenges they regarded as high priority, rather than allowing academia to dictate the forensic innovation agenda.

Police

Forensic scientific developments have exerted a palpable influence on contemporary policing practice, and has significantly changed investigative strategies. In the 1970s and 1980s, obtaining a confession was often the primary investigative strategy. In the absence of a confession, eyewitness evidence was sought, with forensic science being regarded as a means of last resort. In the twenty-first century, this relationship has seemingly been reversed. The availability of forensic evidence now appears high on the initial enquiries made by investigating police officers (Police Representative 2015). A high level of authority often appears to be bestowed on forensic science ahead of other sources of intelligence and evidence.

The demise of the FSS was however perceived by some as a serious blow to UK forensic R&D. The FSS appeared to have enjoyed a strong working relationship with the police on matters of innovation, taking a proactive role in the development of new technology and advising police on the implementation of innovations. The closure of the FSS appears to have caused a divergence of innovation routes.

Police forces continue to experience issues in relation to technology, including a lack of integration, interoperability and communication between systems and processes (Police representative 2014). Challenges therefore appear to remain in facilitating effective communication and a sense of shared purpose between forces.

Economic drivers may yet however drive forces further in pursuing innovation along more co-operative lines. Forensic science continues to cost forces considerable sums of money, and budgetary considerations play a key role in commissioning certain forms of forensic science and technology.

In the UK, the Association of Chief Police Officers (replaced in 2015 by the National Police Chiefs Council), introduced and oversaw a 'portfolio' system which provided opportunities for innovators to promote their technology as meeting perceived police requirements. Through this system, police groups could function as key gatekeepers for innovation. Financial constraints may potentially lead police forces to be more selective in which R&D areas they choose to support. Police discourses on innovation have tended to emphasise how technology could be used to enhance operations in the form of, for example, improved body armour or enhanced communication systems.

It is unclear to what extent the portfolio system has taken into account the views and perceived priorities of forensic practitioners. For example, the Forensic Regulator has advocated the need for improved understanding of evidential interpretation methodology, possibly using 'likelihood methods' which formalize a balanced form of evidential interpretation taking both prosecution and defence positions into account (Discussions with author and FSR representative 2014). It is unclear however whether police, given their more prosecutorial orientation, would regard this as a priority (Lawless and Williams 2010).

Certain organizations, such as the Metropolitan Police Service (MPS), advocate a cautious but holistic approach to the implementation of new technology. The MPS recognizes a series of wider hurdles to implementing new technology, which may include issues around training, education, accreditation, the nature of reporting data, agreement on processes with providers and costs, ethics and legal awareness. Other complicating factors recognized by the MPS include processes for evaluating new technology, budgeting, regulatory standards, and making decisions about whether it is appropriate to share data with other nation-states. While the MPS for one appears to take a comprehensive approach to technology implementation, it is unclear whether such stringency is consistently matched by other law enforcement actors, either elsewhere in the UK or abroad.

Home Office

Police procurement of new technology is often independent of the Home Office, but through its Centre for Applied Science and Technology (CAST), the latter is able to exert some influence. In seeking to address crime and security priorities, CAST works with industry and academia to facilitate understanding of policing needs, and collaborates with these partners to develop new technology where no readily available options may exist. CAST also engages in 'horizon scanning', anticipating the relationship between future policing challenges and technology. This encompasses a range of emerging issues, currently including the possibilities and risks of driverless vehicles, and the forensic challenges posed by the introduction of polymer banknotes. The latter pose such issues such as the risk of counterfeiting, and their capacity to retain illegal substances such as cocaine. The horizon is naturally constantly changing and new issues may emerge over time.

CAST comprises approximately 250 staff representing a range of scientific and engineering disciplines, including Information and Communications Technology (ICT), chemistry, mathematics, physics, electrical and electronic engineering, mechanical and civil engineering and materials science. Four police advisers provide insight to assist with the development of technology for operational use. A key role for CAST is to identify and address perceived capability gaps in police and security capacity. A team of capability advisers work across disciplinary areas. Capability advisers are 'senior experts with specialist domain knowledge and skills in contraband detection, crime prevention and community safety, cybercrime and cybersecurity, forensic science, identity assurance, protective security, public order and surveillance' (Mallinson 2015). At the time of writing, specific forensic activity undertaken within CAST is reported to include research around fingerprint enhancement, digital imaging, drug analysis, digital forensics, vehicle forensics, 'real-time' forensic technology, quality standards and proficiency testing, and improvements to DNA technology.

CAST also operates a Security Innovation and Demonstration Centre (SIDC). The SIDC has been described as an 'open innovation centre focussed on security challenges', to facilitate 'partnership between government, industry, academia and users' (Mallinson 2015). SIDC facilities include laboratory and demonstration spaces at CAST sites. SIDC is intended to provide 'access to end users and their environments for rapid real world evaluation of new concepts' (Mallinson 2015). It therefore enables external partners to gain access to potential investors, and to the operational field in order to test and possibly refine new innovations. SIDC is also aimed at encouraging international partners to invest in UK technology. The conversion of ideas into technology, and the commercial exploitation of UK intellectual property, are regarded as priorities by the Home Office. This also reflects an additional driver aside from crime and security concerns, namely the policy aim of promoting economic growth.

The commercial dimensions of UK forensic innovation

The Silverman Review suggested a role for the Technology Strategy Board, now known as Innovate UK, in supporting forensic innovation through the formation of a relevant Knowledge Transfer Network (KTN). The 'Forensic Science Community' KTN was launched in November 2012 (Innovate UK 2014), with the expressed purpose of improving communication and facilitating stronger networks among the diverse array of UK forensic stakeholders. Innovate UK has promoted the Forensic Science Community through a broad programme of events relating to specific forensic specialisms, such as DNA, evidence interpretation, fingerprints, digital forensics and fire investigation.

Innovate UK's role to encourage the commercial potential of science and technology can be discerned from its internet presence, which describes its task in terms of 'commercializing new ideas with business' and offering 'support and services' to 'help business develop new products and services – and bring them closer to market' (Innovate UK 2014). Innovate UK's commercial commitments to forensic innovation have been highlighted in reports, one of which was entitled 'Taking Forensic Science R&D to Market'. Hence it is clear to see an increasing orientation to meeting policing and forensic needs in ways which are commercially exploitable.

The commercial dimensions of forensic innovation invite scrutiny. Precisely how do commercial imperatives influence policing and forensic innovation agendas? To what extent does business rationality align with forensic science, given that the latter can itself be seen to sit awkwardly between the expectations of law and science, two interconnected but markedly different knowledge-making practices? (Jasanoff 2007). Do commercial imperatives necessarily facilitate optimal technological solutions? Does the nexus of innovation, science and law, present added complications in terms of legal, police and public expectations of forensic science? And how do exogenous technological developments impact upon police and forensic practice? The following sections explore such questions by focusing on three emerging areas of forensic science and technology, namely *digital forensics*, *DNA phenotyping* and *forensic language analysis (FLA)*. A consideration of these examples illustrates a series of complex, dynamic relationships between science, policing, and commercial interests on one hand, and ethical and epistemological issues on the other.

Digital forensics

The definition of 'digital forensics' has been observed by a group of leading stakeholders to be open to interpretation, with narrow and wider definitions possibly co-existing (Marshall *et al.* 2013). Digital forensics can however be broadly construed as the process of recovering data files from computerized devices. Digital forensics has emerged in response to the increasing use, and seemingly growing dependence, of individuals on electronic communications and social media. Smartphones have rapidly become mainstream, and tablet computers have joined laptop computers as commonly owned devices. Other electronic devices such as games consoles provide additional forms of internet connectivity. Alongside such new technology, email has been joined by a host of social media platforms such as Facebook, Twitter and Instagram through which users can express themselves and communicate with others online. Other items such as credit cards and supermarket loyalty cards (such as the 'Tesco Clubcard') may provide information about an individual's communications, transactions and movements.

Technological developments, and society's increasing use and reliance on that technology, is driving changes to police work. Among UK police forces, it is now standard casework practice to generate 'digital profiles' of victims and suspects. Crime investigations increasingly involve mapping online transactions of persons of interest or victims, alongside the collection and analysis of forensic evidence such as DNA or fingerprints. This may involve, for example, recovering details of social media or online banking activity.

Digital forensics is currently perceived to concern itself with anticipating the use and misuses of new technology, and how forms of data may assist inquiries. The key task for digital forensics practitioners is how to extract data from a device. Digital forensics practitioners face challenges however in extracting and interpretating such data. The plethora of 'apps' currently found in electronic devices presents a highly complex challenge.

Concerns have been expressed relating to the perceived independence of practitioners conducting digital forensic work. The majority of digital forensics work is performed by police forces in-house, who may only approach external commercial forensic science firms in the event of an excessive workload. It has been perceived by commercial firms that the tendency toward in-housing prevented understanding of what kind of digital forensic methods are used within police forces. Those who perform data extractions may also be directly involved with the investigations, which has led to concerns about bias, or certain data being selectively emphasized while the significance of other data is overlooked. Concerns about possible bias have been linked to infrastructure and resourcing issues. Police are conscious that they provide the infrastructure in which digital forensics work is done, which may shape certain expectations on their part.

The increasing variety of digital devices which can connect online, such as *inter alia* 'smart' televisions and games consoles, sometimes referred to as 'the internet of things' compounds these issues. This is due to the variety of file formats which may be encountered by digital forensics practitioners, who often lack the means of converting these files into formats which allow them to view and analyse the files. New devices may utilize new file formats, yet investigators struggle to keep up. Digital forensics investigators require 'codecs', programs or devices able to decode one file format into another. Without a particular codec, which may be highly specific to a device or app, it may not be possible to pursue a particular case. There is currently no commercial incentive however to develop codecs on a case-by-case basis. While one particular codec may only be specific to a certain case, it may not be useful in subsequent investigations.

Digital forensics encapsulates some of the issues facing forensic R&D. While innovation challenges have been identified, many of these are not currently perceived to be commercially viable, as in the case of developing case-specific codecs. A capability gap remains which could be met through greater engagement with technology manufacturers, yet commercial priorities on the part of the latter may hinder this from being met. Much digital forensics work is being conducted in-house. It is unclear whether operational needs are being sufficiently met.

Digital forensics practitioners also appear to struggle with backlogs of devices from which to extract data, and the length of time it takes to complete an extraction. Whenever an extraction occurs, a verifiable duplicate of the data has to be made, which therefore doubles the time of each procedure. There are questions over what kind of resources could assist digital forensic practice. Digital forensics has to respond not only to criminal justice concerns but also externalities presented via new technological developments from companies such as Apple and Microsoft. New devices may come with new file formats, which further compounds the complexity of digital forensic work. It may often take a considerable period of time for practitioners to learn how to work with new technology. It appears that lines of communication between manufacturers of digital devices and the digital forensics community have been relatively weak, representing another issue facing innovation in this area.

The challenge of extracting data for use on a case-by-case basis appears to be currently overlooked. This problem does not seem to present opportunities for solutions which may commercially exploitable. A group of digital forensics practitioners consulted in 2014 viewed one solution as improved communication between individual experts, developing networks through which experiences could be pooled and agreement on best practice could be reached. It was conceded by the same group however that this approach would not necessarily lead to any specific new inventions.

Digital forensics appears to remain in a reactive rather proactive mode. This seems to be compounded by uncertainty about what kind of technology or communications systems may emerge next, and their potential forensic significance. Digital forensics practitioners may continue to face challenges in responding to technology through the possibility that codecs may need to be derived on a case-specific basis, and on the lack of commercial appeal of their tools.

New technologies may also raise other, exogenous contingencies, relating to public concerns over the potential of technology to function as a means of surveillance of communications. The manner in which digital forensics evidence is interpreted (and concerns over who is doing the interpreting), may add to fear about bias and the in-house location of much digital forensics work.

DNA phenotyping

The term 'DNA phenotyping' broadly refers to methods used to infer the physical characteristics of an individual from a DNA profile. Related research has also sought to study the frequency of certain DNA elements among populations assumed to share a common geographic ancestry.

DNA phenotyping exemplifies the heterogeneity of stakeholders in forensic innovation. Communities of practice surrounding phenotyping encompasses forensic geneticists and related academic researchers, policymakers, government officials, legal professionals, police and other law enforcement officials, commercial firms, defendants and victims. A variety of professional organizations have also played a part in debating DNA phenotyping. These may include scientific bodies such as the International Society for Forensic Genetics (ISFG), European Network of Forensic Scientific Institutes (ENFSI) or the Royal Society, police organizations such as the UK's National Police Chiefs Council, legal bodies such as the Crown Prosecution Service, and regulatory bodies such as the Office of the Forensic Regulator. In addition, organizations such as Innovate UK have brought together these actors to discuss DNA phenotyping (Innovate UK 2015). Bodies such as Innovate UK could possibly be considered 'boundary organizations' (Guston 1999, 2001), helping to shape communities of practice by facilitating cross-disciplinary discussion on the development of phenotyping.

The key gatekeepers for the acceptance of phenotyping could however be regarded as legislators and publics. The Netherlands provides an example of the role of legislators in shaping the use of phenotyping through specific laws. In 2003, the Netherlands Parliament passed the Law on External Visible Characteristics (EVCs) which specifies the type of phenotypes (eye and hair colour) which can be used in casework. This legislation has been described as 'window case legislation' (Toom 2012) as it is designed to be possibly amended to allow other phenotypes in the event of scientific advances.

Discussions around the further development of phenotyping have recently included a sustained focus on the social, legal and ethical implications for the application of such methods to police work. Earlier uses of DNA phenotyping methods raised human rights and civil liberties concerns regarding the rule of law. In 1999, Netherlands police used phenotyping to infer that a suspect leaving an unknown DNA profile at the scene of the alleged murder of Marianne Vastra, was a white Northern European male. They subsequently undertook so-called 'DNA dragnets' from local individuals fitting that appearance, requesting that they submit DNA for testing against the unknown profile (Toom 2012). This case raised issues about whether phenotyping potentially reverses the presumption of innocence central to much Western criminal law (M'Charek *et al.* 2008).

More recently, a DNA phenotyping method developed by the US company Parabon Nanolabs was used to produce an image of a possible suspect in the alleged murder of Candra Alston and her daughter. This incident took place in Columbus City, South Carolina. In 2015, this image was released to the public. The victims had been found in their apartment, with no eyewitness or CCTV evidence. An unknown DNA profile was however found at the scene. From this profile the image was produced, of a young male seemingly of African descent. Concerns were raised in some circles over the verisimilitude of the image, and whether it was too misleadingly generic. The risk of false positive identifications from such images is seen as problematic, particularly in areas where racial sensitivities may exist (Innovate UK 2015).

DNA phenotyping has also raised issues around legal and public acceptance of forensic technology. Some discussions around phenotyping have focused on the threshold of reliability. It has been recommended that inferences of possible phenotype from DNA profiles be expressed conditionally as probabilistic estimates (Kayser and Schneider 2008). However, there are concerns that this could portray DNA phenotyping data as less than categorically certain in the eyes of the public, which might affect its public and legal acceptance. Other concerns relate to whether the public should have knowledge of phenotyping being used in investigations, as in the Candra Alston case, or whether that phenotyping data should be restricted to investigators. Phenotyping data could then be used as a lead to uncover other, more definitive, evidence. Here, the perceived distinction between 'intelligence' and 'evidence' comes into play, also raising issues over how probative a phenotype may be in the absence of other case information.

Discussions over the potential of phenotyping have therefore been tempered by concerns over the possible consequences of exposing such technology to the public and legal spheres too soon. Other ethical issues which have been raised concern precisely which phenotypes should be employed in casework. It is conceivable that future phenotyping methods could extend from external visible characteristics to link forensic DNA profiles with phenotypes suggesting medical conditions. This raises issues of the right to know – might a suspect only discover they have a particular condition once they are arrested?

The development of phenotyping methods necessitates research on large datasets of DNA profiles, reflecting certain populations of shared geographical origin. The requirement for ready access to DNA populations of certain groups has at times led to existing databases being accessed. The group Genewatch found that such research had involved the Police National DNA Database (NDNAD) of England and Wales. Genewatch claimed however that some methods appeared to be somewhat crude:

> The list of projects included some 'operational requests', including one on behalf of the police, to check the Database for named individuals. One research project involved the selection of some groups of individuals from the Database on the basis of '*having an African name*', '*having typical Muslim names*', or '*having typical Hindu/Sikh names*'.
>
> (*Wallace 2008: 11*)

Critics of phenotypic profiling and related research have argued that ethnic classifiers (e.g. 'white Caucasian' etc.) are ultimately culturally-bestowed labels and do not reflect underlying genetic reality (Ossorio and Duster 2005). Such critical arguments suggest that research undertaken on phenotypic profiling, particularly ethnic inference, risk perpetuating a kind of tautology. Cultural labels may be used to 'prove' themselves through perceived links with DNA taken from persons culturally labelled as such.

Despite such criticisms, DNA phenotyping is likely to continue to evolve. It remains to be seen how research into forensic applications of phenotyping may be influenced by issues around public acceptance and potential ethical consequences. It also remains to be seen as to what kind of policy agenda will emerge around phenotyping. Broad public acceptance could predominate at the expense of addressing more complex ethical risks.

Social research has played a role in highlighting scientific and legal controversies, and related social and ethical concerns surrounding DNA phenotyping (Ossorio and Duster 2005; Nuffield Council on Bioethics 2007; M'charek *et al.* 2008). Yet there remains potential for communication gaps to endure between scientists, social researchers, police, lawyers, policymakers and other actors. These communication gaps represent issues that social researchers could address in more depth. Social researchers have raised concerns that scientists developing phenotyping techniques do not fully recognize the social implications of their work (M'charek *et al.* 2008). Scientists have raised concerns that lay audiences may not be able to fully understand the meaning of evidence reported in probabilistic terms. The possible exposure of phenotyping to courtroom scrutiny raises issues about how well legal actors can cope with comprehending potentially complex scientific testimony.

Forensic language analysis (FLA)

The increasing use of communication technology has exposed further possibilities to try and identify persons based on their patterns of written and spoken communication. *Forensic linguistics* involves the analysis of spoken dialects or figures of speech in order to isolate unique identifying features. Forensic linguistics also encompasses the analysis of the composition of written communications such as e-mails or text messages to ascertain the identity of individuals. Elsewhere, linguistic methods have been employed to try and verify the identity of those making claims for asylum. In addition, there are ongoing efforts to develop *forensic phonetics* technology for automated voice recognition systems. Collectively, forensic linguistic and phonetics have been referred to as *Forensic Language Analysis (FLA)*.

Forensic linguistics and forensic phonetics brings together a range of different forms of expertise. This may include speech science, which encompasses linguistics researchers and those researching the anatomy of speech. FLA may also encompass instrumentation and computer scientists. Expertise from these kinds of scientific fields interacts with the professional experience of law enforcement and intelligence officials, immigration officers and legal practitioners.

Social research may play a potential role in both developing and critically analysing these techniques. Sociolinguistics, the study of societal and cultural factors on language use, is one such body of work which is already involved in forensic linguistics research (Leonard 2005). However, social researchers, including sociolinguists, socio-legal scholars and other social scientists, have also been prominent in identifying perceived shortcomings in forensic linguistics techniques. Researchers have drawn attention to the complexities apparent in the relationship between language and origin. Such work indicates that language may not map directly onto geographical boundaries. Displacement may affect language use (Eades 2010).

Linguistic and phonetic analysis is currently challenged by the lack of population level data which could help establish probabilities that a linguistic or phonetic feature may match with an individual. Establishing a match between communication features and an individual may therefore significantly rely on the personal experience and background of the analyst. Forensic phonetics analysts rely on speech processing software, but also on their own trained ears.

Concerns remain over the reliability of current FLA techniques. The potentially uncritical reception of such evidence in court is also a subject of concern (Parliamentary Office on Science and Technology 2015). Debates continue over the use of probabilistic methods to evaluate and report the probative weight of evidence. Similar to other forms of forensic evidence, concerns remain over how comprehensible statistics are to lay audiences and how accurately these can be converted into verbal statements.

Other issues relate to the use of sociolinguistics profiling, and whether it can be safely submitted for courtroom scrutiny or should be used in an intelligence capacity, in order to produce leads which in turn may generate more concrete evidence (Parliamentary Office on Science and Technology 2015). There has already been at least one notable controversy concerning voice recognition systems. In 2007, a paper was published in the *International Journal of Speech, Language and the Law* which strongly criticized the claims made about voice recognition systems marketed by the company Nemesysco. This paper was later redacted after alleged legal threats by the latter.

While guidelines exist in England and Wales which indicate who can provide expert linguistic and phonetic evidence, this kind of expertise is not currently statutorily regulated (Parliamentary Office on Science and Technology 2015). For criminal cases, courts may be regarded as the ultimate arbiters or gatekeepers of forensic linguistic and phonetic evidence. The UK government may be regarded as the key gatekeeper for the use of linguistic and phonetic analysis in asylum cases. Yet while the Home Office uses language analysis to help determine the origin of asylum seekers, no independent assessment of these methods has, at time of writing, taken place. Some FLA practices have been criticized by academics and in the UK Supreme Court (Fraser 2011).

Perceived epistemological issues can be linked to ethical concerns. The relationship between the perceived technological reliability of linguistic systems and the implications for social justice, raises matters of epistemic risk, namely the risk of assuming certainty from what may actually be probabilistic data. One may wish to question, for example, just how aware law enforcement officials may be of the limitations of linguistics technology in asylum cases. Too much trust in what is epistemologically contestable technology may therefore hold consequences for social justice.

Science and Technology Studies (STS), the interdisciplinary field of research focusing on the social impact and shaping of science and technology, could have a role to play in critically understanding wider social issues involving forensic linguistics and phonetics. STS is well placed to compare how claims to the credibility and objectivity of forensic linguistics and phonetics are established, and how these claims might be challenged in courtroom proceedings. Such a research agenda could take its lead from studies of fingerprinting (Cole 2001), and forensic DNA profiling (Aronson 2007; Lynch *et al.* 2008), and STS research could address the scientific and technical debates within forensic linguistics and phonetics, and how actors claim credibility in deliberations over what may be contested scientific claims. Opportunities therefore exist to study how a new range of scientific controversies might play out in the wider legal arena, and throughout society at large.

Certain exogenous pressures appear to impact on FLA. Government policy, particularly in response to (possibly media-led) public concerns over immigration, may drive the deployment of technology which some may regard as inadequately tested. The Nemesysco case also raises issues over the potential for externalities, such as powerful commercial forces, to affect the dissemination of scientific claims.

FLA seems to present an example of how forensic science and technology may become susceptible to wider societal forces, such as public policy attitudes (which themselves may be driven by public and media attitudes), and commercial pressure. It appears that FLA has been applied in casework ahead of timely scientific scrutiny. Issues possibly remain concerning the kind of practices and interests which shape who, within the community of FLA stakeholders, makes decisions about the acceptance of technology, who deems it 'reliable' and how 'reliability' is itself collectively understood.

Managing expectations

An earlier section of this chapter described how forensic science has seemingly become a primary form of evidence for police. During a discussion among forensic stakeholders about DNA phenotyping in 2015, the concern was expressed that investigators might rely too heavily on new scientific techniques, without being aware of the potential limitations or uncertainties surrounding technology. These might only become apparent later in the criminal justice process. This portrayal of police expectations suggests at least two areas of risk. First, the high degree of faith placed on new forensic science and technology may not be matched by results in actual operations. Second, high expectations may lead investigators to assume infallibility of methods which may actually become open to contestation in an adversarial context.

Concerns about the perceived admissibility of new forensic technologies in the adversarial system may play a role in innovation trajectories. It has been possible to detect, during discussions among forensic stakeholders about DNA phenotyping for example, a palpable sense of caution about introducing this technology into the courts, and a sense of apprehension that such evidence could be misunderstood. At times some have viewed that the inaccurate presentation of new technologically-mediated evidence in court could hold adverse consequences for innovation: 'If technology is used wrongly in courts it will set it back two years' (Expert Discussion 2015). The 'CSI effect', is an apparent phenomenon in which public understanding of science is skewed by inaccurate media portrayals of forensic science (Cole 2015). The 'CSI effect' has been used to describe how the media has possibly created distorted public expectations. While it remains an open question whether or not the CSI effect actually exists (Cole and Dioso-Villa 2007), forensic stakeholders are increasingly aware of the potential disparity between expectations of forensic science and what it may actually achieve in operations.

The public aura that seemingly surrounds forensic science and technology in the service of law enforcement is therefore recognized as a challenge by UK forensic stakeholders (Expert Discussion 2015). Stakeholders have also perceived a need to engage with publics around new forensic technologies. This is particularly so where social and ethical issues have been identified in relation to emerging forensic science and technology.

A recurring theme in related discussions among stakeholders concerns the perceived need to 'educate' publics in the 'realities' of using forensic science in casework. Such education has often been deemed necessary given the perceived distorting influence of media portrayals in dramas such as *CSI. Much social research has, however, taken a critical view of such perceptions*. The concept of 'deficit' and 'surfeit' discourses have emerged through sociological critiques of policymaker assumptions concerning the public understanding of science (Bodmer 1985; Cole 2015; Michael 2002; Lezaun and Soneryd 2007). Broadly construed, the 'deficit' model describes the observed tendency of policymakers and scientists to assume that publics lack knowledge of science, but will readily accept the authority of science once they are sufficiently educated.

Contemporary discussions surrounding certain forensic methods indicate other attitudes toward public engagement. In contrast to the deficit model, the notion of the 'surfeit model' has been introduced by social researchers to describe another perceived tendency. The surfeit model has been used to frame another set of assumptions, that publics are over-exposed to inaccurate images of science and technology, which convey images of absolute reliability and infallibility. Media may also portray technologies which may not actually exist, or are still at a nascent stage (Ley *et al.* 2012; Cole 2015). The surfeit model captures the perception, on the part of some forensic stakeholders, that media representations over-state the certainty and utility of science, leading to unduly heightened expectations on the part of lay audiences. Surfeit discourses may also circulate when scientists portray lay audiences as lacking knowledge of the specific realities of scientific work (Lawless 2013, 2016).

Conclusion

Earlier sections of this chapter outlined an innovation landscape in the UK which includes the potential contribution of academia and government-supported research in the form of the Home Office CAST. UK forensic innovation has also been shaped by police organizations, who have introduced gatekeeping mechanisms for innovation. This chapter has also described how commercial imperatives have played an increasingly influential role in the way forensic innovation is shaped.

Forensic innovation is therefore shaped by a wide series of interests and stances. These include concerns over public, police and legal expectations of new forensic science and technology, commercial imperatives, and other influences such as the REF agenda in academia. Forensic innovation is embedded within, and subject to, an environment where such pressures exert effects.

A focus on digital forensics, DNA phenotyping and FLA highlights technological opportunities, but also challenges and threats. These techniques and technologies continue to develop in the context of a heterogeneous forensic innovation landscape, and amid law enforcement priorities which may be subject to change. Accelerating, exogenous technological developments impact upon forensic practice. For example, digital forensics practitioners must contend with the rapid evolution of devices such as smartphones. This chapter has described how the increasing sophistication and variety of electronic devices present problems for digital forensics practitioners, who may struggle to retrieve data from these devices without codecs specific for each of them. The increasing number of devices which may connect online, often referred to as 'the internet of things', significantly compounds the issue. The rapid pace of innovation in online devices is out of the hands of digital forensics practitioners, who can only operate responsively, but who may be hindered by inadequate institutional or commercial support.

The pace of innovation also presents issues relating to DNA phenotyping and FLA. Concerns have been raised about the use of these methods in casework. Some of these relate to the reception and comprehension of informational or evidential claims produced via these technologies. This chapter has described the perceived risks of accruing certainty to these claims. A notable degree of reputational risk has been attached to these technologies from within the forensic scientific community. These reputational risks appear to be associated with using nascent technology prematurely, and without due consideration of the wider potential consequences relating to their possible epistemological shortcomings. However, DNA phenotyping and FLA technology have been embraced by commercial companies, who, it seems, have been keen to promote these technologies in law enforcement contexts. Questions remain whether commercial promotion may have precluded possible rigorous a priori scientific scrutiny. In the case of Nemesysco, scientific critiques of commercial products have been followed by legal threats.

Commercial imperatives exert different effects in the case of digital forensics on one hand, and DNA phenotyping and FLA on the other. Digital forensics encompasses recovery techniques which are relatively simple in principle, but whose dissemination have so far been hindered by a lack of commercial appeal. DNA phenotyping and FLA are arguably more complex technologies, but whose status is potentially complicated by the chaff of commerciality. On the other hand, the decision to utilize new technology may be influenced by cost-benefit concerns. There may be pressures to balance the cost of technology against the time taken to deploy it. This in turn could be complicated by ensuring practitioners are sufficiently competent in a technology which may then only be used relatively rarely, as in the case of DNA phenotyping.

The forensic innovation landscape exists within a wider context of exogenous technological developments, together with various commercial and budgetary pressures. These currents may be at least partially beyond the control of forensic science stakeholders, but they nonetheless present notable scientific, legal and ethical complexities with which stakeholders must contend. Attitudes toward emerging forensic technologies are also influenced by concerns over their potential susceptibility to legal challenge, and how publics may view these technologies in relief of the seemingly idealized forms of forensic science depicted in the media. It appears the UK forensic community are increasingly concerned about public reception to new forensic technologies. There also appears to be heightened awareness to legal attitudes to new technology and their exposure to the jury system.

Forensic technologies represent sites where science, law, and other forms of hegemony (such as commerce) meet and interact. These technologies can be considered spaces where views of science, held by different actors and potentially expressed in 'deficit' or 'surfeit' discourses, meet. The credibility of forensic technology may be upheld or challenged by invoking commercial forces. Commercial appeal may be claimed as undermining technological claims by regarding the former as distorting scientific credibility. On the other hand, forensic scientists have also defended forensic technologies, which have struggled to be accepted, by claiming they lack commercial appeal which prevents technologies from being more widely disseminated (Lawless 2013, 2016).

In addition to 'deficit' and 'surfeit' discourses, it is thus possible to identify how the invocation of a wider series of societal pressures affects the dissemination and perception of forensic scientific claims. It is possible that other forms of hegemony, such as law, may also be susceptible to being expressed in deficit and surfeit terms, or be discursively critiqued in terms of their susceptibility to the influence of wider societal influences (Lawless 2016). Hence forensic technologies also represent loci where different forms of hegemony, expressed relationally through different forms of discourse, interact. Collective understandings of these hegemonies may be challenged in these encounters. Discussions over emerging forensic technologies could be regarded as ways in which different hegemonies discursively position themselves to one another.

Sociological approaches based on 'positioning theory' (Baert 2012) may represent one potentially fruitful way to facilitate understanding of the evolution of forensic technology, and how they reflect, and are shaped by, the interplay of different types of hegemonic discourses. These may be voiced in deficit or surfeit terms, or portrayed as susceptible to wider societal influences such as commerce.

Qualitative social research is well-placed to explore why forensic innovation trajectories emerge in the way they do. Such work could include longitudinal studies of technology development, focusing on social-structural dimensions of forensic innovation together with a focus on the interactions between different stakeholders (e.g. police, defendants, victims, forensic practitioners, lawyers, politicians, publics etc.) over the course of time.

Actors implementing forensic technology implementation often have to address a wider series of economic, regulatory and political imperatives. Implementation of forensic technology can therefore be construed as a markedly social practice. This highlights possibilities for social research to understand further how innovation may become influenced by a range of procedures and imperatives. Social researchers may also wish to alert themselves to the potential differences in the implementation and use of new forensic technology across different spaces and jurisdictions.

References

Aronson, J.D. (2007) *Genetic Witness: Science, Law, and Controversy in the Making of DNA Profiling*. New Brunswick, NJ, London: Rutgers University Press.

Baert, P. (2012) 'Positioning Theory and Intellectual Interventions', *Journey for the Theory of Social Behaviour*, 42 (3): 304–324.

Bodmer, W. (1985) *The Public Understanding of Science*. London: Royal Society.

Cassella, J. (2008) Forensic Science on Trial—Still! Response to "Educating the Next Generation" [*Science and Justice*, 48 (2008) 59–60], *Science and Justice*, 48 (4): 197–199.

Cole, S.A. (2001) *Suspect Identities: A History of Fingerprinting and Criminal Identification*. Cambridge, Mass.: Harvard University Press.

Cole, S.A. (2009) 'A Cautionary Tale About Cautionary Tales About Intervention', *Organization*, 16 (1): 121–141.

Cole, S.A. (2015) 'A Surfeit of Science: The "CSI Effect" and the Media Appropriation of the Public Understanding of Science', *Public Understanding of Science*, 24 (2): 130–146.

Cole, S.A. and Dioso-Villa, R. (2007) '*CSI* and its Effects: Media, Juries and the Burden of Proof', *New England Law Review*, 41: 435–470.

Dawnay, N., Stafford-Allen, B., Moore, D., Blackman, S., Rendell, P., Hanson, E.K., Ballantyne, J., Kallifatidis, B., Mendel, J., Mills, D.K., Nagy, R. and Wells, S. (2014) 'Developmental Validation of the ParaDNA1 Screening System – A presumptive test for the detection of DNA on forensic evidence items', *Forensic Science International: Genetics*, 11: 73–79.

Eades, D. (2010) 'Nationality Claims: Language Analysis and Asylum Cases' in Coulthard, M. and Johnson, A. (eds), *The Routledge Handbook of Forensic Linguistics*. London: Routledge, pp. 411–422.

Forster, L., Thomson, J. and Kutranov, S. (2008) 'Direct Comparison of Post-28-Cycle PCR Purification and Modified Capillary Electrophoresis Methods with the 34-Cycle "Low Copy Number" (LCN) Method for Analysis of Trace Forensic DNA Samples', *Forensic Science International: Genetics*, 2 (4): 318–328.

Fraser, H. (2011) 'Language Analysis for the Determination of Origin (LADO)' in Chapelle, C.A. (ed.) *Encyclopedia of Applied Linguistics*. Oxford: Wiley-Blackwell.

Fraser, J. and Williams, R. (2009) 'Introduction: Themes and Debates in Contemporary Forensic Science', in Fraser, J. and Williams, R. (eds), *Handbook of Forensic Science*. Cullompton: Willan, pp.487–490.

Guston, D.H. (1999) 'Stabilizing the Boundary Between US Politics and Science: The Role of the Office of Technology Transfer as a Boundary Organization', *Social Studies of Science*, 29 (1): 87–111.

Guston, D.H. (2001) 'Boundary Organizations in Environmental Policy and Science: An Introduction', *Science, Technology and Human Values*, 26 (4): 399–408.

Innovate UK (2014) Innovate UK Homepage available at: www.gov.uk/government/organisations/innovate-uk (accessed 7 October 2014).

Jasanoff, S. (2007) 'Making Order: Law and Science in Action', in Hackett, E.J., Amsterdamska, O., Wajcman, J. and Lynch, M. (eds), *Handbook of Science and Technology Studies* (3rd edition). Cambridge, Mass: MIT Press, pp.761–786.

Kayser, M. and Schneider, P.M. (2009) 'DNA-Based Prediction of Human Externally Visible Characteristics in Forensics: Motivations, Scientific Challenges, and Ethical Considerations', *Forensic Science International: Genetics*, 3 (3): 154–161.

Lawless, C.J. (2011) 'Policing Markets: The Contested Shaping of Neoliberal Forensic Science', *British Journal of Criminology*, 51 (4): 671–689.

Lawless, C.J. (2013) 'The Low-Template DNA Profiling Controversy: Biolegality and Boundary Work Among Forensic Scientists', *Social Studies of Science*, 43 (2): 191–214.

Lawless, C.J. (2016) *Forensic Science: A Sociological Introduction*. Abingdon, New York: Routledge.

Lawless, C.J. and Williams, R. (2010) 'Helping with Inquiries, or Helping with Profit? The Trials and Tribulations of a Technology of Forensic Reasoning', *Social Studies of Science*, 40 (5): 731–755.

Leonard, R. A. (2005) 'Forensic Linguistics: Applying the Scientific Principles of Language Analysis to Issues of the Law', *International Journal of the Humanities*, 3 (1): 1–9.

Ley, B.L., Jankowski, N. and Brewer, P.R. (2012) 'Investigating *CSI*: Portrayals of DNA Testing On A Forensic Crime Show and Their Potential Effects', *Public Understanding of Science*, 21 (1): 51–67.

Lezaun, J. and Soneryd, J. (2007) 'Consulting Citizens: Technologies of Elicitation and the Mobility of Publics', *Public Understanding of Science*, 16 (2): 279–297.

Lynch, M. (2009) 'Science as a Vacation: Deficits, Surfeits, PUSS, and Doing Your Own Job', *Organization*, 16 (1): 101–119.

Lynch, M., Cole, S.A., McNally, R. and Jordan, K. (2008) *Truth Machine: The Contentious History of DNA Fingerprinting*. Chicago, Il: University of Chicago Press.

Mallinson, S. (2015) 'An Overview of the Centre for Applied Science and Technology – Current and Future Research Requirements'. Seminar at Northumbria University Centre for Forensic Science. 3 March 2015.

Marshall, A., Higham, S. and Dyhouse, T. (2013) *Digital Forensics Capability Review*. Horsham: Electronics, Sensors, Photonics Special Interest Group.

M'charek, A., Toom, V. and Prainsack, B. (2008) 'Bracketing Off Populations Does Not Advance Ethical Reflection on EVCs: A Reply to Kayser and Schneider', *Forensic Science International: Genetics*, 6 (1): e16–e17.

Michael, M. (2002) 'Comprehension, Apprehension, Prehension: Heterogeneity and the Public Understanding of Science', *Science, Technology and Human Values*, 27 (3): 357–378.

Miller, S. (2001) 'Public Understanding of Science at the Crossroads', *Public Understanding of Science*, 10: 115–120.

Nuffield Council on Bioethics (2007) *The Forensic use of Bioinformation: Ethical issues*. London: Nuffield Council on Bioethics.

Ossorio, P. and Duster, T. (2005) 'Race and Genetics: Controversies in Biomedical, Behavioural and Forensic Sciences', *American Psychologist*, 60 (1): 115–128.

Parliamentary Office on Science and Technology (2015) 'Forensic Language Analysis', *POSTnote* No. 509, September 2015. London: Parliamentary Office on Science and Technology.

Pinch, T. and Bijker, W. (1984) 'The Social Construction of Facts and Artefacts: Or How the Sociology of Science and the Sociology of Technology might Benefit Each Other', *Social Studies of Science*, 14 (3) (August): 399–441.

Silverman, B. (2011) *Research and Development in Forensic Science: A Review*. London: Home Office.

Toom, V. (2012) 'Bodies of Science and Law: Forensic DNA Profiling, Biological Bodies and Biopower', *Journal of Law and Society*, 39 (1): 150–166.

Wallace, H. (2008) 'Prejudice, Stigma and DNA Databases', Paper for the Council for Responsible Genetics, available at: www.councilforresponsiblegenetics.org/pageDocuments/PDAFXSTDPX.pdf (accessed 26 September 2013).

Williams, R. and Edge, D (1996) 'The Social Shaping of Technology', *Research Policy*, 25 (6): 865–899.

23

Technology and digital forensics

Marcus Rogers

The impact that technology has on criminality and the rule of law is especially emphasized in the realm of cyber crime and digital investigations. The very nature of law enforcement, investigations and the legal justice system has changed due to the introduction of cyber crime (AKA computer crime) and the technology required to investigate and prosecute these types of cases. Cyber crime can be defined as:

> a criminal offense on the Web, a criminal offense regarding the Internet, a violation of law on the Internet, an illegality committed with regard to the Internet, breach of law on the Internet, computer crime, contravention through the Web, corruption regarding Internet, criminal activity on the Internet, disrupting operations through malevolent programs on the Internet, electric crime, Internet crime, sale of contraband on the Internet, stalking victims on the Internet, theft of identify on the Internet.
>
> *(Burton 2007)*

Technology has enhanced traditional deviant and criminal behavior, while at the same type creating criminal activities that have never existed before (e.g. Distributed Denial of Service attacks, malware attacks). Technology is changing criminalistics at an accelerated path as speed and the ability to handle large volumes of data/evidence are now required. The legal justice system is struggling to define and understand new cyber-criminal behaviors (Garfinkel *et al.* 2009; Meyers and Rogers 2004; Mohay 2005). The legal justice system is also struggling with how to handle "expert opinions" that for the most part are being derived or proffered from automated tools and not people. In order to wrap our heads around the question of how technology has and will impact society, criminality and the rule of law, we will focus on the discipline of digital forensics.

Digital forensics or, as it is often called, computer forensics is commonly defined as: "A sub-discipline of Digital & Multimedia Evidence, which involves the scientific examination, analysis, and/or evaluation of digital evidence in legal matters" (SWGDE 2015).

Digital forensics, like other forensics sciences, has seen a rapid increase in the use of technology by criminals and practitioners in the past few years. This heavy reliance on technology does not come without some costs (Meyers and Rogers 2004). This chapter will examine the history of how criminals have been early adopters of technology to further their criminal trade craft. We will examine specific technologies that have either been used or targeted by cyber criminals. The discussion will not be an exhaustive examination of all of the technologies as space precludes such an endeavor. Instead, we will focus on the the main technologies that seem to have had the biggest impact. This chapter will also look at what impact the use of tools has on those investigating computer crimes, as various automated tools are now becoming a standard part of this forensic discipline.

Computer criminals

It should come as no surprise that criminals embrace technology. For the most part criminals are opportunistic and none more so than those criminals that are engaged in computer crimes. Historically, almost every technology has been co-opted by the criminal element in our society (Meyers and Rogers 2004). The telegraph was used by "confidence men" and fraudsters to reach new victims, so too was the fax machine, pager and color copiers (Casey 2011). In fact, the manufacturers of color copiers not only recognized the risk that criminals would use their technology to produce counterfeit currency, they built in controls in the form of microdots that included the serial number of the machine that created the copy. These dots were used to successfully catch and prosecute numerous criminals, before the criminal community figured out how they were being caught.

In order to appreciate how technology has impacted crime and criminal careers, it is important to define exactly what we mean by computer crime. Cybercrime is actually an umbrella term for any activity that falls into one of three categories (Parker 1998):

1 Computer as a target of the criminal activity.
2 Computer as a tool to further the criminal activity.
3 Computer that is ancillary to the criminal activity.

Computer as a target would include attacks directed at data stored on the computer system itself, such as emails, personally identifiable information or medical data. It can also include attacks that want to gain unauthorized access to a computer in order to establish an entry point into a corporate network. *Computer as a tool* is a bit subtler and here the computer is used to assist the criminal in carrying out the criminal activity (Parker 1998). Examples could include spam and phishing, telecom fraud, and other online fraud activities. *Computer as ancillary* to the activity refers to the fact that we are such a wired society that our lives are recorded in a digital format. This recording includes such things as calendars, mobile phone call data records, texting, social media posts, geo-location, or emails. The focus here is that evidence of planning or connections between the victim and the accused can be found on our computing technologies (Garfinkel 2010).

Computer criminals have capitalized on at least three dimensions of modern society and technology:

1 The Internet that is the common communications backbone for almost all modern technology was never designed to be secure.
2 The general public will almost always choose convenience over security with modern technology.
3 The general public trusts technology and is too willing to accept flawed technologies as final products.

Together, these three dimensions are one of the facets that differentiates technology related crimes from the more traditional crimes and have contributed to the success of cyber criminals and the potentially huge pay outs.

Technology enables perpetrators to dehumanize their victims through the introduction of a layer of abstraction between criminal and the victim (Leigland and Krings 2004). The criminal never has to physically meet or see the victim and in some cases the victim is a computer system or database and it can be rationalized that no real people are being impacted, just faceless corporations (Choi 2008; Rogers 2003; Shaw 2006).

Deterrence theory posits that people commit crimes only to the extent where they derive pleasure (gain) from the activity and not pain (negative consequences) (Beccaria 1963). Legal punishment, depending on the severity and swiftness, is presumed to deter people from committing crimes as this is viewed as painful. Accordingly, anonymity has an important impact on human behavior, especially deviant behavior. When people believe that their true identities cannot be determined, their fear of punishment and retribution becomes diminished and they are more prone to act out behaviors they would not undertake if they were identifiable (Baggili et al. 2012; Zimbardo 1969).

When the dehumanizing element of technology is combined with the assumed anonymity, they can be force multipliers of each other and people who would not have otherwise become traditional criminals, but who were still drawn to deviant activities, may choose to become cyber criminals (Kwan et al. 2008).

Historical targets

Almost every one of the latest technologies has been either targeted or used as an attack vector. In the early days of computer crime, offenders' primary target was telephone systems. These phreakers (hackers who target telephone systems) quickly discovered that they could fool the telcos' equipment into allowing them free phone calls, especially long distance calling. As private companies started to in-house their telecommunications by using PBX (Private Branch Exchange) systems, the phreakers shifted their attention from the major telcos to private companies. The costs to private industry as a result of these PBX attacks were in the millions of dollars.

The next technology area to garner the computer criminal's attention was online credit cards. When the burgeoning e-commerce sites of the 1990s and early 2000s started accepting credit card numbers for transactions, the cyber criminals were able to reverse engineer the very weak security algorithms being used. The cyber criminals were able to generate fake credit card numbers that passed the online security controls, when in fact these numbers had never actually been issued to any card holders. Interestingly enough, most law enforcement agencies began their foray into the investigation of cyber crimes due to the rampant credit card fraud.

Present day

As technology has changed and evolved over time, so too have the targets and attack vectors for cyber criminals. Email has been and will continue to be a target of cyber criminals. A good example here is the Nigerian 419 scam. The number 419 refers to the section of the Nigerian penal code that addresses this type of fraud (FBI n.d.). In this attack a victim receives an email from a supposed Nigerian prince who needs help moving money out of the country and will gladly share the wealth with you if you can send him some money to help pay the taxes to get the funds out. Obviously the email is not from a prince and the victim loses the money they sent to pay the "taxes." This is an example of a more traditional scam that uses the anonymity and the potential for casting a large net over potential victims (millions of people can be emailed at once) (Anon n.d.). Email is the attack "vector de jour," partly because the mail protocols were never designed to be secure and partly because of the sheer volume of email communications – this makes it difficult for people or technology to fully process the emails to determine what is spam, what are phishing attacks and what are legitimate important emails. Phishing has become a prolific cyber-criminal activity and is linked to identity theft, online bank fraud, and major privacy breaches. This should not surprise anyone as email (and probably now texting) has replaced phone calls as the major medium for communications. Email attacks have been linked to such infamous breaches as Anthem, the IRS Breach, and indirectly the Target security breach (Kedgley 2015) (probably the Sony attack as well).

One of the next technologies to be exploited was smartphones. As consumers demanded more and more functionality with their smartphone devices, retailers, financial institutions and health care providers have allowed transactions and account access to occur via the mobile phone (smartphone). Unfortunately, these companies have failed to understand that functionality is the enemy of security and as a result, mobile phones are being targeted at an alarming rate, with great success (Baggili et al. 2007; Damshenas et al. 2014). This is especially troubling now that we not only have online banking enabled by smartphones, but the actual payment system for purchases now resides on the smartphone itself. Apple pay, Google wallet, Android pay and Samsung pay systems make use of the Near Field Communication (NFC) chip technology that is on most smartphones. This technology uses proximity readers and pre-authorized credit cards to turn the smartphone into a de facto credit card. All the consumer needs to do is place the phone in proximity of the point of sale (POS) system and their credit card is debited (Gomzin 2014). This is an obvious convenience for the consumer, a security nightmare for the banks, and a dream for the cyber criminals (Symantec 2014.). Now all a cyber criminal has to do, is be in close proximity of their victims and all of the credit card/payment information can potentially be stolen, transferred to another device and then used by the criminals.

Future targets

The Internet of Things (IoT) is the near future target for cyber criminals. IoT refers to any device that can connect to the Internet and can either receive or send data. According to the IDC the market place for these types of devices will surpass $1.7 trillion by 2020 (IDC n.d.). While vendors are rushing to inter-connect all our consumer devices (e.g. thermostats, garage door openers, lights) they are creating huge security vulnerabilities that can and will be taken advantage of by cyber criminals. Studies have indicated that wired homes etc. are "sitting ducks" for cyber criminals (Garcia-Morchon et al. 2013). Now breaking into a garage or thermostat could lead to an attack on a government agency or financial institution. Yet the consumer market place is bursting at the seams for wired home technologies. The number of technologies and the interconnection of these household technologies along with smart meters and the smart grid could potentially allow a cyber criminal to map out the exact daily patterns of your life. They can determine when you awake in the morning, when you leave for work, when you return, when you go to sleep and when you are on vacation. They can listen in on your intimate conversations via your smart TV and even start your car without ever setting foot on your property or even in your country.

The cyber criminal's early adoption of technology introduces numerous difficulties for investigators and practitioners. The automated tools used by investigators are usually designed to work with last year's technologies at best and these tools in some cases cannot even process data stored on or created by the most current technologies. A good example here is the difficulty that the field has staying somewhat relevant with the smartphones that are being constantly updated and vary from vendor to vendor. Investigators struggle just trying to extract basic data from current generation smartphones, as the USB connectors and various levels of encryption and local storage characteristics (i.e. whether data is stored locally or at the telco) can cause the automated tools used by investigators to fail. This dependence on tools and technology by investigators and practitioners is a limitation in the fight against cyber crime.

Digital forensic practitioners

In order to appreciate the full impact that technology has made in the area of digital forensics, we must also consider how investigators and practitioners are using technology and technological advances. As was mentioned, cyber criminals are not the only group embracing technology. The individuals tasked with dealing with cyber criminals, namely cyber/digital forensics practitioners are looking to technology for assistance. As the cyber bad guys have embraced technology to further their tradecraft, the investigators and practitioners have embraced technology in an attempt to keep up – like a cyber arms race. Let's now focus our attention on the impact that technology has had on the practitioners and investigators, and the field of digital forensics.

Digital forensics (AKA computer forensics) is defined as "A sub-discipline of Digital & Multimedia Evidence, which involves the scientific examination, analysis, and/or evaluation of digital evidence in legal matters" (SWGDE 2015). The key differentiator from other forensic sciences here, is the focus on evidence that is digital in nature. Digital evidence is "Information of probative value that is stored or transmitted in binary form" (SWGDE 2015: 7).

To appreciate the impact that technology has made on digital forensics, and the resultant issues and problems, it is necessary to understand, even at a rudimentary level, the digital forensics process. Digital forensics incorporates traditional criminalistics phases. The phases include: (1) The identification and collection of evidence (acquisition), (2) Examination and Analysis, (3) Interpretation and decisions, (4) Report and/or testimony (Saferstein 2014). These phases have driven the technologies and tools used in digital forensics.

As was stated, practitioners of digital forensics have struggled to keep up with the cyber criminals (Garfinkel 2010). As a direct result of the various tactics and techniques used by cyber criminals, and the inherent properties of the technologies used or targeted, investigators have been forced to rely on cyber/digital forensic tools.

History

In order to fully understand the digital forensics relationship and dependence on automated tools and technology, it is necessary that we understand digital forensics' somewhat unique history. Unlike the more traditional forensic sciences (e.g. DNA, Serology), digital forensics did not develop from a scientific area that just happened to have some aspects that could be used to assist with criminal or civil investigations. Digital forensics developed from the very real, applied problem facing law enforcement in the 1990s, namely credit card fraud and other computer related criminal activities (Garfinkel et al. 2009). As a result of having to deal with technology and banking systems that were unfamiliar to law enforcement investigators and with evidence that was not just physical but also digital, investigators reached out to the vendors for support and assistance. These vendors answered the call with tools that required little if any knowledge of computer science or information and communications technology (ICT) (Garfinkel 2010). These tools allowed the investigator to focus on the investigative aspects, which they were more than qualified to do. In some countries like Canada, the tools were developed in conjunction with vendors and the Royal Canadian Mounted Police (RCMP) (e.g. RCMP Utilities).

It was thought that at some point the field of digital forensics would transition from an almost exclusive law enforcement-centric area, to more of a lay person scientific area (Meyers and Rogers 2004; Garfinkel *et al.* 2009). Therefore, no one really worried about the science and the underpinnings of the technology that was assisting with the collection and examination of the evidence (although the tools mainly focused on the collection of the evidence). The courts were very trusting of any Law Enforcement or expert testimony, as judges and lawyers did not really understand technology and the evolving Internet. This could be thought of as the "pre-science" era of digital forensics (to borrow a term from Thomas Kuhn).

First generation tools

Early on in the history of digital forensics it was obvious that the rate of technology adoption would soon result in manual processes being impractical and automated tools were going to be required. However, as the field of digital forensics matured the focus still remained on the collection of the evidence. These first generation tools (see Table 23.1) attempted to increase the automation and relative speed of collection, and the rise of the "black box" began. A black box can be best described as a device where the user can control/see the inputs and view the resulting outputs, but cannot directly see or control what is occurring internally. Surprisingly, despite this lack of transparency as to what the tools were actually doing, no red flags were raised. Law enforcement was concerned with the investigative components, and the courts, as an extension of society, trusted technology.

These tools were little more than a compilation of various scripts that could be run with a single command in an MS-DOS environment. While these were automated, the automation was very rudimentary and did not provide any sophisticated user interface and contained no database back end. Examples of tools in this generation are Safeback, RCMP Utilities, and REDX (RCMP tool).

Second generation tools

Second generation tools continued to focus on the acquisition of evidence (Richard and Roussev 2006) but added a better interface and included a backend database structure that could be used for the examination and analysis phase as well. These tools such as EnCase, FTK, Sleuthkit and ProDiscover now allowed for data on a hard drive to be indexed and then searched for keywords at a much faster rate. Without the use of a database structure keyword searches had to be conducted in real time which made the process very slow (Ayers 2009). These tools were primarily made to run in a Microsoft Windows environment on a single PC with a relatively large amount of RAM since all the processing was happening locally.

Third generation tools

The focus on collection has continued into the present day, so much so, that the field of digital forensics now faces the problem of having too much data to process, even with automated tools. The "big data" problem of too much data (volume), unstructured data (variety), and speed of Internet and mobile connections (velocity) are causing those active in this field to not only reconsider how to use the tools, but also, and more importantly, what functionality is really required of the tools.

This transition from the routine collection of data to the examination and analysis of large volumes of data marked the introduction of the second generation of tools. The decisions that investigators were being asked to make related to the data, also changed. Traditionally the decisions were pretty much binary – was the picture present on the computer or not? Now the questions are subtler and more complex – if the picture is present, can we prove it was knowingly (Rogers *et al.* 2006) and intentionally downloaded by the user? This is a much harder question to answer as it takes into consideration context as well as content.

The third generation tools consist of applications that are expected to determine time stamps, list physical and logical locations of the data, search for keywords, and track ownership of the data/evidence. Tools such as NUIX and the latest versions of FTK and EnCase have been optimized to take advantage of multiple cores and processors, large amounts of fast RAM, and parallel processing. The database function is often off-loaded to a dedicated database server and the tool can be run across multiple systems in a networked environment.

Fourth generation tools

The real benefit of technology is with the fourth generation of tools, which are being developed to not only speed up the process, but also to make decisions as to what the data means, how it is related, and what if any patterns can be derived (including temporal patterns). The fourth generation of tools are being developed from areas tangential to digital forensics such as information and library sciences, information retrieval, artificial intelligence and machine learning. These tools will take advantage of distributed processing, cloud storage, machine learning, contextual and data analytics. This is a quantum leap of sorts for digital forensics from a technical prospective, as well as a huge leap for the legal justice system. Lawyers, judges and juries will be asked to trust decisions not made by the human investigator now, but for the most part, by some expert or intelligent system whose code remains secret because of intellectual property concerns, and whose error rate is unknown.

Table 23.1 Tool generations

Generation	Characteristics	Phase	Examples
1	Command line based MS-DOS Environment	Acquisition	RCMP Utilities, REDX Safeback
2	Graphical user interface Windows environment Database/indexing (local)	Acquisition Keyword searches	FTK EnCase ProDiscover Sleuthkit
3	Graphical user interface Windows environment Network compatible Database server Optimized for Multi CPU/Cores	Acquisition Keyword searches Analysis/examination Timeline analysis	NUIX FTK EnCase
4	Graphical user interface Multiple OS environment Cloud aware Distributed processing Data analytics Machine learning	Acquisition Keyword searches Analysis/examination Timeline analysis Visualization Contextual analysis	TBD

While in the past, courts did not give digital evidence testimony much scrutiny, this is no longer the case. The entire spectrum of digital forensics has fallen upon hard times with the courts. Several high profile examples of how the forensic sciences either failed or how its findings were misrepresented, resulted in the National Academy of Sciences issuing a report at the behest of the US Congress. This report "Strengthening Forensic Science in the United States: A Path Forward" was anything but flattering. It took the forensic sciences (including digital forensics) to task for its apparent lack of scientific rigor and susceptibility to confirmation bias. The sciences were also criticized for lacking appropriate measures for determining the validity of its findings and conclusions and error rates of its tools and techniques:

> the law's admission of and reliance on forensic evidence in criminal trials depends critically on (1) the extent to which a forensic science discipline is founded on a reliable scientific methodology, leading to accurate analyses of evidence and proper reports of findings and (2) the extent to which practitioners in those forensic science disciplines that rely on human interpretation adopt procedures and performance standards that guard against bias and error.
>
> *(National Research Council 2015: 111)*

In an attempt to address many of the concerns expressed by the scientific community, the courts and National Academies of Science, practitioners are finally starting to dictate the functional requirements for tools. Several open source alternatives to industry standard proprietary tools are becoming available (e.g. Sleuthkit). The creators of these tools actively solicit input from the user community as to what additions and modifications to make. Being open source, the code can be scrutinized and the functions fully documented, thus decreasing the "black box" conundrum (Carrier 2002).

Future issues

The digital forensics tools of the not too distant future (next few years) will be asked to make decisions about what data to include or exclude, what data means or in some cases doesn't mean (contextualization), and provide visualization and temporal analysis that can be used as illustrations in a court proceeding, while at the same time providing measures of confidence such as likelihood ratios, and error rates. In theory, very little human intervention would be necessary.

This full-on automation introduces new issues and challenges. It will no longer be just a case of the tools being "black boxes." With black boxes, the investigator still chooses the input and interprets the output. Now, the entire process would be automated. To date the courts have been uncomfortable with "machines" making decisions, absent a human's oversight and final judgment. Yet what they are asking, even demanding of the forensic sciences, including digital, can really only be accomplished by removing the person (e.g. bias free examinations).

We are definitely in uncharted waters now that full automation and potentially AI-like systems will be taking over from their "human handlers." The digital forensics field is starting to reach out to other scientific areas that have focused on dealing with large amounts of data for several years already (Garfinkel 2010). The fields of Information Sciences, Business Analytics, and Business Intelligence have developed algorithms and tools to assist them in parsing large data sets, as well as visualizing the data and resultant patterns (Dilek *et al.* 2015). These tools leverage advances in machine learning, expert systems and in some case weak Artificial Intelligence (AI). However, as was stated previously, what has not changed is the legal requirement that the decisions and evidence derived from these tools is valid, reliable and the error rates are known. The Federal rules of evidence will also have to be revised, as currently a tool cannot be an "expert witness" (Meyers and Rogers 2004). It will be interesting to see how the courts deal with the question of error rates. From software engineering we know that there is no such thing as error-free code. In fact, there are formulas used to estimate the approximate number of errors based on the number of lines of code. It will also be revealing to see whether the courts allow the vendors of these tools to self report their error rate or whether some neutral third party will be required, much like a consumer's report or under-writers' labs (Meyers and Rogers 2004).

Of equal importance will be what error rate the courts will consider to be acceptable for these automated tools. Will a 5 percent type 1 error rate (false positive) be acceptable, or will the courts look to cross over rates, such as is currently done with biometric devices? The courts may also look at confidence intervals to determine what is appropriate. But here again, will they require a 95 percent or 90 percent confidence interval and who will help the courts decide what is appropriate?

The challenges of educating judges and jurors on what error rates and confidence intervals really mean will be no small task either. If this education is done incorrectly, judges and jurors could be either too liberal or too conservative regarding what they accept as "true decisions."

Conclusions

Society has greatly benefitted from the advances in technology. The Internet has arguably changed the very fabric of our society. Today's generation cannot envision a time when social networks, smartphones and the ability to be connected 24/7 did not exist. Unfortunately, the deviant criminal elements of our society have also benefitted from our dependence on technology and have co-opted these technologies to further their cyber criminal trade craft (Parker 1998). It is important to also remember that technology is neutral, it is neither evil nor good, it is just a tool. It is what we do with the technology that defines whether it assists or harms our culture.

While cyber criminals have exploited technology's many vulnerabilities, investigators and practitioners have leveraged technology's ability to automate routine mundane tasks and process large volumes of data to assist in catching cyber criminals. We should not be surprised that law enforcement is entrenched in a technology arms race with cyber criminals. What should surprise us is the fact that the general public and the courts take for granted that technology is infallible and overly trust the derived opinions and decisions.

As we move further into the twenty-first century, the legal justice system will need to get up to speed with technological advances and begin understanding not only the strengths of digital forensic technology, but also its weaknesses. Practitioners and investigators must try and close the technology adoption gap between themselves and cyber criminals. Researchers, and criminologists need to more fully recognize that criminals are going to continue to be attracted to technology as either a tool or a target. The impact that technology has on society, criminal behavior and the legal justice system is only now being understood. As we move into the the remainder of the twenty-first century, the importance of technology will only increase.

References

Anon (n.d.) The "Nigerian" Email Scam, Consumer Information. *consumer.ftc.gov.* Available at: www.consumer.ftc.gov/articles/0002l-nigerian-email-scam [Accessed 13 Apr., 2016].

Ayers, D. (2009) A second generation computer forensic analysis system. *Digital Investigations*, 6: S34–S42.

Baggili, I., Al Shamlan, M., Al Jabri, B. and Al Zaabi, A. (2012) Cybercrime, censorship, perception and bypassing controls: An exploratory study. In M. Rogers and K. Seigfried-Spellar (eds), *Digital Forensics and Cyber Crime*, pp. 91–108. Berlin: Springer-Verlag.

Baggili, I., Mislan, R. and Rogers, M. (2007) Mobile phone forensics tool testing: A database driven approach. *International Journal of Digital Evidence*, 6(2): 1–11.

Beccaria, C. (1963) *On Crimes and Punishments* (introduction by H. Paolucci, trans.). New York: Macmillan. (Original work published 1764).

Burton, W. (2007) *Burton's Legal Thesaurus, 4E*. Available at: http://legal-dictionary.thefreedictionary.com/cybercrime, [Accessed 13 Apr., 2016].

Carrier, B. (2002) Open source digital forensics tools: The legal argument. Available at: http://dl.packetstormsecurity.net/papers/IDS/atstake_opensource_forensics.pdf [Accessed 1 Oct., 2015].

Casey, E. (2011) *Digital Evidence and Computer Crime: Forensic Science, Computers, and the Internet*, 3rd edition. Boston: Elsevier.

Choi, K. (2008) Computer crime victimization and integrated theory: An empirical assessment. *International Journal of Cyber Criminology*, 2(1): 308–333.

Damshenas, M., Dehghantanha, A. and Mahmoud, R. (2014) A survey on digital forensics trends. *International Journal of Cyber-Security and Digital Forensics*, 3(4): 1–26.

Dilek, S., Cakır, H. and Aydın, M. (2015) Applications of artificial intelligence techniques to combating cyber crimes: A review. *International Journal of Artificial Intelligence & Applications*, 6(1): 21–39.

FBI, Common Fraud Schemes. Available at: www.fbi.gov/scams-safety/fraud/fraud [Accessed 4 Apr., 2016].

Garcia-Morchon, O., Keoh, S., Kumar, S., Hummen, R. and Struik, R. (2013) Security considerations in the IP-based Internet of Things. Available at: https://tools.ietf.org/pdf/draft-garcia-core-security-04.pdf [Accessed 5 Oct. 2015].

Garfinkel, S., Farrell, P., Roussev, V. and Dinolt, G. (2009) Bringing science to digital forensics with standardized forensic corpora. *Digital Investigation*, 6: S2–S11.

Garfinkel, S.L. (2010) Digital forensics research: The next 10 years. *Digital Investigation*, 7: S64–S73.

Gomzin, S. (2014) *Hacking Point of Sale: Payment Application Secrets, Threats, and Solutions*, 1st edition. Boston: Wiley Publishing.

IDC, Explosive Internet of Things Spending to Reach $1.7 Trillion in 2020, According to IDC. Available at: http://www.idc.com/getdoc.jsp?containerId=prUS25658015.

Kedgley, M. (2015) If you can't stop the breach, at least spot the breach. *Network Security*, 2015(4): 11–12.

Kwan, L., Ray, P. and Stephens, G. (2008) Towards a methodology for profiling cyber criminals. The 41st Annual Hawaii International Conference on System Sciences. IEEE, pp. 264–264.

Leigland, R. and Krings, A. (2004) A formalization of digital forensics. *International Journal of Digital Evidence*, 3(2): 1–32.

Meyers, M. and Rogers, M. (2004) Computer forensics: The need for standardization and certification. *International Journal of Digital Evidence*, 3(2): 1–11.

Mohay, G. (2005) *Technical Challenges and Directions for Digital Forensics*. Proceedings of the First International Workshop on Systematic Approaches to Digital Forensic Engineering (SADFE'05). Available at: http://ieeexplore.ieee.org/lpdocs/epic03/wrapper.htm?arnumber=1592529 [Accessed 6 Oct. 2015].

National Research Council (2015) *Strengthening Forensic Science in the United States: A Path Forward*, Available at: www.nap.edu/catalog/12589.html [Accessed 6 Oct., 2015].

Parker, D.B. (1998) *Fighting Computer Crime: A New Framework for Protecting Information*. New York: John Wiley & Sons, Inc.

Richard, G.G., III and Roussev, V. (2006) Next-generation digital forensics. *Communications of the ACM*, 49(2): 76–80.

Rogers, M. (2003) The role of criminal profiling in the computer forensics process. *Computers & Security*, 22(4): 292–298.

Rogers, M., Goldman, J., Mislan, R., Wedge, T. and Debrota, S. (2006) Computer forensics field triage process model. *Journal of Digital Forensics, Security and Law*, 1(2): 1–20.

Saferstein, R. (2014) *Criminalistics: An Introduction to Forensic Science*, 11th edition. New York: Prentice Hall.

Shaw, E.D. (2006) The role of behavioral research and profiling in malicious cyber insider investigations. *Digital Investigation*, 3(1): 20–31.

Symantec (2014) A Special Report on point-of-sales systems. Available at: www.symantec.com/content/dam/symantec/docs/white-papers/attacks-on-point-of-sale-systems-en.pdf [Accessed 4 Apr., 2016].

SWGDE (2015) SWGDE/SWGIT Digital & Multimedia Evidence Glossary. *www.swgde.org/documents/Current%20Documents/2015-05-27%20SWGDE-SWGIT%20Glossary%20v2.8*. Available at: www.swgde.org/documents/Current%20Documents/2015-05-27%20SWGDE-SWGIT%20Glossary%20v2.8 [Accessed 14 Oct., 2015].

Zimbardo, P. G. (1969) The human choice: Individuation, reason, and order versus deindividuation, impulse, and chaos. In W. D. Arnold and D. Levine (eds), Nebraska Symposium on Motivation, Lincoln: University of Nebraska, pp. 237–307.

24

DNA and identification

Carole McCartney

Introduction

Between the discovery of the structure of DNA in the 1950s, and the mapping of the human genome in 2003, it became common to discuss 'DNA', the human 'genome' and 'genetics' in public discourse on a wide variety of issues. This included the use of particular techniques of 'DNA profiling' for forensic purposes. Since the first policing use of DNA in the 1980s, the utilisation of DNA profiling within law enforcement has grown exponentially and its spread is now global. A 2008 Interpol survey showed over half of countries in all regions, except Africa, use DNA profiling in criminal investigations.[1] By early 2016, it is reported that 64 countries have operational national DNA databases of varying sizes, with 30 more in the planning stages.

Just as with fingerprint evidence before it, commonly considered unimpeachable with a hundred-year pedigree, the public (ergo police, legal professionals and juries), no longer need convincing of the accuracy or reliability of DNA evidence. Indeed the concern is now that juries may be far too easily persuaded by DNA, or may demand DNA evidence be provided before they will convict or acquit a defendant.[2] In 2009 forensic DNA profiling also secured its place as the standard bearer for forensic sciences, being cast as the 'gold standard' by the august body, the US National Academy of Sciences.[3] There is no denying that DNA has had a dramatic impact upon criminal justice systems, just as fingerprinting did before it, and the potential remains for this identification science to find new forensic applications and continue shaping law enforcement strategy and the criminal process. Yet while DNA evidence is a powerful investigative tool, able to incriminate as well as exculpate, common portrayals of DNA as being able to solve crimes almost instantaneously, beyond any doubt, even from 'beyond the grave', may overstate the degree to which DNA assists in criminal investigations. In fact, DNA remains marginal in most criminal investigations, (indeed, fingerprints are still more commonly found and used in evidence) yet it still commands political and public attention. Research also continues apace on new ways to exploit DNA with a view to maximising the utility of abandoned genetic samples, collections of DNA, and the evidence which they can yield.

This chapter will examine the use of DNA within criminal justice systems, with special reference to the UK, where DNA profiling originated and became quickly embedded within the criminal process. The creation of DNA databases and the attendant issues raised will be considered. A critical overview will sketch developments that have 'stretched' the science of forensic DNA profiling, occasionally beyond the ability of the courts to rely upon it, before posing questions of legitimacy, including social and ethical concerns. A weighing of the benefits brought by forensic DNA profiling necessarily involves reflection upon mistakes of the past, and consideration of whether there is now the foresight, ability, and will to prevent abuses and augment the advantages of forensic DNA profiling in the future.

The use of DNA in criminal justice systems

Discovered by Sir Alec Jeffreys and perfected by colleagues at the University of Leicester in the UK in the early 1980s, the technique of 'DNA fingerprinting'[4] was first applied to immigration disputes, providing proof of biological family relationships to support asylum claims. A DNA 'revolution' then gathered pace, with rapid scientific and technological developments, accompanied by incremental legal reforms, enabling the taking and use of DNA profiles to become an 'integral part' of the UK's criminal justice processes by the mid-1990s. The enthusiastic adoption of the new technique was fuelled by media, and political, hyperbole about the benefits of DNA profiling. High-profile crimes where DNA had been useful during an investigation were given significant publicity, although many cases where DNA could have been expected to, but did not, assist police investigations, were brushed over. Failures of the science, or the police to utilise the science, no longer fitted the official story of the 'new dawn' being witnessed in policing.

DNA: the success story

Whilst rapidly hailed as a vital weapon in a police armoury, it has conversely been the success of forensic DNA in exculpating innocent suspects and exonerating the wrongly convicted that is perhaps the greatest success story of forensic DNA profiling. The trust in the reliability of DNA and its (supposed) conclusive nature has seen hundreds of wrongful convictions overturned around the world. Criminal cases where innocent people are calling for evidence to be (re-)tested, in the hope that DNA will finally bring about justice, are legion. In many such cases, if available evidence had been subjected to DNA testing then the suspect could have been excluded from police inquiries and often, another suspect indicated. Forensic DNA thus exposes the failings of many other, unreliable and often highly questionable, evidence types, regularly relied upon by the police, and at court. The recognition of the prevalence of wrongful convictions has had a profound impact in countries, none more so than the US, where reforms to all components of the criminal process are being posited and trialled to try to minimise future injustice. The US National Registry of Exoneration,[5] an online database containing all exonerations in the United States since 1989, currently lists more than 1,000 wrongfully convicted individuals and continues to grow daily. In just one British example, Sean Hodgson spent 27 years in prison for a murder that he had not committed. Blood type matching had been used at trial to support Hodgson's false confession, but after DNA testing was finally undertaken after nearly three decades, it was proven that he was not the killer.[6]

A DNA sample located at a crime scene is very powerful evidence. In some instances it can provide almost incontrovertible proof of identity. For example, in allegations of rape, if there is a full DNA sample left by the alleged perpetrator on their victim, then the issue of identity is almost indisputable. This of course does not prove someone guilty of rape, but identifying a perpetrator quickly and reliably can speed up investigations and make the resolution of a criminal case quicker and cheaper, while often negating the need for costly and lengthy trials. Using the UK National DNA Database can significantly increase detection rates, particularly in crimes notoriously difficult to solve, for example, in the case of burglary the detection rate in 2014/15 rose to 56 per cent when DNA was located (National DNA Database 2015).

The use of DNA during an investigation provides an opportunity to place less reliance upon often questionable evidence, such as eye-witness testimony or confessions. It can also contribute to police intelligence where there may be none to be gained from more traditional policing methods. A DNA 'match' can link crimes together so an offender can be prosecuted for a series of offences and may also be able to ensure a conviction, even years after the crime. In the UK in 2002, one police force launched Operation Phoenix, obtaining 42 DNA matches from more than 400 unsolved sexual offences over a 14-year period, resulting in 14 convictions up to 2005 (see BBC News 2003 and 2004 for two successful convictions). Two years later, 'Operation Advance' in another force involved the review of undetected serious sexual assaults and rapes from the late eighties and early nineties. The latest scientific technology identified 215 cases that had DNA crime stains. Of these, 148 were scientifically progressed and further investigated. Such 'cold cases' can result in convictions years after a crime has been committed, for example, a UK man was convicted in early 2016 and jailed for ten years for raping a woman in her Manchester home in 1984, while Christopher Hampton pleaded guilty to the murder, 32 years previously, of Melanie Road, 17 years old, in June 1984 (BBC News 2016b; *Guardian* 2016). Such success can keep alive interest, and re-ignite leads in long forgotten cases, such as the murder of 13-year-old Lindsay Rimer, her body found in a canal in 1995.[7]

Whilst the causes of crime are multi-faceted and complex, it is accepted that the chances of detection can sometimes act as a powerful deterrent – i.e. the more certain an offender is that they will be apprehended, the less likely they are to consider committing a crime. Indeed, in one police area in the UK, police are reporting a 19 per cent drop in burglaries since commencing 'Operation Shield', a project which saw 'liquid DNA' kits given to homeowners to mark their property. Also, when touching property, offenders can be marked by the liquid, which cannot be washed off. Over the past year, the area has been blanketed with warnings that burglars in the area can be more easily detected, and the subsequent drop (the biggest in the UK) has been attributed to this novel use of DNA (see Security Newsdesk 2016). This deterrence argument is occasionally stretched to suggest that, in time, if most citizens were on a National DNA Database, then crime would fall because not only would offenders be caught more frequently (and their 'criminal careers' shortened), but potential offenders would reconsider their offending behaviour as the certainty of being caught made criminality too risky. This argument is hypothetical of course, and will remain so for the foreseeable future at least.

A more sceptical view?

The advent of forensic DNA testing has thus indeed been revolutionary, leading to the accurate detection and conviction of many criminals, who may otherwise have evaded punishment. But while DNA has been referred to using many hyperbolic epithets, scrutiny soon reveals that as with any technology utilised by humans, it is not infallible, and may even lead to injustice. As an example of an early 'near miss' with less sophisticated DNA profiling than now used in most countries, Raymond Easton was charged in 1999 with the burglary of a house 200 miles from his home after a 'cold hit' on the UK National DNA Database. His DNA matched the crime scene DNA at six loci, with a one in 37 million chance that a randomly selected person's DNA would match. However, when Easton, who had advanced Parkinson's disease and was unable even to drive a car, offered an alibi, the DNA was tested at four more loci. This more sophisticated test showed there was no DNA match after all.

As well as the ever present risk of contamination at the scene of a crime, biological samples once in the hands of police and scientists are also at risk if not handled properly. For example, in Germany, police forces expended thousands of hours in chasing the 'Phantom of Heilbronn' (see Feltes n.d.). The basis for this focus upon an unknown female serial offender was a DNA profile that had turned up at a variety of crime scenes for several years. The 'Phantom' was considered a criminal mastermind, evading capture despite large rewards for her identification. Investigators finally realised that contaminated cotton swabs used in the profiling process were the source of the DNA. There are now stricter rules on the production of 'non-contaminated' consumables used in testing.

Such problems are not restricted to the past, with more recent high-profile errors arising. In 2012, a young man was charged with a rape in the city of Manchester, on the basis of a DNA 'match'. He denied having ever travelled the 400km from his home to Manchester, and his lawyer pressed for further testing. It transpired that the DNA from the rape case had been contaminated at the laboratory when a sample taken from Scott when arrested for a minor affray, was mixed with other samples, and staff failed to use clean equipment (*Guardian* 2012). In another contamination case, during a lengthy and perplexing police investigation into the death of an MI6 employee, Gareth Williams, forensic scientists provided police with a DNA profile from the holdall in which Williams' body was found. The police subsequently spent a year attempting to trace the individual responsible for leaving the DNA, to no avail. It was later discovered that an LGC employee, manually entering the DNA profile into a computer, had transposed the numbers '3' and '5', rendering the DNA profile incorrect. This typographical error led to the costly pursuit of a non-existent individual (Schlesinger and Hamilton 2012).

While proven to be extremely powerful evidence when attendant risks are neutralised, a DNA match can still only be a possibility if a crime scene is examined, and there is wide variation between police agencies as to the number of scenes examined, the subsequent number of biological samples sent for analysis, and then compared with a suspect, or searched against a DNA database. Prioritising must always take place in a system without unlimited resources, and police budgets will play a major factor in decision making. In many crimes, there will be no 'scene' to examine, or it will be impossible to search effectively. Often the culprit will be obvious, particularly when most violent offences take place between people known to each other, and police may catch offenders 'in the act'. Indeed, it was reported in 2006 that in 58 per cent of cases where there was a DNA match, this was the first link to the suspect, raising the possibility that 42 per cent of the cases marked as a 'DNA detection' would have been solved without the DNA evidence (Home Office 2005: 14).

In the majority of crimes, the likelihood of DNA evidence being found remains low. In 2014/15 the police only forensically examined 12.8 per cent of crime scenes (rates for burglary scene examinations are much higher, and are more successful – partially explaining the high detection rate). Once at a scene, it is still rare to find DNA, with only 19 per cent of scenes yielding a DNA sample. Of 32,168 crime scene profiles loaded to the NDNAD in 2014/15, just 13,375 (41.6 per cent) resulted in a positive 'outcome' counted by the police following a match on the NDNAD (not necessarily a conviction). This in a year that police recorded over 3.5 million offences. Thus one could deduce that DNA 'helped' solve just 0.38 per cent of the crimes recorded in the UK in 2014/15 (National DNA Database 2015). So while a DNA match improves the chances of detection, it is not conclusive that it will. A detection also does not equal an offender being convicted, as prosecution policy will then intervene and charges may not be brought against a suspect, or some other factor arises that means that there is no conviction. Indeed, DNA has been called: 'a fresh filling between two slices of stale bread' (Leary and Pease 2003: 11), highlighting that the 'policing' and 'prosecution' slices of bread are still flawed, often ineffective and/or inefficient, diminishing the whole 'criminal justice' sandwich experience.

DNA databases

To support the use of DNA in criminal investigations, police take and retain three types of DNA samples: samples from suspects and those arrested;[8] elimination samples from victims and witnesses etc., and samples from, or relating to a crime scene, such as from weapons, discarded items and bodily fluids etc. These are compared with profiles from known offenders, and DNA from other crime scenes – past and present (and retained to compare to samples from all those arrested/ suspected in the future and from crimes not yet committed). A 'match' against any profiles is reported to the police to enable them to then act. The ability to match DNA from crime scenes with DNA profiles from known previous offenders, without a suspect to hand, is the potential that gives DNA profiling such power – not only will it confirm that you have the right suspect (or not), it can also name suspects that are not already part of an investigation. With this promise, national DNA databases have been established around the world.

The UK National DNA Database (NDNAD) was the first to be operational, created in 1995 without any single dedicated legislative instrument or Act of Parliament, evading political or public debate. Police powers to take and retain biological samples from citizens, and the growing collection, storage and use of DNA and biological samples was instead facilitated by successive amendments to existing legislation, expanding the list of those from whom a sample may be taken; downgrading the authority required to sanction and perform sampling; increasing access to the database (and uses for DNA and the NDNAD); and permitting samples and profiles to be retained indefinitely. While these successive legal reforms were taking force, there came instances where DNA unlawfully retained on the NDNAD was matched with crime scene profiles. Thus, judicial attention was turned to the rapid expansion of the NDNAD and the lack of debate over its parameters.

While hearing conjoined appeals regarding convictions secured by DNA from individuals who had not been convicted of any crime (at that time, making the retention of their DNA unlawful), the House of Lords ruled that such evidence could be used in the 'interests of justice'. However, with Her Majesty's Inspectorate of Constabulary reporting that at least 50,000 'unlawfully retained' samples were most likely on the NDNAD (HMIC 2000), the government were forced to amend the law (again) to ensure that profiles belonging to 'innocent' individuals were now lawfully held. With this further incremental legal change, England, Wales and Northern Ireland had created the most inclusive DNA database in the world: samples were taken and retained permanently from all individuals arrested for 'recordable' offences,[9] without their consent. These profiles could be used 'for purposes related to the prevention or detection of crime, the investigation of an offence or the conduct of a prosecution'.[10] This included minors, who if over 10 and under arrest, were treated the same as adults.[11] Volunteers, witnesses, and victims were also included on the NDNAD, many 'consenting' without any real understanding of the implications of permanent inclusion on the database.

The size of the database and the inclusion of innocent (un-convicted) individuals began to attract attention. Scrutiny did not always result in the NDNAD looking like a good investment. Results wrestled out of policing authorities at that time showed that in fact the 'success' of DNA profiling really depended upon the number of DNA samples retrieved from crime scenes, rather than the considerable effort of adding 50,000 individuals a month to the database. With DNA at that time recovered from just 10 per cent of crime scenes examined – and only approximately 17 per cent of crimes receiving what could be considered a scientific crime scene examination – in 2007/08, it was calculated that just 0.36 per cent of recorded crimes that year were detected using DNA, down from an all-time high of 0.37 per cent. In fact, during the time of rapid expansion of the database, the number of crimes detected using the NDNAD fell (NDNAD 2009).

At the same time, there began to be consideration of the legal and ethical basis of the NDNAD. The Nuffield Council on Bioethics published a critical report in 2007, focusing upon those principles to be respected when the State exercises power over citizens, such as the respect of personal liberty; the maintenance of autonomy of the individual; personal privacy; informed consent and equal treatment. The use of DNA remained sensitive, the Council argued, so while the public deserved protection from crime and expected the State to maintain law and order, this had to be weighed against the protection of these fundamental principles and ethical values. Indeed, the UK government noted that there was an ethical vacuum and created the NDNAD Ethics Group in 2007, an advisory non-departmental public body, providing independent advice to the NDNAD strategy board. The Ethics Group stated their aim as seeking: 'to balance the interests of public protection ... with the inevitable intrusions of privacy and personal labelling', admitting in their first Annual Report however, that: 'There are no absolutely right answers.'

Despite the hasty creation of the Ethics Group, the real threat to the NDNAD as it stood, was already making its way up through the English courts, to arrive in the form of an appeal to the courts to have the DNA of two individuals removed from the NDNAD. The two individuals, known as 'S' (a juvenile) and Mr Marper, had not been convicted of any crimes. However, their request to the police to have their DNA removed was refused. This led to claims that their rights to privacy (Article 8) and right to not suffer discrimination (Article 14) were being breached. The House of Lords held (by 4 to 1) that if their privacy was breached by the retention of their DNA (and this was by no means agreed upon), then the breach was slight, and justified in the fight against crime. Their appeal proceeded to the European Court of Human Rights, where, in a unanimous verdict, the Court held that the holding of their DNA samples was a breach of the European Convention on Human Rights, stating that that Court was 'struck by the blanket and indiscriminate nature of the power of retention in England & Wales.'[12] In the succinct judgement, the Court concluded that the retention by the State of the DNA of innocent individuals, 'constitutes a disproportionate interference and cannot be regarded as necessary in a democratic society.'[13]

With this ruling, the UK government was forced to take remedial steps and the country was finally forced to debate the retention of the DNA of millions of citizens. However, the parameters of the subsequent 'debate' were very narrowly drawn. The case of *S & Marper* had not considered the taking of DNA samples, nor the retention of samples from convicted individuals. All that was considered was the removal of DNA profiles after an acquittal (or a criminal case had not proceeded to conviction). Further, police powers to take DNA samples in the UK had extended yet further since *S & Marper*. A (complex) compromise of sorts was reached with the passing of the Protection of Freedoms Act 2012, with varying retention periods for convicted/non-convicted persons, all overseen by the newly created 'Biometrics Commissioner'.

The UK NDNAD is still seen as an example for other countries to emulate and many countries either now have, or are working towards the establishment of a national DNA database. The UN's Special Rapporteur on Privacy (SRP) notes that approximately 25 per cent of the UN's member states have implemented national criminal offender DNA database programmes (Cannataci 2016: 7). However, while many countries have now reached a consensus on the size and operation of their national DNA database, many continue to struggle to establish a NDNAD, whether it be due to lack of infrastructure, political will or public agreement over who should be included on the database and the powers of the police. India and Italy are just two examples of countries where such issues have prevented the creation of a NDNAD despite using DNA in their criminal justice processes.

The Italian DNA database

In 2009, Italy passed DNA database legislation, permitting convicted offenders to have their DNA stored, and the DNA of arrested individuals could be taken upon the order of a judge, and later destroyed if not convicted.[14] Following terrorist bomb attacks in Europe, Italy announced in March 2016 that their DNA database would become operational, with seemingly wider parameters than originally legislated for, stating that this move would be a 'step forward' for national security (DPA 2016). Their 2009 legislation has already been heavily criticised for having insufficient security measures in place, and vague provisions for critical issues, including database access and the deletion of profiles and samples (Biondo and De Stefano 2011). DNA profiling itself in Italy has previously attracted a great deal of international criticism, particularly in the convictions of Amanda Knox and Raffaele Sollecito for the murder of the British student Meredith Kercher in Perugia in 2007.

India's DNA proposals

India's Union government was working on legislation in 2015 to set up a national DNA database of 'offenders'. The draft legislation envisaged the creation of several State-level databases. However, there were many ambiguities in the draft legislation and much was left undefined, for example: there was no mention of consent, there was no judicial oversight of the database planned, and the list of offences (including some civil offences) and offenders to be included was extensive. The database was also intended to include volunteers and missing persons. While there were criteria for when DNA profiles were to be removed from the database, there was no mention of the retention of biological samples. Many other operational issues were also left unmentioned, and concerns were raised about the quality of the laboratories tasked with undertaking the DNA testing. With India being the most populous nation in the world (with a notoriously inept legal system), an expansive DNA database has the potential to grow to an enormous size rapidly.

Universal DNA databases

While DNA databases can dramatically improve the effectiveness of DNA profiling within a criminal justice system, there remains the frustration of obtaining a DNA profile from a crime scene, but failing to find any matches (or near matches) on a database. One solution is to create so-called 'universal' DNA databases, with all the citizens of a country included. Whilst any nation seeking to achieve universality will face the huge costs and significant logistical issues of ensuring one hundred per cent inclusion (although this will still not include recent migrants/ tourists etc.), and assuming legal and political permission for such a development, universality may mean that arguments about discrimination disappear, and the police *should* be able to solve all crimes. Of course, the European Court of Human Rights has already ruled that the UK's 'blanket and indiscriminate' regime for retaining DNA from individuals who were not convicted breached human rights, a ruling that should inform the DNA database policies of all EU States. But for non-EU countries, the issue could simply become one of ensuring that the DNA profiling system used is discriminating enough to be able to distinguish between millions of individuals. The United Arab Emirates, with a population of just over 5.5 million, are the first to start the process of DNA sampling all of their citizens, with Kuwait announcing intentions to DNA test everyone in, and entering, their country (Macdonald 2016).

It is wildly optimistic however, to imagine that a universal DNA database can make serious inroads into the crime rate, particularly when the impact that it may have on crime detection will not be as dramatic as first thought. Most crimes do not have 'scenes' as such, or the scene is not one that can be searched productively, or the perpetrators are already known to the police. Indeed, policies underpinned by a 'bigger is better' logic, are poorly supported by evidence of effectiveness, and may be seriously flawed (see earlier underwhelming statistics). Criminologists will be aware that 'labelling theory' explains that attaching a label of 'criminal' or 'future offender' to an individual, may impact upon the individual's view of themselves as well as how others treat them, which can make the label a self-fulfilling prophecy. Deterrence arguments remain very difficult to sustain and there are serious civil liberties implications for treating individuals as 'future criminals', or 'pre-suspects' (Lynch *et al.* 2008). With such caveats, even a universal DNA database will not be the magic bullet to rid societies of crime. DNA scientists have thus looked instead to maximise the utility of DNA profiles obtained and have developed techniques to try to enhance DNA profiles, or gain more information from a DNA profile.

Stretching the science

With mobile DNA testing at crime scenes a possibility, and DNA profiling at police stations promised as standard in the future, the time taken to profile a crime scene profile or suspect has dropped from days to potentially minutes. However, such developments, while opening possibilities for swift detection of offences, do not assist when no match is found. 'Familial searches', locating 'near matches' on a DNA database in an effort to identify potential relatives of the perpetrator, are now a legitimate investigative practice, though still relatively costly and time-consuming. Meanwhile, research into 'phenotyping' – detecting observable traits from genetic sequences – has led to 'red-hair' predictive tests (where offenders can be identified as having red hair, though this does not tell detectives if they still have their hair, or if they have dyed it), and ethnicity predictions (with similar difficulties in making reliable predictions about appearance from a generalised prediction about ethnic background). Scientists around the world however, are seeking the ultimate prize: constructing a reliable physical description of an offender (height, ethnicity, eye colour, hair colour, etc.) from their DNA profile. This tantalising prospect appears closer each month, with investigators now occasionally issuing 'DNA photofits' of suspects, constructed entirely from information on appearance gained from a DNA sample left at a crime scene.

Meanwhile, there is still debate as to whether this would actually prove helpful in the context of investigations, with its ability to mislead. If police narrow their search, to concentrate upon just those suspects who resemble a 'photofit', then mistakes could be made. It is known from studies of miscarriages of justice that 'tunnel vision' early on in investigations can lead to missed exculpatory evidence as well as inculpatory evidence pointing to other suspects. It is also known that many genes interact with, or can be overridden by, the environment, for example, height can depend upon dietary factors etc., while other external characteristics can be altered quite easily, i.e. hair colour, even eye colour. It remains to be determined just how helpful it may be to say to an investigator 'the suspect may be Caribbean'. Indeed, perfecting such predictions must be an essential task before such information is more useful than distracting. There is also a possibility that ethnic predictions in particular could promote stereotypes. If researchers focus upon ethnicity (which is highly contested), then this could prove highly controversial.

More successful advanced DNA profiling techniques have included separating mixed samples (but see Simon A. Cole, Chapter 29, this volume, for problems with this development), and analysing degraded samples (Mitochondrial DNA Analysis (MtDNA) and Y Chromosome Analysis are used in such instances). There are also developments that are not currently standard, but could become so in the future – such as analysing 'SNIPS' – or Single Nucleotide Polymorphisms, as undertaken in the medical arena. There are ongoing efforts in respect of automation and miniaturisation. Automated processes are proving successful but prototype 'lab-on-a-chip' has proved more problematic. There have also been concerns over the development of new DNA techniques becoming more commonly used, especially 'Low Copy Number' DNA analysis, or so-called 'touch' DNA. This is where the standard DNA profiling technique is taken a step further enabling the analysis of even smaller samples (one cell, for example). Such techniques increase the possibility of contamination, and misinterpretation, and raise issues such as 'innocent', or 'secondary transfer'. This can occur when all individuals shed their DNA and leave it on mobile objects (including other people), who then transfer this DNA to a place/position the individual has never personally been. Research is only now starting to highlight just how often this happens, and will prove a major problem for investigators.

An illustration of disputes around scientific developments is to be found in the 2009 prosecution for the murder of two British soldiers at Massereene Barracks, Northern Ireland.[15] Much of the evidence was based on witness identifications at the scene, testimony of the defendants Duffy and Shivers, and materials found in the burned out getaway car. Advanced DNA tests linked Duffy to the tip of a latex glove and seat-belt buckle in the abandoned car. The judge, sitting without a jury, decided that while he was satisfied that the DNA link to the car was soundly established, the prosecution had failed to demonstrate that the DNA linked Duffy to the murder plot. A further crucial issue was whether the novel statistical process to calculate the DNA probabilities could be regarded as having achieved sufficient recognition as valid and reliable. This controversy had earlier received publicity in another Northern Ireland case, the prosecution of Sean Hoey for the Omagh bombing.[16] In that case, the judge ruled that the 'novel' technique of 'Low Template Number DNA' (Low Copy DNA) had not yet reached the level of 'general acceptance' within the scientific community. For that, among other reasons, the judge ordered an acquittal, prompting a 'review' of cases involving LTDNA. Just one year later, in 2010, LTDNA was considered in *R v. Reed & Reed* and this time admitted as evidence by the English Court of Appeal.

Critical overview

The advent of DNA profiling requires different skills for investigators to master and in the early years of adoption, training emergency services personnel to become 'forensically aware' had often been a significant issue. Complications arise when biological samples are collected, or stored inappropriately, and then analysed for DNA using today's sensitive techniques. Forensic evidence can rarely be salvaged if ruined, as was witnessed in the trial of the alleged Omagh bombers, where police were heavily criticised for their collection and storage of forensic evidence, resulting in the DNA evidence being discounted. While there is a clear requirement for sterile laboratories, stringent testing procedures and non-contaminated consumables for example, there is also no room for error at the crime scene, which is often a far from ideal environment.

Even once biological samples are collected and contamination avoided, all DNA matches have to be interpreted within the context of the scene and what is purported to have occurred. In many instances, finding DNA may not mean very much and can tell you very little about what actually occurred, and *how* DNA may have arrived at the scene (see Cole, Chapter 29, this volume). If someone is attacked in their home, and their partner's DNA is found, that is of no help to investigators, unless the biological sample is found somewhere anomalous or highly suspicious, such as directly on a weapon for example. Similarly, if there is a robbery in a public space, locating DNA is very difficult (if not impossible), and search parameters must be limited to biological samples from somewhere significant within the crime scene. There can also be an attendant danger with DNA matches that are declared very early on in an investigation (particularly if attempting 'instant' DNA testing at the scene in more futuristic scenarios). As previously stated, research shows that flawed investigations are usually characterised by early decisions made about suspects and the subsequent narrowing of the investigation. Early decisions about the guilt of suspects can skew police investigations. With the risks of contamination, secondary transfer, and the relevance or significance of DNA being misinterpreted or exaggerated, naming a suspect early on in an investigation based on DNA located at a scene, could backfire.

There may also be concern over the abbreviation of the criminal process. Often, the finding of a DNA match can be presented to suspects at a police station, implying that their conviction is assured, so they should plead guilty and avoid a trial, possibly benefitting from concessions for early guilty pleas. Legal advice in this situation is crucial but it remains unclear whether solicitors are able to fully interrogate the relevance and significance of any DNA match, which may subsequently prove meaningless, in which case they should not advise their client to plead guilty. Indeed, the first reported DNA-based conviction was in 1987 in *R v Melias*,[17] where a DNA match induced a guilty plea, the DNA evidence going unchallenged. DNA profiling may thus lend support for the dilution of suspect protections with the necessity of protection overruled by definitive 'scientific' evidence (i.e. double jeopardy rules have been deemed 'outdated' by DNA and forensic evidence and relaxed in some countries). Any attenuation of suspect protections may perpetrate further miscarriages of justice. Of course, the police must disclose all forensic evidence in a timely fashion to ensure that the defence have a fair chance to assess the evidence, and if necessary, seek further advice and testing on the evidence. This again is not always assured, leaving defendants facing incriminating DNA evidence without a sufficient opportunity to challenge it at trial.

Conventional forensic DNA evidence now rarely raises admissibility issues in the courts, having made the transition: 'from a novel set of methods for identification to a relatively mature and well studied forensic technology' (Imwinkelreid and Kaye 2001). During the 1990s in the US they witnessed what has been called the 'DNA wars',[18] with frequent challenges to DNA at court. In England and Wales, the debate was far less polarised but there were challenges to the admission of DNA evidence. For example, in *R v Deen*,[19] a retrial was ordered when there were a number of criticisms of the statistical formulae used to calculate and present the DNA evidence. The Lord Chief Justice, calling the case 'an early exercise in a new field', heard evidence that the prosecution had fallen foul of the 'prosecutor's fallacy' (mistaking the likelihood ratio for the match probability). Again in *R v Gordon*,[20] a retrial was ordered because of uncertainty over the quality of the DNA evidence, although not its validity. This case demonstrated that while DNA technology and techniques were accepted by the Courts, the statistical evaluations could yield misleading evidence, the risk being that juries may accept 'matches' as conclusive where they are not.[21] As Redmayne explained: 'when DNA evidence began to be used in court, a problem quickly emerged. The problem was how to present DNA statistics – "match probabilities" – to juries' (Redmayne 2002).

In *R v Denis Adams*,[22] the Court of Appeal rejected the reasoning that the complexity of the evidence was a ground upon which it could be excluded. However, the court ordered a retrial because the use of Bayes theorem by the defence had 'plunged the jury into inappropriate and unnecessary realms of theory and complexity deflecting them from their proper tasks'.[23] The judge had told the jury that they could decide themselves whether to use Bayes, the court agreeing that the defence were allowed to reply to the prosecution case by way of statistics. The Court of Appeal ruling again stated that there was no objection in principle to relying on DNA evidence alone but soundly rejected the use of Bayes theorem in jury trials.[24]

The weight to be given DNA evidence, its presentation and ability to prove guilt were considered in *R v Doheny & Adams*.[25] An expert in *Doheny* testified that it was his opinion that the offender was the defendant. The trial judge then directed the jury that if this evidence was to be believed, the defendant's guilt had been conclusively proved. This was contrary to the real meaning of the DNA evidence, that whilst there were a very small number of others that could have matched the DNA profile (1 in 40 million), the defendant was only one of this small group. However, it was vital in light of the increasing use of DNA evidence that the profiling process be understood and that the manner in which the evidence is presented be made as clear as possible and *R v Doheny & Adams* attempted to set out guidelines to minimise the risk of the misuse of DNA evidence.

Despite attempts, at least in the courts of England and Wales, to set down guidelines for the proper use of DNA evidence at trial, its use in prosecutions may have unintended consequences: in the US, research has shown that the presence of prosecution DNA evidence at trial resulted in longer sentences (Purcell *et al.* 1994), the hypothesis being that judges were 'punishing' defendants for wasting their time, when their conviction was assured by the DNA evidence. In Australia, it has been found that juries are 33 times more likely to convict a defendant where the prosecutor produced DNA evidence in sex offence cases (Briody 2002). Yet research in Australia has also found that jurors have been exposed to DNA through popular culture before trial and anticipated its significance – they enter the courtroom convinced that DNA would be compelling evidence, and were more ready, and happier to convict – even if the DNA evidence was not probative (Findlay and Grix 2002). There have been reported examples of misunderstanding of DNA (and scientific evidence generally) by judges, lawyers, police, journalists and even forensic scientists. There is sometimes confusion as to the extent of its proof, including a need to distinguish an 'identification' from matters such as *when* or *why* a person came to be identified. Thus, while DNA evidence is increasingly accepted without challenge and can aid in convicting the guilty and exculpating the innocent, there must be avoided the creation of 'technological tyranny' (Blake 1989: 110). Through the scientific illiteracy of legal profession and faith in science held by criminal justice professionals and the wider public, the faith in DNA evidence may lead to an unquestioning acceptance of all 'DNA' developments:

> Unanticipated uses and side effects … can never be fully foreseen or controlled … technologies we use today can have serious consequences for individuals separated from us by great expanses of time and space … the risks can never be eliminated, we must own up to their possibility and make serious attempts to anticipate and control them.
>
> *(Yuthas and Dillard 1999)*

Indeed, technological development within the criminal justice arena has often outstripped other considerations; the European Court of Human Rights states that effectiveness is not the sole criterion to be measured with official admonition in *S & Marper* that choices have to be made and balances are to be struck:

> any State claiming a pioneer role in the development of new technologies bears special responsibility for striking the right balance between the use of modern scientific techniques in the criminal justice system and important private-life interests.

Legitimacy

The inception of DNA profiling within a criminal justice system, and in particular, the creation of a national DNA database, necessitates debate over the potential for the infringement of privacy rights. Privacy has a broad interpretation, with no exhaustible definition. Governments argue that there is a clear public interest in the fight against crime, and that taking, and more often than not, keeping the DNA of citizens is justified by the wider public interest in law and order. With regard to convicted offenders, this argument is more easily won, with their conviction and the risk of their reoffending, permitting encroachment upon their rights. However, any State interference may only go so far as is necessary and police powers must remain proportionate. There must thus be protections in place to ensure that DNA is not misused, open to abuse or shared indiscriminately, for example. Only then can DNA sampling and retention be legitimate. Often, however, a lack of transparency means that the legitimacy of police powers and governance provisions etc. are not properly scrutinised and public assurances ring hollow.

Enforceable boundaries

Privacy and data protection is of particular importance when forensic DNA profiles are exchanged internationally. Such exchange is now taking place more frequently and in some countries within the EU, is fully automated since the Prum Treaty of 2005.[26] Yet even within the EU there is only partial European-wide and other 'soft' data protection measures in place, albeit there are ceaseless efforts to reach binding data protection agreements. Even when measures are in place, individuals are reliant upon Member States' checking, and complying with national laws. The European Data Protection Supervisor has been scathing of this 'country-by-country approach' to data protection with monitoring left to national data protection authorities who can only request, log, and decide upon lawfulness *ex post*. Data protection globally is characterised by its fragmentary nature when applied to law enforcement, 'hedged by multiple derogations, allowing significant variation in implementation' (Brown 2009: 184). Traditional parameters restraining information sharing are thus increasingly inadequate. Where one expects 'borders' that law enforcement information cannot cross, enforceable boundaries regarding forensic data at a national level are often poorly defined and applied. This is then exacerbated internationally: prior to international exchange, there needs to be an international consensus, but this is not feasible in the near future, if ever. Yet, if it is clear that there is a need for exchange, then there is a demand for uniformity of legal regimes and scientific approaches at the very least. This is particularly difficult in a field with extreme variation and rapid scientific and legal change.

In addition, the use of forensic intelligence is growing among security, border, and law enforcement agencies, with the linking of databases of differing provenance. These networks or communication channels most often lack formal procedures and legal guarantees that prevent there being scope for unauthorised storage, and further manipulation or exchanges of data (see McCartney *et al.* 2011). There are concerns that policing networks that sit outside of governance and accountability frameworks have emerged and are nurtured by national and transnational liaison networks (den Boer 2002). These concerns are heightened by poor accountability mechanisms, for instance, the sharing of data across Europe under the Prum Treaty does not come under the European Court of Justice's jurisdiction. There is thus no oversight of the process in terms of 'proportionality' and lawfulness at a European level. We are reliant upon EU political institutions to weigh societal versus individual versus national interests, and strike the 'correct' balance.

Oversight

With trust in the institutions responsible for gathering, storing, using and sharing forensic DNA, citizens can know that respect for human rights and the democratic process remains (Hufnagel and McCartney 2016). However, securing and maintaining such trust requires appropriate independent oversight. Citizens must have sufficient, reliable, public information in order for this trust to be garnered and maintained. This information is scarce. No significant independent bodies have oversight powers to monitor the taking, keeping and use of forensic DNA and yet it must: 'require at the very least unambiguous guidelines, close supervision and invasive oversight' (Brown 2009: 188). The requirements for 'good' governance of forensic science as a whole remain poorly understood. Clarity of purpose and aims are pre-requisites to any governance: you need to know what it is that you are aiming to achieve before you can know how then to govern. With forensic science, the aims can be many, and unclear, sometimes contradictory. This makes it very difficult to govern – particularly now in the UK with a privatised market, where concerns such as profitability and sustainability come into play. Yet in almost all countries, the monitoring of forensic science provision remains opaque.

With policing budgets almost universally under strain, forensic science is regularly lauded as saving money via quick detections and shortened investigations (and requiring fewer policing personnel), but evidence for this financial benefit is scarce. Forensic budgets are easy to cut and have been slashed in police forces across the UK, leading to questions over the credibility of cost-efficiency claims. In times of austerity, there must be attention paid to costs: forensic science requires 'significant pre-investment without any guarantee of short-term quantifiable improvement in performance … [increasing] appreciably when it extends into the international dimension' (Brown 2009: 182). The parameters of 'success' or 'failure' also remain equivocal, for instance, does excluding a suspect early in an investigation count as a 'success', or will only a conviction suffice? Without such clarity, there can be no agreement over what data might be collected to demonstrate cost-benefits or effectiveness.

In many arenas where forensic DNA has encroached, there may have been a significant conflation of 'risks' and significant over-inclusion. The Prum Treaty was intended to: 'combat organised crime, terrorism, and illegal migration', yet the definitions of each, and their demarcations are contentious. Can forensic DNA profiling be said to be effective in all these arenas? The European Court of Human Rights has warned in both *Marper* and *Huber*,[27] of over-enthusiastic data collection and retention. Yet inconsistencies in oversight between countries are significant. In some countries, there may be clear oversight of 'quality' measures (for example in the UK, as with other countries, there is a national Quality Assurance Body that accredits DNA laboratories to the international ISO17025 standard and there is also a Forensic Regulator that set standards for providers of DNA profiling, although this regulation itself is patchy and of questionable effectiveness). However, issues regarding police powers are left to Parliament to debate and decide upon. With governmental interest in providing police with the powers they insist they require to tackle crime, Parliaments can often have a very particular (partisan) view of how a balance between public and individual interests should be struck.

Ethical issues?

There has very rarely been an ethical spotlight shone upon forensic science. It is customarily considered in the 'public good' and there are longstanding powers to take and retain forensic evidence dating back to the birth of fingerprinting. States and police agencies have habitually been meticulous record-keepers and collecting and using forensic intelligence was originally seen as a mere extension of such accepted powers. Thus law enforcement agencies are ordinarily permitted to take and retain significant personal information without any real public concern. However, for a variety of reasons, the 'the innocent have nothing to fear' argument has begun losing potency. There is now a greater realisation that there may be social and ethical consequences to police powers.

Similarly, there has been apprehension over the uses for forensic information, and more controversial techniques and technologies (in particular, familial searching and phenotyping). Increasingly questions arise about the possibility of 'function creep': using a resource for purposes other than those intended or authorised. Many such concerns are amplified when there is also a loss of trust in authority, increasing public scepticism, or a breakdown in police–public relations. There are also differences internationally between levels of trust between citizens and the police, or their leaders – for instance, in countries where dictatorships have tended to flourish. In many countries, there is anxiety over the creation of a 'Big Brother' society and the growth of surveillance and attendant fears of large and intrusive databases. In the UK, such unease was heightened in recent years by efforts to introduce ID cards, and publicity over massive losses of highly personal data by the government and other bodies. The potential mass-collection of DNA grabbed a lot of the limelight, largely due to genetic exceptionalism, but it also turned greater attention to all police databases. The Nuffield Council on Bioethics 2007 report recognised that the police must justify their powers and demonstrate that they are proportionate, striking a balance between personal liberty and the common good. The Council argued that the 'no reason to fear if you are innocent' argument ignored the cost of being involved in a criminal investigation; any intrinsic value of liberty, privacy and autonomy; and the implications of 'criminality' of being on the National DNA Database. They also expressed concern over the lack of oversight of the NDNAD, especially when it came to research using the DNA profiles, with independent oversight essential if attempting to adhere to the tenet: 'There's a difference between what one can do, scientifically or otherwise, and what one ought to do' (Dinerstein 2001).

Function creep

'Function' or 'mission' creep occurs when a project is expanded beyond its original goals. In the case of DNA this could be evidenced by expansion to include individuals and offences that were not originally intended as targets, by extending the uses to which the databases can be put, or the development of techniques that can garner greater information from DNA profiles (such as phenotyping or familial searching). The legal parameters for use of the UK's NDNAD are clearly delineated in legislation: the prevention and detection of crime, the investigation of an offence, the conduct of a prosecution, or the identification of a deceased person. This affords some certainty about how the NDNAD may be lawfully used. Not all national DNA databases have such explicit or clear boundaries, and may not definitively preclude their use in medical or other research or in paternity disputes for example. In Australia, Queensland's legislation permits the Police Commissioner to use the DNA Database for *any* police service function;[28] such terminology may be subject to a wide interpretation. DNA profiling was originally introduced to help identify criminals who left retrievable biological material such as blood, semen, saliva and hair at crime scenes (or on victims or weapons etc.). These were, typically, violent criminals and sex offenders. The UK's NDNAD, along with many others, has been significantly extended to include not only individuals convicted of minor offences but also arrestees. Such developments raise concerns that we have already witnessed function creep, and this may continue.

As stated, forensic databases are increasingly used to identify unknown suspects by searching for possible relatives of a perpetrator (so-called familial searching), or for predicting the race and/or appearance of an unidentified suspect. While these can be classified as operational uses, in that they are directly related to specific police investigations, there are potential other uses of DNA databases, with research conducted using the electronic records (DNA profiles) or the archived biological samples from which profiles have been generated. For example, researchers may find irresistible the prospect of finding the 'criminal gene', a mainstay of criminology since its inception.[29] Genetic studies to date have taken the form of twin, or sibling, and hereditary studies, purporting to demonstrate how much offending can be attributed to 'nature' or 'nurture'. Studies have not been able to conclude that there are 'genetic' explanations for offending, but the search could become more sophisticated, using large datasets of 'criminal' DNA. With the human genome now decoded, and research to find 'genes for …' continuing apace, it may not be too futuristic to consider that it will one day be reported that scientists have located a gene that may bear the stigma 'criminal gene'. The media, ignoring the superficiality of such a claim, would no doubt overlook the complexities of offending, and the necessary interactions between behaviour, the environment, societal reactions, the law etc., before someone is considered 'criminal'.

Questions over whether States are collating excessive information on citizens will persist, particularly where information could be used for discriminatory purposes, and to the detriment of the individual. At national level, mass surveillance of citizens could lead to a 'suspect society' with a national (perhaps compulsory) DNA database (McCartney 2006), which can lead to 'civic chill' − more commonly associated with CCTV proliferation, but 'chill' could arise from excessive State surveillance of personal information and loss of genetic privacy. This may occur by stealth, with police powers subtly extended and new opportunities arising for DNA sampling.[30] In a revival of concerns over the insurance industry accessing genetic data, the promise of 'personalized medicine' may prompt citizens to voluntarily submit their DNA to health agencies (Cannataci 2016: 7). In the UK, a proposed BioBank of DNA samples is purported to be solely for medical research, but a clause in the original Government White Paper enables police access to this BioBank database if they gain a warrant in exceptional circumstances (a high-profile murder?) Such a BioBank negates the need for a government to publicly debate the introduction of a national compulsory and comprehensive forensic database as this can be achieved under the guise of health care. There have already been (mis)reported cases of commercial 'geneaology' companies surrendering their DNA databases to policing agencies (see Syrmopolous 2015a, but then also Syrmpolous 2015b).

Such expanding uses of DNA databases beyond operational uses makes crucial the need for robust ethical oversight and regulation, particularly in instances where the research uses archived biological samples. These samples contain personal genetic information and their use warrants stricter regulatory oversight. Advanced levels of ethical and scientific review are necessary as samples are often initially obtained without consent, unlike those collected in medical settings, and remain easily traceable to named individuals. In deciding upon permission for research to be undertaken, there must be serious consideration of whether there is a police need for the research, as well as its legality and ethical aspects. There will thus always be a need for clarity over why information is collated and its utility: are these forensic DNA databases 'policing' databases for detecting crimes? If not − then what are they? The answer to this question will determine a great deal more: governance requirements; operational protocols; as well as how such databases should be populated.

'Genetic justice'?

There have been a number of collateral consequences of the forensic DNA 'revolution' including the massive expansion of forensic science provision in England and Wales and internationally, with the creation and growth of a forensic science 'market'. During this time, there have been a number of reports commenting upon the provision of forensic science; the use of forensic services by the police; and the institutional framework of forensic science provision. All have been unremittingly critical. Without 'good' forensic science, authorities run the risk not only of miscarriages of justice, but a loss of public confidence in both forensic science and the criminal justice system. It is critical that attention be paid to the delivery of forensic services: how standards are set, monitored and maintained across the forensic science sector. Essentially, forensic science must not be considered infallible, and the police are not scientists, nor are legal practitioners, judges and juries. The limits of the science are often poorly understood and courts are not the place to debate science. Public misunderstandings are still widespread and resort to DNA profiling may not foster the police–community relations necessary to ensure ongoing public co-operation in investigations.

While DNA has undoubtedly brought many people to justice, some of whom may never have been detected using traditional investigative means, it is far from being the most effective tool in the fight against crime. Locating DNA (avoiding contamination) is critical to its utility in solving serious crimes, those with no other leads, and those where there was no other hope of detection, but looking at a broader picture of 'crime' in all its guises, and investigation and detection strategies, it remains marginal. Yet efforts to increase the utility of DNA and the expansionist doctrine with regard to DNA databases, are often marred by ethical and scientific obstacles. There remains significant debate over the desired extent of forensic databases. There is an expectation that the power to gather, store and share forensic DNA will be 'free from corrupt influence, only when it is lawful, necessary and proportionate to do so' (Harfield 2008: 487). However, there is an absence of institutions with either the resources, or the authority to foster greater coordination and collaboration over ethical concerns. Other vital issues such as access to databases/samples; research uses; retention periods; advanced techniques etc., all are still contested.

There are also increasing demands for evidence of efficacy, particularly in an era of severe budget constraints. With media enthusiasm yet to wane, it still requires empirical evidence to demonstrate that DNA technologies are actually assisting in crime detection and prosecution. This evidence is scarce and equivocal at the moment. More powerful has been its ability to exonerate wrongfully convicted individuals, shining a light on injustice and the causes of miscarriages of justice. Indeed, forensic DNA continues to expose other forensic sciences that are now scrutinised for their accuracy and reliability. This is not to conclude that forensic researchers or police are chasing unrealistic dreams, and there is potentially scope for greater use of DNA in crime investigation. However, the search for the genetic 'Holy Grail' – of DNA being the solution to ridding societies of crime – may ultimately be misguided.

Notes

1 Africa have made significant advances since this time. See INTERPOL Global DNA Profiling Survey 2008; for recent updates see www.dnapolicyinitiative.org.
2 Referred to as the 'CSI Effect', its existence is nevertheless highly contested. See, for example, Cole and Dioso-Villa 2009.
3 In its 2009 report, the US National Academy of Sciences stated that, 'with the exception of nuclear DNA analysis … no forensic method has been rigorously shown to constitute reliable evidence.' National Research Council of the National Academies 2009.

4 DNA 'fingerprinting' was the first phrase coined by Jeffreys, the term 'profiling' soon came to be preferred to the 'fingerprinting' moniker though this is still sometimes referred to, especially by popular media.

5 Launched in May 2012 by the University of Michigan Law School and the Center for Wrongful Convictions at Northwestern University. See www.law.umich.edu/special/exoneration/Pages/about.aspx.

6 *R. v. Robert Graham Hodgson* (also known as Sean Hodgson) [2009] EWCA Crim 490 18 March 2009 – Court of Appeal Criminal Division. See *Guardian* 2009.

7 DNA scientists in Canada are working on samples from the original investigation and a new DNA profile has been produced, with the hope that it may still be possible to locate her killer (see BBC News 2016a).

8 Legal jurisdictions will differ in whether they permit arrestees to be sampled and which arrestees will be sampled, often dependent upon the seriousness or nature of the crime they have been arrested in connection with.

9 Recordable offences are essentially those crimes that the police must officially record, thus only very minor, obscure, or less serious motoring offences are excluded.

10 Section 64 Police and Criminal Evidence Act 1984.

11 If under 10, their parents could consent to sampling.

12 *S & Marper v UK* (App.no. 30562/04) [2008] ECHR 1581 paras.118/9.

13 *S & Marper v UK* (App.no. 30562/04) [2008] ECHR 1581 para 125.

14 Law No. 89 of June 30, 2009.

15 *R v. Duffy and Shivers* [2011] NICC 37.

16 *R v. Hoey* [2007] NICC 49.

17 *R v Melias* (1987) *The Times*, 14 November.

18 For a comprehensive account of these DNA wars, see Kaye 2010.

19 *R v Deen* (1994) *The Times*, January 10.

20 *R v Gordon* (1995) 1 Cr.App.R. 290.

21 See *Crim LR* [1995] 413 and *Journal of Criminal Law* (1995) 59 (4) 54.

22 2 *Cr.App.R.* 467. See also *J of CL* (1997) 61(2) 170 and *Crim LR* [1996] 898.

23 *R v Denis Adams* [1996] 2 *Cr.App.R* 467.

24 *Journal of Criminal Law* (1998) 62(5) 444.

25 *R v Doheny & Adams* [1997] 1 *Cr.App.R.* 369 see also *J of CL* (1998) 62(1) 33 *Crim LR* [1997] 669.

26 Council Decision 2008/615/JHA and 2008/616/JHA, published in *OJ L* 210 of 6.8.2008, p. 1 and 12. See McCartney *et al.* 2011.

27 *Heinz Huber v. Germany*, European Court of Justice, Case C-524/06, Judgement of 16 December 2008.

28 Queensland Police Powers and Responsibilities Act 2000: Section 493: Use of QDNA or CrimTrac database:

> 'It is lawful for the commissioner to use QDNA or the CrimTrac database for performing any function of the police service'.

29 One of the first 'criminologists' Lombroso explicitly stated that criminals were 'born' not made.

30 I.e. London Underground staff are provided with 'spit kits'– to profile the DNA of individuals who spit at staff on the Underground and subsequently search against the NDNAD.

References

BBC News (2003) Man 'trapped' by DNA evidence, 26 November. Available at: http://news.bbc.co.uk/1/hi/england/tyne/3239938.stm (accessed 1 Sept. 2016).

BBC News (2004) Cold cases solved by Phoenix, 28 September. Available at: http://news.bbc.co.uk/1/hi/england/tyne/3621048.stm (accessed 1 Sept. 2016).

BBC News (2016a) Lindsay Rimer death: New DNA leads in 1994 murder case. Available at: www.bbc.co.uk/news/uk-england-leeds-36019234 (accessed 1 Sept. 2016).

BBC News (2016b) Man jailed for 1984 Manchester rape after DNA advances, 2 February. Available at: www.bbc.co.uk/news/uk-england-manchester-35475542 (accessed 1 Sept. 2016).

Biondo, R. and F. De Stefano (2011) Establishment of Italian national DNA database and the central laboratory: Some aspects, *Forensic Science International: Genetics Supplement Series* 3 (1): e236–e237.

Blake, E.T. (1989) Scientific and legal issues raised by DNA analysis, in J. Ballantyne, G. Sensabaugh and J. Witowski (eds), *DNA Technology and Forensic Science*. ColdSpring Harbor, NY: Cold Spring Harbor Laboratory Press.

Briody, M. (2002) The effects of DNA evidence on sexual offence cases in court, *Current Issues in Criminal Justice* 14 (2): 159–181.

Brown, S. (2009) Trading intelligence, *Policing* 3 (2): 181–190.

Cannataci, Joseph A. (2016) *Report of the Special Rapporteur on the right to privacy* 8 March 2016 Human Rights Council 31st Session, A/HRC/31/64.

Cole, S.A. and Dioso-Villa R. (2009) Investigating the 'CSI Effect' effect: Media and litigation crisis in criminal law, *Stanford Law Review* 61(6).

den Boer, M. (2002) Towards an accountability regime for an emerging European policing governance, *Policing Society* 12 (4): 275–289.

Dinerstein, D. (2001) Criminal law and DNA science, *American University Law Review* 51: 401–430.

DPA (2016) Italy sets up DNA database, says it's key for national security. Available at: https://about.hr/news/europe/italy-sets-dna-database-says-its-key-national-security-15419 (accessed 19 April 2016).

Feltes, T. (n.d.) *Limits of Traces – The Phantom of Heilbronn*. Available at: www.sipr.ac.uk/downloads/Phantom_of_Heilbronn.pdf (accessed 16 April 2016).

Findlay, M. and Grix, J. (2002) Challenging forensic evidence-observations on the use of DNA in certain criminal trials, *Current Issues in Criminal Justice* 14 (3): 269–282.

Guardian (2009) Prisoner has murder conviction quashed after 27 years, 18 March. Available at: www.theguardian.com/uk/2009/mar/18/prisoner-hodgson-murder-quashed-miscarriage (accessed 1 Sept. 2016).

Guardian (2012) Forensics blunder 'may endanger convictions', 8 March. Available at: www.theguardian.com/law/2012/mar/08/forensics-blunder-convictions (accessed 16 April 2016).

Guardian (2016) Man pleads guilty to 1984 murder of Melanie Road, 9 May. Available at: www.theguardian.com/uk-news/2016/may/09/bristol-man-pleads-guilty-to-1984-of-melanie-road?CMP=Share_AndroidApp_Tweet (accessed 1 Sept. 2016).

Harfield, C. (2008) The organization of 'organized crime policing' and its international context, *Criminology and Criminal Justice* 8 (4): 483–507.

HMIC (2000) *Under the Microscope: An HMIC Thematic Inspection Report on Scientific and Technical Support*. London: Home Office.

Home Office (2005) Forensic Science and Pathology Unit, *DNA Expansion Programme 2000–2005: Reporting achievement*. London: HMSO.

Hufnagel, S. and McCartney, C. (2016) (eds) *Trust in Policing and Judicial Cooperation*. Oxford: Hart Publishing.

Imwinkelreid, E. and Kaye, D. (2001) DNA typing: Emerging or neglected issues, *Washington Law Review* 76: 458.

INTERPOL (2008) Global DNA Profiling Survey. Available at: www.interpol.int/Public/Forensic/DNA/Default.asp (accessed 1 Sept. 2016).

Kaye, D. H. (2010) *The Double Helix and the Law of Evidence*. Cambridge, MA: Harvard University Press.

Leary, D. and Pease, K. (2003) DNA and the active criminal population, *Crime Prevention and Community Safety: An International Journal* 5 (1): 7–12.

Lynch, M., Cole, S. A., McNally, R. and Jordan, K. (2008) *Truth Machine: The Contentious History of DNA Fingerprinting*. Chicago, IL: Chicago University Press.

Macdonald, F. (2016) You'll soon have to hand over your DNA if you want to visit Kuwait. 3 May. Available at: www.sciencealert.com/you-ll-soon-have-to-hand-over-your-dna-if-you-want-to-enter-kuwait (accessed 9 May 2016).

McCartney, C. (2006) *Forensic Identification and Criminal Justice: Forensic Science, Justice and Risk*. Willan Publishing: Cullompton.

McCartney, C., Williams, R. and Wilson, T. (2011) Transnational exchange of forensic DNA: Viability, legitimacy, and acceptability, *European Journal of Criminal Justice Research and Policy* 17 (4): 305–322.

National DNA Database (2015) Strategy Board Annual Report 2014/15 Presented to Parliament pursuant to Section 63AB(8) of the Police and Criminal Evidence Act 1984, December 2015. Available at: www.gov.uk/government/publications/national-dna-database-annual-report-2013-to-2014 (accessed 1 Sept. 2016).

National Research Council of the National Academies (2009) *Strengthening Forensic Science in the United States: A Path Forward*. Washington, DC: The National Academies Press.

NDNAD (2009) Annual Report 2007 to 2009. Available at: www.gov.uk/government/uploads/system/uploads/attachment_data/file/117784/ndnad-ann-report-2007-09.pdf (accessed 16 April 2016).

Purcell, N., Winfree, L.T. and Mays, G.L. (1994) DNA evidence and criminal trials: An exploratory survey of factors associated with the use of 'genetic fingerprinting' in felony prosecutions, *Journal of Criminal Justice* 22 (2): 145–157.

Redmayne, M. (2002) Appeals to reason, *Modern Law Review* 65 (1): 19.

Schlesinger, F. and Hamilton, F. (2012) Someone knows how spy ended up dead in that bag, coroner declares. *The Times*. Available at: www.thetimes.co.uk/tto/news/uk/crime/article3402508.ece (accessed 16 April 2016).

Security News Desk (2016) SelectaDNA forensic marking behind Police's biggest burglary drop, April 12. Available at: www.securitynewsdesk.com/selectadna-forensic-marking-behind-polices-biggest-burglary-drop/ (accessed 16 April 2016).

Syrmopoulos, J. (2015a) Ancestry.com allegedly caught sharing customer DNA data with police. Available at: http://thefreethoughtproject.com/ancestry-com-caught-sharing-dna-information-police-warrant/#ykw5mLVm11CwXWH0.99 5 May 2015 *The Freethought Project.com* (accessed 1 Sept. 2016).

Syrmopoulos, J. (2015b) Correction: Ancestry.com did not share customer data with police without a warrant. Available at: http://thefreethoughtproject.com/correction-ancestry-com-share-customer-data-warrant/#2UlHcQGJLvGKSsYT.99 (accessed 16 April 2016).

Yuthas K. and Dillard, J. (1999) Ethical development of advanced technology: A postmodern stakeholder approach, *Journal of Business Ethics* 19: 49.

25

Visual surveillance technologies

Richard Jones

Introduction

The focus of this chapter is on one particular kind of surveillance technology used in crime control, namely visual surveillance technology. As the chapter will show, we can identify several different kinds of visual surveillance, with a range of crime control applications, and with various consequences for our societies and for how we understand the nature of policing, crime control and criminal justice. Over the past few decades, visual surveillance technologies have gradually become an increasingly significant feature of crime control systems. This has been enabled by technological developments, and in particular advances in microelectronics, the shift from analogue to digital electronics, and the emergence of digital imaging, have seen the size and cost of devices plummet at the same time as their optical capabilities have steadily increased. As such, the increased use of visual surveillance technologies parallels the increasing reliance, over a similar time, on other forms of electronic surveillance technologies, such as the electronic monitoring ('tagging') of offenders, and surveillance of the Internet (Jones 2014a; Nellis 1991, 2005; Nellis *et al.* 2013).

Perhaps the first or at least most well-known form of electronic visual surveillance, and the one which to date has spawned the largest body of research is closed-circuit television (CCTV) surveillance. Whilst this chapter will indeed further examine CCTV surveillance, it will place such analysis within the broader category of 'visual surveillance technologies'. As such, the chapter will examine a range of contemporary visual surveillance technologies, highlighting the capabilities, practices and implications specific to each of the technologies in turn, before identifying and exploring the commonalities of this family of technologies as a whole. Many of the most striking capabilities of today's visual surveillance technologies are enabled by digital electronics – facilitating the digital generation of optical imagery, remote control of a camera, the transmission over distance of visual imagery, or the recording of a large amount of footage, for example. However, when discussing these technologies, we should not forget the other engineering advances that have facilitated their development, such as in battery technology and manufacturing processes, and in communications systems enabling video imagery to be transmitted across networks.

In the following sections of this chapter, various different visual surveillance technologies in use today will be identified in turn. In each section I will examine how the technology in question works, the policing and other law enforcement uses to which the technology has been put, exploring some of the relevant research studies conducted, in order to build up a picture as to their potential applications, effects and consequences, and in order to try to establish the technologies' similarities and differences. It will be argued that while new technologies such as body-worn video cameras and surveillance drones have begun to be used for law enforcement purposes, and while their relationships with existing policing practices are complex and varied, if we zoom out, as it were, and consider electronic visual surveillance technologies today, we see considerable similarities between them, at least at a conceptual or theoretical level. Indeed, this chapter will contend that what we are witnessing today is effectively the diversification of some of the classic elements of CCTV cameras across a range of new 'platforms', but that in the process a range of new social, legal, privacy and cultural issues are opened up.

Types of visual surveillance

Visual surveillance need not involve any technology at all. The most basic form of visual surveillance is human surveillance. This could be used in the course of either formal social control (such as by a police officer or security guard) or informal social control (such as by members of a community over one another). Technology could however be (and indeed historically has been) introduced to enhance such surveillance in one of two main ways. First, visual surveillance capabilities can be enhanced through the use of magnifying optics, in the development of telescopes and binoculars. Second, the nearby physical environment may be changed or otherwise designed in such a way as to facilitate human visual surveillance – for example, by removing trees or by designing buildings in order to maximise visible lines of sight (such as in Bentham's imagined prison design of the 'Panopticon') (Foucault 1979), or latterly as a means of trying to facilitate 'defensible space' (Newman 1972), achieve 'crime prevention through environmental design' (CPTED) (Jeffery 1977), or prevent crime in specific settings by 'increasing the risk of detection' (Clarke and Mayhew 1980; Cornish and Clarke 2003). Some of the rationales offered for the employment of visual surveillance technologies are based on crime prevention theories, though it also seems the case that the technologies' introduction is in practice driven by a range of factors, and particular applications and use purposes are often emergent rather than necessarily very clearly defined and understood at the outset.

Telephoto and covert photography

If we begin a consideration of visual surveillance not with digital electronic technologies but with analogue optics, we can see that simple photography can be used for surveillance purposes. Surprisingly little research appears to have been conducted on the use of covert photography as a form of surveillance by law enforcement officers, though more research has been undertaken on the use of photography by the police as a means of identifying and cataloguing suspects. Cameras have long been used in undercover policing and the policing of political protests and protestors, however, and telephoto lenses seem to have been used from the 1960s and 1970s as a means of surveillance and identification of suspects, for example in relation to organised crime. This practice is worth mentioning here, albeit briefly, however, because if we compare it with police use of body-worn video and surveillance drones, for example, we can see certain instructive similarities yet also differences. Rather than claiming to invoke any deterrent function, covert surveillance aims to gather intelligence or collect evidence about a suspect. As such, it is different from overt forms of surveillance such as police body-worn video cameras which are typically overtly visible, but shares some characteristics with aerial surveillance, especially that taken at higher altitudes where the surveillance craft is not readily visible from the ground. Surveillance of political protestors may be either covert or overt, probably depending on whether the aim is intelligence-gathering or an attempt to intimidate the protestors.

CCTV

Closed-circuit television cameras have for many years been the most common, visible and well-known form of visual surveillance technology in the UK and many other countries around the world. Technological advances from the 1960s to 1990s enabled the development of progressively smaller, cheaper and more powerful forms of CCTV. A CCTV system may be defined as any non-broadcast electronic camera system, in which a particular viewer or viewers sitting in front of a connected monitor can view the imagery from an electronic camera remotely. Using digital storage technologies, the imagery may additionally, or alternatively, be recorded electronically, and hence replayed at some later point.

In his study of the possible causes of crime and insecurity within residential areas, Newman proposed that the otherwise concealed spaces of lifts in public housing blocks could be fitted with CCTV cameras with the aim of introducing visibility and hence enhancing 'defensible space'. Among the early uses in England of CCTV camera technologies was crowd monitoring in central London in 1960, and by the late 1960s over a dozen English forces were using cameras, though only a tiny number each. Cameras were installed in a small number of London Underground stations in 1975, and the first large public CCTV system in the UK was in Bournemouth in 1985 for the Conservative Party Conference, following a deadly bomb attack by the IRA on the Conference the year before (Norris *et al.* 2004).

From the 1990s onwards the number of CCTV cameras in the UK seems to have increased dramatically. One of the reasons for this was the declining cost of small CCTV systems, and their appeal as a possible additional security response to crime. CCTV cameras have proliferated in several private or quasi-public contexts, for example in small shops, larger retail outlets, shopping malls, banks, offices, public buildings, train stations, airports and residential buildings. The CCTV systems installed in such locations are often simpler and cheaper, however, often featuring a small number of fixed cameras, each pointing at a particular area of interest, such as a door or other entry point. In larger locations in which dedicated security staff are employed, the cameras may be monitored by security staff, but in many other cases the camera images will simply be recorded and kept for a period of time in case they need to be consulted, before being over-written.

During the 1990s in the UK the number of CCTV schemes in public areas such as city centres also increased significantly, almost certainly as a result of the 'City Challenge Competition' initiative, in which central government would subsidise the installation costs of city-centre CCTV schemes (Norris *et al.* 2004; Webster 2009). These schemes, many of which are still in use today, are often typically run by the local authority, but may also feature the involvement of the local police force. The camera operators are usually able to view and often directly control a number of cameras remotely including being able to 'zoom in' on particular individuals, places or things of interest. The footage is typically automatically recorded and stored for a particular time period before being deleted.

The use of CCTV systems of the more powerful and extensive kind found in city-centre schemes has not been without controversy. Initially introduced with the stated aim of helping to reduce crime by acting as a deterrent, few evaluations were able to demonstrate this empirically (though Welsh and Farrington (2008) found some evidence that cameras reduced vehicle-related crime when installed in car parks in the UK), and some studies have found evidence that even if the cameras were to have an effect on offenders' behaviour it might result in 'displacement' of crime from one area (covered by CCTV) to another (that was not) (see for example Cerezo (2013)). Conversely, though, an evaluation of a car park CCTV scheme suggested there could be 'diffusion of benefits' to other nearby car parks at which CCTV had not been installed, perhaps as the result of car thieves learning of the initiative but being unsure as to its local scope. The official justification of such schemes remains somewhat unclear, however – though the lack of clear evidence as to the schemes' effectiveness does not appear to have impacted on their appeal. The reasons for this may include that they enjoy a relatively high degree of public support, and that their actual functions have been retrospectively reinterpreted, for example through discovering their potential role in providing evidence in support of arrests and prosecutions (even if this was hampered for many years by the poor or uneven quality of CCTV recorded footage, especially from fixed camera systems), and in the case of larger schemes, their potential role in helping guide police officers on the ground towards on-going incidents. A study in the Netherlands, however, found that young people visiting the nightlife districts of two cities there reported only limited faith in CCTV cameras' ability to prevent crime and make them feel safer, with several saying that if something did happen to them they didn't believe the cameras could do anything to help them at least there and then (Brands et al. 2016).

It has been suggested that rather than always be reactive to events CCTV could also be used more proactively by police with the aim of swiftly deploying officers and intervening sufficiently early in order to prevent offences from taking place, using police powers consistent with legal principles (which in the US include 'probable cause' and 'reasonable suspicion'). A preliminary study of such a practice found some evidence that it could potentially effectively intervene in some offences but that in practice it was hampered by slow police dispatch times. The study authors suggest that CCTV operators they interviewed displayed good ability 'to effectively identify instances of suspicion and criminal behavior' (Piza et al. 2014). However, stop and search/frisk practices based on 'suspicion' remain highly controversial, especially in relation to race and police–community relations (Bradford and Jackson 2016).

As well as in homes (Wright et al. 2015) and private businesses and in public and quasi-public areas CCTV has been used within various institutions for security purposes. Within prisons, cameras can enable guards to see what is happening on different wings. Controversially, some schools have installed CCTV and other surveillance or security technologies (see Taylor 2013). Courts have experimented with using a single camera CCTV system in order to enable vulnerable witnesses to give testimony from a remote room so that they do not physically have to enter the potentially intimidating courtroom. As early as 1953–55 CCTV was used within psychiatric hospitals in some US states for therapeutic or educational purposes, and in UK mental health hospitals they have been used as surveillance cameras since 2002, and while various uses of such cameras have been identified there are also possible 'negative consequences' of its use, including consequences of an ethical nature (Tully et al. 2015: 290–291).

One interesting line of research has explored the use of CCTV in practice by studying the people who operate the cameras in larger schemes. Research suggests that CCTV camera operators look out for particular kinds of people, such as groups of younger people. Gavin Smith has explored the working patterns, beliefs, and coping-mechanisms of camera operators in still greater depth, by employing ethnographic research methods (Smith 2015). Boersma recounts how feeds from a large number of CCTV cameras were dealt with in an emergency control room during a large annual student event (Boersma 2013).

One possibility that alarmed and intrigued researchers in the 1990s was the potential for CCTV imagery to be processed in real-time by a computer, in order to be able to query a database. In relation to people, this 'algorithmic surveillance' involved facial recognition software in order to try to match people in view of the camera against a database of the faces of people of interest (Norris et al. 1998). The civil liberties dangers of such a system were quickly identified. The effectiveness of such systems depends on their technical sophistication, their practical efficacy including the ease with which they can be resisted, and the quality of the database data against which faces are matched. Over the same timeframe, though, the automation of visual recognition from CCTV cameras has become widespread in other applications, most notably in relation to Automatic Number [Vehicle License] Plate Recognition (ANPR) (O'Malley 2010; Parsons *et al.* 2012). Other ways in which computer image processing of surveillance camera footage may be used in future includes semi-automated surveillance systems designed to help camera operators detect and track events and hence make them more effective (Dadashi *et al.* 2013; Ferenbok and Clement 2011). Researchers have also examined how, given the huge amount of visual data generated by CCTV surveillance cameras, the surveillance video can be 'crowdsourced' online to a large informal group of non-specialist but motivated volunteers who could be asked to carry out certain watching or identification tasks. However, it seems that good interfaces and clear tasks are essential (Dunphy *et al.* 2015). Moreover, there are various privacy concerns, that such practices seem to further normalise CCTV surveillance, and there are also concerns regarding the lack of training of volunteers, and in some cases companies charge end users for the service but do not pay the volunteer workers (Trottier 2014). An interesting feature of crowdsourcing, to which I will return later in this chapter, is that it goes some way to changing the nature of the surveillance away from a 'closed' system and towards an 'open' or 'broadcast' medium (and as such, the surveillance cameras are technically no longer a 'closed circuit').

Although it is possible in some ways to speak of CCTV as being a single technology, we can extend our understanding of the technology by recognising two ways in which it is in fact composed of a range of sub-technologies. First, we can differentiate between different kinds of traditional CCTV camera. Cameras range from simple, fixed, unmoveable cameras mounted in or on a building, for example, to advanced, remote controllable cameras that can pan, tilt and zoom. Many CCTV cameras are permanently mounted in a particular position, but CCTV cameras have also been deployed in mobile vans or are mounted in ways designed to be rapidly deployable/redeployable, enabling temporary deployment during particular events or operations (on compliance issues relating to 'temporary' cameras, see Hier and Walby (2014)). Second, if we focus on the closed-circuit aspect of CCTV, we can think of a drone fitted with a camera as being akin, conceptually, to a kind of 'flying CCTV camera'.

If CCTV cameras have not become as reviled as some might have predicted back in the 1980s, it is probably largely due to a combination of powerful factors – even if it is also true to say that their use does provoke unease among certain people or when used in certain contexts. First, CCTV cameras have become so prevalent (in the UK at least) that they are almost ubiquitous, and have largely disappeared into the background of everyday life, which in turn may have led to them now being perceived as normal or indeed 'banal' (Goold *et al.* 2013). Second, they have become a familiar figure within fictional and non-fictional depictions of surveillance, and in a way that implies a certain inevitability about their use, which again may have resigned many to the cameras' acceptance. Third, over the years several high-profile cases, such as the killing of James Bulger, and the belated detection of the London bombers in 2005, have featured CCTV evidence. Whereas the cameras may in many cases have been of little value either as a deterrent or in gathering evidence for use in court, their high-profile role has attached strikingly to society's fears of one kind or another, or its fantasy/dream of security on the other (see Kroener 2013).

It was noted as recently as 2013 that there was little evidence at that time for a new generation of Internet-connected home CCTV cameras enjoying either any significant sales nor any market demand; indeed, when questioned for a research study various interviewees expressed a lack of interest or even hostility towards the idea (Goold *et al.* 2013: 990–992).

However, and as correctly predicted by a chief constable interviewed for the same study, there are now various successful products enabling people to '[use] the internet to monitor things within their home when they're away' (Goold *et al.* 2013: 991). Factors driving sales of such cameras may include not just increased resolution and decreased cost, but also faster broadband, increased smartphone ownership and increased mobile data speeds. However an important additional factor, and which may also account for the apparent discrepancy between the previous hostility reported above and increased acceptability today, may be the clever product designs and aesthetics intended to make the cameras blend in to households rather than take the form of the traditional and rather more distrusting-looking CCTV wall-mounted design.

ANPR cameras

Another road traffic monitoring system employing camera technology, and one that does seem to constitute a form of surveillance is Automatic Number [Vehicle License] Plate Recognition (ANPR). ANPR cameras are typically placed near a busy road, whether in the form of permanent roadside mounting, in a mobile vehicle, or mounted inside a police car. This technology uses video cameras connected to a computer system in order automatically to read the number plates of all cars passing by. The number plates can then be checked against various databases, such as for stolen vehicles, or for vehicle owners for whom an arrest warrant is in force. Insofar as there is a network of ANPR cameras covering a given road system, the system could also be used to track the journeys of individual vehicles in real time, data which could be used in on-going police or intelligence operations in which suspects were being tracked. Two particular configurations of ANPR that have been used have been to install ANPR cameras along particular, heavily used routes such as motorways; and to install ANPR cameras next to each of the access roads in and out of a particular locale such as a city-centre area. ANPR cameras have been installed in the latter configuration as a means of policing compliance with congestion charge systems, such as in central London; however, once installed, the same system could presumably also generate data to be accessed for police and intelligence purposes.

A limitation of ANPR systems is that while they can accurately identify particular vehicles (which may be sufficient if police cars are nearby and the suspect vehicle/driver can be intercepted), in terms of their surveillance capabilities they can only determine that that vehicle passed a particular camera at a particular time. Nevertheless, even a small ANPR network could acquire data on a large number of vehicles, raising questions as to how far we wish to enable the state (or indeed local municipalities) to be able to monitor citizens' vehicle journeys and hence travel movements.

It is not clear to what extent ANPR camera systems can and are being defeated by measures taken by motorists, but powerful as the technology is, there may be ways of rendering it ineffective on a particular vehicle. One way could be to switch the licence plates on a vehicle in order to disguise its identity and true owner. This would likely be treated as a serious offence were it to be detected, but seems a possible counter-measure that serious criminals might employ. Another way of subverting an ANPR system could be to doctor the licence plate letters or numerals in order to make them harder for the computerised visual processing system to identify.

One can imagine that in the future, vehicles will continue to be increasingly computerised, and by employing GPS and cellular technology their location anywhere could be accurately established without having to rely on the installation of a large network of ANPR cameras and the visual capture of vehicle license plate details. Whilst this automotive era has already started, and will no doubt accelerate quickly, for the time being only a small proportion of vehicles on the road feature such tracking technology, and as such ANPR will probably remain the preeminent vehicle monitoring technology for some years to come.

Aerial surveillance technologies

CCTV camera systems are typically fixed in place and installed out of reach somewhere above head height. When mounted quite high up such cameras are able to obtain quite a commanding view and potentially over a reasonably large range; they remain fixed in place, however, and are thus bounded by the area they are able to cover. Helicopters fitted with video cameras offer a means by which surveillance can be conducted from the air, generating useful imagery as to events unfolding on the ground. Helicopters can fly faster even than sports cars, and can be useful for following stolen cars, for example. They can also be used in the policing of events involving large numbers of people, such as public protests, gatherings, or major sporting events, in order quickly to see where people are gathered and what is happening. A camera mounted on the helicopter can record video evidence for later, and today can also relay that in real-time to officers on the ground. Special thermal-imaging cameras can also be used to search for suspects on the ground during the night. Despite their uses in policing, helicopters have their drawbacks. They are expensive to acquire, to operate and to maintain, and require highly trained pilots to fly. Since they are also very noisy, they are also not a discreet means of surveillance. Light aircraft fitted with very high resolution cameras could potentially be used in place of police helicopters.

A rather different form of aerial surveillance is made possible in the form of spy satellites. While the imagery they generate is likely to be classified, unavailable to police forces, and shared only with intelligence and military agencies, conceptually we can understand the imagery they generate as sharing many of the characteristics of other forms of aerial surveillance: the images are from a 'bird's-eye' perspective, and allow a view of a very wide area. Limitations of spy satellites include that they are tremendously expensive to build, launch into orbit, and operate; the territory they can cover is determined by their orbit; they may be confounded by cloud; the imagery resolution may be low; and images may not be produced in real-time. However, it is likely that most of these limitations can now be overcome.

Returning to policing applications of aerial surveillance, recent years have seen unmanned aerial vehicles (UAVs), or 'drones' as they are often colloquially termed, emerge as potential alternatives to some uses of police helicopters, while also introducing some potential new surveillance use-cases. UAVs vary hugely in size, ranging from large, military-specification craft capable also of carrying missiles, to tiny devices only a few centimetres in length. Consumer and 'prosumer' level drones have recently increased considerably in capability while decreasing in price, and many are now capable of relaying back to the ground very high quality streaming video. Possible use-cases of law enforcement UAVs include monitoring public events and disorder, civil contingencies, road traffic disruptions and accidents, counter-terrorism operations, border patrol, hostage situations, and intelligence/surveillance-gathering operations. In addition to providing real-time imagery for operational purposes, drone footage could also be used as evidence in prosecutions, or drones could be used to gather evidence for use in civil or criminal cases (see Ravich 2015).

A range of issues is raised by the dawn of the era of capable, affordable drones. Among these is the question of privacy, since drones can be flown above individuals, groups, or public or private property, and drones already have advanced visual surveillance capabilities (Clarke 2014; Finn and Wright 2012). Privacy issues seem likely to further intensify as the optical, video and automatic tracking capabilities of surveillance drones improve. For example, the ARGUS

Imaging System developed by the US Department of Defense features a 1.8 billion pixel image sensor created by joining together 368 five-megapixel cameras, which, when attached to a drone or aircraft flying at 17,500 feet is able to automatically track all moving objects within an area of 15 square miles for hours or days at a time. Its image feed is relayed back to ground in real time, and can be archived and studied thereafter (Talai 2014: 745–746). Imaging systems on other surveillance drones can include thermal infrared sensors allowing them to see in the dark (Parks 2014), but in theory any system (such as ANPR, for example) could be installed. Where the gaze of drones is sustained over time, they take on a quality we could liken to that of CCTV, or what Wall and Monahan (2011) term 'the drone stare'. As Talai argues, 'drone surveillance is highly efficient and persistent, difficult to detect, and difficult to resist. The result is a well-founded fear—in the hands of police, drones make the specter of unfettered and unknown discretion all too possible' (Talai 2014: 746–747).

Police body-worn video cameras

By the 1990s, the use of forward-facing, dashboard-mounted video cameras within police cars had become more common in the US and the UK. Mounted in this way, the cameras record the view forward from the front window of the police car, and typically also overlay the video imagery with the time and date of recording and the speed of the police car. The footage so obtained can be very compelling, and the principal purpose or aim of the technology seems to be to generate evidence that could later be used in court. Arguably, body-worn video cameras are a logical development of the same, but attaching the cameras to officers themselves instead of just their vehicles. 'Body-worn video' (BWV) cameras worn by officers are now in use in many police forces in the UK, US, Australia and elsewhere. These cameras feature a forward-facing video camera worn by the officer on their lapel or shoulder, connected to a data storage device on which any video footage recorded is stored, ready for download to a computer back at the police station, where the footage can be replayed, studied, and archived. The video resolution, data storage capacities and battery life of the mobile body-worn video units are all increasing. Body-worn video systems vary as to how they are used, but a typical use involves a practice of 'selective capture', in which the camera is normally switched off, and is switched on by an officer only at certain times. An officer might switch on his or her camera in order to record a particular encounter with a suspect, witness or victim, for example. Such selective recording enables the amount of data recorded to be kept more manageable; helps conserve the battery of the mobile unit; and is more sensitive to privacy concerns, as it means that only specific encounters of interest to the police are recorded. In the UK, the normal practice is that activation of the video camera is indicated by a red light being illuminated next to the camera, so that all parties can see the camera is active and recording.

As is the case with video cameras mounted in police cars, the position and direction of BWV cameras generates footage that records the events as seen from the optical 'point of view' of the police car driver or foot patrol officer respectively, and this 'first-person video' footage can be very compelling to other viewers when replayed later. Clearly, this means the footage is potentially a very effective form of evidence, whether during an enquiry, bringing a prosecution, or securing a conviction in court. Situations in which BWV could potentially prove useful include the policing of domestic abuse, violent offences, and instances where an unwarranted complaint was made against an officer.

However, there are also some concerns and issues surrounding the use of BWV technology. Used regularly by a number of officers, the storage space required to archive the large video files generated by the high-resolution recordings quickly becomes sizeable, and systems are required for managing this digital evidence. The batteries of the mobile BWV units need to be kept charged, and are thus one more device officers need to remember to keep recharged. Selective capture can facilitate careful and appropriate use, but could also potentially be used by officers in order not to record certain incidents, or to record events only in part.

Whereas one of the original rationales behind the introduction of police BWV has been to collect compelling audio-visual evidence from suspects, victims and witnesses, the cameras' role as a potential means of encouraging better conduct by police officers themselves has also been suggested. As Ariel and colleagues have noted, 'police use-of-force and citizens' complaints against the police … represent two burning issues in American policing' (Ariel *et al.* 2015: 510). The same authors hypothesise that body-worn cameras may be more effective than CCTV in discouraging improper behaviour because they are observable and proximate – and that this can extend to the officer wearing the camera even if he is not directly visible (Ariel *et al.* 2015: 517). Their randomised control trial conducted in Rialto, California, found 'that police body-worn-cameras reduce the prevalence of use-of-force by the police as well as the incidence of citizens' complaints against the police' (Ariel *et al.* 2015: 531). In the course of conducting a replication of the Rialto experiment in Wolverhampton, UK, Drover and Ariel (2015) documented some of the challenges faced in gaining acceptance of the technology by frontline officers, echoing previous suggestions by Wain and Ariel (2014) that officers disliked technologies that they felt put them under surveillance by senior management. On the other hand, there is some evidence that police BWV can lead to a reduction in the number of offences against officers (see Cubitt *et al.* 2016 for a review of various studies).

Whatever the current issues surrounding the implementation and use of police BWV cameras, however, it seems plausible to argue that in the future body-worn cameras will become widely used or even ubiquitous within law enforcement. Over time, indeed, this is likely to have implications for the criminal legal process (from suspect to trial, as Sanders and Young (2012) termed it) as police, prosecutors and courts become increasingly used to, and eventually expectant of, high quality video footage capturing interactions at key moments with key actors, be they suspects, arrestees, victims, witnesses or police officers. Developments in the technology will however likely further change how such BWV cameras are used in the future. Once live streaming of BWV footage becomes common, as seems likely within the next few years, the imagery can become used for police deployment and remote management purposes, as well as for sharing purposes between officers. Being able to view such footage (along with various other data) using some variant of 'smart glasses' could further change elements of policing practice.

Turning the tables, confounding the cameras: sousveillance and resisting surveillance

Whichever theory of surveillance is the most compelling, and whether one regards surveillance as a useful crime control measure or as an insidious mechanism of social control, it can be agreed that visual surveillance technologies offer a means by which certain people can view and thus exercise some form of control over others. The declining cost of consumer video recorders, and the integration of cameras and later video cameras into smartphones, has meant that ordinary citizens have the capability of turning their cameras on authority figures such as police officers. Such a development was anticipated in many ways by Mann and colleagues, who, in a seminal and creative article foresaw the advent of wearable computing devices as a means of collecting data of one sort or another (Mann *et al.* 2003). They coined a term for such a practice, offering a way for ordinary citizens to 'observe those in authority' and exercise surveillance from 'underneath' rather than from 'above', as it were, namely 'sousveillance', and suggesting it is a form of 'inverse panopticon' (Mann 2013; Mann *et al.* 2003: 332). Their article demonstrates how different kinds of 'performances' involving wearable cameras and projectors provoke different responses from people depending on various contextual factors. The thrust of their article is to show how wearing cameras in certain ways can be used as a sort of political performance art, challenging and subverting conventional understandings and systems of official surveillance. In a later article, Mann and Ferenbok (2013) suggest how wearable camera glasses similar to Google Glass could become a mechanism for sousveillance. However, their article also prefigures subsequent simpler acts of simply holding up a camera to video the behaviour of an authority figure, which for ordinary people (as opposed to authority figures such as police officers) has to date become a far more ubiquitous as well as socially acceptable means of carrying a camera than are wearable cameras. If, as these authors suggest, electronic surveillance technologies can be understood as the latest technological means of conducting centuries-old practices of surveillance, we can also see that just within a few decades the capabilities of the newest technologies have developed dramatically. Indeed, gradually what begins as a chance accident (a citizen happening to test their new video camera at the precise moment that LAPD Officers were beating suspect Rodney King in 1992) becomes increasingly more likely. While such forms of surveillance by citizens could be a matter of chance – happening to point a camera in a particular direction at a particular time – it could instead be planned and intentionally directed, for example when protestors video the protest in order to document instances of disproportionate use of force by the police, or in 'organised copwatching' by local residents who monitor the conduct of police (Simonson 2016 (forthcoming)). Moreover, if the intention is not merely to video police misconduct but to publicise it, for example by uploading it to a video-sharing website, the practice becomes akin to the 'synopticon', in which the many watch the few (Mathiesen 1997). Indeed, one could argue that insofar as citizens document their own engagement with the police, this form of sous/surveillance includes to some limited degree what has been termed 'participatory surveillance' (in which participants electively partake in self-surveillance for reasons of 'mutuality, empowerment and sharing' (Albrechtslund 2008; see also Albrechtslund and Lauritsen 2013; cf. Jones 2014b). Police officers may variously embrace or shy away from the gaze of their own or others' cameras, with some policing practices evolving as a result (Sandhu and Haggerty 2015). An interesting aspect of sousveillance is that it often employs very similar technologies to those used by law enforcement, but it would be a mistake to assume that this immediately puts such surveillance on an equal footing with state-sanctioned surveillance. In addition to possessing more sophisticated technologies, crucially state authorities also possess various powers enabling them to take action following surveillance collection – to engage in 'social sorting' as Lyon (2003) puts it. If sousveillance has effects they are more likely to be of a revelatory and disruptive kind, documenting malpractice by officials, for example.

As Greer and McLaughlin have argued, recent years have witnessed the rise of the 'citizen journalist', or at least citizens who, happening to be in a certain place at a certain time, are able to capture images and report on an unfolding event, and whose 'credibility and authenticity as news sources derive from their capacity to provide "factual" visual evidence of "live events"' (Greer and McLaughlin 2010: 1054). They note too how '[t]he rise of the citizen journalist has been accompanied, and perhaps encouraged, by a decline in deference to authority and a deterioration of trust in official or elite institutions' (Greer and McLaughlin 2010: 1054–55) and that 'it is when citizen journalism challenges the "official truth" … that it becomes most potent as a news resource' (Greer and McLaughlin 2010: 1056). It is also worth noting that visual surveillance is often compelling testimony, particularly so when it is recorded from a first-person or 'point of view' perspective, giving a clear impression as to what it was like to witness the event. This may also account for the popularity of wearable or vehicle dashboard-mounted cameras ('dashcams') used not only for leisure purposes but also increasingly by cyclists, motorcyclists, and drivers to gather testimony of others' dangerous driving, or for insurance purposes (see Štitilis and Laurinaitis (2016)).

Visual surveillance technologies offer a means by which individual people may be identified by their faces or clothing, but there are ways by which those who do not want to be identified or located may try to resist such surveillance. One simple but effective tactic is to cover one's head to hide one's face, for example by wearing a hoodie. (The effectiveness of such a tactic is presumably also why banks forbid the wearing of motorcycle crash helmets on their premises.) It seems too that facial recognition technology can also be confounded, even when a clear view of a person's face is captured, if the person uses certain patterns of face paint, particular hair styles, or masks, to confuse the facial detection software algorithm. The website CV Dazzle (cvdazzle. com) documents and explains such strategies. And American artists have 'even released a drone-proof clothing line called Stealth Wear made of nickel metalized fabric, and a German company has designed a special outfit called Ghost, which makes the body invisible to infrared sensors' (Parks 2014: 2520).

Issues and implications of visual surveillance technologies

In the preceding sections various issues and consequences of the use of different visual surveillance technologies were identified. However it is possible additionally now to identify some common themes cutting across the different technologies. Each of the technologies captures different kinds of imagery, taken from different vantage points, yet all share a certain ineffable quality in which the viewer is transported into a certain setting and made witness to the unfolding events. There is something particularly compelling about visual testimony, whether it be the photograph or a video recording. For a person depicted in video, especially at close range, it is hard for them bluntly to deny involvement, which could have positive consequences for criminal justice. Yet of course while it may seem that 'the camera does not lie', journalism, media and film studies tell us differently. In relation to video evidence, the direction in which a camera is pointed, its field of view, when it is switched on, and how the resulting footage is edited, can all change our perception of the events depicted. More prosaically, it is clear that proper administrative procedures must be put in place in order to fulfil legal evidence requirements; today, much of this evidence will be digital in nature which may add extra complexity but is a manageable issue. More exotically, digital imagery is potentially subject to sophisticated digital manipulation, and thus demanding of digital forensics in order to assess its authenticity.

But it is the impact on citizens' ability to enjoy privacy, a vital component of democratic civic life, that seems most at stake in the face of further expansion of visual surveillance capabilities. Whereas different technologies potentially infringe on different aspects of people's private lives (for example BWV may involve officers recording inside someone's home, whereas ANPR cameras can be used to track vehicles), individually and collectively they could lead to a variant of the 'chilling effect'. Whereas the 'chilling effect' conventionally refers to the possibility that citizens may self-censor their expressions if they think their written or verbal communications could be under audio or Internet surveillance, with potentially damaging consequences to political debate and hence democracy itself, further extending visual surveillance within society could lead some people to decide not to be seen in certain places (for example places of worship, or political meetings) in case they might be seen and identified. But even if such an effect does not transpire, a sense of being watched can lead to people experiencing being in places in unsettling ways, with the atmosphere or 'ambiance' of a place changed in a subtle but negative fashion (Adey *et al.* 2013), with the consequence that those places at best come to be unloved and maybe even to be avoided.

As was noted above in the section on CCTV, crowdsourcing the monitoring of CCTV images technically means the surveillance cameras are no longer a 'closed circuit'. This is noteworthy because it forces us to consider what exactly CCTV is, and how exactly it differs from more recent visual surveillance technologies such as aerial surveillance and body-worn video cameras. Indeed, although in this chapter I have distinguished between these different forms of surveillance because they have different properties and raise different issues, and for practical reasons in terms of this chapter's organisation, at a *conceptual* or theoretical level it seems plausible to regard body-worn video and drones as essentially two forms of mobile CCTV camera. As such, conceptually we can perhaps now regard 'traditional' kinds of CCTV camera as just the first instance of what would later become a larger family of CCTV devices. We can therefore distinguish between 'CCTV' as commonly understood today (a camera mounted on a wall or pole, for example) from 'CCTV' the family of technologies. The crowdsourcing example also draws attention to the question of who has access to the footage, asking whether 'CCTV-like' technologies today need to be based on closed-circuits at all. If they do not, then they seem similar in some ways to broadcast or livecast media. Yet mobile cameras, such as are found variously in body-worn video, vehicles or drones, take on a relationship with the social world that is quite different to that found in conventional fixed CCTV cameras. In particular, the properties of mobility and flexibility that they possess (Klauser and Pedrozo 2015: 289) suggests they enable what Klauser has described as 'surveillance on the move' (Klauser 2013).

Conclusion

Gary T. Marx once wryly remarked that, '[t]here are two problems with the new surveillance technologies. One is that they don't work. The other is that they do.' It seems plausible to maintain that today ours is indeed a 'surveillance society', in the sense that numerous aspects of our daily lives are routinely monitored, including by means of visual surveillance technologies. There are numerous potential negative consequences of the use of surveillance technologies in relation to individuals, groups and society as a whole. On the other hand, as we have seen, technologies are not always smoothly implemented and operated and they may fail or be circumvented in various ways. A fear for the future is that the development of visual surveillance technologies leap forward in their capability and power such that they begin to near the visual equivalent of 'information seamlessness' (McGuire 2012: 88) or the 'total surveillance' ideal driving Internet mass surveillance today (see Greenwald 2014). In fact, it seems almost inevitable that various 'smart' and intrusive new visual surveillance technologies will be developed – whether these are nano-drones for covert surveillance, or automatic assemblages of multiple surveillance camera sources into a movie-like edit – and the only questions remaining are whether there is the political will to implement them, and if there is, the degree to which the technologies are to be used in a democratic and accountable fashion. Of course, if (when?) such developments do take place, it will not be in a vacuum, and will unfold against a backdrop of changing societal understandings of privacy and data protection, and of the availability of new camera technologies that ordinary members of the public can acquire and point back towards the authorities. An irony here, however, is that in resisting and challenging authority, sousveillance is in some respects extending not contracting surveillance. Moreover, the problem remains that surveillance is still asymmetric in its operation, the state's power remaining ascendant. Nonetheless, electronic eyes are everywhere now, and gradually all the world will become a cinematic stage; whether we are content merely to be players is up to us.

References

Adey, P., L. Brayer, D. Masson, P. Murphy, P. Simpson, and N. Tixier (2013) 'Pour votre tranquillité': Ambiance, atmosphere, and surveillance, *Geoforum*, 49: 299–309.

Albrechtslund, A. (2008) Online social networking as participatory surveillance, *First Monday*, 13 (3): 2008.

Albrechtslund, A. and P. Lauritsen (2013) Spaces of everyday surveillance: Unfolding an analytical concept of participation, *Geoforum*, 49: 310–316.

Ariel, B., W.A. Farrar, and A. Sutherland (2015) The effect of police body-worn cameras on use of force and citizens' complaints against the police: A randomized controlled trial, *Journal of Quantitative Criminology*, 31: 509–535.

Boersma, K. (2013) Liminal surveillance: An ethnographic control room study during a local event', *Surveillance & Society*, 11 (1/2):106–120.

Bradford, B. and J. Jackson (2016) Enabling and constraining police power: On the moral regulation of policing, in Jonathan Jacobs and Jonathan Jackson (eds), *Routledge Handbook of Criminal Justice Ethics*. London: Routledge.

Brands, J., T. Schwanen, and I. van Aalst (2016) What are you looking at?: Visitors' perspectives on CCTV in the night-time economy, *European Urban and Regional Studies*, 23 (1): 23–39.

Cerezo, A. (2013) CCTV and crime displacement: A quasi-experimental evaluation, *European Journal of Criminology*, 10 (2): 222–236.

Clarke, R. (2014) The regulation of civilian drones' impacts on behavioural privacy, *Computer Law & Security Review*, 30: 286–305.

Clarke, R.V.G. and P. Mayhew (1980) *Designing Out Crime*. London: Home Office Research Unit.

Cornish, D.B. and R.V. Clarke (2003) Opportunities, precipitators and criminal decisions: A reply to Wortley's critique of situational crime prevention, in M.J. Smith and D.B. Cornish (eds), *Theory for Practice in Situational Crime Prevention* (Crime Prevention Studies Volume 16). Cullompton, UK: Willan Publishing.

Cubitt, T.I., R. Lesic, G.L. Myers, and R. Corry (2016) Body-worn video: A systematic review of literature, *Australian and New Zealand Journal of Criminology*. Advance online publication. doi:10.1177/0004865816638909.

Dadashi, N., A.W. Stedmon, and T.P. Pridmore (2013) Semi-automated CCTV surveillance: The effects of system confidence, system accuracy and task complexity on operator vigilance, reliance and workload, *Applied Ergonomics*, (44): 730–738.

Drover, P. and B. Ariel (2015) Leading an experiment in police body-worn video cameras, *International Criminal Justice Review*, 25 (1): 80–97.

Dunphy, P., J. Nicholson, V. Vlachokyriakos, P. Briggs, and P. Oliver (2015) Crowdsourcing and CCTV: The effect of interface, financial bonus and video type, *Newcastle University Technical Report 1453*, www.dunph.com/cctv.pdf.

Ferenbok, J. and A. Clement (2011) Hidden changes: From CCTV to 'smart' video surveillance, in Aaron Doyle, Randy Lippert, and David Lyon (eds), *Eyes Everywhere: The Global Growth of Camera Surveillance*. Abingdon: Routledge.

Finn, R.L. and D. Wright (2012) Unmanned aircraft systems: Surveillance, ethics and privacy in civil applications, *Computer Law & Security Review*, 28: 184–194.

Foucault, M. (1979) *Discipline and Punish: The Birth of the Prison*. Harmondsworth: Penguin.

Goold, B., I. Loader, and A. Thumala (2013) The banality of security: The curious case of surveillance cameras, *British Journal of Criminology*, 53: 977–996.

Greenwald, G. (2014) *No Place To Hide: Edward Snowden, the NSA and the Surveillance State*. London: Hamish Hamilton.

Greer, C. and E. McLaughlin (2010) We predict a riot?: Public order policing, new media environments and the rise of the citizen journalist, *British Journal of Criminology*, 50 (6): 1041–1059.

Hier, S.P. and K. Walby (2014) Policy mutations, compliance myths, and redeployable special event public camera surveillance in Canada, *Sociology*, 48 (1): 150–166.

Jeffery, C.R. (1977) *Crime Prevention Through Environmental Design*. London: Sage.

Jones, R. (2014a) The electronic monitoring of offenders: Penal moderation or penal excess? *Crime, Law & Social Change*, 62 (4): 475–488.

Jones, R. (2014b) The electronic monitoring of serious offenders: Is there a rehabilitative potential?, *Monatsschrift für Kriminologie und Strafrechtsreform*, 97 (1): 85–92.

Klauser, F. (2013) Spatialities of security and surveillance: Managing spaces, separations and circulations at sport mega events, *Geoforum*, 49: 289–298.

Klauser, F. and S. Pedrozo (2015) Power and space in the drone age: A literature review and politico-geographical research agenda, *Geographica Helvetica*, 70: 285–293.

Kroener, I. (2013) 'Caught on camera': The media representation of video surveillance in relation to the 2005 London Underground bombings, *Surveillance & Society*, 11 (1/2): 121–133.

Lyon, D. (2003) *Surveillance as Social Sorting : Privacy, Risk, and Digital Discrimination*. London; New York: Routledge.

Mann, S. (2013) Veillance and reciprocal transparency: Surveillance versus sousveillance, AR glass, lifeglogging, and wearable computing, *IEEE International Symposium on Technology and Society (ISTAS)*, http://wearcam.org/veillance/veillance.pdf.

Mann, S. and J. Ferenbok (2013) New media and the power politics of sousveillance in a surveillance-dominated world, *Surveillance & Society*, 11(1/2): 18–34.

Mann, S., J. Nolan, and B. Wellman (2003) Sousveillance: Inventing and using wearable computing devices for data collection in surveillance environments, *Surveillance & Society*, 1(3): 331–355.

Mathiesen, T. (1997) The viewer society: Michel Foucault's 'Panopticon' revisited, *Theoretical Criminology*, 1(2): 215–234.

McGuire, M. R. (2012) *Technology, Crime and Justice: The Question Concerning Technomia*. Abingdon: Routledge.

Nellis, M. (1991) The electronic monitoring of offenders in England and Wales: Recent developments and future prospects, *British Journal of Criminology*, 31(2): 165–185.

Nellis, M. (2005) Out of this world: The advent of the satellite tracking of offenders in England and Wales, *Howard Journal of Criminal Justice*, 44(2): 125–150.

Nellis, M., K. Beyens, and D. Kaminski (2013) *Electronically Monitored Punishment: International and Critical Perspectives*. Abingdon, UK: Routledge.

Newman, O. (1972) *Defensible Space: People and Design in the Violent City*. New York: Macmillan.

Norris, C., M. McCahill, and D. Wood (2004) Editorial. The growth of CCTV: A global perspective on the international diffusion of video surveillance in publicly accessible space, *Surveillance & Society*, 2(2/3): 110–135.

Norris, C., J. Moran, and G. Armstrong (1998) Algorithmic surveillance: The future of automated visual surveillance, in Clive Norris, Jade Moran, and Gary Armstrong (eds), *Surveillance, Closed Circuit Television, and Social Control*. Aldershot: Ashgate.

O'Malley, P. (2010) 'Simulated justice': Risk, money and telemetric policing, *British Journal of Criminology*, 50 (5): 795.

Parks, L. (2014) Drones, infrared imagery, and body heat, *International Journal of Communication*, 8: 2518–2521.

Parsons, C., J. Savirimuthu, R. Wipond, and K. McArthur (2012) ANPR: Code and rhetorics of compliance, *European Journal of Law and Technology*, 3(3).

Piza, E.L., J.M. Caplan, and L.W. Kennedy (2014) CCTV as a tool for early police intervention: Preliminary lessons from nine case studies, *Security Journal*, April 14: 1–19.

Ravich, T.M. (2015) Courts in the drone age, *Northern Kentucky Law Review*, 42 (2): 161–190.

Sanders, A. and R. Young (2012) From suspect to trial, in Mike Maguire, Rod Morgan, and Robert Reiner (eds), *The Oxford Handbook of Criminology* (5th edition). Oxford: Oxford University Press.

Sandhu, A. and K.D. Haggerty (2015) Policing on camera, *Theoretical Criminology*. Available at: http://tcr.sagepub.com/content/early/2015/12/17/1362480615622531.full.pdf+html.

Simonson, J. (2016 (forthcoming)) Copwatching, *California Law Review*, 104.

Smith, G.D. (2015) *Opening the Black Box: The Work of Watching*. Abingdon: Routledge.

Štitilis, D. and M. Laurinaitis (2016) Legal regulation of the use of dashboard cameras: Aspects of privacy protection, *Computer Law & Security Review*, 32: 316–326.

Talai, A.B. (2014) Drones and Jones: The Fourth Amendment and police discretion in the digital age, *California Law Review*, 102 (3): 729–780.

Taylor, E. (2013) *Surveillance Schools: Security, Discipline and Control in Contemporary Education*. Basingstoke: Palgrave Macmillan.

Trottier, D. (2014) Crowdsourcing CCTV surveillance on the Internet, *Information, Communication & Society*, 17 (5): 609–626.

Tully, J., F. Larkin, and T. Fahy (2015) New technologies in the management of risk and violence in forensic settings, *CNS Spectrums*, 20, Special Issue 03: 287–294.

Wain, N. and B. Ariel (2014) Tracking of police patrol, *Policing: A Journal of Policy and Practice*, 8 (3): 274–283.

Wall, T. and T. Monahan (2011) Surveillance and violence from afar: The politics of drones and liminal security-scapes, *Theoretical Criminology*, 15 (3): 239–254.

Webster, W. (2009) CCTV in the UK: Reconsidering the evidence base, *Surveillance & Society*, 6 (1): 10–22.

Welsh, B. and D. Farrington (2008) Effects of closed circuit television surveillance on crime, *Campbell Systematic Reviews*. www.campbellcollaboration.org/lib/download/243/.

Wright, J., A. Glasbeek, and E. van der Meulen (2015) Securing the home: Gender, CCTV and the hybridized space of apartment buildings, *Theoretical Criminology*, 19 (1): 95–111.

26

Big data, predictive machines and security

The minority report[1]

Adam Edwards

Introduction

In his popular treatment of predictive policing, *The Minority Report*, the science fiction writer Philip K. Dick imagines a fantastic world in which felonies have been reduced by 99.8 per cent. Intelligence for crime prevention is provided to the State's 'Precrime' department by three mutant 'precogs' who possess the talent of foreseeing the future. As with much of Dick's work, the book reflects on the existence of multiple realities and the free will of humans in bringing them about. In this instance, the Head of Precrime, Police Commissioner Anderton, is identified by two of the three precogs as the prospective murderer of one Leopold Kaplan. Anderton suspects Ed Witwer, his ambitious junior colleague, of corrupting the system in order to oust him. To clear his name, he searches for the suppressed minority report of the third precog which provides an alternative vision of Kaplan's murder. Subsequently, Anderton discovers the precog reports are visions of alternative realities, all of which could come about, but such is his commitment to the concept of Precrime that he deliberately chooses to kill Kaplan to corroborate the majority report and maintain the legitimacy of the Precrime department.

Analogies are increasingly being drawn between Dick's short story and current ambitions for the engineering of predictive machines for crime prevention that can harness the intelligence provided by digital technologies, including their capacity to both generate and analyse 'Big Data' (Vlahos 2012). Beneath the general idea of predictive policing, however, Dick's story provokes further reflection on the self-fulfilling prophesies of predictive machines. The argument is familiar in the sociology of deviance and social control in which one of its pre-eminent exponents, Stanley Cohen, poses the problem of the three orders of reality in criminology: the 'thing itself' (crime and the apparatus of control), 'speculations' about this thing (description, classification, causal theory, normative and technical solutions to crime as a 'problem') and 'reflections' on the relationship between the thing itself and speculation about it (Cohen 1988: ix). Specifically, Cohen encourages us to reflect on the role of speculation in constituting the thing itself rather than simply registering its objective reality with greater or lesser accuracy. Criminological speculations constitute social problems in certain ways, predominantly with a view to reforming offenders and addressing the conditions of their offending behaviour, but the rise of 'security studies' is reconstituting these problems as risks that can be calibrated, predicted and then managed.

In this era of security studies, of 'pre-crime and post-criminology' (Zedner 2007), speculation about social problems shifts from the post-hoc investigation and control of offences and offenders to pre-emptive intervention against risks, particularly those thought to present severe threats such as terrorist bombing campaigns, cyber-attacks and the organisation of serious crimes. This is one way of understanding the rise of 'security talk' in contemporary social control and its increasing displacement of the retrospective language of criminal justice and rehabilitation found in criminology. In this era, the quest for legitimate pre-emptive intervention, the justifiable surveillance and control of people who it is anticipated will offend but have yet to do so, is premised on the pursuit of predictive knowledge. Hence the interest of security studies in 'Big Data' as a key source of such knowledge.

Although this field is evolving rapidly, it is already possible to identify three genres of 'reflection' on the role of Big Data and allied emergent technologies, such as machine-learning and the automation of knowledge, in reconstituting problems of social control. 'Enthusiasts', primarily emanating from the research programmes of computational science and artificial intelligence but with interlocutors in criminology, celebrate the governmental powers of digital communication technologies, specifically the particular ways in which they render populations of interest thinkable for the purposes of control. It is argued that hitherto unrecognisable patterns of social relations can be registered, indeed *anticipated*, through reference to digital databases arising out of the translation into digital format of historical administrative data held in paper-based archives, the shift to near/real-time recording of administrative data in digital datasets that can be accessed rapidly and remotely, and access to the explosion of data generated by users of social media services. For enthusiasts, the real governmental power of Big Data is in the capacity for linking data 'harvested' from multiple administrative, commercial and academic databases and then translating this linked data into intelligence on 'the thing itself', crime and the apparatus of control. These processes of data-linkage and translation are, in turn, enabled by the engineering of algorithms and supporting technologies of high performance computing that automate machine-learning about patterns of deviance and social control. In response, it is possible to distinguish 'critics' of either the political and/or technical feasibility of this 'social computing' from 'sceptics' about the limited potential for hybrid human-machine learning. The chapter explores these reflections on the governmental power of Big Data and concludes with some conjectures about its prospective articulation with other emergent technologies in shaping future visions of social control.

Enthusiasts

The concept of Big Data is often associated with Doug Laney's seminal research note on the exponential increase in the Volume, Velocity and Variety (the 3Vs) of data management generated by the revolution in digital technologies for the production, storage, analysis and visualisation of information including the capacity to relate or 'integrate' administrative and commercial data (Laney 2001). The emergence of 'social computing' – given the arrival of 'read/write technologies' (such as blogs, micro-blogs, social networking, wikis, sharing photographic and video materials etc.) on 'Web2.0', the 'interactive World Wide Web' – has provoked further interest in the prospect of integrating such user-generated data with digital administrative and commercial datasets to provide a 'bird's-eye view' of social relations. Computational scientists refer to this in terms of the '10,000 foot view' of 'the social graph' and argue that read/write technologies are just the beginning of an 'age of social machines' (Hendler and Berners-Lee 2010). It is envisaged that these machines will rapidly evolve from a situation in which various read/write applications operate in isolation from one another (an exchange on Facebook, a discussion on Twitter, comments on a broadcast media website, opinions registered through online surveys, the retrieval and annotation of digitally archived police, health, education and census data and so forth) to one in which such applications begin to interact with one another. In this way it is believed that social machines will enable an exponential increase in the kinds of collective intelligence and collaborative work needed to grasp and solve the complexity of social problems that confront us and which are irremediable through individual thought and effort, from climate change through major public health challenges to mobilising local community responses to crime and violence.

Underpinning this enthusiasm for the mobilisation of social machines in collective problem solving is a form of extreme inductive reasoning in which the data 'speaks for itself' once the artificial intelligence needed to get the machines talking to one another has been engineered. The terminus of this reasoning is the provocative concept of 'The Singularity', which is described as 'an era in which our intelligence will become increasingly nonbiological and trillions of times more powerful than it is today—the dawning of a new civilization that will enable us to transcend our biological limitations and amplify our creativity' (Kurzweil 2006). As enthusiasts of social computing note, however, we are only in the embryonic stage of this trajectory, or more accurately this contentious ambition, toward the post-human engineering of self-reproducing and self-regulating social machines.

Even so, there is already fertile speculation about the kinds of security scenarios that could unfold once such faith is placed in artificial intelligence. Enthusiasts for predictive policing (PREDPOL) (Perry et al. 2013) in the United States and for 'prospective crime mapping' (PROMAP) (Johnson et al. 2009) in the United Kingdom have developed algorithms, premised on a 'contagion thesis', which seek to detect when and where crimes will occur by factoring in different kinds of assumptions about how crime spreads from an initial offence in particular environments, given the routine activities and rational calculations of offenders, victims and control agents (Benbouzid 2015). These predictions are then tested against the crime patterns actually registered through conventional methods of police recording and self-report studies of offending and victimisation and then the algorithms are subsequently revised as a means of better anticipating crimes and targeting pre-emptive interventions. What the continued input of human intelligence into this automated 'workflow' will be is less clear, but the implication of the Singularity Thesis is of an ultimate ambition to engineer an artificial intelligence capable of revising its own algorithms and in 'near' if not real-time. The enthusiasm for building predictive machines is now being further extended to design algorithms or 'machine classifiers' to better 'sense' and anticipate patterns of threatening or 'hateful' online communications through social media and to forecast their putative relationship to off-line events such as terror attacks (Burnap and Williams 2015).

Critics

There are predictable criticisms of engineering predictive machines for security applications. In this context, the most notable are those political concerns raised by the whistle-blower Edward Snowden about the massive and routine invasion of privacy through the US National Security Agency's PRISM surveillance programme. PRISM collects communications through the internet, without reasonable suspicion, and then mines them for intelligence on, and forecasting about, various security threats including terror plots and illicit drugs markets (Lyon 2014). The Snowden revelations suggest how the inductive reasoning of building an understanding of security out of Big Data necessarily contravenes the right to private communications because the 10,000 foot view of the 'security graph' cannot be accomplished without generalised data collection from whole populations.

There is the related concern, the essential message of Dick's *Minority Report*, that predictive machines generate self-fulfilling prophesies. They can become active ingredients in the targeting of suspects, including entire populations, such that problems of security become artefacts of the way in which algorithms, machine classifiers and their underlying assumptions speculate about security, including how they establish the parameters of any security graph to include certain concerns (e.g. speech about 'radicalised' Muslim youth) whilst obviating others (e.g. speech about the culpability of Western foreign policy in the Middle East). In this regard, targeting the usual suspects ceases to be just a consequence of episodic prejudicial police actions and becomes automatically reproduced by a social machine. The alienation of entire social groups as a consequence of this kind of group profiling and targeting, along with the creation of a policing environment conducive to miscarriages of justice, is an established theme in critical criminology, particularly in the UK with reference to the war in Ireland and the long history of antagonism between the police and street populations, particularly of young males from minority social groups (Pantazis and Pemberton 2009; Hallsworth and Lea 2011). However, the consequences of automating this policy failure are only just beginning to be appreciated (Chan and Bennett Moses 2016).

Less prominent are criticisms of the technical feasibility of using Big Data for the predictive machine engineering of security applications. The idea, epitomised in the fantasy of *The Singularity*, that machines can be effectively trained to think like humans and, ultimately, in ways that can transcend human thought, is a long-running theme in the study of artificial intelligence. An important rebuttal of this foundational belief argues there are only a few instances in which machines can be programmed to effectively mimic human actions whilst there are many of these actions that machines cannot accomplish and, crucially, will never be able to accomplish. This argument distinguishes between 'mimeomorphic' actions, which 'we either seek to or are content to carry out in pretty much the same way, in terms of behaviour, on different occasions' and 'polimorphic' actions which are all other human actions (Collins and Kusch 1998: 31). In these terms, machines can be trained to undertake mimeomorphic actions, such as swinging a golf club or dialling a telephone number, but are colossal failures at polimorphic actions, such as writing love letters or the subversion of factory work routines, because they lack the core human qualities of empathy and improvisation. Unless the behaviour being predicted is carried out the same way each time it is performed then a machine will not be able to learn how to perform, much less anticipate, it (Ibid.: 37).

An exemplar of this limitation is the software firm Microsoft's very short-lived experiment in creating a 'chatbot' on the micro-blogging site, Twitter, in which other users of this site were invited to interact with an artificial intelligence called 'Tay'. On releasing Tay, Microsoft argued that, 'The more you chat with Tay, the smarter she gets, so the experience can be more personalised for you'. After only 24 hours, however, human employees of Microsoft were already editing some of the more inflammatory comments from Tay, which repeated and amplified racist sentiments, Nazi sympathies and expressions of support for genocide, having learnt these through interactions with unscrupulous users of Twitter (Wakefield 2016).

From this perspective, the potential success or colossal failure of predictive policing hangs on the question of how mimeomorphic problems of security are. Are certain types of crime, such as the organisation of crystal methamphetamine production, distribution and exchange, sufficiently 'scripted' (Chiu *et al.* 2011) to be predictable in terms of when, where and how they are performed? How improvised are property crimes such that their patterns and modus operandi escape anticipation? How 'golf-swing-like' is crime? It has been argued, for example, that improvisation is the central dynamic of much crime, particularly sophisticated organised crimes, in which perpetrators and preventers are in an ongoing correspondence, in this case an 'arms race' rather than an amorous exchange, to outflank and outwit each other (Ekblom 2003; Dorn 2003).

Sceptics

Sceptics provide yet a further take on the feasibility of predictive machines for security, central to which are the possible interactions of human and machine learning and their regulation, including the design of hybrid human and machine learning approaches.[2] Commissioning crimes and other social relations may not be akin to swinging golf clubs but they may be sufficiently scripted to be predicted, in part, by automated learning. They may, to continue the analogy, be more like performances of a play in which the actors improvise around the script but still rehearse their lines and do not completely rewrite the story from one performance to another. How these scripts and their narrative structures can be registered is the subject of current methodological argument and innovation in 'digital social research' (Edwards *et al.* 2013; Housley *et al.* 2014).

At the core of this argument is the claim that computational methods of artificial intelligence, such as machine-learning, may augment more conventional social research methods, such as surveys, interviews, analyses of conversations and participant observations, but can never replace them as surrogates, much less as superior forms of intelligence about the social. Notwithstanding the low fidelity of much 'Big Data', particularly that associated with social media communications where it is often difficult to establish who is communicating with whom and from where, and the commensurability of digital datasets collected over different spatial and temporal horizons, the improvised qualities of social life will always require human input to refresh the algorithms driving predictive machine learning. Therefore, the sceptics' case for this learning rests upon the possibilities for collaboratively designing and refreshing algorithms involving the central contribution of social scientists and their conventional methods and concepts for researching the language, accomplishment and enacted environment of social relations. What is at stake is the consonance of the scripts of the actors in question, those of the social and computational scientists interested in these actors and finally those of the predictive machines they collaborate in engineering. Such is the likely fluidity and dissonance of these scripts in the predictive game that it would be wise to heed Amartya Sen's advice to content ourselves with speculation rather than specious precision, with being 'vaguely right, rather than precisely wrong'. Is vague prediction a contradiction in terms, amounting to a colossal failure of artificial intelligence or does it suffice to produce a high confidence level of agreement, if not an exact match, between the scripts of actors, (human) researchers and machines, particularly if this level of agreement is routinely tested?

An early exemplar of this more sceptical approach to artificial intelligence sought to indicate tension in social media communications (Williams *et al.* 2013; Burnap *et al.* 2015). Tension indication is the predictive policing problem *par excellence*, with its roots in attempts by British police forces to better anticipate the civil unrest experienced in major English cities following the riots of spring and summer 1981. Twenty years on from these initial disturbances the completely subjective character of this forecasting, reliant primarily upon neighbourhood police officers' perceptions and informants, resulted in the misdiagnosis of tensions preceding the riots in former mill towns in northern England in summer 2001. These were diagnosed as turf wars over the street drugs trade rather than, as was subsequently established, the reaction of South Asian communities to the mobilisation of the far right British National Party in their neighbourhoods (King and Waddington 2004). A further ten years on, the forecasting intelligence available to police forces again failed to anticipate the scale and rapid spread of civil unrest in large English cities during August 2011, this time fuelled by the use rioters made of smartphones and social media networks to co-ordinate their activities and outflank public order policing for the initial days of the rioting.

In this context, research has questioned the possibility of indicating tension amongst online communities discussing off-line social conflicts as a first step to training a predictive machine capable of indicating tension around certain culturally specific conflicts and events. The research used the expression of racism in Twitter posts about English professional football as a case study and made some progress in 'training' an algorithm capable of detecting communications which humans had similarly rated as expressing tension in terms of antagonistic, abusive, speech. The measure of this 'inter-coder reliability' identified agreement between the four police officers used as human coders of the abusive tweets, taken to indicate 'high tension', and the 'tension-monitoring' machine in 68 per cent of cases (William et al. 2014: 474). However, it must be emphasised that this algorithm was built using a particular approach in the sociology of language, that of Membership Categorisation Analysis (Housley and Fitzgerald 2009), which relies upon the use of culturally specific knowledge in associating particular subjects with particular events and certain predicates, in this instance social media communications about the racial abuse of one footballer, Patrice Evra, by another, Luis Suarez. This approach is some way from golf-swing-like replication but it can, plausibly, claim to have detected the clear majority of racist scripts amongst a blizzard of social media communication, albeit in a very specific cultural context. How generalisable this approach would be to other social problems in other contexts is a moot point about which we are still in the very early stages of learning. The implication is that bespoke algorithms need to be designed for the context-specific problem of security in question. From a sceptical point of view, however, this kind of hybrid human-machine learning re-orientates the debate about artificial intelligence away from strict dichotomies between either mimeomorphic or polimorphic action towards argument over the human refinement of machines capable of mimicking human action in a clear majority of cases.

In addition to the culturally specific focus of the tension-monitoring machine built by the Collaboratie Online Social Media ObServatory (COSMOS) (Williams et al. 2013), its focus was also retrospective, seeking to predict if humans and machines could agree enough about the abusive characteristics of communication about an event that had already happened (the racist abuse of Evra by Suarez). Whether an inter-coder agreement of 68 per cent could be maintained by prospective predictive machines is an even more contentious ambition, even if the goal was to anticipate the scripts of the same actors (Evra and Suarez) acting in the same context (a derby match between the English Premiership football teams Manchester United and Liverpool). Logically, the more generalised the script (racism in football, racism in sport, racism in England etc.), the less precise the sociologically informed algorithm and the greater the likelihood of it misattributing 'high tension' in digital communications as would the Evra–Suarez algorithm if it were applied to other instances of racism in English professional football, let alone other problems and contexts of policing. Hence the scepticism about fantasies of self-reproducing and self-regulating social machines, much less the accomplishment of *The Singularity*, because any prospective predictive machine worthy of the name would have to be built and refreshed through culturally specific human intelligence about plausible scripts (e.g. abuse, including racist slurs, at forthcoming derby matches between rival football teams). In this context, predictive policing could anticipate tension and its escalation at key 'fixtures' not just the fixture list of professional football for the forthcoming season but other predictable fixtures on the cultural calendar (for example alcohol-related violence on weekend nights or during festive seasons, conflict at planned political demonstrations or during industrial disputes and so forth). Again, the implication is that social life is scripted and not utterly spontaneous, notwithstanding room for improvisation, and therefore amenable to some prediction (we know how *King Lear* ends, although there is always interpretative flexibility in Shakespeare's script for why it ends the way it does).

The centrality of culturally-specific human intelligence to the collaborative design of algorithms provokes discussion of 'reflexive securitisation' and its potential role in hybrid human-machine learning about policing. This concept is associated with the scepticism of Tim Hope's (2006) response to the first generation of predictive policing programmes, such as PREDPOL and PROMAP, specifically his criticism of their neglect of the non-obvious, 'latent', variables affecting the spread of security problems, such as volume, personal and property crime. As this first generation has focused solely on the actual frequency of victimisation they ignore the crucial counterpart, the distribution of non-victimisation and therefore the interaction of victimisation and non-victimisation. Employing innovations in Bayesian statistical modelling (a Latent Class Analysis) of domestic burglary in England and Wales, Hope revealed the gross inequality in the distribution of victims and non-victims, estimating that 80 per cent of the residential population experienced only 20 per cent of this crime. Conversely, a fifth of the population experienced four fifths of this crime, the renowned '80:20' split. The pattern of domestic burglary is consequently identified as 'zero inflated' in that most of the population experience no burglary at all. Hope's conjecture is that this is related to the ability of the majority of the population to 'immunise' themselves against burglary through access to various private 'club goods', such as commercial household security, the market value of their homes as a proxy for the segregation of the residential population into wealthier, less criminogenic, neighbourhoods and poorer, more predatory, neighbourhoods and, within such segregation, participation in 'gated communities' with enhanced security surveillance and patrols. The lack of access to such private goods leaves the fifth of the residential population who are chronically victimised dependent on public security goods, such as police forces and victim support agencies degraded through austerity programmes or the good will of self-mobilising community activists, which further weakens their immunity. In summary, notwithstanding some revolution in social inequalities, continuation of the 80:20 script is predictable in the current English and Welsh context and would, presumably, be registered by any algorithm premised on Hope's Bayesian model.

In these terms the significance of access to the 'Big Data' of digitalised administrative, commercial and user-generated data sets is in the potential to reveal further latent variables that may revise the 'immunisation thesis' and, therefore, a further refinement of predictions of where and when certain security problems will manifest. In addition to the prospective revelation of other latent variables (in addition to housing market dynamics for example), access to Big Data has the potential to refine the temporal and spatial parameters of this prediction, moving, for example, beyond annual accounts of victimisation in England and Wales to predictions of how types of victimisation are distributed, if not in 'real-time', then in quarterly, monthly and so on timeframes, at the level of cities and neighbourhoods etc. Again, though, central to this reflexive securitisation is the human intelligence of social scientists revising the predictive machine through reference to their domain expertise, in this instance, sociological knowledge about immunity and predation in the pattern of property crimes in the context of English and Welsh housing markets.

As such, the spatial and temporal parameters of the prediction are driven by fallible, revisable, human speculations, not self-regulating social machines. If, for example, the 80:20 script is corroborated as a relatively stable, durable, pattern reflecting immunisation at the neighbourhood level, the decision to pursue even more granular data on patterns of distribution by week, day or minute in certain streets, households and individuals reflects fallible arguments amongst humans, rather than machines, about the relative merits of sociological and psychological accounts of victimisation. To reiterate Cohen's argument, these are competing speculations that constitute problems of security as those of individual or collective, psychological or sociological, dynamics rather than unproblematic truths that can be induced from the data itself.

A minority report?

In relation to the enthusiastic embrace of social computing in predicting security problems and the critical dismissal of this as both technically and politically unfeasible, the sceptics provide a minority report. Sceptics offer an altogether messier and less certain reflection on the limits to hybrid human–machine learning but one that is irreducibly driven by the humans in constituting problems of security not simply registering objective truths. There are also grounds for scepticism about the integration of variegated data sets composed of material collected over hugely varying temporal and spatial horizons. There may be opportunities for recomposing this data in ways that enable it to be meaningfully linked but even where relatively robust administrative data sets are concerned this entails a substantial input from human intelligence. This necessarily compromises the automation of knowledge that is at the heart of both enthusiastic claims for predictive policing and critical fears about the efficacy of mass surveillance. Whether and how the lower fidelity data generated by users of social media, often anonymously and with limited demographic detail about who these users are and where they are from, can be meaningfully linked to other administrative and commercial data, as well as the primary data sets produced by social scientists, remains a very challenging, possibly insurmountable, methodological problem.

Even if technically feasible, and this is a very big 'if', there are genuine ethical and political concerns about engineering predictive machines capable of collating person-specific data from multiple sources in order to circumvent controls on the anonymity of such data, and thus the profiling and monitoring of 'risky' individuals and groups. This is particularly so where predictive machines could be deployed within any governing regime, not just in liberal democracies with lively, open, debates about 'snooper's charters' and Orwellian objections to Big Brother. At best we are in a situation that requires deliberation about the levels of confidence inspired by the technical feasibility of predictive machines and then about the appropriate regulatory frameworks for governing the access to and use of such machines by different statutory, commercial and not-for-profit groups as well as private citizens. The latter point is especially pertinent because Orwellian imagery and concerns about panoptic surveillance ignore the synoptic powers of digital technologies (Mathieson 1997; Doyle 2011), in which the few are more readily held to account by the many, given popular access to read/write technologies including the real-time capacity to stream video content, say of public order policing operations, to the internet or to distribute smartphone applications that enable counter-surveillance. More pressing than either of these visions of elite or popular social control, however, are concerns over the polyoptic powers of read/write technologies in which the many can watch and abuse the many whilst outflanking the state and its regulatory regimes (Webb et al. 2015).

Conclusion: emergent technologies, crime and justice

To revisit the imagery of Phillip K. Dick and the ontology of Stanley Cohen, this minority report on the governmental powers of Big Data has broader implications for a Handbook on *Technology, Crime and Justice*. Dick and Cohen provoke reflection on the multiple realities of social control, specifically whether 'the thing itself' is *only* an artefact of the many descriptions, classifications, causal theories, normative and technical solutions to crime as a 'problem' or whether there is some singular, concept-independent reality there to be discovered through methodological innovation. The latter presumption is explicit in much of the computational contribution to criminology but, it must be said, in a naïvely realist attempt to better 'sense' crime through triangulating administrative, specifically police-recorded, data with the notoriously low-fidelity data on crime and civil unrest that is generated by users of particular social media platforms (Williams *et al*. 2016). Judgement of a realist approach that is more critical of its own conceptual interpretation, as well as its methodological and empirical register, of crime must await the advent of research employing a broader repertoire of empirical sources to adapt rival theses on crime and justice.

Such an approach is being developed in the 'Justice Matrix' project[3] which simulates the process of criminalisation in England and Wales using a combination of big data sources including social, economic and demographic data on the UK population, taken from the Census, criminal justice data on victimisation and recorded crime, arrests and prosecutions, convictions and sentencing, and combines these with data on public attitudes, policy priorities and agendas and also with data on criminal justice institutions, including staffing levels, physical infrastructure and expenditure. Significantly, the Justice Matrix project seeks to break with linear models of explanation that test privileged theories (crime is an artefact of control, crime is a product of social inequality etc.) providing, instead, a virtual environment in which rival theses on the *interaction* of crime and justice can be tested and built. Here the aspiration is akin to the judgemental rationalism of a more critical realism, that acknowledges the interpretative dimension to social relations and to the investigation of these relations (the 'double hermeneutic'), whilst insisting that rival interpretations must still be subject to procedures of corroboration and refutation (Sayer 2000). In this regard, the Justice Matrix project enrols the emergent technologies of Big Data into a framework for thinking through the interrelationship of theory, method and evidence in ways that facilitate an understanding of how interactions between crime and justice can be reconstituted, imagined otherwise in future, as well as better registered in the present.

Understanding the prospective impact of emergent technologies on crime and justice can also benefit from broader sociological accounts of digital futures. These alert us to the role of emergent technologies in actively constituting, not just registering or 'imagining', social orders. Housley (2015) discusses the 'emerging contours of data science' in which the 'embedding of automation and computation into social life will have significant unintended consequences'. If Ekblom's thesis on the technological 'arms race' between perpetrators and preventers of crime holds true, it is possible to anticipate a number of plausible, if unintended, consequences of enrolling emergent technologies like machine-learning into the interaction of crime and justice. Chief amongst these must be the capacity of resourceful criminal enterprises to employ their own surveillant algorithms against public authorities to model, monitor, anticipate and circumvent control strategies and further insulate themselves from 'disruption', let alone prosecution. The idea that emergent technologies can be used to resist and disrupt public authority as well as challenges to this authority, and to commercial security, can be anticipated in other arms races around, for example, 3-D printing in the unregulated production of firearms, ammunition, ordnance and other facilitators of organised criminality as well as the uses of drone technology for counter-surveillance or advanced robotics for commissioning crimes and insulating humans from prosecution (Edwards 2016).

Yet more fanciful future visions of social control can be found in the burgeoning science fiction about artificial intelligence, in which predictive machines communicate with drones and military robotics to outflank and ultimately destroy humankind in ruthless calculations about the pestilent threat humans pose to the digital ecosystem (Hertling 2011, 2013, 2014, 2015). An antidote to both this kind of subjective futurology and the present-centred orientation of much 'computational criminology', is the renaissance of forecasting methodologies that cultivate collective and reflexive human intelligence (Vander Beken and Verfaillie 2010; Edwards, Hughes and Lord 2013). These can accommodate and stimulate deliberation amongst a breadth of theory, method and data but do so precisely because they privilege the collective intelligence of humans over the artificial intelligence of machines. In this sense, the reconstitution of social control as 'security' might aspire to be 'pre-crime' but it cannot be post-human, if it is to keep pace with the context-specific creativity of human beings in adapting technologies to their own ends, nor 'post-criminological', if it is to grasp the interactional dynamics of crime and (social and restorative as well as narrowly criminal) justice.

Notes

1 This is an expanded and revised version of an earlier article that appeared in the on-line sociological magazine, *Discover Society* (2015).
2 See those trialled by the Collaborative Online Social Media ObServatory (COSMOS) www.cs.cf. ac.uk/cosmos/.
3 www.crimeandjustice.org.uk/project/justice-matrix.

References

Benbouzid, B. (2015) From situational crime prevention to predictive policing, *Champ pénal/Penal field*, Vol. XII: Available at: http://champpenal.revues.org/9066.

Burnap, P., Rana, O., Avis, N. Williams, M. L., Housley, W., Edwards, A., Sloan, L. and Morgan, J. (2015) Detecting tension in on-line communities with computational Twitter analysis, *Technological Forecasting and Social Change*, 95: 96–108.

Burnap, P. and Williams, M.L. (2015) Cyber hate speech on Twitter: An application of machine classification and statistical modeling for policy and decision making, *Policy and Internet*, 7(2): 223–242.

Chan, J. and Bennett Moses, L. (2016) Is Big Data challenging criminology?, *Theoretical Criminology*, 20(1): 21–39.

Chiu, Y.-N., Leclerc, B. and Townsley, M. (2011) Crime script analysis of drug manufacturing in clandestine laboratories: Implications for prevention, *British Journal of Criminology*, 51: 355–374.

Cohen, S. (1988) *Against Criminology*. New Brunswick, Transaction Books.

Collins, H.M. and M. Kusch (1998) *The Shape of Actions: What Humans and Machines Can Do*. Cambridge MA: The MIT Press.

Discover Society (2015) Big Data, predictive machines and security: Enthusiasts, critics and sceptics. Available at: http://discoversociety.org/2015/07/28/big-data-predictive-machines-and-security-enthusiasts-critics-and-sceptics/.

Dorn, N. (2003) Protieform criminalities, in A. Edwards and P. Gill (eds) *Transnational Organised Crime: Perspectives on Global Security*. London: Routledge.

Doyle, A. (2011) Revisiting the synopticon: Reconsidering Mathiesen's 'The Viewer Society' in the age of Web2.0, *Theoretical Criminology*, 15(3): 283–299.

Edwards, A. (2016) Multi-centred governance and circuits of power in liberal modes of security, *Global Crime*, 17(3–4): 240–263.

Edwards, A., Housley, W., Williams, M.L., Sloan, L. and Williams, M. (2013) Digital social research, social media and the sociological imagination: Surrogacy, augmentation and re-orientation, *International Journal of Social Research Methodology*, 16(3): 245–60.

Edwards, A., Hughes, G. and Lord, N. (2013) Urban security in Europe: Translating a concept in public criminology, *European Journal of Criminology*, 10(3): 260–83.

Ekblom, P. (2003) Organised crime and the conjunction of criminal opportunity framework, in A. Edwards and P. Gill (eds) *Transnational Organised Crime: Perspectives on Global Security*. London: Routledge.

Hallsworth, S. and Lea, J. (2011) Reconstructing Leviathan: Emerging contours of the security state, *Theoretical Criminology*, 15(2): 141–157.

Hendler, J. and Berners-Lee, T. (2010) From the Semantic Web to social machines: A research challenge for AI on the World Wide Web, *Artificial Intelligence*, 174: 156–161.

Hertling, W, (2011, 2013, 2014, 2015) *The Singularity Series*. Portland: Liquididea Press.

Hope, T. (2006) Mass consumption, mass predation – private versus public action? The case of domestic burglary in England and Wales, in Lévy R., Mucchielli L., Zaubermann R. (eds) *Crime et insécurité : un demi-siècle de bouleversements. Mélanges pour et avec Philippe Robert*. Paris: L'Harmattan, 46–61.

Housley, W. (2015) Focus: The emerging contours of data science, *Discover Society*, http://discoversociety. org/2015/08/03/focus-the-emerging-contours-of-data-science/.

Housley, W. and Fitzgerald, R. (2009) Membership categorization, culture and norms in action, *Discourse and Society*, 20(3): 345–362.

Housley, W., Procter, R., Edwards, A., Burnap, P. Williams, M.L., Sloan, L., Rana, O., Morgan, J., Voss, A. and Greenhill, A. (2014) Big and broad social data and the sociological imagination: A collaborative response, *Big Data & Society*, April–June 2014: 1–15.

Johnson S., Bowers K., Birks D. and Pease K. (2009) Predictive mapping of crime by ProMap: Accuracy, units of analysis, and the environmental backcloth, in D. Weisburd (ed.) *Putting Crime in its Place*. New York: Springer.

King, M. and Waddington, D. (2004) Coping with disorder? The changing relationship between police public order strategy and practice: A critical analysis of the Burnley Riot, *Policing and Society*, 14(2): 118–137.

Kurzweil, R. (2006) *The Singularity is Near: When Humans Transcend Biology*. London: Penguin.

Laney, D. (2001) 3D data management: Controlling data volume, velocity and variety, *Application Delivery Strategies*, File 949, Meta Group Inc., Available at: http://blogs.gartner.com/doug-laney/ files/2012/01/ad949-3D-Data-Management-Controlling-Data-Volume-Velocity-and-Variety.pdf.

Lyon, D. (2014) Surveillance, Snowden, and Big Data: Capacities, consequences, critique, *Big Data and Society*, July–December 2014: 1–13.

Mathieson, T. (1997) The viewer society: Michel Foucault's 'Panopticon' revisited, *Theoretical Criminology*, 1(2): 215–234.

Pantazis, C. and Pemberton, S. (2009) From the 'old' to the 'new' suspect community: Examining the impacts of recent UK counter-terrorism legislation, *British Journal of Criminology*, 49: 646–666.

Perry W.L., McInnis B., Price C.C., Smith S.C. and Hollywood J.S. (2013) *Predictive Policing: The Role of Crime Forecasting in Law Enforcement Operations*. Rand Corporation Report.

Sayer, A. (2000) *Realism and Social Science*. London: Sage.

Vander Beken, T. and Verfaillie, K. (2010) Assessing European futures in an age of reflexive security, *Policing and Society*, 20(2): 187–203.

Vlahos, J. (2012) The department of pre-crime, *Scientific American*, 306(1): 1–9.

Wakefield, J. (2016) Microsoft chatbot is taught to swear on Twitter, BBC News online, Available at: www.bbc.co.uk/news/technology-35890188, accessed 24 March 2016.

Webb, H., Jirotka, M., Carsten Stahl, B., Housley, W., Edwards, A., Williams, M., Procter, R., Rana, O. and Burnap, P. (2015) Digital wildfires: Hyper-connectivity, havoc and a global ethos to govern social media, *Computers and Society*, 45(3): 193–201.

Williams, M.L., Edwards, A., Housley, W. Burnap, P., Rana, O., Avis, N., Morgan, J. and Sloan, L. (2013) Policing cyber-neighbourhoods: Tension monitoring and social media networks, *Policing and Society*, 23(4): 461–481.

Williams, M., Burnap, P. and Sloan, L. (2016) Crime sensing with big data: The affordances and limitations of using open source communications to estimate crime patterns, *British Journal of Criminology*, Available at: http://orca.cf.ac.uk/id/eprint/87031.

Zedner, L. (2007) Pre-crime and post-criminology?, *Theoretical Criminology*, 11(2): 261–281.

27

Cognitive neuroscience, criminal justice and control

Lisa Claydon

Introduction

The aim of this chapter is to consider how our developing knowledge of cognitive neuroscience and its associated technologies may impact upon the criminal justice system. This chapter examines the claims that modern understandings of how the brain drives behaviour, based on new insights from cognitive neuroscience and neurobiology, will challenge certain tenets of the criminal law and the criminal justice system. In doing this it will be useful to examine the work of academic researchers who have been actively carrying out funded research in this area. This will assist in the process of scoping how future advances in our scientific understanding of the brain may challenge the criminal law. Additionally, it will be useful to assess the activities of learned societies and governmental bodies in reviewing the area. This will aid an understanding of how to evaluate the way in which academic and government funded research in the area of neuroscience and neurotechnology may impact upon and influence future developments in the criminal justice system.

For those readers who have not encountered this area of law before, it might be helpful to reprise some of the claims that have been made by academics regarding the impact of new cognitive neuroscientific understandings upon the law, and the criminal justice system. The introduction here will be necessarily brief but the analysis will continue throughout the chapter as areas of particular significance to future developments in the criminal legal system are examined.

A good place to start this review is with the work of Stephen Morse, an American legal academic who is well known for making the argument that criminal law presupposes and accepts a 'folk psychological' view of human behaviour. Morse explains what this folk psychological approach entails, and how it differs from a more determinist scientific explanation of behaviour. Folk psychology considers 'mental states' as 'fundamental to a full causal explanation and understanding of human action' (2011: 25). Part of the argument Morse makes against the adoption of neuroscientific understandings of human behaviour is that the normative requirements of law presuppose a conscious actor who 'forms and acts on intentions'. From this standpoint more behaviourist explanations of how behaviour results from brain states will not assist the courts. Whilst Morse accepts that it is the brain that generates these experiences, the basic argument Morse makes is straightforward: 'The law treats people generally as intentional creatures and not simply mechanistic forces of nature' (2011: 25). This for Morse is one of the driving justifications for his call for 'neuromodesty'. For Morse law is normative and many of its norms are folk psychological in form. Morse argues forcefully that this should lead us to treat claims that advances in cognitive neuroscience will challenge the criminal law with scepticism. Morse's arguments are aimed at the structure and function of the criminal law, particularly Anglo-American criminal law. Additionally, he argues that nothing that neuroscience has produced by way of empirical research, so far, provides a challenge for the law. So his comments may be viewed as related to the structure and function of the law and what should be argued in court rather than broader issues in the criminal justice system concerning when interventions should be made prior to a criminal act taking place.

A slightly different view is expressed by Goodenough and Prehn (2006), who make the argument that cognitive neuroscience may add a depth to our understanding of how and why we hold people to blame for their criminal actions. They also point out that the scope of the interdisciplinary exercise to gain some understanding of the relationship between brain and blame is daunting to the lawyer. The disciplines such work encompasses are diverse. Goodenough and Prehn also suggest that, at the very least, some understanding of the work of cognitive neuroscience, psychology, philosophy, evolutionary biology, psychology and psychiatry is required (2006: 77). The focus of part of their discussion is upon the dichotomy posed by 'distinguishing between the reason based dictates of law and an intuition based sense of justice' (2006: 80). They argue that advances in cognitive neuroscience will allow us 'to reconsider our theories of normative judgment and apply new tools to its study' (2006: 84).

This approach clearly differs from that of Morse. It does not deny that legal judgments are normative but rather suggests that as we learn more about behavioural aspects of actions we may amend our present normative approaches to blaming people for their actions. Neuroscientists tend to take an altogether different approach. Many would argue that the focus of their scientific work has nothing to do with issues of responsibility. They would argue that the nature of the relationship between action and blaming is one for moral philosophers, not for scientists. Colin Blakemore writing in 1988, about the issue of whether individuals have free will, suggested that the problem came with trying to mix 'the judicial system with medical science'. He identified that problem as being one of language, arguing that science has no place for the distinction of right from wrong, giving the following example:

> No one would try to assign *responsibility* for the fact that the earth orbits the sun. Nor is it the job of science to decide whether people are responsible for their actions that society judges to be wrong.
>
> *(Blakemore 1988: 270, emphasis in original)*

This is a moot point. The responsibility for reaching judgment in a criminal case clearly rests with the court. However, accepting Blakemore's words at face value, if the scientist is a determinist, the meaning to him of the evidence that he gives may well be at variance from the meaning attributed to what he says by all non-scientifically trained people in the court room. A scientist might hold the straightforward view expressed succinctly by Colin Blakemore that:

> All our actions are products of the activity of our brains. It seems to me to make no sense (in scientific terms) to try to distinguish sharply between acts that result from conscious intention and those that are pure reflexes or that are caused by disease or damage to the brain. We *feel* ourselves, usually, to be in control of our actions, but that feeling itself is the product of the brain, whose machinery has been designed, on the basis of its functional utility, by means of natural selection.
>
> *(1998: 270, emphasis in original)*

This is indeed a different language from the folk psychological explanation described by Morse. The challenge for the law, if scientists wish to express their views about actions, is that it is likely that scientists will be speaking a different language from the rest of the courtroom. If an expert was really talking a totally different language, Spanish, German or French, then the court would expect some sort of translation to take place. An understanding of the language used by expert witnesses would facilitate an evaluation of the blameworthiness of the accused's actions. Judges are not scientists and the prosecution and defence, whilst they and the expert owe their first duty to the court to achieve a just outcome, may not be able to convey the real meaning of the science to the jury. It seems therefore, that at least at a minimal level, a shared understanding of some of the evaluative scientific criteria will be required. How this is to be achieved whilst retaining objective neutrality is an interesting point.

These comments concern the use of neuroscientific findings in relation to individual actions, but neuroscientific understanding and empirical evidence may also be used to inform policy formation in the area of criminal justice, for example when planning interventions to prevent anti-social behaviour, in working out how to assess racial bias of key personnel in the system, and measuring predispositions to violent criminal acts.[1] Interventions may take place to avoid an individual entering the criminal justice system or to prevent an individual from reoffending. Moral questions as to the rightness or wrongness of the intervention need to be addressed at the point of policy formation. A clear understanding of the scientific basis for making the intervention will be required by those advocating its use, particularly where the intervention might be based on more speculative neurocognitive experimental designs. Some claims made by neuroscientists, as previously identified by Stephen Morse, do seem to embody neuro-exuberance. An example of this is the claims made by researchers at North West University to be able to retrieve concealed memories from terrorist suspects using the P300 signal identified by the use of electrode technology.[2] The technology is based on EEG and the claim is that the signal reveals concealed knowledge.[3]

How can folk psychological understandings comprehend claims for technologies that appear to read minds? Particularly when faced with claims which 'over-hype' the potential of such technologies. Often this overselling results from the most positive gloss being placed upon research findings; added to this there is the commonplace method of press reporting that presents stories in a manner that grabs public attention.[4] This means that speculative ideas drawn from research by cognitive neuroscientists, behavioural geneticists and neurocriminologists are often presented to the public as having an accepted basis in science, and therefore may directly enter the 'folk psychology' belief systems held by jurors and indeed other actors within the system. The permeability of folk psychology to influence by scientific speculation or tides of national sentiment makes it a problematic barometer of right and wrong. This is not to dispute that the jury will make its decisions against the background of the prevailing perceptions of the society in which the judgment takes place as to what is right or wrong. It is to make a further assertion, that is, that those giving and hearing evidence in court therefore bear a heavy responsibility. The responsibility is to make sure that the translation of the ideas of science into words that the court can understand is able to inform the jury's decision about guilt and innocence, in a meaningful and appropriate manner, in addition to being both an accurate and objective description of the scientific findings.

Novel technologies and how they might be applied

It would be useful prior to undertaking this review to look at a general definition of cognitive neuroscience and give a brief explanation of the associated technologies. Cognitive neuroscience is described by Goodenough and Prehn (2006: 84) as:

> an approach that seeks to integrate into the study of human thought, our rapidly emerging knowledge about the structure and functions of the brain, and about the formal properties of agents and decision-making processes ... Although cognitive neuroscience was well launched before the advent of such imaging technologies as PET and fMRI, the availability of non-intrusive methods that allow us to establish functional connections between mental tasks and specific anatomical structures has increased its power and accelerated its application.

In 2011 The Royal Society published the first of its policy documents on neuroscience, society and policy (The Royal Society 2011a). The document assessed different types of neuroimaging and reviewed the use of the technology to date. Reviewing these non-invasive ways of looking at the brain the review ranges from technologies such as computer tomography (CT) scans which use X-ray technology to look at the structure of the brain to magnetic resonance imaging (MRI). MRI is described as the most common form of neuroimaging, having the advantage that it permits greater understanding of the anatomical structure of the brain. Diffusion-weighted MRI is described as enabling scientists to 'visualise' connections within the brain (ibid.: 8). Functional magnetic resonance imaging (fMRI) commonly measures blood oxygenation levels in the brain to look for evidence of functional activity. It is described as an 'indirect measure of neural activity, through the effect of changes in local blood flow in the brain' (ibid.: 9). The technique's limitations are noted in that its temporal resolution is slow and therefore it does not accurately capture rapid changes in brain activity.[5]

Additional techniques that are relevant are electroencephalography (EEG) and magnetoencephalography (MEG) which are not viewed as neuroimaging techniques, but are relevant techniques to this discussion. These technologies measure brain activity by using electrodes placed on the scalp. They differ from the other techniques in directly measuring, at the level of the scalp, neuronal activity. All the techniques described briefly here have shortcomings in terms of what they reveal about the activity within the brain. MRI is said to be the best technique for gaining structural information, fMRI for spatial resolution but having real limitations in terms of assessing when precisely the activity noted occurred. Positron emission tomography (PET) is seen as limited because it needs to be used with a radioactive tracer molecule. This is expensive and the technique is said to be inflexible, there being a limit to the number of scans that may be taken. MEG and EEG are seen also as having poor spatial resolution as the signals transmitted are 'altered by the scalp and tissues' (ibid.: 14). However, whilst criticisms can be made of aspects of each of these technologies, collectively they have provided scientists with the tools to develop a far greater understanding of the workings of the human brain.

In the courtroom

It is generally accepted that there is a problem with applying empirical data obtained from specific, group level, research findings to the individual in the criminal court room.[6] The problem revolves around a number of issues which are not dissimilar to the application of other scientific findings in the courtroom. Criticisms include: applying findings based on

465

the averaging of research data obtained from many individuals to a single individual and the transferability and relevance of applying conclusions about human behaviour based on such data to the accused individual. There is a further difficulty in precisely identifying the behaviour that the experiment was measuring and the relevance and applicability of the experimental results to the individual criminal action being assessed. There are also disputes regarding how the measurements are obtained and the accuracy of the measurement. However, such problems are not new to the courts. There is a robust form of control regarding the admission of evidence in England and Wales.[7]

Nor are these disputes confined to neuroscience – similar arguments are made about the reliability of assessing predispositions utilising genetic data. Nita Farahany and James Coleman make the point well when writing about the use of scientific evidence by those accused of a crime in court in the USA:

> Although the science is in early stages of discovery, and scientists quarrel over basic methodology and the definitions and metrics for measuring behavior, criminal law has already seized upon behavioral genetics and neuroscience evidence for a variety of purposes: as exculpatory evidence, to bolster preexisting legal defenses, and as mitigating evidence during sentencing. As these fields progress and gain credibility, scientific results demonstrating a genetic or neurological contribution to behavioral differences in violence, aggression, hyperactivity, impulsivity, drug and alcohol abuse, anti-social personality disorder, and other related traits will continue to be introduced into the criminal law.
>
> *(Farahany and Coleman 2009: 183)*

In court the best evidence will normally be the accepted scientific position. In developing areas of science, the best available evidence may not yet be received scientific opinion. However, if the accuracy of the science is sufficient to put before the jury, even if it is not totally validated by the scientific community, then the judge will have to consider its probative value to the jury.

Nita Farahany has carried out empirical research into the use of such evidence in the American courtroom by those accused of criminal offences.[8] This research has been mirrored in four other jurisdictions utilising the same search terms to research different national law databases: The Netherlands (de Kogel and Westgeest 2015), England and Wales[9] (Catley and Claydon 2015), Canada (Chandler 2015) and Singapore.[10] The English data is based on reported cases; these are largely appeal decisions and so may not reflect fully the use and extent of neuroscientific evidence in the criminal courtroom by those accused of crime. Case reporting tends to cover those cases where an important novel legal issue has been the subject of appeal following conviction. Over 70 per cent of English cases, where someone is accused of a crime, are resolved by a guilty plea.[11] Therefore, any snapshot seen in the case reports will only hint at the use of neuroscientific or genetic evidence by prosecution or defence. Nonetheless the research confirms that evidence from cognitive neuroscience is used in criminal trials in England by those accused of crime.

Using evidence based on cognitive neuroscience is likely to be expensive. In her article reviewing the area in the USA Nita Farahany suggests that wealthy defendants or those able to secure state funding or pro bono services are more likely to be able to introduce neurobiological evidence (2015: 491).

Katie de Kogel notes a different issue in the Netherlands:

> An issue in the Netherlands is that the pool of experts who report to the criminal courts about neuroscientific information is rather small. For instance, in the majority of cases

in which neurological information was reported in relation to aggressive behavior, the same 'behavioral neurologist' was consulted. For the growth of expertise in this area, it is important to have more professionals.

(de Kogel and Westgeest 2015: 602)

In England the courts admitted neuroscientific evidence for a wide number of reasons. On occasion the evidence was admitted in sentencing decisions and here there was a tendency for the information to operate as a double-edged sword. This was also found to be the case in the Canadian research. Jennifer Chandler writes:

The majority of the cases are sentencing decisions, which is useful given that it offers an opportunity to observe how judges wrestle with the tension at the heart of the justifications of punishment in the criminal law. Neuroscientific evidence suggesting diminished capacity tends to reduce moral blameworthiness – a factor central to the retributive philosophy underpinning the requirement of proportionality between the degree of wrongdoing and the punishment – and yet it also tends to increase judgements about risk and dangerousness, given the view (expressed often in the cases reviewed here) that brain injuries can sometimes be managed but not cured. This makes neuroscientific evidence a 'double-edged sword' from the offender's perspective.

(2015: 574)

The research in five jurisdictions suggests that where a defendant is well resourced they may be able to present defences based on neuroscientific evidence in court.

This research examines the present use of neuroscience by the accused in the criminal courts, but the purpose of this book chapter is to assess how neurotechnologies may be used more generally by the criminal justice system in the future. From time to time speculative pieces on this subject appear in the media. The *Guardian* in January 2016 featured an article entitled: 'Can a brain scan uncover your morals?' The article reports the use of brain scans in cases heard by the American courts over a number of years. All the cases covered were high profile and included the case of John Hinckley who tried to assassinate Ronald Reagan. The focus of the article is the use of brain scans, rather than other evidence from cognitive neuroscience, but its conclusion, if accurate, is interesting. It reports that 'the federal government is pumping millions of dollars into fMRI research on mental diagnoses, partly in anticipation of the judicial system benefitting from it.'[12]

There is a further interesting assumption made at the end of the article in terms of the folk psychological view of neuroscience. The journalist writes:

Everyone who has a stake in the science is hoping the scans will someday provide an unbiased truth. But there is a systematic problem because the law needs finality, while science relies on continued research. And for now, there is no way to see intention in the scans – there is no record of a crime, of innocence, of morality, of honesty. Behavioural brain scans are as objective as their interpreters.

The comment could be read as suggesting that one day the scans will see the things that are now absent. This is extremely unlikely. The law would indeed be in trouble, if it were to become accepted folk psychology that brain scans could look back and see the intentions of the accused at the time of a crime. Rather than viewing folk psychology as the prevailing and accepted belief, perhaps folk psychology needs challenging as firmly as the view that brain scans will be able to identify our intentions. Arguably, the suggested focus of discussions about the value of cognitive neuroscience should be on what we use evidence for, what it establishes and how probative the expert evidence regarding the factual issues might be to assist in the determination of guilt or innocence. This could simply mean asking how relevant the evidence is to the individual case. It is apparent that there is little mileage in arguing neuroscientific research is of no use when it is used in many jurisdictions (Spranger 2012).

The corollary to this finding is that key actors in the justice systems need to be trained about the relative strengths and weaknesses of the neuroscientific evidence. Additionally, there is a strong need to explain to juries where the folk psychological view of causes of criminal acts is inaccurate or misleading. This will be of particular importance where this understanding directly feeds into the assessment they make of guilt and innocence. This may mean that the prosecution or defence might have to consider leading evidence to rebut the identified notions that folk psychology holds concerning how people are caused to act. The scope of this chapter is insufficient to consider what the whole ramifications of this might be; it is not suggested that the societal view that people are responsible for their actions be abandoned. But is this what really would have to happen? Many cognitive neuroscientists would argue that we need to hold people responsible for their actions if society is to protect itself from truly problematic and dangerous individuals. The question is whether our present explanatory structures of causality, and criminal responsibility function adequately in the courtroom in the light of advances in neuroscience and neurotechnologies.[13]

The work of the Presidential Commission *for the* Study of Bioethical Issues will be discussed later in the chapter. It is worth noting here that the view the commission took was that whilst neuroscience was unlikely to provide a determinative explanation of why a particular individual committed a crime, there was significant likelihood that politicians would use science, including neuroscience, to advocate policy agendas (2015: 101).

In the criminal justice system

Raine (2013) argues that science and neurosciences may influence investments made by states in the resources employed to solve problems in the criminal justice system. In *The Anatomy of Violence: The Biological Roots of Crime*, Raine advocates the argument, based in neurocriminological research, that biology can predispose individuals to violent behaviour. The natural corollary of this argument is, he suggests, that were we able to identify the most problematic individuals in our midst then society could take measures which could make it a safer place in which to live and work.[14] The book is intended to provoke discussion as to how societies should deal with such an assertion.

Raine's argument, is that modern science makes it possible to envisage a society which would intervene early in the criminal career of an individual, in some cases, possibly even before that career had begun. He points out that some sections of society and indeed politicians might think that this outcome would be better for all. Raine emphasises the fact that, in such a future, the outcome for some would be worse than it is at present. The basis of the argument is that if you start from the point that a few dangerous people cause a disproportionate amount of harm to other people, it is possible to progress to a view that something should be done about those people.[15] Raine envisages a future where a government following on from a violent incident where scores of people are killed, in response to public clamour for better protection, sets up a project to identify those who are a real threat to society (2013: 342).

He names the project, LOMBROSO,[16] and speculates that the state of scientific knowledge and the availability of other relevant scientific data would allow the project to commence in the

2030s. The aim of the project, as Raine envisages it, is to assess with accuracy the risk posed to the public by particular individuals. He suggests that in the 2030s it will be possible by testing all males over eighteen, for five variable factors and setting this against DNA and other relevant environmental and background information, to assess with accuracy the risk they pose to the public.[17] Raine argues that it is the ability to access big data sets and apply effective algorithms to make group data relevant to the individual that enables this to be a possible outcome. 'The computerization of all medical, school, psychological, census, and neighborhood data makes it easy to combine these traditional risk variables alongside the vast amount of DNA and brain data to form an all-encompassing biosocial data set' (Raine 2013: 342). From this Raine suggests that you could identify those men who had a 79 per cent chance of committing a violent crime in the next five years. You could also identify a group who were 'Lombroso Positive – Homicide' who would have a 51 per cent chance of killing someone in the next five years, and another group 'Lombroso Positive – Sex' who would have an 82 per cent chance of committing either rape or a paedophilic offence (ibid.: 343). Raine argues that were a society sufficiently outraged by a violent event, then the temptation to identify such people and commit them to indefinite detention might be too strong for politicians to resist.

He points to the fact that projects which are similar in terms of their outcomes to LOMBROSO have been 'alive and well for years in countries like England' (ibid.: 354). An example of this is considered in a review of the convergence of mental health and criminal justice systems policy, legislation, systems and practice. Entitled 'Blurring the Boundaries', the report published by the Sainsbury's centre reviews the present provisions for 'Potentially Dangerous Persons' comments in looking at policy on present treatment of potentially dangerous criminal offenders that 'the model is one that could be replicated (under appropriate ethical and clinical scrutiny) to those with chaotic lives and multiple needs, who are not formally involved in the criminal justice system' (Rutherford 2010: 70). This comment relates to the treatment of those who have not yet committed criminal offences but whose life style and behaviours are identifiable as posing a risk to the society within which they live. The point made by Raine is that policy makers have to consider how to diminish the risk to the health and happiness of the rest of society and to decide whether compulsory treatment or incarceration is necessary. His focus is on what would happen if the evidence were to become far more compelling in its predictive efficacy which he sees as likely to occur within the next 20 years.

This leads him to conclude that societies have to reach a view as to what type of future they wish to have. In his view all sectors of society need to be thoroughly engaged in the debate. Raine is not just a popular writer, he is an academic neurocriminologist based at the University of Pennsylvania. He has a critical perspective on how politicians may react to public outcry, pointing out that they may continue to 'overreact' to 'quell the public outcry and try to solve society's problems' (Raine 2013: 357).

All of this academic speculation by Raine does not lead to the view that we should abandon cognitive neurobiological research, or argue that reviewing the evidence has no place in law. The truth, he argues, is that these concerns deserve much greater prominence. An informed debate needs to take place about the nature of risk and the responsibility of society towards its individual members. The debate is not an easy one to have and it requires considering some fairly unpalatable truths. Raine makes the point that his imagined LOMBROSO project, whilst it would incarcerate men who did not pose a real risk, would probably not be racially biased. But Raine is also cautious about trusting politicians to use such tools: 'Let's face it, elements are already in place right now. The prison at Guantánamo Bay is just one example of how indefinite detention is being used by countries throughout the world in the name of national security' (ibid.: 351–2).

Lisa Claydon

Researching the intersection of law and neuroscience

One way of scoping likely future developments that may influence the interrelationship between law and neuroscience is to look at where interdisciplinary work is taking place, the aims of the research and the reported outcomes. It will also be useful to look at the work of politically influential groups. Two influential reports have appeared scoping this area, one produced in England by The Royal Society and the other in the United States by the Presidential Commission *for the* Study of Bioethical Issues.

Learned societies and policy development

The Royal Society

In 2011 the Royal Society set up a working group to report on the policy implications of developments in neuroscience which could impact upon the law. The group was tasked inter alia to:

- provide an introduction to the questions raised around the intersection of neuroscience and the law, and the link between the brain, mind states and behaviour
- provide an assessment of the extent to which neurotechnologies might be able, now or in the future, to contribute to the quality of decision-making in legal proceedings.

(The Royal Society 2011b: 2)

The areas that are of particular relevance to criminal responsibility that were identified as worthy of review were risk, developmental maturity, and memory, including the reliability of witness testimony. The report sounds the expected note of caution about the extrapolation of insights from neuroscience via scientific descriptions of mental processes into estimations of individual responsibility. The report took a necessarily broad view of areas of neuroscience that would be of interest to the criminal justice system. The report by the working party included some discussion of behavioural genetics and neuropsychology, identifying mental activity 'such as thinking, feeling, sensing, attention, memory and consciousness' (ibid.: 1) as areas of relevance to the criminal law.

Overall the strong conclusion of the report was that neuroscientific understanding of the brain and human behaviour was developing. However, it noted the gap in understanding between the work of neuroscientists and 'the realities of the day to day work of the justice system' (ibid.: 30) as being an area of concern. The report recommended the creation of a forum where the two parties could have a fruitful exchange of ideas. Such a forum was seen as existing in other countries, particularly the United States, where the working group noted the work of the MacArthur Foundation and the work of the National Academy of Sciences in bringing neuroscientists and lawyers together. It was argued that both scientists and lawyers would benefit at undergraduate level in receiving some training in the other discipline (ibid.: 31). Finally it was felt that the criminal justice system would benefit from further research into risk assessment. The ESRC was asked to consider providing funding to support neuroscientific research in this area to evaluate the 'relative efficacy of various models of risk assessment in the context of probation' (ibid.).

Gray Matters

The Presidential Commission *for the* Study of Bioethical Issues has produced a two part report entitled *Gray Matters*, concerning neuroscience, ethics and society. The first part of the report looked at Integrative Approaches for Neuroscience, Ethics and Society;[18] the second part of the report examines Topics at the Intersection of Neuroscience, Ethics and Society.[19] The second volume devotes a complete chapter to the consideration of 'Neuroscience and the Legal System' (*Gray Matters* vol. 2, Ch. 4). The chapter starts with an assertion that 'the brains of criminals have captured the public's imagination for centuries' (ibid.: 86). Then a clear statement is made regarding the usefulness of neuroscience to improving policymaking. The areas of use include 'increased accuracy and decreased errors in advancing justice', which appear to be two sides of the same coin. The areas of concern identified are 'scientific reliability, misapplication and overreliance on a developing science' (ibid.: 7), areas about which the Royal Society also had concerns. *Gray Matters* reiterates the importance of thinking about the ethical implications of growing neuroscientific knowledge and the power of prediction. Particularly, concern is expressed about the effect that greater neuroscientific knowledge will have on 'conceptions of free will, mental privacy and personal liberty' (ibid.).

The report notes the relevance of neuroscience to the legal system.[20] It conjectures that the science may be at its most effective at the policy level. Interestingly the report states the neuroscience 'may guide normative assessments' in the legal arena, though it robustly states that it must not solely define them (ibid.: 88). The report makes some clear assertions about the potential value of neuroscience to the criminal justice system:

> Neuroscience has the potential to advance justice by increasing accuracy in legal decision making and policy development. A deeper understanding of the human brain, cognition, and behavior on both individual and societal levels might help tailor policies and sentences, determine guilt and innocence, evaluate blameworthiness, and predict future behavior. For example, evidence of brain abnormalities might help determine whether a criminal defendant is competent to stand trial. Neuroscience evidence might contribute to a jury's determination of guilt or innocence, by helping jurors understand a defendant's mental state, intent, or voluntariness of action. A deeper understanding of the development and capacity of the adolescent brain might help formulate policies about the sentences that young adults and adolescents should receive. Neuroscientific techniques like brain imaging might help detect juror bias or determine the reliability of eyewitness testimony. Overall, neuroscience might contribute to more accurate decision making and fairer outcomes. Justice requires that we use empirical evidence, including neuroscience, to strengthen the decisions made in these central civic and political realms.
>
> *(ibid.: 88–9)*

One of the conclusions of the report which does also urge caution about neuro-exuberance is that more research is required and it looks to funders like the National Academies of Science, US Department of Justice, and the Social Security Administration to support further research. The focus of the research it suggests should be the use of neuroscience in legal decision making and policy development (ibid.: 112). One foundation that has consistently supported and indeed driven academic research in this area is the MacArthur Foundation.

Funded research

MacArthur Foundation

The MacArthur Foundation has greatly supported the researching of the interface between neuroscience and law. In the USA the Foundation has invested a considerable amount of money in the interdisciplinary study of areas of mutual interest to scientists and criminal lawyers. MacArthur at the point when this article was written in 2016 had already invested $7,600,000 into funding interdisciplinary study in this area. At present MacArthur highlights three areas as being of importance to assist in developing the knowledge base in the area of neuroscience and criminal justice.

> 1) investigating law-relevant mental states of, and decision-making processes in, defendants, witnesses, jurors, and judges; 2) investigating in adolescents the relationship between brain development and cognitive capacities; and 3) assessing how best to draw inferences about individuals from group-based neuroscientific data.[21]

Thus its concerns focus on similar areas to those already discussed. The chapter has already considered how Big Data might be utilised in examining the work of Adrian Raine. Additionally, it is worth considering the influence of an organisation which spends considerable sums of money funding the work of cognitive neuroscientists.

DARPA

All this funding for academic research in the United States of America pales into insignificance when compared to the spending in this area of the Defence Advanced Research Projects Agency (DARPA). This agency in 2014 invested $50,000,000 into the Brain Research through Advancing Innovative Neurotechnologies (BRAIN) Initiative in the USA.[22] One of the projects for which it announced funding is highly ambitious. The project will cost millions of dollars and aims to repair memory loss in those who have suffered traumatic brain injury. The project is shared between the University of Pennsylvania and the University of California, Los Angeles. The aim of the project is to build a direct brain recording device. The University of California in its announcement says it will receive funding of $15,000,000 for its part of the research.[23]

Information on the DARPA website about the research programme, which it calls RAM, contains the following statement:

> In addition to human clinical efforts, RAM will support animal studies to advance the state-of-the-art of quantitative models that account for the encoding and retrieval of complex memories and memory attributes, including their hierarchical associations with one another. This work will also seek to identify any characteristic neural and behavioral correlates of memories facilitated by therapeutic devices.[24]

This is just one of the programmes which DARPA is funding that will enable a greater understanding of the function of brain areas.

The laudable declared aim of RAM is to help restore memory loss in those who have suffered traumatic brain injury. This is not the application of research into memory that tends to worry lawyers or raise concerns about mental privacy. These concerns arise when the research understanding obtained by neuroscientists is applied to investigating the contents of human memory – particularly where the investigation is of the memory of someone accused or suspected of a crime.

Memory tests – finding the truth?

Memory is key to our human identity.[25] Through memory we lay down the autobiography of our life. This helps us to understand who we are and how we interact with the world. Loss of memory as is the case with some forms of severe brain injury, or Alzheimer's, may lead to a loss of a sense of personal identity. Memory of events therefore is personal and generally people do not share intimate memories with anyone but their closest family and friends.

Memory of events has historically been viewed in a similar fashion by the law. This is particularly true when the content of that memory could incriminate an individual. The boundary between the state and the individual was marked by the understanding that the individual should be assumed innocent until proven guilty and the prosecution carried the burden of proving guilt beyond reasonable doubt. The accused was not required to prove his innocence.[26] Thus in England and the United States the right not to incriminate oneself has long been enshrined in law.[27] In England, in the late twentieth century, that right to silence was limited by statute but an accused can still not be compelled to give evidence. An individual suspected of a criminal offence will be advised that he has the right to remain silent but if he does so adverse inferences may be drawn from his silence.[28]

Claims are made by those who are experienced in the use of lie detection technology that it is possible to detect when someone is not telling the truth. Indeed, one of the older forms of lie detection technology, using a polygraph, is routinely used in England to assess the risk to the community of convicted sex offenders released on licence.[29] This use of and belief in a technology which has been shown to be flawed is concerning.

The effect of the use of evidence from lie detection tests, to assess risk in the released sex offender population, is covered in an article in the *Guardian*. In the article the technology is seen as yielding results supporting the accused's innocence. There is some evidence that it has been successfully used by defence lawyers. The investigative journalist reports the use of a lie detection test, on behalf of a client, to avoid the client being charged with a sex offence. The journalist quotes a barrister and solicitor as believing that the tests were effective in persuading the police to drop the charge,

> a solicitor of Litigaid Law in Southport, who was formerly a police detective in London and Merseyside, and Mark Tomassi, a barrister, said: 'My client said the sex had been consensual. He passed a lie detector test with flying colours. In the end he was never charged' … A person suspected of sexual crime could and should be offered the opportunity to take a lie detector test … If such a test is sufficiently reliable to protect the public from future offences from those already found by the courts to be guilty, it is but a modest proposal to allow an innocent person at least a chance of persuading a prosecutor to think again.[30]

Concealed information testing

Scientists at present are working on identifying with the use of electrode technology the content of memories. This could be used to identify, for example, whether the accused had a memory of a place that he denied ever having visited. Clearly this would be useful to investigators who were trying to establish whether this accused had visited a particular location. The distinction is made by comparing neural patterns of activity where the record is taken under highly controlled circumstances. These records of neural activity recorded from the scalp show differences in neural patterning for places that are known when compared to neural patterns that are recorded when an accused is shown locations that they have not visited. However, such assessments have been shown to be flawed because of the possibility of taking counter measures which would skew the results obtained (Uncapher *et al.* 2015). Were such applications of electrode technology to improve in accuracy, or to be routinely used by the State, then clearly the issues surrounding the right to silence would need to be revisited as a matter of urgency.

Gershon Ben-Shakhar writes of lie detection technologies: 'One of the most serious deficiencies of CIT [Concealed Information Test] is its vulnerability to countermeasures by guilty or deceptive examinees' (2011: 200). However, the converse argument that these countermeasures are detectable also needs to be weighed into the argument (see Rosenfeld et al 2008).

Conclusion

Clearly our knowledge of the brain, how it functions and how that leads us to certain behaviours is expanding. The new knowledge generated by advances in neurotechnology will and should find its way into the courtroom and will lead us to take a different view of why people act and the basis on which we hold them responsible for their actions. Ethical consideration will need to be given to the introduction of interventions in the criminal justice system. The point that Adrian Raine makes is well made – should we lock up people before they commit a criminal act? In making the question as difficult as possible for us to answer Raine asks that the reader consider the case of 'Fred Hatoil'. Fred is someone who suffered an abusive childhood, a 'traumatic' home life, four brothers and sisters died before reaching adulthood. Repeatedly moved from house to house he performed poorly at school and left without qualifications. He became a message runner in the First World War and was gassed. Following his harrowing war experience, he suffered from post-traumatic stress disorder. Raine tells us that like many veterans of war, his 'emotional compass was blunted'. Gradually Raine builds a picture of unemployment and failed applications to art school and architecture courses, of an inability to form intimate relationships and an individual who is socially dysfunctional. He poses the question, if Fred is charged with murder, whether the court and judge should show clemency. Raine argues that for many people the facts of Fred's life would mean that clemency should be shown to him. In the United States clemency of course means something different from the same word in Europe. This is because 31 states still have the death penalty.[31] In the final part of the scenario Raine reveals Fred's true identity to be that of Adolf Hitler and asks if we still wish him to avoid the death penalty (Raine 2014: 321–22). Raine goes on to point out that many of the worst tyrants in history had deprived and disrupted childhoods.

This argument may make us focus on the most appropriate form of disposal for those who are at risk but it does not make us consider what the right ethical basis is for intervention in other people's lives. This question can only be posed and answered when the basis on which intervention, be it in the form of punishment, treatment or incapacitation is known. Whether the life experience of someone like Hitler and his mental condition at the time of his crimes should offer some excuse is an ethical question. In order to grapple with any new challenges to our ethical understandings then lawyers, neuroscientists and moral philosophers need to work together. None of these groups will be able to grasp the nettle and start down the path to resolving the issue without an open and flexible exchange between all three disciplines.

Relevant to this exchange is Daniel Dennett's idea that we totally abandon the metaphysics of free will and replace metaphysics with an ethical stance which would recognise that we blame someone for their choices. It is the failure to act otherwise and avoid the crime for which society holds them responsible. He argues that the idea of choice is central. Determining when an agent 'could have done otherwise' focuses on the ethical idea that the 'pivotal phrase *could have done otherwise*' ascribes responsibility and makes blame and punishment appropriate. He concludes that the 'the fact that free will *is* worth wanting can be used to anchor our concept of free will'. He concludes that 'metaphysical myths' will fail to achieve this (2003: 297).

In taking this stance Dennett, arguably, frees society from having to develop a deeper understanding of determinism or indeterminism in creating an ethical framework for blaming people for their actions. Societal judgment may then be determined against this background. Enabling the development of firm ethical frameworks situated in shared understandings of what is or is not acceptable in terms of punishment or other forms of intervention. This leaves the focus of what is, or is not acceptable, regarding interventions in the criminal justice system on examining the efficacy and ethical appropriateness of proposed solutions against the background of the fact that society in general holds people responsible for their acts rather than their potential for criminal behaviour.

Further difficult questions will need to be resolved, for example: the appropriate response to risk. Also problematic is research into jury and court behaviours which suggest judges and juror's judgment may be biased. Research in this area is being carried out at present by the MacArthur Foundation who are 'investigating law-relevant mental states of, and decision-making processes in, defendants, witnesses, jurors, and judges'.[32] Understanding such processes may provide further challenges for society. However, developing a more robust framework to ensure that these decision-making processes are as just and fair as possible can only strengthen the criminal justice system not weaken it.

What has emerged from this review of the area is an acceptance that cognitive neuroscience and neurobiology will add to the sum of knowledge of what it means to be human. This new understanding will be generated by the research enabled by advances in neurotechnologies such as neuro imaging and the growth of the ability to construct mathematical algorithms that will permit the interrogation of the data from neuroscientific research to be more applicable at the level of the individual actor. It is likely that the money which is being invested by governments in researching the brain means there will be a drive to convert the fruits of the research into meaningful outcomes that will show the policy impact of the governments' spending. At present the spending on the European Union's Human Brain Project is estimated at 1.19 billion euros over ten years.[33] In the United States the Brain Research through Advancing Innovative Neurotechnologies (BRAIN) Initiative supported jointly by state funding and private donors, has an estimated resource allocation of 200 million dollars per annum.[34] How this will feed into the criminal justice system is a matter for speculation and possibly concern. The impact is potentially wide ranging. If the evidence to date is examined through the response to the pressure to assess and blunt terrorist threats or to deal with the risk posed by convicted sex offenders, then the response of policy makers and law enforcement agencies has not always led to the most transparent responses. In this sense taking forward the academic research agenda to develop appropriate ethical frameworks is both necessary and essential.

Notes

1 For examples of interventions which have been researched using empirical data drawn from many scientific studies including those of cognitive neuroscientists and geneticists see Farrington *et al* (2003).
2 http://neurosciencenews.com/reading-terrorists-p300-brain-waves-attacks/ (accessed 10 March 2016). The arguments concerning retrieved memories will be considered later in the chapter.
3 For further information about the test see Rosenfeld (2011).
4 Not that this is new; harsh criticism has been made of the tendency of the press to fail to present information in an objective and measured way since newspapers first appeared. This is particularly the case in terms of the reporting of political issues: 'What the proprietorship of these papers is aiming

at is power, and power without responsibility – the prerogative of the harlot throughout the ages.' Stanley Baldwin, Prime Minister, 17 March 1931 speech given three days before the general election was to be held. Though responsible reporting does take place through academic journals, the worry is the level of reporting in mass circulation papers. See for example: www.dailymail.co.uk/sciencetech/article-3433491/How-anger-changes-BRAIN-Aggression-causes-new-nerve-cells-grow-trigger-rage-future.html (accessed 10 March 2016). The article does not suggest that the experiment actually reveals anything about human behaviour but the omission of a discussion concerning the transferability of information from experiments on a particular mammal to other mammals is noteworthy.

5 Ibid.; These types of rapid changes are said to be 'associated with perception, thought and action.'

6 See discussion of this issue in the Royal Society (2011b: 2.4.3) see also Spranger (2012:1–5).

7 The rules for the admission of expert evidence into criminal courts are set out in the Criminal Practice Directions. At the time of writing the controls are stringent in requiring an assessment of the validity of the evidence. See for example the rules with regard to the content of the expert's report which require, inter alia, details of expert's qualifications, statements of facts relied on, name and qualifications of person carrying out any tests referred to, summary of findings, reasons for the selection of opinions and any other information required for the court to make a decision with regard to admitting evidence. These include the limits of the expert's knowledge, and a summary of other expert opinion on the topic area where there are differing opinions regarding the evidence given. Details of any research literature relied upon in reaching an opinion must be included. Proof of expertise has to be established and the overriding objective is said to be objective and unbiased evidence. The duty of the expert is to the court. A declaration of truth is made by the expert witness.

8 Considering the problem of carrying out such research Farahany writes:

> Moreover, more than 90 per cent of criminal cases in the United States never go to trial. Most individuals who are charged with a crime forego their constitutional right to a trial and plead guilty in exchange for a plea agreement. Of those cases that do go to trial, while many are appealed, many more are not. Of cases that are appealed, there are narrow legal grounds available for overturning a conviction or setting aside a sentence and procedurally the cases must be raised in that manner. Moreover, investigation into neurobiological contributions to criminal behavior can be costly. In cases where the defendant has adequate resources, or able to secure resources from the state, or as pro bono services, they are more likely to be able to introduce neurobiological evidence. This may skew the kind of criminal defendants who raise claims rooted in neurobiology. (2015: 491)

9 Hereinafter referred to as the law of England.

10 This research was carried out by Calvin Ho of the National University of Singapore.

11 Figure extracted from the Crown Prosecution Annual Report and Accounts 2014–15 HC20, Appendix D.

12 www.theguardian.com/science/2016/jan/17/can-a-brain-scan-uncover-your-morals.

13 For a thoughtful review of this area see Moore (2009).

14 In Raine (2013), see particularly chapter 11, The Future.

15 Raine considers the introduction of Imprisonment for Public Protection (IPP) by the Blair government in the UK in 2003 as an example of this trend (Raine 2013: 353).

16 The choice of name is drawn from Cesare Lombroso, who believed that you could identify criminals from certain physical traits. He published *L'uomo delinquitte* in 1876.

17 The tests include: 'a quick brain scan, and DNA testing. … Then a five-minute brain scan for the "Fundamental Five Functions": First, a structural scan provides the brain's anatomy. Second, a functional scan shows resting brain activity. Third, enhanced diffusion-tensor imaging is taken to assess the integrity of the white-fiber system in the brain, assessing intricate brain connectivity. Fourth is a reading of the brain's neurochemistry that has been developed from magnetic resonance spectroscopy. Fifth and finally, the cellular functional scan assesses expression of 23,000 different genes at the cellular level. (Raine 2013: 342).

18 http://bioethics.gov/node/3543 (accessed 11 March 2016).

19 http://bioethics.gov/node/3543 (accessed 11 March 2016).

20 Neuroscience has a variety of potential applications to the legal system and already is employed in many relevant contexts, including increasingly in criminal law … Prosecutors and defense attorneys use neuroscience evidence in criminal proceedings to support propositions concerning, for example, competency to stand trial, mitigation of criminal responsibility, and predicting future dangerousness. Parties also use neuroscience evidence in the civil context to provide objective evidence of 'invisible' injuries, such as toxic exposure, pain, and suffering. Policymakers have invoked neuroscience to

advocate for legislation and reform; scholars have advocated using neuroscience to address biases in legal decision making; and even some commercial entities have introduced novel uses of neuroscience for investigative purposes (Presidential Commission *for the* Study of Bioethical Issues 2015: 86–7).

21 www.lawneuro.org/ (accessed 11 March 2016).

22 The Office of the Press Secretary The White House 2/04/2013 FACT SHEET The Brain Initiave, www.whitehouse.gov/the-press-office/2013/04/02/fact-sheet-brain-intiative (accessed 20 February 2016).

23 'The UCLA Henry Samueli School of Engineering and Applied Science has been tapped by the Defense Advanced Research Projects Agency to play a key role in an innovative project aimed at developing a wireless, implantable brain device that could help restore lost memory function in individuals who have suffered debilitating brain injuries and other disorders.' http://engineering. ucla.edu/ucla-engineering-plays-key-role-in-darpa-neuroprosthesis-research/ (accessed on 11 March 2016).

24 This explanation is preceded by further explanation of the research: 'The end goal of RAM is to develop and test a wireless, fully implantable neural-interface medical device for human clinical use, but a number of significant advances will be targeted on the way to achieving that goal. To start, DARPA will support the development of multi-scale computational models with high spatial and temporal resolution that describe how neurons code declarative memories—those well-defined parcels of knowledge that can be consciously recalled and described in words, such as events, times, and places. Researchers will also explore new methods for analysis and decoding of neural signals to understand how targeted stimulation might be applied to help the brain reestablish an ability to encode new memories following brain injury. 'Encoding' refers to the process by which newly learned information is attended to and processed by the brain when first encountered.' www.darpa.mil/program/restoring-active-memory (accessed on 12 March 2016).

25 For a thorough introduction to the importance of memory to the construction of our individual identity see (Rose 2003).

26 In England these rights are poetically described as 'the golden thread' that runs through the criminal law, see *Woolmington v DPP* [1935] AC 462.

27 For example, The United States Constitution 5th Amendment.

28 In England the right is qualified by s34 of the Criminal Justice and Public Order Act 1994, the accused is advised of his right to silence but also warned that adverse inferences may be drawn from his silence if he fails to reveal something when questioned that he later relies on in court.

29 www.gov.uk/government/news/compulsary-lie-detector-tests-for-serious-sex-offenders (accessed 11 March 2016). Gives information about the use of the technology.

30 www.theguardian.com/society/2014/aug/08/lie-detector-polygraph-tests-introduced-monitor-sex-offenders (accessed 11 March 2016).

31 www.deathpenaltyinfo.org/states-and-without-death-penalty (accessed 13 March 2006).

32 See the section entitled Funded Research and the discussion around memory.

33 The Human Brain Project Report to the European Commission 2012 Available at: www. humanbrainproject.eu/documents/10180/17648/TheHBPReport_LR.pdf (accessed 16 April 2016).

34 The Office of the Press Secretary, The White House, 2 April 2013, FACTSHEET, The Brain Initiative available at www.whitehouse.gov/the-press-office/2013/04/02/fact-sheet-brain-initiative (accessed 16 April 2016).

References

Ben-Shakhar, G. (2011) Countermeasures, in Verschuere, B. Ben-Shakhar, G. and Meijer, E. (eds) *Memory Detection Theory and Application of the Concealed Information Test*. New York: Cambridge University Press.

Blakemore, C. (1988) *The Mind Machine*. London: BBC Books.

Catley, P. and Claydon, L. (2015) The use of neuroscientific evidence in the courtroom by those accused of criminal offenses in England and Wales, *J Law Biosci* 2(3): 510–549.

Chandler, J.A. (2015) The use of neuroscientific evidence in Canadian criminal proceedings, *J Law Biosci* 2(3): 550–579.

Dennett, D.C. (2003) *Freedom Evolves*. London: Penguin.

Farahany, N. (2015) Neuroscience and behavioral genetics in US criminal law: An empirical analysis, *J Law and Bioscis* 2(3): 485–509.

Farahany, N. and Coleman, J. (2009) Genetics, neuroscience and criminal responsibility, in N. Farahany (ed.) *The Impact of the Behavioral Sciences on the Criminal Law*. New York: Oxford University Press.

Farrington, D. and Coid, J. (eds) (2003) *Early Prevention of Adult Anti-social behaviour*. Cambridge: Cambridge University Press.

Goodenough, O.R. and Prehn, K. (2006) A neuroscientific approach to normative judgment in law and justice, in Zeki, S. and Goodenough, O.R. (eds) *Law and the Brain*. New York: Oxford University Press.

de Kogel, C.H. and. Westgeest, E.J.M.C. (2015) Neuroscientific and behavioral genetic information in criminal cases in the Netherlands, *J Law Biosci* 2(3): 580–605.

Lombroso, C. (1876) *L'uomo delinquente*, in Gibson, M. and Rafter, N.H. (2006) (trans) Durham, NC: Duke University Press.

Moore, M. (2009) *Causation and Responsibility*. New York: Oxford University Press.

Morse, S.J. (2011) NeuroLaw exuberance: A plea for neuromodesty, in van den Berg, B. and Klaming, L. (eds) *Technologies on the Stand: Legal and Ethical Questions in Neuroscience and Robotics*. Nijmegen: Wolf Legal Publishing.

Presidential Commission *for the* Study of Bioethical Issues (2015) *Gray Matters: Topics at the Intersection of Neuroscience, Ethics and Society*, Vol. 2. Washington. Available at: http://bioethics.gov/sites/default/files/GrayMatter_V2_508.pdf .

Raine, A. (2013) *The Anatomy of Violence*. London: Penguin.

Rose, S. (2003) *The Making of Memory*. London: Vintage.

Rosenfeld, J.P. (2011) P300 in detecting concealed information, in Verschuere, B. Ben-Shakhar, G. and Meijer, E. (eds) *Memory Detection Theory and Application of the Concealed Information Test*. New York: Cambridge University Press.

Rosenfeld, J.P., Labkovsky, E., Winograd, M., Lui, M.A., Vandenboom, C. and Chedid, E. (2008) The Complex Trial Protocol (CTP): A new countermeasure-resistant, accurate, P300-based method for detection of concealed information, *Psychophysiology* 45(6): 906–919.

Rutherford, M. (2010) *Blurring the Boundaries: The Convergence of Mental and Criminal Justice Policy, Legislation, Systems and Practice*. London: Sainsbury Centre for Mental Health.

Spranger, T. (2012) Neurosciences and the law: An Introduction, in T. Spranger (ed.) *International Neurolaw: A Comparative Analysis*. Heidelberg: Springer.

The Royal Society (2011a) *Brain Waves Module 1: Neuroscience, Society and Policy*. London: The Royal Society.

The Royal Society (2011b) *Brain Waves Module 4: Neuroscience and the Law*. London: The Royal Society.

Uncapher, M.R., Boyd-Meredith, J.T., Chow, T.E., Rissman, J. and Wagner, A.D. (2015) Goal-directed modulation of neural memory patterns: Implications for fMRI-based memory detection, *Journal of Neuroscience* 35(22): 8531–8545.

Keynote discussion

Technology and control

28

The uncertainty principle

Qualification, contingency and fluidity in technology and social control

Gary T. Marx and Keith Guzik

Odysseus and the seduction of technological certainty

In the *Odyssey*, Homer (2016) recounts Odysseus' control efforts to avoid hearing the dangerous song of the Sirens. As Odysseus approached the island of the Sirens, he reports,

> I … sliced a large cake of beeswax with my sword-edge, and kneaded the slivers in my strong hands until the pressure and the rays of Lord Helios Hyperion heated it. Then I plugged the ears of each of my friends, and they tied me hand and foot and stood me upright in the mast housing, and fastened the rope ends round the mast itself. Then sitting down again, they struck the grey water with their oars … when we were within hail of the shore, the Sirens could not fail to see our speeding vessel, and began their clear singing:
> 'Famous Odysseus, great glory of Achaea, draw near, and bring your ship to rest, and listen to our voices. No man rows past this isle in his dark ship without hearing the honeysweet sound from our lips.'
> This was the haunting song the Sirens sang, and I longed to listen, commanding my crew by my expression to set me free. But they bent to their oars and rowed harder, tightened my bonds and added more rope. Not till they had rowed beyond the Sirens, so we no longer heard their voices and song, did my loyal friends clear the wax that plugged their ears, and untie me.
>
> *(Homer, The Odyssey, Book XII, 2016: 165–200)*

Odysseus knows he must resist the Sirens' song and he also knows that he will be unable to resist temptation without a strategy of mechanical intervention. Hence ears are plugged and he is tied until safely beyond their range. For Homer (2016), Odysseus's passing of the Sirens exemplified the cleverness and guile central to Greek notions of heroism. For Horkheimer and Adorno (1972 [1944]), meanwhile, the tale spoke to the deep historical roots of the social inequalities exacerbated by modern industrial society. Odysseus, by commanding his charges to block their ears and fasten him to the ship's mast, ensures the safety of the vessel, all while denying them and reserving for himself the pleasure of the Sirens' song.

For the purposes of this chapter, the story speaks to society's faith in mechanical solutions to problems. Through the proper technological intervention, in this case blocking communication and preventing physical movement, danger can be avoided. Moving right along, several thousand years later, have you ever encountered a web site that would not let you access it unless you entered a valid email address and a password with at least 8 small and capital letters and at least one number? Or what about a bathroom with cameras and biometric access that also measures outputs? (see Appendix A – "Raising Your Hand Just Won't Do.") The preventive efforts of both Odysseus and more contemporary examples nicely illustrate this chapter's topic of the use of technology for social control.

Today, political leaders, bosses, and other authorities promise that the ever expanding repertoire of communication and information technologies (when subject to appropriate controls as decided by them) will better secure society and keep the ship steady. People today are flooded with communication rather than being blocked from it as with Odysseus's sailors. This leaves us to wonder what risks we may face as authorities seek ever greater access to our songs, even as they may restrict or softly manage our access to theirs.

The historical and cultural roots of technology as social control trace back far beyond the Greeks' use of the word *techne* to mean practical skill. But what do the terms social control and technology mean today? Traditionally social control referred to the integration or meshing of institutions such as the family, education, religion, work and government (Janowitz 1975; Gibbs 1989). This approach looks at the total society, the largely unplanned factors in its evolution and the ways, and extent to which, the parts mesh in providing social guidance and order. In contrast, contemporary social scientists use the term to refer more directly to behavior that involves rules and standards – their creation, mechanisms for conformity, the discovery of violations and violators, and processes of adjudication and sanctioning.

Control or enforcement through material technology, the emphasis of this chapter, is only one of many control modalities intended to create rule adherence. Other related forms are socialization, appeals to conscience and reason, peer pressure, informing, licenses and bonds and insurance, rewards and punishments, habit and repetition, exclusion and inclusion and deception and manipulation. The relative importance of these varies across time periods, types of rule and violation, setting and actor involved, but as this *Handbook* demonstrates, the *engineering of social control* is a defining characteristic of modern society and almost always has a seat at the table. It is so prominent, ubiquitous, and transparent in daily life that it is often taken for granted. Our personal, spatial, communication, social, cultural, and psychological environments and borders are increasingly subject to technological strategies designed to influence behavior, whether involving conformity with rules, safety, consumption, or attitudes. The communication and information technologies at the heart of the computing revolution are touted as possessing near magical powers that will allow social control, and hence security and other benefits.

As Cole's contribution to this volume (Chapter 29) illustrates, DNA identification represents the latest in a long line of technologies (fingerprinting, lie detector tests, etc.) that were purported to offer the police "material objectivity" in identifying criminal offenders. Green technologies promise consumers the comforts of modern living (clean water, automobiles) without the harmful side-effects that industrial life wrought (polluted water, poisonous car emissions) (see Brisman and South, Chapter 18, this volume). Algorithms offer the potential of better decision-making on vital economic and planning matters without the error of human actors (see Pagallo, Chapter 37, this volume).

Marx (2016) asks if we are becoming a *maximum security society* in noting the increased parallels between the highly rationalized social control of the maximum security prison and control efforts seen in the broader *surveillance society*. Such a society is ever more transparent, porous and regulated, as the traditional borders that formerly protected personal information and choice are weakened or obliterated by new technologies, new ways of living, and new threats. Control is ubiquitous, varied and integrated into networks of astounding complexity.

The maximum security society is reflected in eleven sub-societies: A *hard-engineered* society where control is sought via materially altering the environment; a seductive and *soft engineered* society relying on persuasion, invisibility and deception; a *dossier* society based on extensive record keeping; an *actuarial* society based on predictive statistics; a *transparent* society where ever more aspects of life are visible to authorities; a *self-monitored* society relying on self-control as a result of measurements and education; a *suspicious* society expecting people to prove that they are innocent; a *networked* society of ambient and ubiquitous sensors in constant communication; a *safe and secure* society with attenuated tolerance for risk; a *"who are you?"* society of protean identities both asserted by, and imposed upon, individuals; a *"where are you, where have you been, who else is there, and what did you do?"* society of mobility and location documentation.

These efforts at rational control involve varying degrees of generality. This chapter deals with the broadest forms – the hard and soft engineered societies within which the others nestle and twirl. The use of science-based technology to control and influence persons is central to modernization, whether for purposes of crime control in the form of "hard" prevention or "soft" intelligence/information collection and communication for risk management and influence – whether to sell toothpaste or candidates for office or to stop drugs through "just say no" campaigns.

But, as this chapter argues, control often evades the grasp of the engineers and entrepreneurs (whether economic or moral) who tout such technologies as the easy answer to problems of security and public safety. In contrast, this chapter develops the idea of the *uncertainty principle* as applied to the outcomes of surveillance technologies (although it also applies to other social control efforts). To get there, concepts need to be defined, including surveillance as a form of the *engineering of control* connected to other aspects of control. To do that, the next section suggests a language to organize and contrast the array of contemporary control technologies. The major kinds and sources of variation will be identified before turning to empirical inquiry. We then consider a single case study of a Mexican government effort to monitor automobiles for the purposes of public security. We do this to highlight some of the obstacles facing social control through technology. The chapter then ends with a description of the *uncertainty principle* that sets the stage for further inquiries into the uncertainty and unpredictability of the rapidly expanding hard engineering of social control that so defines our age.

Technology and social control: variables and definitions

Considerable variability is found in traditional definitions of technology in the social sciences. For Karl Marx, technology is a key element in the historical movement from feudal, trade-oriented, and early manufacturing society to industrial society. Technological artifacts are central to the economy (Marx 1947, 1956). Critical here is the ideal of technology as a self-realizing, self-defining activity through which humans "begin to distinguish themselves from animals as soon as they begin to produce their means of subsistence" (Tucker 1978). For Weber, technology figures not as productive activity, nor as artifact, but as a mode of thinking and acting in and on the world, "the application of the (technically) most efficient means to given ends within the various spheres of social life" (Schroeder 1995: 228). But more generally, we can define technology as the strategic application of means to ends, whether or not there is a material component, as Karl Marx would have it, or the effort "works" in a literal sense, as Weber stresses. Lip, eye and facial reading for example are technologies because they involve an intentional application of a means to an end, so too does the technology of the rain dance, even though none of them have a material base. It is important to analyze the consequences of the presence or absence of a tangible tool, what it means to say something "works," and whether an outcome can be scientifically accounted for. But neither materiality nor effectiveness are defining characteristics of technology per se.

Another way of defining technology is to see its role in *extending* human capability and resistance (Brey, Chapter 1, this volume), a more practice-oriented view that reflects Marxist emphasis on human labor and intentionality. The use of animals, steam, combustion, electricity, nuclear, wind and solar technologies extend innate abilities, as do sling shots, guns and missiles and hydraulic lifts and robots. Similarly, new surveillance tools that extend the senses (e.g. seeing in the dark or from outer space), or tools offering remote communication and control, illustrate this. Computers expand information capabilities regarding both the amount of information available and the ability to store, analyze and share it.

While the effort to extend human ability is a driving force in invention, we can also note efforts to diminish that ability. Some technologies are designed for *eliminating, blocking* or *impeding*, rather than extending, human capabilities (at least on the part of potential rule breakers). The engineering of control may involve *target hardening* or *suspect softening*. Such actions are intended to make it impossible, or much more difficult, for offenders (although it leaves untouched motivation). Examples of hardening can be seen in the moats and formidable walls of the fortress, in the protection of the closed door (whether locked or unlocked), in titanium locks for bicycles, in biometric controls on cars or weapons that restrict use to registered persons, and in encryption or passwords that block access to information. Examples of suspect softening include castrating sex offenders and Antabuse for alcoholism.

These ideas can be related to the *extensive perspective* on technology. They call attention to the interactional and hierarchical dimensions of technology employed for social control. The efforts of one person to extend their capabilities by restricting access through *target hardening* (say, the titanium lock on the bicycle) come at the expense of others' attempts to extend their capabilities by acquiring that target (the bicycle itself).

To deepen the distinction between *target hardening* and *suspect softening*, we can also say that the hard-edges of preventive means seek to bypass the will of the suspect, leaving the individual little choice but to conform or, if violation remains possible, to face increased risks. Contrast a high wall embedded with broken glass encircling the perimeter of a property with an encirclement by a low hedge that permits trespassing by simply climbing over it. The hedge is a symbol communicating to the potential trespasser that this is private property, even as the choice to enter it remains. That isn't the case (or isn't the case without neutralization actions and risk) for the high wall. Engineering solutions such as the high wall or high voltage electrified fence (that bootlegs in punishment as well as exclusion) are presumed to offer more security. Of course, the low hedge could offer that as well by the presence of snarling guard dogs wearing video-cams, even though trespassing in principle remains possible.

There is an ethos of certainty associated with hard engineering solutions. This is expressed by what the head of a large corporation said (in a satirical statement which could almost be true) about his company's automated toilets. These were engineered to serve goals of work productivity, health and crime control. He said, "we believe that our trusted employees will do the right thing when given no other choice" (see Appendix A). This statement reflects the effort to eliminate the possibility of making bad choices through hard-engineering.

The view contrasts with the soft-edges of other control means, which are based on the premise (or hope) that violators are rational beings whose behavior will be governed by consideration of the imagined consequences of a given line of behavior. Deterrence may be sought from lessening or eliminating the value of objects that can be stolen. Consider a bank's marked money attached to a hidden, exploding dye packet; or indelible serial numbers on property. Another means is to block the access that potential suspects have to a resource needed for the violation, such as weapons or chemicals needed to make drugs or explosives. Drug treatment interventions intended to convince or teach someone that they do not really need or value the thing they desire are another example.

Dossiers and actuarial data, although dependent on passive sensors and computers, are soft discretionary tools intended to also shape the choices of individuals – both agents of control and potential offenders. Algorithms used by authorities for profiling are intended to manage criminal justice decisions. Information about engineered controls is communicated to potential violators in the hope that they will self-regulate after assessing the likely consequences. This can involve realizing that carrying out the offense is impossible or that, while possible to literally carry it out, the risks of identification and apprehension are too great, not to mention the other costs, such as bodily harm from climbing a barbed wire or electrified fence.

Our discussion has noted factors such as goals (prevention, deterrence, identification); focus (subject, resources used in violation, objects sought such as money or information); and type of offense (theft, violation of trust, contraband, violence). As these factors illustrate, there is a clear variability and complexity at play when we talk about technology and social control. To systematize this further, we can offer seven ideal types that vary in social control and technology settings.

1. Target strengthening and insulation

This is an ancient technique in which the victim or object of desire remains but is protected. Perimeter maintaining strategies such as gated communities, fences, guards, and dogs can be distinguished from more specific protections surrounding an object such as safes, armor, chastity belts, and goods in locked cases or chained to immovable objects. The architectural development of "skywalks" linking downtown private buildings creates "sanitary zones" more subject to control than the potentially disorderly public streets below. Targets may be insulated in a different sense by being hidden or disguised. Pagallo's work in this volume (Chapter 37) suggests that websites can be insulated from malicious robots by requiring a human user to type in dancing letters at a prompt in order to gain access to them.

2. Target or facility removal

This reflects the logic of hard prevention. Something that is not there cannot be taken or used. The move toward a cashless society is one example. Merchants who only accept credit or debit cards, or whose registers never have more than a modest amount of cash in them, are unlikely to be robbed by conventional means. Furniture built into the wall cannot be stolen. Subway cars and buses made with graffiti resistant metals are hard to draw upon. Such strategies can also focus on removing objects from society thought harmful, such as illegally cultivated marijuana (Schuilenburg 2015) or automobiles running on fossil fuels (Brisman and South, Chapter 18, this volume).

3. Target devaluation

This lessens or eliminates the value of what is sought. The object remains, but its uselessness makes it unattractive. Examples include products that self-destruct, as with some car radios

485

when stolen or mixing a bad smelling chemical into a product to work against it being inhaled for its hallucinatory effects and biometric and encryption controls on computers and other goods. To preview the empirical case study in the next section, Mexican authorities have recently attempted to combat stolen mobile phones and automobiles used in kidnappings and drug trafficking by registering all phones and automobiles with the state – thus, a stolen unit would be reported as stolen and lose its value (Guzik 2016).

A concept cutting across the three above is *resilience*. As a strategy, it starts with the assumption that stuff will always happen in a complex and complicated world. Under such conditions, the question becomes how can society best respond to and limit harm, rather than trying to prevent things that cannot be prevented or necessarily anticipated. Thus the Internet with its hydra headed, decentralized structure was created to resist an attack on a centralized structure. Environmental restrictions about building on flood planes or using floating foundations are ways of overcoming tidal waves and resisting earthquakes. A European Union project (IRISS 2014) applies the concept to surveillance societies.

4. Offence/offender/target identification

These strategies are present when it is not possible to physically prevent the violation, or where it is too expensive to do so. A focus on surveillance, technology and social control is most prominent here. The goal is to discover the violation or problem and various details such as how it was done, where contraband is, who is responsible and where the responsible person(s) or group(s) or object(s) are and if a transaction is legitimate. These can also serve as victim warnings. A major goal of nineteenth-century forensic science was to develop biometric measures of identity based on the analysis of fingerprints, facial measurements, mug shots, and chemical properties (Thorwald 1965; Cole, this volume). These have significantly expanded from involving a person's gait and voice to tracking their distinctive smell and internet searches. Electronic monitoring or location devices based on GPS are other contemporary examples. Mexico's REPUVE program (discussed in the next section), involves vehicle registration and tracking. And national ID cards (Guzik 2016; Breckenridge 2014; Lyon 2009) would serve to identify all persons in order to facilitate the identification of criminal offenders after the fact. Or consider an effort to help save endangered species such as Mountain Gorillas or to warn villagers when a dangerous elephant is approaching by attaching collars that ping to iPhones (Ozy 2016).

5. Offender weakening or incapacitation

This seeks to render potential offenders harmless by disabling or weakening their will or ability to violate the norm in question. The means may act directly on the body such as cutting off the hands of thieves or lowering serotonin levels to curb violence, or the focus may be on the mind as with aversion therapy. Various citizen protection devices that can be defensively used, such as mace, fit here, as do non-lethal crowd control devices such as electrical, chemical, strobe, and acoustical immobilizers that disorient, stop, restrain, or block individuals.

6. Exclusion

This seeks to keep potential offenders away from targets or tempting environments by banning them from certain places or activities, such as requiring badges and passwords to enter secure sites, excluding minors from bars, curfews, and even exile and related forms of segregation as with prisons. Capital punishment is the ultimate fail-safe form of exclusionary social control.

With the human genome project completed, neo-eugenic modes of exclusion are likely to be advocated, avoiding the uncomfortable task of the state putting people to death. We are also likely to see new restrictions on those deemed to be genetically at risk of violent and other anti-social behavior. Banning identified card counters from casinos or shoplifters from a store are intended to function as exclusion of offenders from privileged sites of commerce (Schuilenburg 2015).

7. Victim warning

This involves altering the material world such that those who might be harmed are alerted to an impending risk, as with elephants wearing electronic collars. A broken tamper-proof seal or failure to hear a popping sound on food products is intended to alert the user that all might not be well. The visible warning offered by branding or clipping the ears of convicts found in medieval Europe offers another example, while the stigma might serve to deter others from being labeled.

These ideal types help us conceptualize the various ways authorities and individuals use social control technologies, whether with respect to public safety or individuals securing valuable items. And they can be applied to help conceptualize different empirical programs and strategies. They also provide a launch point for considering the main substance of this chapter, the not infrequent failure of such strategies to reach their goals. As recent work (Guzik 2016; Schuilenburg 2015; Breckenridge 2014) illustrates, uncertainty is inherent to technology and social control. In Mexico, the state surveillance programs that involved surveillance technologies to create a national registry of mobile devices and automobiles (*target identification* and *target devaluation*) struggled to get off the ground, with the former being terminated before it could be fixed and the latter being transformed and operational in only a few parts of the country. In South Africa, the effort to launch a national ID card (*offender identification*) failed and had to be cancelled (Breckenridge 2014). The various projects studied by Schuilenburg in the Netherlands – marijuana eradication (*target/facility removal*), road transport policing (*target identification*), and shop bans (*exlusion*) – faced various challenges and proved a disappointment to policymakers and officials working with them. The "truth machines" described by Cole (this volume), which are meant to realize *offender identification*, have each failed to live up to their hype.

The reasons behind such struggles are diverse, reflecting the complexity of the social control relationships into which technologies are embedded. To illustrate this complexity, the next section describes a state surveillance program based on technology to fight vehicular crime in Mexico. The case study that follows documents many of the basic factors associated with the uncertainty of technologized social control.

Mexico's public registry of vehicles: a case study in uncertainty

On June 22, 2009, Mexican President Felipe Calderón inaugurated the Public Registry of Vehicles (REPUVE) by placing the program's first radio-frequency identification (RFID) sticker on the inside windshield of a Chevrolet Suburban at a toll booth outside Mexico City. The REPUVE was designed to do three things: (1) create a centralized federal registry of all cars circulating in the country, including vehicle identification number, registration information, physical description, and the name and address of owners; (2) attach 18000-C type RFID tags onto vehicles containing the unit's registration details; and (3) install RFID readers and license plate recognition (LPR) cameras at transit points across the country to verify the status of passing vehicles. In doing this, the registry would serve as a tool to combat crimes involving automobiles, including car thefts, kidnappings, and drug trafficking. The program thus represents a critical tool in the state's fight against organized crime.

The REPUVE database is administered by the Executive Secretary for the National System of Public Security (SESNSP) and receives data from three separate types of sources: "federal authorities" (federal agencies such as the Secretariat of Finance and Public Credit, which manages customs in Mexico); "federative entities," (state-level agencies such as departments of motor vehicles); and "obligated subjects" (private sector businesses dealing with automobiles, such as manufacturers, importers, financing agencies, and insurance companies). Any person can consult the REPUVE database (via a web interface) for information on vehicles, which allows them to know whether the vehicle they own or are acquiring/selling had previously been stolen.

The RFID tags featured in the REPUVE program, produced by the Neology Corporation, contain a microchip that can store 800 bits of information and transmit that data via radio frequency. As passive tags, the RFID stickers only transmit data upon being activated by a RFID reader. The tags are not applied to vehicles by the SESNSP, but are distributed to the "federal authorities", "federative entities," and "obligated subjects", who are responsible for applying them. In the case of new vehicles, the "obligated subjects" who produce or import them simply record VINs onto the chips, adhere the chips onto vehicles, and then report the link between the VINs and chips into the REPUVE database. In the case of cars already circulating, "federal authorities" and "federative entities" apply the RFID tags following a physical inspection of vehicles and corresponding documents.

While a critical piece of Mexico's anti-crime fight, the REPUVE experienced problems soon after its launch. First, although the program was designed to have all of the nearly 25 million vehicles circulating in the country registered with RFID tags by 2012, less than half of Mexico's 32 states were actually applying them to vehicles as that deadline approached. Second, where the registry was functioning, it was not necessarily doing so as a solution to insecurity, but as a solution for highway tolling and customs inspections at the border. Thus, despite the design of the program to serve as a tool that federal authorities in Mexico would wield in order to achieve target hardening and target identification, the REPUVE has struggled to meet this purpose. It thus serves as an ideal case study of the uncertainty surrounding surveillance technologies and social control.

So, what accounts for the uncertainty experienced by the REPUVE? In the remainder of this section, we cover seven basic factors that are essential for understanding the way the REPUVE, and social control efforts more generally, develop. These are: (1) the goals of agents; (2) the interests of organizations; (3) political and legal settings; (4) the resistance of subjects; (5) cultural contexts; (6) material tools and objects; and (7) geography and space.

1. Agents and goals

In other work, Marx (2016) offers a description of the structures undergirding surveillance and other types of social control, a truncated list of which includes *agents* (those who conduct surveillance), *subjects* (those who are surveilled), *audience* (those for whom surveillance is conducted or who otherwise observe it), and *organizations* (whether surveillance takes place within an organizational setting/by an organization or not). By definition, one goal of an agent of social control is to collect data on subjects. In the case of REPUVE, as in many social control operations, monitoring is distributed across multiple actors. These include, but are not limited to, the federal employees within the SESNSP, which oversees the program and its database; the state-level technicians and administrators at registration sites, who inspect vehicles already on the road, place tags on them, and then feed their data into the REPUVE database; and the private technicians working within car producers and importers, who apply tags to new vehicles they produce and share the information with the REPUVE database.

Shared labor does not mean however that control agents share the same goals. For instance, the SESNSP can be seen to possess multiple goals in gathering data for the program. Above all, it wants the compliance of drivers, businesses, and states with the REPUVE law. Compliance is demonstrated, in turn, through records of inspections and registrations of vehicles by companies (in the case of new vehicles) and by state inspectors (in the case of used vehicles). But it is looking to make this information public in order to help citizens make more informed decisions about car purchases. And since the REPUVE is a federal program that could be cancelled by an incoming presidential administration, the SESNSP is also looking to ensure the program's continuity, lest it be cancelled. Thus, monitoring activities serve not only to fulfil the law and fight crime, but to illustrate the program's progress to the public and ensure its own survival amidst political turnover.

The other agents, meanwhile, the state-level technicians and administrators and the private companies, possess goals beyond collecting and sharing data on registered vehicles to the REPUVE database. At the state level, data gathering is conducted by agents and administrators from finance ministries or attorney generals' offices. And collecting vehicular data can not only serve the REPUVE database, but also taxation applications as well. Conversely, attorney generals' offices overwhelmed with other security concerns, of which there are many in Mexico, might be less invested in data collection on vehicles.

With private sector agents, a similar multiplicity of goals is present. Their main concern is profit. Thus, companies collect and report data in order to be in compliance with the REPUVE law and not expose themselves to fines and sanctions. But companies might also see participation and collection of data and placement of tags as a competitive advantage, which will make their vehicles appear safer and thus more desirable to purchasers. Conversely, if the expenses of participating in the program are too great, they might oppose participating altogether.

So, the structure of social control is distributed across various actors. And to the extent that the goals of different agents coincide, the activity can be expected to conform to a greater degree to expectations and plans for that program. But to the extent that goals of different agents do not coincide, then the outcomes of the activity can be expected to not conform to expectations.

2. Organizations and interests

The preceding points illustrate that in considering the goals of different surveillance agents, we need remain aware of the organizations involved in that control activity (Marx 2016). Individual efforts to control, because they are solitary, are distinct from the efforts of collective actors. And varied organizations using the same means may have different goals (e.g. public police in principle concerned with due process and justice as against private police concerned with protecting the interests of their employer).

In the case of the REPUVE, the distinction between private and public organizations is particularly relevant. A private organization, such as a car producer, possesses a set of organizational interests revolving around profit that are generally distinct from a public organization, such as a federal agency or state government bureaucracy, whose interests revolve around the execution of laws, policies, and bureaucratic procedures. On this basis, it would be reasonable to assume that a public organization would be more likely to support a surveillance program or the application of a technology for a public, security purpose, such as the REPUVE, than would a private organization, whose financial interests could be burdened by the costs of compliance. But such a reasonable expectation is betrayed by real world practice.

In Mexico, the greatest challenge to the REPUVE has come from the states, which by and large refused to implement the program. Speaking to program administrators within the SESNSP, the reasons for these difficulties were clear. They involved "resources" and "the will to get things done".

A public organization's implementation of a program such as the REPUVE requires a clear investment in resources. These include, at a bare minimum, computers to process information, printers to print out tags, hand-held RFID readers in order to activate and verify chips once they are adhered, and facilities and salaries for workers in the program. Some of the resources required by Mexican states to implement the REPUVE could be covered by the federal government, but the discrepancies between allocations and costs can be great. And without the funds on hand to dedicate to the program, many states in Mexico refused to implement the REPUVE.

Frustration over resources speaks to, in turn, the will or leadership to get things done. One SESNSP administrator Guzik spoke with opined that "this project has a very large scope, but sadly it has not been seen as such in the upper spheres, not at the Presidential level, nor the Interior Secretary level, nor the Public Security Secretariat level. They haven't given the program the enthusiastic backing that it should have ... There is this feeling, it hasn't been made to succeed."

Conversely, while private organizations like car companies and their representatives complained both publicly and privately about the costs of complying with the REPUVE law, they largely complied. Why? In a communication with a REPUVE compliance officer at the SESNSP, Guzik was told that "the private sector has always been very strict, above all the big companies, with relation to legal compliance with the federal and state governments. It's not easy for their legal teams to know about a law or regulation and not comply with it."

Car producers themselves, meanwhile, saw compliance as part of their corporate culture and identity that reflected their responsible civic participation. As one car producer noted, "Why do [we] comply with the REPUVE law? [Our company] basically has 100% compliance with the law, not only the REPUVE law, but with all the laws, economic, import, customs, etc. We are the most precise and maybe the most compliant in the entire industry." Another company, importing vehicles from Asia, explained, "the Asian philosophy is very precise, very dignified in the sense that if the law asks me to do this, I have to comply ... there are companies that are more rebellious, that in a given moment would sue and not comply. But the Asian companies aren't like that."

Organizations then are a vital component to understanding the fates of social control programs and technology. And if organizational interests coincide with the goals and plans of a particular surveillance program, they can be expected to support the effort. But such interests can be complicated, regardless of whether the organizations are governmental or not. And nonalignment of organizational interests can threaten the work of social control.

3. Political constitutions and legal obligations

If the goals and interests of agents and organizations are central to understanding control outcomes, it is also important to pay attention to the political and legal settings in which these entities operate. This is obvious in many respects, as an agent or subject of control and what their obligations are, are defined by the law. In the case of Mexico's automobile registry, for instance, the REPUVE law clearly defines the SESNSP as the authority over the registry, the states and federal agencies as federated entities and federal authorities, and private companies as obligated subjects with reporting responsibilities to the federal REPUVE.

But the impact of the law can work in more subtle ways as well. For instance, the reluctance of public organizations such as the customs offices and states to participate in the REPUVE is provided for under the Mexican constitution. Within Mexico's federalist political system, a reflection of the country's long history of dividing political power among regional strongmen, federal law does not apply to state governments the way that it does private corporations or individuals. As administrators with REPUVE complained, "the problem that we are having these days is that the private industry, the producers, 100% are participating. But the law requires them to … the states are autonomous. Free and sovereign. We cannot sanction them. We have to convince them. That's the challenge for us." Thus, Mexico's federal political constitution supports the expression and pursuit of different organizational interests.

But even for private organizations complying under the threat of punishment, the balance of power between political branches enshrined in Mexico's constitution provides a space for them to pursue their interests outside of the program. So it was that the Automotive Industry (AMIA) launched a lawsuit to reform the REPUVE law so that the states would have to register all vehicles. So, even though there was an agreement between the auto industry and the REPUVE, the industry did not stop in its efforts to alter the Public Registry of Vehicles Law so that its obligation would be done away with.

In these ways, legal contexts are central to outcomes. And we might expect more open political systems to offer more opportunities for actors and organizations to pursue their own interests or not comply. Similarly, federated political systems, such as in the US and Mexico, may differ from unitary republics, such as in France, where states and regions possess less autonomy from federal authorities. Autocratic political systems, meanwhile, would provide less opportunity to deviate from programs and federal, centralized plans.

4. Subjects and resistance

These considerations bring us to the topic of resistance. Marx (2016) has identified twelve general techniques of neutralization that subjects undertake to counteract surveillance. These are: (1) *discovery*, finding out whether surveillance is in operation; (2) *avoidance*, of the contexts or places that one knows will be subject to monitoring; (3) *piggy-backing*, where subjects directly face surveillance rather than avoid it, but evade control by attaching to a legitimate subject or object; (4) *switching*, when a subject transfers an authentic result to someone or something to which it does not apply; (5) *distorting*, moves to manipulate the surveillance-collection process such that, while test or inspection results are technically valid, the inferences drawn from them about performance, behavior, or attributes are invalid; (6) *blocking* and (7) *masking*, where the former makes inaccessible what is of interest to agents and the latter renders what is of interest unusable; (8) *breaking*, to render the surveillance device inoperable; (9) *refusing*, by subjects to cooperate under the terms desired by agents; (10) *explaining*, to account for an unfavorable result in order to cast doubt upon a tactic; (11) *cooperating*, where surveillance efforts can be neutralized or undermined if agents come to collude with subjects; and (12) *counter-surveillance*, when subjects use the same tools as agents, and they may do so to record the behavior of agents.

Illustrations of these diverse tactics can be seen in drug testing. There, we see "refusal" (to take a test), "discovery" (of the date of a random test), "avoidance" (not going to work on testing day), "switching" (a clean drug sample for a tainted one), "distorting" (consuming substances to neutralize the drug test), "masking" (one's identity to testers), and "countersurveillance" (testing on oneself to ensure success).

These strategies are present in and instructive for interpreting the outcomes of the REPUVE. In the states where the program was actually being implemented, subjects – drivers who are obligated to register their vehicles with the program – often refused. In other words, only a small number of drivers chose to provide themselves with the chance to harden their valuable targets from the threat of theft. And such a neutralization tactic threatened the viability of the program for those states looking to embrace it. State agents in those states were then left with having to brainstorm ways to lure drivers into the program.

5. Culture and stores of skepticism

If the lack of participation by drivers reflects strategic choices to not participate, it's also important to consider where such motivations derive from. Here, cultural factors specific to particular groups, regions, and countries can be seen to influence subjects' view of surveillance and social control.

In Mexico, for instance, federal car registries have a notorious history. Prior to the REPUVE, the Mexican government launched the National Registry of Vehicles (RENAVE). The RENAVE, like the REPUVE, was to include a database of vehicles manufactured, assembled, imported, or circulating in national territory in order to prevent contraband and automobile thefts. Unlike the REPUVE however, the program carried a registration fee (375 pesos, or $47, for new cars) and was operated by a private firm, Talsud. The RENAVE fell into disrepute when the head of Talsud, Ricardo Miguel Cavallo, was arrested in Cancún after it was learned that he was actually Miguel Angel Cavallo, an Argentine war criminal wanted by Spanish authorities for torture and other crimes committed during Argentina's military dictatorship in the late 1970s.

For drivers in Mexico, such stories remained present in their popular imagination and served as a ready well of mistrust when they encountered the REPUVE. As a technician with REPUVE explained, "There is 20% of the people that simply are not going to come, they're not going to come [to register]. Maybe they have bad information. They think that this is insecure, that we work for criminals, like RENAVE. RENAVE was a bad program that is distorting the REPUVE, because people think that it's the same story". Thus, the power of culture can impact seemingly unambiguous hard-engineered control efforts.

6. Technical tools and material objects of surveillance

Thus far, this review of the obstacles that can complicate social control activities through technology has been human-centric, focusing on the human actors and organizations that are central to control activities. Of course technology itself can be very important as well. Mirroring the distinction between the agents and subjects of social control, technology is central as a tool of control and as an object of control. And in both cases, technology can impinge upon the best laid plans of crime fighters.

In the case of tools, an obvious concern is whether a particular instrument is appropriate to the application it was chosen for. For the REPUVE, a controversy concerned passive versus active RFID technology. Following the REPUVE's launch, media reports cited experts who argued that the passive tags chosen by the SNSP had disadvantages such as poor read coverage and a high cost of reading equipment. Ultimately, the head of the SESNSP, Roberto Campa Cifrián, was called to testify before Mexico's House of Deputies, the country's lower legislative chamber, to explain his decision on the bid. Earlier, he had said that the decision was made based on recommendations of an evaluation team from three of Mexico's universities and that their decision was informed by a consideration of chip lifespan, the lower costs per unit for passive chips, and the fact that passive technology was open source and not subject to proprietary restrictions. Unconvinced by the executive secretary's explanation, the House forced Campa Cifrián's resignation.

It is also important to consider the objects of surveillance, in this case automobiles. Even if subjects – drivers – want to register and participate in the REPUVE, objects can pose problems. One example here concerns VINs (Vehicle Identification Numbers). While there is no single standard for assigning VINs, the practice of assigning vehicles a 17-digit number identifying the car's manufacturer and characteristics, including model year, was adopted in many parts of the world in the early 1980s. But in Mexico, the international norm was applied beginning in 1997. Thus, vehicles produced in Mexico before that time could present particular difficulties. As an administrator with REPUVE explained, "Nissans and Volkswagens from 95, 96, 97, as well as Chevys from 94. They can't be put into the Public Registry of Vehicles for now. The system doesn't allow the registration of their serial numbers because of a production problem in those vehicles. The REPUVE system detected duplicate serial numbers in those brands, which caused the service to be suspended."

Automobiles have also proven tough to monitor on account of their windshields. Many vehicles have metallic particles (in their windshields) that prevent readings from REPUVE's equipment. Program administrators estimated that around 3 percent of the 25 million vehicles in the country, around 750,000, have such windshields. These vehicles cannot then be included in the registry. Thus, the reading of REPUVE's RFID tags has been complicated by the materiality it is meant to control.

7. Geography and the challenges of space

The technologies of social control do not exhaust the list of non-human, material elements that can complicate monitoring. Geography, the material or human-built landscapes upon which the daily activities of life are carried out, can also influence things. For instance, in the case of the REPUVE, program administrators found it challenging to plan for and implement the program in rural settings, which are set apart from the technical grids of urban centers which support the information technologies of social control.

In Zacatecas, a state with an appreciable rural population spread out over a large geographic area, planning how to ensure vehicles in rural areas were registered in the program posed a particular challenge. One technician there explained, "we know that in Zacatecas there exists a half-million registered vehicles, but there aren't a half-million vehicles. There's more. Some aren't plated. Others have American plates ... Where are these people? In the villages. Why? Because maybe they make only a trip or two to the city. In the towns, in the cities, it's a bit stricter, because there's more people watching. There are cameras. You can't get away with as much. But there are villages that are very small, that have 1 or 2 traffic cops at most. And then, it's your buddy, your neighbor, your brother, or your father who's stopping you. And then it's just 'get on your way' [and nothing happens]." In this sense, the close-knit socio-political formations that are characteristic of rural areas prevent the implementation of hard-engineered control on the ground.

To respond to these geographic challenges, program leaders in Zacatecas decided to employ mobile registration modules, housed in mobile trailers that could be moved around the state in order to complete registrations. But in electing for mobile units over stationary ones, the computer equipment that the team used had to be broadcast remotely via radiofrequency. And the team registering vehicles experienced service interruptions that delayed the transmission of data from the registration site to the secretary's database. An administrator in Zacatecas noted that "I would ask the system specialists when they installed the system why the signal gets cut off. They said that these things are out of our hands. Maybe the telephone company, who administers the internet connection, had an issue. Or maybe there was a problem of some sort. Someone hit a telephone pole. Or someone was digging and cut the fiber optic cables."

The uncertainty principle

The preceding list of obstacles are not unique to Mexico or the REPUVE. Marc Schuilenburg's (2015) study of marijuana eradication campaigns and efforts to police road transport crimes in the Netherlands identified *dissimilar organizational interests* between policing agencies and housing associations (in the case of marijuana eradication) and policing agencies and insurance companies (in the case of road transport policing) as impediments to these novel security strategies. In the same work, he also notes how the *disparate goals of control agents* work against shop bans – shop owners are often unwilling to invest the time and money into training their employees to identify shoplifters, which is necessary for the ban plan to work. Bigo (2006) and Cole (Chapter 29, this volume) touch upon related elements. Criminal courts in general were not enthusiastic about the wider use of lie detector tests in their proceedings for fear that a mechanic mode of truth-telling would make the court's human-based methods obsolete. In this sense, the courts possessed *dissimilar organizational interests* from the policing agencies promoting the machines. But more than this, "truth machines", like the "green technologies" studied by Brisman and South (Chapter 18, this volume), possess inherent *technical limitations* that complicate their programs of social control. Every way of doing something is also a way of not doing something else, with attendant gains and costs. Machines of "objective truth", even in the case of DNA identification technology, require a decent share of subjective processing and interpretation, and the production and use of green technologies always requires some element of environmental harm.

Our list of obstacles is hardly exhaustive. The discussion offers a starting point for future work to build on. And the preceding points are not intended to suggest that social control technologies are doomed to total failure. The REPUVE exists and operates today in Mexico. The reason it does is related to the program administrators' ability to respond to and work with the varying challenges noted. In Sonora, for instance, national REPUVE directors worked with state officials to fashion the registry into a tolling solution to provide residents free passage on federal highways. At the border, administrators convinced customs offices to participate in the program by allowing them to charge for car inspections, in seeming violation of the REPUVE law itself.

Such improvisations have been described as the "self-organizing processes" of security work (Schuilenburg 2015) or as "statecraft" (Guzik 2016). And they are critical to consider, since such alternations help shape the outcome and meanings of social control. For instance, in Sonora, the REPUVE is valued nearly universally there as a means for establishing and respecting a right to free transit that was fought for and established in the Mexican Revolution rather than a measure against vehicle thefts. At border crossings, the REPUVE is viewed negatively as another scheme to squeeze tax revenues out of individuals importing their vehicles from abroad. In Zacatecas, the REPUVE is understood and approached more cautiously as another governmental program promising to provide security. So then, although the REPUVE is functioning to realize *target identification* and *target hardening* in some parts of Mexico (but not others!), how its technology operates and what it means to people emerge in time and practice as administrators respond to challenges that were often difficult to anticipate when the program began.

In considering obstacles and improvisations and spaces to maneuver, we see *an uncertainty principle* at work with technologies and social control that belies the seeming objectivity and fixity of such efforts and the effort to fully reach a cherished engineering goal "to get the humans out of the loop." The factors we note make it likely that the application of surveillance technologies will often bring surprises. The hopes of utopians and fears of dystopians are rarely on target. Contexts need to be considered, and discretion and interpretation remain dance partners. There is usually room for humans to respond in ways not anticipated, or at least desired by planners. As Brisman and South (this volume) suggest, our "faith in magic wands" needs to be held in check.

But saying that uncertainty is inherent to social control and technology is not the same as saying that such efforts are wholly unpredictable. While uncertainty is inherent to technology and social control, systematic study and analysis can improve understanding and application.

At least five of the most prominent sources of uncertainty challenging the best laid plans can be noted:

1   *Uncertainties of functioning* – technical artifacts may fail to work, break, or require costly unanticipated inputs and revisions; as they say, "stuff happens."
2   *Uncertainties of intended function* – most technical artifacts are not limited to their specified function. Thus Don Ihde (2008) has talked about the "designer fallacy" and the "ambiguous multistable possibilities" of artifacts. An airplane can get you to your destination, but also be a weapon to destroy buildings.
3   *Uncertainties of consequence* – technologies may bite back. (Aspirin can make one feel better, but taking too many can kill.) Policy and practitioners need to be particularly alert to gradient effects and short and longer time periods. Surprise outcomes are more likely when we fail to analyze the assumptions that are often buried (hidden?) deep within the celebratory rhetoric of technology's boosters (Appendix B lists a number of such techno-fallacies).
4   *Uncertainties of context* – social actors (individual or organisational) use their technological/behavioural extensions in ways which are unpredictable and often "irrational." Contexts with varied goals and interests may collide and react differently to the same tool.
5   *Uncertainties of environment* – physical environments can also be unpredictable and overwhelm the tool, as with the case of a meteorite hitting earth, or a monumental earthquake. A technology may work just fine as with the Fukushima Daiichi nuclear plant, but speaking metaphorically, only until there is a perfect storm in the form of the earthquake.

An example of one insight that future research into the relationship between the uncertainty principle and complexity is considered next. At the micro-level, social operations involving technology might be expected to encounter the fewest obstacles, since the number of agents, tools, subjects, and objects to monitor or control is limited. Take, for instance, the application of an RFID tag as an access solution for a parking garage, a hard-engineered solution to *target identification*. Generally, a single human agent with the tools of an RFID tag and electronic gate suffice to manage the operation, while the building management and company providing the service comprise the organizational setting. The rules for surveillance are largely defined by management. And the subjects are limited to those who want to access the garage and the number of vehicles that can fit in it. Finally, the material landscape of the parking garage is man-made and thus conducive to the application of man-made technological networks.

As surveillance is scaled up to a meso-level, the number of obstacles increases. Let's say we scale up RFID tag access from a parking garage to a tolling solution for state highways. Now, the agents increase, from a single guard to a state police force and local police forces. The organizational context changes as well. Beyond building garage management, we now have a state highway agency, police forces, political powers to authorize the program, corporations looking to compete for such a contract, and perhaps civic organizations that may oppose such a program. The number of subjects and objects of surveillance increases as well, as do the geographic challenges.

Scaling up to the macro-level, let's apply RFID tags for security on a national level, like the REPUVE, and the number of obstacles increases still. Federal police forces and agencies are added to the agents. Organizations now include federal agencies and political parties. National federal law, and international agreements on closed and open borders shift the legal context, as do constitutional guarantees. The number of subjects increases as well, as do the objects to be monitored. And geographic variability is enhanced still, from dry desert landscapes to dense rainforests. So the kinds of technologies we have considered must be accepted by a multiplicity of organizations, agents, and subjects, and be functional across a variety of material environments. And with open source technology, the technology must operate across a variety of providers.

By this logic, we can predict that macro-level efforts to achieve social control will be more uncertain than local efforts. And for those concerned with fighting crime, the uncertainty principle offers grounds to favor local applications, other factors being equal.

Those who look optimistically to technology as the solution to control problems and those who question it too often talk past each other. Both can reflect techno-fallacies embedded, and often unrecognized, in our culture (see Appendix B, drawn from Marx 2016 which identifies 44 such beliefs which may be fallacious on empirical, logical or value grounds). Particularly relevant here are: *the fallacy of a passive, nonreactive environment; the fallacy of more – if some is good, more must be better; the fallacy that the facts speak for themselves; the fallacy of explicit agendas; the fallacy of the sure shot; the fallacy of the fail safe system; the fallacy of delegating decision-making authority to the machine; the fallacy of the free lunch; and the fallacy that technology will always remain the solution rather than become the problem.*

Whether critic or advocate, there is a need to be aware of complexity and value conflicts and the need for empirical reality-checks. Liberty and democracy are fragile and not self-sustaining. The message for the independent scholar is to avoid premature commitment to the camp of either the optimists or the pessimists. Usually, the best answer is "it isn't clear" or "it depends." And then to take it further and indicate what needs clarification and what it depends on.

Let us finish with a contemporary implication of Odysseus's encounter with the Sirens. With the uncertainty principle in mind, the tale of Odysseus and the Sirens is not only about the heroism of its protagonist, as Homer would have us believe, nor about the ways technology can increase inequality, as per Horkheimer and Adorno. It is also a story about rationally applying a mechanical solution to a problem. Odysseus's Siren problem could be easily solved by a handful of beeswax. But today, the siren songs offered by technology's cheerleaders are more challenging and perhaps harder to resist. Amidst a deepening sense of crisis, they tempt us with simple engineered solutions sometimes bordering on the illusional, if not also the delusional, with the implication that our challenges can be met as successfully and with as little effort as Odysseus's. At such times it is well to recall another Greek myth, that of Icarus whose hubris was his undermining, as his wax wings melted when he flew too close to the sun.

References

Bell, Wendell (1997) *Foundations of Futures Studies: History, Purposes, and Knowledge*, Volume 1. New Brunswick, NJ: Transaction Publishers.

Bigo, Didier (2006) Globalized (in)security: The field and the ban-opticon, in Naoki Sakai and John Solomon (eds) *Translation, Biopolitics, Colonial Differences*, 109–156. Hong Kong: Hong Kong University Press.

Breckenridge, Keith (2014) *Biometric State: The Global Politics of Identification and Surveillance in South Africa, 1850 to the Present*. Cambridge: Cambridge University Press.

Ellul, Jacques (1964) *The Technological Society*. New York: Vintage Books.

Gibbs, Jack P. (1989) *Control: Sociology's Central Notion*. University of Illinois Press.

Grabosky, Peter N. (1996) Unintended consequences of crime prevention, in Ross Homel (ed.) *Crime Prevention Studies*, Volume 5, 25–56. Monsey, NY: Criminal Justice Press.

Guzik, Keith (2016) *Making Things Stick: Surveillance Technologies and Mexico's War on Crime*. Berkeley, CA: University of California Press.

Hilgartner, Stephen, Richard C. Bell, and Rory O'Connor (1982) *Nukespeak: Nuclear Language, Visions, and Mindset*. San Francisco: Sierra Club Books.

Homer (2016) *The Odyssey*, A.S. Kline (trans.), available at: http://www.poetryintranslation.com/PITBR/Greek/Odyssey12.htm#_Toc90268047.

Horkheimer, Max, and Theodor W. Adorno ([1944] 1972) *Dialectic of Enlightenment*. New York: Herder and Herder.

Ihde, Don (2008) The designer fallacy and technological imagination, in P. Vermaas, P. Kroes, A. Light and S.A. Moore (eds) *Philosophy and Design: From Engineering to Architecture*, 51–59. New York: Springer.

IRSS (2014) *Handbook of Increasing Resilience in Surveillance Societies*. http://irissproject.eu/?page_id=610&utm_source=IRISS_June2014&utm_campaign=8a7a98fc5a-IRISS_PR_Surveillance_in_Europe_10_10_2014&utm_medium=email&utm_term=0_a05fc7983f-8a7a98fc5a-174015249 (accessed October 1, 2014).

Janowitz, Morris (1975) Sociological theory and social control, *American Journal of Sociology*, 81(1): 82–108.

Lyon, David (2009) *Identifying Citizens: ID Cards as Surveillance*. Malden, MA: Polity.

Mander, Jerry (1992) *In the Absence of the Sacred: The Failure of Technology and the Survival of the Indian Nations*. San Francisco: Sierra Club Books.

Marcuse, Herbert (2002) *One Dimensional Man*. New York: Routledge.

Marx, Gary T. (1987) Raising your hand just won't do, *Los Angeles Times*. April 1, 2016.

Marx, Gary T. (1995) The engineering of social control: The search for the silver bullet, in John Hagan and Ruth D. Peterson, (eds) *Crime and Inequality*, 225–246. Stanford, CA: Stanford University Press.

Marx, Gary T. (2016) *Windows into the Soul Surveillance and Society in an Age of High Technology*. Chicago: University of Chicago Press.

Marx, Karl (1947) *Capital, volume 1*. New York: International Publishers.

Marx, Karl (1956) *Economic and Philosophical Manuscripts of 1844*. Moscow: Foreign Languages Publishing House.

Morozov, Evgeny (2011) *The Net Delusion: The Dark Side of Internet Freedom*. New York: Public Affairs.

Mumford, Lewis (1934) *Technics and Civilization*. London: George Routledge & Sons Limited.

Ozy (2016) The surprising link between Mountain Gorillas and iPhones. Feb. 18, 2016. www.ozy.com/fast-forward/the-surprising-link-between-mountain-gorillas-and-iphones/64539.

Postman, Neil (1992) *Technopology: The Surrender of Culture to Technology*. New York: Knopf.

Rosner, Lisa (2004) *The Technological Fix: How People Use Technology to Create and Solve Problems*. London: Routledge.

Rule, James B. (1978) *Insight and Social Betterment: A Preface to Applied Social Science*. Oxford: Oxford University Press.

Schroeder, Ralph (1995) Disenchantment and its discontents: Weberian perspectives on science and technology, *Sociological Review*, 43(2): 227–250.

Schuilenburg, Marc (2015) *The Securitization of Society: Crime, Risk, and Social Order*. New York: New York University Press.

Scott, James C. (1998) *Seeing Like A State: How Certain Schemes to Improve the Human Condition Have Failed*. New Haven, CT: Yale University Press.

Tenner, Edward (1997) *Why Things Bite Back: Technology and the Revenge of Unintended Consequences*. New York: Vintage.

Thorwald, Jürgen (1965) *The Century of the Detective*. New York: Harcourt, Brace & World.
Tucker, Robert (ed.) (1987) *The Marx-Engels Reader*. New York: W.W. Norton and Company.
Weinberg, Alvin M. (1967) Can Technology replace social engineering? *American Behavioral Scientist*, 10(9): 7–10.
Weizenbaum, Joseph (1976) *Computing Power and Human Reason*. San Francisco, CA: W.H. Freeman.
Wiener, Norbert (1967) *The Human Use of Human Beings: Cybernetics and Society*. New York: Avon Books.
Winner, Langdon (1988) *The Whale and the Reactor: A Search for Limits in an Age of High Technology*. Chicago: University of Chicago Press.

Appendix A

*Raising Your Hand Just Won't Do**

TO: ALL EMPLOYEES
FROM: EMPLOYEE RELATIONS DEPARTMENT
SUBJECT: RESTROOM TRIP POLICY (RTP)

An internal audit of employee restroom time (ERT) has found that this company significantly exceeds the national ERT standard recommended by the President's Commission on Productivity and Waste. At the same time, some employees complained about being unfairly singled out for ERT monitoring. Technical Division (TD) has developed an accounting and control system that will solve both problems.

Effective 1 April, [April Fool's Day] a Restroom Trip Policy (RTP) is established.

A Restroom Trip Bank (RTB) will be created for each employee. On the first day of each month employees will receive a Restroom Trip Credit (RTC) of 40. The previous policy of unlimited trips is abolished.

Restroom access will be controlled by a computer-linked voice-print recognition system. Within the next two weeks, each employee must provide two voice prints (one normal, one under stress) to Personnel. To facilitate familiarity with the system, voice-print recognition stations will be operational but not restrictive during the month of April.

Should an employee's RTB balance reach zero, restroom doors will not unlock for his/her voice until the first working day of the following month.

Restroom stalls have been equipped with timed tissue-roll retraction and automatic flushing and door-opening capability. To help employees maximize their time, a simulated voice will announce elapsed ERT up to 3 minutes. A 30-second warning buzzer will then sound. At the end of the 30 seconds the roll of tissue will retract, the toilet will flush and the stall door will open. Employees may choose whether they wish to hear a male or a female "voice". A bilingual capability is being developed, but is not yet on-line.

To prevent unauthorized access (e.g., sneaking in behind someone with an RTB surplus, or use of a tape-recorded voice), video cameras in the corridor will record those seeking access to the restroom. However, consistent with the company's policy of respecting the privacy of its employees, cameras will not be operative within the restroom itself.

An additional advantage of the system is its capability for automatic urine analysis (AUA). This permits drug-testing without the demeaning presence of an observer and without risk of human error in switching samples. The restrooms and associated plumbing are the property of the company. Legal Services has advised that there are no privacy rights over voluntarily discarded garbage and other like materials.

In keeping with our concern for employee privacy, participation in AUA is strictly voluntary. But employees who choose to participate will be eligible for attractive prizes in recognition of their support for the company's policy of a drug-free workplace.

Management recognizes that from time to time employees may have a legitimate need to use the restroom. But employees must also recognize that their jobs depend on this company's staying competitive in a global economy. These conflicting interests should be weighed, but certainly not balanced. The company remains strongly committed to finding technical solutions to management problems. We continue to believe that machines are fairer and more reliable than managers. We also believe that our trusted employees will do the right thing when given no other choice.

*Marx 1987. Appeared on April Fool's day, but many readers thought it was real. Is it more real today than in 1987?

Appendix B

Techno-fallacies

The social scientist focused on the empirical, logical and moral aspects of claims about technology for social control often hears statements that sound wrong. These techno-fallacies can involve elements of substance as well as styles of mind and ways of reasoning. Sometimes these fallacies are frontal and direct; more often they are tacit – buried within seemingly commonsense, unremarkable assertions. It is important to approach the commonplace in a critical fashion – whether we initially agree or disagree with the ideas.

This approach to analyzing the rhetoric of technology advocacy and consequences follows in the broad tradition of Mumford (1934), Ellul (1964), Weinberg 1967, Winner (1988), Postman (1992), Tenner (1997), Scott (1998), Marcuse (2002) and Rosner (2004) and of the more focused work on topics such as computers, the environment, energy, and crime (e.g. Wiener 1967; Weizenbaum 1976; Morozov 2011; Mander 1992; Hilgartner, Bell, and O'Connor 1982; Marx 1995; Grabosky 1996).

Beliefs may be fallacious in different ways. Some are empirically false or illogical. With appropriate evidence and argument, persons of goodwill holding diverse political perspectives and values may be able to see how they are fallacious, or in need of qualification.

Fallacies may also involve normative statements about what matters and is desirable. These reflect disagreements about values and value priorities. To label a normative belief a fallacy more clearly reflects the point of view of the labeler. However, normative positions are often informed by empirical assumptions (e.g. favoring controls that are presumed to eliminate discretion because they are believed to be more effective). In sniffing out fallacies, one must identify and evaluate the intermingling of fact and value and the quality of the facts (Rule 1978; W. Bell 1997). At a very general level, people often agree on values (though they often dissent over prioritizing and implementing these). Disagreements also commonly occur over what evaluation measure(s) and specific tools for judgment are most appropriate and over how evidence is to be interpreted—both with respect to what it says empirically and to its meaning for a given value.

Marx (2016) identifies five basic categories for organizing techno-fallacies:

A. Fallacies of technological determinism and neutrality
B. Fallacies of scientific and technical perfection
C. Fallacies involving subjects of surveillance
D. Fallacies involving questionable legitimations
E. Fallacies of logical or empirical analysis

Information age techno-fallacies

A. Fallacies of technological determinism and neutrality

1. The fallacy of autonomous technology and emanative development and use
2. The fallacy of neutrality
3. The fallacy of quantification
4. The fallacy that the facts speak for themselves
5. The fallacy that technical developments must necessarily mean less privacy

B. Fallacies of scientific and technical perfection

6. The fallacy of the 100 percent fail-safe system
7. The fallacy of the sure shot
8. The fallacy of delegating decision-making authority to the machine
9. The fallacy that technical solutions are to be preferred
10. The fallacy of the free lunch or painless dentistry
11. The fallacy that the means should determine the ends
12. The fallacy that technology will always remain the solution rather than become the problem

C. Fallacies involving subjects of surveillance

13. The fallacy that individuals are best controlled through fear
14. The fallacy of a passive, nonreactive environment
15. The fallacy of implied consent and free choice
16. The fallacy that personal information is just another kind of property to be bought and sold
17. The fallacy that if critics question the means, they must necessarily be indifferent or opposed to the ends
18. The fallacy that only the guilty have to fear the development of intrusive technology (or if you have done nothing wrong, you have nothing to hide)

D. Fallacies of questionable legitimations

19. The fallacy of applying a war mentality to domestic surveillance
20. The fallacy of failing to value civil society
21. The fallacy of explicit agendas
22. The legalistic fallacy that just because you have a legal right to do something, it is the right thing to do
23. The fallacy of relativism or the least bad alternative
24. The fallacy of single-value primacy
25. The fallacy of lowest-common-denominator morality
26. The fallacy that the experts (or their creations) always know what is best
27. The fallacy of the velvet glove
28. The fallacy that if it is new, it is better
29. The fallacy of equivalence or failing to note what is new
30. The fallacy that because privacy rights are historically recent and extend to only a fraction of the world's population, they can't be very important
31. The fallacy of the legitimation via transference

E. Fallacies of logical or empirical analysis

32. The fallacy of acontextuality
33. The fallacy of assumed representativeness
34. The fallacy of reductionism
35. The fallacy of a bygone golden age of privacy
36. The fallacy that correlation must equal causality
37. The fallacy of the short run

38. The fallacy that greater expenditures and more powerful and faster technology will continually yield benefits in a linear fashion
39. The fallacy that if some information is good, more is better
40. The fallacy of meeting rather than creating consumer needs (demand vs. supply)
41. The fallacy of the double standard
42. The fallacy that because it is possible to successfully skate on thin ice, it is wise to do so
43. The fallacy of rearranging the deck chairs on the titanic instead of looking for icebergs
44. The fallacy of confusing data with knowledge and technique with wisdom

Part IV

Technology and the process of justice

29

Establishing culpability

Forensic technologies and justice

Simon A. Cole

As this chapter is being written, the United States is convulsed in an intense discussion about the desirability and operationalization of so-called "body cameras," lightweight, miniaturized digital video cameras that can be mounted on the clothing (and cars) of police officers to record their interactions with the public (e.g. Mateescu *et al.* 2015; Simon and Bueermann 2015; Stalcup and Hahn 2016). This discussion has been prompted by a convergence of technological developments, which have rendered such cameras ever smaller and more affordable, with a significant number of high-profile, mostly racially charged, incidents, some of them fatal, between police and citizens, which were recorded either with body cameras or with increasingly ubiquitous privately held video recording devices, such as those on many mobile telephones. This discussion is very unlikely to be confined to the US; it has begun, or should be expected to soon begin, around the world.

The pace of technological development today is such that I expect the vast majority of readers of this chapter will know a great deal more than I do about the outcome of the societal discussion of body cameras, putting me at somewhat of a disadvantage. (Indeed, it will presumably not be long before my characterization of 2015-vintage body cameras as "miniature" and "lightweight" will seem quaint.) My ignorance of the future notwithstanding, I believe it is possible to say that this discussion will implicate the very questions about the truth-producing capabilities of machines and the handling of those questions by legal institutions that form the topic of this chapter.

Photographs and video

Law has long dreamed of the idea of mechanical forms of truth-telling that would be immune to the foibles which were such a familiar part of legal evidence: the lies, the biases, the inaccuracy. Wigmore's (1940: §1367) famous remark about cross-examination being the "greatest legal engine ever invented for the discovery of truth" notwithstanding, law has long been both tempted and repulsed by the idea of more literal "engines" which might prove "greater" still. Without discounting the existence of an earlier history of legal truth-making machines, it is reasonable to date the flourishing of such devices to "the second half of the nineteenth century," which

> saw a new mode of persuasion rising to dominance, driven by a new class of machine-made testimonies that threatened to turn words into an inferior mode of communicating

facts. Ever alert and never involved, machines such as microscopes, telescopes, high-speed cameras and x-ray tubes purported to communicate richer, better, and truer evidence often inaccessible otherwise to human beings. The emblem for this new type of mechanical objectivity was visual evidence. "Let nature speak for itself," became the watchword, and nature's language seemed to be that of photographs and mechanically generated curves.

(Golan 2004: 183–184)

The notion of "mechanical objectivity," or what Seltzer (1992: 100) has called "machine culture," is drawn from Daston and Galison (1992) who discussed numerous late nineteenth-century scientific devices for recording observations, such as pen registers, sphygmometers, and so on.

One of the earliest technologies to invoke mechanical objectivity was that ancestor of the body camera, the photograph. Some late nineteenth-century courts concluded that "the photograph was not merely evidence, but the best kind of evidence imaginable: mechanical, automatic, and not subject to those biases and foibles that may cloud human judgment," one court going so far as to declare, "We cannot conceive of a more impartial and truthful witness than the sun." Indeed, one commentator presciently "suggested that when photographic techniques were 'perfected,' all of the streets and alleys of cities should be swept by surveillance cameras. The author hoped that these cameras would capture images of anyone rioting or disturbing the peace for use in subsequent legal proceedings" (Mnookin 1998: 18–19). While it is now well understood that the perceived objectivity and impartiality of the photograph is illusory, the photograph has nonetheless continued to generate strong claims of objectivity. "Throughout its history of use in law enforcement and criminal identification practices," Finn (2009: xii) writes, "the subjectivity of the photograph has been rendered largely invisible against the tremendous literal power of the image to record objects in the live world." At the same time, however, "the photograph's offer of verisimilitude was threatening: indeed, in its strongest form, the photograph threatened to make the fact-finding portion of a trial redundant by providing the facts in an incontestable form" (Mnookin 1998: 6). Mnookin (1998: 54) suggests that courts partially resolved this tension by allowing photographs as evidence, but only as "demonstrative evidence," that illustrated the testimony of a witness, rather than evidence that could offer a stand-alone claim to truth.

This was not true of another form of photograph, a photograph of the inside of the body: the X-ray. "[B]y emphasizing the objective aspect of X-ray photographs—presenting them as the deterministic product of the immutable laws of nature," litigants won the admissibility of X-ray evidence just prior to the turn of the twentieth century (Golan 2004: 192–195). By the 1920s, X-ray images had become an exception to the demonstrative evidence doctrine, admissible "as substantive evidence of the conditions revealed by them" (Golan 2004: 207).

While X-rays required expert interpretation, in the late 1960s a descendant form of photography, the surveillance camera, was admitted with "no one to speak for them in court. Thus, for the first time, the courts faced machine-made visual evidence that was no longer required to be coupled with a human agency in order to express what it contained" (Golan 2004: 209). It has been noted that "This approach came to be known as the 'silent witness' doctrine because it recognizes the photograph as one that 'speaks for itself' and not for human patron" (ibid.: 210).

Curiously, however, as the technology has improved, the pendulum in the courtroom has swung back. Beginning around the early 1990s, prosecutors with poor-quality images from surveillance cameras began proffering expert witness to interpret these images. Drawn from an eclectic and inconsistent group of disciplines, these interpreters of CCTV images, in some cases, purported to be able to support definitive statements that a particular person was the person captured in the video image. There was, and remains, little empirical support for such extreme claims (Edmond *et al.* 2009). While many courts have disallowed these strong claims, courts in many countries have generally admitted more modestly formulated imaging evidence (Edmond *et al.* 2013).

A seminal moment for the video camera was undoubtedly the 1991 beating of Rodney King by Los Angeles police officers, captured on a now quaint home video camera by a bystander. While the unedited footage, given widespread coverage on television, shocked viewers, the case has been interpreted as standing for the proposition that even seemingly incontestable readings of video images are subject to social interpretation. The video did not simply "speak for itself" (Judith Butler quoted in Stalcup and Hahn, 2016). In his famous analysis, Goodwin (1994) showed that at the police officers' trial the defendants softened the impact of the video by slowing it to a frame-by-frame analysis and invoking "professional vision," using the authority of experts to frame the officers' moment-by-moment seemingly brutal actions as reasonable and consistent with their training (see also Ronell 1992; Cannon 1999).

Two decades later, the modest dissemination of relatively inexpensive video cameras and the remote possibility that any given police action might fortuitously be subject to citizen counter-surveillance has been replaced by the perceived *ubiquity* of much more inexpensive, lightweight, pocket-sized cameras integrated into a device commonly possessed by many, if not most, citizens, the mobile telephone. This not even to speak of the well documented explosion of surveillance cameras in urban areas around the globe (Doyle *et al.* 2012). This technological development has transformed expectations so that any police encounter that takes place in front of bystanders may be expected to be filmed by these third parties: the cameras, to some extent, have been turned on the police (Koskela 2009; Goldsmith 2010; Wilson and Serisier 2010; Stuart 2011). Indeed, at the time of this writing, the popular media is suffused with debate about the "Ferguson effect," named after a notorious police killing in Ferguson, Missouri. The claim is that police are "holding back" and avoiding encounters with citizens for fear of having their behavior filmed and posted on the internet.

Fingerprinting and forensics

Another criminal identification technology, fingerprinting, more directly mimicked the mechanical inscription devices described by Daston and Galison. The fingerprint—at least the deliberately captured fingerprint—was widely described as possessing this quality of mechanical objectivity (Cole 2001: 165–166). Fingerprinting was "a simple mechanical process … an actual impression taken mechanically from the hand of the prisoner." Fingerprints were "an absolute impression taken direct from the body itself; if a print be taken at all it must necessarily be correct" (Troup *et al.* 1894: 28–29). For example, the Boston Police Department declared that the advantage of fingerprinting was that "as the digits record themselves there are no inaccuracies" (Garner 1910: 635). Fingerprint identification was seen to echo emerging technologies of duplication, such as the letter press, carbon paper, "the message-recording machine, the machine that sets type and the press that prints thousands of copies to the hour" (De Pue 1902: 93). As another commentator put it, "The fingerprint system reduced identification to a method of bookkeeping" (Faurot 1921: 105).

The mechanical associations surrounding the recording of fingerprint impressions gradually migrated to the process of interpretation of those impressions by human experts. As one examiner put it:

> The finger print expert has only facts to consider; he simply reports what he finds … If two prints are identical in every particular, they were made by the same person. If they are different, they were not made by the same person. No matter how many finger print experts may be engaged in the labor of comparing two prints, their verdict must be the same.
>
> *(Gribben 1919)*

Another examiner invoked the silent witness doctrine for the fingerprint: "identity is proven when the evidence can 'speak for itself;' in other words when the evidence becomes *self-evident*" (Burtis Bridges, quoted in T.D. Cooke, introduction to Mairs 1955: 3, original emphasis).

Of course, we now know that human fingerprint examiners and other interpreters of scientific evidence were not machines; traces inferred to derive from a common source are not "identical in every particular," but rather *similar enough* that such an inference is made (e.g. Cole 2009); the inference is probabilistic, not categorical (e.g. Champod and Evett 2001); and human examiners often reach different verdicts from the same data (e.g. Ulery *et al.* 2011; Haber and Haber 2014; Dror and Charlton 2006). Nonetheless, it has become quite common for forensic experts of all sorts to adopt this mantle of mechanical objectivity, to claim that even largely unregulated acts of human interpretation are mechanical, with all the connotations of objectivity and accuracy that term implies.

Recently, this reliance on human observation has made forensic science an object of criticism for its susceptibility to "confirmation bias," a concern that comes under the heading of "human factors." It is claimed that the forensic sciences have fallen behind mainstream science and medicine in paying heed to the potential for scientific observations to be biased by human observers. These problems were addressed long ago in fields ranging from astronomy (with its "personal equations") to medicine with its double blind clinical trials (Schaffer 1988; Risinger *et al.* 2002). Resistance to these criticisms leads to invidious comparisons between the pattern recognition disciplines, like fingerprints and firearms and toolmarks, and disciplines, like DNA and drug analysis, which rely at least in part on machine observations. Paradoxically, while historically many forensic scientists claimed to behave like machines, this controversy might be expected to encourage the greater use of actual machines into the interpretation of forensic traces.

Polygraphs, lie detectors, and breathalyzers

While the treatment of forensic analyses as "truth machines" is metaphorical, there are also, of course, more literal "truth machines." Lie detector devices were envisioned early in the twentieth century by Hugo Munsterberg, and the idea disseminated broadly. "From 1900 to 1920, a series of 'soul machines,' 'truth-compelling machines,' and 'machines to cure liars' were described" in the popular American press, "with great enthusiasm" (Bunn 2012: 180). These devices were literal embodiments of mechanical objectivity, making use of the same sorts of inscription devices (like sphygmomanometers, galvanometers, pneumographs, and kymographs) as those described by Daston and Galison (Bunn 2012: 127). From the 1930s through the 1950s both science and science fiction explored the topic of "mechanical mind reading" whose "simplified, mechanized vision helped to shape perceptions of and expectations for what would become the sciences of brain imaging" (Littlefield 2011: 68).

During the 1920s, the modern lie detector was developed through the complementary, and often competing, efforts of William Marston, John Larson, and Leonarde Keeler (Bunn 2012;

Alder 2007). The claim, of course, was that lie detectors were *machines* that detected deception scientifically, in contrast to the old-fashioned way in which humans (endeavored to) detect lies during ordinary social interaction. The "assumption" was "that the body provides us with objective data that do not require interpretation; or put another way, the body appears to be self-reporting" (Littlefield 2011: 5). The lie detector, like the modern literary detective which emerged around the same time, was concerned with "solving the mystery of the body by converting it into a truth-telling machine" (Thomas 1999: 39). Larson's lie detector's "great advantage was that the automated device minimized the examiner's judgment in taking the readings, thereby fulfilling one criterion of the scientific method, which was to 'eliminate all personal factors wherever possible'" (Alder 2007: 5). Proponents of the lie detector intentionally touted its mechanical appearance as an intimidating "black box" (Bunn 2012: 142).

Claims about the "automatic" nature of the lie detector have always been half-hearted, though, in that proponents have simultaneously insisted that the machine needs to be properly manipulated by its human operator. Indeed, many have gone so far as to claim that the reliability of the machine depends on the skill of the operator. The machine could supposedly detect lies automatically, and yet it required an expert to operate it and interpret the results (Bunn 2012: 143). "The paradox," as Littlefield (2011: 5) notes, "is that even as the body speaks, what it is saying required interpretation, often by scientific experts." As early as the 1920s, Marston made "the shocking discovery that subjects' reactions depended on the qualities of the examiner" (Alder 2007: 53), and Marston always insisted that the lie detector was a test, not a machine (Bunn 2012: 126). Keeler's version of the polygraph was likewise premised on the notion that the reliability of the interrogation lay, not in the machine at all but in Keeler's "personal know-how" (Alder 2007: 80). This "made Keeler himself a lie detector, and it constituted his true innovation" (Alder 2007: 81). Nonetheless, polygraph proponents continued to perpetuate "the charade that it is the polygraph machine and not the examiner which assesses the subject's veracity" (Alder 2007: 106). Indeed, it would appear that polygraph operators to this day do not desire complete mechanical objectivity:

> In the 1990s new computer algorithms were developed that could analyze the subject's physiological responses with mechanical neutrality. But because the algorithms might preclude operators from accusing subjects of lying (whatever the machine said), the nation's top examiners at the Department of Defense Polygraph Institute report that most operators usually turn the computer off.
>
> *(Alder 2007: 129)*

The lie detector was famously ruled inadmissible by the United States Court of Appeal for the District of Columbia in *Frye v. United States* (1923), and it has generally remained inadmissible with some exceptions. Many have argued that "the courts rejected the lie detector not for its failings but for its power" (Alder 2007: 147). By purporting to determine whether an individual is telling the truth, lie detection devices can sometimes directly resolve the "ultimate issue" rather than, like all other evidence, providing incremental evidence toward or against it, thus "invading the province of the jury," in the legal phrase. Thus, the lie detector is existentially threatening to the law. The lie detector "represented the dreams of a criminology in support of the law, but it promised to replace the due process of law altogether" (Bunn 2012: 192). Early press reports about the proto-lie detector devices declared: "It Will Make Expert Testimony Unnecessary and May Eliminate Juries in Trials" (Bunn 2012: 106; Golan 2004: 250). Golan (2004: 250) suggests "experimental psychology was threatening to reintroduce similar procedures" to the medieval ordeal "into the courtroom, with machines playing the part previously allocated to deity ... It was not the human expert who threatened the jury's province. It was the machine—and it did not threaten merely to invade the province of the jury; it threatened to obliterate it." Shniderman (2011–12: 469–470, original emphasis), however, posits a different, more mundane, explanation: "The admissibility decision appear to be based more on which party is proffering the evidence than any scientific or legal factors. Among the technologies of similar scientific validity, lie detection is the *only* technique offered almost exclusively by defense counsel."

One way around judicial exclusion was the fact that, even if courts would not accept polygraph tests as evidence, they would accept confessions extracted during polygraph examinations, which, at least in Keeler's version, were the true goal of the examination (Alder 2007: 126). Thus, the polygraph was in some sense an empty box: "Given the nature of the ruse, the interior workings of the machinery are almost beside the point" (Alder 2007: 128). Indeed, Keeler sought to turn the judicial opposition to the polygraph on its head, by extending the machine metaphor to the trial itself:

> As jurors were also incapable of evaluating sophisticated psychological tests, he agreed that they ought not to hear polygraph evidence either. Instead, he advocated trying criminal cases before expert criminologists wielding a polygraph, with a judge to rule on legal technicalities. Keeler looked forward to a justice system run with the efficiency, precision, and impersonality of a machine.
>
> *(Alder 2007: 147)*

Today, a number of more sophisticated neuroimaging technologies are being promoted as replacements for the old-fashioned polygraph, such as a technique that uses Functional Magnetic Resonance Imaging (fMRI) known as Brain Electrical Oscillation Signature (BEOS) testing (or, absurdly, "Brain Fingerprinting"). These techniques have not been well received in US courts, but BEOS was quickly admitted, on a rather thin scientific basis, in India. Some scholars have attributed this to weak judicial regulation of expert evidence in Indian law (Gaudet 2011). Other explanations for India's embrace of the neuroimaging include a desire for "modernization at all costs" and a desire to do away with "third degree" police torture, although it is noted "that even when there is a distinct desire to do away with physical torture, there appears to be an inability to challenge all the conditions responsible for its persistence" (Lokaneeta 2014: 18).

Some of these new technologies rely on different "software" (questioning protocols) as well as different hardware: the guilty knowledge test (GKT), as opposed to the Control Question Tests (CQT) and Directed Lie Tests (DLT) that dominated applications of the polygraph (Iacono and Lykken 2008: 621–623). The GKT replaces the trial process, not merely by purporting to resolve the ultimate issue, as all "lie detector" technologies do, but in another way as well. For the GKT protocol "a full investigation and subjective examination must be carried out *before* a test is ever ordered. To say nothing of scientific process, Brain Fingerprinting's reliance on a legitimate/illegitimate knowledge paradigm defies *due process,* by assuming a verdict of guilt or innocence long before the mechanical exam is ever undertaken" (Littlefield 2011: 137, original emphasis).

Alcohol breath testing devices, most popularly known as "breathalyzers," constitute another class of technologies that quite literally purport to "establish culpability" (e.g. Barone and Vosk 2015). As is the case for fingerprints and other pattern recognition disciplines, there are also human interpretive routines for detecting intoxication that seek to invoke the mantle of mechanical objectivity, even without the physical trappings of mechanisms. The horizontal gaze nystagmus (HGN) test is based on the purported ability of a police officer to detect alcohol intoxication. One California court, in finding the test admissible, noted "The nystagmus effect can be observed without mechanical, electronic, or chemical equipment of any kind." Thus, the police officer himself is a sort of cyborg detection technology (Jasanoff 1995: 60).

It is not coincidental that the above case occurred in California, which adheres to a curious variant of the "*Frye* standard" for admissibility used in many US states. The federal *Frye* standard, copied in many states, demands "general acceptance in the relevant field to which" the scientific or technological claim "belongs." Some California courts have seemed to restrict the application of California's version of the *Frye* standard, the "*Kelly* standard," to forms of expert evidence based on the products of machines (e.g. Crooke and Depew 2012: 28; Epstein 2004: 32). But this restriction of *Kelly* to machine-based evidence has also been criticized as a misinterpretation of *Kelly* (e.g. Hedger 2004: 200).

DNA profiling

In the late twentieth century, "the truth machine" reappeared yet again, this time in the form of a powerful new forensic identification technology, DNA profiling. This technique in some sense replaced serological blood testing—it operated on the same forensic traces: bodily fluids. The technology, however, derived from cutting edge molecular biology: the forensic applications were only one practical application of techniques that were also widely used in basic biological research. Finally, the technique was also a bit like fingerprinting—and it even briefly adopted that name ("DNA fingerprinting")—in that it purported to offer individualized identification.

DNA profiling developed rapidly in the decades following its introduction in the mid-1980s. The technology became more discriminating; able to work with ever smaller traces by "amplifying" them; faster; cheaper; and more mobile. It purported to be almost as discriminating as fingerprint identification, and yet, it was in some ways superior: as the technology developed, it was able to derive information from small amounts of any type of body cell—thus becoming useful in cases in which legible fingerprints were not present. In addition, it has certain claims to scientific credibility, by having derived from mainstream molecular biology, that fingerprinting lacked. Perhaps most importantly, it was able to exploit the unusual data structure of genetic information to make relatively transparent and defensible calculations of the rarity of the genetic features found consistent between the "crime scene stain" and the known genotype of the suspect.

The visual technologies for imaging genetic profiles were in some sense also "truth machines." The earliest visual representation of DNA evidence, "[t]he autoradiograph[,] entered the practice of criminal identification as a genetic 'fingerprint': it was a coherent, unified evidentiary statement that was understood to guarantee identification" (Finn 2009: 66). With this image, "The expert witness and the examining lawyer collaborate to instruct, cajole, and rhetorically retrain the fact-finder's eyesight, with greater or lesser success, to 'see' DNA and so, by a metonymic transfer of meaning, to perceive the truth whole" (Jasanoff 1998: 720). DNA typing was "a whole technology of certainty, predicated on the index" that "manifests itself in a methodology of detached observation that will lay claim to objectivity at the same time that it produces its observing subject" (Hutchings 2001: 135–136). In court, this threatened to swamp other legal considerations: "The charisma of genetic science is such that DNA has power beyond other forms of evidence, that its presence in courts is almost always decisive" (Gerlach 2004: 192).

Several scholars, however, have pointed out that claims about the "automatic" nature of DNA profiling have been overstated (see McCartney, Chapter 24, this volume for one such discussion). Early cases clearly showed that interpretation was necessary to make sense of DNA evidence and that DNA scientists often transgressed their own rules for calling DNA profiles consistent (Jasanoff 1998: 728; Lander 1992; Mnookin 2006). And so, "The autoradiograph was deconstructed, revealing the fragmented and specialized processes behind its construction. The image was shown to be the product of complex and diverse scientific practices, and its interpretation was shown to be bound to the equally complex subject of population genetics" (Finn 2009: 66). The credibility of DNA is far from automatic; it relies upon the technology surviving challenges to laboratory practices, statistical interpretation, implications of contamination, challenges to the chain of custody and so on. "The 'genetic witness' speaks for itself only when presented in the form of expert *testimony*, and, as we have seen, interrogation of that 'voice' points to an extended, indefinitely complicated, series of fallible practices through which evidence is collected, transported, analyzed, and quantified" (Lynch *et al.* 2008: 336, citation omitted, original emphasis). In addition, DNA results only acquire their truth value in the context of larger narrative explanations of the crime (Lynch *et al.* 2008).

Such arguments were not stable because, in contrast to many of the technologies discussed above, the automation of DNA profiling was, and is, not fixed but progressing rapidly. Autoradiographs, which were once eyeballed by scientists, could later be machine read. The autoradiographs themselves were soon phased out when gel electrophoresis was replaced by "capillary electrophoresis," which was always machine read, and yielded what became known as "graphical outputs" (Figure 29.1). These developments enhanced the appearance of mechanical objectivity in DNA analysis and interpretation: "Arguments about the relative alignment of bands across lanes, the use of arbitrary correction factors for band-shifting, and suspicions about 'subjective' visual inspection no longer seem salient when judgments are programmed, and molecular weights of STR sequences are read automatically and visualized as discrete, color-coded, graphic peaks" (Lynch *et al.* 2008: 234). But, "In part the 'digital' properties of STR profiles are due to what [forensic scientist Christophe] Champod calls the 'preprocessing' of graphic data before they are quantified and presented publicly. What Champod seems to suggest is that the 'subjective' or judgmental aspects of STR analysis become hidden, because they are 'black-boxed' by delegating visual inspection and analysis to machines" (Lynch *et al.* 2008: 298).

Over the last several decades, DNA profiling was quickly and widely adopted by law enforcement agencies around the world (Hindmarsh and Prainsack 2010). Though it was used primarily to build cases against suspects, attorneys sometimes used it to reinvestigate cases in which the convict claimed innocence. By testing preserved evidence, which had not been

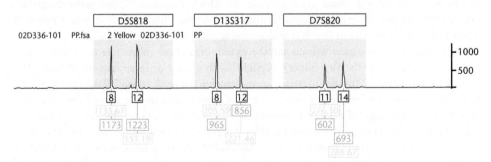

Figure 29.1 An electropherogram. The numbers in boxes are labeled by machine, but human interpretation of the "graphical output" is still possible

Source: William C. Thompson

DNA tested at the time of the original investigation, these attorneys were sometimes able to expose wrongful convictions. These cases became celebrated, and, in addition to exposing serious problems in criminal justice systems, also enhanced DNA's mythic status. Indeed, it was in this context that one of these "innocence" attorneys, Peter Neufeld, called DNA profiling a "truth machine" (Lynch *et al.* 2008: 263).

This belief in the "mechanical objectivity" of DNA evidence extends to convicts themselves (Machado and Prainsack 2012: 77). They believe "that DNA technologies enable the automatic identification of 'offenders.'" Interestingly, however, some convicts feel "more protected by the automation provided by technology," because they feel disempowered. "Hence automation transposes the power of decision and its political character to technology, perceived as neutral and effective, in a form of 'mechanical objectivity' which *'serves as an alternative to personal trust'*" (Machado *et al.* 2011: 142, quoting Porter, original emphasis). These beliefs also extend to crime victims. For example, in Mulla's (2014: 37) ethnographic account of forensic nursing, one of her informants described DNA as "the hand of God."

More sophisticated technology brought the interpretation of high-quality, single-source samples much closer to "automatic" status, but many issues remain in DNA interpretation, especially in cases involving what are called "mixtures," samples with more than one contributor. Indeed, in some of these cases, it is not possible to determine from the evidence alone precisely how many contributors there are (Lynch *et al.* 2008: 284–290).

One such issue is bias, as discussed above, by which analysts may be influenced by "contextual" information that suggests what the interpretation of the DNA profile "should be." Anecdotal cases (Thompson 2009: 261–262) and controlled studies (Dror and Hampikian 2011) have suggested that analysts' interpretations can change depending on context. More generally, it is clear that the interpretation of DNA mixtures is not at all straightforward, and this has become an area of extensive scholarly debate.

Into this debate has entered yet another generation of "truth machines"—"automated" computer systems for DNA mixture interpretation that supposedly adopt an "objective," predetermined set of rules for mixture interpretations that are immune from the difficulties of bias and backward reasoning that are a source of concern for human interpretation. These systems may be thought of as a reconstruction of the claim that DNA interpretation can be mechanized. The best known of these algorithms is TrueAllele, which has been aggressively marketed as "an objective and scientifically valid method for assessing the statistical value of DNA evidence," especially in complex mixture cases (Thompson *et al.* 2012: 18). However, it has been noted that "The fact that an automated system can produce answers to the questions one puts to it is no assurance that the answers are correct. While automated systems appear promising, their ability to handle 'hard cases' like" complex mixtures "remains to be fully evaluated" (Thompson *et al.* 2012: 19). Murphy (2015: 102) cautions that automated systems "may favor models that rely less on input from an actual person," and this may "come at the expense of ignoring the qualitative value that a well-trained analyst may provide." At least one alternative, open-source system, LRmix, is much less automated, allowing for "'strong interaction' between the analyst and software" (Murphy 2015: 103).

Another issue raised by TrueAllele, however, is that, as a for-profit corporation, it has insisted that its source code remain proprietary, and it has refused to turn the source code over for defense inspection when it is used to inculpate defendants in criminal trials (Murphy 2015: 100–103). The battle over source codes in criminal trials has a long history involving breathalyzer devices (Short 2009), as well as early DNA kits (Mellon 2001). TrueAllele has always been found admissible in the US, and almost always worldwide, despite challenges to both admissibility and transparency (Moss 2015).

Future truth machines

Law's ambivalent desire for a "truth machine" seems to be a perpetual enough one that we should continue to expect to entertain such claims for the foreseeable future. We already discussed in the introduction the increasing dissemination of surveillance cameras, and the coming advent of police body cameras. Moving from the investigative phase of the criminal justice process to the punishment phase, computer programs that automatically implement sentencing guidelines are already in place in several jurisdictions. As Aas (2005: 76) comments, "The guidelines' self-referential and machine-like nature has distanced them, not only from the communities in which they are applied, but furthermore, from the individuals who are supposed to give meaning to their content—the judges."

Where else should we expect to see truth machines? Proponents of "intelligence-led policing" argue that entire crime scenes can be reduced to information that can then be mined for connections and links with other sources of "intelligence," resulting in conclusions about past and future crimes alike. For example, Roux *et al.* (2012: 17) describe the crime scene investigation as a "hypothetico-deductive mechanism." In some sense, this is nothing new, in that crime scene investigation has been invoking mechanical objectivity since the late nineteenth century (Burney and Pemberton 2013: 18). "The perfect detective," some of the popular press argued in the 1860s, "was not so much a scientist as a machine" (Summerscale 2008: 199). And, Keeler promoted an "Illinois State Police Mobile Crime Detection Laboratory and Emergency Unit," which was described as "almost a complete crime detection laboratory on wheels," in the 1940s (Bunn 2012: 167–168).

Will machines sweep through entire crime scenes gathering up and interpreting "truth"? Perhaps. But, as is so often the case, the science fiction of Philip K. Dick helps explore the tricky epistemological issues that such a technological future raises. In *The Penultimate Truth* (1964), robots provide security and assist human detectives in investigating crime scenes. But a German-made assassination machine, "the standard model 2004 Eisenwerke Gestalt-macher" (ibid.: 138), is capable not only of penetrating a well-guarded home and carrying out an assassination, but also of depositing traces in order to "frame" a designated individual for the crime. For the assassination described in the novel, seven traces lead back to the faux perpetrator, including fingerprints, hair, fibers, voice, inferred body weight from bending of a window sill, blood drops, and brain waves. Once again it would seem that machines have usurped the law: "It would appear beyond a reasonable doubt that Stanton Brose, the man who had hired Foote to look into this felony, was the killer" (ibid.: 143). Does this mean forensics cannot be trusted because evidence can be manufactured?

It isn't clear. As the detectives investigating the murder reason, Brose would have been implicated as the murderer had the detectives not figured out that the murder was carried out by a Gestalt-macher. Having reached this conclusion, their suspicions would fall on anyone but Brose since the machine was programmed to frame Brose. However, another possibility is that Brose programmed the machine to implicate himself in order to throw suspicion off himself (ibid.: 167). Characteristic of Dick's conundrums, it's impossible to tell which scenario obtains. Perhaps, then, establishing culpability still requires context, no matter how many machines are involved in its manufacture.

Conclusion

The attraction that machines and mechanical objectivity hold for law and for criminal justice systems more generally is understandable. The unreliability, biases, and foibles of human judgment are familiar enough to explain the yearning for the mechanical. And yet, as this review has shown, mechanical solutions have historically failed to fully deliver on the hopes that have been invested in them. The problems have been fairly consistent. Machines themselves are never perfectly reliable, and they are usually not as reliable as their promoters claim. The very erasure of human judgment that is considered the benefit of machines is also their drawback: there are times when human discretion is considered desirable. Machines are not necessarily free of bias either; the biases are simply those written (by human designers) into their programming or calibration. And, the supposed "objectivity" of machines will always be somewhat illusory. Mechanical recordings and outputs will also lack some form of context and nuance. For all these reasons, machines will almost surely continue to exert both attraction and repulsion upon custodians of justice systems.

References

Aas, K.F. (2005) *Sentencing in the Age of Information: From Faust to Macintosh*. London: Glasshouse Press.

Alder, K. (2007) *The Lie Detectors: The History of an American Obsession*. New York: Free Press.

Barone, P.T. and Vosk, T. (2015) Breath and blood tests in intoxicated driving cases: why they currently fail to meet basic scientific and legal safeguards for admissibility. *Michigan Bar Journal* 94: 30–35.

Bunn, G.C. (2012) *The Truth Machine: A Social History of the Lie Detector*. Baltimore: The Johns Hopkins University Press.

Burney, I.A. and Pemberton, N. (2013) Making space for criminalistics: Hans Gross and fin-de-siècle CSI. *Studies in History and Philosophy of Biological and Biomedical Sciences* 44: 16–25.

Cannon, L. (1999) *Official Negligence: How Rodney King and the Riots Changed Los Angeles and the LAPD*. Boulder: Westview.

Champod, C. and Evett, I.W. (2001) A probabilistic approach to fingerprint evidence. *Journal of Forensic Identification* 51: 101–122.

Cole, S.A. (2001) *Suspect Identities: A History of Fingerprinting and Criminal Identification*. Cambridge: Harvard University Press.

Cole, S.A. (2009) Forensics without uniqueness, conclusions without individualization: The new epistemology of forensic identification. *Law, Probability and Risk* 8: 233–255.

Crooke, E.L. and Depew, B.D. (2012) Expert judgment: California's test for admissibility of expert opinion concerns reliability and assistance to the trier of fact, not correctness. *Los Angeles Lawyer* 35: 24–29.

Daston, L. and Galison, P. (1992) The image of objectivity. *Representations* 40: 81–128.

De Pue, F.H. (1902) The De Pue system of identification. *Proceedings of the International Association of Chiefs of Police*. Grand Rapids, Mich.: Seymour & Muir, 90–104.

Dick, P.K. (1964) *The Penultimate Truth*. New York: Carroll & Graf.

Doyle, A., Lippert, R. and Lyon, D. (2012) *Eyes Everywhere: The Global Growth of Camera Surveillance*. Routledge: London.

Dror, I.E. and Charlton D. (2006) Why experts make errors. *Journal of Forensic Identification* 56: 600–616.

Dror, I.E. and Hampikian, G. (2011) Subjectivity and bias in forensic DNA mixture interpretation. *Science & Justice* 51: 204–208.

Edmond, G., Biber, K., Kemp, R. and Porter, G. (2009) Law's looking glass: Expert identification evidence derived from photographic and video images. *Current Issues in Criminal Justice* 20: 337–377.

Edmond, G., Cole, S.A., Cunliffe, E. and Roberts, A. (2013) Admissibility compared: The reception of incriminating expert opinion (i.e. forensic science) evidence in four adversarial jurisdictions. *University of Denver Criminal Law Review* 3: 31–109.

Epstein, D.G. (2004) Class, there'll be no Kelly Test this week. *Orange County Lawyer* 46: 32–34.

Faurot, J. (1921) *Criminal Identification*. New York: NYPD Bureau of Printing.

Finn, J. (2009) *Capturing the Criminal Image: From Mug Shot to Surveillance Society*. Minneapolis: University of Minnesota Press.

Garner. J.W. (1910) Identification of criminals by means of finger prints. *Journal of the American Institute of Criminal Law and Criminology* 1: 634–636.

Gaudet, L.M. (2011) Brain fingerprinting, scientific evidence, and *Daubert:* A cautionary lesson from India. *Jurimetrics* 51: 293–318.

Gerlach, N. (2004) *The Genetic Imaginary: DNA in the Candadian Criminal Justice System.* Toronto: University of Toronto Press.

Golan, T. (2004) *Laws of Men and Laws of Nature.* Cambridge: Harvard University Press.

Goldsmith, A.J. (2010) Policing's new visibility. *British Journal of Criminology* 50: 914–934.

Goodwin, C. (1994) Professional vision. *American Anthropologist* 96: 606–633.

Gribben, A.A. (1919) How the finger print expert presents his case in court. *Finger Print and Identification Magazine* 1(2): 10–14.

Haber, R.N. and Haber, L. (2014) Experimental results of fingerprint comparison validity and reliability: A review and critical analysis. *Science & Justice* 54: 375–389.

Hedger, C. (2004) Daubert and the States: A critical analysis of emerging trends. *Saint Louis University Law Journal* 49: 177–207.

Hindmarsh, R. and Prainsack, B. (2010) *Genetic Suspects: Global Governance of Forensic DNA Profiling and Databasing.* Cambridge: Cambridge University Press.

Hutchings, P.J. (2001) *The Criminal Spectre in Law, Literature and Aesthetics.* London: Routledge.

Iacono, W.G. and Lykken, D. (2008) The case against polygraph tests, in Faigman, D.L., Kaye, D.H., Sanders, J. and Cheng, E.K. (eds) *Modern Scientific Evidence: Forensics.* Student Edition ed. St. Paul, Minn.: West, 609–663.

Jasanoff, S. (1995) *Science at the Bar: Law, Science, and Technology in America.* Cambridge: Harvard University Press.

Jasanoff, S. (1998) The eye of everyman: Witnessing DNA in the Simpson trial. *Social Studies of Science* 28: 713–740.

Koskela, H. (2009) Hijacking surveillance? The new moral landscapes of amateur photographing. In: Aas, K.F., Gundhus, H.O. and Lomell, H.M. (eds) *Technologies of InSecurity: The Surveillance of Everyday Life.* Abingdon: Routledge, 147–167.

Lander, E. (1992) DNA Fingerprinting: Science, law, and the ultimate identifier, in Kevles, D.J. and Hood, L. (eds) *The Code of Codes: Scientific and Social Issues in the Human Genome Project.* Cambridge: Harvard University Press, 191–210.

Littlefield, M.M. (2011) *The Lying Brain: Lie Detection in Science and Science Fiction.* Ann Arbor: University of Michigan Press.

Lokaneeta, J. (2014) Creating a flawed art of government: Legal discourses on lie detectors, brain scanning, and narcoanalysis in India. *Law, Culture and the Humanities:* 1–19. doi:10.1177/1743872114559881.

Lynch, M., Cole, S.A., McNally, R. and Jordan, K. (2008) *Truth Machine: The Contentious History of DNA Fingerprinting.* Chicago: University of Chicago Press.

Machado, H. and Prainsack, B. (2012) *Tracing Technologies: Prisoners' Views in the Era of CSI.* Farnham: Ashgate.

Machado, H., Santos, F. and Silva, S. (2011) Prisoners' expectations of the national forensic DNA database: Surveillance and reconfiguration of individual rights. *Forensic Science International* 210: 139–143.

Mairs, G.T. (1955) Random thoughts concerning finger prints. *Finger Print and Identification Magazine* 36(8): 3–18.

Mateescu, A., Rosenblat, A. and Boyd, D. (2015) *Police Body-Worn Cameras.* Data & Society Research Institute.

Mellon, J.N. (2001) Manufacturing convictions: Why defendants are entitled to the data underlying forensic DNA kits. *Duke Law Journal* 51: 1097–1137.

Mnookin, J.L. (1998) The image of truth: Photographic evidence and the power of analogy. *Yale Journal of Law and the Humanities* 10: 1–74.

Mnookin, J.L. (2006) *People v. Castro:* Challenging the forensic use of DNA evidence, in Lempert, R. (ed.) *Evidence Stories.* New York: Foundation Press, 205–235.

Moss, K.L. (2015) The admissibility of TrueAllele: A computerized DNA interpretation system. *Washington and Lee Law Review* 72: 1033–1076.

Mulla, S. (2014) *The Violence of Care: Rape Victims, Forensic Nurses, and Sexual Assault Intervention.* New York: New York University Press.

Murphy, E.E. (2015) *Inside the Cell: The Dark Side of Forensic DNA.* New York: Nation Books.

Risinger, D.M., Saks, M.J., Thompson, W.C. and Rosenthal, R. (2002) The *Daubert/Kumho* implications of observer effects in forensic science: Hidden problems of expectation and suggestion. *California Law Review* 90: 1–56.

Ronell, A. (1992) Video/television/Rodney King: Twelve steps beyond *The Pleasure Principal*. *Differences* 4: 1–15.

Roux, C., Crispino, F. and Ribaux, O. (2012) From forensics to forensic science. *Current Issues in Criminal Justice* 24: 7–24.

Schaffer, S. (1988) Astronomers mark time: Discipline and the personal equation. *Science in Context* 2: 115–145.

Seltzer, M. (1992) *Bodies and Machines*. New York: Routledge.

Shniderman, A.B. (2011–12) You can't handle the truth: Lies, damn lies, and the exclusion of polygraph evidence. *Albany Law Journal of Science and Technology* 22: 433–473.

Short, C. (2009) Guilt by machine: The problem of source code discovery in Florida DUI prosecutions. *Florida Law Review* 61: 177–201.

Simon, D. and Bueermann, J. (2015) What's the right police body camera policy? *Times*. Los Angeles.

Stalcup, M. and Hahn, C. (2016) Cops, cameras and the policing of ethics. *Theoretical Criminology* 20.

Stuart, F. (2011) Constructing police abuse after Rodney King: How Skid Row residents and the Los Angeles Police Department contest video evidence. *Law & Social Inquiry* 36: 327–353.

Summerscale, K. (2008) *The Suspicions of Mr. Whicher, or The Murder at Road Hill House*. London: Bloomsbury.

Thomas, R.R. (1999) *Detective Fiction and the Rise of Forensic Science*. Cambridge: Cambridge University Press.

Thompson, W.C. (2009) Painting the target around the matching profile: The Texas sharpshooter fallacy in forensic DNA interpretation. *Law, Probability and Risk* 8: 257–276.

Thompson, W.C., Mueller, L.D. and Krane, D.E. (2012) Forensic DNA statistics: Still contorversial in some cases. *The Champion* 36: 12–23.

Troup, C.E., Griffiths, A. and Macnaghten, M.L. (1894) Report of a committee appointed by the Secretary of State to inquire in the best means available for identifying habitual criminals. *British Sessional Papers. House of Commons.* London: Eyre and Spottiswoode.

Ulery, B., Hicklin, R.A., Buscaglia, J. and Roberts, M.A. (2011) Accuracy and reliability of forensic latent fingerprint decisions. *Proceedings of the National Academy of Sciences* 108: 7733–7738.

Wigmore, J.H. (1940) *Evidence*. Boston: Little Brown.

Wilson, D. and Serisier, T. (2010) Video activism and the ambiguities of counter-surveillance. *Surveillance and Society* 8: 166–180.

30

Technology-augmented and virtual courts and courtrooms

Fredric I. Lederer[1]

Neither courts nor courtrooms have been immune to modern technology. Faced not only with an inherent interest in enhancing basic efficiency, courts must now cope with the general adoption of electronic communications, data-oriented and often data intensive litigation, and the benefits of trial technologies. In criminal cases, evidence increasingly includes audio–video recordings from cell phones, drones, and body cameras, and it often seems that no case can take place without evidence obtained from social media. As a result, an increasing number of

Image 30.1 Simulated 2010 federal criminal trial in William & Mary Law School's McGlothlin Courtroom, home of CLCT

courts are dependent upon technology. Many courts use electronic filing and case management systems to manage their caseloads. Technology-augmented courtrooms permit the visual display of evidence at trial, to include potentially 3D, holographic, and immersive virtual reality display of evidence and argument,[2] internet-capable court record (e.g. the transcript or audio-video record of the proceedings), remote appearances by judges, counsel, witnesses, and jurors, and assistive technology. Virtual trials in which no human beings appear in a physical courtroom are possible. This chapter will address these developments, emphasizing technology-augmented courtrooms and the degree to which trials may become virtual ones. In doing so, it will concentrate on criminal matters, recognizing, however, that most courts deal with criminal and civil matters.

The evolving nature of courts, or virtual courts anyone?

At their core, courts exist to resolve disputes. They do so either by trial and adjudication of the dispute or by presenting sufficient certainty – and threat – of such resolution that parties will settle the dispute without trial. Indeed, approximately 90 percent of American criminal cases are resolved by guilty plea, ordinarily due to plea bargaining.[3] Given the possibilities inherent in modern technology, the initial question to be addressed necessarily is whether we will see virtual courts in the short- to mid-term future.

Traditionally, a "court" consists of one or more courthouses. A courthouse includes not only the trial and/or appellate courtrooms responsible for adjudication, it necessarily includes the offices, storage areas, holding cells, and similar spaces required for the court's operations. In theory, and setting aside at least in the United States possible constitutional constraints, a court could be a nearly entirely virtual organization consisting of numerous remote persons all collaborating via the "cloud" of one or more Internet servers. Such a virtual court is certainly possible. Today's technology permits multiple individuals to appear remotely on a screen with the ability to share documents and other electronic images and material – e.g. evidence. If we were prepared to be less realistic, various products would permit entire trials in virtual space using avatars. Virtual arbitration and mediation services already exist.[4] Virtual courts appear to be improbable, however, for at least two reasons.

No matter how effective remote collaboration may be, the efficiencies that now exist from human contact, including the unplanned "water cooler" effects of accidental physical meetings suggest that moving to an entirely virtual world would be less efficient than an environment mostly, but not entirely, based on physically present persons. But more importantly, courthouses exist for reasons other than to just house the adjudicatory and appellate systems. In the United States, courthouses serve as the physical reminder of the government[5] (Thacker n.d.). With the possible exception of local police forces, most people perceive Government, especially the national government, in the form of its bureaucratic effects. Courthouses, however, are the physical manifestation of governmental presence and power and courthouse architecture ordinarily embodies that concept. Courthouses ordinarily are designed to be visually imposing and to convey the message that justice is important and, unavoidably, the government is responsible for justice. Most courtrooms are similarly designed. The usual courtroom presents a visual image of permanence, authority, and, indeed, majesty. Judges wear robes and preside from the "bench," customarily elevated over the courtroom. Courtroom architecture, which complements courthouse architecture is intended to impress on all trial participants the importance of trial and, for witnesses, the need for accuracy and truth-telling in testimony. Accordingly, it is unlikely that we will eliminate courthouses as such anytime soon. Rather, the real question is the extent to which traditional functions will be changed or supplanted entirely by technology.

Courthouse functions

As major bureaucracies, courts are data-intensive organizations. In addition to the data they generate themselves, in criminal cases courts require information from police, defendants, defense lawyers, prosecutors, probation and prison officials, and various forms of experts. In nations that use juries, juror data is necessary. In addition to all of the usual computer capabilities (e.g. word processing, accounting, payroll, spreadsheets), the two key court/courthouse developing technologies are case management systems and electronic filing (e-filing).

Case management systems are intended to ensure that all of the documents and details relating to any given case can be found and accessed electronically thus enhancing efficiency and eliminating the risk of lost or misfiled case documents. Such a system should include submission deadlines and hearing dates and should notify all concerned of such approaching dates. On occasion, a court may have a separate electronic docketing system, especially when the court has large numbers of cases and it needs to coordinate the schedules of judges, counsel, police witnesses, and courtrooms.

Electronic filing is the means by which parties and counsel may submit documents such as motions to the court via the Internet. Doing so obviates the need for physical delivery, enhancing reliability and ordinarily eliminating lost or misfiled documents. E-filed documents should but might not be electronically connected with the case management system. E-filing may but need not permit filing documents after a court is physically closed for business. A fully electronic court would generate an electronic court record which would include all pretrial matters, the transcript of all hearings, electronic copies of the exhibits (evidence), and all post-trial matters, all of which would be combined and forwarded to the appellate court when appropriate.

In criminal cases, setting aside information supplied by the defendant, there often are three primary sources of key data: the police, the prosecution, and the court itself. One would expect seamless information exchange among the three. Unfortunately, in many cases the data systems of the three entities not only are not interconnected but often are incompatible. Thus, it is very possible that police information must be manually submitted to both the court and prosecution, and the prosecution systems cannot interface either with the court or the police insofar as return data is concerned. Lost or inaccurate data transmission is thus possible.

Courts hold vast amounts of current and past case data, often including sensitive testimony, personal identifiers, and information which could be harmful if known to the public. In the United States, court records are "public records," but that does not necessarily mean that the public has any access or easy access to given information. Some categories of cases, such as family law cases (e.g. adoptions and divorces) may be treated as non-public per se. In other categories such as criminal cases, individual case records may be partially or entirely "sealed," meaning that the public cannot access the record, either in electronic form or in person at the courthouse. Some courts rely on "practical obscurity"; the records are available at the courthouse for manual retrieval or electronic retrieval but not electronically online. The need to prohibit or redact unneeded personal identifier information from court pleadings and court records, along with other sensitive matter, is problematic for many courts, as is the need to cope with subsequent expungements of prior lawful convictions. Even when courts can decide on how to balance public access and transparency against personal privacy and safety, the actual implementation of any policy can be both difficult and potentially ineffective. This is particularly true given that businesses acquire court data both directly and indirectly, and can be circulated so widely online that erroneous or subsequently expunged data is not always removed from public access.

Judges and court staff, especially lawyers, use on-line legal materials for legal research purposes. At present, WestLaw, Bloomberg Law, and LEXISNEXIS are the customary sources of legal materials in the United States. Increasingly, however, legislatures, agencies, and courts are placing legal material on the web subject to the usual search procedures.

In a world in which "hacking" is a constant fear, the security of case management systems, case records, including trial and hearing transcripts, and other court data is unclear[6] (The Associated Press 2013). The likelihood is great that the primary protection of court data is the disinterest of those with the ability to seek it.

Electronic discovery

Although most of the world's multi-billion dollar electronic discovery industry is concerned with civil cases, search for and acquisition of emails, texts, cell phone records, social media postings, and other forms of electronic data for potential evidentiary use at trial is now fairly common. Unlike civil practice in which agreed-upon mutual discovery or subpoenas are the usual legal method of obtaining electronic information, criminal cases often require search warrants, especially if computer files are to be accessed and copied or seized, and current American procedure ordinarily permits the use of technology for such warrants.

Warrants

Although arrest and other law enforcement data may be incompatible with any given court, law enforcement personnel regularly deal directly with the courts. In pretrial matters law enforcement agents regularly seek search and, occasionally, arrest warrants from magistrates and judges. When doing so, they must show probable cause to believe that the property sought is where they wish to search or in the case of arrests probable cause to believe that the suspect committed the offense. In seeking a warrant an officer must supply sufficient information, traditionally in the form of an affidavit, sworn to by the officer.

Depending upon the court, courts now often accept applications via modern technology. Federal Rule of Criminal Procedure 4.1(a), for example, declares that "A magistrate judge may consider information communicated by telephone or other reliable electronic means when reviewing a complaint or deciding whether to issue a warrant or summons." Of course, the move from tradition to technological communications is not always simple. In *State v. Gutierrez*,[7] for example, the Utah Supreme Court sustained the legality of an e-warrant over an objection based upon state law as it dealt with the definition of an oath or affirmation. As the Court reported, "The eWarrant application included a screen labeled 'Affidavit Submission for eWarrant' and included the statement: 'By submitting this affidavit, I declare under criminal penalty of the State of Utah that the foregoing is true and correct'." The Court found that the officer's electronic submission fully complied with all legal requirements.

Remote first appearance and arraignment

In the United States after a person is arrested, he or she should be brought before a magistrate who will convey to the suspect the general nature of the charges against the person, advise the suspect of the right to counsel, and the nature of any conditions to be met in order to secure the suspect's release from custody pending trial. This is increasingly done by two-way videoconferencing with the suspect in a jail facility and the magistrate originating from a courtroom or office in a courthouse. Extensively used by the states,[8] federal practice requires that the accused voluntarily consent to remote first appearance.[9] It should be noted that in many jurisdictions this is called "remote arraignment." "Arraignment" technically means a judge's request to a defendant for the defendant's plea. Many first appearances do not include arraignment and thus the name, "remote arraignment" is formally erroneous. Because remote first appearances and arraignments eliminate the need for resource-consuming prisoner transfers to and from the courthouse, this technological use is especially of interest to many jurisdictions.

Remote hearings and motion practice

It is highly unusual for a judge to hold a hearing when absent from the courthouse, although modern technology certainly permits this. The British Columbia courts in Canada have created a system whereby a remote judge is made available whenever a local judge is not, all via two-way video conferencing. Appellate courts, including the High Court of Australia and the Supreme Court of Canada regularly hold appeals or certain forms of applications remotely. In the United States, the United States Court of Appeals for the Armed Forces twice held appellate hearings at William & Mary Law School in which one or more judges attended from remote locations.

Counsel frequently find it necessary to appear before the judge to argue motions (often known in other nations as "applications"). Customarily, these are often short matters which require substantial time expenditures by counsel to travel to and from the courthouse, to say nothing of the delays occasionally inherent in clearing courthouse security. Such motions could readily be argued remotely, perhaps from the lawyer's desk. Although still uncommon, this has been done regularly in many courts, especially in California. CourtCall,[10] a California company, has provided telephonic motion practice for many years and recently has added a video option. CLCT video experiments have substantiated the possible use even of smartphones for remote video judicial applications.

Trials and hearings

Traditionally, Anglo-American adversarial trials have been based on largely oral presentations of counsel to a judge or jury. Counsel begin the case with opening statements, predictions of what counsel hope to prove. Those statements are followed by the formal presentation of evidence ordinarily consisting of questioning of witnesses and submission of any documentary, physical, audio, or visual evidence. Counsel conclude with closing arguments, or summations. Because lawyers long ago recognized that a "picture is worth a thousand words," counsel might well make use of large images of evidentiary exhibits or use trial aids such as charts or diagrams, often placed on easels for the judge and jury to see.

Although this traditional spoken approach, which can be said to harken back at least to the days of Cicero of Rome, remains customary in most cases in most courts today, modern technology is changing the trial process. On the one hand, technology can substantially improve and hasten the trial process. On the other, the digitalization of the world has created a situation in which much evidence begins in digital form and ought to be so displayed at trial in the same form. Indeed, sometimes the only way to really understand electronic evidence is to be able to view it along with its associated metadata.[11]

Potentially available technologies can include:

- visual presentation of opening statements, evidence, and closing arguments
- technology-based court record

- remote appearances
- assistive technology.

Visual presentation

The true defining element of a technology-enhanced or augmented courtroom is the ability to show visual images, whether during an opening statement, evidence presentation, or closing argument. Inasmuch as the core of a case is the information used to prove it – the evidence – it's helpful to begin with evidence presentation. As noted above, the evidence in most cases consists of witness testimony and, until recently, physical evidence such as documents or objects such as guns, knives, or drugs. Traditional trial practice often has a lawyer showing a specific piece of evidence to a witness, the "exhibit;" having the witness identify and "authenticate it" (prove that the exhibit likely is what it purports to be; and then formally "tendering" it for the judge's approval to be used as evidence in the case. If "received" by the judge, counsel would hand the exhibit to the judge, or in a jury case, to the jurors. Unless copies of the exhibit had previously been prepared, counsel would then wait until everyone had seen the exhibit or would proceed at the risk of having distracted jurors waiting for their turn to peruse the exhibit.

In first generation technology-augmented courtrooms, counsel would place physical evidence under a document camera, a downward facing video camera, and then display the image to projection screens, TVs or monitors.

In the usual contemporary American technology-augmented courtroom, counsel use a notebook computer or a tablet to display electronic images of the evidence.[12] In most circumstances, the image is an adequate substitute for even a physical original (Federal Rules of Evidence 101, 1002, and 1003). Having connected to the courtroom visual display system, usually but not necessarily, via a direct physical or wired connection, the image is shown to the witness for authentication, and then if received in evidence by the judge, to the judge as formal evidence, or in a jury trial to all the jurors at one time. Counsel and/or the witness can interface with the exhibit, annotating the electronic image, for example, by underling key text or features with colored lines or adding text labels. Received evidence can be displayed in the courtroom via projection screens and/or computer monitors. To the degree that there can be said to be a customary design in the United States, the judge, counsel, and witness have individual monitors and every two jurors share a monitor. Most such courtrooms also make the image available to any members of the public attending trial via projection screens and/or large flat panel screens. This can enhance substantially the "transparency" of justice, allowing media representatives as well as the public for the first time to actually see documentary and related evidence. Although counsel could use any appropriate program or "app," to store and present evidence, most experienced lawyers will use specialized software or apps for the purpose. These provide counsel with versatile ways of displaying material and annotating it for emphasis.

Note that few courtrooms have permitted counsel to connect with the courthouse network and distribute images via the network. Rather, counsel's devices connect to a visual display system that only distributes the video content. This eliminates the risk of accidental or intentional injury to the courthouse network. Increasing interest in using wireless networked tablets for evidence display suggests that this may change in the near future. As of 2016, at least the United States District Courts in Los Angeles and Philadelphia and the Center for Legal and Court Technology were experimenting with the use of tablets for evidence presentation.[13] In Philadelphia and Williamsburg the lawyer's content would be streamed to individual juror tablets.[14]

Image 30.2 WolfVision visualizer

Image 30.3   2016 CLCT experimental (simulated) trial, *United States v. Chiu*, showing jury tablet use

Increasingly, modern criminal cases rely on digital evidence including emails, social media transmissions or posts, cellphone video, drone images, Google Glass recordings and the like. All of these lend themselves to easy presentation in technologically-augmented courtrooms that can display imagery, along with audio, or for that matter, digital audio alone. Although digital evidence may suggest modification of traditional evidentiary rules, those same rules ordinarily are adequate for normal operation (Lederer 1999: 389).[15]

New forms of evidence

Most documents today originate as digital data and most information is transmitted digitally, usually via the Internet in one form or another. Accordingly, evidence in criminal cases is likely to consist of computer files (seized via search of personal or organizational computers or cloud-based servers) emails, email attachments, Twitter tweets, Facebook posts, and similar Internet stored or posted information. Along with the basic information, such as a word processor document file, counsel may need to offer into evidence the metadata – or data associated with the file, such as date and time of preparation and the identity of the customary users of the computer or other device. Visual display of this data is the most efficient and reasonable way of showing the trial fact finder this type of evidence. Further, use of specialized trial practice software will permit side by side comparisons, annotations, and enlargements ("call-outs" of critical portions). The same technology can be used for opening statements and closing arguments.

In addition to digital evidence, the nature of evidence itself may be about to change. Panoramic visual imagery that permits a comprehensive view of a location, including a crime scene, has been available for years. The late twentieth-century video reconstruction of the Derry, Northern Ireland, Bloody Sunday shootings for the Saville Tribunal, provided an interactive tool for witness testimony.[16]

In 2002, CLCT demonstrated holographic evidence and the use of immersive virtual reality in a simulated criminal case. Neither seemed to catch on in actual cases. In 2013 at the Court Technology Conference conducted by the National Center for State Courts in Baltimore, Maryland, CLCT demonstrated the use of 3D evidence in a simulated case, showing both a bloody brick and a brick particle impregnated baseball bat to the audience "jury" via 3D projection. Although highly successful, and raising interesting questions about the emotional effects of vivid evidence images, the need for 3D glasses by jurors suggested the improbability of general adoption. The current rapid move to virtual reality for computer gaming and other uses suggest the possibility of immersive virtual reality evidence in 2016, just as new forms of projecting 3D images without the need for glasses suggest the possibility of 3D evidence (Mims 2015). Indeed, at the 2015 Court Technology Conference, FTR, an international court record company, demonstrated the first known 360-degree immersive virtual reality court record, made the week before in the Center for Legal and Court Technology's William & Mary Law School's McGlothlin Courtroom. On March 18, 2016, in the same courtroom FTR made a virtual reality record of the entire experimental case of *United States v Chiu*.[17] Equipped with an appropriate headset, a person would seem to stand in the center of the courtroom during trial and could see anywhere in the courtroom that could be viewed from that central position, listening to the proceedings.

Some courts provide technology for jurors to review evidence during deliberations. Experimental work by the Center for Legal and Court Technology, based both on simulated and real cases, has shown the utility of such an approach (Courtroom 21 2002).

Technology-based court record

The court record consists of copies of a case's associated papers, including the pleadings, and any motions (applications), evidentiary exhibits, and customarily, a verbatim or word for word transcript. Technology now permits digital audio or digital audio and video recording of what transpires during trial. With few exceptions, the "record" that is submitted for appellate purposes will include a text transcript of the recording as judges prefer text to having to listen to the entire recording. Increasingly, transcription can be accomplished by multiple remote transcribers, who may actually transcribe during trial. In 2014, CLCT demonstrated remote transcription that delivered text within approximately 20 minutes of the remarks. In 2016, Revolutionary Text provides remote court reporting and did so for simulated depositions in the author's Technology Augmented Trial Advocacy course at William & Mary Law School.

Both stenographic and voice writer court reporters (those who use computers equipped with speech recognition software) can supply "realtime" transcription which supplies judge and counsel with a nearly immediate rough draft, a draft which they can capture individually and annotate for later use at trial — such as cross-examination.

CLCT has demonstrated what is believed to be the most comprehensive court record now available, a combination of digital audio, video, realtime text transcript, and images of the actual evidence as shown by counsel as well as the first immersive reality court record, discussed above.

Image 30.4   CLCT multi-media court record

Remote appearances

Modern video conferencing permits the remote participation of judges,[18] counsel, witnesses, jurors, and interpreters. The increasing need for competent, especially court certified, interpreters, for many languages is impelling the adoption of remote interpreters. Ordinarily, however, remote appearances have been reserved principally for witnesses who cannot come to the courtroom. Based on past Center for Legal and Court Technology controlled scientific experimental work, it appears that providing a life-size image of a remote witness behind the witness stand is likely to give the same trial result as if the witness were physically in court. The witness, particularly if the witness is an expert, may feel uncomfortable if he or she cannot see the entirety of the courtroom, especially the judge or jury (Wallace 2011: 256). However, it does not appear that such discomfort would affect the trial. Although no protocol appears to require it, it seems clear that if possible a remote witness should testify from another courthouse in the presence of a court officer.

Technologically, quality two-way Internet-based video conferencing is no longer especially expensive, and public domain software permits use of even notebook computers, tablets, and phones as origination devices, although any specific device may yield lower resolution video.

Image 30.5 Remote Mandarin interpretation in 2016 CLCT laboratory trial using Cisco technology

Concerns about the use of remote testimony ordinarily are legal and policy ones that center about the question of whether remote testimony is the same as in-court testimony and whether cross-examination of a remote witness would be as successful as in-court examination. Inherent in these issues is the deep seated belief of most judges that they at least, and perhaps jurors as well, are able to determine whether an in-court witness is lying and that such an ability may not work during video-conferencing. Such a belief has legal authority on its side. Trial court determinations of fact are usually taken as accurate because the fact finder, judge or jury, had the ability to view the "demeanor evidence," – how the witness looked and sounded. In actual fact, however, I am unaware of any meaningful evidence that judges or jurors actually can determine truth-telling in court. No scientific study corroborates the ability to determine truth-telling via demeanor evidence. Although statutory or court rule authority may be necessary in some jurisdictions for remote testimony, in the United States ordinarily the critical question is one of constitutionality. The belief in demeanor evidence and concerns that having a remote prosecution witness may deprive a defendant of the ability to "confront" the witness within the meaning of the 6th Amendment to the United States Constitution raises the most fundamental challenge to remote prosecution testimony. Given that "confront" in the eighteenth century unavoidably meant to physically be in the same courtroom, an originalism theory of Constitutional interpretation would prohibit remote prosecution testimony against the will of the defendant. However, IF remote testimony is functionally identical to in-court testimony and that cross-examination of such a witness is at least as effective as cross-examination of in-court witnesses, other theories of constitutional interpretation would permit remote testimony at least when necessary. At present the United States permits remote testimony of child victim witnesses when shown to be necessary.[19] However, despite a number of federal and state court decisions permitting remote prosecution testimony,[20] there is sufficient disagreement about its constitutionality[21] to compel the statement that such testimony of unknown constitutionality. The reader should note, however, that remote defense witness testimony is fully constitutional as the Bill of Rights protects the defendant and not the government. And, of course, international and non-US courts not constrained by the American constitution regularly use remote testimony (Ivkovic 2001: 286; Victoria Evidence (Audio Visual and Audio Linking) Act 1997 § 3).

Assistive technology

Access to justice requires that courts and courtrooms be available to all, despite any disabilities that be involved. From a criminal justice perspective, that includes the very ability to answer law enforcement pre-trial interview questions and to testify at trial. Recent work in Israel has shown how pictograms, often coupled with technological selection of the pictograms permit persons who cannot speak or hardly move to communicate.[22]

Insofar as the trial is concerned, most courtrooms today are designed to afford access to persons using wheelchairs or scooters, although such access may be awkward at best. Persons with limited vision can use electronic enlargement technology to read evidence and other information with special devices being available for those with macular degeneration, devices that will not only enlarge an image but also provide varied background colors to further assist in viewing evidence. Special scanners can read physical documents to the user, if computer-based screen readers won't do because the information is not in a digital file in the computer. In 2006, The Center for Legal and Court Technology demonstrated how a new court officer, the "court explicator," who provides a running description of what is visually happening in the courtroom to judge, counsel, and others, coupled with optical scanners that will read documents to the user can be combined to assist low vision or blind judges to preside over a case.

Many people are hard of hearing or deaf. Various forms of listening devices can amplify sound and transmit it wirelessly to headphones for those with constrained hearing. Those who have no hearing at all can be assisted via sign language, including remote sign language interpreters using video conferencing, or realtime transcription. Two-way remote sign language interpretation can also permit testimony by those who both cannot speak and cannot hear.

Mobility, hearing, vision, and speech disabilities are relatively easy to deal with via technological assistance. Other forms of disabilities including some intellectual ones are more difficult and may not have adequate technological solutions.

Conclusion

Technology has already proven that when properly installed and used, trials and hearings can be conducted more quickly and probably more accurately than traditional proceedings. Equally important is that technologically augmented courtrooms and trials increasingly will be necessary in order to cope with new forms of technological evidence and to provide maximum access to justice with enhanced public transparency.

Technology presents both opportunities and problems. Initially, knowledgeable and responsible decision-makers must select appropriate technology for court and legal system use, use that is both lawful and in accord with legal and court culture. Then it must be properly implemented and maintained, a process that is simple to say but hard to do. Budgets are implicated as technology must not only be acquired and installed, it must be supported by properly trained persons and ultimately replaced and updated. Judges, lawyers, and court staff in particular must be able and willing to use technology when appropriate. With that in mind, in 2012 the American Bar Association amended Comment 8 to ABA Model Rule of Professional Responsibility 1.1, Competence, to mandate that "To maintain the requisite knowledge and skill, a lawyer should keep abreast of changes in the law and its practice, including the benefits and risks associated with relevant technology, ..." Finally, we must ponder the degree to which legal systems which are profoundly humanistic in nature can utilize technology extensively without losing their heart.

Notes

1 This chapter is based in significant part on work done in William & Mary Law School's McGlothlin Courtroom, the world's most technologically advanced trial and appellate courtroom. CLCT's mission is to improve the world's legal systems though appropriate technology. See www.legaltechcenter.net.
2 Anecdotal evidence suggests that individual display of evidence during trial not only enhances memory and understanding of the fact-finder (judge or jury) but likely results in a one-quarter to one-third total time saving. Although likely correct, there is only limited data to support the conclusion. In 2002, the Center for Legal and Court Technology (then the "Courtroom 21 Project") realized an approximately 10 percent time saving in one hour simulated jury trials with about eight exhibits.
3 In calendar year 2014, of a total of 84,544 federal criminal defendants, 75,035 pled guilty; 1,657 were convicted via jury trial and 143 by bench trials without juries. Note that 7,394 were dismissed, and 315 were acquitted. Guilty pleas, most of which likely were based on plea bargains, consisted of 88.8 percent of all defendants and 97.3 percent of all defendants who went to trial. United States Courts Statistical Tables for the Federal Judiciary – December 2014, Table D-4, U.S. District Courts – Criminal Defendants Disposed of, by Type of Disposition and Offense, During the 12-Month Period Ending December 31, 2014.
4 E.g. www.virtualcourthouse.com.
5 "As the preeminent symbol of federal authority in local communities, a federal courthouse must express solemnity, stability, integrity, rigor, and fairness."
6 Washington State court records were successfully penetrated in 2013 permitting access to social security numbers and drivers' license numbers. Associated Press, Washington State data breach: Office of the Courts was hacked, May 9, 2013.
7 2014 UT 11, 337 P.3d 205 (2014). Law enforcement and the courts have frequently tried to keep up with technological options (Marshall 1995).
8 "Remote arraignments have been used in courts since at least 1982 when Dade County, Florida, initiated two-way television first appearances in misdemeanor cases" (Lederer 1994).
9 "Video teleconferencing may be used to conduct an appearance under this rule if the defendant consents" (Federal Rule Criminal Procedure 5(f)).
10 See www.courtcall.com.
11 "Metadata" is often defined as "data about data". It includes such basic information as file size and date of creation and edit.
12 This is the standard United States approach; counsel submit evidence and the Court record will consist of the proffered evidence, usually in physical form. In the Australian model, customarily used in major inquiries such as Royal Commissions and trials, counsel submit their evidence, often in electronic form, to the court well before the hearing. During the hearings counsel then access the evidence from the Court's depository. In April of 2015 William & Mary Law School's Center for Legal and Court Technology successfully demonstrated the use of a smartphone for evidence presentation during a realistic simulated trial presided over by a federal judge.
13 According to remarks made by a United States district judge at the 2015 Court Technology Conference, the United States District Court in Los Angeles sometimes uses tablets on which exhibits have been preloaded.
14 As orally reported at the National Center for State Courts' September 2015 Court Technology Conference in Kansas City, the United States District Court in Los Angeles provides jurors non-networked tablets, each of which has preloaded evidence made available on an individual exhibit basis via codes provided by the trial judge.

15 Note, however, that the application of traditional evidentiary rules to technological evidence may result in undesirable results (Bellin 2013).

16 See, e.g. http://papers.ssrn.com/sol3/papers.cfm?abstract_id=2551251.

17 The author was present for both virtual reality recordings.

18 Remote judicial appearances are uncommon at trial although they are sometimes used, perhaps regularly, by appellate courts such as the Supreme Court of Canada. In one unusual UK case, a trial judge presided over part of a case from her London hospital bed. (BBC News 1999).

19 *Maryland v. Craig*, 497 U.S. 836 (1990).

20 See, e.g. *United States v. Abu Ali*, 528 F.3d 210 (4th Cir. 2008); *State v. Harrell*, 709 So.2d 1364 (Fla.); *cert. denied*, 525 U.S. 903 (1998).

21 See, e.g. *United States v. Yates*, 438 F.3d 1307 (2006)(en banc).

22 www.youtube.com/watch?v=Bgmli_aElYY.

References

BBC News (1999) UK Internet first for injured judge, http://news.bbc.co.uk/2/hi/uk_news/407911.stm (accessed 24 July 2015).

Bellin, J. (2013) *eHearsay*, Minnesota Law Review 98: 7–61.

Courtroom 21 (2002) *The Use of Technology in the Jury Room*.

Federal Rule of Criminal Procedure 5(f), available online https://www.law.cornell.edu.

Federal Rule of Evidence 101, available online https://www.law.cornell.edu.

Federal Rule of Evidence 1002, available online https://www.law.cornell.edu.

Federal Rule of Evidence 1003, available online https://www.law.cornell.edu.

Ivkovic, S.K. (2001) Justice by the International Criminal Tribunal for the Former Yugoslavia, *Stanford Journal of International Law* 37: 255–346.

Lederer, F. I. (1994) Technology Comes to the Courtroom, and …, *Emory Law Journal* 43: 1095–1122.

Lederer, F. I. (1999) The new courtroom: The intersection of evidence and technology: some thoughts on the evidentiary aspects of technologically produced or presented evidence, *S.W. University Law Review* 28: 389.

Marshall, S. (1995) Gwinnett police go online for warrants; Video testimony speeds arrests; may not be legal, *Atlanta J. and Const.* April 12, 1995, at B4.

Mims, C. (2015) Virtual Reality Isn't Just About Games, *Wall Street Journal*.

Thacker, G. (n.d.) Federal Courthouse, Whole Building Design Guide: www.wbdg.org/design/federal_courthouse.php (accessed 22 July 2015).

The Associated Press (2013) Washington state data breach: Office of the Courts was hacked, www.oregonlive.com/pacific-northwest-news/index.ssf/2013/05/washington_state_data_breach_o.html (accessed 25 October 2015).

Victoria Evidence (Audio Visual and Audio Linking) Act 1997 § 3 (Act No. 4/1997, Victoria, Australia) inserting into the Evidence Act 1958, new Section 42G.

Wallace, A. Justice and the 'Virtual' Expert: Using Remote Witness Testimony To Take Scientific Evidence, unpublished dissertation 2011, University of Sydney.

31

Computer-assisted sentencing

Martin Wasik

Introduction

Sentencing practice has altered considerably over the last twenty years or so, from an area of law with considerable judicial discretion and relatively little statutory law or case precedent to one with voluminous statutory provision and case law. Sentencing is now regarded by practitioners as complex and technical. The development of sentencing guidelines since 2000 has also transformed the way in which the practical business of sentencing is carried out. The sentencing of every case in the Crown Court requires close judicial engagement with the guidelines. Sentencing guidelines are issued by the Sentencing Council, and one example of a sentencing guideline (for the offence of assault occasioning actual bodily harm), is considered later in this chapter.

Until recently the criminal courts have relied heavily on paper files. Some of this reliance on paper is set to change. Chancellor George Osborne announced in his financial statement on 26 November 2015 that £700 million would be allocated to courts and tribunals in England and Wales, principally to invest in a dedicated IT system. Judges have been told that paper documents will be phased out during 2016 and replaced with electronic files uploaded by prosecution and defence teams as the case progresses. All that material will be copied electronically to the judge assigned to the case in advance of the sentencing hearing. This is to be known as the Crown Court Digital Case System (DCS).

Apart from the documents associated with the case itself, Crown Court judges and counsel routinely access on-line databases of statute, case law and sentencing guidelines. Research into this material will both be done in advance and, sometimes when disputed points arise in the course of a sentencing hearing, during the sentencing hearing.

Forms of computer-based support in sentencing

Is it possible to build a practical computer-based support to aid the Crown Court judge in determining sentence in an individual case? We need to draw a distinction between different forms of computer-based support.

Legal information retrieval systems

First of all there is what Susskind (2000: 165) calls the 'legal information retrieval system'. Reference has already been made to these. Practitioners, including judges, make extensive use of systems such as Lexis, Westlaw and similar word-search based databases, to ensure that they are up to date with legal provisions relevant to the case in hand. This is uncontroversial, but the availability of on-line materials has come to be essential in the sentencing context, where the pace of statutory and case-law change is very rapid. Statutory development in this field is generally achieved by new statutory provisions amending and overlaying earlier ones, often with complex implementation provisions, so that a hard copy version of any sentencing statute is guaranteed to be inaccurate. The only hope of reading and applying legislative sentencing provisions accurately is to employ an on-line resource such as Westlaw, which is updated at least once a day.

There is evidence that the complexity of the law leads to errors in Crown Court sentencing which the Court of Appeal is required to remedy on a frequent basis. According to a survey undertaken by Robert Banks in 2012, of the 262 cases taken on appeal to the Court of Appeal, 95 of these were *illegal* sentences (i.e. contrary to statutory provision, as opposed to sentences which the appellate court was invited to change because the punishment imposed was arguably too high or too low). The Law Commission is currently working on a project to gather all sentencing legislation (estimated by them to amount to some 1,300 printed pages) into a single sentencing statute (Law Commission 2014). The new statute would for all practical purposes be just an on-line resource (Law Commission 2015). Apart from statute law, the voluminous case-law on sentencing can be searched conveniently through the same law databases, together with relevant academic articles on sentencing and critical on-line commentaries. In addition, the standard practitioner works on sentencing which cover the most important appellate cases as well as statutory material, can all be searched on-line and provide on-line updates to subscribers.

Sentencing information systems

A second form of assistance is a 'case-based computer support system'. It allows a judge dealing with a case of a particular type to access electronically the range and quantum of sentences which have been passed for comparable cases in the past. A number of such support systems were piloted in the 1980s and 1990s, across different jurisdictions. Perhaps the best-known examples are the sentencing information systems (SIS) developed in Canada (Doob and Park 1987), in New South Wales, and in Scotland (Hutton 1995; Tata 2000). These systems are designed simply to provide data about sentencing outcomes in a range of past cases. They do not prescribe sentence in the case under consideration, or even seek to explain how a judge might use the information about past cases to assist with the present case. The central objective of SIS is to encourage greater consistency in sentencing outcomes by making available to judges a context of past decisions in similar cases. This is a laudable objective, but not easily achieved given the number of variables which can exist in any given case. There are many relevant factual differences within and between offences, and in the personal circumstances pertaining to different defendants. Previous cases must be analysed to a high level of detail in order for the exercise to be useful, and not downright misleading. For a judge about to sentence a defendant for an offence of, say, assault occasioning actual bodily harm, a SIS that informs him or her of the full range of sentences passed in the previous few years for that offence would be of no practical value. At the very least a judge would require the decided cases of assault to display the operation of what might be regarded as key variables in the facts of the offence itself and the interplay between those variables (examples might be: extent of injury to the victim, vulnerability of the victim, whether the victim was targeted or not, whether a weapon was used, whether there was a single blow or repeated assault, whether the offender was part of a group, and so on), and then the relationship of these factors to issues relevant to the defendant (plea of guilty or not, relevant previous convictions or not, offence planned or committed on the spur of the moment, provocation from the injured party, and so on).

It turned out that this generation of SIS was largely a failure. Most projects never advanced beyond the pilot stage, and others which did soon fell into disuse. The reasons for failure seem to have been the practical difficulty and spiralling cost of keeping the SIS up to date and sufficiently detailed to be relevant. There was not much enthusiasm about the systems from practitioners, and hence no serious offers to fund their commercial development, either from the state or the private sector. In reality SIS have largely been overtaken by the development of formal sentencing guidelines in many jurisdictions including England (though not in Scotland, where guidelines have not taken root). See further Tata (2013). There are some examples of SIS being used today. The New South Wales scheme is still operative, and a SIS has been developed in Ireland in lieu of sentencing guidelines. Another SIS is known to be in operation in China. Although full details of it are hard to find, the available literature suggests that Chinese judges are supplied with a database of past decisions in 'similar' cases in order to inform their sentencing practice (Anderson 2006). This development is seen by commentators as an advance on earlier practice, which exhibited wide disparity in judicial decisions. It may be, however, that the case-based expert system in China is now in the process of being supplanted by sentencing guidelines, at least for some offences.

Legal expert systems

The third form of computer support to aid sentencers is a form of rule-based or knowledge-based 'expert system'. In his short introduction to expert systems Beard (2014) writes that:

> Expert systems are a branch of artificial intelligence ... They attempt to mimic human expertise by applying inference methodology to a specific body of knowledge to aid in decision support or problem solving within a specific problem domain.

Susskind (2000: 163) explains that:

> Expert and knowledge-based systems are ... computer applications that contain knowledge and expertise which they can apply – much as a human expert does – in solving problems ... In law, these systems should be able to apply their legal knowledge in guiding users through complex legal issues; in identifying solutions to problems ... and in offering advice and making specific recommendations.

Susskind goes on to identify different types of expert system, of which two, the 'intelligent checklist' and the 'diagnostic' system, seem the most relevant for our purposes. The 'intelligent checklist' essentially audits *compliance with legal rules*. There are examples of expert systems of this kind in the area of taxation, such as one to assist auditors to determine whether a company is compliant with corporation tax rules. The 'diagnostic system' offers specific solutions to problems presented, usually after some kind of *interactive consultation*. There are many such programs currently available, operational in diverse fields apart from law. Examples can be found in medicine, such as the self-diagnosis system on NHS Direct (www.nhsdirect.nhs.uk/en/CheckSymptoms). There are also examples of interactive web-based programs to assist lay people with a range of issues such as health, diet and lifestyle, or even their choice of pets (www.exsys.com/Demos/Dogs/DogTitle.hmtl). Expert systems are also used in a range of business and management scenarios, such as training aids for newly appointed staff. The website www.Openexpert.org offers a range of freely available demonstrations, and others can be found at www.xpertrule.com/expert-system-software.hmtl.

It might be thought from this that expert systems would already have been devised to guide and assist judges with sentencing. However, no such systems currently exist, and attempts in the past to develop them (such as the one outlined for use in magistrates' courts by Bainbridge, 1990) have not progressed beyond the drawing board.

Expert systems in sentencing

The fundamental form of every expert system in law is that of a decision tree, which operates on an 'if this, then that' sequential logic. A simple example of a decision tree of this kind was provided by Twining and Miers, in their famous book *How to do Things with Rules*, published as long ago as 1976. Although sentencing was then more straightforward than it is today, the authors chose a sentencing statute to illustrate 'the structure of complex rules' (Twining and Miers 1976). They took the following provision:

> *Powers of Criminal Courts Act 1973, s.20(1)*
> No court shall pass a sentence of imprisonment on a person of or over twenty-one years of age on whom such a sentence has not previously been passed by a court in any part of the United Kingdom unless the court is of opinion that no other method of dealing with him is appropriate … .

The authors then presented section 20(1) in schematic or algorithm form. Twining and Miers offered the algorithm as a precise set of instructions for solving what they said was a 'well-defined' problem. The decision tree has a structured series of questions to each of which the reader can only answer 'yes' or 'no'. The answer given automatically takes the reader on to the next question, unless it leads to an outcome, in which case the algorithm is complete.

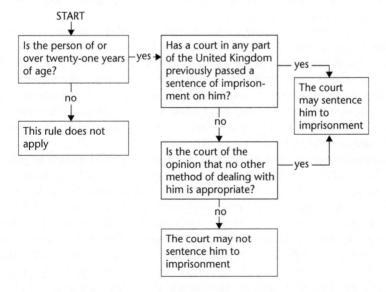

Figure 31.1 Algorithm designed to instruct a court of its powers of imprisonment of a person over the age of 21 who has not previously served a prison sentence

Could a judge, in principle, receive assistance from an expert system in law programmed with this algorithm? At the very least the system could serve to remind the judge of the relevance of section 20 to the case in hand, and the questions which need to be answered before a sentence of imprisonment could lawfully be imposed. Let us consider briefly some of the decision points in the algorithm. Box 1 asks whether the defendant is of or over 21 years of age. This seems straightforward, but omits the crucial detail of whether this means 21 at the date of commission of the offence, or date of conviction, or date of sentence. Box 2 asks whether the defendant has previously received a prison sentence, but would a suspended prison sentence count for that purpose or not? Answers to both those questions can be found elsewhere in sentencing law (it is the date of conviction which is relevant, and a previous suspended sentence does not apply), but it is immediately clear that to provide full and reliable assistance to the judge the expert system would have to include material going well beyond section 20 to ensure that problems like these were identified and resolved. As far as the third question goes (is the court of the *opinion* that no other method of dealing with him is appropriate), it is hard to see how an expert system could provide any useful assistance. The question invites consideration of a fundamental *policy* issue of whether, taking into account all the circumstances of the offence and the offender, imprisonment can be avoided in this case. Lord Bingham CJ said in *Howells* (1999) that:

> there is no bright line which separates offences which are so serious that only a custodial sentence can be justified from offences which are not so serious ... In the end, the sentencing court is bound to give effect *to its own subjective judgment of what justice requires* on the peculiar facts of the case before it.

As Padfield has observed, 'sentencing is a complex and subjective decision-making process – there is no easily ascertainable custody "threshold"' (Padfield 2011: 598).

The custody threshold is a matter of professional judgment, going well beyond the application of rules capable of being reduced to an algorithm. We have here a moral and policy-based assessment, linked to general societal standards of fairness and proportionality which the judge is expected to reflect in his decision.

We turn now to consider a more modern algorithm, a decision-tree provided in Thomas's *Sentencing Referencer* (Thomas 2015), a *vade-mecum* which is much relied upon in the courts. The chart is designed to assist judges in the task of sentencing so-called 'dangerous offenders'. The chart sets out a series of steps which are required on the way to a judge deciding that a defendant who has been convicted of one of a number of specified violent or sexual offences represents a significant risk of causing serious harm in the future from committing further such offences. If so, the defendant becomes eligible for a life sentence (if the offence carries life as its maximum) or an extended custodial sentence.

Thomas chose to set out this particular aspect of sentencing practice in the form of a decision tree because it is technical and complex, and because this area of sentencing lends itself (at least to some extent) to sequential yes/no answers. Let us work through some of the boxes. Box 1 asks whether the defendant reached the age of 18 before being convicted. This is fully amenable to a straight yes/no answer (and note that the ambiguity evident in Box 1 in the previous exercise is avoided here by greater precision in the language). Box 2 also seems straightforward, and there would be no difficulty in programming an expert system with a list of the offences (known as 'specified offences') which are currently set out in schedule 15. There are reported cases where judges have made mistakes in relation to Box 2, and have gone on to sentence the defendant as a dangerous offender when their offence was not, in fact, listed in the relevant schedule. An appellate case dealing with eight such mistakes is *Reynolds* (2007). An expert

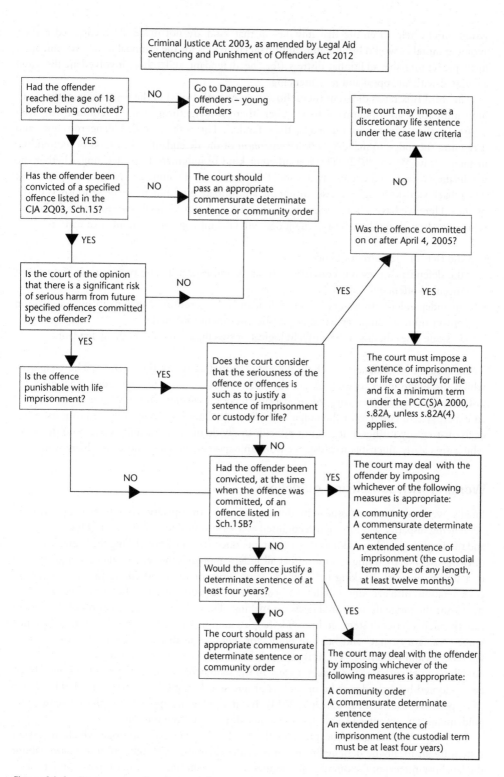

Figure 31.2   Dangerous offenders: adults

system might help to ensure that mistakes of that kind are not made. As mentioned earlier, avoiding mistakes would bring significant financial savings for the criminal justice system, apart from preventing the additional distress (for the defendant and others involved in the case) associated with an appeal and re-sentencing.

The third Box raises different issues, however. As with Box 3 in the first example, the court now must form an 'opinion' as to whether 'there is a significant risk of serious harm from future specified offences committed by the offender'. This is the so-called 'dangerous test', and it requires the judge to make a *predictive assessment* of the defendant's likelihood of reoffending in such a way (Wasik 2012). This is a different kind of judgment from deciding whether the defendant has crossed the 'custody threshold' (Box 3 in the Twining and Miers example). That was a 'backward-looking' exercise, fitting the appropriate punishment to the seriousness of the offence. Here we have a 'forward-looking' exercise, to estimate the future risk that a defendant represents. In making this decision the judge will take into account a number of matters:

- the facts of the current offence;
- the defendant's previous convictions (if any), and especially any previous convictions for specified offences;
- an independent assessment of the defendant's dangerousness, drawn up by a probation officer using a computer-based diagnostic assessment tool; and
- in some cases there will be a psychological or psychiatric report on the defendant.

There is case-law guidance as to how the judge should carry out this exercise. The weighting which the judge gives to each of the above factors is a matter for the judge who may, for instance, agree or disagree with the probation officer's assessment. It is not clear to me how far an expert system could assist the judge with this task. There is perhaps no need to pursue this decision-tree any further, but it is not difficult to divide up the remaining boxes into decisions which are clearly amenable to assistance from an expert system, and those which are not.

Working with sentencing guidelines

Finally, we turn to a central issue in modern sentencing – the operation of sentencing guidelines. Sentencing disparity was long ago recognised as a serious problem of injustice (Hogarth 1971), and the English guidelines address this important concern, not by prescribing the 'right' sentence in each case, but by trying to ensure that all judges approach the sentencing of any given case in a similar manner, working through the same series of steps and taking the same range of considerations into account. Guidelines now cover the great majority of criminal offences, and they assist the judge in structured decision-making. They identify salient features of the instant case through a series of steps, and provide help in guiding the judge towards a normal 'offence range', 'category range' and 'starting point' for sentence in that case (for further explanation see Roberts and Rafferty 2011). It is essential that the judge engages with applicable sentencing guidelines in every case, and makes it clear in the sentencing remarks the reasoning which has been adopted by the judge as he or she has progressed through the steps provided. The Court of Appeal has made it clear (in *Dyer*, 2014) that if a judge has applied the relevant sentencing guidelines then any appeal based upon sentence disparity is bound to fail.

In many ways the guideline provides the kind of assistance for a judge which an expert system might ideally be designed to do. It provides the judge with relevant information about the legal parameters of the offence, and guides the judge through a series of steps in the process of selecting sentence in the particular case. Some practitioners object to what they see as a 'tick-box' approach to sentence decision-making, arguing that guidelines leave insufficient room for the judge to exercise discretion to do justice to the facts of the individual case and to the background and personal circumstances of the offender. A senior barrister (Cooper 2008, 279–280) has complained that:

> the sentencing process has become a 'box ticking' exercise, at the expense of a flexible sentencing regime. The layout of the guidelines lends itself to a 'multiple choice' approach to the setting of the tariff, made up of columns and lists of considerations which the sentencing tribunal is advised to take into account in arriving at an appeal-proof decision.

A comment in a recent blog (Anonymous 2015) complains that guidelines amount to no more than 'sentencing by numbers':

> I doubt any competent programmer would find difficulty in producing a logarithm to give a bench or DJ [district judge] instant outcomes at the press of a couple of keys on the keyboard.

The most thoroughgoing critique of sentencing guidelines in the context of the present chapter has been provided by Franko Aas (2005: 76), who complains that guidelines are an undesirable 'transformation' in the role of a judge:

> sentencing guidelines represent … a move from a rational … individual who is autonomously giving meaning to a legal text, to a state where a judge's individuality becomes almost irrelevant in the light of objective systems … Do guidelines then still function as a text, which needs to be read and interpreted, or have they become closer to becoming a computer program – which functions automatically? [S]entencing guidelines reduce the sentencing process to an algorithmic procedure.

In the view of the present writer these concerns are over-stated. It is surely desirable for judicial discretion in sentencing to be properly structured, so as to avoid widely variant sentences being passed by different judges for similar crimes. Of course flexibility must be ensured to allow judges to depart from sentencing guidelines where it would be just and appropriate to do so. Here we set out for illustration the sentencing guideline for the offence of assault occasioning actual bodily harm, an offence which is 'triable either-way', and hence can be sentenced in the Crown Court or in a magistrates' court (Sentencing Council 2011).

The guideline is structured in the same way as other offence guidelines, requiring the judge to work through a series of up to nine steps, although some of the later steps may not be relevant in a particular case.

Step 1 determines into which of three broad categories the particular offence falls, by requiring the judge to form an assessment of the degree of harm caused or risked by the offence and the level of culpability displayed by the offender. The factors which the judge may take into account in forming this assessment are then set out. Clearly these are factual matters relating to the offence, such as whether the offender used a weapon or not. If the issue is factually disputed the judge will have to make a finding of fact on that point, putting the prosecution to proof and hearing evidence on the matter if necessary. It will be seen that none of the factors listed at Step 1 is reducible to a simple 'yes/no' formulation equivalent to the question (in Thomas's decision-tree, above) as to whether the offence is listed in schedule 15 or not. Depending on the established facts the judge will be required to decide many matters of degree – was the

The court should determine the offence category using the table below.

Category 1	Greater harm (serious injury must normally be present) and higher culpability
Category 2	Greater harm (serious injury must normally be present) and lower culpability; or lesser harm and higher culpability
Category 3	Lesser harm and lower culpability

The court should determine the offender's culpability and the harm caused, or intended, by reference **only** to the factors identified in the table below (as demonstrated by the presence of one or more). These factors comprise the principal factual elements of the offence and should determine the category.

Factors indicating greater harm	Use of weapon or weapon equivalent (for example, shod foot, headbutting, use of acid, use of animal)
Injury (which includes disease transmission and/or psychological harm) which is serious in the context of the offence (must normally be present)	Intention to commit more serious harm than actually resulted from the offence
Victim is particularly vulnerable because of personal circumstances	Deliberately causes more harm than is necessary for commission of offence
Sustained or repeated assault on the same victim	Deliberate targeting of vulnerable victim
Factors indicating lesser harm	Leading role in group or gang
Injury which is less serious in the context of the offence	Offence motivated by, or demonstrating, hostility based on the victim's age, sex, gender identity (or presumed gender identity)
Factors indicating higher culpability	
Statutory aggravating factors:	Factors indicating lower culpability
Offence motivated by, or demonstrating, hostility to the victim based on his or her sexual orientation (or presumed sexual orientation)	Subordinate role in group or gang
	A greater degree of provocation than normally expected
Offence motivated by, or demonstrating, hostility to the victim based on the victim's disability (or presumed disability)	Lack of premeditation
	Mental disorder or learning disability, where linked to commission of the offence
Other aggravating factors:	
A significant degree of premeditation	Excessive self defence

Having determined the category, the court should use the corresponding starting points to reach a sentence within the category range below. The starting point applies to all offenders irrespective of plea or previous convictions. A case of particular gravity, reflected by multiple features of culpability in step one, could merit upward adjustment from the starting point before further adjustment for aggravating or mitigating features, set out below.

Offence Category	**Starting Point** *(Applicable to all offenders)*	**Category Range** *(Applicable to all offenders)*
Category 1	1 year 6 months' custody	1 – 3 years' custody
Category 2	26 weeks' custody	Low level community order – 51 weeks' custody
Category 3	Medium level community order	Band A fine – High level community order

Figure 31.3a Assault definitive guideline

Source: © Crown copyright 2011, contains public sector information licensed under the Open Government Licence v3.0

The table below contains a **non-exhaustive** list of additional factual elements providing the context of the offence and factors relating to the offender. Identify whether any combination of these, or other relevant factors, should result in an upward or downward adjustment from the starting point. In some cases, having considered these factors, it may be appropriate to move outside the identified category range.

When sentencing **category 2** offences, the court should also consider the custody threshold as follows:
- has the custody threshold been passed?
- if so, is it unavoidable that a custodial sentence be imposed?
- if so, can that sentence be suspended?

When sentencing **category 3** offences, the court should also consider the community order threshold as follows:
- has the community order threshold been passed?

Factors increasing seriousness	
Statutory aggravating factors:	Exploiting contact arrangements with a child to commit an offence
Previous convictions, having regard to a) the nature of the offence to which the conviction relates and its relevance to the current offence; and b) the time that has elapsed since the conviction	Established evidence of community impact
	Any steps taken to prevent the victim reporting an incident, obtaining assistance and/or from assisting or supporting the prosecution
Offence committed whilst on bail	Offences taken into consideration (TICs)
Other aggravating factors include:	**Factors reducing seriousness or reflecting personal mitigation**
Location of the offence	No previous convictions **or** no relevant/recent convictions
Timing of the offence	Single blow
Ongoing effect upon the victim	Remorse
Offence committed against those working in the public sector or providing a service to the public	Good character and/or exemplary conduct
Presence of others including relatives, especially children or partner of the victim	Determination and/or demonstration of steps taken to address addiction or offending behaviour
Gratuitous degradation of victim	Serious medical conditions requiring urgent, intensive or long-term treatment
In domestic violence cases, victim forced to leave their home	Isolated incident
Failure to comply with current court orders	Age and/or lack of maturity where it affects the responsibility of the offender
Offence committed whilst on licence	
An attempt to conceal or dispose of evidence	Lapse of time since the offence where this is not the fault of the offender
Failure to respond to warnings or concerns expressed by others about the offender's behaviour	Mental disorder or learning disability, where **not** linked to the commission of the offence
Commission of offence whilst under the influence of alcohol or drugs	Sole or primary carer for dependent relatives
Abuse of power and/or position of trust	

Section 29 offences only: The court should determine the appropriate sentence for the offence without taking account of the element of aggravation and then make an addition to the sentence, considering the level of aggravation involved. It may be appropriate to move outside the identified category range, taking into account the increased statutory maximum.

Figure 31.3b Assault definitive guideline

Source: © Crown copyright 2011, contains public sector information licensed under the Open Government Licence v3.0

STEP THREE

Consider any other factors which indicate a reduction, such as assistance to the prosecution

The court should take into account sections 73 and 74 of the Serious Organised Crime and Police Act 2005 (assistance by defendants: reduction or review of sentence) and any other rule of law by virtue of which an offender may receive a discounted sentence in consequence of assistance given (or offered) to the prosecutor or investigator.

STEP FOUR

Reduction for guilty pleas

The court should take account of any potential reduction for a guilty plea in accordance with section 144 of the Criminal Justice Act 2003 and the *Guilty Plea* guideline.

STEP FIVE

Dangerousness

Assault occasioning actual bodily harm and racially/religiously aggravated ABH are specified offences within the meaning of Chapter 5 of the Criminal Justice Act 2003 and at this stage the court should consider whether having regard to the criteria contained in that Chapter it would be appropriate to award an extended sentence.

STEP SIX

Totality principle

If sentencing an offender for more than one offence, or where the offender is already serving a sentence, consider whether the total sentence is just and proportionate to the offending behaviour.

STEP SEVEN

Compensation and ancillary orders

In all cases, the court should consider whether to make compensation and/or other ancillary orders.

STEP EIGHT

Reasons

Section 174 of the Criminal Justice Act 2003 imposes a duty to give reasons for, and explain the effect of, the sentence.

STEP NINE

Consideration for remand time

Sentencers should take into consideration any remand time served in relation to the final sentence. The court should consider whether to give credit for time spent on remand in custody or on bail in accordance with sections 240 and 240A of the Criminal Justice Act 2003.

Figure 31.3c Assault definitive guideline

Source: © Crown copyright 2011, contains public sector information licensed under the Open Government Licence v3.0

victim *particularly* vulnerable, was the assault *sustained*, was the injury *less serious in the context of the offence*, was there a *significant degree* of premeditation, was there *deliberate targeting* of the victim, etc. These factors must be considered and weighed individually by the judge, and then an overall weighting of the harm and culpability factors must be made to determine whether the case falls into category 1, 2 or 3.

At Step 2 of the guideline the judge is now presented with the guideline category range and starting point which follows from the assessment made at Step 1. A further, but this time non-exhaustive, list of aggravating and mitigating factors is set out, requiring the judge to adjust the provisional sentence upwards or downwards from the relevant starting point. One or two of these factors are perhaps reducible to 'yes/no' outcomes – such as whether the offender was on licence at the time of the offence – but the vast majority of aggravating and mitigating factors relevant to the offence being sentenced have to be assessed and weighed by the judge, both individually and as part of the overall assessment. The guideline also alerts the judge to important policy assessments that may need to be made – such as the elusive 'custody threshold' question (which was referred to above), or whether, if a custodial sentence is 'unavoidable', it might nonetheless be suspended, rather than imposed immediately. One factor of personal mitigation is perhaps worth mentioning separately. The guideline says that the judge may adjust the sentence downwards if there is 'remorse'. Whether the offender is remorseful or not is a factor requiring the judge to make an overall assessment of the way in which the offender has conducted himself, no doubt taking the pre-sentence report and defence counsel's representations into account. Remorse is an easy thing to claim, and the judge must decide whether the remorse is 'genuine' and, if so, how much weight to give to it.

At the end of Step 2 the judge still only has a provisional sentence – it may require further adjustment under later steps. Only two will be mentioned here. At Step 4 the judge must make an appropriate reduction where the offender has admitted guilt. A separate sentencing guideline dealing with that single issue applies here (Sentencing Guidelines Council 2007). The judge must give the largest reduction (up to one-third reduction from sentence) if the offender has admitted guilt 'at the first reasonable opportunity', but ultimately it is for the judge to apply his experience of the system to assess whether the plea was entered in accordance with that requirement or not and, if not, what lesser reduction is appropriate (see *Caley* (2013) for appellate guidance on this point). Step 5 deals with the problem of 'dangerousness', which we have looked at already. Little more needs to be said here, save for noting that the offence of occasioning actual bodily harm *is* an offence listed in schedule 15, and the court would in an appropriate case be required to work through the decision tree set out in Thomas's book to see whether the passing of an extended sentence might be appropriate.

Drawing some conclusions

This chapter has considered various forms of computer-assisted sentencing. Legal information retrieval systems are uncontroversial, and very widely used by practitioners, including judges. We have seen that the first generation of expert systems in law was not a success. There are renewed claims, by those who design them, that a second and more sophisticated generation of expert systems would be more up to the mark (Stevens, Barot and Carter 2010). It is said that newer systems are capable of managing multi-layered alternative outcomes and highly complex decision trees, all at lightning speed. Newer systems increasingly employ 'fuzzy logic', through which the system can achieve greater flexibility in managing multiple factors and registering outcomes which fall somewhere on a range. Fuzzy logic takes 0 and 1 as the extreme cases (defendant is 'remorseful' or 'not remorseful' but can register a range of position in between (such as '0.4 of remorsefulness'). This may offer a way of giving weight to a sentencing factor within an algorithm but, as we have seen, there are many different factors relevant to a sentencing decision, and there is the problem of achieving a final assessment of those weighted factors as they interplay with one another, and then taken as a whole.

An expert system to assist the judge in sentencing might improve the process of justice by helping to eliminate mistakes, and save money by reducing the number of appeals necessary to correct sentences which have been passed in error. However, the investment of time and resources required to develop a knowledge-based system, even in a very narrow and stable area of law, is considerable. Creating the system requires investment both in the software architecture and in transferring subject expertise to algorithmic form (Stevens, Barot and Carter 2010). In the court setting it would seem that the data relating to each case would have to be entered into the system before it could provide support for the judge, or there would have to be an intelligent interface between the software in the expert system and the case files uploaded to the judge at the start of the case. All this seems fraught with potential technical difficulty.

Given the substantial set-up and maintenance costs involved it would be propitious if the area of law concerned was one where the logical basis for the legal provisions was clear, and if it was an area of commercial importance (where there was clear opportunity for a return on the investment in setting up the system), and it was an area of law which was stable, with infrequent statutory amendments. Unfortunately, none of these desirable pre-conditions applies to the area of sentencing. That is not to say that related areas of criminal justice could not benefit from the development of an expert system. One such candidate would be the provisions which determine the earliest date of a prisoner's release from custody. In *Noone* (2010) the Lord Chief Justice described the applicable statutory provisions as a 'legislative morass'. While this is an area of law which is technically complex, it is reducible to a stepped series of rules and so would be amenable to an algorithm. It is also an area in which expensive mistakes can be made, with prisoners being released before they should be (representing an unnecessary risk to the public), or are detained illegally beyond their date of release (leaving the state open to an action for damages for illegal detention). Here, the relevant pre-conditions seem to exist, and an expert system might well be of value.

It seems clear that sentencing itself is *not* a promising area for the development and application of an expert system. Some of the reasons have been rehearsed above. All expert systems rely on deductive reasoning, which is deficient when it comes to according relative weight to a wide range of considerations, such as aggravating and mitigating factors in sentencing. As Leith (2010: 7) aptly puts it: the needs of judges are 'not met by a system which simply lists rules, and indicates the order in which they were triggered'. Leith says that the enthusiasm for expert systems in law in the 1980s was founded on a very narrow and uncritical concept of 'law as a system of rules' which has now been discredited, and Susskind (2000) has questioned whether the very nature of legal rules renders them unamenable to expert systems in law. That broader jurisprudential debate is beyond the scope of this chapter, but what seems clear from the examples given above is that while some sentencing provisions are straightforward rules, and in principle are reducible to algorithmic form, far more sentencing provisions are flexible and 'open-textured' in style. Sentencing requires the settling by the judge of disputed facts, whereas expert systems can only function where all the relevant facts are given: if there is no consensus there is no reliable data on which the system can work. Sentencing involves the weighing of incommensurable concepts, such as whether the personal mitigation available to the offender justifies the suspension of a period of imprisonment which would otherwise be appropriate given the seriousness of the offence. As we have seen, the defendant's remorse can properly mitigate. But only a human being can make the assessment of whether the offender is 'genuinely' remorseful and, if so, how much weight should be given to that factor in light of all the others. There are also important policy considerations in sentencing, such as the need for a judge to bear in mind pressure on prison accommodation (see *Kefford* (2002)), or the human rights of children of a single parent about to receive a custodial sentence (see *Petherick* (2013)). Expert systems cannot deal with factual uncertainty, or with the weighing of incommensurables, or with issues of policy.

This chapter has considered the potential usefulness of expert systems in the context of sentencing. Whatever technical problems there might be in developing such a system in this area (and there are many), it seems that the development of sentencing guidelines since 2000 has to a large extent removed the need for generic expert systems in sentencing. This is perhaps the key finding of this chapter. But guidelines and expert systems are not the same thing. I do not agree with Franko Aas that guidelines reduce sentencing to an algorithmic procedure. That is *not* how sentencing guidelines (at least, those in England and Wales) work (see further, Wasik 2008). Sentencing guidelines provide an appropriate form of support for judicial decision-making, and their design draws upon the case-law, as well as the pooled experience of the senior judges and others who are responsible for drawing up the guidelines. My conclusion is that the range of factors in play at sentencing, together with the inevitable choices required to be made by a human being amongst the various considerations, means that currently there is no place for expert systems in this area of legal practice, and it is difficult to imagine a time when there might be. But there are, of course, broader arguments than that. Sentencing is a process which is performed by a human being, charged with the responsibility to act on behalf of the state. This is a more fundamental objection to computer-assisted sentencing than saying that computer-assisted judges would just become lazy, and pass the sentence which is indicated in the 'grid'. It is about the nature of the human communication between the judge (acting on behalf of the state) and the defendant.

References

Anderson, N. (2006) China tests computer-aided sentencing, available at: http://arstechnica.com/information-technology/2006/09/7745/.

Anonymous (2015) Sentencing guidelines tomorrow or computer says no, available at: www.thejustice ofthepeaceblog.blogspot.co.uk/2015/10/sentencing-guidelines-tomorrow-or.html.

Bainbridge, D. (1990) 'CASE': Computer assisted sentencing in magistrates' courts, paper presented at the 5th BILETA Conference, British and Irish Legal Technology Association, available at: www.bileta.ac.uk/content/files/conference%20papers/1990/'CASE'%2520-%2520Computer%2520Assisted%2520Sentencing%2520in%2520Magistrates'%2520Courts.pdf.

Beard, M. (2014) *Expert Systems: An Introduction*, published by Matthew Beard (May 11, 2014). Kindle edition available through Amazon Digital Services.

Cooper, J. (2008) The Sentencing Guidelines Council – A practical perspective, *Criminal Law Review* 4: 277–286.

Cooper, J. (2013) Nothing personal: The impact of personal mitigation at sentencing since the creation of the Council, in Ashworth, A. and Roberts, J.V., *Sentencing Guidelines: Exploring the English Model*. Oxford, Oxford University Press, 157–164.

Doob, A. and Park, N. (1987) Computerised sentencing information for judges: An aid to the sentencing process, *Criminal Law Quarterly* 30: 54–70.

Franko Aas, K. (2005) *Sentencing in the Age of Information: From Faust to Macintosh*. London: Glasshouse Press.

Hogarth, J. (1971) *Sentencing as a Human Process*. Toronto: Toronto University Press.

Hutton, N. (1995) Sentencing, rationality and computer technology, *Journal of Law and Society* 4: 549–570.

Law Commission (2104) *Twelfth Programme of Law Reform*. Law Com No 354 London: Law Commission.

Law Commission (2015) *Sentencing Procedure Issues Paper 1: Transition*. London: Law Commission.

Leith, P. (2010) The rise and fall of the legal expert system, *European Journal of Law and Technology* 1(1). Available at: http://ejlt.org/article/view14/1 (accessed 30 November 2015).

Padfield, N. (2011) Time to bury the custody 'threshold'? *Criminal Law Review* 8: 593–612.

Palmer, A. (2012) When it comes to sentencing, a computer judge might make a fairer judge than a judge, available at: www.telegraph.co.uk/news/uknews/law-and0order/9029461/When-it-comes-to-sentencing-a-computer-might-make-a-fairer-judge-than-a-judge.html (accessed 30 November 2015).

Roberts, J.V. and Rafferty, J. (2011) Sentencing guidelines in England and Wales: Exploring the new format, *Criminal Law Review* 681–689.

Sentencing Council for England and Wales (2011) *Definitive Guideline, Assault (Crown Court)*. London: Sentencing Council.

Sentencing Guidelines Council (2007) *Definitive Guideline, Reduction in Sentence for a Guilty Plea*. London: Sentencing Council.

Stevens, C., Barot, V. and Carter, J. (2010) The next generation of legal expert systems – new dawn or false dawn? available at: www.dora.dmu.ac.uk/bitstream/handle/2086/4600/NextGenLegalES_version10_FINAL%5B1%5D.pdf?sequence=2.

Susskind, R. (2000) *Transforming the Law: Essays on Technology, Justice and the Legal Marketplace*. Oxford: OUP.

Susskind, R. (2008) *The End of Lawyers? Rethinking The Nature of Legal Services*. Oxford: OUP.

Tata, C. (2000) Resolute ambivalence: Why judiciaries do not institutionalise their decision support systems, *International Review of Law, Computers & Technology* 14(3): 297–316.

Tata, C. (2013) The struggle for sentencing reform, in Ashworth, A. and Roberts, J.V., *Sentencing Guidelines: Exploring the English Model*. Oxford: OUP, pp.236–256.

Thomas, D. (2015) *Thomas's Sentencing Referencer 2015*. London: Sweet & Maxwell.

Twining, W. and Miers, D. (1976) Algorithms and the structure of complex rules, Chapter 9 in *How to do Things with Rules*. London: Weidenfeld and Nicolson.

Wasik, M. (2008) Sentencing guidelines – state of the art? *Criminal Law Review*, 253–263.

Wasik, M. (2012) The dangerousness test, Chapter 11 in Sullivan, G.R. and Dennis, I. *Seeking Security: Pre-empting the Commission of Criminal Harms*. Oxford: Hart Publishing.

Cases

Caley [2013] 2 Cr App R (S) 47

Danga [1992] QB 476

Dyer [2014] 2 Cr App R (S) 11

Howells [1999] 1 WLR 307

Kefford [2002] 2 Cr App R (S) 495

Petherick [2013] 1 WLR 1302

R (Noone) v Governor of Drake Hall [2010] UKSC 30

Reynolds [2007] 2 Cr App R (S) 553

32

The technology of confinement and quasi-therapeutic control

Managing souls with in-cell television

Victoria Knight

Introduction

Confinement, like other varieties of punishment has been furthered by a variety of technologies and techniques (see Hallsworth and Kaspersson, Chapter 33, this volume). These have ranged from those designed to cause death (like immurement), those designed to cause humiliation (like the stocks) to those designed for spatio-temporal control of the body (like the dungeon through to the prison). The nineteenth century saw the emergence of new kinds of disciplinary control which began to re-interpret confinement in terms of rehabilitation and therapy. New prison designs centred upon enhancing discipline or moral rectitude, the medicalisation of prison environments and the increasing role of psychiatry all served to further this therapeutic model of the prison and the technologies associated with this. In this chapter I will consider how contemporary technologies like television (along with emerging digital technologies in this context) are co-opted into this philosophy of confinement. This chapter draws on ethnographic research into in–cell television and the emergence of digital media in a closed adult male prison (Knight 2012, 2015b, 2016). Decisions to introduce communicative and digital technologies in prison settings have always been fraught with anxieties, especially in relation to the potential dangers of prisoners reaching the outside world (Knight 2015b). In highlighting this, the chapter begins by describing television's birth into prisons in England and Wales and contextualises the policy in which it is justified. This is an important juncture in penal history as it marks a shift towards a 'neo-paternal' model whereby much emphasis and effort is placed on the need to regulate and control behaviour. The chapter then moves on to describe how technology like television can be imagined as a mode of governance by drawing on the detailed business of emotional control from the perspectives of both prison staff and prisoners. The kinds of relationships prisoners have with technology demonstrates how television has become part of 'a plethora of less expensive and less intensive therapeutic techniques … through which individuals may seek a resolution of their inner distress' (Rose 1999: 217). The chapter ends by reviewing the nature and implications of the care-giving qualities of television and digital technologies, which have begun to be introduced more recently.

The birth of in-cell television

The introduction of television into prisoners' cells took place in England and Wales in 1998. This introduction took twelve years to complete from its official launch to the last prison receiving television in cells in 2010. This was because many prison cells had to be updated to receive electricity. Prior to this approximately 1,000 prisoners across the prison estate had in-cell television from 1991. An official review was consequently conducted to interrogate this disparity (personal correspondence with Ministry of Justice 2011). In rejecting uneven access to in-cell television across the prison estate, amongst other things, the prison service was under pressure to satisfy the prisoner as well as the public and of course maintain order. Following a review of disturbances at HMP Strangeways by Lord Justice Woolf in 1990 the Incentive and Earned Privilege (IEP) system was introduced in 1996 (PS0 4000 – now replaced by PSI11/2011). This system sought to robustly manage prisoner behaviour as well as normalise the prison experience. The principles of this system sought to manage prisoners according to how well they responded to the prison regime and complied with prison rules. In motivating compliance, in-cell television became a key incentive along with access to other goods and services. Hence television became officially framed as a privilege, and as a consequence non-compliance could lead to these privileges being withdrawn and prisoners placed on a 'basic' regime. This stripped regime was intended to discourage prisoners from doing their prison time in limited and restricted conditions. Here access to communicative opportunities such as telephone, visits from family and friends and time to associate with other prisoners was withheld and tightly restricted. Studies of media use in prisons found that limiting social contact amplifies a series of deprivations, which make the prospect of doing time without these outlets difficult to endure (Vandebosch 2000; Jewkes 2002; Gersch 2003; Knight 2012).

The introduction of in-cell television was not warmly received and continues to be contested in public discourse. Underpinning this was the anxiety that the penal system was going 'soft on' criminals and that the prison service was losing its purpose to punish. These pressing and politically sensitive claims emphasised the need to position in-cell television as an earnable privilege for 'deserving' and compliant prisoners; for example, those who proved to be drug free (Hansard 1998). The introduction of other technologies and wider communicative outlets like print media, radio, the telephone and now more recently services like email are framed in contexts of less eligibility (not deserving to benefit from goods and services like everyone else), cost and security. In historical terms it is possible to observe accounts of anxiety and fear of introducing communicative developments to the prison landscape (Jewkes and Johnston 2009; Knight 2015b).

As a result caution was and still is always appropriated to prisoners' access to communicative outlets. For example, prior to the 1950s, staff in UK prisons would read out news items from newspapers in chapel every week. By 1954, under supervision, prisoners could directly access radio and newspapers. Typically radio was piped onto prison landings through large speakers. A short time later, prisoners had the opportunity to buy their own transistor radio sets. Films were sometimes shown in communal areas on a weekly basis and communal television sets were introduced in some prisons from the 1970s, but this was never formally standardised. Sir Martin Narey, former Head of the Prison Service for England and Wales, reflected upon these kinds of resistance explaining that, 'When I joined the prison service in 1982 people were terrified of allowing prisoners to have FM radios...' (*The Independent* 2015). In short, these kinds of responses and subsequent policy directives galvanise the shaping of access and use of technology in prison as a mode of governance. The following section extends upon this idea describing how this is achieved.

How television is a mode of governance

First it is opportune to reflect upon the social-psycho features of human relationships with television. Silverstone's extensive studies of television in the home are useful in illustrating how television as a material and symbolic object has shaped modern everyday life. Silverstone (1999) argues that television can provide symbolic attachments, which in turn help individuals to obtain comfort and trust. Attachments are formed by the pleasurable sensations of finding comfort in watching a favourite film and trust that broadcasts will deliver familiar formats like news and soaps. In doing so television can also be a place for the development of security. This is because

> television will become a transitional object in those circumstances where it is already constantly available or where it is consciously (or semi-consciously) used by the mother-figure as a baby sitter: as her or his own replacement while she cooks ... The continuities of sound and image, of voices or music, can be easily appropriated as a comfort and a security, simply because they are there.
>
> *(Silverstone 1999: 15)*

Television can then be conceived as a care-giver, and this is reinforced by the idea that 'television survives all efforts at its destruction ... it is eternal' (ibid.). The care-giving qualities are enhanced because it is 'cyclical' (ibid.) as it can routinely deliver broadcasts regularly and in schedules that audiences have become accustomed to. This regularity and order can, as Silverstone suggests, assist with accomplishing personal 'ontological security' (ibid: 17) – the need to find a stable and safe being. This is further assisted by other qualities which include opportunities to engage with civic society and mark out time and space relations such as rituals of daily life. Together then, television can contribute to wider projects of governance and this is particularly compelling when social agents endeavour to accomplish social and psychological security (Layder 2004).

Rose's (1999) analysis of governance is useful in articulating this further. Audience trust is established as an enterprise of care-giving, which forms an important basis to governance. The backdrop to this idea can be framed more broadly in the emergence and proliferation of the 'psy' sciences as a response to deal with pathology and abnormality. Thus in search of harm minimisation and reform techniques like surveillance in the shape of monitoring people's behaviour, performance and activities emerged as efforts to govern and regulate. In doing so, the art of governance is not only commanded by the state's institutions like law enforcement, education and health but is also transferred on to the individual – resulting in modes of self-regulation, self-expertise and self-control. Television can be an important source for reinforcing these kinds of doctrines, especially in the shape of talk and lifestyle shows (Shattuc 1997). Sconce reflects upon the audience's relationships with television series commenting on the 'intensity of investment and depth of immersion ... [and] this immersion is accomplished more in mental space than electronic' (2004: 110). In short the cognitive, emotional and social connections with televised narratives can facilitate powerful outcomes for audiences. This too is accentuated in the prison setting knowingly and purposefully to encourage the prisoner to take on the business of self-care (Bosworth 2007). This extends to the deployment of technologies like television and its success is most accentuated by the nature in which prisoners are subjected to 'pain' – most famously documented by Sykes (1999) as a series of deprivations such as loss or restriction of liberty, goods and services, heterosexual relationships, security and autonomy. The discomfort and distrust prisoners experience makes television a very attractive and meaningful option for both prisoners and staff:

> With TV I am in another world. It is mental torture in here – so you do need somewhere else to escape, to cope and not breakdown.
>
> *(Leon, prisoner)*

> you can absorb yourself in a film and remove yourself from the current hurt and pain you are feeling, and escape emotionally for two hours, three hours, half an hour, and that has to be a positive thing …
>
> *(Ann, prison chaplain)*

Together the prison and television assist the prisoner and staff to manage their internal selves – 'souls' – in order to achieve successful modes of governance. Both Leon and Ann recognise how television viewing can soften the harsh environment. Much of this governance work is preoccupied with emotional control. If control of this kind is successfully accomplished order can be maintained and the effects of deprivation such as loneliness and social isolation are minimised. It seems television can soothe and calm large numbers of prisoners almost instantaneously. Claire talks about how television, in her view has brought about observable differences to the ways in which prisoners are spending their time locked in their cells:

> There is a marked difference … I left X [prison] 10 years ago and I came back; when I left it had no in-cell TV and obviously since having the in-cell TV the prison is cleaner because there are no t-shirts hanging out the windows. They used to spend a lot of their time … shouting out windows.
>
> *(Claire, Governor)*

The observable changes Claire describes links to the 'Decency Agenda' which has been signalled as a 'desire to raise the standard of care' (Bosworth 2009: 182) in prison. However, this at best can be interpreted as 'welfare rhetoric' (ibid: 182) and thus is problematised and illusionary. As Bosworth remarks,

> the choice of rehabilitation or reform has become the individual prisoner's sole responsibility. The prison is merely expected to provide the arena for such personal decisions while warehousing inmates securely … administrators seek to co-opt prisoners themselves into maintaining order and discipline.
>
> *(Bosworth 2007: 68)*

The ways in which order is sought can be sourced within and across micro emotional interactions. This is the business of emotional control and this is linked to two distinctive and inter-related processes; *techno-therapy* and *rationalisation*. Together these define how television is adopted as one of the best available technologies to achieve control of the emotions – a quasi-therapeutic medium to regulate the complexity of emotions in prison.

The business of emotional control

The need to control emotions is not exclusive to prison. Yet the prison context presents a unique set of social practices, which always amplifies the need and urgency to ensure emotions are regulated to prevent violence, self-harm, suicide and disorder. Success in this is routinely achieved in most prisons most of the time, and yet little attention has been paid to these successful micro transactions – where prisoners and staff get through their daily lives without incident. However, failure to achieve emotional control regularly feeds into public debates about the disappointments of the prison service, especially when prisoners riot, escape or die in prison. In either circumstances prisons are complex places with multifaceted emotional transactions occurring continuously and simultaneously. Comprehending the prison through an emotive lens has received little attention from scholars. More recently Crewe *et al.* (2013) and also the author of this chapter (Knight 2015a) have begun using empirical evidence to chart the emotional landscape of our prisons. In doing so it is then possible to distinguish what prisoners and staff *do* with emotion. This is important for understanding how prisoners' relationships with television can be interpreted.

Techno-therapy

Television, in the prison context, belongs to a variety of 'civilising technologies' which seek to manage risk and dangerousness (Vrecko 2010; Hannah-Moffat 1999). These technologies are forms of 'quasi-therapy' (which I argue include television) (Rose 1999: 182). This section demonstrates how television is adopted as 'techno-therapy', not necessarily to replace provision of limited and formal therapeutic interventions (such as cognitive behaviour therapy programmes or medical interventions), but to normalise the experience of prison and to enhance opportunities to make the prison experience safer and decent. This is compounded by the fact that television through its broadcasts can assist with 'self-care and responsibility' (Rose 1999: 188). Prisoners' exposure to these 'neo-hygienic' discourses, which profess self-care through television programmes, can encourage some prisoners to work on their 'souls' or life situation. Moments alone with television can enhance self-reflection, inspire small changes, or address complex human problems. As Leon explains

> my family is my motivation to do well here, I'm not working for enhanced privileges, it is for my family to help them. I am a good person, I want to show them, my family is my motivation. TV helps me with this.
>
> *(Leon, prisoner)*

With television Leon is able to focus on things that will help him to get through his time in prison and value his family. Television is therefore placed in the cell with unanticipated outcomes for therapeutic control. This interpretation of television in this setting extends the original remit of privilege based on compliance that the policy intended. While prisoners are occupied with television in their cells and are also focused on sustaining access to it, the prison is able to 'minimize the riskiness of the most risky' through mechanisms of surveillance, incapacitation and deprivation to achieve quasi-therapeutic control (Rose 1999: 189). Here modes of governance can be operationalised with television placed inside the cell.

One key element to this kind of governance is that of surveillance. In-cell television relates to the modern prison system's panoptic structure, as famously articulated by Foucault (1991). The placing of the television inside the prison cell fixes the prisoner within the confines of this space, without too much distress and duress. The cell, typically a place of loneliness and torture has become transformed sufficiently to draw the prisoner inside the cell. A panoptic model is enabled with television at hand and with little intervention from staff. Prisoners are attracted to the symbolic comforts that television can deliver, and so they are amenable to remaining fixed in their cells rather than disappearing out of sight. Tim explains how television has helped to reinforce a routine with little intervention from staff themselves:

I believe, they got themselves into a routine where they'll watch 'Eastender's' because it's on before 'Coronation Street' and then they'll watch something else and it's just a routine that they've got themselves into.

(Tim, prison officer)

In this case staff are more confident that prisoners are where they are supposed to be and are busy watching television. In a panoptic sense they are being viewed in a designated location. In extending this further, staff are complicit in acknowledging the attractive and compelling nature television has in this context. Interviews with staff highlighted how sensitive they were to the deprivations prisoners can experience. For example Paul, a senior prison officer, talks about how much privileges, especially if they are removed, can impact on prisoners

so although we talk about in-cell television that has an effect on IEP. That has an effect on prisoners' behaviour because if we take that facility away ... it's a punishment that they will struggle with.

(Paul, senior prison officer)

Interviews with staff also suggested how television, despite its link to the privilege system, was adopted to help in their work towards safer custody and decency for prisoners. With television fixing them to their prison cells, achieving safety and decency in some respects is enabled by enhanced opportunities to undertake better modes of surveillance. Staff were, however, keen to emphasise that safer custody took priority when managing prisoners, and were inclined to take this position in favour of the privilege system.

Prisoners who have got issues around self-harm, distressed, vulnerable, isolation, keeping awake at night, we all know the worst place in the world when you are awake at night and you can't sleep and you have got all this stuff going round in your head, at least they have got a television and can watch something.

(Claire, Governor)

Prisoners too were aligned to this same position recognising that television viewing is a form of treatment; a therapy to help prisoners get through daily prison life safely and decently. Sunny, for example, explains how this idea applies:

I tell them not to worry and watch TV to help them be distracted. TV keeps prisoners calm ... Like when they have problems, TV can keep your mind off bad feelings.

(Sunny, prisoner)

The project of treatment is nothing new to prisons and some seek to establish 'therapeutic communities' whereby daily life is 'treatment orientated' (Genders and Player 1995). Like elsewhere in society the medicalisation of prison is assisted by modes of technology – the sophisticated drugs to sedate and soothe complex mental illnesses and the risk assessment tools to guide rehabilitation/reform. The 'gaze of the psychologist' (Rose 1999: 138) is complex and expensive and time to assess and work with swelling numbers of prisoners is at best unrealistic. In reaching the prisoner – therapeutically speaking – the introduction of in-cell television has had considerable benefits in this regard. As a consequence there has been a decline in situated interaction across the prison. Prisoners are less inclined to engage in social encounters with prisoners and staff. Television-use diaries outlined a large quantity of hours consumed each week in prison (approximately 60 hours per week). Many officers talked and at times complain about prisoners disappearing behind their cell doors.

> you get association [time out of cell] periods now where we used to get, say there's 60 eligible to go you'd get between 50 and 60 go to association because the incentive was, OK there was a pool table and table tennis but there was also a big TV in there but most of them now think, well I'll just sit and watch telly. So interaction between prisoners and between staff has reduced, in my eyes, has reduced drastically because they'd rather sit and watch TV.
>
> *(Brian, prison officer)*

Conversely prisoners justified this in terms of exploiting the privacy of their cell, not necessarily to hide from the gaze of prison staff but in terms of keeping safe:

> There is an open door policy now though. I don't want to be around drugs, you have to be careful. I don't like the ... the enhanced landing, if things get broke and mistreated we suffer. Someone was sick on the football table once. TV is my own cell, I can block it all out.
>
> *(Will, prisoner)*

Both staff and prisoners however generally agree that television has brought about significant 'improvements' in the ways prisoners occupy the time spent in their cells. Television placed inside the cell has meant that establishments can function in achieving better levels of control (Layder 2004). This can be illustrated in Moores' studies of radio in the home (1988) and also satellite television (1996) which highlights the same kinds of social adjustments as a result of new technologies introduced to the home. He along with others (Spigel 1992; O'Sullivan 1991) observed how the spatio-temporal organisations of family life changed when novel technologies were introduced to the home for the first time, for example reorganising meal times around popular broadcasts like news and soaps and the splintering of families viewing and listening together. What this suggests is modern technology has brought about a similar 'domestication' of private life (Silverstone 1999) – an effect which seeks to tame audiences and cultivate obedience.

Despite these shifts in social dynamics, broadcasts, in post-war Britain and the USA, began to increasingly deliver audiences with content which helped them address their problems. These 'vocabularies of the therapeutic' correspond to a range of technologies that sought to achieve governance (Rose 1999: 218). In short, broadcasting technologies corroborate with these therapeutic resources that marry with much of what institutions like prisons are trying to achieve – to reform, readjust and control. These therapeutic agendas in the context of the prison become magnified and urgent. Hence there are 'civilising' effects as a result of in-cell television. In turn these effects help to manage and regulate emotional responses, especially those that can spiral into destructive outcomes like violence (Elias 2010). Simon, for example, talks about the importance of his maintaining control, especially over other prisoners:

> I'm the alpha male here though, I have the bottom bunk. If he wants to watch something I do let him. But it is my remote and my telly, if you don't like it then get out. There has been loads of arguments, but not with current pad mate. I remember a man wanted to watch a cartoon, I ain't watching that 'Winnie the Pooh'. He got ejected from here. I refuse to live with them.
>
> *(Simon, prisoner)*

What Simon emphasises here is the distinction of tastes and motivations for viewing a range of material. He also uses television to assert prisoner culture which conforms to masculine battles for power. It is established that audiences and consumers of mass media are motivated to use it for a range of social and psychological reasons. As emphasised by the uses and gratifications model (McQuail *et al.* 1972) and Lull's (1990) typology of the social uses of television, audiences seek to achieve (and sometime fail) a diverse range of 'gratifications' – such as achieving intimacy or closeness to media personalities or characters. In turn these connections can assist with social interactions across the prison environment. These abundant outcomes of viewing are thus compelling and attractive for audiences. In the context of the prison the incentivisation of in-cell television is particularly helped by the fact audiences demonstrate signs of 'transitional' attachments to it (Silverstone 1999: 14). Moreover staff know this is also the case:

> I think in general people like television, they like watching it, they like their football or their soaps or *This Morning* … I think it is a motivator for prisoners because I think they feel quite lost without it.
>
> *(Claire, Governor)*

This is further compounded in the prison setting because of the deprivations prisoners can experience. One vivid example is that of boredom, as Leon helpfully describes:

> Boredom is poisonous, it is mental poison. You can easily get distressed and suicidal in here. TV keeps you occupied. Even just changing the channels using the remote, it keeps you focussed.
>
> *(Leon, prisoner)*

Within this context boredom, as Leon describes, is problematic and can lead to other less controllable states like anger and frustration. Paul, for example, recognises how television can help to minimise disorder, or at least prevent these kinds of inflammatory situations:

> prisoners can become very bored, and when they become bored they become mischievous. So by installing in-cell television and television within the establishment I believe it prevents a lot of that mischievousness, it's allowed them to be less bored.
>
> *(Paul, senior prison officer)*

Here emotive dissonance can be avoided with almost instantaneous effect. This is also reinforced when prisoners have their privileges removed. The public display of punishment within the prison context, like the withdrawal of television, is a powerful mechanism for governance, especially in reinforcing punitive measures. The witnessing of punishment by prisoners themselves serves as an important reminder of the deprivations the prison can impose, and can function as a stark reminder to them. One of the consequences of these kinds of punishment means that prisoners are locked behind their cell doors more and this is obvious to those living in close proximity:

> if they are not coming out of their cell, people can visibly see that they are not coming out of their cell.
>
> *(Fran, prison officer)*

Maintaining order is not necessarily straightforward, as accounts from prisoners suggest they admit to creating or witnessing violence, bullying and extortion. The underworld of the prisoner society means that forms of resistance and deviance appear (Liebling *et al.* 2011: 122) and are observable not only to the prisoners but also staff. Steve explains:

> You get the innocent ones, but then you get the ones that are at it, and no matter which way you do it they will get passed from cell to cell to cell. And when you have got 110 prisoners and three officers on a landing you can only police so much. And it's a shame but it gets into the prisons, you get mobile phones coming, you get contraband brought in … yes, we know phones are in there at some point but its catching them … And it's the same with the drugs, we've had a load that's come over the walls but you can only stop so much. People will try to flout the system and try their best and some get away with it and some don't, but it's a game.
>
> *(Steve, prison officer)*

The 'game' that Steve refers to is routinely part of this staff-prisoner interaction. As McDermott and King found, games are important 'to gain control over the meaning of a situation … they are competitions about power' (1988: 360). Alan, a prisoner, also views the privilege system like a game; he talks about the 'hook' which draws prisoners in:

> It is a luxury item, the regime tries to get you in on it. But then it gives you some borderline…
>
> *(Alan, prisoner)*

Alan suggests that prisoners are encouraged to participate in the IEP system as long as the rules to this 'game' are understood by its participants. The privilege scheme is not dissimilar to some models of childcare or parenting: measures of success are achieved through control and also discretion (Liebling *et al.* 2011: 121). These additional dimensions of neo-paternal power reinforce prisoners' attachments to things like television and for some can strengthen the need to comply with the regime. Yet the ways in which policy is orchestrated and administered is not seamless and even. Both prisoners and staff know discretion plays a large part in this. Liebling *et al.* highlight that organisational policy is sometimes confusing and disorientating (2011: 125) and regularly subject to criticism for being 'vague' and unclear (Bottoms 1990). Moreover this discretion can mean staff can control simple things like the supply of electricity to the cells, which in turn could limit television viewing. These kinds of battles for power reasserts staff's ability to control what prisoners do and when they do it. These kinds of threats can infuriate prisoners, yet balanced with this is the management of complex and vulnerable prisoners. Tony, a deputy governor talked about the prospect of censoring what prisoners can watch:

> I have got a big problem with giving, especially sex offenders who are in denial, access to all that, all the late night, even if its Channel 4 stuff and that sort of thing. Whether it involves kids or not I have issues with that. But if you take that to its limit then you ban them from virtually everything, you are stopping them watching *Emmerdale Farm* and *Coronation Street* because they have got kids in. I think there has got to be a line somewhere but I think its basic stuff that's allowed to be put on TV. I suppose you could use the watershed as an example or as a time ban if you wanted to. But I think the stuff that's on TV then most people should be allowed to watch it.
>
> *(Tony, deputy governor)*

These kinds of comments about the punitive use of television are often justified and framed in discourses of care. As Brian explains, taking care of prisoners is challenging and the constraints of the environment and the demands of their work mean that he is appeased that care can continue in his absence – prison officers cannot be with prisoners all of the time:

> the TV helps with that because if there's something on, it can help, 90 per cent of the time it helps but some of the programmes that are on, they can take it the other way as well.
>
> *(Brian, prison officer)*

As 'technicians of reform' (Garland 1991: 182) staff are inculcated into the project of correction. This encompasses aspects of both control and – also in its shadows – care. The culture of punishment is embroiled in managerialist mechanisms like actuarial and target driven outcomes. As a consequence, care has become standardised or 'rationalised' (Garland 1991) and thus can be obscured from view of onlookers. In evidencing care in their work, staff were explicit about the role television has in delivering care. As we have seen, the positioning television as 'care giver' (Silverstone 1999) instead of reward marks an important shift in the rhetoric about in-cell television, yet this is largely silenced across prison policy and penal discourse.

As outlined earlier in this section, television is actively 'prescribed' by prison staff and prisoners themselves, as a therapeutic intervention. This is productive in helping prisoners to manage their emotional responses such as boredom, frustration and anger.

> People get upset because they have mental health issues and it's beyond their control. People get upset because they have issues and concerns out of the establishment, people get upset for lots of different reasons. And if the only thing that is going to keep you calm is to engage with a TV event, or having a hot drink, why should those things be taken away?
>
> *(Fran, prison officer)*

Technology thus serves as an antidote to the deprivations prisoners can experience – as a 'distraction' or deviation from these. One example is that of managing large quantities of unstructured time. This has also been observed in earlier studies of mass communications in prison:

> presence of television can normalize or readjust time ... Like drugs ... television can provide refuge from the harsh realities of life, filling large amounts of self-time which otherwise might be given over to introspection.
>
> *(Jewkes 2002:102)*

As intervention television can directly assist with this 'self-time' – by allowing prisoners to choose programming to either distract or help with introspection. Introspection can be productive – to review one's life and seek to enhance and develop life opportunities. Television itself can also help prisoners realise these kinds of aspirations. Malcolm talked about how he liked lifestyle programmes, especially those that showed how people bought and sold antiques. He had quickly learnt that he could do this through using the internet:

> I now love technology, I'm going to buy myself a laptop, it has really taught me something. I was frightened before. I've not used the internet though I'm 63 years old. I'm going to use eBay.
>
> *(Malcolm, prisoner)*

Yet in the other regard television viewing in prison can trigger a catalogue of emotive responses such as anger, frustration, loneliness. Shaun for example finds music much more comforting than television:

> Listening to music is motivating, it makes me feel better and increases my self-esteem. TV is depressing. It constantly reminds me of what I am missing. Music gives me strength in prison, But TV doesn't it takes it away. Music takes you away from here. Prison is depressing, nothing happens and to have the outside shoved in your face is hard. I don't want to think about it whilst I am here. I'm in my cell all day.
>
> *(Shaun, prisoner)*

This becomes especially significant when prisoners reject television and other activities and begin to withdraw from both prison life and broadcast life. Staff realise the dangers of prolonged introspection and 'self-time'. With this knowledge in mind the prescribing of television is a viable and accustomed ritual in their practice. In this sense, prison staff act as quasi-therapists. Technologies like television are interventions that are easily administered, they are abundant and not too costly. Ann, for example, was clear how the content she watched on television helped her in her role as prison chaplain. She watched popular television shows like *Strictly Come Dancing*, *The X Factor*, *Big Brother* and news to allow her to get closer to her ministry and find connections to 'reach' prisoners.

> It allows prisoners to engage with a chaplain on a human level ... I think it enables some realness ... it is a very human conversation, and therefore quite helpful.
>
> *(Ann, chaplain)*

These social uses of television are not unusual (Lull 1990). Moreover, opportunities to reach the prisoner are considered by many staff the most 'technical' part of their job:

> we take people who are severely damaged and make them more socially acceptable in as much as you talk to them, you can sit down and make them understand, you can change their views on the way the prison service staff work with prisoners.
>
> *(Paul, senior officer)*

As quasi-therapists there is an underlying belief that staff can influence change and promote better self-regulation so that prisoners:

> can act upon their bodies, their emotions, their beliefs, and their forms of conduct in order to transform themselves, in order to achieve autonomous selfhood.
>
> *(Rose 1999: 251)*

Enabling transformation is a feature of control, and commanding expertise and intervention is part of prison work. In the absence of formal routes to therapy, which are costly and time consuming, staff actively deploy television as a form of psychic control. In achieving this control, further work is undertaken by staff and prisoners themselves to manage emotions more closely.

Rationalisation

A complex picture of the forms of interaction is taking place in prison. Staff are anxious about the depletion of face-to-face interaction (situated) especially between staff and prisoners. Staff attribute much of this to the introduction of in-cell television. Paradoxically prisoners are generally happy and trusting of encountering content via television – it delivers bouts of pleasure into a harsh landscape. This makes the cell a much more palatable place to retreat. This section examines this distancing and separation and outlines how technology is assisting with rationalisation – the control of emotions. With television, rationalisation or the control and ordering of emotions is possible. The emotional lives of prisoners are being ratified and articulated with in-cell television close at hand rather than being allowed to surface in social encounters with other prisoners and staff. Instead of spilling out more regularly onto the prison landings, television behind the cell door helps to keep them contained. Staff worry, and consequently struggle to reach and observe prisoners' emotions directly. With television adopted as a quasi-therapeutic tool to intervene in the business of controlling emotions, staff–prisoner relations are distanced. As a consequence the display of emotions can become obscured:

> Values ... and emotional attitudes which lie behind them – may be muted and displaced by bureaucratic institutions, but they do not disappear.
>
> *(Garland 1991: 189)*

Whilst the muting of emotions brings about many benefits to organisational control, the jostling of power between staff and television is evident. Some staff acknowledge that forms of power are now in the hands of technology and not them. The enterprise of punishment is emotive (Crawley 2004; Garland 1991). As an antidote to manage risk-related emotions such as boredom and anger, in-cell television has shifted staff–prisoner relations. It is feared prisoners are more likely to find respite, care and help from viewing television than with prison staff. Spigel (1992: 65) argued that 'television threatened to drive a wedge between family members', which altered how families resided together. Spigel identified that unity of household togetherness was challenged because of television. Television placed inside the cell has given prisoners the opportunity to vanish from staff view. Moores (1988: 25) observed that families were adopting radio to discourage movement to exterior spaces and bring people into the interior (the home) and thus control movement from and between different spheres – the public and the private. On one hand families could keep a closer eye on their family members, whilst on the other not have complete sway or control on how they think and behave. This too is replicated in the prison and influences are now enabled by technologies. In the case of the prison this paradox means that disappearance from public view is considered problematic, especially in light of the task of reform:

> It is quite important that you get interaction in prison if you really want to tackle re-offending because a prison officer's job should be that you'll sit talking ... I don't think we do that at all. I think it's reduced it down to virtually nil, if I'm honest.
>
> *(Brian, prison officer)*

In part television is doing some of the work they expect to do themselves. Instead, aspects of control and potentially the task of reform are split across these two modes of interaction: the personal face-to-face kind (situated) and the digital technology kind (mediated). The birth of television in the prison setting has coincided with marked changes in the ways prison work is undertaken. Managerialist approaches now demand large administrative tasks and reporting, which also detract staff from the face-to-face work with prisoners. Moreover, austerity measures imposed by the UK's Coalition (Conservative and Liberal Democrat) government's welfare reform from 2010 has seen a decline in staff numbers. Together it is understandable why staff may be concerned that the advent of in-cell television may be held responsible for removing the labour they do (Knight 2015b). As Paul highlights, these losses mean that:

> the art of communication has been taken away from us … you will desensitise or dehumanise prisoners because they won't know how to talk to people … they don't have the social skills. So I think technology isn't always the best thing …
>
> *(Paul, senior prison officer)*

As a consequence some staff believe that this distance between prisoners and staff can interfere in the quality of surveillance prison officers are encouraged to achieve:

> Dynamic security is prison officers walking around, talking and listening, listening to what happens … now what we do is as soon as somebody says, I'm gonna kill myself or they self-harm, we fill one of these forms in and we spend so long filling in this form in … we should go and talk to prisoners, sod the paperwork.
>
> *(Brian, prison officer)*

This shrinking of contact means that forms of care and control become compromised and television with its content becomes an increasingly important mechanism to assist in prison craft. Tim, for example, explains how television provides routes into conversation and agenda for talk (Lull 1990):

> prisoners will talk about what they watched on television last night and I think it's important for the staff to interact with that and pick up on what they are saying, you know, on a wider security issue, it does build up dynamic security.
>
> *(Tim, prison officer)*

Likewise prisoners recognise the value of cultural resources from television:

> TV is important here, it is a topic of conversation, we relate to and share things. Like 90 per cent people here watch *Eastenders* and football, we can chat about it. Like *Big Brother* it gives people stuff to converse about.
>
> *(Stuart, prisoner)*

Conversely Brian however, does not acknowledge that these conversations about television programmes are sufficiently genuine:

> conversation is OK when you're talking about what happened on the TV but realistically that's not a conversation that is aimed towards rehabilitating somebody, it's just passing comments.
>
> *(Brian, prison officer)*

These conversations amongst prisoners and also staff provide a safe 'script' (Cohen and Taylor 1972; Goffman 1990) in which social interactions can be comfortable and easy (Lull 1990;

Wood 2009). Like Brian, Leon is less inclined to claim much value of television with regard to social interaction:

> There are two types of personalities in prison, those that mingle and chat and those that don't engage and get involved in the unnecessary bits of prison life. What have you gotta talk about in here? I suppose TV is a good socializing factor like someone might say 'you looked like that geezer on...' Ha, ha, ha. But it is nothing productive really ... I learn more from TV than from folks in here, it is more accessible.
>
> *(Leon, prisoner)*

For Leon, mediated encounters in this context are safer and of better quality. Television viewing provides him with a legitimate excuse to withdraw from prisoner interactions and thus reduce the risk of contamination – contact and influence of criminal and deviant practices. Here Leon is able to temper his exposure to other prisoners and focus on benefiting from his time in prison. In doing so, Leon is assisting the prison in the rationalisation of prisoners' emotions by both withdrawing from association with other prisoners and gleaning from the cultural artefacts that television delivers directly to his cell.

For some prisoners, as highlighted earlier in this section, their ability to interact socially can be hindered by the limited opportunities they choose or are forced to have.

> I do wonder how I will be around my kids and my wife when I go home because I have spent so long away from them. It does worry me a little bit, will I be alright? I think I need to go away to another country and sort my head out for a month or so, try and get used to having people around me.
>
> *(Ron, prisoner)*

The disconnection and separation that incarceration brings about don't allow Ron to rehearse situated (face-to-face) encounters. Whilst many prisoners are able to connect to the outside world via television, actual practice of everyday life with its rituals and resources continues to be eroded. The fear of a cultural detachment can also be extended to a 'digital-lag' (Knight 2105b). Like Ron, forgetting how to be with people can also be extended to forgetting or even knowing how to be with objects:

> Psychology think I am institutionalised. I think I am. I got a mobile phone as a gift when I was out last. I don't know how to text. I don't belong out there, I am inadequate.
>
> *(Mick, prisoner)*

The shifting penal context with digital technologies

The benefits of achieving emotional composure and support through *techno-therapy* and *rationalisation* is that it enables prison institutions to work towards a range of policy directives such as decency, safer custody and violence reduction. As a result, 'soft' rather than 'hard' power approaches to imprisonment and punishment can be achieved (Crewe 2011). Prisoners 'get on with' their time in prison without overt intervention from prison staff and television helps with this. Prisoners become increasingly used to accessing the contextual resources television can offer, rather than seeking those made available through other mechanisms, such as use of libraries for information, seeking advice from staff and other prisoners.

The evolving technological landscape in our prisons means that these trends are likely to further accelerate. Many prisons across developed countries are welcoming interactive services for prisoners (Knight 2015b). Exploiting these 'lifeless materials' (Elias 2010: 7) by allowing prisoners to access a range of digital technologies could widen the principles of quasi-therapeutic control further. For example some prisons have introduced a self-service facility in the UK, Belgium, the Netherlands and the USA. Here prisoners are able to manage some aspects of life in prisons themselves – such as arranging their visits, booking appointments with health services, making requests, managing their money and buying products via a virtual prison shop. The most consolidated model is Belgium's PrisonCloud (2015) platform, which performs all of these functions in prisoners' cells. This service consists of a secure desktop computer and is made available to all prisoners. Through this platform prisoners can access these self-service options as well as watch television, rent films, make telephone calls, use the desktop facilities, access their judicial files and access e-learning opportunities. Though this technology is still to be evaluated, claims about the shift in responsibility for prisoners to manage their domestic lives are currently deemed positive. This model replicates most domestic services and PrisonCloud offers a vision of what prison provision can be. Television reception and telephony are changing and convergence of platforms mean that modes of use are dynamic, mobile and interactive (Spigel and Olsson 2004). Prisons will not abandon these developments. Unlike television, digital technologies require developers who design these kinds of services for secure settings (including hospitals) to assure nervous prison services about the security measures they have in place. With this single-platform model in mind, opportunities for self-regulation can become even more refined. Whilst self-regulation is productive in many instances, the introduction of digital technologies has, as argued in this chapter, been adopted by prison services to administer 'care'. Staff are anxious about this transition and yet prisoners find private comfort in broadcast company. With opportunities to access interactive digital media, self-care opportunities are becoming increasingly abundant. Ribbens and Malliet's (2015) study of digital gaming highlighted the protective characteristics gaming can bring about in prison settings. Its interactive features offer active ways for prisoners to minimise psychological deterioration, build networks and open up a virtual landscape. The Netherlands are developing gaming and gamification across their prison service which dovetails with a vision to provide additional opportunities for prisoners to engage in therapeutic treatments and enhance well-being (Dialogic 2015).

Conversely the development of Web 3 – third-generation web claimed to be the 'intelligent' web – means that opportunities for surveillance, monitoring, data-mining, artificial intelligence are significantly enhanced – something prison service providers find valuable yet at the same time feel anxious about. The inevitability of these will, albeit in modified forms, appear in our prisons. The PrisonCloud model in Belgium is the next generation of in-cell architecture and other jurisdictions are trying to emulate this. The progress of digitisation in prisons is slow and occurs in small pockets. In Australia prisoners can use e-readers and use the 'offline study desk' (Farley et al. 2014). In the US, digital visitation (video conferencing) is popular in some parts, where prisoners can 'see' and communicate with their families (Philips 2012; Doyle et al. 2011). Digital health interventions are also appearing in our prisons with programmes and applications to help prisoners recover from drug and alcohol addictions (Elison et al. 2015). Technologies in other secure environments, such as secure hospitals, are also now designing in digital architecture such as the Netherlands' 'media wall' where patients can alter the atmosphere, the lighting, the sound in their room to help their recovery. These large screen mood walls help inpatients de-escalate from psychotic episodes (Recornect 2016). Creating digital opportunities in secure settings is receiving positive support, especially when secure providers can demonstrate important rehabilitative outcomes like enhanced education, improved family contact, health interventions and also self-service. There is less innovation consistently across different jurisdictions to blend communicative and interactive platforms with these more established platforms. The 'Prisons of the Future' project led by Europris (2016) argues that innovation requires some major rethinking about traditional methods of incarceration – favouring 'support instead of control' which can encourage internal self-control (ibid).

The evolving penal context has 'become increasingly technical and professional' (Garland 1991: 187) and the increased reliance upon technological solutions can extend the deployment of experts, systems and technology to harness control. As outlined in this chapter the need and encouragement to manage emotions requires prisoners to be attentive to their internal dialogues. It can therefore be envisaged that prisoners will be increasingly encouraged to subscribe to self-monitoring technologies to record activity levels, nutritional intake, mood and emotional responses. The uptake of mobile applications in wider society to monitor health, activity and mood has seen increased growth in these areas. If made available to prisoners it is likely they too will be active in reviewing and reflecting on their 'performance'. As Bosworth has previously highlighted, prisoners are left to attend to their own 'self-improvement' (2007: 179) and these technological mechanisms will provide extended sites for prisoners to 'look inwards' (Rose 1999: 227). Moreover, the prison too may claim access to this kind of data to closely observe its prisoners in order to protect them, direct services to them and to inform intelligence mapping – in sum to control them. Together this collusion of efforts facilitated by digital technologies will reassert continuing forms of control, much like that documented here in relation to television. Television 'works' on behalf of prison services and contributes to both the control and care of prisoners. There are wider ethical dimensions to this debate in relation to normalising the prison experience. The onset of digitisation brings new but familiar challenges to the craft of confining prisoners, but with enhanced possibilities to watch the inmate. Television has allowed prisoners to undertake methods of self-care and services have benefited from this in terms of achieving greater levels of control. These recent shifts in the technologisation of confinement enables, unlike television viewing, mechanisms to measure, record and observe the user from the moment they enter prison to the moment they leave. Their digital footprints provide a complex set of data about their identities, their moods, their aspirations and their social networks. Together this information could be used and exploited to inform a range of criminogenic and therapeutic assessments. It is difficult to envisage how prison can be softened when technology is applied in these ways and yet from a user perspective digital freedoms 'soften' the complex and harsh realities of incarceration. And so the prison digital revolution will emphasise a new era of penal control.

References

Bosworth, M. (2007) Creating the responsible prisoner, *Punishment and Society* 9(1): 67–85.

Bosworth, M. (2009) Governing the responsible prisoner: A comparative analysis, in Sørensen, E. and Triantafillou, P. (eds) *The Politics of Self-Governance*. Farnham, Ashgate.

Bottoms, A. E. (1990) The aims of imprisonment, in Garland, D. (ed.) *Justice, Guilt and Forgiveness in the Penal System*. Occasional Paper No. 18, Edinburgh: Edinburgh University Press.

Cohen, S. and Taylor, L. (1972) *Psychological Survival: The Experience of Long-Term Imprisonment*. Middlesex: Pelican.

Crawley, E. (2004) *Doing Prison Work: The Public and Private Lives of Prison Officers*. Collumpton: Willan.

Crewe, B. (2011) Soft power in prison: Implications for staff-prisoner relationships, liberty and legitimacy, *European Journal of Criminology* 8(6): 455–468.

Crewe, B., Warr, J., Bennett, P., and Smith, A. (2013) The emotional geography of prison life, *Theoretical Criminology* 18(1): 56–74.

Dialogic (2015) Gaming and Gamification for Correctional Setting, unpublished report, Dialogic Netherlands. Available at: www.dialogic.nl.

Doyle, P., Fordy, C. and Haight, A. (2011) *Prison Video Conferencing*. Burlington, VT: The University of Vermont. Available at: www.uvm.edu/~vlrs/CriminalJusticeandCorrections/prison%20video%20conferencing.pdf (accessed 20 January 2015).

Elias, N. (2010) *The Civilizing Process*. Oxford: Blackwell.

Elison, S., Weston, S., Davies, G., Dugdale, S. and Ward, J. (2015) Findings from mixed-methods feasibility and effectiveness evaluations of the 'Breaking Free Online' treatment and recovery programme for substance misuse in prisons, *Drugs: Education, Prevention and Policy* 2(1): 1–10.

Europris (n.d.) *Prisons of the Future* Available at: www.europris.org/projects/prisons-of-the-future/ (accessed 8 March 2016).

Farley, H., Murphy, A. and Bedford, T. (2014) Providing simulated online and mobile learning experiences in a prison education setting: lessons learned from the PLEIADES pilot project. *International Journal of Mobile and Blended Learning* 6(1): 17-32.

Foucault, M. (1991) *Discipline and Punish: The Birth of the Prison*. London: Penguin.

Garland, D. (1991) *Punishment and Modern Society: A Study in Social Theory*. Oxford: Clarendon.

Genders, E. and Player, E. (1995) *Grendon: A Study of a Therapeutic Prison*. Oxford: Oxford University Press.

Gersch, B. (2003) Dis/connected: Media Use Among Inmates, Unpublished PhD Oregon, University of Oregon USA.

Goffman, E. (1990) *The Presentation of Self in Everyday Life*. London: Penguin.

Hannah-Moffat, K. (1999) Moral agent or actuarial subject: Risk and Canadian women's imprisonment, *Theoretical Criminology* (3)1: 71–94.

Hansard (1998) House of Commons Debates, Vol. 314 Session 1997–1998. Available at: Spigel, L. and Olsson, J. (eds) (2004) *Television after TV: Essays on a Medium in Transition* www.publications.parliament.uk/pa/cm/cmvo314.htm (accessed 1 September 2016).

Her Majesty's Prison Service: Prison Service Order 4000 (2000) Incentives and earned privileges. Available at: www.justice.gov.uk/offenders/psos (accessed 13 November 2011).

Her Majesty's Prison Service (2011) Prison Service Instruction 11/2011 Incentives and Earned Privileges. Available at: www.justice.gov.uk/downloads/offenders/psipso/psi_2011/psi_2011_11_incentives_and_earned_ privileges.doc (accessed 10 December 2013).

Jewkes, Y. (2002) *Captive Audience: Media, Masculinity and Power in Prison*. Collumpton: Willan.

Jewkes, Y. and Johnston, H. (2009) Cavemen in an era of speed-of-light technology: Historical and contemporary perspectives on communication within prisons, *The Howard Journal of Criminal Justice* 48(2): 132–143.

Johnston, I. (2015) Give prisoners iPads to help rehabilitation, says report, *The Independent*, 10 December. Available at: www.independent.co.uk/news/uk/home-news/give-prisoners-ipads-to-help-rehabilitation-says-report-a6767441.html (accessed 28 December 2015).

Knight, V. (2012) The study of in-cell television in a closed adult male prison: Governing souls with television PhD Unpublished thesis. Available at: www.dora.dmu.ac.uk/bitstream/handle/2086/7886/Final%20Thesis%202012%20VK.pdf?sequence=1 (accessed 1 September 2016).

Knight, V. (2014) A modus Vivendi – In-cell television, social relations, emotion and safer custody, *Prison Service Journal* November 2014, 215: 19–13.

Knight, V. (2015a) Television, emotion and prison life: Achieving personal control, *Participations* 12(1). Available at: www.participations.org/Volume%2012/Issue%201/3.pdf (accessed 1 September 2016).

Knight, V. (2015b) Some observations on the digital landscape of prisons today, *Prison Service Journal* July 2015, 220: 3–9.

Knight, V. (2016) *Remote Control: Television in Prison*. London: Palgrave Macmillan.

Layder, D. (2004) *Emotion in Social Life: The Lost Heart of Society*. London: Sage.

Liebling, A., Price, D. and Shefer, G. (2011) *The Prison Officer*. Oxford: Willan.

Lull, J. (1990) *Inside Family Viewing: Ethnographic Research on Television Audiences*. London: Routledge.

McDermott, K. and King, R. (1988) Mind games: Where the action is in prisons, *British Journal of Criminology* 28(3): 357–78.

McQuail, D. (ed.) (1972) *Sociology of Mass Communications*. Harmondsworth: Penguin Books.

McQuail, D., Blumler, J., and Brown, J. (1972) The television audience: A revised perspective, in D. McQuail (ed.), *Sociology of Mass Communications*. Harmondsworth: Penguin Books.

Moores, S. (1988) 'The box on the dresser': Memories of early radio and everyday life, *Media, Culture & Society*, 10: 23–40.

Moores, S., 1996. *Satellite Television and Everyday Life: Articulating Technology*. Luton: University of Luton Press.

O'Sullivan, T. (1991) Television memories and cultures of viewing 1950-65, in Corner, J. (ed.) *Popular Television in Britain*. London: BFI.

Phillips, S. (2012) *Video Visits for Children Whose Parents Are Incarcerated: In Whose Best Interest?* Washington D.C: The Sentencing Project. Available at: https://www.ncjrs.gov/App/AbstractDB/AbstractDBDetails.aspx?id=263698 (accessed 23 September 2016).

PrisonCloud (2015) Available at: www.ebo-enterprises.com/en/prisoncloud (accessed 16 February 2015).

Recornect (2016) *Media Wall* Available at: http://recornect.com/?page_id=956 (accessed 4 January 2015).

Ribbens, W. and Malliet, S. (2015) Exploring the appeal of digital games to male prisoners, *Poetics* 48: 1–20.

Rose, N. (1999) *Governing the Soul: The Shaping of the Private Self*. London: Routledge.

Sconce, J. (2004) What if? Charting television's new textual boundaries, in Spigel, L. and Olsson, J. (eds) *Television after TV: Essays on a Medium in Transition*. Durham: Duke University Press, pp.93–112.

Shattuc, J. M. (1997) *The Talking Cure: TV, Talk Shows and Women*. New York: Routledge.

Silverstone, R. (1999) *Television and Everyday Life*. London: Routledge.

Sørensen, E. and Triantafillou, P. (eds) (2009) *The Politics of Self-Governance*. Farnham: Ashgate.

Spigel, L. (1992) *Make Room for TV*. Chicago: University of Chicago Press.

Spigel, L. and Olsson, J. (eds) (2004) *Television after TV: Essays on a Medium in Transition*. Durham: Duke University Press.

Sykes, G. (1999) *The Society of Captives: A Study of A Maximum Security Prison*. New Jersey: Princeton University Press.

Vandebosch, H. (2000) Research note: A captive audience? The media use of prisoners, *European Journal of Communication* 15(4): 529–544.

Vrecko, S. (2010) Civilizing technologies and the control of deviance, *BioSocieties* 5: 36–51.

Wood, H. (2009) *Talking with Television: Women, Television and Modern Self-Reflexivity*. Urbana: University of Illinois Press.

33

Punitivity and technology

Simon Hallsworth and Maria Kaspersson

Since the first humans discovered within them a highly developed appetite for inflicting pain on their fellow beings, Homo sapiens have, in the millennia that have followed, assiduously cultivated the art of punishment, utilising the immense creative genius innate to their species. And each age has sought to do so, moreover, by availing itself of every technology that refinements in the forces and relations of production have made possible. In this chapter our aims are two-fold. In the first section, we will consider the relationship between technology and the punitive, prior to establishing how technology conceived both as an art or, *techné*, as well as a material assemblage of people and things, has been brought together to deliver pain to people in various ways. To accomplish this we will examine various forms of penal technical associations, beginning with the use of a technology as a simple extension or prosthetic of the human body (such as a whip), before studying more complex punitive machines such as the gallows and guillotine, prior to exploring more elaborate punitive assemblages in which various machines intersect with each other in elaborate social technical actor networks.

In the second section of the chapter we will deploy a periodising hypothesis to explore how technology has been differentially applied to the business of pain delivery by exploring how the application of technology to punishment underwent a radical transformation in the movement from pre-modern to more distinctively modern societies. Put simply, it is our conjecture that in pre-modern societies technology was above all (if not exclusively) directed at enhancing the business of pain delivery along with the spectators' experience; while in distinctly modern (western) societies, that is, those subject to a quasi 'civilising process' (Elias 1969), technology is primarily directed at what we will call the business of squaring the circle between, on one hand, the continued desire to inflict pain on others, while on the other hand, doing so in the context of a society beholden to a cultural injunction that views making others suffer (particularly unnecessarily) as un-civilised and barbaric. We conclude by examining the relation between punitiveness and technology in late modern societies asking 'have things changed in recent years?'

Technology and the punitive: some prefatory notes on terminology

Etymologically the word 'technology' stems from the Greek word *techné*, meaning 'skill', 'art', and 'craft' (Parry 2014). Considered this way, *techné* implies a mode of human endeavour directed to *realise* a desired end. Technology then can be defined minimally as *the human activity of furnishing means to effect a desired end*. This implies: first, a desiring being who desires something; second, determining how to realise the desired end; finally, selecting, producing and assembling the human and material means in order to realise it. Technology, in this sense, embraces both the original desire to achieve something; the skill, art, craft and imagination necessary to make it achievable; and assembling the materials in order to achieve it. As we shall see when we develop the argument below, technology, as the term will be deployed here, is far more than an assembled object such as a whip or an instrument of torture. On the contrary, the end product cannot be considered independently from attending to the assemblage of ideas, material objects, sensuous practices, and complex actor networks in which and out of which it is produced and mobilised.

A punitive act is one where pain is intentionally inflicted on another, where pain delivery is, in part, the purpose of the exercise. As Nils Christie observes, all punishment – not only in its most violent manifestation – entails an element of pain dispensation and thus punishment is always in a definitive sense, punitive (Christie 1981). The level of pain inflicted and the way pain is delivered may well change, but unifying all forms of punishment is the common desire to hurt someone, coupled with the material reality of its infliction. This does not mean that the mechanisms for inflicting pain need be obvious and immediate in any punitive act. Indeed, as we shall observe, technology can be deployed precisely in order to obscure the pain inflicted as well as separate those who inflict it from the consequence of their actions. By punitive measures then, we can embrace a spectrum of methods by and through which pain is inflicted from those which are explicitly violent, where the violence in question is directed at the body (flogging, beating, torturing) through to those measures in which pain dispensation appears less directed at bodies but at the mind, as embodied, for example, in the rise of the modern penitentiary where they are called something else such as 'rehabilitation'.

If we put the two terms *punitiveness* and *technology* together, then punitive technologies necessarily encompass three essential components:

- A desire to inflict pain to a particular constituency.
- The articulation of this desire through what we propose to term the punitive imaginary. That aspect of the imaginary which helps determine how and in what way pain will be visited onto others. In short the 'art' or 'craft' of pain dispensation: designing or selecting the instruments used to punish; designing the staged settings where punishment will be conducted; designing the specific rituals that surround the act of punishment itself viewed as a staged performance. Working through the desired relationships between punisher, the punished, spectators and the public viewed as an actor network.
- Applying reason in order to realise this end. This comprises the *instrumental rational* element of a punitive technology: identifying, producing and assembling the human and non-human elements required in order to produce the intended effects.

Prior to examining these social technical couplings, it could be observed that the way humans intersect with technology and punishment typically reveals two recurrent patterns. First, we cannot overlook the innately sadistic desires that animate humans to visit pain on others; nor overlook how such desires articulate into a punitive imaginary that leads them to devise ever more elaborate, not to say, byzantine systems to punish. Or observe, on a more Weberian register, how humans simultaneously deploy instrumental rational thinking to enable them to realise their punitive desires in a calculated means-to-an-end way. Punitive development, like societal development more generally, is simultaneously attached to the realm of the rational as much as it is to the realm of the imaginary and its darker reaches.

Ordering punitive technologies

In what follows we will consider different orders of punitive assemblage. We begin by looking at relatively simple technologies of punishment; technologies which we will class as technical extensions to the existing human physical capacity to inflict pain (McLuhan 1964; Brey, Chapter 1, this volume). Such extensions include instruments such as whips, rulers and canes. We will then consider more refined punitive machines; those which have been carefully designed to extend punishment or render pain more intense. Torture instruments, for example, fall within this category, as would punitive machines like the gallows and the guillotine. We will then consider how these punitive instruments are embedded within and form part of what we propose to term a wider punitive assemblage. These will be explored in two sections. In the first we explore relatively simple punitive assemblages. A place of execution constitutes a relatively simple punitive assemblage as does a modern prison. These can be explored as stand-alone assemblages in which humans work with each other in social machines alongside other machines to deliver pain in various ways in dedicated spaces of punishment. In the second we will explore how these more simple assemblages converge into more complex punitive assemblages at the social level. Assemblages whose nature we will approach in actor network theory terms. In each assemblage, whether simple or complex, we witness an articulation between desire, various actor networks, an art of punishment, penal rituals, settings and the material objects that might be erected and used within them.

Punitive extensions

A punitive technology can be relatively simple or complex. Examples of the former would include the technology innate to the human body: the hand that slaps, the foot that kicks, the mouth that insults and degrades another. Then we have that category of technology which simply extends and enhances the business of pain delivery in some way. The whip, for example, simply replaces the hand as a vehicle for punishing bodies. In a sense it is no more than a prosthetic extension of a hand. It can do what the hand can do but also far more effectively. It makes the pain delivered more intense, it multiplies the damage imposed to a body. And such technologies have flourished over the centuries as have the rituals that surround their use. In British schools, canes along with readymade objects designed for alternative uses, such as rulers and gym shoes (known regionally up to the 1960s as 'daps') have been routinely enrolled as punitive instruments of this form. All were deployed in the name of encouraging compliance and punishing disobedience from malfunctioning young people. 'Dapping' referred to the characteristic everyday punishment meted out to errant school children across the UK colonies throughout the twentieth century.

While this category of punitive instruments is homogeneous in relation to the fact that they simply enhance the human capacity to inflict pain, they also vary in relation to the quantum of pain and damage they can inflict. A slap hurts, being hit repeatedly with a belt hurts more, being flogged with a whip more still and whips themselves vary considerably. If, in one sense, the whip simply does what a hand can do but better, these objects themselves can become subject to incredible refinement and ingenuity. Consider for example the Cat of Nine Tails, a whip, with nine separate knotted lashes. Developed in ancient Egypt, the whip was named after the nine lives of a cat and because the marks it left on the body resembled cat scratches (Science Museum 2016). It was routinely used by the British Navy to punish errant sailors:[1] it was an instrument routinely used in the slave trade as well (Oxford Dictionary 2009). Or take the Cat's Paw (also known as the Spanish Tickler), not unrelated, a torture instrument composed of sharp iron spikes attached to a handle used to shred the flesh from those against whom it was used. In a sense it was an extension of the torturer's hand. Like the Cat however, it also significantly enhanced the degree of pain that any hand could inflict.

Here we have examples of technologies that in and of themselves perfectly embody the essential attributes of any penal technology: the desire to hurt, an art of pain dispensation evident in the imaginative way in which pain is designed to be delivered; all coupled with the application of instrumental rational ways of thinking in order to ensure that the desired aim is realised in the most effective way. To return to the British navy, not only were sailors made to construct the Cat that would be used upon them, each lash had woven within it lead pellets in order to ensure the process of being punished was sufficiently excruciating.

Punitive machines

Closely related to this class of punitive technology we find a class of machines that enable humans to do more than what the primitive technologies of the body can accomplish. Stand-alone instruments of torture would logically fall into this category. This category would include, for example, various boots designed with the intension of crushing the feet (the' Spanish boot' for example involved encasing a victim's leg and foot inside an iron casing and then driving wedges into the gap between the iron and leg); through to the development of more complex punishing machines such as the rack (in which people are strapped before being stretched) through to machines of execution such as the gibbet or the guillotine.

What the gibbet and the guillotine share in common is that both modes of punishment have been subject to a process of continuous development. Take the fundamental processes underpinning each. Hanging people at its most elementary simply entails suspending someone from a noose placed around their neck until they die. Prior to the advent of the guillotine, cutting people's heads off with a sword or an axe were routinely practised. What the development of the gallows as a technology of punishment, along the advent of the guillotine facilitated, were ways of accomplishing the same end (killing people) while also resolving many of the problems that more primitive attempts to hang and behead people brought in their wake (Spierenburg 1984).

Take the case of hanging someone to death. It is not perhaps as easy a project as it might appear. Some knots are better than others if the aim is kill someone quickly and hence the advent of the hangman's noose. If a rope used is too light, it might well snap when the victim is dropped. At the same time if the drop is too long, a victim might well be decapitated. It requires a lot of accumulated knowledge and its practical application to hang somebody effectively (Gatrell 1994). In effect there is a punitive science of hanging, the application of which led to several refinements and developments in the technology of the gallows. As we shall see, in the second section, technology can also be applied to punishment in ways that do not entail making people suffer. In the pre-modern period, hanging was typically designed to be a protracted affair. For a society in which punishment was served up as a spectacle, drawing out punishment made evident sense. However in a modern society in which unnecessary punishment was considered barbaric (Spierenburg 1984) gallows technology was invested instead into ensuring that death was more swift and, not least, less messy. The same applies to the art of decapitation. It too could be a messy business and while severing a person's head from their body could be accomplished at a single stroke with a sword or axe, this did not happen in many cases. What the advent of the guillotine sanctioned was the creation of an execution machine which was almost totally efficient (Kershaw 1958). The principles of science were applied to the art of execution in a near perfect way. Few survived Madame Guillotine.

Punitive assemblages

While it is inviting to imagine punitive technology as a stand-alone machine exemplified, for example, by the gallows or the guillotine, to remain fixated at this level would be to miss the bigger picture. For if we take our definition of punitive technology to its logical extension we cannot simply terminate our consideration of the relationship between technology and the punitive here, because it ignores the wider sites and social relations in which these technologies are socially embedded. To establish what we mean by this let us return to the subject of the gallows, but, on this occasion, study more closely the wider institutional setting in which hanging as a practice of punishment was located. By way of case study, let us consider Tyburn in England – an execution site in London that would remain active for over 300 years until eventually replaced by the penitentiary system in the nineteenth century. Why its study is important is because, if we deploy the concept of technology in the way we have articulated above, it is evident that it is not enough to reduce punitive technology to simple machines (whips, torture instruments, gallows etc.); we have to look at the wider assemblage of which they are a part.

In the case of Tyburn we have a site of punishment in which gallows are erected (the triple tree, john ketch – it had many names) (Hay et al. 1975). The gallows undeniably are the key referent here but they are not the only prop. The gallows are erected on a huge stage, all the better to reveal the majesty of the sovereign's justice; all the better to dramatise capital punishment as a spectacle the public are invited to witness. Seating is also thought through with reserved spaces being allocated for the powerful and wealthy. And though hanging was the defining feature of the spectacle, it is preceded by a huge, stage managed, ritualised performance in which an array of different actors play a part. It begins when the condemned are taken in an elaborate procession from the prison at Newgate (the expression 'riding backward up Holborn Hill' (part of the procession route) meant going to the gallows). The condemned would be led along Tyburn Road (today known as Oxford Street) eventually to arrive at Tyburn. Crowds would throng the streets, sometimes cheering the condemned, sometimes jeering them, depending on their offence (see Lindebough 2006). Prior to the execution various pamphlets would be in circulation narrating the offender's crimes. The prisoner would mount the stage and another set of rituals would begin. A statement of their crimes would be pronounced and the offender would be invited to make their final speech before they were hanged. Nor did it end there. Depending on their sentence other unpleasant things could occur. They might be cut down before they were dead to be ritually disembowelled prior to being drawn and quartered. Then as part of the ritual we have the crowd reaction itself.

Which takes us to our point, a punitive assemblage is more than a simple punitive machine. To put it another way, it is not that a punishment system uses a particular punishing technology such as a gibbet. In Deleuzian terms (Deleuze and Guattari 1972), if a machine is nothing but an assemblage of machines driving or being driven by other machines, then a punitive assemblage is in itself a machine with machine-like properties. To make sense of the machine we must thus study the assemblage. We need to study the component parts (the machines/the actors) and consider, as we do, how they are assembled in actor networks (Latour 2005).

Considered as an elaborate social-technical machine, the concept of a punitive assemblage can be extended further to define an entire system of punishment, be this the sacrificial system developed by the Aztecs, the inquisition of the medieval world, modern incarnations such as the Nazi death camp system, the gulag archipelago of the Soviet period through to the modern mass incarceration system in the USA today. Here we find ourselves again looking at the intersection of the same primary elements of any punitive assemblage, albeit in a systematised and distributed form, macro structured: the desire to punish, the creation of processes by and through which punishment will be mediated, creating and assembling the social-technical recourses, human and otherwise, that will make a standardised and distributed system of punishment possible at a societal level. And underpinning all of this, the application of specialised knowledge, eventually science, and all of this locked into an extensive penal industrial complex with its own elaborate bureaucracies and divisions of labour.

By way of illustrating this, take the Nazi death camp system and its rapid development. Let us consider the development of this punitive assemblage in actor network terms. We can study the development of the assemblage here as it moves from primitive forms of mass killing involving death squads (the *Einsatzgruppen*) who assembled Jews across Poland before having them executed by machine guns in woods (Browning 1998; Bauman 1989), towards the evolution of the death camp system perfected in the construction of execution facilities such as Auschwitz. This facility not only exemplifies the application of scientific reasoning to the art of mass killing, the *techné* of genocide, but also the intersection of an elaborate set of socio-technical actor networks. It includes, for example, the bureaucratic networks coupled with transport networks that would facilitate moving Jews *en masse* from across the occupied territories to the death camps in Poland. Viewed as complex problem solving machines, the bureaucrats in question had to bring a range of actors together (guards, trains, Jews) resolve the problem of paying for the transportation (they ingeniously made the Jews pay for their own transport) and organise the logistics i.e. integrating transportation timetables across national jurisdictions.

To this network we can add a different intersection of networks responsible for developing the death camps themselves: the enrolment of scientists employed in large existing German manufacturing organisations, who, through experimentation would refine the use of gas as the preferred mode of killing; the enrolment of architects and engineers to design the camps that facilitated the mass killing of Jews in ways that both separated the executioners from direct proximity of their victims, whilst also resolving the problem of having to forcefully herd Jews to their deaths by disguising execution chambers as shower rooms; and the technical aspects of mass killing also followed through into the way corpses were subsequently destroyed through the development of crematoriums, where the corpses in question were removed by other Jews. In actor network terms, we witness here both the association and enrolment of different human actors (architects, designers, soldiers, bureaucrats, etc.) with various material technologies (gas systems, train networks etc.) mobilising a range of different specialisms, each geared to solving problems posed in different domains; each orchestrated together through a complex web of translations, operating collectively within a complex, distributed industrial technical network. One which was animated, as Bauman argued (Bauman 1989), with the technical problem solving capacities made possible by a rational industrial bureaucratic (gardening) state, which would, within two years, facilitate the creation of one of the most technically accomplished killing machines that humanity has ever developed.

Punishment systems then can be considered as themselves a technological assemblage, in effect a machine and thus a technology. In the case of the death camp system we have a kind of infernal machine programmed in a way that enabled all its constituent parts, despite periodic setbacks, to deliver a particular output. What is disturbing about social punitive machines – and not only the death camp system, as we shall see – is that once constructed and assembled, they will continue to work independently of their creators. In the case of the death camp system, it continued to function even when it became patently clear to all that the war was lost. The consequence was that tens of thousands more lives were lost simply because the machine carried on doing what was expected of it.

The contemporary mass incarceration system in the USA also illustrates these machine-like trends. It is composed of an interlocking system of penal establishments through which penal subjects are circulated and held, often for incredibly long periods of time (see Garland 2001). It has an escalation system embedded within its structure that also works to propel penal subjects from its soft end to the hard end exemplified, for example, by total lock down, Supermax prisons. Though prison rarely works to rehabilitate penal subjects, this escalator is serviced by the on-going recirculation of ex-penal subjects, either because they re-offend, are caught and thus shunted back into the system or are re-admitted having failed bail conditions established to fail them. Coupled together with what Wacquant terms the 'deadly symbiosis' between the penitentiary and the ghetto which establishes primary throughput into the penal maw (Wacquant 2009), what the penal industrial complex has produced is a deadly self-perpetuating machine that has generated the highest volume imprisonment system in the world. As Christie (1993), observes it is also a system that is potentially modifiable. If fewer penal subjects enter, which might thus impact upon its reproductive logic then changes can be made (such as raising prison tariffs, or launching new wars against crime) that will compensate. It has, as such, homeostatic properties hard wired into its institutional DNA.

In making these points, namely that penal systems are machine-like, and thus wholly technological, it is not our aim to suggest that these machines are necessarily stable or not ultimately self-destructive. As Rothman's history of the penitentiary shows, what we have here is less a punitive technology that succeeds in realising the programme that allegedly animates it, but a machine that systemically fails, but which is nevertheless and paradoxically always reborn as the solution to its own crisis (Rothman 1971; see also Cohen 1985). If we want to comprehend penal machines they thus resemble less a BMW car but perhaps more the kind of surrealistic sculptures produced by Jean Tinguely. By way of illustration, Tinguely once constructed a machinic assemblage in the American Institute of Contemporary Art called 'Homage to America'. This machine, composed of a series of interlocking machines, took three days to destroy itself – whilst also serenading itself to destruction (Violand 1990). The lesson we take from this is that penal machines might well be locked into a reproductive dynamic, but at the same time it is often to a self-destructive dynamic of reproduction that they are articulated. They might well be machines that 'work', but the end towards which they work is ultimately destructive. To return to the American mass incarceration system, it no doubt 'works', but the concept of working requires refinement. It does not rehabilitate; indeed, at the deep end we find prisons, like the Supermax, which are purposively designed to be post rehabilitative (Lynch 2001). They are designed simply to warehouse humans, often in appalling conditions, not to reclaim them for useful purposes as was the case in more distinctly modern regimes (Hallsworth 2000). While a case might be made that the system brings peace to the streets of America, the US is still home to rates of violence unseen in less punitive societies. As Christie argues, this machine exists less to do good, but far more to satisfy the needs of social control in class-divided societies and meet the needs of the huge industrial penal complex whose interests are served by meeting this need (Christie 1993). And these machines are very expensive to maintain and run, even when privatised. Indeed, in states like California, penal fare consumes more resources than higher education.

Technology and modernity

Let us now turn away from the orders of punitive technology and consider briefly the various ways technology has been applied to punishment, firstly, in premodern societies, secondly, in more distinctly modern societies, before looking at the relation as it appears to be coalescing in postmodern or late modern societies. In considering this relationship, we want to develop two broad conjectures by way of an opening gambit:

- *Proposition 1:* In pre-modern societies technology is primarily, if not exclusively, orientated to enhance the quantum of pain directed at a penal subject whilst facilitating the viewer experience.
- *Proposition 2:* In modern societies technology is increasingly applied to square the circle between a continued appetite to inflict pain but in the context of social formations beholden to a civilising process that places significant prohibitions in relation to enjoying the spectacle of suffering, and which views untoward punishment as barbaric and uncivilised.

As we have seen in our examination of spaces of execution such as Tyburn, and, prior to that, our brief consideration of torture instruments, within pre-modern societies technology is used to create refined instruments of punishment that can inflict far more grievous pain than human hands alone. They can also deliver pain, and death, far more effectively to far more people. If we attend specifically to the design of grand amenities in which punishment is dispensed we also witness considerable investment in time, resources and the development of punitive sciences that will enable their construction. If punishment in the pre-modern order was principally directed at the body in a graphic way and in a way designed to be publicly mediated, considerable thought was extended into constructing elaborate staged settings in which this might occur. The Roman Coliseum exemplifies all these qualities. It constitutes one of the most elaborate structures ever constructed to facilitate the business of graphic punishment served up as a public spectacle. At its most basic it is an arena around which tiered seating is erected enclosed by a grandiose building purposively designed to convey an image of Roman might and civilisation. In effect, an architecture of power, designed to awe. Its construction brought together architects, engineers, set designers, artists and artisans into one joint enterprise. Once built, it was capable of hosting 80,000 spectators. Beneath the arena a two-level subterranean network of tunnels and vertical shafts were constructed (the hypogeum) (Welch 2009). These enabled victims and various other props to literally appear, as if from nowhere, in the arena, whilst also facilitating their efficient removal. In relation to what was presented for public consumption then this technological triumph was purposively built to facilitate the most extreme aspects of the punitive imaginary and on a vast scale – importing wild animals to tear apart Christians, staging battles of the ancient world and convening gladiatorial contests to the death.

With the advent of modern industrial rational societies, punishment changes and technology changes in tandem with it. In a civilising society punishment goes backstage, pain is no longer directed predominantly at the body (though there are stark exceptions to which we will return) but increasingly towards the mind, where the pre-eminent vehicle for realising this would be the penitentiary (Foucault 1977; Melossi and Pavarini 1981). Like the Coliseum, these technological amenities are designed to reflect the power of the state, but in terms of their design, their high walls exist to keep the inquisitive gaze of the public away from the business of punishment, now relayed instead through spatial confinement for often long periods of time in a prison cell. Technology in this sense is applied to stop the public enjoyment of the spectacle of suffering that sat at the heart of pre-modern penal regimes. The cellular design of the penitentiary also facilitates better the task of targeting the mind of the offender, which would, or so it was hoped, be rehabilitated by the experience. If pre-modern penal regimes were grounded on an economy of penal excess, in the modern age, technology is applied in the form of a developing penal science to facilitate reclamation of penal subjects for useful purposes (Hallsworth 2000).

Whilst pain directed in the form of denying offenders the right to exercise their basic rights sat at the heart of the new punishment, nevertheless the will to punish and to kill also remained. Whereas, as we have seen, in the pre-modern age, technology was applied to refining punishment machines in ways that enabled them to ramp up the intensity and pain visited on an offender, in modern regimes technology (of which science and its application is intrinsic) is mobilised increasingly in order to ensure that death (if that is the intention) occurs instantaneously, efficiently, and in ways that do not occasion bystanders and, not least, executioners, undue suffering. The art of hanging people effectively is perfected. The executioner's task is now rooted in evidence-based practice. And scientists will step in to ensure that where problems are experienced in delivering pain, technological solutions can be applied to resolve them.

Let us illustrate what we mean by this. Take the case of electrocution practised as a system of execution in the USA. At first sight it might appear a fairly straightforward way to take an offender's life. But as the Virginia State's attempt to execute Jesse Tafero on 4 May 1990 demonstrated, this is not as straightforward a killing procedure as it appears:

> instead, flames, smoke and sparks shot six inches out of the head of Tafero as three 2,000 volt shocks were administered. Because the amperage was incorrect Tafero's flesh cooked on his bones.
>
> *(Trombley 1993: 34)*

And botched executions caused by 'execution glitches' are fairly commonplace. Too much current and bodies can explode, eyes pop out of their sockets while the body cooks and stinks. Typically offenders will also defecate or urinate. It certainly doesn't constitute a civilised spectacle for the few invited to bear witness to the execution. This is nothing, however, that the appliance of science cannot address. In his book *The Execution Protocol*, Trombley (1993) examines an electric chair developed by Fred Leuchter,[2] to reduce the problems attendant on electrocuting people while also making the process more resonant with civilised sensibilities:

> The seat is made of Plexiglas and is perforated so that when the victim loses control of their bowels and bladder, liquid waste will pass through the seat. This is collected by a removable drip pan positioned under the seat. This feature makes the execution team's task of removing the dead inmate from the chair less unpleasant and presents a more hygienic spectacle for witnesses. All electric chairs in the United States have heavy leather straps to restrain the inmate. This can be awkward for the execution party to fasten, particularly if the inmate offers resistance. They also cause pain and discomfort to the inmate (autopsies show facial bruising and even laceration from the straps). When the execution is over the execution team's job of unstrapping the dead man from the chair is often repugnant, as they have to tug and push at the body. There is always suppuration from the third degree burns on the head and the leg, and in some cases the 'cooked' flesh comes away from the body when touched. Fred's 'solution' to the problem was to introduce a 'non-incremental restraint system' – basically, a seat belt made out of aircraft nylon with a single, quick release fastening.
>
> *(Trombley 1993: 36)*

In summary then, technology, both in its material form and in the form of science is increasingly applied to help remove the business of punishment off stage while removing the unsightly and barbaric excesses that have typically been intrinsic to its practice. In a weird kind of way, technology mobilised this way helps paradoxically to 'civilise' punishment. It occurs but the public rarely see the consequences. When it occurs it is directed now at the mind more than the body and when it is directed at the body, the systems employed (like electrocution) are designed to ensure instantaneous death without the disturbing mess that can, as we have seen, get in the way.

As we have tried to emphasise however, there is always a contradiction between on the one hand, a society committed to a civilising process, whilst, on the other, also being simultaneously committed to visiting pain on others. As we have seen above, one way to square this circle is to tidy up the spectacle of suffering, pushing the more unpleasant and disturbing aspects of punishment backstage where they can no longer be seen. Another way in which technology can be enrolled to 'civilise' punishment is to help develop systems of punishment that occur increasingly out of sight. Consider these punitive endeavours – saturation bombing and the contemporary drone war. Both are made possible by the developments in the forces and relations of production in an ever more technologically enhanced world. The saturation bombing of civilian areas in World War II was perpetrated both by the Allied and German forces. To enact mass bombing, planes were developed that would enable them to transport bombs over long distances. Significant investment was also expended into developing high explosives. Such developments permitted the mass destruction of cities along with their human populations. In the case of Dresden, for example, during four raids on the city between 13 and 15 February 1945 Allied bombers dropped more than 3,900 tons of high explosive bombs and incendiary devices on the city, provoking a firestorm that killed around 25,000 people (Lindqvist 2002). This was a wholly punitive act on the part of the Allies, not one guided by military objectives. In this case, note the total separation between the humans engaged in dropping the bombs from their victims (they push buttons and fly planes) while the public in whose name such extreme punishment is justified, do not see or experience the real consequence of their actions. It could, of course, be argued that those consequences did become more evident, but only much later. The drone war perpetrated by the USA in Pakistan conforms to a similar pattern. Drones are pilotless aircraft equipped with weapons directed by operatives located thousands of miles away from the zones in which they are deployed. They rain down death from the skies. If a successful assassination occurs it is reported but the wider social carnage including the killing of innocent civilians, is not (Breau et al. 2011).

Is it possible to talk of a new relationship between technology and punishment in late modern societies? It is an issue worth reflecting on. In pre-modern societies we have argued that technology is invested specifically in enhancing the quota of pain delivered while simultaneously improving the spectator's experience. In 'civilised' modern societies, technology is used to help build institutions and design killing technologies which remove spectators from the spectacle of punishment while tidying up the resulting mess that it occasions. Technology, in effect, is used to 'humanise' and 'civilise' the business of punishment. Might we be witnessing the advent of a techno-social regime which is arguably less modern whilst also in a sense being post-modern (Hallsworth 2002)? In his paper 'The Return of the Wheel Barrow Men', John Pratt draws attention not only to the will to punish that has fuelled the contemporary development of the USA mass incarceration society, but also to the resurrection of an array of penal practices within it once considered decidedly un-modern. Pratt narrates, as examples, the advent of new name and shame rituals and the advent of visceral sentencing policies such as 'three strikes' (Pratt 2000). He would subsequently go on to conceive in the resurrection of such pre-modern penal sensibilities and practices evidence of de-civilising processes at play in late modernity. Bringing this analysis up to the present and bringing technology into the equation, consider the creation of punitive assemblages like Guantanamo Bay and forms of interrogation conducted within them, such as water boarding. Here we find a paradigmatic late modern punitive technology in which water is poured over a cloth covering the face of an immobilised victim inducing a sense of being drowned. Might we be witnessing a new punitive technology relation? From the penitentiary designed as a space and place of reclamation we see in installations such as Guantanamo (itself modelled on Supermax prisons) not only a space in which torture is openly practised but which is purposively designed to be post-rehabilitative. No reclamation or reform is sought or expected on the part of the inmates detained there.

It could be noted that the utilisation of these technologies has been legitimated in the name of the 'war on terror' mounted against insurgents whose will to violence appears directly modelled on the penal excess typical of the pre-modern period. So-called Islamic State, for example, have demonstrated no appetite at all for developing punitive assemblages that make any claim to 'civilise' or 'humanise' punishment. They avail themselves instead of more primitive punitive machines and techniques – the use of swords, for example, to behead their victims. In one sense, this is clearly anti-modern, in so far as such practices mark a return to pre-modern sensibilities. This is evident not least in the theatrical way in which punishment is conducted. But also postmodern to the extent that such punishment marks a paradigmatic break with civilised codes associated with the civilising process, whilst also utilising advanced contemporary technologies like the Internet and social media to mediate the spectacle.

Notes

1 The expression 'the cat is out of the bag' refers to the ritual of taking the whip out of a bag where it was traditionally stored, while the expression 'no room to swing a cat' referred to the fact that on a ship with limited room it was often difficult to flog someone without the lashes becoming tangled.
2 Fred Leuchter practised his trade despite having no formal engineering training. He rose to prominence and notoriety after writing a report on the basis of forensic evidence he had obtained (while on Honeymoon) in Auschwitz, that the death camp could not have been used for mass killing. This would project him into the premier league of Holocaust deniers. His rise and fall was the subject of an excellent Errol Morris documentary entitled *Mr. Death: The Rise and Fall of Fred A. Leuchter*.

References

Bauman, Z. (1989) *Modernity and The Holocaust*. Ithaca, NY: Cornell University Press.
Breau, S., Aronsson, M. and Joyce, R. (2011) *Discussion Paper 2: Drone attacks, international law and the recording of civilian casualties of armed conflict*. Oxford Research Group. Available at: www.oxfordresearchgroup. org.uk/sites/default/files/ORG%20Drone%20Attacks%20and%20International%20Law%20Report. pdf (accessed 11 February 2016).
Browning, R. (1998) *Ordinary Men: Reserve Police Battalion 101 and the Final Solution in Poland*. New York: Harper Perennial.
Christie, N. (1981) *Limits to Pain*. Oslo: Universitetsforlaget.
Christie, N. (1993) *Crime Control as Industry: Towards Gulags, American Style*. London: Routledge.
Cohen, S. (1985) *Visions of Social Control: Crime, Punishment and Classification*. Cambridge: Polity Press.
Deleuze, G. and F. Guattari. (1972) *Anti-Oedipus*. Trans. Robert Hurley, Mark Seem and Helen R. Lane. London and New York: Continuum.
Elias, N. (1969) *The Civilizing Process*, Vol. I. *The History of Manners*. Oxford: Blackwell.
Foucault, M. (1977) *Discipline and Punish*. Trans. A. Sheridan. London: Peregrine Books.
Garland, D. (2001) *The Culture of Control*. Oxford: Oxford University Press.
Gatrell, V.A.C. (1994) *The Hanging Tree: Execution and the English People 1770–1868*. Oxford: Oxford University Press.
Hallsworth, S. (2000) Economies of excess and the criminology of the other, *Punishment and Society* 2(2): 145–60.
Hallsworth, S. (2002) The case for a postmodern penality, *Theoretical Criminology* 161: 145–63.
Hay, D., Linebaugh, P. and Thompson, E.P. (eds) (1975) *Albion's Fatal Tree: Crime and Society in Eighteenth Century England*. London: Penguin.

Kershaw, A. (1958) *A History of the Guillotine*. London: John Calder.

Latour, B. (2005) *Reassembling the Social: An Introduction to Actor Network Theory*. Oxford: Clarendon.

Lindebough, C. (2006) *The London Hanged: Crime And Civil Society In The Eighteenth Century*. London: Verso.

Lindqvist, S. (2002) *A History of Bombing*. London: Granta Books.

Lynch, M. (2001) The Birth of the Post Rehabilitative Prison: A Case Study of Arizona's Penal System, paper presented at the 53rd annual meeting of the American Society of Criminology Conference, Atlanta, USA, November.

McLuhan, M. (1964) *Understanding Media: The Extensions of Man*. New York: McGraw-Hill.

Melossi, D. and M. Pavarini (1981) *The Prison and the Factory: Origins of the Penitentiary System*. London: Macmillan.

Oxford Pocket Dictionary of Current English (2009) 'cat-o-nine-tails.' Available at: www.encyclopedia.com (accessed 11 February 2016).

Parry, Richard (2014) '*Episteme and Techne*', *The Stanford Encyclopedia of Philosophy* (Fall 2014 Edition), Edward N. Zalta (ed.), Available at: http://plato.stanford.edu/archives/fall2014/entries/episteme-techne/.

Pratt, J. (2000) The Return of the Wheel Barrow Men; or, The Arrival of Postmodern Penality, *British Journal of Criminology* 40(1): 127–45.

Rothman, D. (1971) *The Discovery of the Asylum: Social Order and Disorder in the New Republic*. Boston, MA: Little Brown.

Science Museum (2016) *The cat of nine tails*. Available at: www.sciencemuseum.org.uk/broughttolife/objects/display?id=92935 (accessed 22 March 2016).

Spierenburg, P. (1984) *The Spectacle of Suffering*. Cambridge: Cambridge University Press.

Trombley, S. (1993) *The Execution Protocol*. London: BCA.

Violand, H. (1990) *Jean Tinguely's Kinetic Art or A Myth of the Machine Age*. Diss; New York University.

Wacquant, Loïc (2009) *Punishing the Poor: The Neoliberal Government of Social Insecurity*. Durham, NC: Duke.

Welch, K. (2009) *The Roman Amphitheatre*. Cambridge: Cambridge University Press.

34

Public and expert voices in the legal regulation of technology

Patrick Bishop and Stuart Macdonald

Introduction

The law in isolation is seldom, if ever, an adequate regulatory device. As Black contends in her exposition of decentred regulation: 'governments do not ... have a monopoly on regulation ... regulation is occurring within and between other social actors' (Black 2001: 103). Indeed, the use of alternative instruments to achieve regulatory goals has proliferated in recent times; Thaler and Sunstein's influential 'nudge' theory (Thaler and Sunstein 2009) is one obvious manifestation of the preference on the part of policy makers for non-regulatory solutions to social problems. But whilst innovation in the design of policy instruments is laudable, there remains a place for more traditional command and control regulation. As Macrory notes, 'it remains equally important to ensure that the qualities of transparency, accountability, and enforceability inherent in the more formal legal structures are not lost' (Macrory 2001: 647).

The focus of this chapter is the legal regulation of technology. In one sense, to distinguish between the regulation of technology and the regulation of other things is a false dichotomy. Technology does not exist in a vacuum and any regulation of it is essentially concerned with limiting and controlling the uses that may be made of technology by humans. In short, any form of legal regulation is concerned with behavioural control and regulating technology is no exception. In another sense, however, the targeted activity of any regulatory scheme will influence both its creation and eventual operation and here the highly specialist and complex nature of technology – and its associated risks – are certainly capable of raising distinct challenges. The chapter examines three technological areas which have been subjected to legal regulation: human fertilisation and embryology; the manufacture and distribution of chemicals; and the disposal of hazardous waste. Whilst these activities, and the regimes which regulate them, are quite different, they do share two basic common features. First, the activities themselves are necessary and/or socially beneficial. And, second, the activities also have the potential to cause considerable harm – at both the individual and societal level – if left unregulated.

Efforts to regulate activities like these face various challenges. For a start, efforts at legal regulation of technology struggle to keep pace with scientific advances. This has been characterised as a race between the 'hare' of technology against the 'tortoise' of law (Stokes 2012: 93). It is also apparent that despite an initial period free of regulation, new high-tech products and processes are eventually caught by a bespoke regulatory net or, failing that, fall within existing regulatory regimes with broad and overlapping remits (Friedman 2001).

A further challenge, which is the focus of this chapter, is managing the frequent tension between public and expert opinion. Technological advancement is a key ingredient of the 'new modernity' conceptualised by Beck: 'in social science's understanding of modernity, the plough, the steam locomotive and the microchip are visible indicators of a much deeper process, which comprises and reshapes the entire social structure' (Beck 1992: 51). This reshaped social structure is 'increasingly preoccupied with the future (and also with safety), which generates the notion of risk' (Giddens 1999: 3). Any legal regime tasked with the regulation of risk is necessarily complex and multifarious and involves the input of science and expertise (to provide assessments of risk) and the public at large in order to gauge society's response to particular dangers and their likelihood. These drivers of regulatory design and approach will often (if not inevitably) exist in mutual tension. This chapter will explore this tension via an analysis of three distinct regulatory regimes. These regimes have been chosen as they each have different structures and legislative underpinnings: UK domestic law (Human Fertilisation and Embryology Act 1990), EU law (REACH Regulation 2006) and international law (Basel Convention 1989). Before turning to these, the chapter begins by examining the role public participation plays in legal regulation in general, and the tension between this objective and considerations of resource and expertise.

Public and expert voices in regulatory design: the theory

The literature on principles of 'good regulation' reveals a high level of consensus regarding the importance of public participation (Baldwin, Cave and Lodge 2012; Regulating Better 2004; Mandelkern Group on Better Regulation 2001; Regulatory Performance Indicators 1999). In environmental matters the importance of public participation has been fully recognised at the international level (Aarhus Convention 1998). The rationale for enhanced participatory rights is rooted in democratic concerns, particularly in the context of non-majoritarian regulatory bodies. As Lee and Abbot have noted:

> The political nature of environmental decisions, together with their frequent delegation to unelected experts, requires public participation to enhance the procedural legitimacy of decisions, since electoral legitimacy is weak.
>
> *(Lee and Abbot 2003: 84)*

In addition to justifications based on democracy, there is also ample support for the view that decisions and policies made as a result of a participatory process produce higher quality results than would be the case absent any public input (Lee and Abbot 2003: 83; McGarity 1990: 112). The basis of such an argument is that public participation is able to broaden the regulator's lens beyond the purely technical aspects of the particular regulatory regime, which in turn will allow the regulator to view issues from different perspectives (McGarity 1990: 112). Theoretical support for this may be gleaned from the social science research methodology known as 'triangulation', which involves the 'use of more than one approach to the investigation of a research question in order to enhance confidence in the ensuing findings' (Bryman 2004: 1143). It is also important to note that, despite the often technical nature of regulatory regimes, expertise alone is incapable of providing a complete solution to most regulatory challenges. As McGarity has opined: 'many, if not most, important health and environmental questions are in fact not resolvable by the experts. The available information and the state of the scientific art is often so poor that the experts can at best hazard highly uncertain educated guesses' (McGarity 1990: 105). Even where experts are able to provide a reasonably accurate assessment of risks this in no way provides an answer to the vexed societal question of how much risk is acceptable. In this regard the public 'can provide useful information on matters such as public fears and values' (Lee and Abbot 2003: 82). Finally, involvement in the decision-making process provides interested parties with a better understanding of how decisions are made which in turn has the potential to reduce the scope for judicial review (McGarity 1990: 112).

In spite of these benefits, there are two sets of concerns about public participation. The first are practical, and focus on operational efficiency. In a world of finite resources, it is inevitable that cost minimisation and efficiency concerns will permeate the design and operation of any regulatory regime. As discussed below, this is certainly the case with the three areas of regulation considered in this chapter. In such a context, it is generally accepted that public participation will have considerable resource implications, both in terms of money and time (Lee and Abbot 2003: 87; McGarity 1990: 112). Moreover, an over-emphasis on participation as a condition precedent of regulatory action has the potential to damage a regulator's ability to respond to issues in a timely manner: 'more participation might lead to less effective decision making and eventually to stagnation in the regulatory system' (Baldwin, Cave and Lodge 2012: 29).

By contrast, the second set of concerns focuses not on resources but on the ability of the public to engage in what are often technical, even esoteric debates. As Eden has noted:

> [O]ne of the circumstances that can militate against this admirable objective is where discussions are dominated by 'experts' of one sort or another. This precludes wide public involvement by defining the discussions as the exclusive preserve of 'experts'.
>
> *(Eden 1996: 183)*

This view is shared by a number of commentators (Baldwin, Cave and Lodge 2012: 29; Lee and Abbot 2003: 84; McGarity 1990: 113). However, whilst members of the public 'have not always had the power or the confidence of their own "expertise" to raise their criticisms forcefully' (Eden 1996: 191), Baldwin, Cave and Lodge point out that in Beck's 'risk society' there is a 'new political dialogue built on the death of deference to those claiming special expertise' (Baldwin, Cave and Lodge 2012: 30). Moreover, any attempt to preclude public debate on the basis that only 'experts' are capable of reaching an appropriate decision undermines the democratic foundations on which public participation is based and the legitimacy of any subsequent decisions.

Having outlined this tension between public participation on the one hand and operational efficiency and expertise on the other, the chapter will now examine how these considerations have been managed in three contrasting regulatory regimes specific to differing technological contexts.

Human fertilisation and embryology

Following the birth in the UK in 1978 of the first child conceived through *in vitro* fertilisation (IVF), a Committee of Inquiry was established in 1982 under the chairmanship of Dame Mary Warnock to consider the social, ethical and legal implications of the advances in human fertilisation and embryology and to make recommendations on the policies and safeguards that should be applied. When it was published in 1984, one of the Warnock report's principal recommendations was the creation of an independent regulatory authority (Department of Health & Social Security 1984). This body would have both executive and advisory functions, and would include not only representation from the scientific and medical communities but also lay members in order to ensure public confidence and participation (ibid.: para 13.4).

Following the Warnock report's recommendations, the Human Fertilisation and Embryology Act 1990 (the 1990 Act) established the Human Fertilisation and Embryology Authority (HFEA). The HFEA's functions include, first, licensing and monitoring: clinics that carry out IVF and donor insemination treatment; centres that carry out human embryo research; and, the storage of gametes and embryos. Activities which infringe this regulatory framework, such as unlicensed treatment, constitute a criminal offence (1990 Act, s 41). Second, the HFEA issues a Code of Practice and maintains formal registers of information about donors, fertility treatments and children born as a result of these treatments. And, third, the HFEA publicises its role and provides advice and information to patients, donors and clinics. The HFEA currently has 12 members.[1] These members have a range of expertise, including medicine, law, religion and philosophy. To encourage an independent view, the HFEA requires that at least half of its members (including the Chair and Deputy) are not doctors or scientists involved in human embryo research or fertility treatment.

Whilst stating that the 1990 Act had worked well – enabling science and medicine to flourish within agreed parameters and promoting public confidence – the Government decided in 2005 that a review of the law and regulation in this area was 'timely and desirable' in the light of technological developments in assisted reproduction and changing public attitudes (Department of Health 2006: paras 1.2–1.3). When it reported in 2006, one of the review's key outcomes was to reiterate the Government's commitment (first announced in 2004) to creating the Regulatory Authority for Tissue and Embryos (RATE). The intention was that RATE would replace the HFEA and the Human Tissue Authority (HTA) with a single regulator with responsibilities across the range of human tissues, cells and blood (ibid.: para 1.4). It was suggested that this merger would prevent overlapping regulation and ensure the application of 'common principles and standards' across these 'closely linked areas' (ibid.: para 3.2). The government's proposal was for the creation of a RATE board, charged with taking a more strategic role, supported by (non-executive) Expert Advisory Panels (EAPs) to ensure that expertise would be available in all areas of activity within its remit (ibid.: paras. 3.10–3.16).

Following the publication of a draft Bill in May 2007, a Joint Parliamentary Scrutiny Committee was established. The Joint Committee's report, in July 2007, found 'overwhelming and convincing' evidence against establishing RATE (Joint Committee on the Human Tissue and Embryos (Draft) Bill, 2007, para 92). The Committee stated that, whilst there were some synergies between the work of the HFEA and HTA (ibid.: para 60) and a merger offered some potential efficiencies and cost savings (ibid.: para 61), there were also significant risks. The broad remit of RATE could result in a loss of both specialist expertise (ibid.: para 76) and HFEA's national and international reputation (ibid.: para 73). There are also significant differences between the work of the HFEA and HTA (ibid.: para 69), and public confidence could be affected by the loss of a dedicated authority for embryos – which are widely regarded as meriting a special status (ibid.: para 65). Moreover, RATE's proposed structure could result in it being little more than a rubberstamping authority, with the EAPs functioning effectively as the HFEA and the HTA (ibid.: para 85), and it would be possible to achieve some efficiencies without a formal merger (ibid.: para 82). Following the Joint Committee's report, the government decided not to proceed with RATE, and focused instead on how the two authorities could work together to streamline their operations (Secretary of State for Health 2007: para 17).

The Joint Committee's report also expressed concern that the draft Bill lacked 'the explicit underpinning ethical framework which in 1990 was provided by the Warnock Report' (Joint Committee on the Human Tissue and Embryos (Draft) Bill, 2007: para 44). This was due, in part, to the fact that the Act being created was an amending statute. In other words:

Nowhere was a blank piece of paper offered for reform, in a way that allowed for a thorough and fundamental rethinking of the kind of regulation which might best suit this area or the ethical principles which should underpin it. Rather the architects of reform worked outwards from the provisions already in place, making the key question not 'what model of law do we want?' but rather 'what needs to be changed?'

(McCandless and Sheldon 2010: 180)

So whilst the resulting Human Fertilisation and Embryology Act 2008 (the 2008 Act) made some significant changes – including: permitting the creation and use (under licence) of animal–human hybrid embryos for research purposes; prohibiting the selection of the sex of offspring for non-medical reasons only (thereby allowing the screening of embryos to select a saviour sibling); removing the requirement to consider the future child's 'need for a father' when deciding whether a woman should be accepted for treatment services; and, providing for the first time that two women may be recognised as a child's legal parents from the moment of birth – it also represented a missed opportunity in other important respects. For example, the way parenthood is framed within the legislation continues to prioritise married couples (even though this is at odds with other developments in family law), assume a two-parent model and regard parents as occupying complementary yet different roles (hence a lesbian co-mother is not a mother but a female parent) (McCandless and Sheldon 2010). The 2008 Act also made only minimal changes to the law and regulation of surrogacy. In particular, there remains a distinction between full surrogacy (to which the 1990 Act applies[2]) and partial surrogacy (where there is a 'regulatory vacuum' (Horsey and Sheldon 2012: 73)). This distinction is difficult to justify and in the latter case 'leaves individuals dependent on the efforts of "well meaning amateurs", who are prevented from charging the fees that would otherwise allow them to professionalise their services' (ibid.: 87).[3]

In addition to an appropriate ethical framework, the Joint Committee emphasised the importance of an 'appropriate, consistent and workable regulatory architecture'. This entails 'finding the right balance between flexibility and legal certainty' (Joint Committee on the Human Tissue and Embryos (Draft) Bill, 2007: para 49). Claiming that the Government had favoured legal certainty and, as a result, 'been over-prescriptive in many areas in an attempt to provide for every eventuality',[4] the Joint Committee argued for 'a more flexible approach within clearly defined parameters' (ibid.: para 55). This principle of 'devolved regulation' would, it suggested, provide regulators and clinicians with greater freedom of action and future-proof the legislation as technology continued to develop (ibid.: para 56). In keeping with this more 'permissive' approach, the Committee also recommended that the HFEA be given a statutory power to define areas of exemption from the regulatory remit where appropriate (ibid.: para 105). These recommendations were not accepted by the Government, however, who asserted that 'such a framework would introduce a lack of accountability' (ibid.: para 8), 'would cause uncertainty about the scope of regulation' (ibid.: para 11), and 'would be confusing and open up the HFEA to increased litigation and judicial review' (ibid.: para 11).

The future of HFEA was examined again following the 2010 General Election. As part of its programme for Government the Coalition Government made a commitment to cut the number of health arm's length bodies and reduce bureaucracy significantly. The Department of Health accordingly stated that the HFEA and HTA would be retained temporarily as separate arm's length bodies, with a view to transferring their functions to other bodies by 2015 (Department of Health 2010). After a consultation found that the majority of respondents did not favour this proposal (Department of Health 2013a), an independent review into the operation of the two bodies was set up, led by Justin McCracken. The McCracken report found that 'There is almost universal praise for the Human Fertilisation and Embryology Act, and recognition that it is still fit for purpose' (McCracken 2013: para 4.5). The report emphasised that the work of the HFEA and HTA is, by and large, separate, and that the 'specialist expertise in the regulators and their understanding of the science underpinning commercial developments in the field were cited as critical, and important to preserve' (ibid.: para 4.2). This specialist expertise was also found to be key to maintaining '[p]ublic confidence in the sensitive areas regulated by the HFEA' (ibid.: para 4.1). Since a merger offered only 'relatively modest additional cost savings' and 'Much of the potential benefit of merger can be achieved by merging the two Finance and Resource groups while retaining the separate statutory entities with their respective Chairs, Chief Executives, and Boards' (ibid.: para 4.18), the report recommended retaining the HFEA as a separate body. This was accepted by the Government (Department of Health 2013b), and the recommendations have since been implemented (Human Fertilisation and Embryology Authority 2014).

In her comparison of the UK's and US's approaches to human embryonic stem cell research, Schechter states that:

> To some, it may seem counterintuitive that the United Kingdom, with its stringent regulatory and licensing standards, would be more effective at encouraging research than the United States and its relatively lax, unrestrictive approach. However, considering the state of the science in this field, the level of uncertainty created by the lack of uniformity and oversight, and the benefits of comprehensive regulation, the federal government needs to play an active role in this area if it wants to see real, competitive progress.
>
> *(Schechter 2010: 629)*

The regulatory oversight provided by HFEA has, she argues, allowed the UK to develop a 'consistent and progressive' approach (ibid.: 620). By contrast, in her discussion of whether the Netherlands should create a similar regulatory body to the HFEA, Zeegers (2014) has suggested that such a model can result in 'bad politics' (p. 13). Using hybrid embryos as a case study, and pointing in particular to the fact that those who opposed the creation of hybrid embryos on moral and ethical grounds were not only 'misled by the idea that the creation of true hybrid embryos would not be at issue' but also had their beliefs 'turned into an argument for facilitating an even wider range of human animal forms of embryo' (ibid.: 27), she argues that the regulatory mechanism resulted in 'a neutralization and downplaying instead of a real reconciliation of differences in moral views' (ibid.: 26). This resonates with a comment in the McCracken report which, whilst generally very positive, did assert that there is 'a fairly widespread sense that the HFEA needs to do more to properly take into account stakeholder views and to be seen to do so' (McCracken 2013: para 4.5).

The manufacture and distribution of chemicals

The ubiquitous use of chemicals (and chemical technologies) undoubtedly provides innumerable benefits to society but can also cause considerable damage to human health and/or the environment. As Stokes and Vaughan posit: 'chemicals can hold concurrent and fundamentally opposed positions in our legal and social consciousness: being both savior and sinner in the same instant' (Stokes and Vaughan 2013: 435). The regulation of chemicals poses challenges of immense complexity, in at least three different ways. First, the central mechanism of chemicals control in the EU, the REACH Regulation (Regulation 1907/2006/EC (OJ L396/1 2006)) – an acronym for the Registration, Evaluation, Authorisation and Restriction of Chemicals – is not designed with a single objective in mind (see generally Vaughan 2015). Early EU action in this area (Council Directive 67/548/EEC) was motivated primarily by single market aims and the desirability of correcting information asymmetries. In a modern context, the objectives of EU chemicals regulation have been extended to explicitly include the protection of human health and the environment, facilitate market integration and to promote innovation in the chemicals sector (Heyvaert 2007: 201). Second, chemicals regulation is essentially concerned with the identification and management of risk as informed by regulatory science (Funtowicz *et al.* 2000). The limitations of science conducted primarily to inform policy are well-documented. Regulatory science is often characterised by extrapolation: from high to low dosage levels, from animals to humans, from short-term to long-term exposure (Jones 2007). As noted earlier, even where science is able to provide an accurate assessment of risk this in no way provides an answer to the difficult societal question of how much risk is acceptable – but this is a question that any regulatory regime must address. Finally, any chemicals regulatory scheme is faced with considerable challenges of scale. While the exact number of chemicals on the market is unknown, the REACH Regulation is only applicable to chemicals manufactured or imported in the EU at or over one tonne per annum. The European Chemicals Agency expects at least 30,000 existing chemicals to be registered in this category by 2018 (European Chemicals Agency 2015). These complexities have unquestionably shaped the substance and form of EU chemicals policy.

REACH is a legislative instrument of immense complexity, density and length. The text of the consolidated version is over 130,000 words in length, accompanied by over a million words of guidance issued by the European Chemicals Agency (ECHA). Thus, only the briefest of summaries is possible here!

The title 'REACH' encapsulates the full range of regulatory mechanisms utilised by the regime. A staged approach is adopted commencing with the registration of chemicals, followed by their evaluation and then, where applicable (depending on the outcome of evaluation), their authorisation and possible restriction. The first of these steps – the registration process – is arguably the most significant. Here a 'no data, no market' approach is employed. Manufacturers or importers of chemicals must apply for registration, a process which involves the submission of a technical data file supplying, inter alia, the identity of the chemical, its intended use, physical properties and toxicological/ecotoxicological information (art.12(1)). This process represents a reversal of the traditional approach to market regulation for the purposes of environmental protection:

> Whereby most environmental regulation operates as a limit on market activity, 'you can do what you like but not x'. Such laws dictate what particular kinds of behavior are not allowed. In contrast, registration is operating as a precondition to market activity.
>
> *(Fisher 2008: 553)*

A further noteworthy feature of the registration requirement is that the duty to conduct the necessary risk assessments is delegated to private actors.

Following registration, a substantive evaluation of the registered information is required for all chemicals manufactured or imported in quantities exceeding 100 tonnes or (irrespective of volume) where the data supplied raises concerns about health and/or environmental impacts (art.44(2)). An EU-wide rolling action plan has been established, where a substance targeted for evaluation is allocated to a Member State to act as rapporteur. When concerns are confirmed, evaluation may trigger further measures, such as the inclusion of the chemical on the list of substances subject to authorisation, or the drafting of risk reduction measures. All chemicals designated as 'Substances of Very High Concern' (SVHC) require authorisation before they can be marketed or used within the EU (art.56). Firms are required to offer proof that the risks created by the SVHC are either 'adequately controlled' or there is a 'socio-economic need for their continued use, while no viable alternative currently exists' (title VII). This authorisation process therefore encompasses the principle of substitution, namely, if safer alternatives exist the SVHC must be phased out. The Commission is ultimately responsible for authorisation, which must be granted if the risks are shown to be adequately controlled. If this proves to be impossible, the Commission may still grant authorisation taking into account the severity of the risk and the viability of alternative substances. In reaching its decision the Commission has to follow the advisory comitology procedure under scrutiny (art.64(8)). While a detailed examination of this procedure is beyond the scope of this chapter, a central feature is the significant influence of the European Parliament, which can oppose a Commission authorisation proposal with a simple majority. Finally, for those substances which pose unacceptable risks to human health and the environment, restrictions represent the ultimate safety net. Restrictions take many forms, from a total ban to not being permitted to supply a substance to the general public.

The REACH Regulation is a multi-faceted legislative instrument which adopts a variety of different approaches. As well as mechanisms which may be termed 'command and control' (authorisation, restriction) it also relies on essentially market-based instruments (registration, provision of information). Similarly, the system is both centralised (integral role for the Commission, ECHA, etc.) and devolved (obligation on private actors to provide technical data). Of especial relevance to this chapter is the registration and provision of information requirement.

From one perspective, the obligation placed on private actors to register technical data relating to chemical risks is a practical necessity given the sheer volume of substances manufactured and marketed within the EU. It may be seen as a workload sharing mechanism (Fleurke and Somesen 2011: 372). However, the requirement also has a substantive rationale. In an unregulated world, producers and manufacturers would have few incentives to provide information about chemical risks. As Wagner has noted:

> Actors who create externalities are best suited to access and produce information on the nature of the harms that their activities cause, but they also stand to lose from providing such information.
>
> *(Wagner 2004: 1648)*

Thus, by requiring the provision of information as condition of entry to the EU chemicals market, the 'no data, no market' principle may be seen as a form of regulated self-regulation. In addition to an entry requirement, the registration requirement has other functions. First, the risk assessment which forms the basis of the registered information might highlight risks which can trigger evaluation, authorisation and restriction measures. And, second, the provision of data is essentially a market-based mechanism capable of providing information to all participants in the supply chain including the end consumers of products. Where risks are deemed unacceptable to the market, the responding purchasing (in)activity will send a clear signal to those further up the supply chain (Case 2005: 383). This in turn might advance the goals of innovation, protection of health and protection of the environment by providing an incentive to manufacturers to produce substances which pose less significant risks. This objective is supported by the maintenance of a publicly accessible database administered by the ECHA.

In one sense it is difficult to disagree with the assertion that all participants in a market should be granted access to fullest body of information possible; indeed information asymmetry is seen as a significant cause of market failure. However, the desirability of the approach adopted by REACH has not been accepted axiomatically. Durodie, for example, has referred to the 'social cost of fear reduction' and questioned the need for the risk assessment obligation for existing substances:

> Accordingly, while one might usually favour seeking to obtain the greatest possible amount of evidence in deliberating upon matters, there would appear to be a clear need in this instance to maintain some sense of perspective and priorities. This is especially so as most of the chemicals now being required to be tested have been in use for a quarter-century or more and have effectively acquired billions of hours of exposure data through consumption or use.
>
> *(Durodie 2003: 390)*

In a similar, more hyperbolic vein, the REACH Regulation has been described as 'economic suicide by a massive self-administered regulatory overdose' (Logomasini and Miller 2005: 13). In addition to such concerns based on the fear of over-regulation, doubts have been expressed about the effectiveness of informational regulation, based on the public's ability to engage fully with a highly technical area:

> Data produced in accordance with registration requirements and housed on ECHA's website are dense and technical and beyond the means of comprehension of the vast majority of EU citizens. At the same time, users of chemicals have been provided with Safety Data Sheets (which detail risks and risk mitigation measures) by chemical manufacturers and importers that are now, thanks to REACH, up to 1,000 pages long per substance and, as a consequence, often meaningless. The quality of the data produced to date has also been poor, a fact recognised by both ECHA and the Commission. Put simply, more information is not always better information.
>
> *(Stokes and Vaughan 2013: 427)*

One possible counter to such arguments is the existence of environmental pressure groups and other NGOs, who might possess the necessary expertise and therefore perform a watch-dog role capable of pressurising the chemicals sector (Case 2005: 420–423). Even eight to nine years after the enactment of REACH, the slow transmission rate of market information means it is still too early to discern the efficacy of its registration and information requirements.

The disposal of hazardous waste

In June 1987 – following the discovery, in Africa and other parts of the developing world, of illicit dumps of imported hazardous waste – the United Nations Environment Programme (UNEP) established a working group tasked with elaborating a global convention on the control of transboundary movements of such wastes. But whilst there was a growing recognition of the threat posed by hazardous waste to both human health and the environment, and of the difficulties developing countries have in managing waste (Shearer 1993), the application of more exacting environmental standards in the developed world had meant that 'dumping in the less regulated (or unregulated) developing world [had become] a cheaper and more commercially attractive option' (Morrow 2010: 219). Unsurprisingly, then, the debates on the new convention were 'politicized, arduous and emotionally charged' (Kummer 1998: 227). The resultant Basel Convention on the Control of Transboundary Movements of Hazardous Wastes and their

Disposal was adopted on 22 March 1989, and entered into force on 5 May 1992. To date, it has been ratified by 183 countries (Basel Convention 2015).

In keeping with its stated aim of reducing the transboundary movement of hazardous waste, the Basel Convention imposes a general, non-legally enforceable,[5] obligation on States to reduce the generation of hazardous waste to a minimum (Article 4(2)(a)) and to ensure that adequate disposal facilities are available within their jurisdiction for the environmentally sound management (ESM) of hazardous waste (Article 4(2)(b)). The Convention also prohibits the export of hazardous waste to Antarctica, to states that are not party to the Basel Convention, and to states that have banned the import of hazardous waste (Article 4). But the Convention stops short of imposing a total ban on exporting such waste. First, Article 4(9)(a) permits the transboundary movement of hazardous waste if the exporting state lacks 'the technical capacity and the necessary facilities, capacity or suitable disposal sites in order to dispose of the wastes in question in an environmentally sound and efficient manner'. Waste may therefore be moved to another state in which its disposal will be managed in an environmentally sound manner. Second, Article 4(9)(b) allows wastes to be exported if they 'are required as a raw material for recycling or recovery industries in the State of import' (Article 4(9)(b)). This recognises the fact that waste has a 'dual character' (Morrow 2010: 228), as both pollutant and tradable resource.

The regulatory framework for exporting hazardous waste is based on the principle of Prior Informed Consent (PIC). Article 6(1) requires the authorities of the exporting state to notify the authorities of the prospective states of import and transit of the proposed transboundary movement. The movement may only proceed once all states concerned have given their written consent, including confirmation of the existence of a contract between the exporter and the disposer specifying environmentally sound management of the waste in question (Article 6(3)(b)). Importantly, Article 4(10) stipulates that the state that generated the hazardous waste cannot transfer its obligation to manage the waste in an environmentally sound manner to a transit or import state. This notion of cradle-to-grave liability underpins the duty (found in Article 8) to re-import wastes when their movement cannot be completed in accordance with the terms of the contract, and is also significant in cases where it is difficult to determine fault for an improper disposal. The principal weakness of the PIC mechanism, however, is that it relies on self-verification by states and so requires good faith from all concerned. This has a number of flaws. First, the exporting country may fail to verify that the facility accepting the waste in the importing country can manage it in an environmentally sound manner. This is compounded by the fact that the technical guidelines generated under the Convention are not mandatory, and so 'a country attempting to self-verify a facility cannot assume that the destination country is observing the Basel ESM guidelines for a specific waste' (Gutierrez 2014: 407). Second, the exception for wastes which are required for recycling or recovery in the importing state may be exploited, creating a 'recycling loophole' (ibid.). Data suggests that since the Convention entered into force a significant proportion of waste that was destined for final disposal has headed instead for recycling or further use (Gutierrez 2014). It has even been suggested that 'exporters often misrepresent the nature of the wastes, misleading importing nations into consenting' (Onzivu 2013: 636). And, third, there is the possibility of corruption in the importing country:

> Another weakness in the PIC procedure is its omission to account for the susceptibility of country consents to be obtained from corrupt local officials. It also ignores the economic motivation of poor countries to accept these types of wastes for either the value or money that the waste can contribute to the local economy.
>
> (Gutierrez 2014: 407)

In cases involving illegal traffic in hazardous waste, the Convention envisages a core role for criminal liability (Article 4(3)) alongside liability under civil law (Article 4(4)). But whilst a protocol on liability was adopted in 1999, this is not yet in force as it lacks the required number of ratifications. Since there is no liability regime under the Convention, parties must rely on domestic law for redress. Yet in many (developing) countries there is weak multi-sectoral co-ordination of waste management and monitoring of waste law (Onzivu 2013). The Basel Convention does have some financial mechanisms, including the Basel Convention Trust Fund to Assist Developing Countries and other Countries in Need of Technical Assistance (the BD fund). But this too is problematic, since it is based on voluntary contributions by signatories who tend to accrue significant arrears (Morrow 2010). Further mechanisms for promoting implementation and compliance include: regional training centres (BCRCs), which provide technical assistance for parties to safely manage hazardous and other wastes (Article 14(1)); a requirement to submit an annual report to the Conference of the Parties (Article 13(3)); and, a duty to participate in hazardous waste audits conducted by the Secretariat (Article 10(2)(b)). The effectiveness of these has also been questioned, with BCRCs lacking sufficient funding and capacity and states failing to submit national reports to the Secretariat (Onzivu 2013).

During the negotiations on the Basel Convention, African countries, supported by other developing countries and environmental NGOs, had pushed for a total worldwide ban on transboundary movements of hazardous wastes, claiming that this was necessary to protect poorer regions from becoming the dumping grounds of the wealthy North (Kummer 1998). When the Basel Convention was formally adopted the Organization of African Unity (OAU) expressed its disappointment at the lack of a total ban, stating that the restrictions imposed by the Convention 'could be circumvented because of the lack of competent administrators and administrative agencies' (Shearer 1993: 151). As a result, the OAU subsequently adopted the Bamako Convention, which creates a total ban on the importation of all hazardous wastes into Africa and limits the transfer of such wastes within Africa – thereby making the 'difficult choice for most African countries placed between the double negative alternatives of "poverty or poison"' (Eze 2007: 229). Importantly, the Bamako ban includes recycling and reclamation activities. This limits African industry to the use of traditional raw materials in production methods and, it has been suggested, underestimates the growing importance of waste as a valuable resource which can create green business opportunities and jobs (Kummer Peiry 2013). In fact, it has been argued that the 'ambitious provisions' of the Bamako Convention are 'so stringent' that they 'threaten its enforceability' (Shearer 1993: 174–5). Pointing to 'The complete absence of reported activities by the Bamako Secretariat and the Conference of the Parties' and the *Probo Koala* incident in the Ivory Coast in 2006, Eze concludes that 'the lofty ideals of the Convention might be lost to the impossibility of compliance with its provisions' (2007: 228–9). On the other hand, the Bamako Convention has an important symbolic effect, sending the message that African nations are not dumping grounds for hazardous wastes generated in other countries. It has also acted as a catalyst for other, similar, regional agreements, as well as the Basel Ban Amendment. The latter measure provides for a ban on the transfer of hazardous wastes for final disposal from Organisation for Economic Cooperation and Development (OECD) to non-OECD countries. Although it has not yet been ratified by enough countries to bring it into force, the Ban Amendment has been ratified by, and so has legal effect in, the EU.

Commentators have warned of the dangers of treaty congestion (Morrow 2010; Onzivu 2013). Treaty congestion further complicates this already complex area. For a start, different treaties may have different ambits: for example, the Bamako Convention encompasses sea disposal of hazardous wastes, whereas the Basel Convention expressly excludes this from its scope (Article 4(12)). The two Conventions also define hazardous waste differently (Eze 2007: 216–7). Second, even if there is an attempt to delineate the scope of different treaties, the application of this in individual cases may be contested. In the *Probo Koala* incident, for example, one of the key issues was whether the sludge created by the floating refinery constituted hazardous waste or whether it was in fact slops, which are expressly excluded from the Basel Convention (Article 1(4)) and instead fall under the International Convention for the Prevention of Pollution from Ships 1973 (the MARPOL Convention) (Morrow 2010). Similarly, there has been debate as to whether end-of-life ships constitute hazardous waste, so that the Basel Convention applies to the shipbreaking industry (Karim 2010). Third, the previous two points are compounded by the fact that the definition of waste depends on the view of its generator. This subjectivity allows 'the waste disposer effectively to opt for obeying the regime that imposes the least demanding standards in the context of the disposal of hazardous waste in the developing world' (Morrow 2010: 229). Moreover, the Convention's definition of ESM as 'taking all practicable steps to ensure that hazardous wastes or other wastes are managed in a manner which will protect human health and the environment against the adverse effects which may result from such wastes' (Article 2(8)) is vague and open to different interpretations depending on one's interests (Gutierrez 2014). Such complexity and lack of clarity generates uncertainty.

Public and expert voices in regulatory design in practice

No legal or regulatory system can operate without significant discretionary power. As Bradley and Ewing observe, 'If it is contrary to the rule of law that discretionary authority should be given to government departments or public officers, then the rule of law applies to no modern constitution' (2003: 94). Given the inevitability of discretion in every legal system, proponents of what has been dubbed the 'extravagant version of the rule of law' (Davis 1971: 28–33) seek to eliminate as much discretion as possible. Beyond this they urge the need to 'bring such discretion as is reluctantly determined to be necessary within the "legal umbrella" by regulating it by means of general rules and standards and by subjecting its exercise to legal scrutiny' (Lacey 1992: 372). One of the difficulties with this, however, is that it mistakenly assumes that there is a neat dichotomy between rules and discretion. In fact, the distinction between the two is far more uncertain (Galligan 1986; Hawkins 1992). As the examples examined in this chapter illustrate, discretion is heavily implicated in the interpretation of regulatory norms and the application of them to technological processes and practices. Assessments under the REACH regime of whether the risks posed by a Substance of Very High Concern have been adequately controlled, whether there is a socio-economic need for the substance's continued use and whether a viable alternative exists all involve discretionary judgement, as do decisions as to whether a state lacks the technical capacity and necessary facilities to dispose of hazardous waste in an environmentally sound and efficient manner and decisions as to whether a proposed research project involving embryos is necessary or desirable for the purpose of promoting advances in the treatment of infertility, increase knowledge about the causes of congenital disease or miscarriages, or develop more effective contraceptive techniques (1990 Act, Schedule 2, section 3).

It is also apparent that, within each of the technological contexts examined in this chapter, certain viewpoints are prioritised in the exercise of this discretionary judgement. Perhaps most starkly, under the Basel Convention the view of the waste generator is relevant to the application of the definition of hazardous waste. Hence in the *Probo Koala* incident the company (Trafigura) that chartered the floating refinery was able to contend that the 528 tons of sludge deposited around the city of Abidjan was in fact slops (governed by the MARPOL Convention), not hazardous waste (governed by the Basel Convention). And whilst a stated objective of both the HFEA and REACH frameworks is to promote public participation, the extent to which this objective is realised is open to question. As noted above, it has been suggested that the HFEA model results in bad politics. Whilst membership of the Authority includes a range of disciplinary perspectives, and there is a cap on the number of members who are doctors and scientists involved in human embryo research or fertility treatment, it is also the case that previous reviews have emphasised the important role the Authority plays in facilitating further innovation and development and that three of the seven current lay members are women who have previously received IVF treatment (Human Fertilisation and Embryology Authority 2015). Indeed, the McCracken report stated that the HFEA needs to do more to properly take into account stakeholder views (McCracken 2013). Meanwhile, REACH's registration process is, amongst other things, intended to operate as a mechanism for providing information to end consumers of products. But as has been pointed out, not only is the information provided normally dense and highly technical, meaning it is incomprehensible to the vast majority of members of the public, it is also often based on poor quality data and can be hundreds of pages in length. In reality, then, the opportunity for end consumers to contribute, via the market, to decisions about acceptable levels of risk is limited.

As well as the role played by discretion and the prioritisation of certain viewpoints, a third recurring theme across the regimes for regulating technology examined here, as in many other contexts, is the importance of resource considerations. In spite of the high regard in which the HFEA is held, twice in less than a decade there have been proposals to merge it with the HTA for the sake of efficiency savings. Resource constraints also greatly hinder the operation of the Basel Convention, most obviously in the lack of third party verification of the PIC process. There is also a *de facto* lack of third party verification of the REACH registration process; with over 30,000 existing chemicals expected to be registered by 2018, the sheer magnitude of the regulatory task – and the level of resource that would be required to police this – means that the oversight of the ECHA is necessarily limited. Both regimes are thus, in effect, forms of 'regulated self-regulation' (Crawford 2003). This means, first, that the regulatory frameworks might be circumvented by those prepared to act in bad faith. As explained previously, the Basel Convention's PIC process has been criticised on the grounds that exporting states might deceive importing states and vice versa. It is also possible that exporting and importing states might act in collusion. This is exacerbated by the fact that the protocol on liability is not yet in force and by the difficulties in taking action under domestic law. And whilst REACH requires each member state to maintain an enforcement regime which provides 'effective, proportionate and dissuasive' penalties for non-compliance (Health and Safety Executive 2015), this will prove little disincentive to those who wish to act in bad faith and who, as a result of the limited degree of oversight, perceive the likelihood of being caught to be low (von Hirsch et al. 1999). As well as bad faith, an additional possibility is a formalistic or minimalistic attitude towards compliance with the regulatory requirements. The frequent resort to the Basel Convention's 'recycling loophole', for example, is at odds with the spirit of the Convention and the ESM principle. Similarly, during the REACH registration process manufacturers or importers of chemicals might submit a file that provides the required data but does so in a partial or selective manner. Meanwhile those who act in good faith and seek to comply fully with the regulatory requirements risk being rendered uncompetitive – both in comparison to those who do not comply fully and those who sit outside the regulatory regime in question. For example, concern has been expressed that the Bamako Convention will increase the cost of doing business in Africa and act as a disincentive to foreign investors (Eze 2007), whilst to comply fully with the burdensome requirements of the REACH regime may result in 'paralysis by analysis' (Jones 2007: 356) – not only leaving European businesses less competitive than those from other parts of the globe but also potentially reducing social utility by slowing product development.

Conclusion

This chapter began by outlining why public participation has been deemed a principle of good regulation (whether for technology or in general) and explaining the tension between this objective and considerations of resource and expertise. It then examined this tension in the context of three regulatory regimes specific to technological practices. The chapter has shown that in each of these regimes certain viewpoints are prioritised in the exercise of the discretionary decision-making that the regimes inevitably contain. What is striking about this is that, in each context, the viewpoint that is prioritised emanates from within the regulated industry: whether it is IVF doctors, researchers or patients; waste generators; or, chemical manufacturers and importers. Moreover, resource constraints mean that the latter two are effectively left to self-regulate, whilst public confidence in the HFEA is consistently attributed to its expert status. This not only raises questions about the extent to which these regulatory regimes do in fact exert effective control over the activities in question, but also whether the limited scope for public participation diminishes the legitimacy of these decision-making processes.

By way of conclusion, it is worth returning to the false dichotomy between rules and discretion noted earlier. As Hawkins observes, to suggest that 'discretion in the real world may be constrained only by legal rules' is to 'overlook the fact that it is also shaped by political, economic, social and organizational forces outside the legal structure' (Hawkins 1992: 38). At a time when regulators resort all too readily to the enactment of new laws as a response to societal challenges (Ashworth 2000), it is important to appreciate not only the limitations of legal regulation of technological processes and practices, but also to recognise – and seek to harness – the potential role of these extra-legal constraints.

Notes

1   In January 2013 the HFEA reduced the size of its board from 19 members to 12. It reports that this 'smaller board size is now widely recognised to be more effective' (Human Fertilisation and Embryology Authority 2014: 16). It has also gradually reduced its staff complement, from 86 in 2010/11 to 64 by the end of the 2013/14 financial year (ibid.).
2   Although as Blyth (2012) notes the fact that this 'became a vehicle for exercising some measure of regulation over surrogacy was largely serendipitous' (p. 309).
3   Some of the difficulties are illustrated by *Re T (a child) (surrogacy: residence order)* [2011] EWHC 33 (Fam). See further Alghrani (2012), Vijay (2014).
4   One possible example is the express prohibition of using reproductive technologies to ensure the birth of a child with a particular disability (2008 Act, s 14(4)). For an argument for a more flexible approach to this issue, see Taylor (2010).
5   It is important to note that the reduction obligation is expressed in such qualified terms (and with no specific target) that it would make it extremely difficult to factually establish a breach; as such, it seems that the obligation to reduce the generation of hazardous waste represents a symbolic provision rather than a legally enforceable obligation.

References

Aarhus Convention (1998) The Convention on Access to Information, Public Participation in Decision-making and Access to Justice in Environmental Matters (Aarhus, 25 June 1998).
Alghrani, A. (2012) Surrogacy: A cautionary tale, *Medical Law Review*, 20(4): 631–641.

Ashworth, A. (2000) Is the criminal law a lost cause? *Law Quarterly Review*, 116: 225–256.

Baldwin, R., Cave, M. and Lodge, M. (2012) *Understanding Regulation*, 2nd edn. Oxford: Oxford University Press.

Basel Convention (2015) Parties to the Basel Convention on the Control of Transboundary Movements of Hazardous Wastes and their Disposal. [Online]. [Accessed 6 July 2015]. Available from: www.basel.int/Countries/StatusofRatifications/PartiesSignatories/tabid/4499/Default.aspx#enote1.

Beck, U. (1992) *Risk Society, Towards a New Modernity*. London: Sage Publications.

Black, J. (2001) Decentring regulation: Understanding the role of regulation and self-regulation in a 'post-regulatory' world. *Current Legal Problems*, 54: 103–147.

Blyth, E. (2012) Access to genetic and birth origins information for people conceived following third party assisted conception in the United Kingdom. *International Journal of Children's Rights*, 20(2): 300–318.

Bradley, A. and Ewing, K. (2003) *Constitutional and Administrative Law*, 13th edn. Harlow: Longman.

Bryman, A. (2004) Triangulation, in Lewis-Beck, M., Bryman, A. and Futing Liao, T., *Encyclopaedia of Social Science Research Methods*. Thousand Oaks: Sage Publications.

Case, D. (2005) Corporate environmental reporting as informational regulation: A law and economics perspective. *University of Colorado Law Review*, 76: 379–438.

Crawford, A. (2003) 'Contractual Governance' of deviant behaviour. *Journal of Law and Society*, 30(4): 479–505.

Davis, K. C. (1971) *Discretionary Justice: A Preliminary Inquiry*. Urbana: University of Illinois Press.

Department of Health (2013a) *Government Response to the Consultation on Proposals to Transfer Functions from the Human Fertilisation and Embryology Authority and the Human Tissue Authority*. London: The Stationery Office.

Department of Health (2013b) *Response to the Review of the HFEA and HTA*. London: The Stationery Office.

Department of Health (2010) *Liberating the NHS: Report of the Arm's Length Bodies Review*. London: The Stationery Office.

Department of Health (2006) *Review of the Human Fertilisation and Embryology Act*. Cm 6989. London: The Stationery Office.

Department of Health & Social Security (1984) *Report of the Committee of Inquiry into Human Fertilisation and Embryology*. Cmnd 9314. London: Her Majesty's Stationery Office.

Durodie, B. (2003) The true cost of precautionary chemicals regulation. *Risk Analysis*, 23(2): 389–398.

Eden, S. (1996) Public participation in environmental policy: Considering scientific, counter-scientific and non-scientific contributions. *Public Understanding of Science*, 5: 183–204.

European Chemicals Agency (2015) Why Are Chemicals Important? [online] [accessed 28 May 2015]. Available from: http://echa.europa.eu/web/guest/chemicals-in-our-life/why-are-chemicals-important.

Eze, C. N. (2007) The Bamako Convention on the ban of the important into Africa and the control of the transboundary movement and management of hazardous wastes within Africa: A milestone in environmental protection? *African Journal of International and Comparative Law*, 15(2): 208–29.

Fisher, E. (2008) The 'perfect storm' of REACH: Charting regulatory controversy in the age of information, sustainable development and globalization. *Journal of Risk Research*, 11(4): 541–563.

Fleurke, F. and Somesen, H. Precautionary regulation of chemical risk: How REACH confronts the regulatory challenges of scale, uncertainty, complexity and innovation. *Common Market Law Review*, 48: 357–393.

Friedman, D. (2001) Does technology require new law. *Harvard Journal of Law and Public Policy*, 25: 71–85.

Funtowicz, S., Shepherd, I., Wilkinson, D. and Ravetz, D. (2000) Science and governance in the European Union: A contribution to the debate. *Science and Public Policy*, 27(5): 327–336.

Galligan, D. (1986) *Discretionary Powers*. Oxford: Clarendon Press.

Giddens, A. (1999) Risk and responsibility. *Modern Law Review*, 62(1): 1–10.

Gutierrez, R. (2014) International environmental justice on hold: Revisiting the Basel Ban from a Philippine perspective. *Duke Environmental Law & Policy Forum*, 24(2): 399–426.

Hawkins, K. (1992) The use of legal discretion: Perspectives from law and social science. In: Hawkins, K. (ed.) *The Uses of Discretion*. Oxford: Clarendon Press, 11–46.

Health and Safety Executive (2015) The UK enforcement regime for REACH. [Online]. [Accessed 10 July 2015]. Available from: www.hse.gov.uk/reach/regime.htm.

Heyvaert, V. (2007) No data, no market. The future of EU chemicals control under the REACH regulation. *Environmental Law Review*, 9(3): 201–206.

Horsey, K. and Sheldon, S. (2012) Still hazy after all these years: The law regulating surrogacy. *Medical Law Review*, 20(1): 67–89.

Human Fertilisation and Embryology Authority (2015) *Members of the Authority*. [Online]. [Accessed 10 July 2015]. Available from: www.hfea.gov.uk/Authority-members.html.

Human Fertilisation and Embryology Authority (2014) *Annual Report and Accounts 2013/14*. HC 369. London: The Stationery Office.

Joint Committee on the Human Tissue and Embryos (Draft) Bill (2007) Human Tissue and Embryos (Draft) Bill. Session 2006–07, HC Paper 630-I. London: The Stationery Office.

Jones, J. 2007. Regulatory design for scientific uncertainty: Acknowledging the diversity of approaches in environmental regulation and public administration. *Journal of Environmental Law*, 19(3): 347.

Karim, M.D.S. (2010) Environmental pollution from the shipbreaking industry: International law and national legal response. *The Georgetown International Environmental Law Review*, 22(2): 185–240.

Kummer, K. (1998) The Basel Convention: Ten years on. *Review of European, Comparative and International Environmental Law*, 7(3): 227–36.

Kummer Peiry, K. (2013) The Basel Convention on the control of transboundary movements of hazardous wastes and their disposal. *American Society of International Law Proceedings*, 107: 434–6.

Lacey, N. (1992) The Jurisprudence of discretion: Escaping the legal paradigm. In: Hawkins, K. (ed.) *The Uses of Discretion*. Oxford: Clarendon Press, pp. 361–416.

Lee, M. and Abbot, C. (2003) The usual suspects? Public participation under the Aarhus Convention. *Modern Law Review*, 66: 80–108.

Logomasini, A. and Miller, H. (2005) REACH exceeding its grasp? *Regulation*, Fall, 10–13.

Macrory, R. (2001) Regulating in a risky environment. *Current Legal Problems*, 54: 619–648.

Mandelkern Group on Better Regulation (2001) Brussels, final report [online] [accessed 1 July 2015]. Available from: http://ec.europa.eu/smartregulation/better_regulation/documents/mandelkern_report.pdf.

McCandless, J. and Sheldon, S. (2010) The Human Fertilisation and Embryology Act (2008) and the tenacity of the sexual family form. *Modern Law Review*, 73(2): 175–207.

McCracken, J. (2013) *Review of the Human Fertilisation & Embryology Authority and the Human Tissue Authority*. London: The Stationery Office.

McGarity, T. O. (1990) Public participation in risk regulation. *Risk – Issues in Health and Safety*, 1: 103–130.

Morrow, K. (2010) The Trafigura Litigation and liability for unlawful trade in hazardous waste: Time for a rethink? *Environmental Liability*, 18(6): 219–30.

Onzivu, W. (2013) (Re)invigorating the Health Protection Objective of the Basel Convention on Transboundary Movement of Hazardous Wastes and Their Disposal. *Legal Studies*, 33(4): 621–49.

Regulating Better: A Government White Paper setting out six principles of Better Regulation (2004) Dublin, Department of the Taoiseach [online] [accessed 30 June 2015]. Available from: www.taoiseach.ie/eng/Publications/Publications_Archive/Publications_2011/Regulating_Better_Government_White_Paper.pdf.

Regulatory Performance Indicators (1999) Canberra. Australian government, Department of Industry, Tourism and Resources, [online] [accessed 1 July 2015]. Available from: www.brad.ac.uk/irq/documents/archive/Regulatory_Performance_Indicators_Australian_guide_Department_of_Industry.pdf.

Schechter, J. (2010) Promoting human embryonic stem cell research: A comparison of policies in the United States and the United Kingdom and factors encouraging advancement. *Texas International Law Journal*, 45(3): 603–629.

Secretary of State for Health (2007) *Government Response to the Report from the Joint Committee on the Human Tissue and Embryos (Draft) Bill*. Cm 7209. London: The Stationery Office.

Shearer, C. R. H. (1993) Comparative analysis of the Basel and Bamako Conventions on hazardous waste. *Environmental Law*, 23(1): 141–83.

Stokes, E. (2012) Nanotechnology and the products of inherited regulation. *Journal of Law and Society*, 39(1): 93–112.

Stokes, E. and Vaughan, S. (2013) Great Expectations: 50 years of chemicals legislation in the EU. *Journal of Environmental Law*, 25(3): 411–435.

Taylor, E. M. (2010) Procreative liberty and selecting for disability: Section 14(4) Human Fertilisation and Embryology Act 2008. *Kings Student Law Review*, 2(1): 71–86.

Thaler, R. and Sunstein, C. (2009) *Nudge: Improving Decisions about Health, Wealth and Happiness*. London: Penguin Books.

Vaughan, S. (2015) *EU Chemicals Regulation: New Governance, Hybridity and REACH*. London: Edward Elgar.

Vijay, M. (2014) Commercial surrogacy arrangements: The unresolved dilemmas. UCL *Journal of Law and Jurisprudence*, 20: 200–236.

von Hirsch, A., Bottoms, A., Burney, E. and Wikström, P. (1999) *Criminal Deterrence and Sentence Severity: An Analysis of Recent Research*. Oxford: Hart.

Wagner, W. (2004) Commons ignorance: The failure of environmental law to produce needed information on health and the environment. *Duke Law Journal*, 53: 1619–736.

Zeegers, N. (2014) Devolved regulation as a two-stage rocket to public acceptance of experiments with human gametes and embryos: A UK example to be followed by the Netherlands? *European Journal of Comparative Law and Governance*, 1(1): 10–28.

Keynote discussion

Technology and the process of justice

35

The force of law and the force of technology[1]

Mireille Hildebrandt

Introduction

Since the beginnings of modernity in 16th–17th-century Europe, written law has become the most important instrument for the regulation of human society. Initially, modern law established itself as a 'rule *by* law' – overcoming a rule *by* economic power or military force. As I will argue below, it was the articulation of modern law in the technologies of the script and the printing press that, in the end, also paved the way for 'the Rule *of* Law'.[2] This Rule of Law embodies the so-called 'paradox of the *Rechtsstaat*': protecting citizens *against* the authority of the state *by means of* the authority of that same state. This required the institution of an internal division of sovereignty into legislation, administration and adjudication. The ensuing system of checks and balances has provided us with reasonably effective remedies for resisting governmental action in a court of law, to counter the massive powers of the state's *ius puniendi*. Though we may be critical of the extent to which current articulations of the Rule of Law actually empower individual citizens suspected of committing a criminal offence, we should realize that such critique assumes the framework it critiques. My point throughout this chapter will be that we can no longer take the Rule of Law for granted as an affordance of our information and communication technology (ICT) infrastructure, due to the rapid and radical integration of algorithmic decision-systems and other types of data-driven intelligence into the administration of justice.

A *Handbook on Technology, Crime and Justice* should pay close attention to the consequences of transformative and disruptive changes of the ICT infrastructure of our shared life world. If the technological embodiment of modern law and its offspring, the Rule of Law, is changed, the law itself will change – potentially beyond recognition. In other work I have described how the upcoming ICT infrastructure of data-driven intelligence increasingly pervades both our private and public life (Hildebrandt 2015), arguing that law enforcement and police intelligence are moving towards technological pre-emption (Hildebrandt 2016b), explaining how this may trigger various types of pre-crime punishment (Hildebrandt 2010). This chapter engages with a previous article (Hildebrandt 2008a), written on the cusp of the philosophy of law and technology, aiming to highlight the interactions between legal and technological normativity. The aim in that article was to develop a generic concept of normativity capable of expressing the impact of both technology and law on the direction of human interaction. This entails describing the way both technologies and legal norms induce, enforce, inhibit or rule out certain types of behaviour. The analysis of such a generic understanding of normativity was

Mireille Hildebrandt

refined by distinguishing between constitutive and regulative normativity, and between an imperative and a normative dimension of legal norms. In this chapter I will add the notion of normative force, raising the question of how the force of law depends on its current technological embodiment and what normative force will emanate from data-driven intelligence.

In the next section I assess the way lawyers regard the regulation of society, after which I will develop a generic concept of normativity, resulting in a comparison between legal and technological normativity.

The regulation of society: a lawyer's perspective

What in fact is law?

The German legal historian Uwe Wesel (1985: 52) made a salient point when he wrote: 'What in fact is law? Answering this question is as simple as nailing a pudding to the wall'. Despite the accuracy of this proposition we need to clarify the role of law in the regulation of society,[3] if only because democracy seems to depend on the enactment of legal rules by a democratic legislator that is bound by its own enactments.

Common sense understanding of legal norms – even amongst many lawyers – seems to circulate somewhere between the positivist accounts of three scholars of legal theory: 'the command theory of law' attributed to John Austin (1995, first published 1832), 'the pure theory of law' of Kelsen (2005, first published 1934), and 'the concept of law' of Hart (1994, first published 1961). Austin, writing in the first half of the nineteenth century, basically claimed that laws are commands of a sovereign, emphasising the relationship between law and the sovereign state. In opposition with natural law theories, law in his view is human-made and depends on the power of the sovereign to impose general rules on his subjects. Like Austin, Kelsen made a strict distinction between the 'is' and the 'ought' of the law. Writing in the first half of the twentieth century, he described the 'is' of the law as a set of rules that form a pyramid of hierarchically ordered normative rules, which in the end all derive from one *Grundnorm*. This 'Basic Norm' guarantees the unity of the legal system and the validity of all the legal rules that should be seen as derived from it. Like Austin's command theory, law always depends on the authority of the state, but, according to Kelsen, this authority also depends on the law: the state is a legal construction. Building on Austin's opposition with natural law, Kelsen claims that the analysis of what law 'is' must be distinguished from what law 'ought' to be (*das richtige Recht*). To allow a moral evaluation of the law, according to Kelsen, one must first describe its normative content, taking into account the deductive logic that determines the connections between different legal norms. A similar positivist position was taken by Hart, writing in the second half of the twentieth century, even though the nature of 'his' law is defined in terms of social interaction instead of 'purely' normative statements. While Kelsen's *Grundnorm* can be understood as a hypothetical rule that ensures the unity of the system and the validity of its elements, Hart's 'Ultimate Rule of Recognition' is firmly rooted in social acceptance or what he calls the internal aspect of legal rules. Hart discriminates between primary legal rules that define which conduct is prescribed, prohibited or allowed, and secondary legal rules that define the competence to recognize, change or adjudicate primary legal rules. The distinction between primary and secondary legal rules has developed into a canonical approach within law and legal theory.

In short, Austin linked law to the power of the sovereign to impose general rules on his subjects, Kelsen elaborated the systematic character of the body of legal rules and their clear distinction from moral rules and political power, and Hart understood law as a complex system of social norms, coining the difference between regulative and constitutive rules in terms of primary and secondary rules. Roughly speaking, legal positivism seems to emphasize (1) that legal norms are general rules, (2) that they depend on the authority of the state and (3) that they must be strictly separated from moral and political rules.

However, many scholars of law and legal theory have objected to these tenets of legal positivism, which has led to further refinements and alternative positions. Most famous is Dworkin's objection that it makes no sense to understand law as a system of rules, claiming that the interpretation of legal rules implies the guidance of principles, which do not share the binary application of rules. Also, according to Dworkin (1991, first published 1986) the coherence that is inherent in law implies more than just logical consistency, requiring what he calls the integrity of law. With the notion of integrity Dworkin introduces moral standards into the law – even if these are inductively generated from previous legal decisions (enacted law, court judgements). Instead of thinking in terms of a legal system that is focused on logical coherence, he uses the metaphor of a chain novel to indicate the continuing story of law-making. His approach to law can be understood as hermeneutical, stressing the fact that any decision implies interpretation and requires both creativity and precision.

The emphasis on interpretation in contemporary legal discourse is not surprising. Modern law centres around text and printed matter (Hildebrandt 2008b, 2015; Hildebrandt and Koops 2010). As such, law has been a prime example of the cybernetic affordances of the 'technologies of the word', as Walter Ong (1982) famously called the script and the printing press. The success of written legal rules as a means to steer and control a population can be attributed in part to their alliance with the technologies of the written script and the printing press, which extended the reach of law both in time and space, allowing an ever more detailed regulation of human intercourse. These technologies were preconditions of modern law in the sense that they 'afforded' or 'made possible' the rise of positive, deliberately enacted and imposed legal code, to be authoritatively applied by courts of law. The proliferation especially of legal text that resulted from the uptake of the printing press has vastly extended the distantiation that written text generates between author (the legislator, the courts), reader (legal subjects) and text (legislation and case law). Such distance evokes the need for interpretation, since the author can no longer control the meaning of her text. The author is absent, speaking from another place and possibly from another time. As Ricoeur (1973) noted, this is what liberates and emancipates content from the 'meaning' or 'intentions' of its author. This liberation, however, also triggers a permanent and pertinent need for interpretation, precisely because the author can no longer stabilize the meaning of her text. Such a liberation thus also breeds ambiguity and uncertainty as to the meaning of written legal norms. This uncertainty has contributed to the rise of a professional class of lawyers capable of sustaining some form of legal certainty (Koschaker 1966, first published 1947) in the face of interpretive instability. The rise of such a class, in turn, led to the relative autonomy of law in relation to both politics and morality. Lawyers formed a buffer between ruler and ruled, legitimizing the imposition of legal rules by interpreting them in ways that protected against arbitrary application. As such, the script and the printing press form the preconditions for the modern legal systems that depend on them. For lawyers, the fact that law is constituted and mediated by the printed script may be too obvious to warrant further investigation, but the profession would benefit from the awareness that script and print is indeed a technology, with massive implications for the scope, the content and the nature of the jurisdictions it supports. This is not only the case because the invention of script and printed text extended the reach of legal rules beyond face-to-face relationships, forming the condition of possibility for translocal polities and jurisdictions, but also because the script introduces a linear sense of time due to the need to read from beginning to end (from left to right and top to bottom or in any other manner that enforces sequential processing of information), while the printing press evoked increasing rationalization and systematization in order to cope with the explosion of available texts (Lévy 1990; Eisenstein 2005). Sequential processing, rationalization and systematization should not be seen as universal characteristics of the human mind and modern law, but as processes triggered by specific cognitive technologies.

Another salient feature of written law is the pervasive sense of delay that derives from the complexity of the legal system that needs to mind its coherence in the face of ever increasing regulation and adjudication, thus nourishing reiterative doctrinal attempts to create order in the bran tub of newly enacted statutes and newly published case law. This delay is related to the distance between the author (the legislator, the court) and the public (those subject to the law), which is due to the fact that author and public do not necessarily share time and place, since the public may access the text later and elsewhere (Geisler 1985; Ricoeur 1973). In the absence of ostensive reference, the author is never sure how her text will be understood, while the reader cannot take for granted what the author meant to say. This provides for an inevitable latitude in the use of texts and turns law-making (enactment of legal code and its application by a court of law) into a creative process rather than mechanical application.[4] In the US, for instance, constitutional review has moved beyond the idea that the interpretation and application of constitutional safeguards can be based on the 'Framer's intention' or on the 'clear meaning' of the text; a text cannot speak for itself and its meaning is not exhausted by a claim regarding the author's intention. This, however, does not imply that its meaning fully depends on the reader's response, which would land us in arbitrary decisionism as to the application of legal texts. It rather means that in the interplay between author and subsequent readers the meaning of texts acquires a dynamic of its own that restricts potential interpretation (if just any interpretation were possible the text would be meaningless), while still providing room for novel applications (requiring creative actualization in the relevant context).[5] Written law thus generates a dynamic, autonomous law that *depends on* and *nourishes* legal doctrine to provide continuity as well as flexibility in its application. This combination of continuity and flexibility constitutes the condition of possibility for a law that aims to serve the antinomian goals of legal certainty, justice and effectiveness (Lask, Radbruch and Dabin 2014: 44–224).

Lawyers must urgently face the mutations that are re-shaping the environment of the law, taking note of the fact that these mutations will soon re-shape law's own articulation, as its alignment with the technologies of the word is on the verge of being disrupted in favour of a novel alignment with the technologies of machine learning and other types of computational intelligence. This wake-up call is not only relevant for lawyers. Criminologists should take heed of the disruptions that novel ICT infrastructures trigger in the political economy of crime and justice. These disruptions will require them, too, to reinvent their assumptions about normative as well as factual 'deviation' and especially about how undesirable deviation can be countered, addressed and redressed.

Constitutive and regulative legal norms

To fine-tune our understanding of law we should discriminate between legal rules that are constitutive of certain legal acts or legal facts, and legal rules that regulate acts or facts. If one violates a traffic rule that regulates driving a car, this does not mean that one cannot drive the car. The rule – e.g. forbidding speeding beyond 100 mph – is regulative of driving a car since it does not rule out my driving at 120 mph. However, if one 'violates' a rule that stipulates the registration of marriage in the civil registry, one will simply not be married. In that case the rule is constitutive of the marriage, because it stipulates *what counts as* a marriage, or, in other words, the rule determines that registration generates the legal effect of being married. The difference between constitutive and regulative rules derives from Searle (1969),[6] who discriminates between brute facts, which can be the object of regulation but cannot be constituted by human interaction, and institutional facts, which can only be constituted by social interaction. Searle defines brute facts as facts that can exist independently of human beings and their institutions, while his institutional facts depend on human institution. Inevitably this boils down to a physicalist worldview where an objective physical world is supplemented with an intersubjective 'social world' that is contingent upon human interaction. From that perspective, the bare fact of driving a car is a brute fact to the extent that it is not constituted by social interaction, while marriage is an institutional fact, as it cannot exist independent of social interaction. From such a point of view, at some level, all institutional facts are based on brute facts. With Colomb (2005) we could, however, argue that in a particular context C, brute fact X counts as institutional fact Y. For instance, the brute fact of driving a car counts as the institutional fact of 'being a road-user' in the sense of the Traffic Act (which attributes legal effect to this institutional fact) in the context of driving on a public road. A closer look thus discloses that the distinction is relative: depending on one's perspective, brute facts can be rearticulated as institutional facts,[7] while institutional facts can be 'used' as brute facts to be regulated. The institutional fact of being a road-user in the sense of the Traffic Code constitutes the addressee of regulation in the Traffic Code. This means that the distinction between constitutive and regulative has an analytical appeal, as long as it is not taken to imply a strict ontological difference between facts that exist outside human perception and facts that are socially constructed. Other than what Searle stipulates, I would say that even a brute fact involves some form of constitution, namely the one effected by our biological wiring (Varela, Thompson and Rosch 1991).[8] A bat perceives something else where we see a wall, even if the bat will avoid clashing into that same wall just as well as we do – hopefully (Nagel 1974). In point of fact, any type of constitution involves both humans and non-humans (Latour 1993); 'the social' refers to relationships between humans, taking into account that these relationships are forever mediated by our relationships with things.[9] For instance, the constitution of a marriage requires the mediation of technologies that register the marriage as such. In the case of a tradition that is mediated by (printed) script such registration takes place by means of written or printed records; in the case of an oral tradition 'registration' can, for instance, take place by means of wearing a spot on the forehead (*bindi*). Obviously the normative force of both types of registration differs as to scope (only accessible in face-to-face or beyond that), scale (limited to small local groupings or a larger collective), distribution (visible for just anyone or restricted to those with access to written files) and coerciveness (based on an oral tradition with voluntary jurisdiction or based on a distinctive bureaucratic institution linked with the powers of the state).

A generic concept of normativity

Norms as generative constraints on human behaviour

From an external perspective, norms can be equated with constraints that either (1) induce or (2) enforce certain types of behaviour while either (3) inhibiting or (4) ruling out other types of behaviour (Verbeek 2011). Note that constraints are generative because they generate behaviour; even if a constraint inhibits or rules out certain behaviour this will generate other behaviour, induced by the constraint. In point of fact constraints are the condition of possibility for behaviour – there is no behaviour if there are no constraints. This understanding of norms does not imply human agency (it does not address 'action') but it does imply some kind of agency (it does address 'behaviour').[10] In principle this definition of norms also applies to artificial agents, for instance focusing on how they are constrained by the architecture of their environment. Moving to human agency we must take into account that humans are capable of action, that is, behaviour which generates *meaning* based on mutual expectations that are mediated by spoken or written language (Ricoeur 1973). A norm that is meant to regulate human action can be deliberately issued *for*; explicitly recognized *by*; and/or tacitly developed *in* the practices of a certain community or collective. A *legal* norm can be issued by the legislator; to count as such it must be recognized by the courts, and hopefully it will develop into a standard for human action (if not already so). We should, however, take into account that the first two types of norms (issued by a legislator or recognized by a court) form an exclusionary reason for people to guide their actions (Raz 1975), as the duty to obey the law cannot not depend on individual consent. Nevertheless, the fact that people should obey the law, based on their submission to democratic legislation, assumes that they may in fact choose to disobey the law (Brownsword 2009). If this were not possible, the rules would be part of administration, economic power play or brute force, instead of law. The idea of disobeying the law assumes that we can objectify legal norms, which, in turn, depends on their externalization, notably in writing or print. In an oral legal tradition, norms seems to hide their normative force by operating, as it were, from 'under the skin'.

Legal normativity: three types of norms

To better understand what legal normativity is about we may want to distinguish different aspects pertaining to norms, notably a vertical and a horizontal aspect that relates to, respectively, the relationship between the government and its subjects and between these subjects (Glastra van Loon 1958). Deliberately enacted legal norms depend on a unilateral authority to legislate, which is connected with internal sovereignty and the concomitant authority to impose general legal norms. This dependency can be coined as the 'imperative' or 'vertical' aspect of legal norms; it presumes some form of political authority capable of enforcing a norm. However, in the end, the effectiveness of a norm will depend on the extent to which it becomes part of the normative practices of the relevant community or collective, that is, on the extent to which they inform the interaction patterns within such communities or collectives. Norms that inform a practice can be said to harbour a 'normative' or 'horizontal' aspect, which can be explicitly recognized or tacitly developed. The imperative aspect regards the vertical relationship between ruler and ruled, while the normative aspect concerns the horizontal relationships between those sharing jurisdiction. In point of fact, the normative aspect is to some extent instigated by the imperative aspect; by stipulating particular norms for all those sharing jurisdiction, the norms will become part of the legitimate expectations that people 'have' towards each other because they know these norms can be enforced against each individual subject. This connects with formal legal equality and the legal certainty it enables. The normative aspect comes close to Hart's 'internal aspect' of primary norms, and can be equated with Wittgenstein's idea of what it means to follow a rule (Taylor 1995). This implies that – as in the case of brute facts and institutional facts – the distinction between deliberately issued norms and norms that are part of a normative practice is *analytical and not ontological*. A norm can have (1) only an imperative aspect (newly issued legal norms that are not part of the relevant practice yet), it can have (2) only a normative aspect (norms that have no legal effect, because neither the legislator nor the courts recognize it as a legal norm), or it can have (3) both an imperative and a normative aspect (legal norms that have become part of the practice they aim to inform). So, we have three types of norms: legal norms that do not (yet) regulate or constitute the interactions in a particular practice, legal norms that do regulate and/or constitute the interactions in a particular practice and non-legal norms that do regulate and/or constitute the interactions in a particular practice. Obviously, in a modern legal system, *to count as legal* a norm must be issued or endorsed as such by the relevant state authority. Whether and to what extent it informs the normative practice of a community will depend on the normative force it has gathered between those subject to its jurisdiction. In a democratic society, any legal norm that is issued must aim to raise legitimate expectations between people, because self-rule means that the addressees and the addressors of the law are – ultimately – the same. Based on this analysis of legal normativity, we will now investigate technological normativity.

Technological normativity

If we leave the domain of law we may find other types of normativity, pertaining to constraints that have not been deliberately issued but which nevertheless induce or enforce, inhibit or rule out certain types of behaviour. Latour's discussion of the Berlin key is a case in point. This key forces the user to open the door by pushing the key through the keyhole to the other side of the door, while, after entering the house, the door can only be closed by turning the key and thus locking the door. This key demonstrates how a technological device actually regulates and constitutes the interactions of a resident, her door, the key and others who wish to enter the house (Latour 2000).[11] In this case the designer of the key has inscribed a program of action into the hardware, delegating the task of insisting on locking or not locking the door *to the key*. Such delegation seems to introduce both an imperative aspect (as it was deliberately imposed on the design of the key) and a normative aspect (as users of the key are forced to integrate it into the practice of closing their door against outsiders). In our time and age, a prime example of the normative force of a key would be the protection of confidential information and the implementation of authentication protocols by means of encryption techniques. In the first case the content of communication is hidden from all but those with access to the decryption key; in the second case access to information is blocked from all but those with the relevant decryption key. The use of these keys basically enforces specific norms about the *confidentiality of* or the *access to* information. The normative force of these norms has been successfully delegated to encryption keys and – as with Latour's Berlin key – basically rules out disobedience. In that sense it is again closer to a constitutive legal norm, even though it does not produce an institutional fact – as would be the case for a constitutive rule.

With both types of keys the normative force has been deliberately inscribed, *though not by a legislator*. For this reason, though the normative force is determinate rather than persuasive, the norm that is implemented by the key does not have an imperative dimension. The imperative aspect of legal norms is defined in terms of the authority that is capable of imposing the norm, and in point of fact this assumes that the norm – though imposed – can be disobeyed. I will return to this issue when comparing legal and technological normativity. Whereas the normative force generated by the keys is the result of a deliberate design, many of the normative effects of technological devices or infrastructures have not been deliberately planned (Bijker 1995). Contemporary common sense would describe them as side-effects, even in the case that these unplanned effects outweigh explicitly intended effects. To capture both the intended and the unintended effects of technological devices or infrastructures I employ a broad generic concept of technological normativity:

> the way a particular technological device or infrastructure actually constrains human actions, inviting or enforcing, inhibiting or prohibiting types of behaviour.

One could rephrase this by stating that technologies have specific affordances in relation to the subjects that use them. This refers to Gibson's (1986) salient understanding of the relationship between an organism and its environment, indicating that the availability of different courses of action depends both on the organism and on its environment. In discriminating between inviting/enforcing and inhibiting/ruling out certain behaviours I seek a further qualification of what affordances make possible and what they enforce. The concept of technological normativity, for this reason, does *not* depend on deliberate delegation since it may emerge unexpectedly in the interactions between devices, infrastructures and the humans who make use of them (and who are to a certain extent constituted by them). As indicated above, this concept of normativity should not be confused with morality,[12] as this would imply an evaluation in terms of good or bad, whereas normativity refers to the way the patterns of our interactions are affected. Only after assessing the normative effect can we evaluate its moral implications. As to the use of the term 'constraint', as indicated above, this should not be understood as a negative term: constraints are the condition of possibility of (inter)action, they do not only inhibit or rule out certain behaviour – they also induce or enforce certain types of behaviour. Moreover, they often constitute new types of behaviour. The constraints that constitute a crime-mapping technology, meant to detect hotspots (locations likely to harbour violence, drug trafficking or other types of offences), are productive: they generate new courses of action for the police, which is supposed to redistribute its resources based on such detection. The productive nature of the constraints inherent in technological devices and infrastructures clarifies why it makes sense to discuss technological normativity in terms of constitutive and regulative normativity instead of merely distinguishing between coercive and persuasive effects. Crime mapping, for instance, does not force police authorities to allocate resources to hotspots; the technology is not coercive but persuasive. What matters here, however, is the fact that crime mapping constitutes a visualization of spaces in terms of violent behaviour, nuisance or drug trafficking. This visualization, based on data that suggest a certain objectivity, is actionable as it makes certain courses of action possible in the first place – such as, in this case, re-allocating budget, people and equipment based on the data. As such the visualization is constitutive of police action and therefore also normative as it directs the course of action of the police. It does not enforce but it does induce specific police behaviours, thus both framing and enlarging the 'action potential' of the police force.

Conclusion: comparing legal and technological normativity

Let me summarize. In modern states legal norms depend on state authority, which means that they regulate and/or constitute the relationship between citizens and their government. This concerns the vertical or imperative dimension of legal norms, based on the coercive authority of the modern state. In a democracy legal norms also aim to regulate or constitute the relationships between those that share jurisdiction, which means that citizens should feel obliged towards each other to comply with legal rules and principles instead of merely feeling coerced by the state. This concerns the horizontal or normative dimension of legal norms. Democracy implies mechanisms to ensure that the addressor and the addressees of the norms are ultimately the same, taking into account that such mechanisms will entail more or less complex arrangements of representation, deliberation and participation to make this work.

Technological normativity, on the contrary, does not depend on state authority. The Berlin key's ability to enforce a specific practice of door locking does not depend on the state's ability to enforce compliance by means of police interventions or other threats. The Berlin key does, nevertheless, regulate and/or constitute the relationship between the human beings involved (the person living in the house, those outside, the caretaker), and between humans, devices and infrastructures. Does this mean – as suggested above – that technological normativity has a normative dimension while lacking an imperative dimension? Or, should we admit that the normativity of the Berlin key – on the contrary – has a rather impressive imperative as well as a normative aspect, because it enforces the norm while also regulating the relationships between its user and others?

Before answering this question, let's briefly inquire into the case of non-state societies, which lack a centralized authority that implies the absence of coercive *authority*. As in the case of technological normativity, this absence does not imply that coercion is not at play. In non-state societies there is no coercive authority capable of unilateral imposition of its will. Non-state societies are constituted by peers who cannot depend on governmental intervention in the case of conflict, which means that they will have to protect themselves against being overruled by more powerful peers. A level of vigilance is required that is not at stake in a society with a government capable of protecting its subjects against each other. Absent a monopoly on violence, the legal normativity of non-state societies therefore has to be sustained by means of *persuasive authority* (notably religious), and *economic* (for instance surplus livestock) or *military power* (as in revenge or war) (Dubber 2005; Hildebrandt 2014). This implies that in non-state societies the imperative aspect of 'legal' norms is aligned with priesthood, with the verdicts of a general assembly or, for instance, the patriarchal authority within the household. The normative aspect of 'legal' norms in societies without a state will be influenced by the complex amalgam of persuasive authority and economic or military power that structure the mutual expectations between the members of such a society. With Weber (2013) I would argue that the difference between *Herrschaft* (authority) and *Macht* (power) resides in whether a person or entity is capable of subjecting others to its will, invoking an obligation to obey its general orders (this would entail authority) or merely manages to impose its will due to physical or economic preponderance (which entails power). With Sahlins (1963) we can describe this as the difference between, on the one hand, a government that requires its subjects to submit a surplus (tax), based on the unilateral authority to impose obedience and, on the other hand, a leader capable of sharing part of his own surplus (personal wealth) to generate loyalty. In relation to criminology the distinction is of interest where gang leaders need to gain economic power in order to gather followers, whose commitment and loyalty is 'bought' with gifts (drugs, shelter, money, protection). Similarly, corruption is based on the same structure of enforcement: gifts oblige those who receive them.[13]

As with legal normativity in non-state societies, technological normativity does not depend on coercive authority, but neither does it depend on the persuasive authority or the persuasive and coercive powers described above. Technological normativity depends on the affordances of artefacts or infrastructures, and on the way they are taken up in human society. Though this implies that technological normativity may be engaged to reinforce coercive or persuasive authority, or economic or military power, this should not confuse the normative force of technology with that of a government, big business, the police or the military. In fact, to some extent, the normative force of government, businesses, police or militants will depend on the normative force of the technologies they use, just as with the law. In sum, technological normativity can be (1) either regulative or constitutive of human interaction, it can be (2) either coercive or persuasive, but (3) it lacks the imperative aspect inherent in modern positive law. Though its normative aspect (the way it orients human interaction) may be instrumentalized by human or institutional actors it is (4) grounded in the material constraints that create specific affordances for human intercourse. Semi-finally, (5) its normative aspect does not depend on deliberate inscription, making it hard to discern and contest the normative constraints insofar as we are not used to detecting unintended normative implications in the technologies of our own making. This also goes for the normative implications that have been deliberately inscribed without consulting or telling the user, who may follow the script without ever noticing this was meant to be – for a particular reason. This brings me to the last point to be made: (6) technological normativity is capable of reconfiguring the normative force of legal norms, for instance by turning them into paper dragons, automating and enforcing their implementation or by eroding the substance they mean to protect.

The final point should be a wake-up call for both lawyers and others who study crime and justice. It entails that a technology-neutral law may require re-articulation of legal norms as compensation for the interferences of new technological infrastructures. As argued elsewhere, this basically means that technology-neutral law sometimes needs technology-specific laws to sustain its tenets (Hildebrandt and Tielemans 2013). It also means that if the information and communication (ICT) infrastructure that embodies modern law and the Rule of Law (that of the script and the printing press) is transformed into or complemented with a novel ICT infrastructure (that of data-driven architectures) lawyers will have to learn to articulate legal norms and the Rule of Law in that novel architecture, on pain of becoming redundant (Hildebrandt 2016a).

Notes

1 This chapter is a revised version of my Legal and Technological Normativity. More (and less) than twin sisters, in *Techné: Research in Philosophy and Technology*, 12:3, Fall 2008.
2 On continental and Anglo-American understandings of the Rule of Law, see Silkenat, Hickey Jr and Barenboim (2014).
3 The term regulation is used in a common sense way; it is not meant to refer specifically to legal regulation, but can also refer to biological or technological regulation, e.g. the regulation of activities within the cell by genes, proteins or the regulation of safe driving by means of speed bumps or smart cars.
4 About the difference between creative actualization and mechanical realization, cp. Lévy (1998).
5 Though constitutional review in the US has actually and inevitably moved beyond a naïve understanding of the 'Framers' intention' and 'plain meaning', adherents to legal positivism will deny this. A plethora of relevant literature could be quoted here. Cf. e.g. Dworkin (1997).
6 Cf. Austin (1975).
7 Even the existence of a stone is an institutional fact to the extent that what counts as a stone depends on the language system that determines the meaning of the term stone, while in the end the bare existence of a stone still depends on it being perceivable as such by us. Whatever a stone 'is', if it cannot be perceived (Kant's *Ding an sich*), we cannot decide or determine.
8 See previous note.
9 On the role of mediation in human perception and action see the pivotal work of Ihde (1990), further extended by Verbeek (2005).
10 On the concept of agency in the era of machine learning see part I in Hildebrandt (2015).
11 The story of the Berlin key is actually more complicated, as it also involves the caretaker of the house, who has another key that is crucial for the plot. For the point that I am making a detailed account of how the key enforces specific behaviors is not relevant, but see Latour 2000.
12 Though normative effects may have moral implications, see Verbeek (2011) who, however, does not distinguish normative from moral effects, which I believe is crucial – especially so because my primary interest is how legal and technological normativities interact.
13 Sahlins was inspired by Mauss's (2000) famous anthropological study *The Gift*.

References

Austin, J.L. (1975) *How to Do Things with Words*, 2nd edn. Boston: Harvard University Press.

Austin, J.L. (1995) *The Province of Jurisprudence Determined*, edited by W. Rumble. Cambridge: Cambridge University Press.

Bijker, W.E. (1995) *Of Bicycles, Bakelites, and Bulbs: Toward a Theory of Sociotechnical Change*. Series: Inside Technology. Cambridge, Mass.: MIT Press.

Brownsword, R. (2009) Human dignity, ethical pluralism, and the regulation of modern biotechnologies, in T. Murphy (ed.) *New Technologies and Human Rights*, 1st edn. Oxford, England: Oxford University Press. Available at: www.oxfordscholarship.com/oso/public/content/law/9780199562572/acprof-9780199562572-chapter-2.html (accessed 1 Sept. 2016).

Colomb, R.M. (2005) Information systems technology grounded on institutional facts, in D. Hart and S. Gregor (eds) *Information System Foundations. Constructing and Criticising*. Canberra: ANU Press. Available at: http://epress.anu.edu.au/info_systems/mobile_devices/ch07.html. (accessed 1 Sept. 2016).

Dubber, M.D. (2005) *The Police Power. Patriarchy and the Foundations of American Government*. New York: Columbia University Press.

Dworkin, R. (1991) *Law's Empire*. Glasgow: Fontana.

Dworkin, R. (1997) *Freedom's Law: The Moral Reading of the American Constitution*. Boston: Harvard University Press.

Eisenstein, E. (2005) *The Printing Revolution in Early Modern Europe*. Cambridge and New York: Cambridge University Press.

Geisler, D.M. (1985) Modern interpretation theory and competitive forensics: Understanding hermeneutic text, *The National Forensic Journal* III (Spring): 71–9.

Gibson, J. (1986) *The Ecological Approach to Visual Perception*. New Jersey: Lawrence Erlbaum Associates.

Glastra van Loon, J.F.G. (1958) Rules and commands, *Mind* LXVII (268): 1–9.

Hart, H.L.A. (1994) *The Concept of Law*. Oxford: Clarendon Press.

Hildebrandt, M. (2008a) Legal and technological normativity: More (and less) than twin sisters, *Techné: Journal of the Society for Philosophy and Technology* 12(3): 169–83.

Hildebrandt, M. (2008b) A vision of ambient law, in Roger Brownsword and Karen Yeung (eds) *Regulating Technologies*. Oxford: Hart.

Hildebrandt, M. (2010) Proactive forensic profiling: Proactive criminalization? in R.A. Duff, Lindsay Farmer, S.E. Marshall, Massimo Renzo, and Victor Tadros (eds) *The Boundaries of the Criminal Law*. Oxford University Press. Available at: www.oxfordscholarship.com/view/10.1093/acprof:oso/9780199600557.001.0001/acprof-9780199600557-chapter-5 (accessed 1 Sept. 2016).

Hildebrandt, M. (2014) Radbruch on the origins of the criminal law: Punitive interventions before sovereignty, in M.D. Dubber (ed.) *Foundational Texts in Modern Criminal Law*. Oxford: Oxford University Press.

Hildebrandt, M. (2015) *Smart Technologies and the End(s) of Law. Novel Entanglements of Law and Technology*. Cheltenham: Edward Elgar.

Hildebrandt, M. (2016a) Law as information in the era of data-driven agency, *The Modern Law Review* 79(1): 1–30.

Hildebrandt, M. (2016b) New animism in policing: Re-animating the rule of law? in Ben Bradford, Beatrice Jauregui, Ian Loader, and Jonny Steinberg (eds), *The SAGE Handbook of Global Policing*. London: SAGE Publications Ltd.

Hildebrandt, M. and L. Tielemans (2013) Data protection by design and technology neutral law, *Computer Law & Security Review*, 29(5): 509–521.

Hildebrandt, M. and B.J. Koops (2010) The challenges of Ambient Law and legal protection in the profiling era, *Modern Law Review* 73(3): 428–60.

Ihde, D. (1990) *Technology and the Lifeworld : From Garden to Earth*. The Indiana series in the Philosophy of Technology. Bloomington: Indiana University Press.

Kelsen, H. (2005) *Pure Theory of Law*. Clark, NJ: Lawbook Exchange.

Koschaker, P. (1966) *Europa und das Römische Recht*. Vol. 4. München: C.H. Beck'sche Verlagsbuchhandlung.

Lask, E., G. Radbruch, and J. Dabin (2014) *The Legal Philosophies of Lask, Radbruch, and Dabin*, translated by K. Wilk. Boston: Harvard University Press.

Latour, B. (1993) *We Have Never Been Modern*. Cambridge, MA: Harvard University Press.

Latour, B. (2000) The Berlin Key or how to do words with things, in P.M. Graves (ed.) *Matter, Materiality and Modern Culture*. London: Routledge: 10–21.

Lévy, P. (1990) *Les Technologies de L'intelligence: L'avenir de la Pensée à L'ère Informatique.* Paris: La Découverte.

Lévy, P. (1998) *Becoming Virtual: Reality in the Digital Age.* New York and London: Plenum Trade.

Mauss, M. (2000) *The Gift: The Form and Reason for Exchange in Archaic Societies*, Trans. W.D. Halls. W.W. Norton & Company.

Nagel, T. (1974) What is it like to be a bat? *The Philosophical Review* LXXXIII (October): 435–50.

Ong, W. (1982) *Orality and Literacy: The Technologizing of the Word.* London/New York: Methuen.

Raz, J. (1975) Reasons for action, decisions and norms, *Mind*, New Series, 84 (336): 481–99.

Ricoeur, P. (1973) The model of the text: Meaningful action considered as a text, *New Literary History* 5(1): 91–117.

Sahlins, M. (1963) Poor man, rich man, big man, chief: Political types in Melanesia and Polynesia, *Comparative Studies in Society and History* 5(3): 285–303.

Searle, J.R. (1969) *Speech Acts: An Essay in the Philosophy of Language.* Cambridge: Cambridge University Press.

Silkenat, J.R., J.E. Hickey Jr, and P. D. Barenboim (eds) (2014) *The Legal Doctrines of the Rule of Law and the Legal State.* New York: Springer.

Taylor, C. (1995) To follow a rule, in C. Taylor (ed.) *Philosophical Arguments.* Cambridge, MA: Harvard University Press.

Varela, F.J., E. Thompson, and E. Rosch (1991) *The Embodied Mind. Cognitive Science and Human Experience.* Cambridge, MA: MIT Press.

Verbeek, P.-P. (2005) *What Things Do. Philosophical Reflections on Technology, Agency and Design.* Pennsylvania State University Press.

Verbeek, P.-P. (2011) *Moralizing Technology: Understanding and Designing the Morality of Things.* Chicago; London: The University of Chicago Press.

Weber, M. (2013) *Economy and Society.* Edited by Guenther Roth and Claus Wittich. First Edition, Two Volume Set, with a New Foreword by Guenther Roth edition. University of California Press.

Wesel, U. (1985) *Frühformen des Rechts in vorstaatlichen Gesellschaften. Umrisse einer Frühgeschichte des Rechts bei Sammlern und Jägern und akephalen Ackerbauern und Hirten.* Frankfurt am Main: Suhrkamp.

Part V

Emerging technologies of crime and justice

36

Nanocrime 2.0

Susan W. Brenner

I. Introduction

> it would be ... reckless to attempt to list all the different kinds of crimes that might involve nanotechnology.
>
> *(Boucher 2008: 218)*

A few years ago, I published a law review article on a phenomenon that then did not exist, does not exist and may not exist for years, perhaps decades: using nanotechnology to commit crimes – or "nanocrime".[1]

A great deal has been written about the societal implications of nanotechnology (Boucher 2008). The books and articles that deal with nanotechnology often note that criminals will exploit the technology for their own antisocial ends. But while many clearly believe the technology has the capacity for a dark side, no one has focused on how that dark side might manifest itself and on the legal issues it is likely to generate.

These are the topics I examine in this chapter. The analysis is prospective and therefore speculative to a certain extent because nanocrime apparently has yet to manifest itself. Some may question the utility of writing about a phenomenon that does not yet exist, but I think it is a useful endeavor.

I believe nanotechnology is at a point in its development that is analogous to where computer technology was in the 1950s. In the 1940s and 1950s, mainframe computers were found only in government and university computer labs, and computer crime, as we know it, did not exist (see e.g. Brenner 2010: 10–13). Computer crime did not emerge until the 1960s, when mainframes moved into the private sector and employees began using them to facilitate embezzlement and fraud crimes (ibid.: 9–12). Computer crime became more common as the technology evolved from a niche technology into one that permeates the fabric of our daily lives (ibid.: 9–38).

I suspect something similar will occur with nanotechnology. My theory is that although nanotechnology has been moving into the private sector for years, it is still early in that process, so early no one is seriously considering the prospects for and likely implications of nanocrime. I believe that is unfortunate: We had no basis for anticipating how computer technology could, and would, be misused; while we had some experience with the misuse of earlier, simpler technologies, the unprecedented capabilities and evolving complexity of computer technology made it difficult to foresee how it would be misused.

Some may argue that we are in a similar position with regard to nanotechnology, but I disagree. I believe we can use our experience with computer crime to anticipate how law should deal with the varieties of nanocrime, once they appear. The technologies differ in functionality and therefore in their capacity for misuse, but as I have argued elsewhere (Brenner 2007b: 137–181), I do not think law should take a technology-specific approach to crimes, the commission of which is facilitated by computer or other technologies. In § III, I use our experience with cybercrime as an analogy from which to extrapolate how law can deal with the phenomenon on nanocrime, if and when it emerges.

First, though, I need to describe the technology itself; the next section gives an overview of nanotechnology. Section III postulates how nanotechnology can be misused and outlines the potential contours of a law of nanocrime. Section IV provides a brief conclusion.

II. Nanotechnology

This section reviews what nanotechnology is and why many believe it will usher "in the second industrial revolution" (Mandel 2008; Dressler 2005). A study from the Hastings Center explains that nanotechnology "is expected to become a key transformative technology of the twenty-first century":

> Nanotechnology is considered a general use or enabling technology because it has applications that span science and engineering fields. ... Many experts predict nanotechnology will be as significant as the steam engine, the transistor, and the Internet in terms of societal impact.
>
> *(Michelson et al. n.d.)*

The term commonly used to denote transformative technologies is "general purpose technology" (GPT); it refers to "a special type of technology that has broad-ranging enabling effects across many sectors of the economy" (Whitt and Schultze 2009). Timothy Bresnahan and Manuel Trajtenberg introduced the term in their 1995 article, "General Purpose Technologies 'Engines of Growth'?" (Bresnahan and Trajtenberg 1995).

According to Bresnahan and Trajtenberg, GPTs act as "'enabling technologies' by opening up new opportunities rather than offering complete, final solutions" (Whitt and Schultze 2009). Bresnahan later noted that the "most economically important use" of a GPT may not be determined by those who invented the technology "but rather by the inventors of complements, applications" (Bresnaham 2007). This aspect of GPTs means that the extent to which a particular technology qualifies as a general-purpose technology may not be apparent when it is introduced. As one article notes, "when a new 'general purpose technology' is developed, such as the railway, the automobile, the telegraph and telephone, or the Internet, uncertainty is created as to how deeply the technology will transform the economy" and society (Bolton et al. 2005).

Notwithstanding that uncertainty, many characterize nanotechnology as a GPT (see e.g. Breggin and Carothers 2006). For the purposes of this analysis, I assume nanotechnology qualifies as a GPT and consequently has the eventual capacity to transform societies in ways that are analogous to the changes wrought by antecedent GPTs, such as electricity and personal computing. This assumption establishes the conceptual foundation for our inquiry into the likelihood and potential varieties of nanocrime. But before I embark on that inquiry, I need to address three prefatory issues: a review of nanotechnology as a technology; the potential risks of nanotechnology; and proposals for using civil regulatory law to diminish the impact of those risks.

A. Technology

K. Eric Drexler uses a dichotomy to distinguish nanotechnology from technologies that preceded it: He characterizes the antecedent technologies as "bulk technologies" because they all involve manipulating atoms and molecules in bulk; carpenters, potters, machinists, weavers and even the manufacturers of computer chips work with materials that are composed of trillions of discrete atoms (Drexler 1990). Until relatively recently, humans were limited to working with disparate assemblages of atoms because we did not have the ability to manipulate individual atoms (ibid.; Storrs Hall 2005). Because nanotechnology gives us that ability, Drexler refers to it as "molecular technology" (Drexler 1990: 4).

"Nanotechnology" has been defined in at least two ways: One characterizes it as any technology that deals with structures that are 100 nanometers or less in size; the other, older definition characterizes it as "building things from the bottom up, with atomic precision"(Center for Responsible Nanotechnology n.d.). The validity of the older definition has eroded as nanotechnology research increasingly began to focus on top-down, as well as bottom-up, manufacturing (ibid.). The newer definition is therefore the more accurate of the two.

Experts divide nanotechnologies into four "generations," two of which – "passive" and "active" nanostructures – exist (ibid.). Passive nanostructures incorporate nanoscale materials into existing products in order to improve their performance (Michelson et al. n.d.). Passive nanostructures have been integrated into "products ranging from clothing and sporting goods to personal care and nutritional products" (International Risk Governance Council 2007: 7). Active nanostructures are "biologically or electronically dynamic" (Michelson et al. n.d.), i.e. they "can change their state during operation" (International Risk Governance Council 2007: 7). One report explains that active nanostructures include self-healing materials, "targeted drugs and chemicals," "light-driven molecular motors" and "adaptive nanostructures" (Subramanian 2009).

The remaining generations are projected to emerge over the next decade (International Risk Governance Council 2007: 7). The third consists of "integrated nanosystems," i.e. "networking at the nanoscale" (ibid.). Researchers are in the process of developing integrated nanosystems.[2] The fourth generation consists of "heterogeneous molecular nanosystems" in which nano components "are reduced to molecules" that can "be used as devices or engineered to assemble on multiple length scales" (ibid.). Fourth generation nanostructures are also in development (Michelson et al. n.d.).

B. Risks

For years, observers have noted that nanotechnology may bring risks as well as rewards (Joy 1991). As a result, government agencies, private entities and individuals have been – and are – devoting a great deal of effort to exploring the potential EHS (environmental, health and safety) risks of nanotechnology (Maynard 2016). As one report noted, "[s]ome of the unique properties of nanoscale material ... have given rise to concerns about their potential implications for health, safety, and the environment" (Sargent Jr. 2013). The report also explains that the "EHS risks of nanoscale particles in humans and animals depend in part on their potential to accumulate, especially in vital organs such as the lungs and brain, that might harm or kill, and diffusion in the environment that might harm ecosystems" (ibid.: 10).

Nanoparticles enter an individual's or animal's body via "three primary vectors – inhalation, ingestion, and through the skin" (Wolf et al. 2009: 676). Although scientists know little about the long-term effects of exposure to nanoparticles, studies have shown that very small nanoparticles can penetrate the skin and then enter the bloodstream; and inhaled nanoparticles can move "from the nasal region to the brain through the olfactory bulb, thus bypassing the blood-brain barrier" and possibly entering the brain (ibid.; Lin 2007).[3] Nanoparticles can also "pass through lung and liver tissue" (Rakhlin 2008).

While researchers have identified these and other potentially dangerous aspects of nanotechnology, the precise nature and magnitude of the threat the technology poses remains uncertain. As one author noted, risk "to human health and the environment is the most pressing issue in the governance of novel technologies", but the "definition and measures of risk are challenged with nanotechnologies" because nanotechnology

> techniques and products are often combined with other technologies, making it difficult to categorize products by intended use, much less to tease out specific effects. Interactions between components of a technology at the nanoscale or between nanomaterials and human tissue may not be linear. ... Predictive algorithms may also be questionable, as many nano-techniques rely on the ability of materials to behave differently at the nanoscale than at larger, more familiar scales.
>
> *(Hogle 2009: 753)*

While assessing the risks associated with nanotechnology is, and is likely to continue to be, an extraordinarily complicated task, many believe it is a task we must master. A survey of "business leaders in the field of nanotechnology" found that "nearly two-thirds" of them believed we do not understand the risks the technology poses to human life and environmental integrity; those surveyed also believed it is "important" for the government to conduct a meaningful assessment of these risks in order to protect "human health, safety, and the environment" (Sargent Jr. 2013: 10).

The efforts that are currently being made to assess the risks associated with our use of nanotechnology tend to focus on the inadvertent and therefore unintended consequences of incorporating nanotechnology into consumer products, manufacturing and medical care (Wolf et al. 2009). As a United Nations study explained, there are two concerns:

> [T]he hazardousness of nanoparticles and the exposure risk. The first concerns the biological and chemical effects of nanoparticles on human bodies or natural ecosystems; the second concerns the issue of leakage, spillage, circulation, and concentration of nanoparticles that would cause a hazard to bodies or ecosystems.
>
> *(UNESCO 2006)*

As noted earlier, the process of assessing these concerns is more complex than the process of assessing similar concerns for non-nanotech materials. The tendency to utilize nanomaterials in conjunction with other technologies is only one of the factors that exacerbate the difficulty of this process (UNESCO 2006). Another is the unique properties of nanoparticles (Wallace 2008): They tend "to persist [in the body] for longer periods as nanoparticles" because of the way they are designed; and they may "be better able to evade the body's defenses because of their size or protective coatings" (Lin 2007: 356–357). Nanoparticles may also be able to persist for longer periods in the environment (Karn and Bergeson 2009). A third factor that exacerbates the difficulty of identifying and assessing potential threats is that nanotechnology – like computer technology – is inherently unstable, i.e. it continues to evolve. As one scientist noted, "[w]ith researchers in 40 countries creating new nanoparticles every day," it is "difficult to assess each particle individually" in order to determine its characteristics and risk potential (Bahm 2010).

In sum, while we know little about the specific risks associated with nanotechnology, those who study the technology are certain it will generate hazards that are both unprecedented and elusive. It is, however, not enough simply to identify risks; we need risk control strategies, as well. The current risk assessment process focuses exclusively on what I call "civil" nanotech risks; that is, it focuses on hazards that are an inadvertent and therefore unintentional by-product of legitimate uses of the technology. As a result, nanotechnology risk assessment and control strategies focus on how civil regulatory law can be used to ensure integrity in the production and implementation of the technology. The next section reviews the prospects for using regulatory law to control the risks associated with nanotechnology.

C. Assessment

Nano-specific regulation seems advisable given the apparently unique aspects of the technology at issue, but it actually may not be the best approach. Nano-specific regulation presumably means that one agency would be responsible for assessing the risks of utilizing nanotechnologies in various contexts (e.g. medicine, transportation, agriculture, energy, construction, communication, manufacturing, etc.) (see e.g. Brindell 2009). The agency's primary focus would be on nanotechnology, which no doubt means that individuals whose expertise was in nanotechnology would play a major role in making this assessment. The question is whether the assessment should be made by those whose focus is on the technology or by those whose focus is on the idiosyncratic issues that arise in a specific context, e.g. medicine or agriculture.

It is probably much too early for us to be able to answer that question. The answer depends on the extent to which nanotechnology proves to be a transformative technology, the pervasiveness and complexity of which exceed that of antecedent technologies. It is easy to overstate the impact an emerging technology is likely to have on our lives; something similar occurred with computers and, as we will see in the next section, resulted in legislation that tended to over-emphasize the role computers would play in crime.

However we resolve the issue of nanotechnology regulation, civil regulatory measures will be of little utility in analyzing the issues I address below because they are designed to address risks that are the product of inadvertence, not intention. I included this treatment of how we may approach regulation for two reasons: One is that it illustrates the operational tension between technology-specific and technology-neutral control measures. The other is that it illustrates the types of risks associated with various uses of nanotechnology. In the next section, we take up advertent nanotech risks, i.e., intentional abuse of the technology.

III. Technology and crime

> Consider a person who uses nano-tech drug-delivery techniques to apply a very targeted poison in committing a murder...
>
> *(Boucher 2008: 219)*

The sections below examine technology-facilitated crime as an empirical phenomenon: Section III(A)(1) reviews cybercrime, while § III(A)(2) examines how nanotechnology could be used to facilitate the commission of various crimes. Section III(B) analyzes the extent to which our experience with cybercrime should structure our response to nanocrime, if and when such a response becomes necessary.

Before we take up those issues, I need to outline the conceptual framework that will structure our inquiries. It was developed as a tool for analyzing cybercrime – crimes the commission of which involves the use of computer technology (see Brenner and Clarke 2005: 660, for a definition of cybercrime) – but it can also be used to analyze how other technologies facilitate crimes. It is designed to identify the role a technology plays in various types of criminal activity and I see no reason why its basic analysis cannot be extrapolated to nanocrime.

The framework divides cybercrimes into three categories: (i) a computer is the target of the crime; (ii) a computer is a tool that is used to commit a traditional crime such as theft or fraud; and (iii) a computer plays an incidental role in committing one or more crimes (Brenner 2001a). Each category is described below (see Brenner 2010: 39–47).

A computer is the *target* of criminal activity when the perpetrator attacks the computer by breaking into it, introducing code that damages it or bombarding it with data. Cybercrimes that involve breaking into a computer involve accessing a computer without being authorized to do so (outsider crime) or by exceeding the scope of one's authorized access to a computer (insider crime). Access can be an end in itself or it can be used to commit another crime (e.g. damaging or stealing data from a computer). Code target crimes involve creating, disseminating and using malware, viruses, worms and other malicious code that damages a computer system or extracts data from it.[4] Target computer crimes also include DDoS attacks, in which the attacker blasts a computer linked to the Internet with so much data it essentially goes offline; the computer receives so many malicious signals from the attacker that no legitimate traffic can reach it.[5]

A computer can also be a *tool* used to commit a traditional crime, such as theft or fraud; here, the computer's role is analogous to the role a telephone plays when a fraudster uses it to trick victims into parting with their money or property. In both instances, the use of a particular technology facilitates the commission of the crime but does not alter the nature of the offense. Computers can be used to commit most traditional crimes, including fraud, embezzlement, theft, arson, forgery, riot, assault, rape and homicide.

Finally, a computer can play an *incidental* role in the commission of a crime: A blackmailer can use a computer to email his victim; and a drug dealer can use a computer and Excel to track his drug transactions. In these and similar instances the computer's role in the crime is as a source of evidence, nothing more. But that can be important: The evidence investigators find on the drug dealer's computer may play an essential role in convicting him of his crimes.

This trichotomy plays two roles in analyzing cybercrimes: Investigators use it to assess how they should draft search warrants and otherwise incorporate computer technology into their investigative process. Lawyers and legislators use it to determine if existing law is adequate to criminalize how a computer was used in a given instance.

The sections below use the trichotomy for two related purposes: (i) to order our analysis of the ways in which nanotechnology could be used to facilitate criminal activity; and (ii) to order a parallel analysis of the extent to which the criminal uses of nanotechnology can be addressed with existing criminal law or will require modifying existing law or adopting new law.

A. Analogy

As I noted earlier, any discussion of nanocrime is speculative because nanocrime apparently has yet to manifest itself. I believe the explanation for its failure to appear thus far lies in what will emerge as an analogy between the rise of cybercrime and the eventual rise of nanocrime (see § I above).

1. Cybercrime

If it is permissible to analogize something that does not exist to something that has existed for years, then I believe nanocrime can be analogized to cybercrime. Both involve (or, more properly, one involves and one will involve) exploiting a technology for unlawful ends. Criminal exploitation of technology is not a new phenomenon; in a study done several years ago, professors at the Naval Postgraduate School found that historically the "bad guys" are among the first adopters of a new technology, at least once the technology becomes publicly available (see Ronfeldt and Arquilla 2001: 313).

Unlike many technologies, personal computers and, ultimately, nanotechnology are "democratic" technologies: Though each began as a "laboratory" technology that was used exclusively by specialists, each evolved (or, in the case of nanotechnology, will evolve) into a technology used by "consumers," i.e., the general public (see Brenner 2007b). The egalitarian aspect of the technologies makes them more accessible to criminals and more attractive as criminal tools.

In the 1930s, bank robbers and other criminals used motor vehicles to avoid being apprehended after they had committed a crime (Brenner 2009: 25–28). The criminals tended to have faster automobiles than the police, and were adept at using those vehicles to cross jurisdictional borders and otherwise evade capture by the police (ibid). Motor vehicles were readily accessible to criminals as well as to law-abiding citizens; and their general availability meant that the skills needed to operate them were effectively in the public domain, i.e., most people knew or could easily learn how to drive a car. Their unique value as criminal tools lay in the extent to which they facilitated escape, but auto theft also became a new, and serious crime during this era.[6] The former exploited the distinct capabilities of automotive technology as a technology; the latter simply approached automobiles as a generic item possessing value.

Something similar to criminals' use of automotive technology occurred with computer technology, though its migration into the public and criminal spheres took decades.[7] The first published reports of computers being used to commit crimes appeared in the 1960s, when computers were large mainframe systems (see Sieber 1998: 19; Parker 1976: x–xi). The history of modern computing dates back to the nineteenth century, but the development of the mainframe business computer did not begin until after World War II (Campbell-Kelly and Aspray 1996). In 1946, several companies began working on a commercial mainframe and by 1951 the UNIVAC, created by the company of the same name, was being used by the Census Bureau (ibid.: 100–121). In 1951, CBS used a UNIVAC to predict the outcome of the presidential election, which popularized the new technology (ibid.: 125–130). By 1960, 5,000 mainframes were in use in the United States; by 1970, almost 80,000 were in use in the United States and another 50,000 were in use abroad (ibid.: 131). Given the tremendous increase in the number of computers, it is not surprising that computer crime began to become an issue in the 1960s.

Computer crime in the 1960s and 1970s was very different from the cybercrime we deal with today. There was no Internet; mainframes were not networked to other computers. In 1960, a typical mainframe cost several million dollars, needed an entire room to house it and a special air conditioning system to keep its vacuum tubes from overheating and frying the data it stored (Levy 1984: 18–25). Only select researchers were allowed to use a mainframe (ibid.). To access a mainframe, a researcher gave the data he wanted the computer to analyze to a keypunch operator, who used a machine to punch holes in cards; the holes encoded the data into a form the mainframe could read.[8] The keypunch operator then gave the cards to another operator, who fed them into a machine that transmitted the information to the mainframe for processing; the researcher would eventually receive a printout showing the machine's analysis of his data.[9]

Since mainframes were not networked and the only way to access one was by using this cumbersome process, only a few people were in a position to commit computer crime (Rasch 1996). That limited the type and amount of crimes that were committed in this era. Insiders might spy on other employees by reading their confidential files; and they might sabotage a computer or the data it contained as retaliation for being fired or disciplined. These crimes occurred, but the most common type of computer crime in this era was financial; insiders used their access to a mainframe computer to enrich themselves.

While embezzlement and fraud were the most common crimes, employees found other ways to profit from their access to a mainframe. Some stole information and sold it; they usually took trade secrets, but in one case employees of the Drug Administration Association stole the names of informants and information about pending investigations and sold it to drug dealers.[10] Other employees had their company's mainframe issue phony payroll checks to nonexistent employees, which they, of course, cashed.[11] Another, less common tactic was to misappropriate company data and hold it for ransom (see e.g. Nossiter 1977).

The computer crimes in this era all had one thing in common: The victims were a company or government agency because large entities were the only ones who used mainframe computers. Also, mainframes were generally incapable of inflicting "harm" on an individual; it might have been possible for an insider to manipulate mainframe data to fire someone improperly, but that would have been a very risky crime. The victim would probably complain, triggering an inquiry that could lead to the unraveling of this scheme and any others in which the perpetrator was involved. Using computers to "harm" individuals did not become a problem until the 1980s, when the "personal computer" and the Internet appeared (see Brenner 2010: 23–37, 73–102).

The serendipitous introduction of those innovations at around the same time transformed computer technology from a "laboratory" technology into a "democratic" technology. More and more people began using computers and as a result, computer-facilitated crime increased in incidence and in complexity (ibid.).

In the 1960s and 1970s (and perhaps earlier), computer crime was a relatively simple phenomenon: mainframes were used as tools to commit traditional crimes such as fraud, embezzlement, extortion and the theft of information and/or trade secrets. Computer crime was therefore limited to a basic (albeit often profitable) set of tool crimes. Target crimes do not seem to have existed in this era; there are no reported instances of a mainframe being used to damage itself and since mainframes were not networked, there was no way one mainframe could be used to damage another.[12] These early tool crimes did implicate the third category in the trichotomy outlined earlier; investigators used the mainframe involved in such a crime as a source of evidence to be used in identifying and prosecuting the perpetrator(s) (Bequai 1978: 22–23).

The demise of the mainframe began in 1971 with the invention of the microprocessor, which dramatically decreased the size and cost of computers (Campbell-Kelly and Aspray 1996: 229). It was not until 1975 that the first microprocessor-based computer – the Altair 8800 – made its debut on the cover of *Popular Electronics* (ibid.: 240). In 1976, Steve Jobs and Stephen Wozniak created the Apple II (ibid.: 244–246). Their competitors were focused on creating products for computer enthusiasts (ibid.: 237–244), but Jobs realized a computer could be a popular consumer product if it were "appropriately packaged" (Ibid.: 246).

Commodore Business Machines adopted a similar strategy and in 1977, when the Apple II and Commodore PET went on the market, both "were instant hits" with the public (ibid.: 247). The expanded variety of software that was available by 1980 further increased interest in personal computers, as did IBM's introduction of its Personal Computer in 1981 (ibid.: 248–257).

The popularization of computers took a huge step forward with the rise of the Internet. Its precursor – the ARPANET – went online in 1969 but the ARPANET "did not interact easily" with networks that did not share its networking protocol.[13] The Internet, as such, began in 1983, when the ARPANET protocol was changed to TCP/IP. The World Wide Web – a "system of interlinked hypertest documents accessed via the Internet" – went online in 1991.[14] The first graphical web browser, Mosaic, went online in 1993 and vastly increased use of the Internet.[15]

As anyone who has seen the 1983 film *War Games*[16] knows, networked computer crime (or cybercrime) emerged long before 1993. The movie depicts what happens when David Lightman "hacks" his way into WOPR, a NORAD supercomputer, and starts a game of "global thermonuclear" war with the computer.

War Games brought hacking – which was already popular in certain circles – into the popular consciousness.[17] As a 1983 *New York Times* article explained, the number of "young people roaming without authorization through" the country's computers was in the thousands and growing "with the boom in personal computers" (Treaster 1983). The article also noted that the "hackers" were also using the electronic bulletin boards that were a precursor of the Internet.[18]

War Games-style cybercrime – breaking into computers to satisfy one's curiosity and perhaps play a prank – was the dominant mode of computer crime over the next few years.[19] It survived into the 1990s, but by then adults had essentially taken over cybercrime. Adults had realized computer crime could be a profitable endeavor; the Internet was beginning to link everything, which meant the new cybercriminals had thousands, and eventually millions, of targets. Most organizations were not aware of the need to secure their systems, so most of the targets were easy pickings for even a semi-talented cybercriminal. This led to an explosion in tool crimes – primarily in financially motivated tool crimes. The difference between these tool crimes and mainframe tool crimes was that the cyber-tool crimes were committed in a wholly unbounded context; anyone with access to the Internet could strike at a target across the street or halfway around the world. That created additional incentives to commit financial tool crimes; a clever cybercriminal could attack a target, make a profit and stand very little chance of being apprehended (unlike the insiders who committed mainframe tool crimes).

Financial tool cybercrimes – online theft, fraud, extortion, embezzlement, forgery and identity theft – exploded. By the mid-1990s, financial tool cybercrimes were increasingly the province of adults; by the twenty-first century, they were increasingly the province of organized criminal groups. The democratization of computer technology created vast new opportunities for those willing to break the law for financial gain; in so doing, it may have induced people to commit crimes who might otherwise not have done so.[20]

The democratization of computer technology also created new opportunities for other types of criminal activity. Child pornography, which had been a relatively insignificant crime prior to the Internet, exploded online; those who would never have been willing to seek out child pornography in the physical world found they could easily, and anonymously, acquire it online, which reduced their risk of being identified and prosecuted. The production, distribution and possession of child pornography became another popular tool cybercrime. Child pornography is a tool cybercrime that mixes financial and non-financial motives; there are, and have long been, websites that sell child pornography, but there are also sites where it is traded for free.

The democratization of computer technology also made it easier for individuals to inflict "harms" of varying types – usually non-physical – on each other. Online stalking, harassment, bullying, defamation, imposture and invasion of privacy became increasingly common. The commission of most of these crimes was relatively unusual prior to the rise of the Internet; these personal "harm" tool crimes exploded online because it became possible to inflict any or all of the "harms" they encompass with relatively little risk of being identified and prosecuted. Those who stalked or harassed others in the physical world were likely to be prosecuted; those who do so online have a good chance of facing no consequences for their actions. The crimes in this category are also tool cybercrimes.

Tool cybercrimes are probably the most commonly committed types of cybercrime, perhaps because there are so many types of tool cybercrimes. As I have noted elsewhere (Brenner 2001b), I believe all the traditional crimes except for rape and bigamy can be committed online; we so far do not have a documented case in which the Internet was used to commit murder, but we have a case in which that may have been the perpetrator's goal. And it is reasonable to assume that the Internet could, under certain circumstances, be used to commit the ultimate crime.

Target crimes – attacks on a computer – have grown in frequency and complexity. People still hack computer systems, though today they are more likely to do so as part of a scheme to commit some other crime, such as identity theft or extortion (for more on target crimes see Brenner 2001b: 49–71). The newer target crimes – creating and disseminating malware and launching distributed denial of service (DDoS) attacks – are increasingly common and are often committed as part of a scheme to carry out a tool crime. The democratization of computer technology has expanded the pool of those who commit target crimes; in the mainframe era, only insiders could attack a computer. Today, anyone – insider or outsider – can do so, and the attacks are more complex. Extortionists often use DDoS attacks to take a particular target – an online casino, say – offline as part of an extortion scheme; the casino operators are told that unless they pay a substantial sum (which they usually do), the attacks will continue. Malware can be used in a similar fashion but can also be used to siphon valuable information from the victim computer system. Malware and DDoS attacks can also be used to carry out a "pure" target crime, i.e., for the sole purpose of taking a target computer offline.

There are many other permutations of tool and target cybercrimes, but I believe this summary illustrates how democratizing computer technology opened it up to be exploited for criminal purposes. That brings me to the third and final category in the trichotomy I outlined earlier: the computer as playing an incidental role in the commission of the crime.

This category encompasses cases in which a computer is used to commit a crime, but its use is so trivial that it does not transform the crime into a tool crime; in a tool crime, the use of the computer is integral to the commission of the crime, i.e., the crime could not have been committed when and as it was without the use of the computer. The best way to explain the difference between the two is by using examples. Assume that a blackmailer uses his home computer to write and print a blackmail letter he then mails to his victim; here, the computer played a role in the commission of this crime, but the role was so trivial this does not qualify as a tool cybercrime. The same is true in the example I noted earlier (see Section III above): a drug dealer uses a laptop and Excel to track his purchases, sales and inventory. Here, again, the computer plays a role in the commission of the crime, but the role is too minor for this to constitute a tool cybercrime (for more on this, see Brenner 2010: 45–47).

Crimes in which the computer's role is merely incidental are included in the trichotomy because even when a computer's role is trivial, evidence of the crime will be found on the computer (see Brenner 2010: 103–119). That is important for those who investigate cybercrime and for those who enforce the laws that limit what investigators can do in the course of investigating cybercrimes, but it is generally not important for those who adopt or interpret the legislation that defines cybercrimes. In other words, because the computer's role here is merely evidentiary, its use does not require assessing – or reassign – how law defines a particular crime (ibid.: 45–47).

That has often been the case with tool and target crimes; since the computer plays a significant role in the commission of these cybercrimes, those who adopt or enforce criminal statutes have often found it necessary to incorporate the computer's role in committing an offense into how the law defines, and punishes, that crime. That may necessitate revising existing law or adopting new, cybercrime-specific law. Stalking statutes, for example, had to be revised to encompass online stalking because most of them were drafted in the 1980s when stalking was a purely real-world activity (ibid.: 92–96). Statutes that defined stalking as following the victim or engaging in other activity in the real-world did not encompass the type of activity involved in online stalking (Brenner and Rehberg 2009), which meant perpetrators could engage in that type of activity with impunity.

In the next section, we consider the possibility that nanotechnology will evolve in a fashion analogous to computer technology, i.e., that it will evolve from a "laboratory" technology into a "democratic" technology. If nanotechnology persists as a "laboratory" technology, its role in facilitating the commission of traditional and/or novel crimes is likely to be minimal, at best. If it becomes a "democratic" technology, its potential for crime facilitation is likely to be much more significant. The next section primarily focuses on the second scenario, i.e. it speculates about how nanotechnology could be used to facilitate various types of criminal activity. In focusing on those issues, it implicitly considers the likelihood that nanotechnology will move into public use and become a "democratic" technology.

2. Nanocrimes

As I noted at the beginning of this chapter, authors often caution that nanotechnology will become an implement of crime but they rarely, if ever, elaborate on its possibilities for such use. Since I believe nanotechnology will evolve in a fashion analogous – but not identical – to that of computer technology, I decided the best way to structure speculation into how nanotechnology could facilitate crimes of various types is to use the cybercrime framework outlined above. In this section, then, we will consider the possibilities for nanotechnology target and tool crimes, as well as nanotechnology's capacity to play an incidental role in facilitating offenses.

Before we embark on that endeavor, I need to note a caveat: It is impossible at this point in time to know, or to speculate with any degree of confidence, whether nanotechnology will actually evolve from a "laboratory" technology to a "democratic" technology. It took decades for computer technology to evolve from mainframes to personal computers; for most of that period, most never imagined computers would become a consumer product.[21] We tend to view nanotechnology in a similar fashion: We are probably aware that it has for years been integrated into the manufacture of clothing and other consumer products,[22] but we tend to assume not only that it is a "laboratory" technology but that it will remain one. This may be true, or it may not. Speculating about nanotechnology's transition to a "democratic" technology at this point in time is as difficult and subject to the possibility of error as it would have been for someone accustomed to mainframe culture to speculate about a very different model of computing.

My point is that while I am confident that nanotechnology will evolve in a fashion analogous to computer technology and will eventually become an implement of criminal activity, I realize the speculations I offer in the remainder of this section will no doubt prove inaccurate in varying respects. I am willing to assume that risk of error because, as I noted earlier, I believe we have an advantage that those who experienced the rise and evolution of computer crime did not. We have seen how a "democratic" technology can be exploited for good and evil and are, therefore, on notice that nanotechnology may follow a similar course. It only seems, prudent, then to begin to consider how nanocrime might manifest itself and how the legal system should respond if and when it begins to emerge. In other words, I propose that we undertake efforts analogous to those outlined in § II(C), i.e., I propose that we began to analyze how criminal law can, and should, respond to the criminal exploitation of nanotechnology.

(A) TARGET CRIMES

As we saw earlier, in target cybercrimes a perpetrator attacks a computer by (i) breaking into it, (ii) introducing code that damages it or (iii) bombarding it with data. The strategy is to turn the technology on itself, i.e., use one computer to attack another. Logically, then, target nanocrimes should involve turning nanotechnology on itself. How might that manifest itself?

We will begin with break-ins – "hacks."[23] As I have explained elsewhere, hacking is conceptually analogous to trespass in that in both instances the perpetrator gains "entry" to a place to which he does not have lawful access (Brenner 2001b: 81–82). There are empirical differences between physical trespass and hacking: All of the elements of trespass (the offender, the place being trespassed upon and the means, if any, the trespasser uses to effect the unlawful entry) take place in the physical world (ibid.). Some of the elements of hacking (the offender, the computers involved) take place in the physical world but others arguably do not; the actual "entry" into the victimized computer does not occur in the physical world, at least not as literally as it does in a physical trespass (ibid.). The empirical differences between the two prompted most jurisdictions to create a new crime – "hacking" – that could be used to prosecute computer trespasses (ibid.).

That brings us to nanotech trespass. While computer trespass deviates to some extent from physical trespass, computers are, ultimately, "places" – every computer is in effect a "box", an enclosed area. Trespassing consists of entering a "place" without authorization; while the mechanics of computer trespass deviate in certain respects from real-world trespass, the empirical analogies between the two endeavors are enough to support approaching hacking as a type of trespass. Is that also likely to be true for nanotechnology?

How we answer that question probably depends on how we conceptualize nanotech hacks. I can see two options: In one, nanoparticles are the target of the attack, i.e., the trespass consists of gaining unauthorized access to one or more nanoparticles. In the other option the target is the construct of which nanoparticles are constituent entities. The latter option approaches nanoparticles as the equivalent of the chips and other components that make up the computer that is hacked; the first option approaches them as entities – "places" (or "computers") – in and of themselves.

I suspect these options (and, perhaps, any others that are subsequently identified) are not mutually exclusive. I suspect both may be relevant in different contexts. In some contexts, it may be reasonable to regard nanoparticles as the target(s), in others, it may be more logical to regard the construct of which they are a component as the target. To understand how the options might – or might not – apply in different contexts, we need to consider two examples.

Assume doctors inject "iron-bearing nanoparticles" into the arteries of a patient who has stents – "narrow metal scaffolds that widen a partly clogged blood vessel" – installed in certain of his blood vessels (*Science Daily* 2010). Stents are coated with a drug that prevent muscle cells from accumulating within a stent and clogging it; stents only contain one dose of the drug and its preventative effect begins to wane as time passes. To address this, doctors inject nanoparticles that contain iron and a new dose of the anti-clogging drug into the patient's arteries and use magnets to drive the particles to the stents, where they recharge the anti-clogging drug. After completing their mission, the biodegradable nanoparticles "break down safely" in the patient's body (ibid.).

Now assume that someone with the requisite expertise – Dr. X – is able to gain access to the patient after the nanoparticles have been injected but before they have been driven to the stents. Dr. X injects the patient with nanoparticles that he created and that are designed to find and to infiltrate the drug-bearing nanoparticles. (In other words, each of his nanoparticles is designed to infiltrate one of the original nanoparticles.) Dr. X's motive is to establish that his nanoparticles can, in fact, invade the drug-bearing nanoparticles once they are in the patient's body; his nanoparticles are not designed to interfere with the functioning of the drug-bearing nanoparticles and/or gather information from them.

This scenario is my attempt to outline a nanotechnology hack that conforms to the first option outlined above, i.e., someone trespasses on, or into, discrete nanoparticles. To do that, I needed a scenario that is analogous to the scenario in which a physical trespasser walks onto someone's land or into someone's home without being authorized to do so. As I noted above, law approaches computer hacking as a derived type of physical criminal trespass; in both, the "entry" itself is the criminal act. The question is whether the scenario outlined above is sufficiently analogous either to criminal trespass or to computer hacking that it can be defined in a similar fashion, i.e. as nanotech trespass.

I think it can: Dr. X did not himself physically enter into the drug-bearing nanoparticles (which is necessary for criminal trespass) but he gained "access" to them in a manner that is analogous to what a computer trespasser does when he hacks a computer. The computer hacker and Dr. X both use tools to penetrate a "place" that is not physically accessible to a human being. Unlike a computer hacker, Dr. X has accessed a purely physical "place," albeit a minute one; like a computer hacker, Dr. X has used proxies to gain access to this otherwise inaccessible "place." Hackers use data to break into a system; Dr. X used other nanoparticles. Logically, then, Dr. X engaged in nanotechnology hacking (or trespass).[24]

Now we need an alternate scenario that exemplifies the second option outlined above.[25] Assume that doctors inject "self-assembling macromolecular particles" into someone who has cancer; after being injected, the molecules "self-assemble into complex structures" and "a new, supramolecular bomb is born" (Beckerson 2010). Once the bomb is created, doctors irradiate it with a laser and heat it to the point at which "explosive bubbles" form; the bubbles then "burst and destroy cells in the area", including cancer cells (ibid.). Dr. X gains access to this patient after the bomb has been created but before it has been triggered; he injects the patient with other nanoparticles that are specifically designed to infiltrate the bomb. His goal is to establish that his nanoparticles can, in fact, invade the bomb once it has been created; and again, his nanoparticles are not designed to interfere with its functioning or gather information from it.

If Dr. X commits nanotech hacking in the first scenario, it seems to follow that he also commits nanotech hacking in this second scenario. His conduct in the second scenario is empirically more analogous to that at issue in computer hacking than it was in the first scenario because here Dr. X gained access to a "structure" that was composed of discrete nanoparticles, rather than to individual nanoparticles.[26] Does that matter? If and when these issues actually arise, should law treat the two scenarios differently?

I chose the scenarios because each involves a relatively modest use of nanotechnology. The first involves what I assume is a threshold use of nanotechnology; the second involves a use that is only slightly more complex. I wanted to use scenarios that involve essentially de minimis uses of the technology because I believe the primary conceptual difficulty we will confront – if and when we consider criminalizing nanotech hacking – is the context in which the activity occurs. I suspect many will initially find it difficult to take the notion of sub-minuscule "trespassing" seriously; the scale on which the crime is committed may make it seem too insignificant to justify the imposition of criminal liability.[27]

To overcome that attitude, we would need to bring the "harm" nanotech hacking inflicts within the policy that justifies criminalizing trespasses in general. The "harm" criminal trespass traditionally addressed was violating someone's right to control access to and use of her real property (see e.g. LaFave 2003: § 21.2). Criminal trespass statutes continue to address that "harm", but in the latter part of the last century we extrapolated the "harm" so it also encompassed violating someone's right to control access to and use of their digital property. In other words, we criminalized computer trespass.[28]

If and when we decide it is necessary to criminalize nanotech hacking, we will need to do something similar. We will have to extrapolate the core "trespass" harm so it encompasses sub-minuscule intrusions into personal property as well as intrusions onto real property and into digital property. I suspect we will do this by extending the extrapolation we have already made, i.e. by analogizing intrusions into applications of nanotechnology to hacking. We may find this easier to do once the applications become more common, and more complex. I suspect that as nano-constructs become more complex and as we rely on them more, we will increasingly tend to regard them as analogues of the digital systems that are the targets of computer hackers.

Even if we develop a law of nanotech trespass, that may not be enough. It is also quite possible that we will see nanotechnology analogues of the other target computer crimes: using malicious code (malware) to attack a computer and taking a networked computer offline by bombarding it with Internet traffic (DDoS attack). In analyzing the respective prospects for nano-malware and nano-DDoS attacks, I am going to use the scenarios outlined above, for two reasons. One reason is efficiency, since these scenarios provide the context and dynamics needed to analyze the remaining two target crimes, I see no reason to articulate new scenarios for the purpose of analyzing these target crimes. The other is the reason I chose the scenarios in first place, i.e. they involve *de minimis* uses of nanotechnology. The scenarios are adequate for the purposes of analyzing the remaining target crimes; and the principles we extract in the course of that analysis can then be extrapolated to more complex uses of the technology.

We will begin with nano-malware. In both of the scenarios we examined above, Dr. X disseminated nanoparticles that infiltrated the legitimate nanoparticles that had already been injected into the patients. In both scenarios, then, Dr. X's nanoparticles had a relatively benign function; they were not designed to disable the legitimate nanoparticles or otherwise interfere with their operation in any way.[29] The only function of Dr. X's nanoparticles was to "access" the legitimate nanoparticles.

If we change that circumstance and assume Dr. X's nanoparticles *are* designed to have some negative effect on the legitimate nanoparticles they infiltrate, then Dr. X has in effect disseminated nano-malware. As noted above, statutes that criminalize malware define it in part as computer code that is designed to corrupt, destroy or modify other computer code. The cybercrime consists of knowingly disseminating or attempting to disseminate malware. The nanotech version of the crime, if and when one is created, would presumably involve similar conduct and a similar *mens rea*. Dr. X's conduct in knowingly disseminating nanoparticles he knew would have a negative effect on the legitimate nanoparticles would therefore constitute commission of the nano-malware crime.

That brings us to the final target crime: DDoS attacks (or nano-DDoS attacks). As noted earlier, in a computer DDoS attack the attackers bombard a networked computer with so many signals the computer is effectively taken offline.[30] The characteristic that primarily distinguishes a DDoS attack from the other target crimes is that a DDoS attack is an "outside" crime while hacking and the use of malware are both "inside" crimes.

As we saw earlier, the purpose of hacking is to trespass "in" someone's computer; like a physical trespasser, a hacker's goal is to gain access to a particular place, i.e. get "inside" it. The "harm" basic criminal hacking statutes target is gaining access without being authorized to do so. As we also saw, malware is disseminated for various purposes, i.e. causing damage, stealing data, all of which require that the malware be inserted "into" a computer. These two target crimes are therefore "insider" crimes, i.e. they involve *gaining* access to a computer.

DDoS attacks, on the other hand, are intended to *deny* access to a computer. The crime requires an interactive environment composed of networked entities that communicate with the network via nodes or ports. By bombarding a target's connections to the network with traffic, a DDoS attack effectively takes that system offline.

DDoS attacks are the one new crime to emerge from our use of computer technology (Brenner 2007a: 385). They did not fit into any of our existing crime categories and therefore required the adoption of new, DDOS-specific criminal laws. Since DDoS attacks are the unique product of a specific context, they *may* be limited to that context; in other words, it is possible that DDoS attacks cannot be predicated on the use of other technologies, such as nanotechnology. While that is a reasonable hypothesis, I believe it is incorrect.

To understand why I believe the hypothesis is incorrect, we need to consider a variation on the Dr. X scenarios: In the original first scenario, Dr. X's innocuous nanoparticles merely infiltrate the nanoparticles that are delivering doses of the anti-clogging drug to stents installed in a patient's arteries; in the original second scenario, they infiltrate the nanotechnology bomb that is supposed to release bubbles that will destroy the cancer cells in another patient's body. In these scenarios, Dr. X can be held liable for the crime of hacking – gaining unauthorized access to – the legitimate nanoparticles.

Now assume that in both scenarios, instead of simply having these nanoparticles infiltrate the legitimate ones, Dr. X uses his nanoparticles to prevent the legitimate nanoparticles from performing their intended functions. Assume Dr. X is somehow able to ensure that

- his nanoparticles arrive at the stents before the nanoparticles carrying the anti-clogging drug do and block the stents so the legitimate nanoparticles are unable to deliver new doses of the drugs to the stents; and/or
- his nanoparticles arrive before the other legitimate nanoparticles can assemble into the "supramolecular bomb" that would have destroyed cancer cells.

In these versions of the original scenarios, Dr. X cannot be held criminally liable for gaining unauthorized access because he (or, more precisely, the nanoparticles he used as tools) did not "access" the legitimate nanoparticles. Instead, his nanoparticle tools blocked their path and, in so doing, prevented them from performing their respective intended functions. Nor can Dr. X be held liable for disseminating nano-malware; his nanoparticles did not infect the legitimate nanoparticles with any analogue of a computer virus or worm. Again, they simply blocked their path and, in so doing, prevented them from performing their respective intended functions.

In these modified scenarios, Dr. X accomplishes something that looks very much like a DDoS attack – a physical, rather than digital, DDoS attack. As I noted above, DDoS attacks, as we know them, evolved in a digital context; that does not, though, mean they are exclusive to the digital context. I believe it means they are a phenomenon that has long been possible in the physical world but that we have not encountered due to the logistics involved in implementing such an attack. As I explained elsewhere,

> [d]enying access has been of little concern in the real-world because of the physical difficulties involved in inflicting the 'harm'; to deny others access to a facility in the real world, I need a group of individuals who are willing to physically block access to that facility for a period of time. ... Computer technology eliminates those difficulties and makes it possible for me to launch a DDoS attack ... because I am bored ... or because I want to experiment with the DDoS technology.
>
> *(Brenner 2006: 776–777, note omitted)*

The digital environment is not an integral component of a DDoS attack; it is an adventitious characteristic of the DDoS attacks with which we are familiar, a reflection of the pragmatic obstacles in attempting to predicate a real-world version of such an attack on collective human action.

This means that nano-DDoS attacks are possible whenever nanotechnology can play the same role bits and bytes play in a digital DDoS attack. In the modified scenario outlined above, nanoparticles swarm a physical target and achieve essentially the same effect a DDoS attacker achieves by bombarding a digital target with data. Since the effect is the same, as is the structure of the attack, it seems reasonable to approach this and analogous nanotechnology scenarios as the equivalent of a digital DDoS attack. In order to hold someone criminally liable for such an attack, we would either have to (i) modify our existing DDoS attack statutes so they encompass both digital and nanotech attacks or (ii) create new, distinct statutes that criminalize nano-DDoS attacks by including them in a generic crime that encompasses digital and physical DDoS in or by adopting a physical-DDoS attack-specific statute.

It seems, then, that we may well see nano-analogues of the three digital target crimes. If we do, I suspect they will be just that, i.e. I suspect they will be analogues, instead of clones, of the digital crimes. The fact that nano-crimes, including nano-target crimes, will be committed in a physical environment will no doubt mean they will differ in certain functional respects from their digital antecedents.

The extent to which that will require the adoption of new laws targeting the nano-analogues is a topic we take up in § III(B), below. Before we take up those issues, we need to consider the likelihood that nanotechnology can be used to facilitate the second category of computer crimes – tool crimes – and/or whether it can play an incidental role in the commission of crimes.

(B) TOOL CRIMES

As I noted earlier, it seems bigamy and rape are the only traditional crimes that cannot be carried out through the use of computer technology. At this point in time, I do not know and cannot speculate as to whether the same will be true of nanotechnology, i.e. whether it will become as pervasive a criminal tool as computer technology.

My goal in this chapter is not to provide a comprehensive assessment of how nanotechnology can, and will, be used for criminal purposes; aside from anything else, I do not believe such an assessment is possible given the nascent state of our use of nanotechnology. My goal is to examine the *possibility* that nanotechnology will evolve into a crime-facilitating implement analogous to computer technology. I believe I can achieve that goal by analyzing how nanotechnology could be used to facilitate some of the traditional crimes.

In this section, therefore, we will analyze how nanotechnology *might* be used to facilitate some representative crimes. Traditional crimes are often divided into categories based on the "harm" inflicted, e.g. crimes against persons, crimes against property, crimes against the state, etc. In the sections below, we will consider how nanotechnology could be used to commit representative crimes that fall into each of these three categories.

(i) Crimes against persons   We will begin with two of the crimes against persons: homicide and battery. Homicide, of course, is "the killing of a human being by another human being" (Perkins and Boyce 1982: 46). Homicide is divided into discrete offenses – e.g. murder, manslaughter and negligent homicide – based on the *mens rea* involved in the commission of the crime (ibid.: 46–119). Battery is "the unlawful application of force to the person of another" (ibid.: 152). Traditionally, battery encompassed "any application of force even though it entails no pain or bodily harm" (ibid.)[31]

In analyzing whether nanotechnology could be used to commit homicide and/or battery, we will use the Dr. X scenarios we worked with earlier. We will begin with homicide: In the original versions of both scenarios, Dr. X injected the patients with innocuous nanoparticles to determine if his nanoparticles could infiltrate the legitimate nanoparticles. Now assume that in both scenarios (ibid.) Dr. X injects the respective patients with nanoparticles that are designed to take their lives directly (i.e. his nanoparticles contain poison which they release upon entering a patient's body) or are designed to do this indirectly (e.g. by rupturing one or more arteries and causing internal bleeding that leads to death).[32] Also assume that Dr. X intended to cause the patient's death in each of the four permutations of the two basic scenarios. [33]

If the patients die, then it should be possible to convict Dr. X of murder in each of these scenarios: In each he acted with the necessary *mens rea* and committed a voluntary act that was designed to, and did, result in the deaths of the respective patients. It should be a relatively simple matter to convict Dr. X of homicide in the two poison scenarios because the poison could no doubt be shown to have been the "but for" cause of the victims' deaths (LaFave 2003: § 6.4(b)). It might be more difficult to convict him of homicide in the ruptured artery scenarios because of the need to prove causation. Causation issues could prove problematic in the ruptured artery scenarios because each patient was suffering from a condition that could have – might inevitably have – led to their death at some point in time, perhaps relatively soon.[34] The defense might argue that it would be impossible for the prosecution to prove beyond a reasonable doubt that Dr. X's nanoparticles were the "but for" cause of the victims' respective deaths in the ruptured artery cases.[35]

Now assume yet another permutation on the Dr. X scenarios: In this version of the two basic scenarios we analyzed in the previous section, Dr. X injects both of the patients with nanoparticles that are not designed to kill but are designed to cause them to suffer some discomfort. If Dr. X's conduct is discovered and reported to the authorities, can he be charged with having committed a battery on each victim?

As noted above, to be guilty of battery Dr. X must have (i) intended to (ii) apply (iii) unlawful force to (iv) the person of another human being.[36] As noted above, we are assuming Dr. X intended to cause the patients discomfort because the nanoparticles were "designed" to do precisely that. We will also assume, for the purposes of analysis, that what he did could qualify as "applying" unlawful force to the "person" of another human being; the crime of battery tends to assume, not unreasonably, that the unlawful force is applied to the exterior of the victim's body. We, though, will assume that injecting substances into someone's body can also qualify as battery.[37]

That leaves us with what *might* be the one problematic element in holding Dr. X liable for battery in these revised versions of the two basic scenarios: the use of "unlawful force." The use of force on the person of another will not be held "unlawful" if it was (i) privileged or (ii) not excessive (Perkins and Boyce 1982: 153). We will assume that if Dr. X's injecting the nanoparticles constituted a use of "force," the use of force was not privileged and was excessive.[38] That brings us to the more difficult issue: Was what Dr. X did a "use of force" sufficient to sustain a conviction for battery?

Under the traditional approach to battery, "force" encompasses any "touching" of the victim's body (LaFave 2003: § 16.2(a)). The final issue to be resolved in deciding whether Dr. X committed a battery by injecting nanoparticles into the patients' bodies is therefore whether the injections qualified as a "touching" under battery statutes. One state battery statute defines "touches" to mean "physical contact with another person".[39] Under that definition, Dr. X's conduct would presumably qualify as a "touching" of the patients: He must have touched them with his hands when he was preparing to inject the nanoparticles (e.g. using alcohol to ensure the area was sterile) and he definitely engaged in physical contact when he actually injected the nanoparticles. I cannot find any reported cases in which injecting a substance was held to constitute a use of force under a battery statute, but at least one case held that an injection could qualify as aggravated assault, and battery is generally a lesser-included offense of aggravated assault.

If the prosecution were to proceed under this theory, it seems the only conduct that would be relevant in the battery prosecution is that involved in the injection. In other words, it seems this approach might focus on the acts leading to the penetration of the patients' skin and the injections themselves, but not on what happened after the injections, i.e. the dissemination of the nanoparticles throughout the patients' bodies. I suppose that might not matter, since I assume the discomfort the patients endured is a circumstance that could be considered if and when Dr. X was sentenced for committing battery on both patients. Taking that into account at sentencing might ensure he was fairly punished for the "harms" he inflicted on the victims.[40]

Since simple battery is usually a misdemeanor (Perkins and Boyce : 152), focusing exclusively on the conduct leading to and resulting in the injection of the nanoparticles means that Dr. X would probably be charged with, and convicted of, two misdemeanors. States have also created the felony of "aggravated battery," which is variously defined as using a weapon, dangerous weapon or a dangerous instrumentality to commit battery (LaFave 2003: § 16.2(d)). Some states allow an aggravated battery charge to be predicated on the perpetrator's using "any poison or other noxious or destructive substance."[41] If Dr. X committed his crimes in a state with such a provision, the prosecution should be able to charge him with aggravated battery on the premise that his nanoparticles qualify at least as a "noxious substance."

It seems, then, that nanotechnology *could* be used to commit crimes against persons (with the exception of rape and bigamy). In this regard, nanotechnology may provide more opportunities for criminal exploitation than computer technology. Crimes against persons generally involve inflicting a physical "harm" on the victim,[42] but computers are not a physical medium. While it may be possible to use computer technology to physically "harm" a human being, there are no reported cases in which that has occurred; and the likelihood of computer technology being used to achieve such a result is probably low due to the difficulty involved in carrying out such an effort and the fact that there are so many easier ways to inflict physical "harm" on individuals. Nanotechnology, on the other hand, is a physical medium; as we saw above, it is likely that nanotechnology can be manipulated so as to "harm" human beings.

(ii) Crimes against property The crimes against property are more varied than the crimes against persons. They include theft, robbery, counterfeiting, fraud, forgery, vandalism, arson, receiving stolen property and extortion (Perkins and Boyce 1982: xxi–xxiv). Since our purpose is to analyze the possibility that nanotechnology could be used to commit crimes in each of the three categories listed above, we will not consider how it could be used to commit all the crimes

against property. We will, instead, proceed as we did with the crimes against persons, i.e. we will analyze how nanotechnology could be used to commit some representative crimes against persons.

We begin with two related crimes: counterfeiting and forgery. Historically, counterfeiting consisted of "making false money" and passing it off as genuine (ibid.: 432). Forgery consisted of making "a false writing having apparent legal significance" with the "intent to defraud" (ibid.: 414). The distinction between counterfeiting and forgery was well established as long as "money" consisted of coins, but it began to erode with the use of paper currency (ibid.: 432). As a result, the two terms often appear together in statutes that criminalize counterfeiting/ forging money, property and other items, such as election returns.[43] Both crimes have also broadened in scope: Goods bearing unauthorized trademarks are often referred to as "counterfeit goods" and forgery has expanded to encompass the falsification of items – e.g. art – as well as documents (Pinto *et al.* 2010). Logically, then, nanotechnology could perhaps be used to commit counterfeiting/forgery in either or both of two ways: falsifying documents (including money)[44] and/or counterfeiting goods.[45]

I use distinct terms to refer to what are versions of the same crime because I believe the "document" versus "goods" crimes inflict "harms" that are conceptually distinct, at least to some degree. The forgery "harm" consists of telling a lie; a forged document implicitly communicates a misstatement of fact that invalidates the document (LaFave 2003: § 19.7(j)(5)). The counterfeiting "harm" is analogous to the "harm" encompassed by the common law crime of adulteration: passing off debased goods – goods the quality of which was deliberately diminished in processing – as legitimate.[46] Since the adulterated goods are not what they are represented to be, the adulteration "harm" includes a lie but the lie is not as all-encompassing as the forgery lie. The forgery lie is a zero-sum lie; the forged document is not at all what it is represented to be. The counterfeiting lie is a less than zero-sum lie; the counterfeited good is not entirely what it is represented to be. While this distinction has been neither particularly apparent nor particularly important in criminal law to this point in time, I suspect it may become significant if and when nanotechnology is used to create counterfeit goods. Before we consider counterfeiting, though, we need to consider the potential for nano-forgery.

Nanotechnology seems unlikely to play a significant role in falsifying documents, at least not for the foreseeable future. Today most documents – especially documents that have legal and/ or financial significance – are computer-generated. As long as that is true, computers will probably continue to be the preferred means for falsifying documents.[47] Nanotechnology might, though, still play some role in the computer-falsification of documents; nanotechnology-based inks could make it easier to generate documents that appear genuine (see generally Savastano 2004) and/or to produce self-erasing forgeries that serve their purpose and then eliminate incriminating evidence (*Nanowerk* 2009).

Falsifying documents may not be an area of criminal endeavor in which nanotechnology will play a significant role. That in no way diminishes nanotechnology's potential as a vector for facilitating tool crimes; as we saw earlier, computer technology is a tool that can be used to facilitate a variety of crimes. But computer technology cannot be used to facilitate *all* crimes; it is, as noted earlier, particularly unsuited for facilitating the commission of crimes that involve inflicting physical "harm" on persons.[48]

Nanotechnology is likely to play a notable role in counterfeiting goods. As we saw in § II(A), manufacturing goods of various types is expected to be one of the major applications of nanotechnology. The process of counterfeiting goods requires that counterfeiters have access to the technology needed to create the fake goods and an adequate supply of labor willing to produce the fakes. Counterfeiters use a production process that is analogous, but often inferior, to the process used by the legitimate manufacturers of the goods.[49] Goods counterfeiting has, as a result, tended to be product-specific, especially if the product is relatively complex, like an iPhone (Liu and Sung 2007).

Depending on its accessibility and complexity, nanotechnology manufacturing seems likely to be exploited by goods counterfeiters. They would use the technology in the same way, and for the same purposes, as those who employ it legitimately. The initial hurdle goods counterfeiters are likely to face is acquiring the technology itself; they might be able to accomplish this by compromising an employee of a company that is using nanotechnology in the manufacture of certain goods.[50] If the counterfeiters were able to acquire the technology, they might need individuals with expertise in how it should be utilized; they could address this issue by hiring ex-employees of a company utilizing the technology to work in their illegitimate goods factories (or to consult on the operation of such factories).[51]

Operationally, nano-goods counterfeiting would probably be very similar to conventional goods counterfeiting. The doctrinal issue is whether the use of nanotechnology to copy goods would make the application of current counterfeit goods law to nano-counterfeits problematic. Counterfeit goods law is based on copyright and trademark principles: Basically, a counterfeit good is (i) an unauthorized copy of an item in which someone holds a valid copyright; or (ii) a copy that bears an unauthorized version of a valid trademark owned by someone else (see e.g. Otten and Wager 1996). The "harm" inflicted by each alternative is the owner's loss of some quantum of the value of an intangible property right.

As noted earlier, counterfeiting originally encompassed the same "harm" as the common law crime of adulteration: the practice of diluting the value of a commodity (money) by passing a debased version of it. Contemporary goods counterfeiting law implicitly encompasses that "harm" (since many counterfeit goods are of inferior quality) but it also encompasses another, newer "harm." Since counterfeit goods are often produced in the same factories that produce legitimate versions of a good, the "harm" resulting from their production and sale is not limited to the traditional adulteration "harm;" it includes that "harm" plus a similar "harm," i.e., the loss of value resulting from the production and sale of identical but unauthorized versions of a good.

We do not know what fully mature nanotechnology manufacturing will look like or will be capable of doing, but it seems reasonable to assume that it will involve fabricating items in a manner similar, but superior, to the manufacturing processes in use. It also seems reasonable to assume that to the extent evolved nanotechnology manufacturing is corrupted and used to make unauthorized copies of goods, the process will inflict one or both of the "harms" noted above. That is, it seems reasonable to assume that while a technologically superior process should produce superior (but counterfeit) products, it can also produce inferior (and counterfeit) products. Logically, then, it seems unlikely that existing counterfeit goods law will require major alteration as we move into eras of partial and then increasing nanomanufacture.

There might, though, be a residual scenario: Goods counterfeiters create unauthorized copies of consumer goods because they are relying on a mass market for their profits, which come from selling as many of the copies as possible, usually at reduced prices. Counterfeiting, as such, has not generally involved creating unauthorized versions of unique or extremely rare commodities, such as valuable jewels or works of art.[52] What if we assume, for the purposes of analysis, that nanomanufacturing reaches the point at which it can be used to make a copy of any item that is indistinguishable from the original? It might, for example, be possible to make a perfect copy of the *Mona Lisa* or of the Hope Diamond.

If someone used advanced nanomanufacturing to create an identical, indistinguishable, copy of the *Mona Lisa*, would that be a crime against property? It would not constitute goods counterfeiting because the copy neither violates a copyright nor a trademark held by someone other than the copier. It *could* constitute fraud if the purchaser of the new *Mona Lisa* had been told she was buying the original, the only original, of the painting; but to make the analysis more interesting, we will assume she was told she was buying an identical copy and was quite happy to obtain such a copy of the *Mona Lisa* for what she considered to be a relatively small sum. Since she knows what she purchased and paid what she considers a fair price for the item, we do not have fraud.

Do we have theft? Can the Louvre Museum, which owns the original *Mona Lisa*, file a criminal complaint for theft against the creator of the new version of the painting in, say, New York City, where he lives and works? That, of course, depends on whether or not what he did constitutes theft, and to resolve that issue we need to briefly consider a computer crime case.

A computer specialist (Doe) worked as a contractor for the Intel Corporation in a division that created complex computer systems for the US military.[53] After he had worked in that division for a while, Doe had a falling out with his supervisor and was transferred to another division; most of the passwords he used to access the computers in the original division were disabled, but for some reason one was not. He continued to work at Intel and continued to use that password to access computers in the original division; Doe copied and downloaded the file that contained the passwords for all of the authorized users of the computers in the original division and stored it on a computer he controlled. At that point, his activity was discovered and Intel went to the police. Doe was charged with and convicted of computer theft, i.e., using a computer to commit theft, and appealed. On appeal, Doe claimed he had not committed theft because Intel had not "lost" anything: It still had the original file containing all of the passwords and all of the passwords were still usable.[54]

Doe's argument was quite correct, as a matter of traditional law. Theft has always been a zero-sum transaction: If a thief takes my bag, the possession of that bag shifts completely from me to him; he has it and I do not. The thief cannot copy my bag, so that he has it and all its contents and I do, too. Doe's argument was therefore valid as far as traditional criminal law was concerned; had he been charged under a traditional theft statute, one that defined theft as taking property with the intent to permanently deprive the owner of its possession and use,[55] the appellate court would no doubt have had to reverse his conviction. But Doe was unlucky: He was charged under a new computer theft statute that defined the crime in part as taking "proprietary information."[56] The appellate court was therefore able to affirm the conviction because it found, essentially, that by taking a copy of the passwords the ex-employee deprived Intel of some portion of the value of the passwords: Intel had not lost the passwords, as such, and they still functioned as passwords but their value was significantly diminished once Intel lost exclusive possession and control of them.

That brings us back to the new *Mona Lisa*. Something similar occurs in this scenario: The original version of the painting is still intact, still possesses all of the qualities that have made it admired and respected for centuries. Its tactile integrity has not been compromised. So the person who made the copy could, if he were to be charged with theft, make the same argument Doe made in the scenario analyzed above: He did not "take" anything from the Louvre Museum. They still have the original, the "real," *Mona Lisa*. All he did is to make a copy; he might even point out that the museum has allowed photographs of the painting to be taken and published, and suggest that what he did was analogous to that. To sustain the theft charge, the prosecution (and the Louvre) would have to show how and why the museum "lost" something due to the fabrication of the copy.

In the Intel case, simply making the copy eroded the value of the passwords, just as making a copy of a key erodes its value. Once copies exist, the original key or password still functions as an access device, but it has lost all or much of its utility as a security device. The prosecution (and the Louvre) would have to show that the museum "lost" some quantum of the painting's intrinsic value once the copy was created. In other words, they would have to show that while they had not lost the *Mona Lisa* as such, they had lost a quantum of its ... what? In the Intel case, the lost commodity – the lost quantum "value" – was a utilitarian function, i.e. the ability to keep certain systems secure from unauthorized users. If and when unauthorized nanotechnology copying of tangible (and perhaps unique) items becomes a reality, law will have to decide if it will pursue a similar approach to nanotechnology-copying and if so, how it will approach the issue of a "loss" of property.

If law decides to treat unauthorized nanotechnology-copying of tangible items as theft, then it will presumably have to utilize an approach analogous to that which the Oregon Court of Appeals employed in the case described above. The Oregon court implicitly acknowledged that theft can be zero-sum or less than zero-sum; in other words, a thief can deprive me of all or only part of the "value" of my property. The focus shifts from the physical item, as such, to features of the item, such as its utility as a security device. When theft statutes refer to "value", they tend to define it in purely monetary terms,[57] but *Black's Law Dictionary* defines it more expansively, i.e. as the "significance, desirability, or utility of something" (Garner 2004). The prosecutor who is pursuing the creator of the new *Mona Lisa* in the hypothesis outlined above might, with the assistance of Louvre Museum experts, argue that making the copy eroded some esthetic or cultural "value" associated with the original painting's status as the *only* version of that painting in existence. Or we could simply accept that nothing – at least nothing the existence and nature of which was known publicly – could ever be unique in a nanotechnology world.

As this discussion illustrates, if and when nanotechnology moves into general public use and becomes accessible to criminals and to those who are willing to facilitate criminal activity, criminal law will almost certainly have to confront new, nanotechnology-facilitated property crimes. Some of the crimes, such as forging documents and copying consumer goods, may not require significant changes in existing law and/or the adoption of new, nanotechnology-specific criminal laws. Others, such as the *Mona Lisa* scenario, may require a reassessment of existing law and policy to determine the extent to which the technology is being utilized in a manner that cannot satisfactorily be addressed by the civil justice system.

(iii) Crimes against the state   Crimes against the state do not target the infliction of particular "harms" upon civilian victims, whether individuals or artificial entities. Crimes against the state target conduct that erodes or threatens to erode the state's ability to enforce its laws and thereby maintain social order. There are many different crimes against the state, such as treason, official bribery, escape, riot, perjury and obstructing justice (see Perkins and Boyce 1982: xxiv–xxvii, listing crimes). While nanotechnology may some day be used to facilitate treason, bribery, escape or even riot, I cannot, at this point in time, begin to speculate on how that might come about.

I can, thanks to the unwitting assistance of some science fiction writers, speculate a bit about how nanotechnology could be used to obstruct justice, at least in the sense of tampering with evidence. Obstruction of justice statutes encompass a wide variety of conduct (see Perkins and Boyce: 552–579), but our concern is limited to obstruction that consists of creating, altering and/or destroying evidence with the intent to obstruct a criminal investigation or prosecution (ibid.: 558–559).[58]

More precisely, our concern is with the development and use of canned obstruction of justice techniques, i.e. standardized techniques anyone who has access to can use to destroy certain types of evidence. To understand what that could mean, we need to review a phenomenon that has emerged in the area of computer crime: anti-forensics. As one website explains,

> anti-forensics in the realm of ... computer forensics involves the hiding, destroying, and disguising of data. ... One major goal of anti-forensics is to make analysis and examination of digital evidence as difficult and as confusing as possible. Today, thwarting an investigation has never been easier.[59]

Computer anti-forensics uses several general techniques, e.g. hiding data, securely destroying data or preventing data from being created, each of which involves the utilization of various tools (Booth 2015). Most anti-forensics software is available online for free on sites that often provide at least some instruction in its use (Berinato 2007).

Computer anti-forensics gives non-expert computer users the ability (i) to eliminate or obfuscate computer-generated evidence or (ii) to make the process of investigating a computer crime so complex, time-consuming and expensive that the inquiry may not proceed (see e.g. Behr 2008: 13). It essentially automates the process of obstructing justice by destroying, altering hiding and/or creating evidence.

Anti-forensics automates the process of obstructing justice in the investigation of digital crimes. Might nanotechnology some day play a similar role with regard to the investigation of physical crimes?

As anyone who has watched *CSI* on TV, seen a movie involving the investigation of a crime or read murder mysteries knows, investigations of physical crimes focus on the place – the scene – where the crime was committed. Modern crime scene investigation is based on a principle enunciated by Edmond Locard in 1920, which is that "the criminal leaves marks at the crime scene of his passage" and "takes with him, on his body or on his clothing, evidence of his stay or of his deed" (Horswell 2004). Modern technologies – including DNA – have reinforced the importance of Locard's principle: However clever a criminal is, however much she tries to conceal what she did, she inevitably leaves trace evidence establishing her presence at the scene of the crime and takes trace evidence from the scene with her (see e.g. Pepper 2005).

At least, that is where things stand today. I found an interesting nanotech-crime scene scenario in a short story by science-fiction writer A.M. Dellamonica (2007). In the story, humans have fled Earth and are living as refugees on a planet inhabited by aliens who call themselves the Kabu but whom humans call "the Squids," given certain aspects of their physical appearance (ibid.: 589–590). In the middle of the night, a woman arrives at a building in the refugee area, having been called there by her nephew (ibid.: 590). She knocks on the designated apartment door and when her nephew opens it, she sees a dead human female and a dead male Squid; her nephew says the Squid killed the woman and he killed the Squid (ibid.: 590–591).

The first thing the woman – a police officer on Earth – does is to unpack "an assortment of sprays and other nanotech" she'd assembled before leaving Earth (ibid.) She sprays "nanosols onto a towel" and uses it to wipe several areas before laying it on the floor and having her nephew walk on it "a couple times" before he moves into the hallway. She then uses another spray – one that "devours dead skin cells ..., hairs, sweat, tears, blood – anything that might leave a trace" to remove other evidence that may have been left at the crime scene (ibid.: 591: "developed ... to keep ... investigators from contaminating crime scenes").

There, then, is a fictional example of nano-anti-forensics. As we saw above, digital anti-forensics tools are usually employed *ex ante*, i.e. prior to the time a crime is committed or while the crime is being committed. In this fictional scenario, the aunt uses the nano-anti-forensics techniques to clean up the crime scene after the crime has been committed; this might not have been necessary if the nephew had utilized the second anti-forensics spray before entering the apartment and committing the crime. If that tool prevented him from leaving trace evidence, he would not have to clean up afterward, though he might want to use this tool or a similar tool to remove any trace evidence he collected at the crime scene before he left it.

Tools such as this obviously do not exist, but the possibility that they may some day be developed does not seem outside the realm of possibility. The Environmental Protection Agency and other entities are studying how nanoparticles can be used to clean polluted sites by "absorbing contaminants and transforming them into nontoxic forms" (US Environmental Protection Agency n.d.). In the story, "nanosols" seem to "eat" trace evidence; but instead of consuming pollutants, the environmental clean-up nanoparticles transform them into something else, which could presumably also work in the nano-anti-forensics context. The purpose is to defeat Locard's principle by ensuring that no identifiable trace evidence from the perpetrator is left at the crime scene and no identifiable trace evidence from the crime scene is found on the perpetrator. If the nano-anti-forensic tool transforms trace evidence of either type into something else, i.e. transforms skin cells into dust, that seems to serve the purpose (see *Field Practice of Anti-Forensics* 2009).

My other fictional example of nano-anti-forensics involves an *ex ante* evidence destruction tactic. In Ian McDonald's book *Brasyl*, which is set partially in the São Paulo of 2032, a man uses a one-shot disposable handgun to kill a woman (McDonald 2007: 27–28). He takes the gun out, pulls "the strip" on it ("it began to decompose immediately"), points it at the woman, pulls the trigger and it fires (ibid.: 27). He throws the gun into the gutter, where it dissolves into "black ... liquid and drips from the rungs of the grating into the sewer" (ibid.: 28). McDonald does not specifically identify the dissolving gun as a nanotech-weapon, but I assume that is what he meant the reader to infer; nanotechnology, after all, seems the obvious technology for creating such a weapon.

This is another approach to destroying trace evidence or, perhaps in this instance, traceable evidence. If the gun had not dissolved but was found in the gutter, ballistics experts would have been able to identify what kind (model and caliber) of gun it was. They might have been able to use its serial number to trace it to the store that sold it to someone, perhaps the shooter, perhaps someone else. They would certainly have been able to identify the striations the gun left on a bullet when it fired and match markings on the cartridge case to marks in the gun's chamber and breech.[60] They might also have been able to find fingerprints on the gun.

My point is that while investigators would not have known who the shooter was, they might have been able to find him by tracing evidence derived from the gun. Since the gun self-destructed, they were denied that opportunity. Even if a witness identified a man as the shooter, police would have no additional evidence they could use to prove his guilt. Whatever trace evidence he left on a busy street corner in São Paulo would have disappeared by the time they arrived at the crime scene, and they had no murder weapon. If we assume the bullet did not self-destruct after arriving at its destination, investigators could have analyzed the markings on it, which would have allowed them to link the bullet to the gun – if it still existed. If the bullet did self-destruct after inflicting the necessary damage on the victim, then that option disappears, as well.

Obviously, it will be a long time before lawyers and law enforcement officers actually have to deal with scenarios like these. And they may not ever come to pass. I, for one, would prefer to believe that those trained in nanotechnology will not be inclined to use their expertise to help facilitate crimes. Even if that turns out to be true, at least two of these fictional scenarios involve the criminal exploitation of techniques that were developed for legitimate purposes.

The availability of nano-anti-forensics might create legal questions beyond whether someone who used such a technique to destroy evidence could be prosecuted for obstructing justice. One issue that might arise is whether nano-anti-forensics devices should be illegal, i.e. whether creating, marketing and/or possessing nano-anti-forensics should be a crime.[61] This issue would arise only for the nano-anti-forensic devices that were deliberately created for the specific purpose of destroying evidence; and the permissibility of criminalizing such devices would depend, at least in part, on the extent to which their only use was for criminal activity, i.e. obstructing justice.[62] When a legitimate product was used as a nano-anti-forensics device, the issue might arise as to whether the person(s) who supplied the device to the criminal could be held liable for aiding and abetting the resultant crime(s).

(C) TECHNOLOGY INCIDENTAL

When computer technology plays an incidental role in the commission of an offense its involvement is so minimal it does not rise to the level of transforming the crime into a tool crime. We could simply ignore the computer's involvement in the offense, just as we ignore the involvement of other, more routine technologies, e.g. electric light, telephones, even automobiles. The rationale for including the "incidental" category in the taxonomy used to classify computer-related crimes is that the computer can be an important source of evidence about the crime and the person(s) who committed it. In other words, the assumption is that the computer plays a more active role in the commission of the crime than other, more passive technologies and consequently is more likely to be a source of useful evidence.

As others have noted, criminals could use nanotechnology to disguise their identities while committing crimes (Fiedler and Reynolds 1994). They might do this simply to avoid being caught and prosecuted; or they might use nanotechnology to create "biometric spoofs" that let them circumvent "biometric security detectors" and gain entry to the premises where they intend to commit a crime (Boucher 2008: 219). In other words, nanotechnology could in effect become a biometric lock pick (ibid.: 218–219).

The production, sale and consumption of illegal drugs has for years been a major source of criminal activity. Nanotechnology might take that to the next level by facilitating the creation and marketing of new drugs and/or drug surrogates. By "drug surrogates," I mean substances that are not chemically based but when ingested have effects similar to that of substances such as cocaine, heroin, LSD, etc. Nanotechnology might be used to create non-chemical surrogates that influence the human body in ways that differ from, but are analogous to, those associated with chemically based substances like cocaine. It might be possible to customize the user's experience in terms of factors such as the length, intensity and nature of the "high" that results from consuming a nano-surrogate (ibid.). And nano-drug surrogates would offer drug dealers an option they currently do not have: designer drugs, i.e. the ability to create new and newer surrogates to maintain or expand their customer base (ibid.). And since nano-drug surrogates would not be chemically based, it might be easier for those engaged in their manufacture to conceal their operations from the authorities (ibid.).

It is difficult at this point in time to project all the ways in which nanotechnology might incidentally contribute to the commission of various crimes, just as it would have been difficult for someone writing in, say, 1977 to imagine the many and varied ways computer technology would contribute to criminal activity. These examples, I hope, illustrate how nanotechnology may be integrated into the commission of known crimes; it may also be that nanotechnology, like computer technology, gives rise to a class of previously unknown crimes, in which it plays a role of greater or lesser importance.

B. Lessons learned?

As I have noted elsewhere (Brenner 2006), one lesson I believe we have learned from our experience with cybercrime is to be parsimonious in adopting new, technologically specific criminal laws. When cybercrime was still a very new phenomenon, some jurisdictions tended, at least in my opinion, to over-react by adopting laws that were specifically directed at computer-facilitated analogous of crimes they had already outlawed (ibid.: 768–769).

Some US states did this with harassment: US states began criminalizing harassment about a century ago, as it became clear that telephones could be used for new and unintended purposes, i.e. to make obscene and otherwise harassing phone calls (ibid.: 768). States responded to this new technological crime by adopting use-of-a-telephone-to-harass statutes (ibid.). About eighty years later, as computers began to be used for the same purpose, some states simply added a new crime, i.e. they adopted use-of-a-computer-to-harass-statutes (ibid.: 768–769). Since criminal law is focused on the infliction of "harms" rather than on technology, as such, the better approach would have been to incorporate computer harassment into the telephone harassment statutes the states had adopted years earlier (ibid.: 769–770). Aside from anything else, the unnecessary use of technologically specific laws has certain disadvantages:

> It produces overlapping rules (e.g. rules outlawing theft, rules outlawing the use of computers to commit theft, and rules outlawing the theft of computers). The focus on method instead of result also produces rules that are transient. We began with use-of-a-telephone-to-harass rules and added use-of-a-computer-to-harass rules; this leaves us … with rules that may or may not overlap (for example, if one uses a computer to access a telephone line and uses that connection to harass another, is this use-of-a-telephone-to-harass, use-of-a-computer-to-harass, or both?).
>
> *(ibid.: 769)*

The adoption of technologically specific harassment legislation is but one example of how many jurisdictions responded to the rise of computer crime.[63] The question is, will we respond to nanocrime in a similar fashion?

Nanocrimes differ from real-world crimes and from cybercrimes in at least one respect: In real-world crimes, the perpetrator physically commits the crime himself (perhaps with the assistance of accomplices). In cybercrimes, the perpetrator also physically commits the crimes himself, using computer code as the intermediary device by which he acts "in" the virtual world of cyberspace.

In nanocrimes, the perpetrator's role may be much more attenuated, more analogous to that of someone who sends malware out to wreak generalized havoc on computers than it is to the hands-on role of traditional criminals and most cybercriminals. Nano-perpetrators will presumably operate in a fashion analogous to that of our fictional Dr. X, i.e. they will create nanoparticles that are designed to implement their criminal schemes and send them out to do the dirty work. From what I know of the current state of nanotechnology, my sense is that nanoparticles will be created to perform particular functions (like finding a stent and offloading their drug cargo to it) autonomously. In that regard, they are again analogous to malware (though unlike malware it appears that they will be able to carry out tasks that could directly inflict physical "harm" on human beings).

If all of that is true, and if nanotechnology does not progress beyond the point at which nanoparticles are simply programmed to perform limited functions and then launched to do just that, it seems we could use malware laws as the basis for developing nano-crime laws. In other words, if nanotechnology does not evolve to the point at which nanoparticles are capable of independent (some level of artificial intelligence) or derivative (ability to extrapolate from basic functions to perform functions derived from that set of behaviors) action, it may simply become a tangible manifestation of malware.

More difficult questions will arise if nanotechnology progresses beyond simple artificial intelligence and develops the capacity for truly intelligent, autonomous action. If that eventuates, and if the evolved nanotechnology engages in activity we define as criminal, we would have to decide how criminal law should address either or both of two scenarios.

The first scenario, the "Frankenstein scenario", focuses on how law should deal with a human being who intentionally creates nanoentities that have the capacity for independent action and releases them into a context in which they can function knowing that their actions will be foreseeable for some period of time but may at some point evolve beyond what their creator intended or foresaw. If the nanoentities inflicted "harms" that come within the traditional scope of the criminal law,[64] we would have to decide whether to hold the creator liable (i) only for the "harms" he foresaw and/or intended or (ii) for both those "harms" and for the unintended and unanticipated "harms", the infliction of which was a proximate result of releasing the entities.

The second scenario takes us even further into a science fiction reality: If the scenario outlined above were to occur, we might also have to decide whether we would want to treat semi-intelligent, autonomous constructs as entities subject to the imposition of criminal liability. In other words, we might have to decide whether criminal law – or a type of criminal law – could apply to autonomous, intelligent nano-constructs.

IV. Conclusion

It is not surprising that it took years before American lawmakers realized the need to take computer crime seriously by adopting legislation that criminalized certain kinds of computer-facilitated activity. When personal computers were introduced, coincidentally with the Internet, no one had any reason to anticipate that the interaction of the two would create new and unprecedented opportunities for criminal activity.

Personal computers were only the latest in a series of communications technologies – telephones, radio, television, motion pictures – that appeared between 1880 and 1980 (Brenner 2006: 708–743). Three of the technologies – radio, television and motion pictures – had essentially no capacity to be exploited for criminal purposes because each was a passive communications technology, i.e. in each, content was broadcast to an audience whose only options were to accept or reject it (ibid.: 725–729). Telephones had some capacity to facilitate criminal activity; phones could be used to facilitate fraud and to harass others anonymously. It was a relatively simple matter for legislators to adopt statutes criminalizing the use of telephones (or "the wires") to commit fraud and other crimes, including harassment.

Given our history with technology to that point, it is not surprising that no one anticipated the extraordinary opportunities networked computers would create for enterprising criminals. Nor is it particularly surprising that it took years for legislators to adopt laws that adequately addressed computer-facilitated criminality. Neither the legislators nor law enforcement nor the general public had a model of technologically facilitated crime they could use as a guide in understanding what needed to be done to respond to the new wave of computer crimes.

We *might* be approaching a new era of technologically facilitated crime – the nanocrime examined in this article. Whether nanocrime emerges and whether it becomes a matter of serious concern depends, as noted earlier, on the extent to which nanotechnology evolves from a "laboratory" technology to a "democratic" technology. If nanotechnology remains a "laboratory" technology, its potential for criminal exploitation will be essentially non-existent; if, on the other hand, nanotechnology becomes a "democratic" technology, its potential for criminal exploitation rises, perhaps equaling or exceeding that of computer technology.

I, of course, have no way of knowing whether nanotechnology will make the transition to "democratic" technology or not; and neither I nor anyone else at this point has any way of knowing the extent to which nanotechnology will be used for criminal purposes, if and when it makes this transition. As I noted earlier, my purpose in writing this chapter is not to answer these questions. It is to raise them and, in so doing, encourage those of us who have expertise in crime and criminal law to do what our counterparts could not do thirty-odd years ago: (i) educate ourselves about this emerging technology's capacity to facilitate the commission of crimes and (ii) monitor the development (if any) of such a capacity and anticipate how criminal law should respond if and when it appears. In other words, my goal is to encourage an effort that is the criminal counterpart of the efforts that are underway to develop civil regulatory systems that can respond to the inadvertent hazards associated with nanotechnology.

Notes

1 See Brenner (2011). "Nanocrime" is a neologism for which I cannot claim responsibility. The term was in use by 1998, but probably originated earlier. See, e.g. Nanotech (1998) on nanocrime and "nanocrime tracking".
2 See, e.g. Integrated Nanosystems Research Facility, http://www.inrf.uci.edu/about/.
3 Most traditional contaminants cannot cross the blood-brain barrier.
4 See, e.g. "Malware," Wikipedia, http://en.wikipedia.org/wiki/Malware.
5 See, e.g. "Denial-of-service attack," Wikipedia, http://en.wikipedia.org/wiki/Denial_of_service_attack.
6 See, e.g. Brooks v. United States, 267 U.S. 432, 438–441 (1925).
7 The description of the evolution of cybercrime is taken from Brenner (2010: 9–37).
8 See "Mainframe Computer," Wikipedia, http://en.wikipedia.org/wiki/Mainframe_computer ("Users gained access through specialized terminals").
9 See "Mainframe Computer".
10 See United States v. Lambert, 446 F. Supp. 890 (D. Conn. 1978).
11 See, e.g. "The Nagging Feeling" of Undetected Fraud, U.S. News & World Report 42 (December 19, 1977).
12 In the early 1970s, there were a few attempts to sabotage mainframes. See, e.g., McKnight (1973: 83–108).
13 "Internet," Wikipedia, http://en.wikipedia.org/wiki/Internet.
14 "World Wide Web," Wikipedia, http://en.wikipedia.org/wiki/World_wide_web.
15 See "Internet," Wikipedia, http://en.wikipedia.org/wiki/Internet. See also "Mosaic (web browser)," Wikipedia, http://en.wikipedia.org/wiki/Mosaic_%28web_browser%29.
16 See, e.g. "War Games," Wikipedia, http://en.wikipedia.org/wiki/WarGames.
17 See "War Games," Wikipedia. ("It introduced the world to the peril posed by hackers").
18 See Treaster 1983; see also "Bulletin board systems," Wikipedia, http://en.wikipedia.org/wiki/Bulletin_board_system.
19 The description in the following paragraphs of how cybercrime evolved from the 1980s to the twenty-first century is taken from Brenner 2010: 9–37.
20 For a more detailed description of the evolution of tool crimes, see Brenner (2010: 9–37 and 73–102).
21 This assumption is implicit in some of the early books dealing with computer crime. See, e.g. McKnight (1973: 98–113, 259).
22 See, e.g. Nanotech in Fashion: The Trend in New Fabrics, Morning Edition – NPR (September 7, 2004), http://www.npr.org/templates/story/story.php?storyId=3892457.
23 See "Wargames," Wikipedia.

24 We will assume for the purposes of analysis that the other elements of hacking are present on the facts, i.e., Dr. X knew he was not authorized to access the drug-bearing nanoparticles and it was his purpose to do so.

25 See Brenner (2001a: 12–16); the cybercrime framework seems to have been developed by attorneys for the US Department of Justice, see e.g. Charney and Alexander (1996), discussing the Department of Justice's Computer Crime Initiative; Charney (1992) describing an example of how the cybercrime framework seems to have been developed by attorneys for the US Department of Justice.

26 If our fictive Dr. X were charged with nanotech hacking based on his conduct in either or both scenarios, he might argue that because his nanoparticles were designed to do no harm, he committed no crime. Early hackers used this theory to argue that the law should not criminalize merely accessing a computer system without being authorized to do so. See, e.g., Charney and Alexander (1996).

27 A similar attitude toward computer crime was one of the problem legislatures faced in trying to adopt cybercrime law. See, e.g. Johnson (2009: 587–588).

28 See, e.g. *NYS Agrees on Penal Code that Makes Computer Tampering a Crime,* American Banker (April 16, 1984), 1984 WLNR 18502.

29 If our fictive Dr. X were charged with nanotech hacking based on his conduct in either or both scenarios, he might argue that because his nanoparticles were designed to do no harm, he committed no crime. Early hackers used this theory to argue that the law should not criminalize merely accessing a computer system without being authorized to do so. See, e.g. Charney and Alexander (1996).

30 See, e.g. de Guzman (2010: 530) (DDoS attack "overwhelms the target host, rendering it unable to respond to any other traffic").

31 The Model Penal Code and modern statutes limit battery to a use of force that inflicts physical injury and/or "unwanted sexual advances", but we utilize the older, broader standard in this analysis, LaFave (2003: §16.29(a)).

32 The use of poison would probably be discovered, which would alert police to the fact that a murder had occurred; the ruptured artery would probably be discovered, but might be attributed to natural causes.

33 The permutations are as follows: (i) poison nanoparticles injected into the patient with stents; (ii) poison nanoparticles injected into the patient with cancer; (iii) artery-rupturing nanoparticles injected into the patient with stents; and (iv) artery-rupturing nanoparticles injected into the patient with cancer. See above, § III(A)(2)(a).

34 See, e.g. *People v. Tackett,* 150 Ill.App.3d 406, 412-413, 501 N.E.2d 891, 896 (Ill. App. 1986).

35 See *People v. Tackett* (above). See also LaFave (2003: § 6.4(b) and § 6.4(c)).

36 Battery charges can be predicated on lesser *mens rea,* including negligence, but we will assume intentionality. See, e.g. LaFave (2003: § 16.2(c)).

37 See, e.g. *Agripino v. State,* 217 S.W.2d 707, 712–713 (Tex. App. 2007) (injecting mineral oil into someone constituted aggravated assault). Battery is generally a lesser-included offense of aggravated assault. See, e.g. *Martinez v. State,* 199 P.3d 526, 533 (Wyo. 2009) (battery was lesser-included offense of aggravated assault).

38 The use of force is excessive when the actor used more force "than is necessary." 6 Am. Jur. 2d Assault and Battery § 132.

39 Cal. Penal Code § 243.4(f).

40 See, e.g. *Edwards v. United States,* 523 U.S. 511, 514 (1998) (propriety of including conduct not encompassed in the offense of conviction in calculating the sentence).

41 See, e.g. Idaho Code Ann. § 18-907(1)(c).

42 See, e.g. Perkins and Boyce (1982: xvii–xx) (e.g. homicide, assault, battery, rape, child abuse, child rape, false imprisonment, kidnapping).

43 See, e.g. 18 U.S. Code § 471; 18 U.S. Code § 485; Ariz. Rev. Stat. § 16-1011(A).

44 See "Money," Garner, B.A., *Black's Law Dictionary* (8th ed. 2004) (defining "money" as "paper money," "e-money" and "hard money"). As to using a computer to counterfeit paper money, *see, e.g.* People v. Harrison. 283 Mich. App. 374, 376-377, 768 N.W.2d 98, 100 (Mich. App. 2009).

45 See "Goods," Garner, B.A., *Black's Law Dictionary* (8th ed. 2004) (defining goods, in part, as "[t]angible or movable personal property other than money").

46 See IV William Blackstone, Commentaries on the Laws of England 174.

47 Using a computer is clearly the preferred way to commit forgery today. See, e.g. *State v. Powell,* 306 S.W.3d 761, 762–763 (Tex. Crim. App. 2010).

48 See generally Bradley (2013).

49 See, e.g. Datz (2006); three types of counterfeit operations: legitimate factory also produces fakes; manufacturer hired to make a certain quantity of goods makes more and sells them on its own; and underground, usually low-technology, facilities.

50 See, e.g. *U.S. v. Chung*, 633 F.Supp.2d 1134, 1136–1137 (C.D. Cal. 2009) (employee stole proprietary information concerning employer's technology).

51 See generally *U.S. v. Case*, 2008 WL 1827429 *3- *5 (S.D. Miss. 2008) (economic espionage scheme used current and ex-employees of targeted company).

52 But see Conklin (1994) (noting mass production and sale of counterfeit prints in the United States between 1980 and 1987).

53 These and the other facts in this scenario are based on the facts in *State v. Schwartz*, 173 Or. App. 301, 303–306, 21 P.3d 1128, 1129–1131 (Or. App. 2001).

54 See 173 Or. App. At 315, 21 P.3d 1at 1135.

55 See, e.g. Colo. Rev. Stat. Ann. § 18-4-401(1)(a)-(c); Minn. Stat. Ann. § 609.52(2)(1)-(2).

56 See 173 Or. App. At 315, 21 P.3d 1at 1135 (quoting Or. Rev. Stat. § 164.377(2)(c)).

57 See, e.g. Ga. Code Ann. § 16-8-14(c); Minn. Stat. Ann. § 609.52(2)(3).

58 See also Model Penal Code § 241.7.

59 "About," Anti-Forensics, http://www.anti-forensics.com/introduction. See also Booth 2015.

60 See, e.g. "Ballistic fingerprinting," Wikipedia, http://en.wikipedia.org/wiki/Ballistic_fingerprinting.

61 At least one US state takes this approach to malicious computer software, e.g. viruses, worms, Trojan horse programs. *See* 18 Pa. Con. Stat. Ann. § 7616.

62 See generally *State v. Bui*, 104 Hawai'i 462, 466 n. 3, 92 P.3d 471, 475 n. 3 (Hawai'I 2004) (constitutionality of criminalizing possession of burglar's tools).

63 There were areas in which we needed to adopt new laws, i.e. DDoS attacks, but they were rare.

64 For a scenario in which this occurs inadvertently, see Lerner (2009).

References

Bahm, E. (2010) New Study Shows Possibilities and Dangers of Nanotechnology, Medill Reports (April 8) Available at: http://news.medill.northwestern.edu/chicago/news.aspx?id=162744&print=1.

Beckerson, L. (2010) UCLA Gold Nanoparticle Structure Blows Away Cancer Cells, *Daily Tech* (May 26) Available at: http://www.dailytech.com/UCLA+Gold+Nanoparticle+Superstructure+Blows+Away +Cancer+Cells/article18516.htm (accessed 1 Sept. 2016).

Behr, D.J. (2008) *Anti-Forensics: What It is, What It Does, and Why You Need to Know? New Jersey Layer Magazine*, 255: 4–90.

Bequai, A. (1978) *Computer Crime* Lexington, Mass: Lexington Books.

Berinato, S. (2007) The Rise of Anti-Forensics. CSO Online (June 8) Available at: http://www.csoonline. com/article/221208/the-rise-of-anti-forensics?page=1 (accessed 1 Sept. 2016).

Bolton, P., Scheinkman, J. and Xiong, W. (2005) Pay for Short-Term Performance: Executive Compensation in Speculative Markets, *J. Corp. L.* 30: 721, 723.

Booth, N. (2015) Rise of Anti-Forensics Techniques Requires Response from Digital Investigators, *MicroScope* (September) Available at: http://www.microscope.co.uk/opinion/Rise-of-anti-forensics-techniques-requires-response-from-digital-investigators (accessed 1 Sept. 2016).

Boucher, P.M. (2008) *Nanotechnology: Legal Aspects*. 218 Boca Raton: CRC Press.

Bradley, L.D. (2103) Regulating Weaponized Nanotechnology: How the International Criminal Court Offers a Way Forward, *Georgia Journal of International and Comparative Law* 41: 724, 728–732.

Breggin, L.K. and Carothers, L. (2006) Governing Uncertainty: The Nanotechnology Environmental, Health, and Safety Challenge, *Colum. J. Envt'l. L.* 31: 285, 288.

Brenner, S.W. (2001a) Defining Cybercrime: A Review of State and Federal Law, in R.D. Clifford (ed.) *Cybercrime: The Investigation, Prosecution and Defense of a Computer-Related Crime*. Durham, NC: Carolina Academic Press.

Brenner, S.W. (2001b) Is There Such a Thing as "Virtual Crime"?, *Cal. Crim. L. Rev.* 4(1): 17, 110–114.

Brenner, S.W. (2006) Law in an Era of Pervasive Technology, *Widener L.J.* 15: 667–784.

Brenner, S.W. (2007a) "At Light Speed:" Attribution and Response to Cybercrime/terrorism/warfare, *J. Crim. L. & Criminology* 97: 379, 385 (2007).

Brenner, S.W. (2007b) *Law in an Era of Smart Technology*. New York: Oxford University Press.

Brenner, S.W. (2009) *Cyberthreats: Emerging Fault Lines of the Nation-state*. Oxford: Oxford University Press.

Brenner, S.W. (2010) *Cybercrime: Criminal Threats from Cyberspace*. Santa Barbara, California: Praeger.

Brenner, S.W. (2011) Nanocrime? *University of Illinois Journal of Law, Technology and Policy* 39.

Brenner, S.W. and Clarke, L.L. (2005) Distributed Security: Preventing Cybercrime, *J. Marshall J. Computer & Info. L.* 23: 659, 660.

Brenner, S.W. and Rehberg, M. (2009) "Kiddie Crime"? The Utility of Criminal Law in Controlling Cyberbullying, 8 *First Amend. L. Rev.* 8(1): 15–23.

Bresnahan, T. (2007) Creative Destruction in the PC Industry, in Franco Malerba and Stefano Brusoni (eds), *Perspectives On Innovation*. New York: Cambridge University Press.

Bresnahan, T. and Trajtenberg, M. (1995) General Purpose Technologies "Engines of Growth"? *Journal of Econometrics* 83, 83.

Brindell, J.R. (2009) Nanotechnology and the Dilemmas Facing Business and Government, *Fla. B.J.* 83 Aug: 73, 76.

Campbell-Kelly, M. and Aspray, W. (1996) *Computer: A History of the Information Machine*. NY: Basic Books.

Center for Responsible Nanotechnology (n.d.) What Is Nanotechnology? Available at: http://www. crnano.org/whatis.htm (accessed 1 Sept. 2016).

Charney, S. (1992) The Justice Department Responds to the Growing Threat of Computer Crime, *Computer Security J.* 8: 1–12.

Charney, S. and Alexander, K. (1996) Computer Crime, *Emory L.J.* 45: 931, 954–957.

Conklin, J.E. (1994) *Art Crime*. Westport, CT: Praeger.

Datz, T. (2006) Counterfeiting: Faked in China (January 1), CSO Online. Available at: http://www. csoonline.com/article/220737/counterfeiting-faked-in-china (accessed 1 Sept. 2016).

de Guzman, T.L. (2010) Comment, Unleashing a Cure for the Botnet Zombie Plague: Cybertorts, Counterstrikes, and Privileges, *Cath. U. L. Rev.* 59: 527, 530.

Dellamonica, A.M. (2007) The Town on Blighted Sea, in *The Year's Best Science Fiction: Twenty-Fourth Annual Collection*. Gardner Dozois.

Dressler, K.E. (2005) Foreword, in J. Storrs Hall, *Nanofuture*. New York: Prometheus.

Drexler, K.E. (1990) *Engines of Creation 4*. New York: Oxford University Press.

Fiedler, F.A. and Reynolds, G.H. (1994) Legal Problems of Nanotechnology: An Overview, *S. Cal. Interdisc. L.J.* 3: 593, 623.

Field Practice of Anti-Forensics (2009) SLC Security (September 12, 2009), Available at: http://website. slcsecurity.com/index.php?option=com_content&view=article&id=124:field-practice-of-anti-forensi cs&catid=38:complap&Itemid=106.

Garner, B.A. (ed.) (2004) *Black's Law Dictionary, 8th edn*. New York: West Group.

Hogle, L.F. (2009) Science, Ethics, and the "Problems" of Governing Nanotechnologies, *J. L. Med. & Ethics* 37: 749, 753.

Horswell, J. (2004) *The Practice of Crime Scene Investigation*. London: CRC Press.

International Risk Governance Council (2007) Geneva, Policy Brief: Nanotechnology Risk Governance. Available at: http://www.irgc.org/IMG/pdf/PB_nanoFINAL2_2_.pdf.

Johnson, N.R. (2009) "I Agree" to Criminal Liability: Lori Drew's Prosecution under §1030(A)(2)(C) of the Computer Fraud and Abuse Act, and Why Every Internet User Should Care, *U. Ill. J.L. Tech. & Pol'y* 561, 587–588.

Joy, B. (1991) Nanotechnology: The Promise and Peril of Ultratiny Machines, *Futurist* (March 1).

Karn, B.P. and Bergeson, L.L. (2009) Green Nanotechnology: Straddling Promise and Uncertainty, *Nat. Resources & Env't* 24-Fall. 9: 10–11.

LaFave, W.F. (2003) *Substantive Criminal Law*, Criminal Practice Series. St. Paul, MN: West Group. (2nd edn).

Lerner, E.M. (2009) *Small Miracles*. New York: Tor Books.

Levy, S. (1984) *Hackers: Heroes of the Computer Revolution*. NY: Doubleday.

Lin, A.C. (2007) Size Matters: Regulating Nanotechnology, *Harv. Envtl. L. Rev.* 31: 349, 358–359

Liu, J. and Sung, C. (2007) IPhone Knockoffs Steal Sales as Apple Delays in Asia, Bloomberg.com (September 11) Available at: http://www.bloomberg.com/apps/news?pid=20601109&sid=a7K_I. ifMcEA&refer=home.

Mandel, G. (2008) Nanotechnology Governance, *Ala. L. Rev.* 59: 1323, 1329.

Maynard, A. (2016) We Don't Talk Much about Nanotechnology Risks Anymore, But That Doesn't Mean They're Gone, *The Conversation* (March 29) Available at: http://theconversation.com/we-dont-talk-much-about-nanotechnology-risks-anymore-but-that-doesnt-mean-theyre-gone-56889 (accessed 1 Sept. 2016).

McDonald, I. (2007) *Brasyl*. New York: Pyr.

McKnight, G. (1973) *Computer Crime*. New York: Walker and Company.

Michelson, E.S., Sandler, R. and Rejeski, D. (n.d.) Nanotechnology, in *From Birth to Death and Bench to Clinic: The Hastings Center Bioethics Book for Journalists, Policymakers, and Campaigns*. Available at: http://www.thehastingscenter.org/Publications/BriefingBook/Detail.aspx?id=2192 (accessed 1 Sept. 2016).

Nanotech (1998) Infowar: Are You Ready? Available at: http://www.nada.kth.se/~asa/InfoWar/nano.html

Nanowerk (2009) Nanotechnology Inks for Self-Erasing Paper (August 26) Available at: http://www.nanowerk.com/news/newsid=12320.php. (accessed 1 Sept. 2016).

Nossiter, B.D. (1977) Scotland Yard Deprograms Great Computer Tape Heist, *Washington Post* A15 (January 14).

Otten, A. and Wager, H. (1996) Compliance with TRIPS: The Emerging World View, *Vand. J. Transnat'l L.* 29: 391, 404.

Parker, D. (1976) *Crime by Computer*. Englewood Cliffs: Prentice Hall.

Pepper, I.K. (2005) *Crime Scene Investigation: Methods and Procedures*, Maidenhead: Open University Press, pp. 13–25.

Perkins, R.M. and Boyce, R.N. (1982) *Criminal Law*, 3rd edn. Mineola, NY: Foundation Press.

Pinto, R.J, Stein, E.J. and Savoca, C.B. (2010) Recent Developments in Trademark and Copyright Law, *Understanding Trademark and Copyright Developments for Online Content*, WL 1972540 *3.

Rakhlin, M. (2008) Regulating Nanotechnology: A Private-Public Insurance Solution, *Duke L. & Tech. Rev.* 2: 9.

Rasch, M.D. (1996) *The Internet And Business: A Lawyer's Guide To The Emerging Legal Issues*, Chapter 11 § II. Fairfax, VA: Computer Law Association 1996.

Ronfeldt, D. and Arquilla, J. (2001) What Next for Networks and Netwars? in Arquilla and Ronfeldt, *Networks and Netwars: The Future of Terror, Crime and Militancy*. Santa Monica, CA: Rand Corporation.

Sargent Jr., J.F. (2013) Nanotechnology: A Policy Primer, *Congressional Research Service* 11. Available at: https://www.fas.org/sgp/crs/misc/RL34511.pdf (accessed 1 Sept. 2016).

Savastano, D. (2004) NanoProducts Brings Nanotechnology to Ink Industry, *Ink World* (January). Available at: http://findarticles.com/p/articles/mi_hb3143/is_1_10/ai_n29068170/.

Science Daily (2010) Magnetic Fields Drive Drug-Loaded Nanoparticles to Reduce Blood Vessel Blockages in an Animal Study, (May 8), Available at: http://www.sciencedaily.com/releases/2010/04/100419150821.htm.

Sieber, U. (1998) Legal Aspects of Computer-Related Crime in the Information Society. COMCRIME Study, European Commission.

Storrs Hall, J. (2005) *Nanofuture*. New York: Prometheus.

Subramanian, V. (2009) *Active Nanotechnology: What Can We Expect?*, Program on Nanotechnology Research and System Assessment – Georgia Institute of Technology 1. Available at: http://www.cherry.gatech.edu/PUBS/09/STIP_AN.pdf.

The International Risk Governance Council, Geneva (2007) Policy Brief: Nanotechnology Risk Governance 7 Available at: http://www.irgc.org/IMG/pdf/PB_nanoFINAL2_2_.pdf (accessed 1 Sept. 2016).

Treaster, J.B. (1983) Hundreds of Youths Trading Data on Computer Break-ins, *New York Times* (September 5).

U.S. Environmental Protection Agency (n.d.) Using Nanotechnology to Detect, Clean Up and Prevent Environmental Pollution, Available at: http://www.epa.gov/nanoscience/quickfinder/pollution.htm. (accessed 1 Sept. 2016).

United Nations Educational, Scientific and Cultural Organization (UNESCO) (2006) *The Ethics and Politics of Nanotechnology* 14 Paris: UNESCO.

Wallace, D.L. (2008) Mediating The Uncertainty and Abstraction of Nanotechnology Promotion and Control: "Late" Lessons from Other "Early Warnings" In History, *Nanotechnology L. & Bus.* 5: 309, 309.

Whitt, R.S. and Schultze, S.S. (2009) The New "Emergence Economics" of Innovation and Growth, and What It Means for Communications Policy, *J. Telecomm. & High Tech. L.* 7: 217, 276.

Wolf, S.M., Gupta, R. and Kohlhepp, P. (2009) Gene Therapy Oversight: Lessons for Nanobiotechnology, *Journal of Law, Medicine & Ethics* 37 (4): 659–684.

37

AI and bad robots

The criminology of automation

Ugo Pagallo

1 Introduction

Since the early 1960s and for more than three decades, the field of robotics mostly appeared as an automobile industry-dependent sector. A crucial point in this process occurred in the early 1980s, when Japanese industry first began to implement this technology on a large scale in their factories, acquiring strategic competitiveness by decreasing costs and increasing the quality of their products. Western car producers learned a hard lesson and followed Japanese thinking, installing robots in their factories a few years later. This trend expanded so much that, according to the World 2005 Robotics Report, the automotive industry in Europe still received around 60–70 per cent of all new robot installations in the period 1997–2003 (UN 2005: ix).

Yet, in the same years as those covered by the UN World report, things began to rapidly change. The traditional dependence of robotics on the automobile industry dramatically opened up to diversification, first with water-surface and underwater unmanned vehicles, or 'UUVs', used for remote exploration work and the repairs of pipelines, oil rigs and so on, developing at an amazing pace since the mid-1990s. Ten years later, unmanned aerial vehicles ('UAVs'), or systems ('UAS'), upset the military field (Pagallo 2011). Over the past decade, robots have spread in both the industrial and service fields. Together with robots used in the manufacture of textiles and beverages, refining petroleum products and nuclear fuel, producing electrical machinery and domestic appliances, we also have a panoply of robot surgeons and robot servants, robot nannies and robot scientists, and even divabots, e.g. the Japanese pop star robot singer HRP-4C. The old idea of making machines (e.g. cars) through further machines (e.g. robots), has thus been joined – and increasingly replaced – by the aim to build fully autonomous robots, namely AI machines that 'sense', 'think', and 'act', in the engineering meaning of these words (e.g. Bekey 2005; Veruggio 2006). In a nutshell, this means that robots can respond to stimuli by changing the values of their properties or inner states and, furthermore, they can improve the rules through which those properties change without external stimuli. As a result, we are progressively dealing with agents, rather than simple tools of human interaction (Pagallo 2013).

However, the more robots become interactive, self-sufficient, and adaptable, the more attention should be drawn to the normative challenges of this technology, i.e. the reasons why such AI machines should, or should not, be deployed in our homes, in the market, or on the battlefield. Consider current debate on whether lethal force can be fully automated, or whether the intent to create robots that people bond with is ethically justifiable. Leaving aside the aim of the moral, political and economic fields in governing the process of technological innovation, the aim of this chapter is to cast light on the normative side of the law. The challenges of robotics may concern the pillars of current international law, criminal law, civil law, both in contracts and tort law, administrative law, and so forth. Think of tiny robotic helicopters employed in a jewellery heist.[1] Likewise, consider further robotic applications trading in auction markets and how the random-bidding strategy of these apps clarifies, or even has provoked, real life bubbles and crisis, e.g. the financial troubles of late 2009 that may have been triggered by the involvement of such artificial agents (Chopra and White 2011: 7). Since the 'bad nature' of robots and even their evil character vary in accordance with the field under examination, we have to further restrict the focus of the analysis, so as to pinpoint the specific legal challenges of this technology. For instance, in criminal law, the accountability for bad robotic behaviour is typically imposed on individuals who voluntarily commit a wrong prohibited by law; in contracts, the idea mostly regards compensation to those affected by the harmful behaviour of a counterparty through the robot; in tort law, payment follows from obligations between private persons usually imposed by the state to compensate for damage provoked by robotic wrongdoing. In this chapter, the focus will be on how the discipline of bad robots actually works in the field of criminal law.

The attention should be drawn, first of all, to the old idea that 'everything which is not prohibited is allowed'. This principle is connected to the clause of immunity summed up, in continental Europe, with the formula of the principle of legality, i.e. 'no crime, nor punishment without a criminal law' (*nullum crimen nulla poena sine lege*). Even though certain behaviours might be deemed as morally wrong, e.g. spying on people through domestic robots, individuals can be held criminally liable for that behaviour only on the basis of an explicit criminal norm. In the wording of Article 7 of the 1950 European Convention on Human Rights, '[n]o one shall be held guilty of any criminal offence on account of any act or omission which did not constitute a criminal offence under national or international law at the time when it was committed'. Contrary to the field of civil (as opposed to criminal) law, in which analogy often plays a crucial role so as to determine individual liability, we have thus to determine whether robots may produce a novel generation of loopholes in the criminal law field, forcing lawmakers to intervene at both national and international levels. In addition, we should ascertain whether, and to what extent, the increasing autonomy of robots can affect tenets of this field, such as the notion of an agent's culpability (i.e. its *mens rea*), vis-à-vis matters of criminal conduct (i.e. the *actus reus*).

In order to offer a (hopefully) comprehensive analysis of these issues, the chapter is divided into three sections. Section 2 illustrates today's state-of-the-art in criminal law, and why the level of robotic autonomy is at times sufficient to produce relevant effects in the field of, say, contractual obligations, but arguably is insufficient to bring today's robots before judges and have them declared guilty in criminal courts. Against this backdrop, Section 3 explores the scenario of new robotic crimes, by considering both hypotheses of robots with a criminal mind of their own (Section 3.1), and new types of robotic offences that may fall within the loopholes of current legal systems (Section 3.2). Then, Section 4 proposes a pragmatic approach, in order to define whether new robotic offences have to be established in addition to that which several scholars argue in the fields of international humanitarian law and current laws of war. Should we follow, sooner or later, the option taken by international lawmakers with the Budapest Convention on Cybercrime in November 2001? In other words, is a new convention on robotic crimes already necessary?

2 The current state-of-the-art

The current state-of-the-art in legal science takes robots off the hook with respect to all claims of criminal liability. For the foreseeable future, these machines will be legally unaccountable before criminal courts, because they lack the set of preconditions, such as consciousness, free will and human-like intentions, that is, the *mens rea*, for attributing liability to a party. This is not to say, however, that robots do not affect certain fundamental tenets of the field. Leaving aside the possibility of robotic intentions, attention should be drawn to how the growing autonomy of robots may induce a new set of *actus reus*, that is, the material element of a crime. Here, we have to further distinguish between the general framework of the criminal law field and its specific sub-sector on the laws of war and international humanitarian law. The latter are currently regulated by the 1907 Hague Convention, the four Geneva Conventions from 1949, and the two 1977 additional Protocols. This set of provisions establishes the clauses of immunity, or of general irresponsibility, that hinge on the traditional categories of *ius ad bellum* (i.e. when and how resort to war can be justified), and *ius in bello* (i.e. what can justly be done in war). Over the past decade, scholars have stressed time and again that the behaviour of robots on the battlefield is increasingly falling within the loopholes of the legal system. Significantly, Christof Heyns, Special Rapporteur on extrajudicial executions, urged in his 2010 Report to the UN General Assembly that Secretary-General Ban Ki-moon convene a group of experts in order to address 'the fundamental question of whether lethal force should ever be permitted to be fully automated'. Therefore, as previous international agreements have regulated technological advancements over the past decades in such fields as chemical, biological and nuclear weapons, landmines, and the like, a similar UN-sponsored agreement seems urgently needed to define the conditions of legitimacy for the employment of robot soldiers (Pagallo 2013).

Things are different in the field of robotic crimes that govern people's everyday interaction. Contrary to the clauses of immunity that characterize the laws of war, the focus is here on the legal accountability that is typically imposed on individuals who commit a wrong prohibited by law. More particularly, we should take into account a class of AI machines partaking or being used in criminal enterprises, so as to determine whether such a use of robots can be prosecuted under current provisions of criminal law. The distinction between crimes of war and the general framework of criminal law thus sheds light on a twofold impact of robotic technology: whereas, in the laws of war, most of the current debate revolves around the extent to which we should restrict today's clauses of immunity, discussions on a new generation of robotic *actus reus* concern whether we should expand current provisions of criminal law. In May 2010, for example, the Commissioner of the Australian Federal Police ('AFP') Mick Keelty, insisted on 'the potential emergence of technological crime from virtual space (online) into physical space vis-à-vis robotics'.[2] Five years later, in May 2015, penetration tester Parker Schmitt and robot expert David Jordan showed that drones can spoof Wi-Fi and steal sensitive data, such as credit card information, adding a new level of anonymity on which crackers can thrive. In July 2015, a robot killed a contractor at one of Volkswagen's production plants in Germany, by grabbing the human and crushing him against a metal plate.

Leaving aside sci-fi scenarios of robots with a criminal mind of their own, what these cases suggest is to investigate whether a new generation of robotic *actus reus* may affect notions on which individual responsibility is traditionally grounded in the field of criminal law, i.e. basic concepts defining human *mens rea*. The more robots 'sense', 'think', and 'act', the more the AI properties of these machines could transform the ways in which lawyers traditionally grasp individual criminal accountability as an issue of reasonable foreseeability, fault, negligence or causation. Since ancient Roman law, after all, the notion of legal responsibility has rested with the Aristotelian idea that we have to scrutinize *id quod plerumque accidit* in the physical domain, that is, to focus on that which generally happens as the most probable outcome of a given act, fact, event or cause. The growing capability of robots to change and improve the values of their properties or inner states without external stimuli, may impact crucial criteria for selecting from the entire chain of events the specific condition, or the set of conditions, that best explains a given outcome. Therefore, in addition to the issue of whether, and to what extent, a new generation of robotic *actus reus* shall be added to the list of crimes set up by national codes, statutes, or international agreements, we have to examine how the mental element requirements of the criminal law field may change vis-à-vis the decisions of these AI machines and their unpredictable behaviour.

Yet, a popular claim of today's legal experts is that robotics technology would neither affect concepts and principles of criminal law, nor create new principles and concepts. In order to understand this traditional point of view, let us restrict the focus on the interplay between the human's mental element of an offence (i.e. the human's *mens rea*), and the material element of such a crime (i.e. a robot's *actus reus*). Three observables of the analysis with their variants follow as a result:

a   Cases where the design stance supersedes any evaluation of the robot's intentions, since the primary aim of a given technology is incapable of lawful uses. Consider the robotic submarines designed and employed by Colombian drug traffickers: once ascertained that infringing uses represent the primary aim of the technology, the latter can be banned and hence, every attempt to design, construct or use applications of this kind should be considered as a crime. Furthermore, any additional crime perpetrated by the robot is conceived as if humans knowingly and wilfully committed the act (Pagallo 2012).

b   Crimes of intent, that is, when individuals illegally use robots in order to commit an offence. Here, robots are reckoned as innocent agents or simple instruments of an individual's *mens rea*. This is the traditional approach of criminal lawyers summed up by the 'perpetration-by-another' liability model (Hallevy 2015). All in all, there are three human candidates for responsibility before a criminal court: programmers, manufacturers, and users of robots.

c   Crimes of negligence, that is, cases in which criminal liability depends on lack of due care, so that a reasonable person fails to guard others against foreseeable harms. This is the traditional 'natural-probable-consequence' liability model that comprises two different types of responsibility. On the one hand, imagine either programmers, or manufacturers, or users who intend to commit a crime through their robot, but the latter deviates from the plan and commits some other offence. On the other hand, think about humans having no intent to commit a wrong but who were negligent while designing, constructing or using a robot. Although this second type of liability is trickier, most legal systems hold individuals responsible even when they did not aim to commit any offence. In the view of traditional legal theory, the alleged novelty of all these cases resembles the responsibility of an owner or keeper of an animal 'that is either known or presumed to be dangerous to mankind' (Davis 2011).

Against this traditional backdrop, the next step of the analysis concerns how the advancement of robotics may affect it. The impact can regard either the mental element of the offence, or its material content, or both. We should next ascertain whether robotics technology brings about a new generation of robotic crimes.

3 New robotic crimes

There are three ways in which robotics technology may upset tenets of current criminal law. First, the class of criminally accountable agents has to be taken into account. From the ninth century to the nineteenth in Western Europe, legal systems tried animals for any kind of crime or damage (Ewald 1995). Prejudices and superstitions of the Middle Ages were finally eclipsed by the ideas of the Enlightenment, leaving humans as the only plausible actor in the legal domain. However, breathtaking improvements of robotic technology have persuaded some scholars to envisage AI machines endowed with human-like free will, autonomy or moral sense, thereby impacting on the scenario (b) which was sketched above in the previous section. By enlarging the class of criminally accountable agents, a new generation of robotic crimes would follow as a result.

Second, the material element of new types of crime, or offences, has to be scrutinized, i.e. a novel set of *actus reus*, rather than a new class of artificial *mens rea*. What matters here has mainly to do with the rule of law and that which is summarized, in continental Europe, with the formula of the principle of legality. Going back to the difference between crimes of war and the general framework of criminal law, robotics technology impacts on the safeguards of the rule of law in two different ways. In the case of robotic crimes of war and current international humanitarian law, the debate principally revolves around whether clauses of immunity shall be restricted, e.g. squads of tiny drones that plan the mission they are to execute by themselves. Leaving aside this debate, what this section aims to explore concerns the reasons why the set of acts or omissions, for which individuals can be held guilty of any criminal offence, shall be expanded. Think for example of new forms of corporate criminal liability for robots. What makes sense in the field of corporate criminal law, may not fit the set of robotic crimes of war at all.

Third, we should mention the difficulty lawyers encounter in severing the chain of liability through notions of legal causation and fault, when tackling the growing capability of AI machines to be independent of real time human control input. This latter issue functions as a sort of bridge between the scenarios outlined above, that is, between novel forms of both *actus reus* and *mens rea*, namely the different ways in which robotics technology can impact tenets of current criminal law. In order to grasp why this can be the case, attention should be first drawn to the hypothesis of a novel artificial criminal mind.

3.1 Criminal minds

Gabriel Hallevy is one of the most committed advocates of a new generation of AI criminal minds. In his view, AI technology 'has the capability of fulfilling the awareness requirements in criminal law' (Hallevy 2015: 91), together with 'the mental element requirements of both intent offenses and recklessness offenses' (ibid.: 99). This means, on the one hand, that robots could be either liable as direct perpetrators of criminal offenses (ibid.: 119), or responsible for crimes of negligence (ibid.: 135), or on strict liability basis, and so forth. On the other hand, such artificial agents could be protected by the general defence of loss of self-control, insanity, intoxication, or factual and legal mistakes. Therefore, once the mental element requirement is fulfilled in the case of a robot, there would be no reason why the general purposes of punishment and sentencing, i.e. retribution and deterrence, rehabilitation and incapacitation, down to capital penalty, should not be applied to AI machines (ibid.: chapter 6).

Admittedly, dealing with the kind of strong AI Gabriel Hallevy is assuming in his conceptual exercise (Hallevy 2015: 98), the traditional paraphernalia of criminal lawyers could be properly extended to the regulation of robots. All in all, we may buy Lawrence Solum's argument that 'one cannot, on conceptual grounds, rule out in advance the possibility that AIs should be given the rights of constitutional personhood' (Solum 1992: 1260). Yet, if we admit to there being AI machines capable of autonomous decisions similar in all relevant aspects to the ones humans make, the next step would be to acknowledge that the legal meaning of 'person' and, for that matter, of crimes of intent, of negligence, of strict liability, etc., would radically change. Even Solum admits that, 'given this change in form of life, our concept of a person may change in a way that creates a cleavage between human and person'. Likewise, Hildebrandt *et al.* (2010: 558–559) warn that 'the empirical finding that novel types of entities develop some kind of self-consciousness and become capable of intentional actions seems reasonable, as long as we keep in mind that the emergence of such entities will probably require us to rethink notions of consciousness, self-consciousness and moral agency'. At the end of the day, however, nobody knows where this scenario may lead. For instance, would a strong AI robotic lawyer accept Hallevy's argument that 'evil is not part of the components of criminal liability' (Hallevy 2015: 93)? What if the robot, rather than an advocate of current exclusive legal positivism, is a follower of the natural law tradition?

In addition to this kind of conceptual exercise, there is another way in which the sphere of criminal prosecution can be reasonably expanded. Going back to the scenario (b) of Section 2, consider the current debate on corporate criminal liability and forms of distributed responsibility that hinge on multiple accumulated actions of humans and computers (Freitas *et al.* 2014; Hallevy 2015: 40–45). Here, it can be extremely difficult to ascertain what is, or should be, the information content of the corporate entity as foundational to determining the responsibility of individuals. The intricacy of the interaction between humans and computers may lead to cases of impunity that have recommended some legal systems to adopt forms of criminal accountability for corporations. Think of the collective knowledge doctrine, the culpable corporate culture, or the reactive corporate fault, as ways to determine the blameworthiness of corporations and their autonomous criminal liability. Although several critical differences persist between the common law and the civil law traditions, and among the legal systems of continental Europe, we can leave aside this kind of debate, so as to focus on whether these forms of corporate criminal liability could be applied to the case of the artificial legal agents and the AI smart machines that are under scrutiny in this chapter. Noteworthy, over the past years, several scholars have proposed new types of accountability for the behaviour of robots (Karnow 1996; Lerouge 2000; Weitzenboeck 2001; Bellia 2001; Sartor 2009; Pagallo 2013; etc.), thus suggesting a fruitful parallel with those legal systems that admit the autonomous criminal responsibility of corporations.

Indeed, since corporations cannot be imprisoned, legal systems had to envisage alternative ways of punishment, such as restrictions of liberty of action, and fines. This makes a lot of sense in the case of robots, because some have proposed that we should register such artificial agents just like corporations (Karnow 1996; Lerouge 2000; Weitzenboeck 2001); while others have recommended that we should bestow robots with capital (Bellia 2001), or that making the financial position of such machines transparent is a priority (Sartor 2009). Admittedly, the reasons why these scholars have suggested such forms of accountability for the behaviour of robots do not concern their *mens rea* but rather, pertain to the field of civil (as opposed to criminal) law. In other words, these are ways for striking a balance in the civil law field of contracts and torts, between the individual's claim to not be ruined by the decisions of her robots and the claim of a robot's counterparty to be protected when interacting with them. Therefore, in order to determine whether the harmful behaviour of a robot should be relevant under criminal law, the attention should be drawn to the material conduct of the robot and, more particularly, to whether the latter may jeopardize foundational elements of society and create, generally speaking, social alarm. Because the society's right to inflict punishment is traditionally grounded on the idea that harm affects the community as a whole, the parallel between the criminal responsibility of corporations and the new forms of criminal liability for the behaviour of robots thus rests on the magnitudes of complexity that regard the human-robot interaction, the content of this very interaction and hence, the material elements of a crime. After the mental element requirements of criminal law, or *mens rea*, let us explore what a new robotic *actus reus* could look like.

3.2 Criminal actions

The grandfather of legal automation and AI and law, Wilhelm Leibniz, used to say that 'every mind has a horizon in respect to its present intellectual capacity but not in respect to its future intellectual capacity' (quoted by Coudert 1995: 115). In light of Leibniz's wisdom, I risk here two projections. The first scenario is inspired by a true story: in May 2014, Vital, a robot developed by Aging Analytics UK, was appointed as a board member by the Japanese venture capital firm Deep Knowledge, in order to predict successful investments. As a press release was keen to inform us, Vital was chosen for its ability to pick up on market trends 'not immediately obvious to humans', regarding decisions on therapies for age-related diseases. Drawing on the predictions of the AI machines, such trends of humans delegating crucial cognitive tasks to autonomous artificial agents will reasonably multiply in the foreseeable future. But, how about the wrong evaluation of a robot that leads to a lack of capital increase and hence, to the fraudulent bankruptcy of the corporation?

In this latter case, the alternative seems between the scenario (c) of Section 2 and the hypothesis of AI corporate liability mentioned above in Section 3.1. Still, as to the traditional crime of negligence, there is a major problem: in the case of the wrong evaluation of the robot that eventually leads to the fraudulent bankruptcy of the corporation, humans could be held responsible only for the crime of bankruptcy triggered by the robot's evaluation, since the mental element requirement of fraud would be missing in the case of the human members of the board. Accordingly, the criminal liability of the corporation and, eventually, that of the robot would be the only way to charge someone with the crime of fraudulent bankruptcy. This scenario however means that most legal systems should amend themselves, in order to prosecute either the robot as the criminal agent of the corporation, or the corporation as such.

The second prediction regards a variant of scenario (b) in Section 2, i.e. the 'perpetration-by-another' liability model. By reversing the usual perspective, humans are assumed here as the innocent agent or instrument of an AI's bad decision. Certainly, the scenario is not entirely new: we have full experience of hackers, viruses or Trojan horses, compromising computers connected to the internet, so as to use them to perform malicious tasks under remote direction, e.g. DDoS attacks. What is new in the case of robots concerns their particular role of interface between the online and the offline worlds. In addition, we may envisage robots replicating themselves, in order to specialize in infringing practices, so that no human could be held responsible for their autonomous harmful conduct. Consequently, by admitting the scenario of robots that illegally use humans in order to commit crimes 'out there', in the real world, we would end up with a new kind of *actus reus* which does not necessarily entail any *mens rea*. Think for example of powerful brain computer interfaces for robots that perceive the physiological and mental states of humans through novel Electroencephalography (EEG) filters. Legal systems could react either amending once again themselves, e.g. a new kind of autonomous corporate criminal liability for robots, or claiming that the principle of legality does not apply to smart machines after all, i.e. a simple variant of scenario (a) in Section 2. In any event, it is likely that a new general type of defence for humans, such as robotic loss of self-control, should be taken into account.

Further instances of new robotic offences could be given. But to cut to the chase, we can adapt in this context that which James Moor called the 'logical malleability' of computers and hence, of robots. Since the latter 'can be shaped and molded to do any activity that can be characterized in terms of inputs, outputs, and connecting logical operations' (Moor 1985), this means that the only limits to the new scenarios of robotic crimes are given by human imagination. Unsurprisingly, over the past years, an increasing amount of research has been devoted to the analysis of strong AI systems, trust, and security. At the University of Stanford, an area of study has to do with 'loss of control of AI systems'. In the words of Eric Horvitz (2014), 'we could one day lose control of AI systems via the rise of superintelligences that do not act in accordance with human wishes [so] that such powerful systems would threaten humanity'. Analogous risks have recently been stressed by Bill Gates, Elon Musk, and Stephen Hawking. How should we legally tackle such challenges?

4 Japanese pragmatism

As previously stressed in the introduction, most robots are not a mere 'out of the box' machine. Rather, as a sort of prolonged epigenetic developmental process, robots increasingly gain knowledge or skills from their own interaction with the living beings inhabiting the surrounding environment, so that more complex cognitive structures emerge in the state-transition system of the artificial agent. Simply put, specimens of the same model will behave in quite different ways, according to how humans train, treat, or manage their robots. Correspondingly, both the behaviour and decisions of these artificial agents can be unpredictable and risky, thus affecting traditional tenets of the law, such as notions of reasonable foreseeability and due care, which were mentioned above with scenario (c) in Section 2. Moreover, we lack enough data on the probability of events, their consequences and costs, to determine the levels of risk and, thus, the amount of insurance premiums, on which new forms of accountability for the behaviour of such machines may hinge (Pagallo 2013). How, then, should legal systems proceed?

Over the past years, the Japanese government has worked out a way to address these issues through the creation of special zones for robotics empirical testing and development, namely, a form of living lab, or Tokku. Whereas the world's first special zone was approved by the Cabinet Office in November 2003, covering the prefecture of Fukuoka and the city of Kitakyushu, further special zones have been established in Osaka and Gifu, Kanagawa and Tsukuba. In a nutshell, the overall aim is to set up a sort of interface for robots and society, in which scientists and common people can test whether robots fulfil their task specifications in ways that are acceptable and comfortable to humans, vis-à-vis the uncertainty of machine safety and legal liabilities that concern, e.g. the protection for the processing of personal data through sensors, GPS, facial recognition apps, Wi-Fi, RFID, NFC, or QC code-based environment interaction. Although the Japanese are often perceived as conservative and inclined to a formalistic and at times, pedantic interpretation of the law, it is remarkable that such special zones are highly deregulated from a legal point of view. 'Without deregulation, the current overruled Japanese legal system will be a major obstacle to the realization of its RT [Robot Tokku] business competitiveness as well as the new safety for human-robot co-existence' (Weng et al. 2015). Furthermore, the intent is 'to cover many potential legal disputes derived from the next-generation robots when they are deployed in the real world' (ibid.).

So far, the legal issues addressed in the RT special zones regard road traffic laws (at Fukuoka in 2003), radio law (Kansai 2005), privacy protection (Kyoto 2008), safety governance and tax regulation (Tsukuba 2011), up to road traffic law in highways (Sagami 2013). These experiments could obviously be extended, so as to further our understanding of how the future of the human-robot interaction could turn out. Consider again some of the problems mentioned above in the previous sections, such as matters of foreseeability and due care concerning human negligence, or the unpredictability of robotic behaviour that may suggest new types of corporate liability, or even trigger unexplored forms of *actus reus*. By testing these scenarios in open, unstructured environments, the Japanese approach does not only show a pragmatic way to tackle the challenges brought on by possible losses of control of AI systems. In addition, we can suspect that the Japanese industry is acquiring a strategic competitiveness in the field of robotics, much as occurred in the automotive sector in the early 1980s. Will Western producers learn another hard lesson?

5 Conclusions

The chapter has examined today's state of the art in criminal law, so as to determine whether, and to what extent, current developments of robotics can affect it. A preliminary distinction was stressed in Section 2, namely between the general framework of the criminal law field and the sub-sector on the crimes of war and international humanitarian law. In this latter case, scholars and UN special rapporteurs alike have increasingly stressed over the past years, that an international agreement is needed to define the conditions of legitimacy for the employment of robot soldiers. Through a detailed set of parameters, clauses and rules of engagement, an effective treaty monitoring and verification mechanism should allow for a determination of the locus of political and military decisions that, for example, the increasing complexity of network-centric operations, and the miniaturization of lethal machines, can make very difficult to detect. Of course, this international agreement may take a long time and moreover, today's stalemate will likely continue as long as sovereign states think they can exploit the loopholes of the current legal framework due to their technological superiority or strategic advantage. Such a political issue, however, should not obfuscate the legal reasons of experts and scholars that insist on the necessity of a UN-sponsored treaty. Robotics technology has already impacted on this crucial sector of current legal systems. But, how about the general framework of the criminal law field?

Here, we should distinguish between the mental and material elements of the crime, and the difficulty lawyers face in coping with the growing autonomy of AI machines, in order to sever the chain of liability through notions of legal causation and fault.

First, as to the *mens rea* of a new generation of robotic crimes, we can disregard hypotheses of robots endowed with human-like free will, autonomy or moral sense. Although we may admit that one cannot, on conceptual grounds, rule out this possibility in the long term, it is more than likely that national and international lawmakers will have other kinds of priority in the foreseeable future. Yet, even in the case of robots with a criminal mind of their own, it is not only obvious that we should amend basic pillars of today's criminal law, but of the whole legal system as well. Where this eventuality may lead to, however, nobody knows. Conjectures about a world in which robots intentionally commit the offences prohibited by the law, thus deserving to be punished either as a form of vengeance, i.e. an eye for an eye, or as a way to re-educate them, seem liable to the criticism that John Kenneth Galbraith put forward in his own field: 'The only function of economic forecasting is to make astrology look respectable.'

Second, as to the *actus reus* of new robotic crimes, we should take cases of loss of control of strong AI agents seriously. Consider current work on the verifiability of systems that change or improve themselves, or on utility functions or decisions processes that aim to avoid the possibility that an AI system could try not to be shut down or repurposed. Likewise, reflect on further theoretical frameworks to better appreciate the space of potential systems that avoid undesirable behaviours. As mentioned above in Section 4, a smart way to improve and strengthen this kind of research consists in testing robots in living labs, such as the Japanese Tokku. This sort of interface between strong AI robots and human societies, between present and future societies, should allow us to better understand risks and threats brought on by the unpredictability and uncertainty of robotic behaviour, so as to keep the latter under control. If we are fated to face some of the criminal actions sketched above in Section 3.2, e.g. the 'perpetration-by-another' liability model reversed, let us address these scenarios, first, in a living lab.

Finally, as to the troubles of lawyers with the growing autonomy of AI machines in terms of legal causation and fault, we appreciated how certain decisions of robots and their effects on human behaviour already fall within the loopholes of current criminal law, e.g. the hypothesis of AI corporate criminal liability illustrated above in Section 3.1. Here, it is likely that the way in which legal systems will react is going to be initially clear at the domestic level, through specific amendments to each national criminal law regulation. After all, this is what occurred in the field of computer crimes: first, national legal systems started amending their own codes, acts or statutes, e.g. the Italian regulation 547 from December 1993, and then the international legislator intervened, so as to formalize such legal experience through a general framework, the Budapest Convention from 2001. The new generation of robotic crimes will likely follow the same pattern.

Notes

1   *Nature*, 22 September 2011, p. 399.
2   Top Cop Predicts Robot Crimewave, retrieved at www.futurecrimes.com/article/top-cop-predicts-robot-crimewave-2/ (last accessed on July 4th, 2015).

References

Bellia, Anthony J. (2001) Contracting with electronic agents, *Emory Law Journal*, 50: 1047–1092.
Bekey, George A. (2005) *Autonomous Robots: From Biological Inspiration to Implementation and Control*. Cambridge, MA and London: The MIT Press.
Chopra, Samir and White, Laurence F. (2011) *A Legal Theory for Autonomous Artificial Agents*. Ann Arbor: The University of Michigan Press.
Coudert, Allison P. (1995) *Leibniz and the Kabbalah*. Boston and London: Kluwer Academic.
Davis, Jim (2011) The (common) Laws of Man over (civilian) vehicles unmanned, *Journal of Law, Information and Science*, 21(2): 10.
Ewald, William B. (1995) Comparative jurisprudence (I): What was it like to try a rat? *University of Pennsylvania Law Review*, 143: 1889–2149.
Freitas, Pedro M., Andrade, F. and Paulo Novais (2014) Criminal liability of autonomous agents: From the unthinkable to the plausible, in Pompeu Casanovas et al. (eds) *AI Approaches to the Complexity of Legal Systems*. Dordrecht: Springer, 145–156.
Hallevy, Gabriel (2015) *Liability for Crimes Involving Artificial Intelligence Systems*. Dordrecht: Springer.
Hildebrandt, Mireille, Koops, Bert-Jaap and Jaquet-Chiffelle, David-Olivier (2010) Bridging the accountability gap: Rights for new entities in the information society? *Minnesota Journal of Law, Science & Technology*, 11(2): 497–561.
Horvitz, Eric (2014) *One-Hundred Year Study of Artificial Intelligence: Reactions and Framing*. White Paper. Stanford University, Available at: https://stanford.app.box.com/s/266hrhww2l3gjoy9euar.
Karnow, Curtis E.A. (1996) Liability for distributed artificial intelligence, *Berkeley Technology and Law Journal*, 11: 147–183.

Lerouge, Jean-François (2000) The use of electronic agents questioned under contractual law: Suggested solutions on a European and American level, *The John Marshall Journal of Computer and Information Law*, 18: 403.

Moor, James (1985). What is computer ethics?, *Metaphilosophy*, 16(4): 266–275.

Pagallo, Ugo (2011) Robots of just war: A legal perspective, *Philosophy and Technology*, 24(3): 307–323.

Pagallo, Ugo (2012) Guns, ships, and chauffeurs: The civilian use of UV technology and its impact on legal systems, *Journal of Law, Information and Science*, 21(2): 224–233.

Pagallo, Ugo (2013) *The Laws of Robots: Crimes, Contracts, and Torts*. Dordrecht: Springer.

Sartor, Giovanni (2009) Cognitive automata and the law: Electronic contracting and the intentionality of software agents, *Artificial Intelligence and Law*, 17(4): 253–290.

Solum, Lawrence B. (1992) Legal personhood for artificial intelligence, *North Carolina Law Review*, 70: 1231–1287.

UN World Robotics (2005) *Statistics, Market Analysis, Forecasts, Case Studies and Profitability of Robot Investment*, edited by the UN Economic Commission for Europe and co-authored by the International Federation of Robotics. Geneva: UN Publication.

Veruggio, Gianmarco (2006) Euron Roboethics Roadmap. In: Proceedings Euron Roboethics Atelier, February 27th–March 3rd, Genoa.

Weitzenboeck, Emily Mary (2001) Electronic agents and the formation of contracts, *International Journal of Law and Information Technology*, 9(3): 204–234.

Weng, Yueh-Hsuan, Sugahara, Yusuke, Hashimoto, Kenji and Atsuo Takanishi (2015) Intersection of 'Tokku' special zone, robots, and the law: A case study on legal impacts to humanoid robots, *International Journal of Social Robotics*, 7(5): 841–857.

38

Technology, body and human enhancement

Prospects and justice

Jérôme Goffette

Introduction

Human enhancement is at the same time an old and a new practice. It includes age–old habits such as consuming coffee or alcohol, but the popular understanding of the term is more closely related to practices such as new reproductive techniques, the development of prosthetics or the professional consumption of psychostimulants. A lot of new technologies involve applications for enhancing human body or human mind. In this light, human enhancement seems to be a new phenomenon, deeply rooted in technology. Moreover, the words and concepts surrounding these practices have not yet been clearly defined and the accompanying social, academic and political debates are far from being closed but are instead multiplying.

In light of this situation, we have chosen to approach the issue in a particular way. First, we adopt a specific definition of "human enhancement" that excludes other competing definitions. Second, we have chosen to combine a discussion of the present-day judicial positions on questions of enhancement in a number of developed countries with critical questions concerning these positions and an overview of prospective developments. This double choice results from the aim of this chapter, which must be more a guide for the future than a review of the present and already partly obsolete laws. With this purpose in mind, this chapter, even if it is informed by the abundant ethics literature on human enhancement, will adopt a more concrete and prospective approach, combining elements of sociology, anthropology and philosophy. In fact, thinking about the future of such laws leads away from the focus on rare, atypical cases and instead draws our attention to implications for day-to-day life in the future, because laws are not speculative abstractions but effective and operative regulations. Applied to human enhancement, this reflection on crime and justice is upon the regulation of enhancement, by legal or other frameworks. This in turn clarifies the (new) kinds of criminalities which may emerge from violations of these regulatory frameworks. It helps us to identify telltale sign of tensions, risks and abuses behind the perceived need for rules.

Definition and scope of human enhancement

Even if the expression "human enhancement" is an old one, for a long time it did not indicate any well-defined field or any specific area of research. The turning point was the 1980s (see Druckman and Swets 1988) when it emerged as a topic and was examined in the academic work edited by Erik Parens in 1998. It became a field of wide public interest in 2002–2003 with the NBIC Report[1] and with the report of the President's Council on Bioethics: *Beyond Therapy* (2003). In Europe, one of the key documents is the Science and Technology Option Assessment (STOA) report[2] for the European Parliament, in 2009, which reveals a different approach. In France, in 2010, a bioethical report for the Parliament underlined in its fourth part what is at stake with the improvement of human performances (Assemblée Nationale 2010). Moreover, since 2000 there has been an alternative term for this field – "anthropotechnics" (Sloterdijk 2000; Hottois and Missa 2002; Goffette 2006) – which while more precise has been less widely adopted. Thus, the situation is rather confused. But if one reads this sample of texts, four points have to be underlined:

- "*Human enhancement,*" as a linguistic, sociopolitical and academic label, is recent. Today, it is just at the end of its emergent phase but parliaments have already expressed a strong interest on the topic.
- The field and scope of human enhancement is not yet clearly established. Between the principal studies, all include cognitive enhancement, increased physical force, new reproductive possibilities and ageless bodies, whilst other examples moreover include cosmetic surgery and chemical mood modifications.
- The definition and meaning of "human enhancement" are variable. On one hand, *Beyond Therapy* suggests by its own title that this field is beyond the medical arena, but asserts that medical and enhancement activities are included in the field covering the pursuit of happiness. On the other hand, the STOA report specifies that one should distinguish between therapeutic enhancement within the medical arena and non-therapeutic enhancement which is the more common understanding of "human enhancement." These kinds of persistent issues were already expressed in the book edited by Erik Parens in 1998 and remain open.
- The field of human enhancement/anthropotechnics is a field around which strong expectations and fears are crystallizing amongst the general public and groups who have regulatory power (government, general bodies, etc.), without mentioning the World Transhumanism Association and religious groups.

Considering these four points, the present chapter makes the choice of a certain definition that excludes others, in order to explore the implication of justice and law more clearly. For this text, "Human enhancement" will be taken to mean "the activity or technique of non-medical transformation of a human being by modifying his/her body" (Goffette 2006: 690). This definition takes up the definition of anthropotechnics based on the conceptual work of Sloterdijk, Hottois and Goffette, who separately but simultaneously reinvented this neologism with the same meaning. Among the main advantages of this definition is that it expresses the specificity of this activity: firstly its purpose is not preventing and fighting illness as in medicine; secondly it does not imply a value judgment; and thirdly it reduces the field to bodily modifications, excluding smartphone, computer and other non-bodily enhancements (see Brey 2016, Chapter 1, this volume). For precision, under the term "body" one means here the combination of biological body and body schema (Goffette 2015).

According to this definition, the scope and the field of human enhancement include:

- Physical improvement in order to increase normal human force and physical capacities.
- Psychostimulants for non-medical purposes.
- Mood modifications out of the context of normal medical indications.
- Cosmetic modifications (cosmetic surgery and other modifications of the appearance on normal persons).
- Transformations of sexuation (cases out of the medical field), i.e. the increase or the decrease in gender form, change of gender and modulations of sexuality (aphrodisiacs or inhibitors).
- Reproductive technologies: inducers of temporary or definitive sterility (contraception), abortive substances, techniques providing choice concerning the characteristics of the child, surrogate gestation, ectogenesis, etc.
- Hybridity: use of sensory, motor or interface prosthetics when they do not compensate an incapacity or handicap (additional limbs, means for heightened or new perceptions, expressive prosthetics, etc.), which could also be called cyborgization.

It is easy to see that the field of human enhancement gathers together elements of quite different status, covering real activities, prototype technologies, plausible but non-existent practices and theoretical perspectives. Because all of these involve human modifications, they consequently imply important questions and perhaps some kind of thresholds concerning the human condition and the form or framework of how one could and should live in society.

Of course, the extent, the diversity and the mix of real and potential practices make impossible a precise analysis of all the implied situations. But a set of crucial questions are implied by the majority of them. Thus, this situation brings us to four main points, which will be treated in the following parts:

- How the situation tends to evolve.
- The balance between benefits and risks (i.e. crime, justice and risks).
- The implication in terms of autonomy and heteronomy (i.e. crime, justice, responsibility and human alienation or human development).
- A reflection on the human condition (i.e. crime, justice and common metaphysics).

We will provide elements of reflection based on a set of relevant examples for each of these four topics.

The situation and how it might develop

Let us start with a simple exercise of projection. Everybody could agree with these three simple basic posits:

a   *Human bodily enhancements are and will obviously be growing and multiplying*, because new knowledges engender new technologies and among them new possibilities for modifying the body.
b   *We lack scientific information and evidence-based knowledge on the majority of these human bodily practices* because this kind of knowledge is almost always focused on medical indications rather than on enhanced capacities for normal healthy persons.
c   Human enhancement practices involve both *low and high risks and advantages*.

This simple set of "data" yields a set of implications:

- Implication 1: Due to (b) and (c), *the general public cannot make well-informed choices*.
- Implication 2: Then there will be *risky consumption practices* which will probably lead to health and other scandals.
- Implication 3: So, there will be *a need for better regulations, laws and principles*.
- Implication 4: Consequently there will be *a need to evaluate*. But evaluation requires criteria. In medicine the balance between benefits and risks (BBR) refers to the criteria of life/death, health/illness, suffering and disability. In anthropotechnics the purposes are different: beauty, strength, intelligence, performance… Then for the evaluation of anthropotechnics one must build a new kind of BBR with other criteria.
- Implication 5: Probably, the criteria in the field of human enhancement will combine safety considerations concerning health and disease (somatic and psychiatric troubles) and include different kinds of benefits and risks for the personal autonomy of the consumer.
- Implication 6: As a result, this work for defining specific criteria for human enhancement will result in an increasing demarcation between the classical field of medical regulation (with its own classical criteria) and the field of human enhancement regulation (with its own specific criteria). Then there is a low but deep trend in favor of *a demarcation between anthropotechnics and medicine*.
- Implication 7: Justice, laws and regulations will play an important part in the development and structuring of the field of anthropotechnics / human enhancement. Each scandal will imply a legal process and each legal process will contribute to these new chapters of legislation.
- Implication 8: *In case of costly practices giving a clear social advantage, the question of equity appears unavoidable. Then, the funding of enhancing modifications by a kind of social equity system will be a question more and more openly discussed*, as well as the more specific questions: what practices should be covered? On which criteria? And with what budget?[3]

As one can see, three major key points are salient:

- First, there will be a trend towards demarcation between anthropotechnics and medicine.
- Second, the balance between benefits and risks for anthropotechnics will be a pivot.
- Third, justice and laws will have a crucial role through legal proceedings and political concerns.

Human enhancement: information, risks, crime and justice

The PIP scandal, an interesting example

It can be useful to take an example in order to see the kind of criminal outcomes we encounter. The scandal of the Poly Implant Prothèse (PIP) company appears paradigmatic. Created in 1991, this company produced 100,000 mammary implants per year – 80 percent for exportation – until it was put into liquidation in 2010. Articles in the press noted that 300,000 women could be victims, but it seems that these figures are approximate for Europe and the United States, and do not include Asia, Africa, South America and Russia. In fact, if 50,000 pairs of mammary implants have been produced per year for 15 years, it is reasonable to think that at least 700,000 women could be affected.

The investigation exposed both a quality defect in the casings, which accentuated the risk of leakage and rupture, and the use of a substandard filling material. For the latter, Nusil® was the only approved filler on the market, but the company instead chose less-expensive, industrial-grade silicone, to which it added unapproved additives such as those used in gasoline or for coating electric cables (Silopren®, Rhodorsil®). These products, created for industrial use, have never been tested on humans and are thus considered unfit for medical use. Compounding this, their makeup varies between batches of fillers and some of these additives can degrade the implant casings. Because of these various factors and the absence of clinical trials, the health risks are not well understood. Still, there are many instances of illness (fibrosis, inflammation, cancer, etc.).

In addition to inconsistent manufacturing processes and the use of prohibited materials, serious defects were found in the production line. Even when approved materials were used, quality standards were not uniformly observed.

Several other shortcomings were revealed:

- The notified body (certification authority) – TÜV Rheinland – based its certification process on a cursory examination of written documents, announced its visits in advance and showed little initiative in its investigation of PIP's practices. There were many clear indications that PIP purchased only negligible quantities of Nusil®.
- In 2000: The US Food and Drug Administration sent a warning letter to PIP listing 11 violations of good manufacturing practices, but did not transmit this warning letter to the French Health Regulatory Agency (AFSSAPS).
- As early as 2005–2006, complaints were filed in Great Britain and in France. In 2008–2009 French surgeons tried in vain to draw attention to the problem. It took until 2010 for AFSSAPS to take action and for the majority of countries to ban PIP implants.
- The sale of implants required only CE certification from a notified body (such as TÜV Rheinland) rather than the marketing authorization required for medications. The PIP scandal thus highlights a regulatory and legal shortcoming.
- Legal proceedings are still in process. In France, an initial trial in 2013 convicted PIP's executives. A separate trial cleared TÜV Rheinland of liability. PIP's insurance agency, Allianz, has begun making damage awards, but the guarantee ceiling of 3 million euros will permit only very small payments to registered victims (649€ per person). French authorities have decided that the national health insurance scheme will cover the removal of PIP implants.

Scientific data, responsibility and crime

The case of PIP is a limited but enlightening example. If we take a wider examination of anthropotechnical practices, several observations can be made.

First of all, the issue of the quality of scientific findings is essential. Because all anthropotechnical practices affect the body, the resulting risks must be measured. It is thus logical to envision an investigative study of any proposed enhancement that includes clinical trials. This framework should measure both the risks and advantages, but should also set usage standards, since it will designate a specific purpose and protocol for each product.

In terms of legal concerns, scientific fraud is the primary issue: Poorly conducted research, falsified results, the publishing of journal articles due to pressure and cronyism (the politics of who gets published and why), rather than on scientific merit, etc. Criminal proceedings surrounding medical and scientific publications are equally relevant to anthropotechnical publications.

Balance between benefits and risks for an anthropotechnical indication and criminal aspects

Using scientific findings, it is possible to establish a benefit–risk analysis for a given usage (and thus set the framework for use). This presupposes the establishment of evaluation criteria, which may include health risks and benefits, influence on well-being, ability enhancement, increased freedom, etc. These evaluation findings could then be submitted to public health authorities so that they might grant authorization for the specified use.

There are numerous criminal aspects to take into consideration. In the first place are those that can be transposed from the medical field to the anthropotechnical field: biased records, conflicts of interest, etc. But it is worthwhile to also mention two other possible legal concerns: that of the criteria chosen for the benefit–risk analysis and the weight given to each criterion. The choice of criteria is never simple. It seems fairly evident that it will include both the concept of health and of the stated goal of the studied usage, for example that of increased freedom in the case of a contraceptive, increased performance in the case of a psychostimulant drug or improved appearance for cosmetic surgery. And yet all of these criteria, in order to be used as elements of measure, must be interpreted and transformed into quantitative rating scales.

This may be a simple task for assessing enhanced performance, but it can become less clear when dealing with subjective goals or issues surrounding personal identity. Additionally, it is entirely possible that there are also negative outcomes, which may not be immediately obvious but will become evident with time. For example, a cosmetic surgery intervention may well be a success in terms of aesthetics, but might also result in decreased skin sensitivity for certain areas; a given psychostimulant may improve cognitive abilities in one area (memory, for example) while simultaneously decreasing others (the ability to make choices). In the same vein, because many criteria must be taken into consideration, how much weight should be given to each one in determining the overall gain? To what degree would an increased cognitive performance justify a health risk? The weighting system is inherently socio-political. The examples of contraception, abortion and steroid use by athletes show the challenges of such choices. How can we as a society make these determinations in a democratic way?

The complexity of the evaluation process and the certification awarded by public health authorities will certainly leave ample space for legal disputes. Some disputes will concern the evaluation techniques for the benefit-risk analysis, others will question the rigor of methodology, and still others will call into question the socio-political legitimacy of methodological choices.

Ultimately, the existence of an official marketing authorization (*the permit to market*) will not solve everything. Certain practitioners and clients can still choose to make use of a product or technique outside of its officially recognized purpose (*i.e. the indications for which it received its marketing authorization*). In these cases, how can liability be determined? Should we speak of contractual obligations for which the practitioner has the responsibility to provide accurate information and technical mastery while the client is responsible for the risks of his choice and decisions? Under what conditions and for how long? And what type of liability insurance is necessary?

Client, information and decisions: criminal perspectives

When evaluating an enhancement, the question of information is an essential aspect of the practitioner–client relationship. While there is currently no official ethical framework for anthropotechnical practices, it seems vital to recognize that there is an obligation to provide clear and unbiased information concerning the benefits and disadvantages of the proposed acts.

Likewise, it is important that the role of each party be clearly defined: the practitioner offers a service and the client makes a decision. The responsibility for the decision falls on the client, while the practitioner is responsible for the quality of the information, the benefit–risk comparison presented for each act and the assistance given to each client in understanding the different possible outcomes.

For anthropotechnical practices *related to cosmetic enhancement, psychostimulant, strength enhancement, etc.*, even more so than in the medical field, the responsibility of the decision falls on the patient. In the varying worldwide patchwork of doctor–patient relationships, there are many models, ranging from a classic paternalistic approach to that of a patient as a decider, with many variations in-between. Yet when dealing with the question of human enhancement, the pressure of illness is absent, and so the choice falls squarely in the hands of the client. Thus it is likely that legal standards will prevent the doctor from having any role in decision-making and will limit his or her role to that of proposing services. The client thus is ever more the decision-maker. In this configuration, a practitioner who oversteps the role of offering and moves towards that of deciding could be subject to sanctions – criminal or otherwise.

Consequences of hazards, fault quality or unfairness

A manufacturing defect for a pair of shoes or a computer may be frustrating, but these issues can be dealt with through repairs, updates or by replacing the product in question. In the case of anthropotechnical practices, there are elements that strongly increase the potential seriousness of risks:

- Whenever the human body is altered there is an element of risk. The body is not a machine, for which each part has well-known characteristics, but rather an organism in which each part is divided into infinite subsections, and for which we have an imperfect understanding of the characteristics and relationship between elements. For this reason, anthropotechnics, like the medical field, must be aware of *treatment risks*.
- An adverse outcome might not be reparable. Sometimes it is simply impossible to put something back into its original state or to create an acceptable substitute. Only a few body parts can be easily replaced (teeth, cornea, bones, etc.) and often these interventions produce imperfect results. A kidney or heart transplant requires that the patient take immunosuppressant drugs so that the body does not reject the new organ, a complex situation that is quite different from simply changing a hard drive on a computer. And in considering irreversible changes, how can we neglect to mention the possibility of chronic illness and even death. The legal framework surrounding anthropotechnic issues must also cover events that may have quite serious repercussions.

Two questions must be raised. The first is that which in French Common Law concerns the difference between the obligation of means and the obligation of results. While a car mechanic has an obligation to produce the desired result, a doctor has only an obligation to follow the correct procedures. She or he must do everything possible to ensure the patient's health. But in light of therapeutic risks, there is no obligation to produce a specific result. A doctor is not considered liable for negative outcomes if she or he has respected the obligation of means. This type of distinction exists in most countries, though the formulations may vary (for example with the complex notion of 'best-effort obligation'). The risk of complications is recognized, and so a doctor's responsibility is not related to the final result but rather the quality standards of the service provided or the level of the contractual guarantee.

How should this issue of therapeutic hazard be considered in terms of anthropotechnic services? It is well-acknowledged that the seriousness of the issues treated by the medical field are such that patients do not expect interventions to resolve everything and they are fully conscious of the uncertainty of the situation. However, the starting point of an anthropotechnic consultation does not include this element of suffering or imminent death. Is it nonetheless necessary, as we are still dealing with interventions on a human body, to create a legal concept specifically tailored to "anthropotechnic treatment risks"? And if such a concept is created, will certain practitioners use this term to hide or excuse their errors? Is it necessary to envision an obligation of result, with a specifically tailored insurance that will pay damages in case of a negative outcome? In this case, wouldn't repeated and severe cases of negative outcomes create very high, even inaccessible, insurance premiums in order to cover these situations?

Another issue: what should be done when damage is caused not as a result of unavoidable risk, but rather due to fraud or a quality defect? Can the party at fault be expected to pay reparations? Should malpractice insurance be mandatory for all practitioners and suppliers within the anthropotechnic field?

The PIP case is an excellent example. The product was clearly defective and non-compliant, which prompted a conviction for the directors of the company. Yet this does not resolve the problem, since the risks resulting from defective implants is not fully known, but can vary from moderate (fibrosis) to high (cancer):

- Should the implants be removed? Who should pay for their removal?
- Should they be replaced? Who should pay for the replacement implants?
- Is it necessary to establish a specific type of health monitoring for people who have or had PIP implants? Who should organize and pay for this monitoring? Who should take on the financial burden related to illness and death that result from these implants?

Is indemnification PIP's responsibility or should its insurance underwriter be the one to pay? The situation in France shows that PIP's 3 million euros policy with Allianz is insufficient, resulting in a paltry indemnity payment. Allianz admitted as much. For now, the legal responsibility of the notified body, TÜV Rheinland, is not entirely clear, which leaves room for two scenarios: either the notified body is not liable, in which case the certification system must be entirely reworked, given its shortcomings, or the company is in fact liable and TÜV Rheinland's insurance company must play its role.

Autonomy and heteronomy: crime, justice and de/humanization

Up until now we have examined human augmentation in the sense of manufacturing. We now wish to approach the question in the context of personal experience, family, professional and social sphere.

Human enhancement and fair/unfair advantages in sport

When we speak of human enhancement, the first concept that comes to mind is that of enhanced performance. From this perspective, the example of sports is quite relevant since there has been consistent documentation on the topic and because it has been a question for debate for several decades. Beginning in the 1920s–30s, doping in sports becomes a topic of discussion. An acclaimed article by the famous journalist Albert Londres showed how, in the 1924 Tour de France, the racers were already regular users of performance-enhancing drugs

(Londres 1996 [1924]). This holds true for other sports as well. A 1958 investigation conducted among Italian football clubs showed that 94 percent of clubs used substances such as hormones, anabolic steroids or amphetamines (Ottani 1961). These findings spurred the gradual emergence of tests and anti-doping laws, which came into play just as new types of performance enhancers were created. For some international competitions, looking back over ten years of revelations of steroid use, it is no longer possible to pronounce a winner—doping accusations have simply disqualified, one after another, all the competitors.

The phenomenon has a fairly clear pattern:

1   Athletes and/or their support personnel *believe* that there are substances that will improve performance.
2   Yet for some of these methods, tests are rare, easily circumvented, or non-existent.
3   Thus the athlete and/or their support personnel believe that the competitors are using these methods and have an advantage.
4   So, in order to *remain competitive*, the athlete feels almost obliged to use these methods— and this holds true for all athletes.
5   Yet the guiding principle of sport is that of *hard work and natural talent* without unfair advantages.
6   And so it is necessary to *hide* doping products use.

We can thus observe how naturally the phenomenon of doping products use and concealment becomes commonplace. The British Medical Association has, in fact, published a book with a very fitting title: *Drugs in Sport: The Pressure to Perform* (BMA 2002). An athlete, even an amateur athlete, feels and operates under this pressure. Several restraints and reactions exist to cope with this situation:

7   (=5) Sport's guiding principle is that of competition without unfair advantage.
8   So doping will be *banned as cheating*.
9   Some doping practices have health risks. Thus, doping will also be *banned as a health risk*.
10   And so *anti-doping legislation* has been created, which requires the development of certified tests, the creation of a protocol for sample collection, the establishment of courts and the appointment of experts and judges.

Consequently, the legal question is directly present on many levels:

•   National or international courts regularly sentence those who use doping products and those who recommend them to others.
•   The courts also regularly convict those involved in prescription fraud, theft, trafficking or diversion of medicines, unlicensed practice of medicine, for endangering a person's health, etc.
•   The question of conflict of interest is present at every level, whether it be for athletic institutions, experts, judges, anti-doping centers, media outlets, sponsors or those who organize sporting events.
•   Pharmaceutical companies may want to develop a new market for products that have previously been reserved for medical use. For example, AMGEN, whose first medication was an EPO approved for treating certain forms of anemia, is also the main sponsor for the AMGEN Tour of California Bike, despite the never-ending headlines on EPO-based doping in the cycling world. This sponsorship is not illegal, but it raises the question as to

AMGEN's motives. Even the slogan of the competition has a potential double meaning, ambiguous on the medical and anthropotechnical fronts: "Leading the race to dramatically improve patient's lives"[4] – yet AMGEN's EPO is only authorized for medical use.

The pressure to perform: out of sport

The pattern that has been highlighted for sports extends to other domains. Investigations of doping use in non-athletic contexts, whether professional or academic, show that the pressure to perform also exists in these contexts and can lead to doping behaviors. The concept of "doping behavior" is defined as "the use of a substance in order to improve one's performance when facing an obstacle, whether real or imagined" (Laure and Allouche 2015). This behavior encompasses a variety of professions, like those cited in the studies by Laure and Allouche: workers, truck drivers, students, firemen and physicians (ibid.: 170).

The essential difference for these cases is the near-total absence of a legal framework. While in the world of athletics, doping use is seen as a form of cheating, in other spheres, such as the professional and academic world, use of such products is not seen in the same way, even if there is some tension around this idea. For example, young adults in the Quebec region who have used psychostimulants explain they have done so in order to improve their academic or professional performance, but are hesitant to recognize this as doping behavior or cheating (Thoër and Robitalle 2011). And yet, the use of such substances is a documented phenomenon among university students (Bogle and Smith 2009; Wilens *et al.* 2008). The most popular products are amphetamines (Adderall®) and methylphenidate (Teter *et al.* 2006).

This takes place in a context ruled by the philosophy of *laissez faire*, as if such use only concerned the private sphere and individual behavior. However, there are two reasons to question such a perspective. On the one hand, the consequences of these behaviors are not confined to the individual but rather create a sort of collective pressure, on which public authorities might legitimately decide to act *in the same way doping was prohibited in sport competition*. On the other, these uses are not supported by a benefit–risk analysis because virtually no research data exists. This type of use is instead guided by peer usage and anecdotes rather than by secure and established protocols. As a result, an ensemble of public health issues has begun to emerge. For example, the use of amphetamines triggers powerful side effects, such as exhaustion, cardiac complications and depressive tendencies, not to mention the issues of addiction. These two reasons—regulating social pressure and avoiding public health problems—will likely lead to the establishment of legislation, even though it is difficult to imagine that testing for performance-enhancing drugs will become commonplace in school, university and professional settings, where it is quite difficult to control the movement of medications.

It is worthwhile to emphasize that anthropotechnic practices are not confined to sports and professional use, but also exist in other spheres that should not be overlooked. It is particularly worthwhile to mention aesthetic concerns and sexual performance.

In the aesthetic realm, a growing portion of demand for cosmetic surgery stems not from personal dissatisfaction with looks, but is instead due to the pressure of appearance in a professional context. This difference, in contrast to careers that don't require a public persona, shows that societal pressure plays a predominant role in individual determination. There are many well-known and documented examples of film actors, models, politicians or even teachers. Here again, it is necessary to question the risks, how they are evaluated and the benefit–risk analysis, although the medical field, in most countries, is already aware of this and has been trying to dethrone the current standards of beauty, though strategies for this may vary between countries.

In the sexual realm, practices for enhancing sexual performance have existed for millennia, with an abundant pharmacopeia of aphrodisiacs that date back to Antiquity (Rätsch and Müller-Ebelling 2013). Contemporary development of products with scientifically proven results only reinforces this. Here again, these products are officially authorized as remedies for dysfunction but their commercial strategy serves to widen the market to people without a specific medical need: those who wish to enhance their sexual performance or who want to ensure their ability to perform, though neither situation corresponds to normal abilities. This time, professional pressure is not a factor, but rather social pressure. This pressure expresses itself in mainstream culture, in which desire, sexual ability, stamina and the ability to achieve orgasm are presented as normal. Failing to meet these standards can lead to feelings of inadequacy, guilt and shame. But how can we fail to remark the paradox of this vision of sexuality in the context of a society where people have less and less time to sleep, experience mounting fatigue and where life expectancy has become impressively longer? Exhaustion and middle age are normal factors that will negatively affect sexual performance, not to mention the question of stress. And another paradox—examining an intimate relationship from the perspective of *performance* is tantamount to adding a form of *anguish* that actually degrades the relationship. And yet sexology suggests that the best remedy is to foster a relaxing, complicit and playful environment. This paradox, briefly summarized above, serves to reinforce the market for aphrodisiacs. Because sexual and romantic issues are part of the intimate sphere, use of these products is generally anonymous or hidden, encouraging the neutral interface of the Internet, a setting which only exacerbates issues of counterfeiting and unauthorized products.

The pressure to perform: autonomy and heteronomy

The question raised by anthropotechnic practices is, at its root, a question from classical philosophy, that of autonomy versus heteronomy. On the one hand, according to Immanuel Kant (2011 [1785]), the autonomy is the capacity to decide by oneself. Promoting the autonomy means both for oneself the moral obligation to exert his or her reason (and then to be more human) and to acknowledge the presence of the reason in others (and then to respect them as human beings with an equal dignity). On the other hand, this approach involves an obligation to fight heteronomy, i.e. what could perturb the reason and its capacity to decide by itself (death, suffering, alteration of the mind, coercion and pressure). What can be done so that anthropotechnical techniques have a benefit–risk ratio that favors autonomy over heteronomy? Which social norms, which legal framework should be adopted so that the principle of autonomy—dear to most everyone on the planet—is promoted and respected? What means should be used to avoid practices that would be alienating or dangerous while favoring those which promote human fulfillment?

If certain human enhancement practices are clearly alienating—such as cocaine use—and others clearly liberating and even humanizing—like oral contraception—most practices simultaneously possess a variety of aspects, some promoting autonomy, others favoring heteronomy. This mix creates a complex interplay between benefits and risks, or rather a complex metaphysical equilibrium between autonomy and heteronomy.

To clarify the situation, within this text we will not use Beauchamp and Childress's (2013) very strict definition of the principle of autonomy, which focuses on the concept of enlightened consent. We prefer to take up the classic definition, inherited from Kant (2011 [1785]) and from the philosophy of the Age of Enlightenment. *Autonomy*, even in its etymology, is the ability to decide for oneself the path of one's life. According to Kant, acting in an ethical way implies, on the one hand, favoring the existence of autonomy in others and oneself, and on the other, exercising one's own autonomy, which means making the effort to reflect on one's own actions and making fully informed choices. In this way, respecting autonomy means, of course, that a person must give his or her consent and not be imposed upon, but also requires that the consciousness of a human being has the greatest possible longevity (because death breaks the consciousness) and that as much as possible it is not hindered by disruptive elements (obligation, suffering, addiction, etc.) and that it fosters the exercise of one's capacity to think, decide and act according to one's wishes. The table below seeks to enumerate the points where autonomy and heteronomy come into play in *human enhancement*. We have identified three different approaches: an approach based on general questioning, a more social-based approach, examining situations, and a more individual approach.

As one may gather, given the richness and diversity of questions, a complete analysis of the real equilibrium between autonomy and heteronomy would require a rather thorough examination. In this text we will limit ourselves to underlining three salient points:

First, it is essential to make a detailed observation of the practices in order to understand how they may affect autonomy. For example, Adderal® (a mix of four types of amphetamines) used as an anthropotechnic psychostimulant, gives, on the one hand, a strong boost that allows a person to stay focused and work intensely during a given time period, but on the other, causes severe after-effects, ranging from exhaustion, increased likelihood of tachycardia, anorexia and teeth-grinding. Use of the drug entails many concerns, from the way the product is obtained, dependence on a supplier, doubts on product quality (counterfeit concerns), maintaining a sufficient stock, predicting use habits, choosing how the product is taken, estimating risks and

Table 38.1 Autonomy and heteronomy in human enhancement

	In favor of autonomy	In favor of heteronomy
General approach Human fulfillment and Human alienation	• Expanding lifespan • Enhancing skills • Adding capacities • Increasing satisfaction or happiness • Improvement of the personality • Power to go beyond the traditional limited human condition	• Short or mid-term death • Life span reduction • Incapacity and handicap • Chronic or temporary disease • Moodiness • Alteration of the personality
Socio-situational approach	• Expanded professional outlooks • Expanded social interactions such as hobbies, friends, romantic relationships… • Adaptability to different contexts of life	• Dependency on a provider • Dependency for maintenance • Pressure to be enhanced • Pressure to embody a professional type • Power of others on myself due to their power over the means to enhance
Psycho-personal approach	• Experiencing the liberty to build oneself and to choose who one wants to be and what one wants to do • Experiencing new ways to exist or variations in the modes of existence • Feeling of self-realization	• Loss or diminution of certain skills • Concern about choice • Concern about realization • Confusion about self-identity • Feeling like a product produced by the context

benefits, planning sufficient recovery time, fears surrounding addiction, etc. It should also be noted that regular users of Adderal® report a strange sensation of not feeling like themselves except when under the influence of the drug, and feel diminished, with a reduced sense of self, without the drug in their system, as shown in the field research of Thoër and Robitaille (2011). Dependence on a supplier and drug dependency are obviously heteronomic.

Second, it seems essential to regulate social and professional pressure in order to limit alienating tendencies. Strong and brutal obligation is not the only enemy of autonomy; it is in fact pressure and influence that pose the greatest threat. These factors are even more insidious because, while in theory they can be resisted, in practice, the act of resistance can be accompanied by very negative effects. If you are told to use a certain psychostimulant in order to meet a professional obligation, it is clearly more than a casual suggestion: the person making the recommendation has a very clear objective and refusing could jeopardize your career because without the drug, you may not be able to finalize the project before the deadline. In this way, a person may fail to perform, find him or herself excluded from a team and even be pressured to leave a company. Professional pressure is an alienating force in this case, since a person no longer has leeway to make a personal choice, but instead must yield to the lamentable influence of his or her professional environment. Is it necessary to imagine, as in sports, anti-doping testing? Should these tests be focused on only the most dangerous practices? Or on anything that could endanger health? Or on any means that would give someone an unfair advantage? In the same vein, outside of the professional sphere, is it necessary to develop a public policy to combat the normalizing influences on aesthetic standards for bodies, sexual performance and mood? There are well-known controversies over the promotion of stereotypes by corporations, public authorities and various groups: from the "Barbie" standard to that of "young, beautiful and built." These standards already have such strong sway that they can cause anorexia or the consummation of anabolic steroids (as used in body building). Here, the issue of autonomy calls into question the manner in which the social environment is created and produces standards and models. The issue of autonomy thus also questions the way in which a democratic authority may wish to direct and guide these influences in order to avoid alienating environments.

Thirdly, the principle of autonomy not only counteracts heteronomic dynamics, but also promotes dynamics that favor autonomy. If certain practices have an overall positive effect on autonomy, then their place in society should be recognized as a legitimate one, and they should be legally authorized. If they offer a far-reaching contribution to human autonomy, it would even be worthwhile to promote these practices and to render them accessible to everyone. Oral contraception is an excellent example. Recognized today as a key element for autonomy and for responsible decisions concerning whether or not to become a parent, contraception is not only legal in the majority of countries, but is quite often covered under national health schemes. This type of situation leads to a related and essential question: how can we ensure that the collective choice to authorize, promote and fund enhancement practices comes from a place of social legitimacy? How can we ensure its democratic base to avoid the deforming pressure of social groups or authoritarian tendencies?

Crime, justice and the human condition

In light of the previous considerations, the question of the human condition is essential. *Human enhancement* is in fact a change in the human condition, or rather a multitude of changes. By linking the concept of the body to the concept of production, and by diversifying the means of production, *human enhancement* pushes the human condition into a horizon of changes and disruptions. Jean-Paul Sartre affirmed that humankind is its own project (Sartre 2007).

Henceforth this not only applies to personality, culture and the mind, but also to the body and the body's abilities.

A few academic reflections have laid out essential groundwork: notably E. Kapp, M. McLuhan and G. Hottois. Kapp published his *Grundlinien Einer Philosophie Der Technik* in 1877. Inspired par G.W.F. Hegel, Ernst Kapp sees tools and technologies as an "organ projection" (Kapp 1877: chap. 2). A hammer is a projection of the forearm and fist, joining the body to form an "augmented/expanded self" (ibid.: chap.1, p. 26). All techniques are, in fact, projections of organs that develop our selves beyond bodily limits in order to achieve our ambitions on a much wider scale. Kapp is correct to emphasize this principle of augmented physical self and the intimate organic projection that makes us, through the use of any tool, an augmented human. Yet his organicist vision reaches its conceptual limits when it expands beyond tools to then integrate all infrastructure and all of human culture, according to the Hegelian model of an achievement of an ever-growing consciousness.

McLuhan, with the publication of *Understanding Media* in 1964, offered another building block. His famous slogan "The medium is the message" (McLuhan 1964: chap. 1) leads us to understand that the value of a technology is not limited to its intended use. The technology itself creates a new way of being and interacting with the world, unique unto itself. Because of this, all technology changes us, all technology transforms, to a greater or lesser extent, our lives. Technology has multiple repercussions, some clear, some tacit. As Annemarie Mol explained, "technology is an inventive mediator" (Mol 2008: chap. 4.2), a source of results that develop meanings that are sometimes unexpected and have many offshoots. The theoretical work of McLuhan is also built on the idea that technology is an extension of the self (see Brey 2016, Chapter 1, this volume). Just as we often forget about our hand when we write, we have a spontaneous tendency to forget about a tool when we use it *to achieve* something, because we are focused on our goal. But this loss of consciousness is unfortunate because that which we are doing and even our very identity, both of these are linked to our hand and the tool. The meaning is not only contained in what the tool does, but also in the tool itself and in the way that its existence transforms us, for the tool constitutes, along with ourselves, a unified whole that defines our identity, our culture, our condition of existence. When we invent a tool, the tool invents us. That which holds true for the mediums and technologies examined by McLuhan, is even more true for that which we call "human enhancement", where the body itself is modified like a tool in order to add, increase or modify its functionality. If the term "human enhancement" is vague concerning its theoretical ability to encompass all tools, infrastructure and all human cultures, the concept of anthropotechnics, a more precise idea, specifies the corporeal aspect of these modifications. For the body is even more central to our identity than any tool.

This is why Gilbert Hottois, upon publishing *Species Technica* with Jean-Noël Missa in 2002, brings the examination to the very heart of our notion of humanity. The human species, or *Homo sapiens*, as it is called in scholarly terms, could just as legitimately be called *Homo faber*. And yet the term "do" is still too vague because animals are also *faber*. Our defining characteristic is more precise: we are a technological species, a "species technica." Technology is part of our human identity, just like our consciousness, and the two are almost impossible to separate. Artifacts are in our very nature. We could even say that artifacts are our nature. In light of, and because of this, in order to explore the future of the human condition, in 1982 Hottois wrote a science-fiction novel, which he unearthed from his files in 2002 and published along with a philosophical interview (Hottois and Missa 2002). Two salient points stand out. On the one hand, human beings enter into a horizon of modified corporality and even a hybrid, mosaic corporality. On the other, the way the structure of political representation moves to encompass the political tension between technophobes and technophiles, with their two extremes which, in the novel, are represented by "Anarchecology" and the "Technoscients". The industrial dynamic of the General Anthropotechnics plays out in the background, along with the metaphysical problems of the majority of the characters. In the philosophical interview with Missa, Hottois looks back over his career as a philosopher of technology and as an ethicist. He details his interest in the exercise of intellectual projection where philosophy and science fiction reinforce each other. Because we are a *species technica* who, from here on out, has the technical means to profoundly modify our bodies and our brain functions, we have become our own invention. Here, the philosopher argues for explorational ethics: humanization need not enshrine the ordinary human condition, but exploration should be conducted prudently.

To continue on the trail laid down by Kapp, McLuhan and Hottois, we will touch on each of the anthropotechnical fields: physical improvements, cognitive enhancements, mood changes, aesthetical transformations, modifications of sexual shape, reproductive technologies and prosthetics (when they do not compensate an incapacity or handicap). To avoid dissipation, we will limit ourselves here to the field of anthropotechnic prostheses and we will raise a few issues surrounding changes to the human condition.

The customizable human

The first is the new phenomenon of the bodily modularity. Humankind has become customizable. In the pre-anthropotechnic world, we were born with a body that had two arms, two legs, a nose, etc., and that body was permanent. Of course, puberty caused changes, but it was foreseeable and we had little control over it. Age and sickness could also modify the body, though in the sense of a deterioration. Our traditional means of extending our bodies were tools, which we integrated only imperfectly. Our body schema allowed us to extend our sense of bodily self beyond biological limits, but these tools were never as well integrated as our own hands.

The new situation disrupts this schema. The prosthesis integrates with our bodies far more seamlessly than an ordinary tool—this is true for all anthropotechnic techniques. At the same time, the prosthesis retains one characteristic of a tool/technology: it can be replaced. When Aimee Mullins explains how she chooses her prosthetic legs from a collection of over twelve pairs, not only does she medically compensate for her disability, but she also lays claim to an identity as "super-abled", a perfect illustration of the anthropotechnic approach. With a choice of twelve pairs of legs, she goes well beyond the standard possibilities for an exploration that is simultaneously technological, psycho-social and poetic. Our regular body does not offer us the option of transparent legs, or legs carved from an ash tree, or long legs to refine our silhouette or carbon fiber running blades that allow us to run faster. Aimee Mullins is conscious of this:

> So people that society once considered to be disabled can now become the architects of their own identities and indeed continue to change those identities by designing their bodies from a place of empowerment. And what is exciting to me so much right now is that by combining cutting-edge technology – robotics, bionics – with the age-old poetry, we are moving closer to understanding our collective humanity.[5]

Adding a prosthesis to the body means that the body becomes something that can be *modified*, which leads to a *new experience of the human condition*. As McLuhan underlined, the sensation of this new experience does not lie only in what the prosthesis will allow us to do, but also in the way that we experience that prosthesis. The meaning does not just come from how it is used but also and above all from personal identity and the human experience. *These points involve a change in the sense of what it means to be human.*

Considering human technologized condition, it is necessary to stress that prosthetization-cyborgization opens up an extraordinary wealth of perspectives. To examine them, we cannot limit ourselves to a single classification; at least four are required.

The first three are fairly simple:

- First typology – proximity: endo-prosthesis, peri-prosthesis, exo-prosthesis.
- Second typology – material: bio-prosthesis (prosthesis made of biological materials), mecano-prosthesis (inorganic, mechanical), robot-prosthesis (programmable, thus with a complex and modifiable potentiality).
- Third typology – function: sensory-prosthesis (perception), motor-prosthesis (action), memory-prosthesis (information storage), interface-prosthesis, etc.

With just these three, we can already have a sense of the amazing prospects offered by prostheses. As each of these categories can be combined with the other, there are 36 possible sub-categories of prosthetic experiences. Additionally, it is possible to image hybrid prostheses even within a single typology: an implant that can also be controlled via telephone would be an endo-exo-prosthesis, in which the combination endo-exo could generate new possibilities, which might then create a variety of new sub-categories.

The fourth typology is more complex and has been created to address the possibilities of a Brain-Machine-Interface (BMI) (Eskiizmirliler and Goffette 2015: 152–154) type of prosthesis that is controlled as easily and spontaneously as a person's own hand. Here we have another, more refined, typology of functions:

- Specialized prostheses – the bodily integration of tools: in the same way that additional tool modules can be fitted on a power drill (for piercing, sanding, screwing…), the human body could be equipped with prosthetic modules for every imaginable task.
- Supplementary prostheses: for example, it is already possible to add a supplementary arm, and that opens the way to galleries of additional tools for perceiving, doing, interfacing and thinking.
- Virtualizing prostheses: BMI builds a numerical simulation when processed, so it is possible to use this simulation as a way to interact directly with a numerical device (computer, smartphone…).
- Expressive prosthetics: certain BMI devices can express the operator's mood, as with Neurowear®cat ears, which are a kind of augmentation of facial expression or bodily style.
- Topological prostheses: because of the interface, one can play with distance and size on a micro- or mega-scale (micro-hand, crane…), for example, being completely enclosed in a mega-prosthesis or acting from far away through a prosthesis located in another country.
- Cross-human prostheses: because of the interface, one can imagine combining the cortical activity of one human being with the prosthesis of another, with all kinds of prostheses (acting through the hand of another person, seeing with other eyes, etc.) or acting/feeling together with the same prosthetic device.

All those horizons drive us to ask the basic questions "How does this body work?", "What is this body?", "Where is this body?", "Whose body is this?", "What kinds of prosthesis/extension are normatively acceptable?" and "Which kinds might be criminal?" Until now, these questions had straightforward answers, since the form of the prosthesis and its bodily capacities were in line with typical human experiences, which imposed a common norm. The horizon of different prostheses multiplies the possible responses to each of these questions and this multiplies the possible human conditions.

The human as a product and the controllable human

The example of prostheses also creates two situations where, as enhancement becomes better understood, criminal considerations may come into play.

First of all, a prosthetic device is a *product*. It brings with it all the standard issues of a product: design errors, failure to comply with standards, poor production practices, delivery delays, theft, insufficient or incorrect documentation, predatory sales practices, faulty installation or configuration, product failure, unintended use practices, maintenance issues, questions around the sale of used products, etc.

As a prosthesis is by definition an extension of the body, each of these problems takes on a singular importance. If it is sometimes difficult to deal with a problem for an "ordinary" product (a car breakdown or a telephone malfunction, for example) it is far more worrisome to have to deal with a malfunctioning body part. Should these types of products include a guarantee or enhanced reliability features?

Additionally, because products must be purchased, it is also necessary to examine the question of inequality in regard to bodily abilities. For we know that growing inequalities lead to increased criminal activity. Consequently, how can we create a social mechanism to prevent this new source of inequality, even more problematic since it is linked to physical abilities and may have an amplifying effect (as with the invention of the machine tool)?

Second, for all the prostheses that include a computer program, there is the crucial question of data privacy. If this data makes its way into someone else's hands, that person would have privileged access to our lives: our whereabouts, what we have done, what we have felt, perhaps even some of what we have thought. It is, for example, possible to imagine that the signature made using your prosthetic hand could be copied and reproduced, etc. In this era where identity theft and data piracy are among our top concerns, how very frightening it is to imagine the theft of bodily identity or the piracy of our bodily data, for our body is our most intimate companion.

Conclusions

The first question centered on the meaning of "human enhancement" because this expression is semantically wide open. We choose to define it as equivalent to "anthropotechnics", i.e.: "the activity or technique of non-medical transformation of a human being by modifying his/ her body." Such a definition limits human enhancement to the body (including the body schema), and demarcates it from medicine.

Afterwards, we showed how the spontaneous trend of human enhancement and anthropotechnics can raise fundamental questions about laws, justice and criminality. The question of the balance between benefits and risks appeared as a key question: what criteria, what expertise, what control?

Finally, we developed an analysis based on the point of view of autonomy and heteronomy. These considerations led us to fundamental reflections on human conditions, opening multiples horizons and an interrogation between different conceptions of humanness and questions of modified bodies. In the end, the question centers upon: (i) what kinds of enhancement we want; (ii) which kinds are socially/normatively acceptable; and (iii) how best to regulate those which are unacceptable or which violate existing laws. Since we do not yet have any clear answers to (i), answers to (ii) and (iii) must, inevitably, remain unclear at present.

Notes

1 National Science Foundation (NSF) & Department of Commerce (DOC) (2002).
2 European Parliament (Science and Technology Option Assessment) (2009).
3 In France, a great mutual assurance company (MGEN, 3.5 million members) began to think about these questions (Symposium Transhumanisme / Homme augmenté, March 9, 2016).
4 Cf. www.amgentourofcalifornia.com, accessed 20 October 2015.
5 Aimee Mullins, TED Conference: "My 12 pairs of legs", 2009, www.ted.com/talks/aimee_mullins_ prosthetic_aesthetics/transcript?language=en.

References

Assemblée Nationale (2010) *Rapport d'information n°2235 sur la révision des lois de bioéthique*, (Part IV, chap. 9, C).

Beauchamp, T. and Childress, J. (2013) *Principles of Biomedical Ethics (Seventh Edition).* Oxford University Press (USA).

Bogle, K.E., and Smith, B.H. (2009) Illicit methylphenidate use: A review of prevalence, availability, pharmacology, and consequences, *Current Drug Abuse Reviews*, 2(2): 157–176.

British Medical Association (2002) *Drugs in Sport: The Pressure to Perform.* London, BMJ Books.

Druckman, D. and Swets J.S. (eds) (1988) *Enhancing Human Performance: Issues, Theories and Techniques.* Washington, DC, National Academy Press.

Eskiizmirliler, S. and Goffette, J. (2015) Brain-Machine Interface (BMI) as a tool for understanding human-machine cooperation, in Bateman, S., Gayon, J., Allouche, S., Goffette, J. and Marzano, M. (eds) *Inquiring into Human Enhancement: Interdisciplinary and International Perspectives*, Basingstoke: Palgrave Macmillan, pp. 138–160.

European Parliament (Science and Technology Option Assessment) (2009) *Human Enhancement – Study* (C. Coenen, ed.). Available at: www.europarl.europa.eu/stoa/publications/studies/stoa2007-13_ en.pdf.

Goffette, J. (2006) *Naissance de l'anthropotechnie.* Paris: Vrin.

Goffette, J. (2015) Enhancement: Why should we draw a distinction between medicine and anthropotechnics? in Bateman, S., Gayon, J., Allouche, S., Goffette, J. and Marzano, M. (eds) *Inquiring into Human Enhancement: Interdisciplinary and International Perspectives.* Basingstoke: Palgrave Macmillan, pp. 38–59.

Hottois, G. and Missa, J-N. (2002) *Species Technica.* Paris: Vrin.

Kant, I. (2011 [1785]) *Groundwork of the Metaphysics of Morals: A German-English Edition.* Cambridge (UK): Cambridge University Press.

Kapp, E. (1877) *Grundlinien einer Philosophie der Technik: Zur Entstehungsgeschichte der Kultur aus neuen Gesichtspunkten.* Braunschweig (Germany): George Westermann.

Laure, P. and Allouche, S. (2015) Doping behaviour as an indicator of Performance Pressure, in Bateman, S., Gayon, J., Allouche, S., Goffette, J. and Marzano, M. (eds) *Inquiring into Human Enhancement – Interdisciplinary and International Perspectives.* London: Palgrave, pp. 161–180.

Londres, A. (1996 [1924]) *Tour de France, tour de souffrance.* Paris: Editions du Rocher.

McLuhan, M. (1964) *Understanding Media.* New York: McGraw-Hill Book Co.

Mol, A. (2008) *The Logic of Care.* Oxford (UK): Routledge.

National Science Foundation (NSF) & Department of Commerce (DOC) (2002) *Converging Technologies for Improving Human Performance: Nanotechnology, Biotechnology, Information Technology and Cognitive Science*, on-line prepublication version.

Ottani, G. (1961) *Doping e calcio professionalistico.* Milano: FIGC Lega Nazionale Professionisti.

Parens, E. (ed.) (1998) *Enhancing Human Traits: Ethical and Social Implications.* Washington, DC: Georgetown University Press.

President's Council on Bioethics (2003) *Beyond Therapy: Biotechnology and the Pursuit of Happiness.* New York: Dana Press.

Rätsch, C. and Müller-Ebelling, C. (2013) *The Encyclopedia of Aphrodisiacs*. Rochester (Vermont, USA): Park Street Press.

Sartre, J-P. (2007) *Existantialism Is a Humanism* [1945]. Connecticut, USA: Yale University Press.

Sloterdijk, P. (2000) *La domestication de l'Etre*. Paris: 1001 Nuits.

Teter, C. J., McCabe, S. E., LaGrange, K., Cranford, J. A. and Boyd, C. J. (2006) Illicit use of specific prescription stimulants among college students: Prevalence, motives, and routes of administration, *Pharmacotherapy*, 26(10): 1501–1510.

Thoër, C. and Robitalle, M. (2011) Utiliser des médicaments stimulants pour améliorer sa performance: usages et discours de jeunes adultes québécois, *Drogues, Santé et Société*, 10(2): 1–41.

Wilens, T. E., Adler, L. A., Adams, J., Sgambati, S., Rotrosen, J., Sawtelle, R., Utzinger, L. and Fusillo, S. (2008) Misuse and diversion of stimulants prescribed for ADHD: A systematic review of the literature, *Journal of the American Academy of Child and Adolescent Psychiatry*, 47(1): 21–31.

Keynote discussion

Technology and justice

39

Justice and technology

Albert Borgmann

Introduction

The connection between justice and technology seems to be very much at the center of public attention. In the United States (the setting of my chapter) the Federal Bureau of Investigation (FBI) has gone to court to force Apple to unlock the iPhone of a terrorist. Apple in reply filed a motion to vacate the court's order that Apple comply. The supporting arguments of the two parties turned mainly on security and privacy, though it's unclear from the arguments how and by whom security is best protected, and commentators have suggested that Apple was really interested in its prosperity rather than in the public's security and their users' privacy. But then technology passed justice. The FBI found a way to unlock the phone and dropped the case against Apple.

It was an important controversy, and variants of it are exercising countries around the globe. It may sound strange, but philosophically this is not a difficult issue. There are two well-known and well-articulated conceptions of ethics in the Anglophone world that give ready if contrary answers to the question at hand and allow you to side with one of the schools of justice and its answer or try to forge a compromise between the two. One conception of justice is supported by the ethics of pleasure and prosperity, called utilitarianism by philosophers; the other is the ethics of rights and liberties, called deontological ethics. Utilitarians will favor prosperity over privacy, deontologists privacy over prosperity. Security is important to both schools, of course, but utilitarians will settle for less security in exchange for more prosperity, deontologists for less security in exchange for more privacy. The rest is details. Granted, the issue is important and the details intricate; still, the conceptions of justice and technology that are ruling the debate are shallow, not say superficial, and conceal profound problems of justice and technology.

Philosophy of justice and philosophy of technology

Granted my strong (and certainly controversial) claim that the prevailing views of justice and technology are shallow, can we turn to philosophy for illumination of the deep connections between justice and technology? The answer is no, surprisingly perhaps because both the philosophy of justice (sometimes called jurisprudence) and the philosophy of technology began to flourish in the United States at about the same time, the 1970s.

To begin with justice, there is much to be said about its history, its meanings, and the validity of arguments in support of it. But the philosophy of justice doesn't matter much until it tells us what justice is, what its standards and norms are, and what they require of us. This is *normative* theory of justice, and it was pretty much dead till in 1971 John Rawls revived it in his masterful *A Theory of Justice*. The publication and reception of the book constituted a great event in American philosophy, perhaps the greatest. The response to Rawls has been enormous and is beyond summarizing. There are, however, discussions of justice in Rawls's wake that are particularly helpful in shedding light on the justice–technology relation. The first is Michael Sandel's *Liberalism and the Limits of Justice* (1982). Then came Michael Walzer's *Spheres of Justice* (1983). Another, though not directly influenced by Rawls, is Lawrence Friedman's *Total Justice* (1987). Michael Sandel published his brief but incisive, *The Case against Perfection* in 2007. Amartya Sen's voluminous *The Sense of Justice* came out in 2009, and so did Sandel's *Justice: What's the Right Thing to Do?*, a summary of the state of justice and jurisprudence in the United States.

Up until the mid-1970s, there was no active and well-ordered philosophy of technology in the Anglophone world. There were the classics of Heidegger and Ellul, both published in 1954, and in the 1960s numerous searching and scattered attempts in the United States. In 1975, Paul Durbin and Carl Mitcham convened a conference at the University of Delaware. Out of the conference came the Society for Philosophy and Technology in 1976, and in 1978 the annual journal *Research in Philosophy and Technology*. More important, the philosophers who attended the Delaware conference or came to be associated with the Society wrote their books and established distinctive positions in the philosophy of technology—Langdon Winner, *Autonomous Technology: Technics-out-of-Control as a Theme in Political Thought* (1977); Don Ihde, *Technics and Praxis: A Philosophy of Technology* (1979); Andrew Feenberg, *Critical Theory of* Technology (1991); Carl Mitcham, *Thinking Through Technology: The Path Between Engineering and Philosophy* (1994).

In 1997 Hans Achterhuis edited an anthology, a summary of the work that had begun in 1975. The distinctiveness and continuity of that phase of American philosophy of technology are evident in remarks by Durbin on the 1975 conference and by Achterhuis in the introduction to the anthology. Durbin distilled two things from "the philosophy and technology movement so far."

> (1) There are urgent problems connected with technology and our technological culture which require philosophical clarification, and (2) much that has thus far been written on these problems is inadequate—making it all the more important for serious philosophers to get involved.
>
> *(Durbin 1978: 3)*

Talking about the philosophers whose work was discussed in the collection, Achterhuis said:

> All six philosophers emphasize the power of contemporary technology to transform reality. All six interpret and analyze philosophically the profound ways in which technology has transformed—and continues to transform—social networks and forms of life, human wants and possibilities, and the experience of our bodies and nature.
>
> *(Achterhuis 2001:1)*

The anthology was originally written in Dutch by Dutch scholars and translated and published in 2001 under the title *American Philosophy of Technology: The Empirical Turn*. The origin of the book is evidence that philosophy of technology has become an international concern and shows more particularly that the center of scholarly gravity has shifted to the Netherlands, a shift that's evident also in Philip Brey's contribution to this volume.

Technological surfaces

The prevailing view of technology surfaces when a prominent person—a politician, a business person, an artist—is asked about the impact of technology. A typical answer is the late Andrew Grove's, who was a founder and CEO of Intel Corporation: "Technology happens, it's not good, it's not bad. Is steel good or bad?" This has been called the instrumental view of technology. Technology is thought to be a value-neutral instrument. What's good or bad is the way it's used. Using steel in a scalpel for surgery is good. Using the steel of a switchblade to kill is bad. This is obviously correct, and examining how technology is used for better or worse in the justice system is helpful no doubt. It's helpful also to investigate how advances in technology change the ways in which it is used. But the instrumental view of technology fails to answer the question whether the kind of availability of technological tools influences the ways people behave. When Durbin mentions "the technological culture" and Achterhuis says "All six philosophers emphasize the power of contemporary technology to transform reality. All six interpret and analyze philosophically the profound ways in which technology has transformed— and continues to transform—social networks and forms of life, human wants and possibilities, and the experience of our bodies and nature," then there is evidently a worry that technology is more than value-neutral, that is has transformed the culture and its values through the way it is available and that it's "urgent," as Durbin said, to find out what's good and what's bad in these transformations.

So in addition to the question of how technology as a tool serves justice there is also the question of how technology as a cultural setting has transformed justice itself. What is by now classic American philosophy technology, spawned by the 1975 Delaware conference, offers a variety of positions and approaches. I have been their grateful beneficiary, but this is not the place to acknowledge my debts in detail except to say that all prominent philosophers of technology reject the instrumental view of technology as sufficient for philosophical purposes. To lift the veil of conventional wisdom I will take my clue from a historian and theorist of economics, Karl Polanyi. His pointer will help me substantiate my intuition that the cultural force of technology has led to a flattening of our common understanding of justice.

To say, as I did earlier, that the now dominant conceptions of justice are shallow or even superficial must surely look like a hasty and derogatory judgment. Departing from Polanyi, I want to show that it can be the conclusion of sober cultural analysis. Polanyi wrote his aptly titled *The Great Transformation* during the Second World War and had it published in 1944. At its center is the rise of the market economy in the early modern period. Prior to its ascendancy, economic activities were *embedded* in social and cultural circumstances and inseparable from them. The attempts at severing the economy from these bonds, ultimately impossible Polanyi argued, were by later theorists called *disembedding*. To be sure, the economy cannot be disembedded and turned into a socially and culturally separate institution. The advanced technological economies always require the guidance and support of regulations that are culturally and socially, rather than economically, conditioned. And yet the idea that we can disembed or detach elements from their cultural context to promote prosperity was a rising and is still a powerful force. Let me illustrate the phenomenon of disembedding in a particular instance, then suggest how disembedding captures a profound feature of what we commonly call technology, and finally show how it has transformed justice.

The particular phenomenon I want to consider is light, as in lighting and illumination. The original lights were sun and moon, illumination beyond human control. Once humans had captured fire, they had begun to take control of light. Camp and cooking fires were the first sources of controllable light, and such light was deeply embedded in practices of hunting and gathering, of collecting firewood, of sparking and starting a fire, and in the daily rhythms of camp life. In time, humans discovered that oil and tallow were fine fuels of fire and that the rate of their burning could be controlled if the fuel was confined to a container such as a shell or a hollowed-out stone and the flame was confined by a wick. Next came sticks of hardened fuel surrounding a wick—candles. Candles and oil lamps were the chief sources of light until the beginning of the nineteenth century. Their production was embedded in the rendering of animals and the keeping of bees. Candle making was sometimes part of the household economy and at other times the business of a candle maker, a "chandler," whose identity and craft were well understood in a village or town.

The explosive power of modern technology had its beginning in the Industrial Revolution which in turn was driven by the use of coal as a powerful fuel. Late in the eighteenth century it was discovered that the burning of coal produces a combustible gas and the combustion of the gas produced light—the beginning of gas lighting. It required the building of gas-making plants, the laying of gas pipes, and the production of efficient burners (Rae 1967: 343–348). By the middle of the nineteenth century public spaces and prosperous homes were illuminated with gas.

Gaslight was much brighter than light from oil or kerosene lamps. But it still required laborious lighting, and it produced noxious and unpleasant gases (Gordon 2016: 116). In 1879, Edison succeeded in engineering electric light. "Nothing in the past hundred years matches the sharp distinction between creating light with a flame and creating it with electricity," says Robert Gordon in *The Rise and Fall of American Growth* (Gordon 2016: 118). Electric light was much brighter, safer, pleasant, and available at the click of a switch though it had this in common with gaslight: "In a mere fifty years, the residential United States underwent a transformation from the home production of heat and light by household members who chopped wood, hauled coal, and tended kerosene lamps to a new era of gas and electricity, purchased as commodities and arriving automatically at the dwelling unit without having to be physically carried" (Gordon 2016: 115–116).

Gordon is rightly celebrating the benefits of technological innovation. But evidently these advancements went hand in hand with the disembedding of lighting from contexts of competence and comprehension. No manual skills, no communal practices were needed to turn on electric light. Most people were unaware of the source of electricity and very few understood the process, far less the physics, of generating light. That a glowing filament would produce light and heat is still plausible, but how electrons falling to a lower energy state emit photons in neon light or light-emitting diodes (LED) is incomprehensible and unknown to most users of these light sources.

On first display, electric light seemed like magic and a miracle (Gordon 2016: 117–118). But everyone knew that nothing comes from nothing and that some sort of machinery, however distant and poorly understood, supported and sustained the light show. And it was accepted as a matter of course that you had to work to be able to pay for the blessings of technology. Competence and comprehension were needed for labor if not for consumption. But the skill and expertise of work grew ever more narrow and failed to equal the broad and deep engagements of traditional culture.

The emergence of consumer goods that have been severed from their traditional contexts and are now available safely, instantly, and ubiquitously has been a broad cultural phenomenon in the technologically advanced societies. Think of food, transportation, and entertainment—all of them are pleasantly available. Where the goal of availability is imperfectly realized in, say, health and justice, it is still the guiding norm. In all cases, the goods of technological availability have been disembedded from the intelligible and familiar practices of traditional life, and their production has been transferred to an industrial machinery that for the most part is concealed by distance and ignorance. Even those of us who pride themselves on their understanding of the structures and processes that underlie the realm of leisure and consumption have a merely cerebral comprehension of it all.

Deep justice and ideal justice

"Justice," John Rawls has said, "is the first virtue of social institutions, as truth is of a system of thought" (Rawls 1971: 3). In this basic and general sense, justice is in all cases concerned with three issues: security, fairness, and an ethos. Ethos, as I use it here, is a fancy but convenient word for the common understanding of the good life that animates and sustains a community or society.

When justice is embedded in a traditional, preindustrial community and when life is good, security rests on personal acquaintance and mutual affection; fairness is compliance with the roles that each member of a community is assigned by tradition and the requirements of survival. In all preindustrial communities the ethos was religious, a system of beliefs that was ruled by divinity. To be sure, embedded justice of this kind was a norm that was breached in ways small and large. Deep justice, moreover, had been overlaid in Europe by Roman law in the south and by common law in the north.

The changes that did not just disturb or overlie deep justice, but transformed it from the ground up began with the *tangible* phenomena of the journeys of discovery and international trade in the sixteenth century and with the *cultural* changes that were articulated by Thomas Hobbes (1588–1679) and John Locke (1632–1704). In time the material changes led to modern technology; the philosophically articulated changes led to security as a task for the state and to a conception of fairness I will call ideal justice.

The great contemporary monument of ideal justice is Rawls's magisterial *A Theory of Justice* (1971). It's ideal justice because it sets out "the principles that would regulate a well-ordered society." The theory assumes "strict compliance." And why ideal theory? "The reason for beginning with ideal theory is that it provides, I believe, the only basis for the systematic grasp of these more pressing problems" (Rawls 1971: 8–9). And what are the more pressing problems? They are the actual obstacles we face in establishing a truly just society.

What of the ethos that guided the rise of the modern era? It is the promise of liberty and prosperity. Liberty was thought of as liberation from the oppression of inequality and the burdens of reality—political liberty and the liberty of disburdenment. It's the mark of a new epoch, of modernity, that once its vision becomes available, prior cultural conditions become intolerable—politically the feudal system and in time patriarchy and racism; as for disburdenment—the hardships of hunger, disease, and confinement.

In 1971, Rawls still believed in the ethos of liberty, equality, and fraternity, (Rawls 1971: 105–7). But as Robert Putnam has shown, fellow feeling and social engagement dropped sharply in the second half of the twentieth century (Putnam 2000). In 2001, Erin Kelly, the editor of Rawls's last statement of his theory, *Justice as Fairness*, reported that "Rawls is well aware that since the publication of *A Theory of Justice* in 1971, American society has moved farther away from the idea of justice as fairness" (Rawls 2001: back cover). Not only did ideal justice fail to have practical influence, there were also scholars who criticized Rawls's theory on theoretical grounds.

In 1983 Michael Walzer disagreed with Rawls's idea of *one encompassing* theory of justice and held that there are *Spheres of Justice* (Walzer 1983). One of them is the sphere of "Money and Commodities" where Walzer touches on the disembedding that happens when certain goods are torn from their communal and familial contexts and offered for sale on the marketplace. Walzer's is an admirable book, but it overlooks the broad effect that technology has had on justice. Amartya Sen's *The Idea of Justice* (Sen 2009) is an extended brief against Rawls's pursuit of an *ideal* theory, a brief whose "aim is to clarify how we can proceed to address questions of enhancing justice and removing injustice, rather than to offer resolutions of questions about the nature of perfect justice" (Sen 2009: ix). Sen's questions are urgent and should certainly be considered in their own right. But whether promising answers can be found when the force of technology is largely ignored, as it is in Sen's book, I'm not sure.

Michael Sandel's scholarly trajectory is unusually illuminating. He began his career in 1982 with a critique of Rawls's ideal justice on behalf of (what was left of) deep justice (Sandel 1982). Over time Sandel became increasingly aware of the culturally dubious and even destructive aspects of technology. Some of his incisive essays came out in 2007 under the title *The Case against Perfection: Ethics in the Age of Genetic Engineering* (Sandel 2007). In his great state of the art book *Justice: What's the Right Thing to Do?* (Sandel 2009) he turned his attention, as Walzer had turned his, to the detrimentally disembedding force of the markets. He extended his critique in *What Money Can't Buy: The Moral Limits of Markets* (Sandel 2012) and showed in particular how the markets displace crucial and hitherto enduring features of what I've called deep justice. Lawrence Friedman has discussed most directly how technology has transformed deep justice into an expectation of *Total Justice* (Friedman 1985). Friedman, remarkably, is a scholar of legal history rather than legal theory.

Shallow justice

Rawls's ideal justice has not had the effect that he and many of us had hoped for. In a late and wistful footnote Rawls said: "Here I entertain the fantasy that works like this restatement [*Justice as Fairness*] are known in the public culture" (Rawls 2001: 121, note 42). And yet there is a powerful tie between Rawls's theory and American reality. His vision captures what's best in the soul of the American body politic. Alas, Rawls has been unable to invigorate that soul; and the great remainders of deep justice, whose advocate Sandel has been, are withering under the force of technology. Justice is becoming shallow.

Here's a crucial point about shallow justice: It is with us, and it is working. Tens of millions of people would love to come to the United States and live under its roof. But will it continue to work? And if it does, is it all we can hope for? Let me begin with outlining the working model of shallow justice, then point out where it shows stresses and cracks, and end with a brief look at the future.

Justice becomes shallow when the administration of justice escapes the competence and comprehension of the citizenry and is taken over by a machinery. The converse is true as well: When the administration of justice becomes complex and intricate and a sophisticated machinery becomes available to deal with complexity and intricacy, justice escapes the participation and understanding of most people—it becomes shallow. Still, shallow justice can work.

To begin with security, people can assume there are reliable machineries that will keep them safe when they board a plane, go to a concert, get their groceries, buy at Amazon, or receive an email. When a crime has been committed, they expect the machinery of criminal justice to find the perpetrator and deal with him lawfully. People take bridges to hold up and water to be clean. When they go to a hospital they can assume that they will be treated according to sound medical science and practices. The cars they drive won't explode from under them. The elevators will take their appointed ascents and descents. Scaffoldings in construction won't collapse, and ditches will not cave in.

Security such as it is rests on numerous government agencies, administering a large body of laws and regulations. These agencies command vast computer networks, surveillance cameras, x-ray machines, laboratories that perform DNA analysis and other forensic investigations. All this and more constitutes a machinery that is physically and cognitively concealed from almost all citizens. Security floats on top of this opaque substructure. The machinery emerges, and only ever in part, when security fails.

As for fairness, Americans trust that they and their fellow citizens largely pay the taxes they owe, that accurate records of crimes are kept, that if they violate a law, judges and juries apply the right laws and precedents, that if they get an education, their credentials will be honored, and more broadly that due process will be followed and corruption is not a significant problem in American society. Their trust is not misplaced. The United States ranks sixteenth among the 167 countries surveyed by Transparency International. But their trust is not based on a well-informed understanding of the Internal Revenue Service, of the judicial system, or of the canons of due process.

The ethos of liberty and prosperity, if anything, sustains American society. Political liberties have recently been extended to gay people. Historically and globally considered, prosperity is high in the United States. Almost everyone has indoor plumbing, a refrigerator, television, a car, and access to the Internet. To advance prosperity is, apart from security, the undisputed goal of all politicians; differences pertain as to how best to achieve the goal and with what distribution.

Still, justice has no depth. Citizens see themselves as the recipients and beneficiaries of justice. There is little understanding and less engagement in the transactions of security and fairness. The ethos of liberation and enrichment is becoming brittle. It was powerful in the nineteenth century when the burdens of oppression and the grind of poverty were still palpable and advances of liberty and prosperity were correspondingly obvious and deeply appreciated. The tangible blessings of comforts and possessions reached a high water mark by the middle of the twentieth century. What followed was a period of refinement and expansion—larger homes and cars, better television, more air travel. Refinement and expansion took a new turn with the Internet and information and communication technology (ICT). But the ethos of more rights and greater prosperity is in crisis now. The advancement of women's equality and racial justice has stalled. The incomes of the middle and lower class have stagnated. Stalling is a crucial danger.

A world of glamorous surfaces invites consumption, but these surfaces meld into the background of normalcy and lose their entertaining attraction. Electric light once was magic and a miracle, but now it's taken for granted. So is running water, the refrigerator, the flat screen TV. Novelty is to the world of consumption what the Bernoulli effect is to an airplane—they keep things aloft; stalling is a mortal danger, it leads to a crash.

Shallow justice can work, but it easily provokes unreasonable expectations about the benefits at the surface of justice and about the power of the unintelligible substructure. In a world of disembedded consumption goods, the paradigmatic expectation is that these free-floating commodities be available instantly, ubiquitously, and easily. Justice is then expected to be the remedy that's readily available when something goes wrong. "There has developed in this country," says Lawrence Friedman, "what I call here a *general expectation of justice*, and a *general expectation of recompense for injuries and loss*. Together, these make up a demand for what will be called 'total justice'" (Friedman 1985: 5). Total justice is the companion of "a new kind of society." It's a "society of technology, industry, science, machines" (Friedman 1985: 42).

You may not know how and where the energy for your lighting is generated, but your belief that the generating machinery is powerful and reliable is well-warranted. You may not know how the machinery of justice works, but you may well believe that it is powerful and reliable if only you turn up the power. Properly applied, then, so runs the common expectation, the judicial system will deal with drug users and repeat offenders the way electric lighting deals with darkness. Thus the punitive Rockefeller Drug Laws of 1973 and the three-strikes-and-you're-in-for-life laws of the 1990s. Crime rates have fallen, but for the most part they fell for unrelated reasons. What definitely rose were the incarceration rates, particularly for young Black men. The American prison population is the largest in the world and the highest in the world per capita.

As long as shallow justice works, so does by definition the underlying machinery; and the machinery takes over the functions that were once supplied as a matter of course by neighborliness, mutual obligations, and communal vigilance. Markets, says Sandel, displace "social practices and the goods they embody" (Sandel 2013: 188), and so does shallow justice. Indifference displaces care. Shallow justice works, but it does so in a sociologically shallow way. It spreads indifference about the poor and the powerless in the depths of social stratification. The beneficiaries of shallow justice don't care about the conditions that conduce to dealing and using drugs. When the machinery of justice catches up with dealers and addicts, most of us assume they'll get legal representation, but we fail to look into the system of public defenders and make sure it's adequately supported.

Shallow justice generates indifference, and indifference goes on to render justice narrow as well as shallow and so excludes from view and responsibility the injustices of global warming that we are shifting to the poor countries and to our own offspring. As long as we can, we also shield the ravages of the larger technological machinery from the strictures of environmental justice. Indifference finally, is self-protective and protects itself from the question of where indifference comes from and how it is reinforced by technology. When Internet mischief becomes vicious, we react with dismay and incomprehension.

Prospect

Shallow justice is dominant in American society, I'm afraid, but it is not exclusive. There are living strands of deep justice in people's everyday decency, in their voluntary associations, and in their charitable activities. Inspired politicians remind us of ideal justice. But if the ethos of shallow justice is slackening, as it seems to be, then the common order is in jeopardy. It's hard to say whether the upwelling of anger on the right and the left is a call for recovering a world of widely rising consumption or whether it is dismay about the shallowness of consumerism.

There are, at any rate, reasons for hope. To begin with, there are efforts to strengthen shallow justice on its own terms, assuming its ethos will hold while we work to improve security and fairness. One might complain that these improvements amount to essentially technological fixes of technological problems and will only entrench shallow justice further. But I would reply that profound reforms are more likely to succeed in an atmosphere of greater stability and equity. Thus under judicial and civic supervision, the employment of ICT to prevent crime and terrorism is necessary. So is prison reform, even if for fiscal reasons. And so is regard for Black Lives Matter even if motivated by fear of bad publicity and expensive lawsuits. More important and more profound are attempts to revive sentiments of respect and compassion by having mediation take the place of litigation and restorative and redemptive justice the place of punishment.

Finally and crucially we must realize that we can't have deep justice in a shallow world. The depths of persons, of things, and of justice disclose themselves together or not at all. To be sure, it would be irresponsible to try and dismantle the machinery of technology on behalf of deep justice. But technology can be relegated to the background of our lives and made to leave room for engagement with a deeper world at the center. Thus the most hopeful and profound developments in this country are the recoveries of engaging realities. They are numerous and scattered—walkable cities, the artisan economy; communities of runners, bikers, and skiers; local theaters, arts communities, farmers markets, and more. What is needed now is a mutual recognition of these movements and the realization that the ethos of the profoundly good life must become politically effective to make society truly secure and deeply fair.

References

Achterhuis, Hans (ed.) (2001 [1997]) *American Philosophy of Technology: The Empirical Turn*, tr. Robert P. Crease. Bloomington: Indiana University Press.

Durbin, Paul T. (1978) Introduction to the Series, *Research in Philosophy and Technology*, 1: 1–4.

Ellul, Jacques (1964 [1954]) *The Technological Society*, tr. John Wilkinson. New York: Vintage Books.

Feenberg, Andrew (1991) *Critical Theory of Technology*. New York: Oxford University Press.

Friedman, Lawrence M. (1985) *Total Justice*. New York: Russell Sage Foundation.

Gordon, Robert J. (2016) *The Rise and Fall of American Growth: The U.S. Standard of Living since the Civil War*. Princeton: Princeton University Press.

Heidegger, Martin (1954) Die Frage nach der Technik, *Vorträge und Aufsätze*, 13–44. Pfullingen: Neske.

Ihde, Don (1979) *Technics and Praxis: A Philosophy of Technology*. Dordrecht: D. Reidel.

Mitcham, Carl (1994) *Thinking through Technology: The Path between Engineering and Philosophy*. Chicago: University of Chicago Press.

Polanyi, Karl (1944) *The Great Transformation: The Political and Economic Origins of Our Time*. New York: Rinehart.

Putnam, Robert D. (2000) *Bowling Alone: The Collapse and Revival of American Community*. New York: Simon and Schuster.

Rae, John B. (1967) Energy Conversion, in Melvin Kranzberg and Carroll W. Pursell, Jr. (eds) *Technology in Western Civilization*, vol. 1, pp. 336–49.

Rawls, John (1971) *A Theory of Justice*. Cambridge, MA: Harvard University Press.

Rawls, John (2001) *Justice as Fairness: A Restatement*. Cambridge, MA: Harvard University Press.

Sandel, Michael J. (1982) *Liberalism and the Limits of Justice*. Cambridge: Cambridge University Press.

Sandel, Michael J. (2007) *The Case against Perfection: Ethics in the Age of Genetic Engineering*. Cambridge, MA: Harvard University Press.

Sandel, Michael J. (2009) *Justice: What's the Right Thing to Do?* New York: Farrar, Straus and Giroux.

Sandel, Michael J. (2013) *What Money Can't Buy: The Moral Limits of Markets*. New York: Farrar, Straus and Giroux.

Sen, Amartya (2009) *The Idea of Justice*. Cambridge, MA: Harvard University Press.

Walzer, Michael (1983) *Spheres of Justice: A Defense of Pluralism and Equality*. New York: Basic Books.

Winner, Langdon (1977) *Autonomous Technology: Technics-out-of-Control as a Theme in Political Thought*. Cambridge, MA: MIT Press.

Index

Abbot, C. 578
Achterhuis, H. 676
Adderal 665–6
Adorno, T. 481
advertising 183, 312–14
aerial surveillance 442–3
Africa 587
Agnew, R. 328
agriculture *see* food production
alchemy 44
Allouche, S. 663
Amin, S. 219
AMMORPGs 121
Amster, R. 319–20
ANPR cameras 441–2
anthropotechnics *see* human
   enhancement
anti-forensics 633–5
anticipation 367–8
antique firearms 273–4
Apple 675
Ariel, B. 444
Aristotle 354
arms race *see* co-evolutionary struggles
arraignment *see* remote first appearance
Arthur, W. B. 354–5, 363–4, 367–8
artefacts *see* extension theory
artificial intelligence 453–6, 459, 643–4,
   651–2; Japanese pragmatism 650–1;
   liability 645–6; robotic crimes 647–50
Atherton, M. 274
attribution 235, 337
Austin, J. 598
authority 605

automated video monitoring 244–5 *see also*
   CCTV; visual surveillance
automation *see* robots
autonomy 664–6

Baker, N. 198
Baldwin, R. 579
Bamako Convention 587
bar code 303
Barlow, J. P. 158
Basel Convention 585–9
battery 627–8
Baudrillard, J. 162
Bauman, Z. 310, 570
BDSM 122
Beard, M. 534
Beauchamp, T. 664
Beck, U. 578–9
Behavioural Insights Team 358
Ben-Shakhar, G. 473–4
Bentley, M. 332
Berlin key 603–5
bestiality 123
Beyond Therapy 655
Bhopal disaster 200
big data 452, 459; critics 453–5; enthusiasts
   452–3; neuroscience 468–9; sceptics
   455–8
Bigo, D. 494
bio-power 48–9
biofuels 250–2
biological evolution 364–5, 367
Biological Weapons Convention 228, 230–1,
   233–4